Smart Innovation, Systems and Technologies

Volume 252

The Smart Innovation, Systems and Technologies book series encompasses the topics of knowledge, intelligence, innovation and sustainability. The aim of the series is to make available a platform for the publication of books on all aspects of single and multi-disciplinary research on these themes in order to make the latest results available in a readily-accessible form. Volumes on interdisciplinary research combining two or more of these areas is particularly sought.

The series covers systems and paradigms that employ knowledge and intelligence in a broad sense. Its scope is systems having embedded knowledge and intelligence, which may be applied to the solution of world problems in industry, the environment and the community. It also focusses on the knowledge-transfer methodologies and innovation strategies employed to make this happen effectively. The combination of intelligent systems tools and a broad range of applications introduces a need for a synergy of disciplines from science, technology, business and the humanities. The series will include conference proceedings, edited collections, monographs, handbooks, reference books, and other relevant types of book in areas of science and technology where smart systems and technologies can offer innovative solutions.

High quality content is an essential feature for all book proposals accepted for the series. It is expected that editors of all accepted volumes will ensure that contributions are subjected to an appropriate level of reviewing process and adhere to KES quality principles.

Indexed by SCOPUS, EI Compendex, INSPEC, WTI Frankfurt eG, zbMATH, Japanese Science and Technology Agency (JST), SCImago, DBLP.

All books published in the series are submitted for consideration in Web of Science.

More information about this series at http://www.springer.com/series/8767

Álvaro Rocha · Paulo Carlos López-López ·
Juan Pablo Salgado-Guerrero
Editors

Communication, Smart Technologies and Innovation for Society

Proceedings of CITIS 2021

Springer

Editors
Álvaro Rocha
ISEG
Universidade de Lisboa
Lisbon, Portugal

Paulo Carlos López-López
University of Santiago de Compostela
Santiago de Compostela, Spain

Juan Pablo Salgado-Guerrero
Universidad Politécnica Salesiana
Cuenca, Ecuador

ISSN 2190-3018 ISSN 2190-3026 (electronic)
Smart Innovation, Systems and Technologies
ISBN 978-981-16-4125-1 ISBN 978-981-16-4126-8 (eBook)
https://doi.org/10.1007/978-981-16-4126-8

This Springer imprint is published by the registered company Springer Nature Singapore Pte Ltd.
The registered company address is: 152 Beach Road, #21-01/04 Gateway East, Singapore 189721, Singapore

Preface

This book is composed of the papers accepted for presentation and discussion at the 7th International Conference on Science, Technology and Innovation for Society (CITIS 2021). CITIS is an international conference on engineering at the service of social transformation that has been organized since 2014 by Universidad Politécnica Salesiana, Ecuador. The main goal of the conference is to propose a platform so that engineering professionals, academics, and students can critically and objectively rethink the theoretical and practical problems of the greatest impact on modern society through an engineering perspective. Due to the pandemic affecting humanity since the beginning of 2020, the seventh edition of CITIS took place as a virtual edition hosted from Guayaquil on May 26th–28th, 2021.

The Program Committee of CITIS 2021 was composed of a multidisciplinary group of 112 experts from 17 countries, who had the responsibility for evaluating, in a 'double-blind review' process, the 120 papers received in the conference. The main topics covered in this edition were focused on five thematic axes: science, technology, and innovation for the industry, mobility, environmental sustainability, information, and telecommunications. The accepted contributions recreate recent research trends in the fields of software engineering, big data analysis, cloud computing, data engineering, data management and data mining, machine learning, deep learning, artificial intelligence, smart systems, robotics and automation, mechatronic design, and industrial processes design.

CITIS 2021 also included several workshop sessions, and several keynote speeches were given on cutting-edge topics related to the thematic axes of the conference. In addition, parallel working sessions were held to strengthen entrepreneurship and innovation skills for engineering students.

The 68 papers accepted for presentation and discussion at the conference are published by Springer (this book) and will be submitted for indexing by ISI, EI-Compendex, SCOPUS, DBLP, and/or Google Scholar, among others.

We acknowledge all of those that contributed to the staging of CITIS 2021 (authors, committees, workshop instructors, and our attendees). We deeply appreciate their involvement and support that was crucial for the success of CITIS 2021.

Guayaquil, Ecuador
May 2021

Álvaro Rocha
Paulo Carlos López-López
Juan Pablo Salgado-Guerrero

Contents

About the Editors

Álvaro Rocha holds the title of Honorary Professor, and holds a D.Sc. in Information Science, Ph.D. in Information Systems and Technologies, M.Sc. in Information Management, and BCs in Computer Science. He is a professor of Information Systems at the University of Lisbon—ISEG and a researcher at the ADVANCE (the ISEG Centre for Advanced Research in Management. He is also President of AISTI (the Iberian Association for Information Systems and Technologies), Chair of the IEEE Portugal Section Systems, Man, and Cybernetics Society Chapter, and Editor-in-Chief of both JISEM (Journal of Information Systems Engineering and Management) and RISTI (Iberian Journal of Information Systems and Technologies). Moreover, he has served as Vice-Chair of Experts for the European Commission's Horizon 2020 program, and as an Expert at the COST—intergovernmental framework for European Cooperation in Science and Technology, at the Government of Italy's Ministry of Universities and Research, at the Government of Latvia's Ministry of Finance, at the Government of Mexico's National Council of Science and Technology, and at the Government of Polish's National Science Centre.

Paulo Carlos López-López is professor and researcher in the Department of Political Science and Sociology at the University of Santiago de Compostela (USC), Ph.D. in Communication, journalist and political scientist. He is a member of the International Research Network on Communication Management, the Political Research Team of the USC and the Iberian Association of Information Systems and Technologies. He has published more than sixty scientific papers on topics such as communication, social media and political science. He was director of the group "Media, Applied Technologies and Communication" and director of a postgraduate degree in Political Communication in Ecuador. He was awarded the Drago 2019 research award for his study on transparency. Work experience on electoral campaigns, organizational communication, Press office and parliament management. He has an H Index of 7 and is an accredited professor in the category Contracted Doctor by the Quality Department of the Galician University System (ACSUG).

Juan Pablo Salgado-Guerrero Vice-Rector of Research at Universidad Politécnica Salesiana de Ecuador PhD. in Economics and Innovation for Sustainable Development at Università degli Studi di Ferrara, Italy. Recognition for the best doctoral thesis on June 4th, 2018, during the "IUSS Day" ceremony. Master's Degree in Innovation for Business Development, Tecnológico de Monterrey. Fourth level diploma in Strategic Marketing Management, Universidad Técnica Particular de Loja. Cooperative Learning and Educational Technology at the Salesian-style University Universidad Católica de Brasilia (Brazil)-Universidad Politécnica Salesiana de Ecuador. Mechanical Engineer with mention in Machine Design at Universidad Politécnica Salesiana de Ecuador. Industrial Technologist in the specialty of Industrial Mechanics at Universidad Politécnica Salesiana de Ecuador. He was Secretary of Liaison with the Internationalization Collective at Universidad Politécnica Salesiana de Ecuador. He was Secretary of Human and Economic Development of the Municipality of Cuenca, Ecuador; Management of the City of Sciences. Coordination and Public Policy of the City Council in the areas of Social and Economic Development, Culture, Education, Health, Tourism. Executive Representative of the Mayor and President of the Directories of Public Companies. Advisor to the Minister of Electricity and Renewable Energy. Member of the Trust "Por Cuenca" with private companies and the Chamber of Industries of the city of Cuenca since March 2020. Delegate of the Rector of the Universidad Politécnica Salesiana de Ecuador in the Consortium of Universities of the City of Cuenca for the Cuenca del Futuro Project. Representative of the Rector and Coordinator of the University Council of Library of the Universidad Politécnica Salesiana de Ecuador. Member of the Executive Council and General Coordinator of the I International Congress "University, Development and Cooperation". Member of the executive council of the III International Congress "University, Development and Cooperation". Co-organizer of the I Engineering Congress "Engineering Education in Ecuador". Member of the Council of Publications at Universidad Politécnica Salesiana de Ecuador. Member of the Board of Directors of the REDU (Network of Ecuadorian Universities for Research). President of the Scientific Committee of the International Congress on Scientific Research and Production and Publishing. Delegate of the Mayor of Cuenca to the Chair of Citizenship, Chair of the Accredited Universities of Cuenca. Main Member of the Network of International Relations and Cooperation of the Ecuadorian Universities. Manager of the Research Agreements with the Universities of Ferrara, Pavia and Florence. (Italy). Financial Management with International Banks and Financial Research Organizations and agreements with the University of Economics of Vienna—Austria. Member of the Advisory Mission for the formation of the Salesian Universities in India—Province of Tamil Nadu-Chennai. Member of the Basin Committee of the Alliance for Entrepreneurship and Innovation. Member of the Dialogue Table of the Project of Academic Cooperation with Ecuadorian Universities and the European Union VERTEBRALCUE.

Chapter 1
Comparative Analysis of the Dispersion Parameters of a Microstrip Antenna for the 57–64, 64–71 and 71–76 GHz Bands

Lenin Aucatoma, Evelyn Chalapud, Verónica Soria, and Germán Arévalo

Abstract This article details the design and simulation of rectangular patch-type microstrip antennas that operate in millimeter wave band. Three resonance frequencies are used: 60, 67 and 74 GHz. The simulation was made in Ansys HFSS software to achieve the results of reflection coefficient (S_{11}), voltage standing wave ratio (VSWR) and radiation pattern. The modeling of each antenna was made with the mathematical expressions that define a microstrip antenna. First, three antennas were created using only the dimensions obtained with mathematical calculus, but the results of S_{11} and VSWR were not optimal (for example: VSWR was <1 in each case). For this reason, in each model, it was necessary to make changes to all parameters, achieving a reflection coefficient in values between −20 and −30 dB, a VSWR between 1 and 2, and establishing the characteristics of each gap for each dimension of the antennas designed mathematically versus antennas designed with adjustment factors. Results indicate that there is a proportional and not proportional correlation between the frequency and these gaps, which denotes a real difference in the behavior of each parameter in the mathematical designs versus the fitted designs.

1.1 Introduction

Currently, there is a big demand for high quality and more effective technology products. The deployment technology for the fifth generation of wireless communications (5G) has marked the route to achieve a completely wireless world. To improve broadband services, millimeter waves (mmW) are used, in order to increase the capacity of the wireless channels since this type of waves support data speeds of 10 Gb/s and higher [1].

In the design of antennas, an important characteristic is the dispersion parameters that describe the linear electrical behavior of the systems, which vary with the frequency used [2]. In this context, new types of antennas are required to match

L. Aucatoma (✉) · E. Chalapud · V. Soria · G. Arévalo
Universidad Politécnica Salesiana, Quito, Ecuador
e-mail: laucatoma@ups.edu.ec

Á. Rocha et al. (eds.), *Communication, Smart Technologies and Innovation for Society*,
Smart Innovation, Systems and Technologies 252,
https://doi.org/10.1007/978-981-16-4126-8_1

the characteristics of the millimeter wave band and the new users' needs. In [3], a comparative analysis of patch antennas, with microstrip technology, was carried out in the X band at a frequency of 10.5 GHz. In [4], a characterization of millimeter waves was carried out to determine their possible application in 5G at a frequency of 2.45 GHz. All these studies focus in one frequency of the left extreme of the EHF band.

In this research work, three antennas consisting of a patch integrated with a transmission line, a dielectric substrate and in the lower part of the structure a conductive ground plane, are designed in the bands of 57–64, 64–71 and 71–76 GHz: We aimed to visualize if there is a linear behavior in the changes suffered by the dispersion parameters and the dimensions of the antenna in relation to the resonance frequencies. With this purpose, we compared the models of each antenna designed with the mathematical formulas of a microstrip and the necessary adjustment made to reach the objectives set for the dispersion parameters in each frequency.

The remaining of this paper is organized as follows. Section 1.2 describes the methods employed for designing the antennas. Section 1.3 presents the main results obtained through this research and an exhaustive analysis of them. Finally, Sect. 1.4 concludes the paper.

1.2 Methods and Design of the Antennas

The rectangular patch-type microstrip antenna was designed at three resonance frequencies: 60, 67 and 74 GHz. These designs were simulated in Ansys HFSS software to obtain the behavior of the antenna in each frequency and recollect the data of dispersion parameters to make a comparative analysis for both mathematical (theoretical) design and adjustment design. With these results, we seek to detect minimum and maximums gaps while varying the resonance frequency. In each simulation, we aimed to achieve a value of −20 dB or less for the reflection coefficient (S_{11}), besides to a value between 1 and 1.5 for the VSWR, and a radiation pattern without a significant number of nulls.

1.2.1 Rectangular Patch-Type Microstrip Antenna Design

In Fig. 1.1, it is shown the analyzed model consists of the following elements: a patch (rectangular patch-type), a transmission line, a dielectric substrate and a ground plane. For the patch and transmission line, when designing for the frequency of 60 GHz, the employed material was iron, for the frequency of 67 GHz, the material was aluminum and for 74 GHz, the material was graphite. The ground and substrate of the three antennas were composed by copper and FR4-epoxy, respectively. These materials were chosen after we made an adjustment to the antennas and achieved the planned objectives for S_{11} and VSWR parameters.

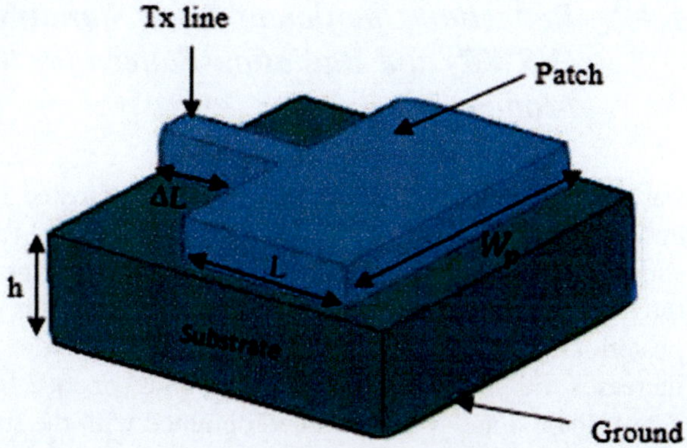

Fig. 1.1 Structure of the rectangular patch-type microstrip antenna

Table 1.1 lists the geometrical parameters for each antenna designed at three central frequencies. The dimensions presented correspond to adjusted values made to the mathematical dimensions for each antenna in each frequency. These models obtained the best results for the S_{11} and VSWR parameters. The dimensions of the fitted models have a crescent gap with the dimensions obtained with the mathematical formulas. For the mathematical modeling were used the formulas described in [5].

Table 1.1 Measurements of the antenna parameters with an operating frequency of 60, 67 and 74 GHz obtained by adjustment

Parameter		60 GHz	67 GHz	74 GHz
		(mm)		
Tx line	ΔL	0.62	0.61	0.61
Patch	W_p	1.52	1.62	1.72
	L	0.84	0.74	0.64
	W_g	1.72	1.82	1.92
	L_g	1.72	1.82	1.92
Ground	x	1.72	1.82	1.92
	y	1.72	1.82	1.92
	z	0.011	0.021	0.022
Substrate	h	0.5	0.32	0.33
	x	1.72	1.82	1.92
	y	1.72	1.82	1.92

1.2.2 Reflection Coefficient (S_{11}), Standing Wave Ratio (VSWR) and Radiation Pattern for the Adjustment Models

For the dispersion parameter S_{11}, in each adjusted model, it we obtained values between −20 and −30 dB. For VSWR, we obtained values from 1 to 1.5. Figure 1.2 shows the 2D radiation pattern in the H plane for the designs with adjustment, for the resonance frequencies of 60, 67 and 74 GHz. This graph allows to identify main, posterior, lateral and null lobes, respectively. Notice that in the radiation pattern, increases the posterior lobules while the resonance frequency increases. Table 1.2 details the S_{11} and VSWR valued obtained with the simulation.

1.3 Analysis and Discussion of Results

In this section, we will analyze the impact of the resonance frequency in each parameter of antenna design, making a comparison between the models of antenna obtained using the formulas specified in [5] versus the correspondent adjustments made in each model. These adjustments consisted in a variation of dimensions to compare the increase or decrease of these values with respect to the formula's results. In each design we have established a correlation between the impact of resonance

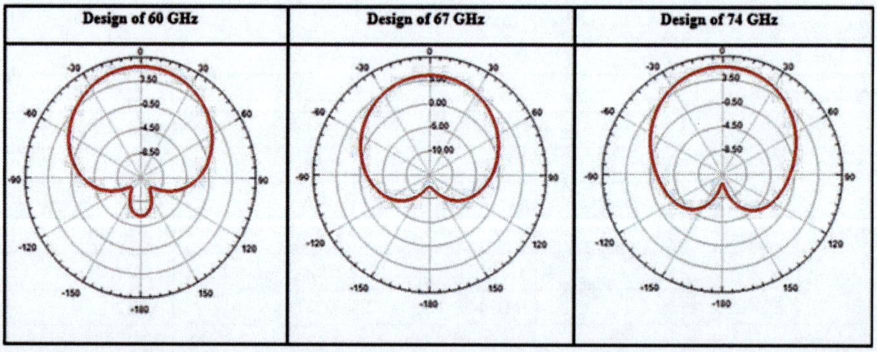

Fig. 1.2 2D radiation pattern of the adjusted designs at the center frequencies of 60, 67 and 74 GHz

Table 1.2 Results obtained from the dispersion parameter and the standing wave ratio (VSWR) for the adjusted design of antennas operating at frequencies of 60, 67 and 74 GHz

Frequency (GHz)	S_{11} (dB)	VSWR
60	−24.4067	1.0472
67,0125	−23.9816	1.0999
74	−24.4920	1.0370

Table 1.3 Results of W_p for three central frequencies

Frequency (GHz)	Patch width (W_p) (mm)		Gap (%)
	Mathematic (M)	Adjustment (A)	M versus A
60	1.5215	1.52	0.1
67	1.3625	1.62	15.90
74	1.234	1.72	28.26

frequency in each parameter, verifying if the correlation persisted when the necessary adjustments were made in each antenna.

1.3.1 Impact of Resonant Frequency on the Patch Width

Table 1.3 presents the results of both theoretical and adjusted values of the patch width (W_p). For theoretical calculations, W_p decreases while the resonance frequency (f_0) increases, and in the adjustment, W_p increases while f_0 increases. The mathematical calculation was made with (1.1).

$$W_p = \frac{c}{2 f_0 \left(\frac{\epsilon_r + 1}{2}\right)} \tag{1.1}$$

where ϵ_r is the relative dielectric constant of the substrate and c is the velocity of light $\left(3 \times 10^8 \text{ m/s}\right)$.

1.3.2 Impact of Substrate Thickness (h) on VSWR and S_{11}

Table 1.4 shows the values of $h = a, h < a, h \ll a$, and $h > a$, where a corresponds to the maximum height of the substrate. For calculation using $h = a$, $h < a$, two constant values were taken: ground width with a value of 0.01 mm and W_p with a value of 0.023 mm. The value of h was calculated using the Eq. (1.2). Finally, the expected values of S_{11} and VSWR were obtained using $h > a$.

For $h \ll a$, the value of ground width is 0.01 mm, substrate width is 0.001 mm and W_p is 0.0002 mm. Very small values were taken as a reference to validate the ratio of the substrate thickness. With the variation of h for h = a, y h < a, we observe while the frequency increases, the thickness of the substrate decreases, obtaining an inverse proportional relation. When $h = a$, $h < a$, $h \ll a$, the peak of S_{11} and VSWR curves have a gap with the value expected of frequency.

$$h \leq \frac{0.3\lambda}{2\pi \sqrt{\epsilon_r}} \tag{1.2}$$

Table 1.4 Results for h obtained by mathematical formulas (theoretical)

Frequency (GHz)	Values of h		S_{11} (dB)	VSWR
60	$h = a$	0.1138	−1.4504	21.5871
	$h < a$	0.1128	−5.5899	10.1416
	$h \ll a$	0.001	−1.2859	22.6283
	$h > a$	0.5	−25.1842	0.9574
67	$h = a$	0.1019	−5.5488	10.2015
	$h < a$	0.1009	−5.6602	10.0403
	$h \ll a$	0.001	−1.5037	21.2754
	$h > a$	0.32	−29.6794	0.5702
74	$h = a$	0.0231	−10.6881	5.2273
	$h < a$	0.0221	−11.3319	4.8335
	$h \ll a$	0.001	−1.6246	20.6073
	$h > a$	0.33	−34.1434	0.3410

1.3.3 Impact of h and W_p on the Effective Dielectric Constant (ϵ_{eff})

Table 1.5 presents the results of the effective dielectric constant (ϵ_{eff}). For the calculation of this parameter, we needed to adjust the value of h in 0.5 mm for 60 GHz, 0.32 mm for 67 GHz and 0.33 mm for 74 GHz. At this point, we were forced to make the first adjustment to obtain the results.

With the applied adjustment, we observe a crescent relation in the effective dielectric constant, in relation with each frequency, and a crescent gap between mathematical and adjustment values, while the resonance frequency is growing. The value of ϵ_{eff} was calculated using Eq. (1.3)

$$\epsilon_{\text{eff}} = \frac{\epsilon_r + 1}{2} + \frac{\epsilon_r - 1}{2}\left[1 + \frac{10h}{W_p}\right]^{-1/2} \tag{1.3}$$

Table 1.5 Theoretical values versus adjust values of ϵ_{eff}

Frequency (GHz)	Effective dielectric constant (ϵ_{eff}) (mm)		Gap (%)
	Mathematic (M)	Adjustment (A)	M versus A
60	3.52113	3.5208	0.01
67	3.6290	3.6856	1.54
74	3.5869	3.6951	2.93

Table 1.6 Effective length (L_e) with mathematical values and values obtained by adjustment

Frequency (GHz)	Effective length (L_e) (mm)		Gap (%)
	Mathematical (M)	Adjustment (A)	M versus A
60	1.3323	1.3324	0.01
67	1.1752	1.1662	0.77
74	1.0703	1.0545	1.48

1.3.4 Impact of f_o on Effective Length (L_e)

In Table 1.6, it can be appreciated the values for the effective length (L_e) with mathematical values and values obtained by adjustment.

While f_o rises, L_e decreases. In this parameter, there is a proportional relation between the values obtained with mathematical formulas and results obtained after we made the adjustment to the antenna's dimensions. For the calculation of mathematical values, it was used (1.4).

$$L_e = \frac{c}{2 * f_o\left(\sqrt{\epsilon_{\text{eff}}}\right)} \tag{1.4}$$

1.3.5 Impact of Substrate Thickness (h) and Effective Dielectric Constant (ϵ_{eff}) on Length Extension (ΔL)

In Table 1.7, it is presented the mathematical results of ΔL versus values obtained by adjustment. The theoretical values were obtained with (1.5).

$$\Delta L = 2\frac{h}{\sqrt{\epsilon_{\text{eff}}}} \tag{1.5}$$

where ϵ_{eff} is the effective dielectric constant and h is the substrate thickness.

Table 1.7 Length extension (ΔL), mathematical values versus adjustment values

Frequency (GHz)	Length extensión (ΔL) (mm)		Gap (%)
	Mathematical (M)	Adjustment (A)	M versus A
60	0.53292	0.53294	0.0004
67	0.3359	0.3334	0.74
74	0.3485	0.3433	1.49

Table 1.8 Mathematical values versus adjustment values for length of the patch (*L*)

Frequency (GHz)	Patch length (*L*) (mm)		Gap (%)
	Mathematical (M)	Adjustment (A)	M versus A
60	0.26646	0.26652	0.023
67	0.5034	0.4994	0.79
74	0.3733	0.3679	1.45

1.3.6 Impact ΔL and L_e on the Patch Length (*L*)

Table 1.8 presents the mathematical results for L versus the values obtained with the adjustment. The gap between mathematical values versus adjustment increases in direct relation to the increase of frequency. The calculation of the mathematical values was made with (1.6), where L_e is the effective length and ΔL is the length extension.

$$L = L_e - 2\Delta L \tag{1.6}$$

1.3.7 Impact of *h* and *L* on Substrate and Ground Length (L_g)

For the calculation of values of L_g, shown in Table 1.9, it was used Eq. (1.7), where h is thickness of the substrate and L is the length of the patch.

$$L_g = 6h + L \tag{1.7}$$

Table 1.9 Results of the length of the substrate and ground (L_g) with mathematical values and values obtained by adjustment

Frequency (GHz)	Length of substrate and ground (L_g) (mm)		Gap (%)
	Mathematical (M)	Adjustment (A)	M versus A
60	3.26646	3.26652	0.002
67	2.4234	2.4194	0.17
74	2.3533	2.3479	0.23

Table 1.10 Mathematical values versus adjustment values of width of substrate and ground

Frequency (GHz)	Width of substrate and ground (W_g) (mm)		Gap (%)
	Mathematical (M)	Adjustment (A)	M versus A
60	4.5215	4.52	0.04
67	3.2825	3.54	7.27
74	3.214	3.7	13.14

1.3.8 *Impact of Substrate Thickness (h) and Patch Width (W_p) on Substrate and Soil Width (W_g)*

The results detailed in Table 1.10, show that, regarding the mathematical values, when f_o increases, the width of the substrate and the ground decreases. On the contrary, with values obtained by adjustment, when the resonance frequency is between 60 and 67 GHz the width of substrate and ground goes down and when the resonance frequency is between 67 and 74 GHz the width of substrate and ground rises. In this parameter, we visualize the first important increases of gap while the frequency increase. To calculate the width of substrate and ground, was it used the expression (1.8).

$$W_g = 6h + W_p \tag{1.8}$$

1.4 Conclusions

To obtain the objective values of S_{11} and VSWR, the key adjustment was the variation of the width of substrate and ground and the width of the patch. Respect to the gap of W_g, it has a factor of approximately 7% of increase, while the frequency increases, and the gap of W_p has a factor of close to 15% of increase.

When the resonance frequency increases, the gap between the values of mathematical dimensions and adjustment dimensions of the antenna increases in a proportion of 0.7%, except for the values of the patch width and the values of the width of the substrate and ground.

It was evident that the values obtained by adjustment resulted in reflection coefficients with return losses below −20 dB, improving the designs by 21.47% respect to the mathematical designs, and concerning the VSWR, we achieved a deviation of 0.061% respect to the mathematical designs.

References

1. J. García Barranco, *Tecnologías 3G, 4G y 5G: Una Perspectiva Económica y Social de la Carrera por la Innovación de las Redes de Banda Ancha* (2019), pp. 1–45
2. R. Domínguez Vásquez, *Análisis y Diseño de Una Antena de Parche Microstrip* (2018), pp. 1–199
3. G. Pazmiño Rodríguez, *Análisis Comparativo de Antenas Tipo Patch con Tecnología Microstrip en la Banda X* (2018)
4. S. Núñez, C. Javier, *Caracterización de las Ondas Milimétricas Para Determinar su Posible Aplicación en la Quinta Generación de Comunicaciones Inalámbricas* (2018)
5. U. Farooq, G.M. Rather, Design and analysis of dual band microstrip antenna for millimeter wave communication applications. Int. J. Comput. Digit. Syst. **90**(4), 607–614 (2020). https://doi.org/10.12785/ijcds/090408

Chapter 2
Data Augmentation Techniques Applied to Improve a Vitiligo Database

Juan Chica, Christian Salamea, Erick Narvaez, and David Romero

Abstract Vitiligo is a pathology that causes the apparition of white patches in the skin, and there are multiple works where different techniques to diagnose, measure the quantity of the pathology, or to identify the existence of vitiligo. In this paper, we proposed a data augmentation technique based on superpixels to increase and optimize a dataset with 4723 images of skin with vitiligo, skin without vitiligo, background, and some other textures that could be used in real applications. Moreover, we also include binary vitiligo images where white represents pixels with vitiligo and black the other elements, and these images could be used in order to contribute to the research field of the vitiligo pathology.

2.1 Introduction

Vitiligo is a skin disorder that causes the apparition of white patches due to the absence or destruction of the melanin in the skin. This pathology affects about 1% of the world's population and without distinction of age, sex, or race [1, 2]. This pathology could be psychologically devastating to the person who has it. Negative feelings are anxiety, depression, and could even trigger suicidal thoughts [3, 4].

In the literature, there are works that propose techniques based on images of the vitiligo pathology to diagnose, measure objectively the repigmentation, or to identify whether or not is vitiligo [5–7]. The images that are used in the experiments have different conditions (Internet, hospitals, etc.) depending on the research field. In the Web, there are very few vitiligo image databases that can be used in scientific research: vitiligo image from Kaggle [8] and vitiligo images from DermNet NZ [9] are some of the ones that are available.

J. Chica
Universidad Politécnica Salesiana, Cuenca, Ecuador

C. Salamea (✉) · E. Narvaez · D. Romero
Grupo de Investigación en Interacción, Robótica y Automática, Universidad Politécnica Salesiana, Cuenca, Ecuador
e-mail: csalamea@ups.edu.ec

Á. Rocha et al. (eds.), *Communication, Smart Technologies and Innovation for Society*, Smart Innovation, Systems and Technologies 252,
https://doi.org/10.1007/978-981-16-4126-8_2

On the other hand, due to their great capabilities, deep learning (DL) models have been used to process medical images [10–12], but to prevent overfitting DL models generally require large amounts of training samples. Data augmentation is a technique that allows to generate new samples by applying different transformations (images are shifted, zoomed in/out, rotated, flipped, and/or distorted) over an original dataset in order to prevent model overfitting. In recent years, new approaches using generative adversarial networks (GANs) or superpixels segmentation have been proposed for data augmentation [13–15], and superpixels approach was shown to improve performance and reduce computational cost during image processing.

Therefore, this paper has two goals: first, shown a technique that could be used for data augmentation based on superpixels segmentation and padding. And second, present a database composed of different kind of images which contain skin with vitiligo, skin without vitiligo, background, and some other textures (that could be present in real clinical applications). We also included binarized images obtained by applying the proposed technique [10] where the pixels with vitiligo are labeled as 1 and healthy skin, textures, and background pixels are labeled as 0. In order to contribute to the research field of this pathology and subsequently postulate a normalized size database of images from where new discoveries can be obtained, the database was constructed from large areas body images such as the back, chest, hands, legs, and feet. From them, smaller normalized size images were obtained using superpixels segmentation. Thus, the new normalized images are mostly composed of similar pixels (e.g., normalized images contain only healthy skin or only skin with vitiligo or only textures and healthy skin, etc.) which allow to separate images composed of vitiligo from those that do not have it.

This paper is organized as follows: In Sect. 2.2, we briefly introduce existing image superpixels segmentation algorithms. In Sect. 2.3, the methodology used to construct the database is described. In Sect. 2.4, the proposed data augmentation technique is described, and the description of the presented dataset is shown, and finally, in Sect. 2.5, the conclusions are shown.

The vitiligo database is not freely accessible due to privacy policy in the data collection, but it can be used for research purposes. In order to obtain more information about the database, contact csalamea@ups.edu.ec.

2.2 Theoretical Background

In the area of image processing, it is very important to extract the most relevant information (feature extraction) in an image. Segmentation is a technique that allows us to perform this task and consists in splitting an image in a group of objects or pixels. Thereby, segmentation consists of assigning a label to each pixel according to its characteristics (color, texture, etc.). In 2003, Ren and Malik introduced the concept of superpixels [16] that consists of group up pixels which have similar properties, reducing the size of input features in the subsequent algorithms and allows algorithms to be more feasible computationally. Thus, in order for superpixels to be useful, certain requirements must be considered, such as simple to use, production of high-quality segmentation, it must be fast, consider the ability to control the size, number, and compactness of the segments (superpixel) [17]. In this section, we briefly introduce different algorithms of superpixels segmentation based on the categorization done by Achanta [18] according to their high-level approach.

2.2.1 *Watershed*

Watershed segmentation is a region-based in watershed. There are three kinds of points to consider: points corresponding to local minimum, points in which, if a drop of water is placed, with certainty it falls at a single minimum and points at which water would fall with equal probability at more than one minimum [19, 20]. Also, this technique uses images morphology and requires selecting at least one marker or seed point in the interior of each object in the image. These markers could be chosen manually or automatically using previous knowledge about the objects in the image. The basic idea of this algorithm is relatively simple, suppose there is a hole in each regional minimum and then the entire topography (3D image) is flooded from below by letting water rise through the holes at a uniform rate. When the water in different zones is about to merge, a dam is constructed to avoid the merge. The flooding continues until only the tops of the dams are visible. The dam boundaries correspond to the divide lines of the watersheds thus giving continuous borders between regions. The principal objective of the segmentation algorithms based on watershed is to find the watershed lines.

2.2.2 *Graph-Based Algorithms*

In these algorithms, the image is processed as a graph, and each pixel is treated as a node in the graph, also there is some connectivity between pixels and some connectivity properties like distances, similarity functions, etc. Based on these properties, the algorithm iterates in the graph dropping the less similar connections and maximizing the strongest. The superpixels are extracted minimizing a cost function defined on the graph. Algorithms based on this approach are Felzenszwalb and Huttenlocher [21] which use an agglomerative clustering of pixel nodes on a graph, normalized cuts [20] which uses contour, and texture cues to recursively partition a graph among others.

2.2.3 Clustering-Based Algorithms

These algorithms are based on clustering such as *k*-means. Here, the initialization is in function of information about color, spatial information, or depth, and it is initialized by a seed of pixels. Among all the algorithms of this type, the simple linear iterative clustering (SLIC) outperforms existing segmentation methods [17] and is a simple and efficient algorithm. In SLIC, two parameters are taken: region size that is the nominal size of the superpixels and regular size that is strength of the spatial regularization besides each pixel is associated to a feature vector and *k*-means clustering is applied on those. SLIC algorithm partitions an image in homogeneous regions starting from the image divided into a grid with step region size. Then, using the center of each grid, the *k*-means are initialized. The last step is performed using the Lloyd algorithm; thus, optimizing the centers in the *k*-means, we obtain the segmented images.

2.3 Material and Methods

The procedure to construct the database was: first, configure the camera, instruct the patient to not make sudden movements (this can generate distortion in the focus of the lens and produce an erroneous image of the injured area), and place it in a position which is possible to observe the lesion. Then, the photograph was taken. It should be noted that, the skin area to be photographed was selected by dermatology specialists in order to the images contain skin with vitiligo and skin without vitiligo. The camera used is a Nikon D7000 configured with exposure time 1/250 s, ISO-100 (ISO speed), auto-capture using flash, priority to quality, and f/5.5 (aperture F). These parameters allow us to control certain factors that directly or indirectly affect the demarcation of the area affected by vitiligo, such as variations in brightness, camera angle, and skin tone [22].

Sixty images of skin were selected (average size of 1200×1000 pixels) to construct the database. The main idea is to divide the images into small images (as regular as possible in size) which are mostly composed of similar pixels (e.g., normalized images contain only healthy skin or only skin with vitiligo or only textures and healthy skin, etc.). Thus, we test three techniques based on superpixels segmentation: SLIC, Felzenszwalb, watershed, and the techniques with the best performance were used.

2.3.1 Patients

The study protocol has been approved by the Research Coordination Center of the Jose Carrasco Arteaga Hospital in Cuenca, Ecuador. Twenty patients from this institution were selected from whom the photographs where obtained to be used in the database (three photographs were obtained from each patient). The inclusion criteria of patients who participated in the study are described below:

- Signed letter of informed consent.
- Type II or type III skin according to the Fitzpatrick scale [23].
- Ages between 18 and 55 years.
- Did not have any type of burn (sunburn and phototherapy) near the photographed area.

2.4 Database Description

2.4.1 Data Augmentation

In this subsection, we show the results of the experiments using the superpixels techniques. In the SLIC segmentation, we obtained an average of 23 segments per image and a total of 1475 segments of all the 60 images. An example of the segmentation using SLIC is shown in Fig. 2.1.

On the other hand, by implementing the Felzenszwalb segmentation technique, a total of 6028 segments and an average of 100 segments per image were obtained. An example of the segmentation using Felzenszwalb technique is shown in Fig. 2.2.

Finally, through the implementation of the watershed segmentation technique, we obtained an average of 248 segments per image making and a total of 14,900 segments. It should be noted that the segments were obtained using automatic seed

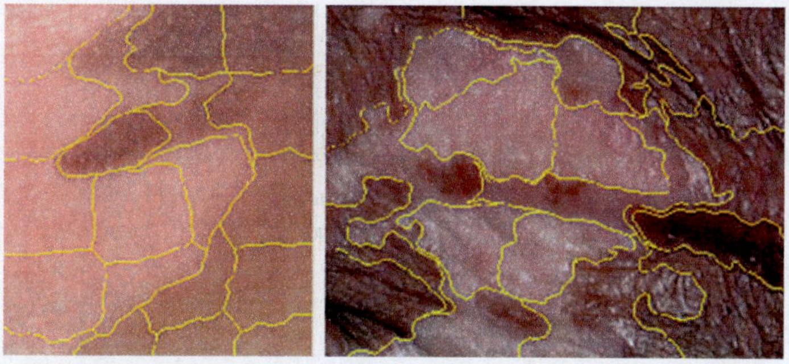

Fig. 2.1 SLIC segmentation in vitiligo images

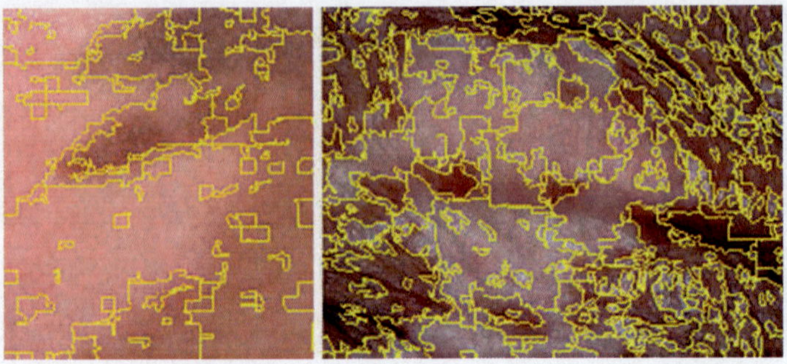

Fig. 2.2 Felzenszwalb segmentation in vitiligo images

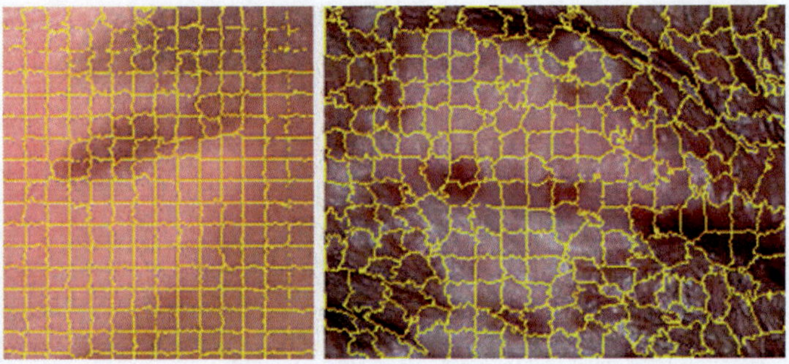

Fig. 2.3 Watershed segmentation in vitiligo images

points based on the relation of the interaction between light and skin described in [10]. An example of the segmentation using watershed technique is shown in Fig. 2.3.

It can be noted that both in Felzenswalb algorithm (Fig. 2.2) and in SLIC algorithm (Fig. 2.1), the size of the segments is highly variable; some segments are bigger in width, other bigger in height, etc. On the other hand, using watershed segmentation, we obtain segments more regular and that all the segments contain majority pixels of similar characteristics (only healthy skin, only background, etc.). Thus, watershed segmentation is selected to obtain images of similar pixels.

In order to develop the data augmentation, we propose to use the images obtained from the watershed segmentation and then add zero padding (add zeros around the segments), resulting into a normalized image of 100×100 pixels where the segment obtained from watershed is in the center. Due to the zero padding, we discard the images where the segments represent less than the 50% of the pixels in the final image (other 50% are zeros), resulting in a total of 4723 images. In Fig. 2.4, an example of images after data augmentation is shown.

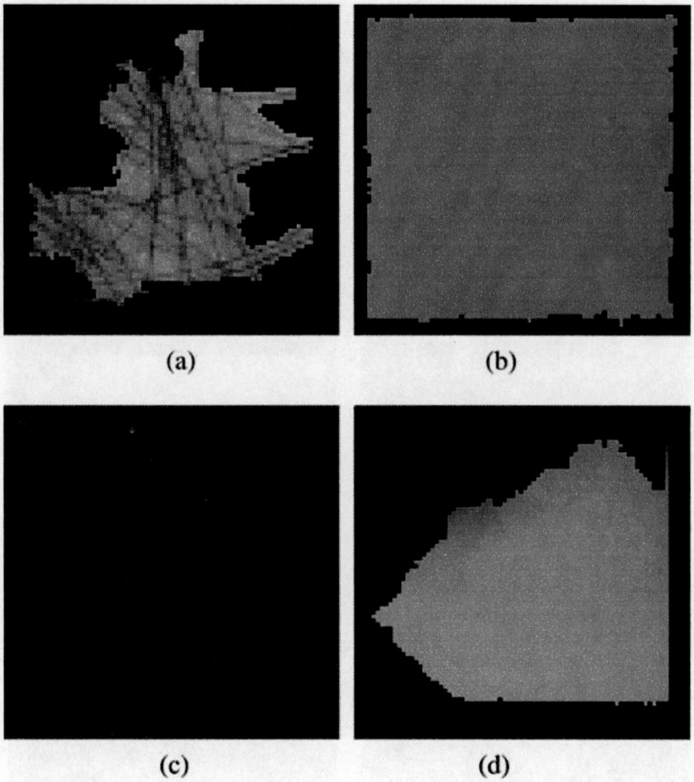

Fig. 2.4 Example of images in the database: **a** skin with hair, **b** normal skin, **c** black background, **d** different textures

2.4.2 Database Description

Vitiligo database contains RGB information of different kind of images such as skin with vitiligo, skin without vitiligo, background, and other textures (cloth) which represent conditions that could be present in real applications, making a total of 4723 images of 100 × 100 pixels in PNG format. In Table 2.1, a summary of the characteristics of the vitiligo database is shown.

Table 2.1 Vitiligo database description

Database size	9446 (4723 RGB and 4723 binarized images)
Images format	PNG
Images size	100 × 100 pixels
Space color	RGB and binarized images
Images types	Skin with vitiligo, skin without vitiligo, background, Textures

On the other hand, an example of the different types of images that compose the database can be observed in Fig. 2.4. Additionally, we present binary images of the database which are segmented on: skin with vitiligo and skin without vitiligo. The proceeding to obtain these images is: firstly, remove pixels which did not contain skin, and then we use the method proposed in [10] to segment the skin, resulting in images where the area affected by vitiligo is white (Fig. 2.5) and the remaining elements is black (including pixels removed). Thus, by each one of the 4723 images there is a corresponding binary image, adding a total of 9446 images to the vitiligo database.

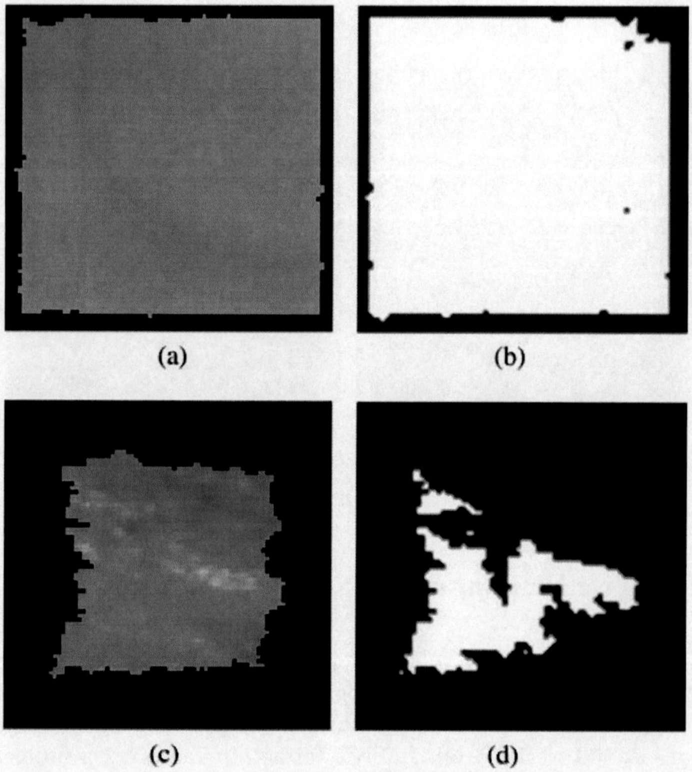

Fig. 2.5 Example of images in the database: **a** skin with hair, **b** normal skin, **c** black background, **d** different textures

2.5 Conclusions

We used three different techniques to segment the images, the watershed is the one that obtains the most regular segments with respect to the size. We think that using techniques like superpixels to data augmentation could prevent the overfitting in deep learning models.

We provide a dataset of medical images of the vitiligo pathology and the different elements that could exist in real applications like skin (healthy, with vitiligo, with hair), background, and different textures (cloth). Moreover, we also include binary images where white represents pixels with vitiligo, and black represents the other elements in the image. The main idea is that the scientific community is able to access and use this database to increase research regarding vitiligo pathology and help improve the field of dermatology. Due to the fact that the quantity of images would be limited in the generality of the databases, we proposed a data augmentation technique using superpixels. We also think that this work could be a starting point toward similar works.

Acknowledgements The authors would like to thank the patients who participated in this study. Authors greatly acknowledge the Universidad Politecnica Salesiana and the José Carrasco Arteaga Hospital by the openness and support to carry out this research.

Author Contributions JC and CS conceived this research and designed methodology. JC collect the database. EN and DR performed the experiments and process the database. JC and CS wrote the paper and participated in the revisions of it. All authors read and approved the final version of the manuscript.

Conflicts of Interest The authors declare that they have no potential conflict of interest in relation to the study in this paper.

References

1. R. Yaghoobi, M. Omidian, N. Bagherani, Vitiligo: a review of the published work: vitiligo. J. Dermatol. **38**(5), 419–431 (2011)
2. A. Alikhan, L.M. Felsten, M. Daly, V. Petronic-Rosic, Vitiligo: a comprehensive overview. J. Am. Acad. Dermatol. **65**(3), 473–491 (2011)
3. G. Schmid-Ott et al., Stigmatization experience, coping and sense of coherence in vitiligo patients. J. Eur. Acad. Dermatol. Venereol. JEADV **21**(4), 456–461 (2007)
4. D. Parsad, S. Dogra, A.J. Kanwar, Quality of life in patients with vitiligo. Health Qual. Life Outcomes **1**, 58 (2003)
5. M.H.A. Fadzil, S. Norashikin, H.H. Suraiya, H. Nugroho, Independent component analysis for assessing therapeutic response in vitiligo skin disorder. J. Med. Eng. Technol. **33**(2), 101–109 (2009)
6. A.F.M. Hani, H. Nugroho, N. Shamsudin, R. Baba, Melanin determination using optimised inverse Monte Carlo for skin—light interaction, in *4th International Conference on Intelligent and Advanced Systems* (*ICIAS*), vol. 1 (2012), pp. 314–318

7. A. Nurhudatiana, A computer-aided diagnosis system for vitiligo assessment: a segmentation algorithm, in *Intelligence in the Era of Big Data*, vol. 516, ed. by R. Intan, C.-H. Chi, H.N. Palit, L.W. Santoso (Springer, Berlin, Heidelberg, 2015), pp. 323–331
8. Vitiligo Images, https://kaggle.com/shaikhshahid/vitiligo-images. Last accessed 03 Mar 2021
9. Vitiligo images | DermNet NZ, https://dermnetnz.org/topics/vitiligo-images/. Last accessed 03 Mar 2021
10. J.F. Chica, S. Zaputt, J. Encalada, C. Salamea, M. Montalvo, Objective assessment of skin repigmentation using a multilayer perceptron. J. Med. Signals Sens. **9**(2), 88 (2019)
11. P. Dutta, P. Upadhyay, M. De, R.G. Khalkar, Medical image analysis using deep convolutional neural networks: CNN architectures and transfer learning, in *2020 International Conference on Inventive Computation Technologies* (*ICICT*) (2020), pp. 175–180
12. Q. Li, W. Cai, X. Wang, Y. Zhou, D.D. Feng, M. Chen, Medical image classification with convolutional neural network, in *2014 13th International Conference on Control Automation Robotics Vision* (*ICARCV*) (2014), pp. 844–848
13. A. Antoniou, A. Storkey, H. Edwards, *Data Augmentation Generative Adversarial Networks*. ArXiv171104340 Cs Stat (2018)
14. C. Qi, J. Chen, G. Xu, Z. Xu, T. Lukasiewicz, Y. Liu, *SAG-GAN: Semi-Supervised Attention-Guided GANs for Data Augmentation on Medical Images*, ArXiv201107534 Cs Eess (2020)
15. Y. Zhang et al., *SPDA: Superpixel-Based Data Augmentation for Biomedical Image Segmentation*, ArXiv190300035 Cs (2019)
16. X. Ren, J. Malik, Learning a classification model for segmentation, in *Proceedings Ninth IEEE International Conference on Computer Vision*, vol. 1 (Nice, France, 2003), pp. 10–17
17. R. Achanta, A. Shaji, K. Smith, A. Lucchi, F. Pascal, S. Sabine, SLIC_superpixels, EPFL technical report 149300 (2010)
18. R. Achanta, A. Shaji, K. Smith, A. Lucchi, P. Fua, S. Süsstrunk, SLIC superpixels compared to state-of-the-art superpixel methods. IEEE Trans. Pattern Anal. Mach. Intell. **34**(11), 2274–2282 (2012)
19. D. Stutz, A. Hermans, B. Leibe, Superpixels: An evaluation of the state-of-the-art. Comput. Vis. Image Underst. **166**, 1–27 (2018)
20. J. Shi, J. Malik, Normalized cuts and image segmentation. IEEE Trans. Pattern Anal. Mach. Intell. **22**(8), 888–905 (2000)
21. P.F. Felzenszwalb, D.P. Huttenlocher, Efficient graph-based image segmentation. Int. J. Comput. Vis **59**(2), 167–181 (2004)
22. I. Hamzavi, H. Jain, D. McLead, J. Shapiro, H. Zeng, H. Lui, Parametric modeling of narrow-band UV-B phototherapy for vitiligo using a novel quantitative tool: the vitiligo area scoring index. Arch. Dermatol., 677–683 (2004)
23. T.B. Fitzpatrick, The validity and practicality of sun-reactive skin types I through VI. Arch. Dermatol. **124**(6), 869–871 (1988)

Chapter 3
Bacteria Identification in Machangara River Water Capable of Metabolizing Emerging Estrone Pollutant

Gabriela Méndez, Karla Vásquez, and Elena Coyago

Abstract Estrone (E1) is an important emerging pollutant due to its significant prevalence in the environment and the association with estrogenic activity, causing ecological effects such as the feminization of aquatic organisms. Although E1 is degraded in wastewater treatment, the elimination is not complete; therefore, biological treatments could be an alternative. Bacteria such as *Nitrosomonas europea, Pseudomonas* spp., *Sphingomonas* spp., and *Sphingobacterium* spp. under aerobic conditions oxidize E1 because of their ability to use estrogens as a source of carbon and energy, producing smaller and less toxic products. For this reason, this study aimed to isolate and identify bacteria from wastewater of two areas with this pollutant in the Machángara river located in Quito-Ecuador, capable of using the emerging pollutant estrone as the only carbon source. *Pseudomonas putida* was identified in this river as a microorganism that uses estrone as the only carbon source, obtaining an estrone consumption of 100% after 120 h of exposure with 5% v/v of inoculum. These results can be used as the basis for future biodegradation assays of this pollutant in water sources.

3.1 Introduction

Knowing the chemical and ecological cycles of water is essential to determine its quality, considering it is a vital source for preserving all living beings. Contaminants in water sources are very diverse in origin and composition including emerging contaminants. These contaminants are any synthetic or natural chemical or any microorganism that are not commonly monitored or regulated in the environment and with ecological and human potential to cause harm, including health effects [1]. Emerging pollutants are a global problem due to the lack of a broad knowledge of their effects, little regulation of these pollutants, or regulatory processes that are still ongoing [2].

G. Méndez (✉) · K. Vásquez · E. Coyago
Universidad Politécnica Salesiana, Quito, Ecuador
e-mail: gmendez@ups.edu.ec

Á. Rocha et al. (eds.), *Communication, Smart Technologies and Innovation for Society*, Smart Innovation, Systems and Technologies 252,
https://doi.org/10.1007/978-981-16-4126-8_3

When present in surface and subsurface waters and even in drinking water [2], emerging contaminants represent a risk to the health of humans and aquatic organisms. These pollutants have as their primary source: urban, agricultural, and industrial wastewater and waste from wastewater treatment [3].

The most studied emerging pollutants are industrial, natural, synthetic estrogens, phytoestrogens, phytosterols, pesticides, pharmaceutical products, and personal care products. The concentrations of these pollutants can vary from 1 ng/L up to values higher than 0.03 mg/L. Estrone (E1) is one of the concurrent organic pollutants in contaminated waters in quantities up to 0.0001 mg/L, being human urine the primary source of these natural and synthetic estrogens. Some studies indicate that estrone at 0.1–1 ng/L can cause reproductive disorders as a sex reversal in small male fish in aquatic species [4, 5].

In Ecuador, rivers such as the Guayllabamba, Machángara, and Esmeraldas present emerging pollutants such as acesulfame (ACS), carbamazepine (CBZ) and its metabolites, the antibiotic sulfamethoxazole (SMX), the psychiatric drug known as duloxetine (Prozac, FLX), venlafaxine (VNF), steroid estrogens (NF) 17a-ethinylestradiol (EE2), 17 b-ethinylestradiol (E2) and estrone (E1) [6]. Estrone in the Machángara river has been identified with concentrations between 7.4 and 11.4 μg/L, showing the highest values compared with other American countries as Brazil with concentrations of 0.003 μg/L in wastewater and 0.7 μg/L in freshwater [7, 8]. This contaminant has affected the aquatic environment, altering the reproductive development of fauna [9].

Estrone has been incompletely degraded in purification plants by treatments such as nanofiltration, oxidation, and activated carbon absorption. So, the research for a biological alternative with the identification of microorganisms capable of using estrone as a carbon source could be an alternative for the degradation of this pollutant [10].

Microbial biodegradation is one of the most suitable technologies for eliminating environmental estrogens, such as estrone, due to its high efficiency, low cost, and because it does not cause secondary contamination. Some strains that degrade estrogens have been identified in other countries such as the genera *Esfingomonas, Nitrosomonas,* and *Rodococo* [11].

There are few studies in Latin America related to the concentration of emerging pollutants as studies carry out by Peña-Guzman et al. and Reichert et al. [7, 8]; however, there is no information about microorganisms capable of degrading estrone for a bioremediation application. This research aimed at the isolation and molecular identification of wastewater bacteria from the Machángara River located in Quito-Ecuador, capable of using estrone as the only carbon source.

3.2 Materials and Methods

3.2.1 Reagents

Pure estrone (99% concentration) from Sigma-Aldrich (USA) was used.

The broth culture used for the isolation of bacteria was basal mineral medium (MBM) composed of 3.5 g of K_2HPO_4, 1.5 g of KH_2PO_4, 0.5 g of $(NH_4)_2SO_4$. 0.5 g of NaCl, 0.15 g of $MgSO_4(7H_2O)$ and 1 mL of trace elements [12].

3.2.2 Sampling Collection Area

Sampling areas were selected based on the study of Voloshenko-Rossin et al. [6], which demonstrated the existence of estrone in these areas. Sampling was carried out in September 2018 at two points on the Machángara River in Quito-Ecuador. The first area was in Pusuquí (coordinates: 0°, 5′ 20.3″ S; 78°, 29′ 11.86″ W) at 2644 masl, and the second area was in Pomasqui (coordinates: 0°, 3′ 5442″ S; 78°, 27′, 6432″ W) at 2439 masl. The water samples were collected in 1 L sterile amber glass bottles, occupying three-quarters of its volume. Samples were transported in a cooler at a temperature of 4 °C and were sent to external laboratories to quantify physical, chemical, and microbiological parameters.

3.2.3 Physical, Chemical, and Microbiological Analyzes

The water collected was transferred to the "Soil and Water Laboratory of the Salesian Polytechnic University" in Cayambe. Biological oxygen demand (BOD), chemical oxygen demand (COD), pH, conductivity, total nitrogen, quantification of total and fecal coliforms were determined as established by the APHA protocol of standard methods for drinking and wastewater analysis [13].

To evaluate water sample quality, the data obtained were compared with the permitted limits established by Ecuadorian law in the Secondary Text of the Legislation of the Ministry of the Environment (TULSMA), Book IV, Annex 1, Tables 3 and 12 [14].

3.2.4 Isolation and Identification of Bacteria Strains

This process consisted of a previous acclimatization stage, which included a phase of exposure of the microorganisms to estrone. Two Erlenmeyer's were autoclaved with 100 mL of mineral basal medium (MBM), in which a stock solution of estrone

(100 mg/L) was added in a 1:10 relation. Water samples were inoculated at 2% v/v and incubated for 15 days, 26 °C and 150 rpm until turbidity of the medium was observed [12]. Finally, several dilutions of acclimated samples were planted with the plate count method in Petri dishes with MBM and estrone (100 mg/L) at 28 °C. Colonies were replicated until pure cultures were obtained for their identification.

Preliminary identification of the strains was carried out by visualization of the macroscopic characteristics of the colonies, considering parameters such as color, shape, and elevation. Gram staining, catalase, and oxidase biochemical tests were performed with colonies of different morphology.

Finally, DNA extraction was, followed by the protocol indicated by the provider of the Thermo Fisher Scientific "Pure link Genomic DNA Mini kit", and DNA was verified in the samples obtained by electrophoresis at 80 V for 45 min. The gel obtained was analyzed in the MicroQBis Photo processor.

After DNA extraction, PCR was performed with an initial denaturation at 95 °C for 5 min and at 95 °C for 30 s, with a melting temperature of 56 °C for 30 s, an extension of 72 °C for 1 min, and a final extension of 72 °C for 10 min (9, 17). In addition, the Thermo Fisher Scientific GoTaq® Green Master Mix was used, with a final volume of 25 μL, with primer forward 27 F (5′-AGA GTTTGATCCTGGCTCAG-3′) and primer reverse 1492R (5′-TACCTTGTTACGATT-3′).

Subsequently, the DNA extracted was quantified using Thermo Fisher Scientific's Qubit 2.0 Fluorometer to determine whether the appropriate concentration was obtained for sequencing. Sequencing was performed using Illumina technology at Molecular Research DNA LAB (MRDNAlab) in Texas, USA.

3.2.5 Batch Assays for Evaluation Estrone Consumption

The identified bacteria were inoculated in MBM with the estrone pollutant (25 mg/L) at 28 ± 2 °C for 24 h in darkness for preparing the inoculum. At 24 h, several colonies were taken, and a bacterial suspension was prepared at a 0.5 McFarland scale (1.5×10^8 CFU/mL). For the batch assays, nine Erlenmeyer's of 250 ml were prepared, with MBM and estrone at a concentration of 25 mg/L, and the identified bacteria (*Pseudomonas putida*) was added to a concentration of 2 and 5%. It was incubated at 28 ± 2 °C in darkness, with the agitation of 150 rpm for 120 h. The microorganism growth and estrone concentration were evaluated over time to determine if the bacterium grows in the presence of this contaminant.

For determining the estrone concentration, UV–Vis spectroscopy was used at a wavelength of 282.5 nm at Jasco V-730 spectrophotometer (Thermo Fisher Scientific Inc. Waltham, MA, USA). Before quantifying samples, a calibration curve was made from a stock solution of 40 mg/L estrone dissolved in methanol and the following standards: 30, 20, 10, and 5 mg/L [12].

The percentage of estrone consumption at the time of each assay was calculated by using Eq. (3.1):

$$= (\text{Initial concentration} - \text{final concentration})/(\text{Initial concentration}) \times 100\% \quad (3.1)$$

For the quantification of bacteria growth, a calibration curve was performed considering the colony forming units per milliliter (CFU/mL) and the optical density measured at an absorbance of 625 nm. Samples from each treatment were measured using the Jasco V-730 spectrophotometer (Thermo Fisher Scientific Inc. Waltham, MA, USA), and in addition, each sample was counted using the pour-plate technique.

The statistical analysis included a completely randomized design with three treatments and three replications for each treatment with three controls. The statistical analysis was performed according to ANOVA with 5% Tukey, using the InfoStat software version 2008.

3.3 Results and Discussion

3.3.1 *Physical, Chemical, and Microbiological Analysis*

Physical and chemical characterization of polluted water shown in Table 3.1 refers to the water quality of the samples used for microbial isolation. In terms of BOD, COD, and pH analyzes, it was observed that the two water samples met the levels allowed by the Ecuadorian laws [13]. The results obtained were like those presented by [6], who obtained pH values of 7.6. The two samples have a slight restriction categorization degree, regarding the conductivity of water, as shown in Table 3.1.

Table 3.1 Average values of physical, chemical, and microbiological parameters of contaminated waters

Analysis	Units	Area 1	Area 2	Maximum permissible limits
COD	mg/L	160.0	112.0	250.0[a]
BOD	mg/L	75.6	62.4	100.0[a]
pH	Adimensional	8.1	7.9	6.5–9.0[a]
Conductivity	µS/cm	886	688.3	≤70 µS/cm: no restriction on water use 700–3000 µS/cm light restriction 3000 µS/cm moderate >3000 µS/cm severe restriction[a]
Total nitrogen	mg/L	47.2	30.5	15.0[a]
Total coliforms	nmp/100 mL	8,400,000	16,000,000	3000[a]
Fecal coliforms	(mp/100 mL	4,000,000	1,800,000	600[a]

[a]Ecuadorian legislation, secondary text of the legislation of the Ministry of the Environment (TULSMA), Book IV, Annex 1

This value was much higher than those obtained in previous years, where the values observed were around 0.7 mS [6].

In the two contaminated water samples, total nitrogen exceeded the allowable limits in the Ecuadorian law (15 mg/L). These results were similar to those reported by [6], where nitrogen values of 31.3 mg/L were obtained.

Thus, the chemical-physical parameters evaluated in this study are associated with changes in water characteristics, which could help the growth of pathogenic microorganisms responsible for alterations in the health of living beings [15].

For total coliforms and fecal coliforms, the limits established in the regulations were exceeded in both sampling areas, maybe because most of the wastewater from Quito is discharged directly to the Machángara river [16]. Total coliforms are used to indicate other pathogens such as viruses, protozoa, helminths, tenias, and intestinal worms. The predominant Fecal coliform is *Escherichia coli*, currently known by the World Health Organization (WHO) as the best indicator of bacteria to monitor fecal contamination of drinking water. Some strains of *E. coli* can cause quite severe diseases and eventually death [17].

3.3.2 Isolation and Identification of Strains

Once water samples were added in the MBM culture medium with estrone as the only carbon source, only one type of colony could be identified, showing that this bacterium was the only one that uses estrone for its growth (see Table 3.2).

Gram stain of the bacteria showed a gram-negative bacillus [18]. In addition, the oxidase test showed a positive result, which concluded that it is a facultative aerobic microorganism [19]; and the catalase test with a positive result indicated that this microorganism could catalyze hydrogen peroxide into water and gaseous oxygen [20]. Based on the results obtained and, on the information presented in the literature, it was determined that the isolated organism belongs to the genre *Pseudomonas* spp., which is a gram-negative, aerobic, mobile bacterium [21, 22].

Table 3.2 Macroscopic, microscopic, and biochemical characteristics of the isolated bacterium

Macroscopic characteristics	Photo of the isolated strain	Microscopic characteristics	Catalase test	Oxidase test
Creamy white colonies with margin		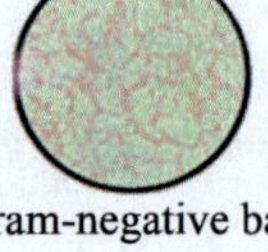 Gram-negative bacillus	 Positive	 Positive

As a result of the molecular identification, it was determined that the bacterium isolated from the Machángara river corresponded to *Pseudomonas putida*, with an identity percentage of 99.758% (Accession number: MW691111). According to [23], *Pseudomonas putida* is a rapidly growing bacterium found in most temperate soils and aquatic habitats where oxygen and high amounts of organic matter are present, related to contaminated water; therefore, it is considered a microorganism that indicates contamination and it is widely used in bioremediation processes of xenobiotic compounds [15].

3.3.3 Evaluation of Estrone Consumption as a Carbon Source

A spectrophotometric calibration curve was performed to calculate *Pseudomonas putida* estrone consumption, as seen in Fig. 3.1. This curve showed an R2 of 0.9998, which implies a good correlation between the absorbance and the concentration of the standard.

In batch assays, *Pseudomonas putida* consumes 43.6% of estrone at 120 h, with 2% v/v of inoculum, whereas when applying 5% v/v, consumption of 100% was achieved (see Table 3.3); thus, the consumption was attributed to the microorganism since the estrone concentration did not decrease in the controls.

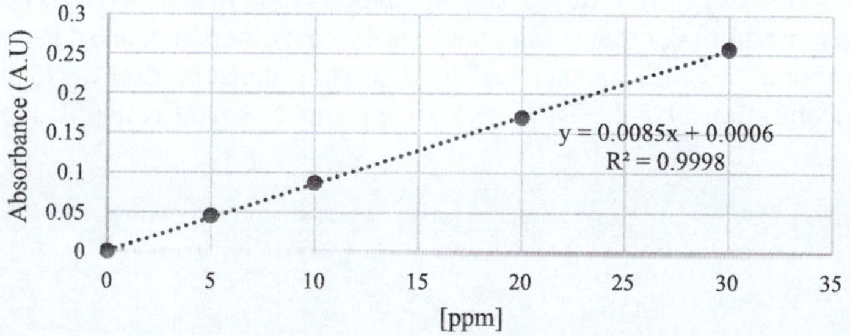

Fig. 3.1 Estrone calibration curve with standards of 30, 20, 10, and 5 ppm at 282.5 nm

Table 3.3 Average values for estrone consumption

<table>
<tr><th>28 °C</th><th>Treatments</th><th>Range</th><th>% estrone</th><th>p-value</th><th>CV</th><th>Significance</th></tr>
<tr><td rowspan="3">120 h and 25 ppm</td><td>T1 Pseudomonas putida 2%</td><td>B</td><td>43.6</td><td rowspan="3"><0.0001</td><td rowspan="3">0.46</td><td rowspan="3">$p < 0.05$</td></tr>
<tr><td>T2 Pseudomonas putida 5%</td><td>A</td><td>100</td></tr>
<tr><td>T3 control</td><td>C</td><td>0</td></tr>
</table>

Results indicate microorganism capacity to use estrone as a carbon source, which implies an important use in bioremediation treatments. These results are like other studies that have identified that the *Pseudomonas* spp. can develop in a medium with estrone as the only carbon source, as observed by [24], who identified *Pseudomonas putida* SJTE-1 with the ability to use estrone (E1) with a 90% consumption and a 5% inoculum concentration. Another study identified the consumption of E1 and EE2 at 55% with 2% v/v of *Pseudomonas aeuroginosa* [4]. Finally, a study found that microorganisms such as *Pseudomonas taiwanensis, Rhizobium radiobacter, Sphingopyxis alaskensis* can use E1, E2, and EE2 as the only carbon and energy sources for their growth [25].

3.3.4 Bacteria Quantification

The results showed that *Pseudomonas putida* at 5% started to consume the estrone at 24 h, which agreed with the increase in the number of CFU/mL. In contrast, at 72 h ended using the estrone, which is proved by the decrease in CFU/mL. In addition, in *Pseudomonas putida* at 2%, it was observed that when the number of microorganisms decreased by about 100 h, the highest estrone consumption was obtained (43.6%), as shown in Fig. 3.2.

According to [24], the period of estrone consumption by Pseudomonas putida SJTE-1 strain was 72 h and 120 h with an initial concentration of 1 × 108 cells/mL, for the consumption of 10 mg/L of estrone, which coincides with the analysis performed in this study, in addition, other studies have shown that others species of Pseudomonas as Pseudomonas aeruginosa that had total estrone consumption at 75 h with an initial concentration of 3.2 ± 0.2 mg/L of E1 and 1 × 109 cells/mL [4]. Finally,

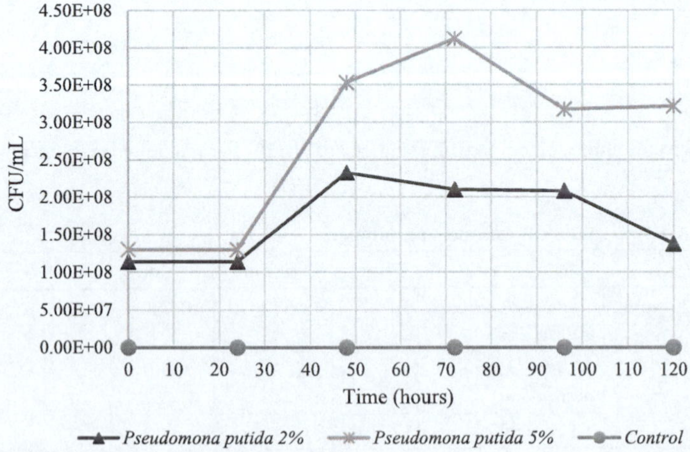

Fig. 3.2 *Pseudomonas putida* growth kinetics

Pseudomonas taiwanensis strain reduced E1–E2 over a 20-day period, where cell counts reached ~6.0 × 108 cells/mL [25]. The above may serve as an indication to say that *Pseudomonas putida* is more effective for estrone consumption.

3.4 Conclusions

Estrone is an emerging pollutant that can cause different adverse effects on ecosystems. On the Machángara river, some microorganisms consume estrone, which was determined by the growth of colonies in a minimal medium with estrone as the only carbon source, and in the batch assays of this study. *Pseudomonas putida* under aerobic conditions was identified as capable of consuming estrone by 100%, using an inoculum of 5% v/v with an initial concentration of 25 ppm. As a micro-contaminant of wastewater, estrone can generate sub-products that can be harmful in aquatic environments; therefore, *Pseudomonas putida* identified in this study may consume pollutants that could help reduce adverse effects on the ecosystem.

References

1. S. Vaz, What are emerging pollutants (EPs) and their fate on the environment, in *Analytical Chemistry Applied to Emerging Pollutants* (2018), pp. 1–13. https://doi.org/10.1007/978-3-319-74403-2_1
2. A.A. Khan, R. Ahmad, I. Ahmad, Density functional theory study of emerging pollutants removal from water by covalent triazine based framework. J Mol Liq 309 (2020) https://doi.org/10.1016/j.molliq.2020.113008
3. S. Mokrane, D. Harik, H. Cabana, *Identification of Emerging Contaminants in Drinking Waters*, pp. 785–787 (2018). https://doi.org/10.1007/978-3-319-70548-4_230
4. B. Pauwels, K. Wille, H. Noppe, H. De Brabander, T. Van De Wiele, W. Verstraete, N. Boon, 17A-ethinylestradiol cometabolism by bacteria degrading estrone, 17B-estradiol and estriol. Biodegradation **19**, 683–693 (2008). https://doi.org/10.1007/s10532-007-9173-z
5. A. Pratush, Q. Yang, T. Peng, T. Huang, Z. Hu, Identification of non-accumulating intermediate compounds during estrone (E1) metabolism by a newly isolated microbial strain BH2-1 from mangrove sediments of the South China Sea. Environ. Sci. Pollut. Res. **27**, 5097–5107 (2020). https://doi.org/10.1007/s11356-019-06894-1
6. A. Voloshenko-Rossin, G. Gasser, K. Cohen, J. Gun, L. Cumbal-Flores, W. Parra-Morales, F. Sarabia, F. Ojeda, O. Lev, Emerging pollutants in the Esmeraldas watershed in Ecuador: discharge and attenuation of emerging organic pollutants along the San Pedro–Guayllabamba–Esmeraldas rivers. Env. Sci. Process. Impacts **17**, 41–53 (2015). https://doi.org/10.1039/C4EM00394B
7. C. Peña-Guzmán, S. Ulloa-Sánchez, K. Mora, R. Helena-Bustos, E. Lopez-Barrera, J. Alvarez, M. Rodriguez-Pinzón, Emerging pollutants in the urban water cycle in Latin America: a review of the current literature. J. Environ. Manage. **237**, 408–423 (2019). https://doi.org/10.1016/j.jenvman.2019.02.100
8. G. Reichert, S. Hilgert, S. Fuchs, J.C.R. Azevedo, Emerging contaminants and antibiotic resistance in the different environmental matrices of Latin America. Environ. Pollut. 255 (2019). https://doi.org/10.1016/j.envpol.2019.113140

9. T.K.A. Tran, R.M.K. Yu, R. Islam, T.H.T. Nguyen, T.L.H. Bui, R.Y.C. Kong, W.A. O'Connor, F.D.L. Leusch, M. Andrew-Priestley, G.R. MacFarlane, The utility of vitellogenin as a biomarker of estrogenic endocrine disrupting chemicals in molluscs. Environ. Pollut. **248**, 1067–1078 (2019). https://doi.org/10.1016/j.envpol.2019.02.056
10. D.P. Moser, B.P. Hedlund, M.R. Rosen, S.M. Blunt, M.J. Benotti, Reversible reduction of estrone to 17β-estradiol by *Rhizobium*, *Sphingopyxis*, and *Pseudomonas* and Isolates from the Las Vegas Wash. J. Env. Qual. **46**, 281 (2017). https://doi.org/10.2134/jeq2016.08.0286
11. Z. Li, R. Nandakumar, N. Madayiputhiya, X. Li, Proteomic analysis of 17β-estradiol degradation by Stenotrophomonas maltophilia. Environ. Sci. Technol. **46**, 5947–5955 (2012). https://doi.org/10.1021/es300273k
12. S. Weber, P. Leuschner, P. Kämpfer, W. Dott, J. Hollender, Degradation of estradiol and ethinyl estradiol by activated sludge and by a defined mixed culture. Appl. Microbiol. Biotechnol. **67**, 106–112 (2005). https://doi.org/10.1007/s00253-004-1693-4
13. APHA, *5210 Biological Oxygen Demand (Bod)**, pp. 2–13 (2012)
14. TULAS, *Norma de Calidad Ambiental y de descarga de Efluentes: Recurso Agua.* TULAS Texto unificado de legislación secundaria del Ministerio del Ambiente 8–9 (2011)
15. S. Ríos, R. Agudelo, L. Gutiérrez, *Patógenos e Indicadores Microbiológicos de Calidad del Agua Para Consumo Humano* (2017). https://doi.org/10.17533/udea.rfnsp.v35n2a08
16. D. Ortega-Paredes, P. Barba, S. Mena-López, N. Espinel, V. Crespo, J. Zurita, High quantities of multidrug-resistant Escherichia coli are present in the Machángara urban river in Quito, Ecuador. J. Water Health **18**, 67–76 (2020). https://doi.org/10.2166/wh.2019.195
17. R. Clarke, D. Peyton, M.G. Healy, O. Fenton, E. Cummins, A quantitative microbial risk assessment model for total coliforms and *E. coli* in surface runoff following application of biosolids to grassland. Environ. Pollut. **224**, 739–750 (2017). https://doi.org/10.1016/j.envpol.2016.12.025
18. L. Yang, X.H. Wang, S. Cui, Y.X. Ren, J. Yu, N. Chen, Q. Xiao, L.K. Guo, R.H. Wang, Simultaneous removal of nitrogen and phosphorous by heterotrophic nitrification-aerobic denitrification of a metal resistant bacterium Pseudomonas putida strain NP5. Bioresour. Technol. 285 (2019). https://doi.org/10.1016/j.biortech.2019.121360
19. B.E. Pierce, M.J. Leboffe, *A Photographic Atlas Microbiology Laboratory*
20. B. Serra, J. Zhang, M.D. Morales, A.G.V. de Prada, A.J. Reviejo, J.M. Pingarrón, A rapid method for detection of catalase-positive and catalase-negative bacteria based on monitoring of hydrogen peroxide evolution at a composite peroxidase biosensor. Talanta **75**, 1134–1139 (2008). https://doi.org/10.1016/j.talanta.2008.01.009
21. I. Altinok, S. Kayis, E. Capkin, Pseudomonas putida infection in rainbow trout. Aquaculture **261**, 850–855 (2006). https://doi.org/10.1016/j.aquaculture.2006.09.009
22. M. Tzirita, S. Papanikolaou, B. Quilty, Enhanced fat degradation following the addition of a Pseudomonas species to a bioaugmentation product used in grease traps. J. Environ. Sci. (China) **77**, 174–188 (2019). https://doi.org/10.1016/j.jes.2018.07.008
23. W. Xiang, S. Chen, D. Tian, C. Huang, T. Gao, *Pseudomonas hutmensis* sp. nov., a New Fluorescent Member of Pseudomonas putida Group. Curr. Microbiol. **76**, 872–878 (2019). https://doi.org/10.1007/s00284-019-01701-z
24. P. Wang, D. Zheng, R. Liang, Isolation and characterization of an estrogen-degrading *Pseudomonas putida* strain SJTE-1. 3 Biotech 9:0 (2019). https://doi.org/10.1007/s13205-018-1537-z
25. D.P. Moser, B.P. Hedlund, M.R. Rosen, S.M. Blunt, M.J. Benotti, Reversible reduction of estrone to 17β-estradiol by, and isolates from the Las Vegas Wash. J. Env. Qual. **46**, 281 (2017). https://doi.org/10.2134/jeq2016.08.0286

Chapter 4
Biotransformation of Hyodeoxycholic and Deoxycholic Bile Acids

María Elena Maldonado Rodríguez, Mishell Enith Araujo Guayasamín, Andrea Alejandra Ruíz Moreno, and Stefania Costa

Abstract Biotransformations involve the use of microorganisms that allow to modify organic compounds by generating different molecules with possible pharmacological, nutritional, or cosmetic properties. The aim of this research was to evaluate the biotransformations of bile acids (hyodeoxycholic acid, HA, and deoxycholic acid, DA), using microorganisms isolated from slaughtering houses of Ecuador. The biotransformations were analyzed by thin-layer chromatography (TLC) and nuclear magnetic resonance (NMR). The genus identification of the bacteria has been carried out by biochemical tests. The results showed that the more efficient bacteria for the biotransformation of HA were identified as *Vibrio alginolyticus, Burk holderia pseudomallei,* and *Vibrio parahaemolyticus.* For DA biotransformations, the most efficient bacteria were *Proteus penneri*, *Burkholdeira pseudomallei*, *Pasteurella multocida,* and *Pseudomona aeruginosa.* The low biotransformation capacity found in fungi may be due to the difference in metabolic pathways between the two types of microorganisms and that the best source for obtaining bacteria with biotransformation capacity (100%) of HA was the container of intestines, whereas for DA the best sampling area was the slaughtering area, isolating bacteria with a biotransformation capacity of 80%.

4.1 Introduction

Bile acids (BA) are the most important end products in the metabolism of cholesterol [1, 2]. These are characterized by a nucleus of 24 carbon atoms, and the differences of these structures are the result of the presence of a variable number of hydroxyl groups in different positions of this kind of molecules [3]. The two most important

M. E. M. Rodríguez (✉) · M. E. A. Guayasamín · A. A. R. Moreno
Carrera de Ingeniería en Biotecnología, Universidad Politécnica Salesiana, Av. Isabel La Católica N. 23-52 y Madrid, Quito, Ecuador
e-mail: mmaldonado@ups.edu.ec

S. Costa
Dipartimento di Scienze Della Vita e Biotecnologie, Università di Ferrara, Via L. Borsari 46, 44121 Ferrara, Italy

Á. Rocha et al. (eds.), *Communication, Smart Technologies and Innovation for Society*, Smart Innovation, Systems and Technologies 252,
https://doi.org/10.1007/978-981-16-4126-8_4

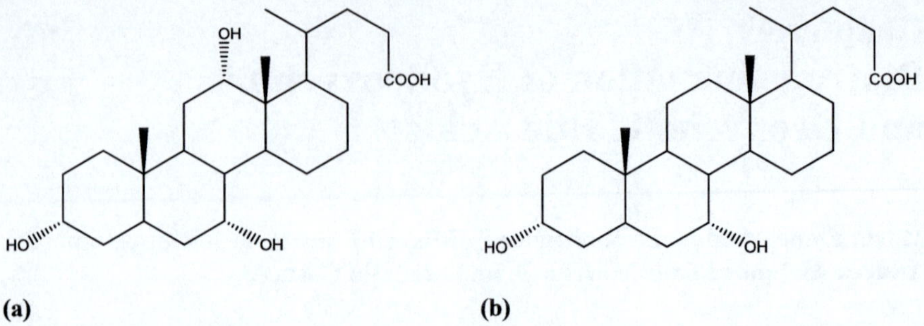

Fig. 4.1 Colic acid (**a**) and chenodeoxycholic acid (**b**)

primary bile acids are colic acid (Fig. 1a) and chenodeoxycholic acid (Fig. 1b). The conversion of cholesterol to primary bile acids requires multiple steps; however, the limiting reaction appears to be the cholesterol conversion to 7-hydrocholesterol [4]. Bile acids are often conjugated with glycine and taurine, forming the group of secondary bile acids (e.g., taurocholic and glycolic acids, derived from colic acid). Bile salts are sodium or potassium salts of secondary bile acids, which are important components of bile and participate in the emulsion of fats during the digestion, favoring the absorption of fat-soluble vitamins [5].

Bile acids and their salts are important in the pharmaceutical industry as some of them (cholic acid and chenodeoxycholic acid) are used as starting products for the synthesis of drugs for different pathologies as for gastrointestinal disorders as well as for the malfunction of the gallbladder [6]. This is the case of the Cuban pharmaceutical industry, which since the 60 s has worked on production lines for obtaining bile acids and their saline derivatives from bovine bile [6], making this kind of steroid molecules very important in the pharmaceutical industry. In fact, they represent the second category (more than 300 steroid drugs are known) on the market after antibiotics [7].

The research group of the University of Ferrara for many years has been looking for new microorganisms able to biotransform bile acids to obtain ursodeoxycholic acid, in order to avoid the steps of chemical synthesis of this pharmaceutical active ingredient, which involves the use of reagents and reaction conditions that have a high environmental impact. Ursodeoxycholic acid is used in the treatment of sandy gallstones.

The aim in the pharmaceutical industry in recent decades is to obtain the active ingredient replacing chemical synthesis with greener processes by using microbial biocatalysts. In order to obtain new steroid molecules through biotransformation processes, it is possible to isolate the microorganisms from the environment or from the processing areas where this type of molecules is present [7]. Various biotransformation processes have been applied on an industrial scale in the production of steroid drugs, using different types of raw materials as waste materials in view of the principles of the circular economy, green chemistry, and sustainability [8]. For this reason, the identification of microorganisms able to biotransform bile acids is an interesting topic for the academia and the industry.

Thus, the objective of this research was to evaluate the biotransformation ability of hyodeoxycholic (HA) and deoxycholic (DA) acids, using microorganisms isolated from slaughtering houses in Ecuador.

4.2 Materials and Methods

4.2.1 Sample Collection

Samples were collected in Latacunga and Cayambe slaughtering houses (89 and 78 km from Quito, respectively), where animals are slaughtered for human consumption with no strict hygiene and environmental controls. The sampling was carried out in the slaughtering areas where the viscera were collected, especially the gallbladder, which by being non-sterile areas favor the presence of microorganisms able of metabolizing steroid compounds.

For the sample collection, a sterile cotton swab was passed over the surface and then was placed inside the sterile physiological solution. Samples have been kept at 4 °C during their transportation to the city of Quito and have been kept at 0 °C in the laboratory until the microorganisms have been isolated. Each sample was assigned a code for its identification.

4.2.2 Biotransformation of Bile Acids (Microorganism Monitoring)

Cultures have been prepared in liquid plate count broth (PCB) medium (10 ml in 50 ml Erlenmeyer flasks) for bacteria, using a sterile loop-isolated bacterium as an inoculum. After 48 h of growth at 30 °C with continuous agitation (100–110 rpm), the various substrates (HA or DA) have been inoculated as solutions of the sodium salts of bile acids (HA and DA) (1 g/L, 100 μl of a 100 ppm solution of each bile acid). After having administered the substrate, the cultures have been maintained for another 48 h at 30 °C with continuous agitation. Samples of 200 μl have been taken every 24 h, acidified with HCl until pH 2, and extracted with ethyl acetate (200 μl). The samples thus obtained have been centrifuged at 6000 rpm for one minute. The organic phase was analyzed by thin-layer chromatography (TLC) using a mixture of organic compounds as a mobile phase according to the polarity of the molecules, thus determining the mobile phase suitable for the research. The following mobile phases have been used: Mobile phase 1, ethyl acetate/glacial acetic acid 50:2 (50 ml ethyl acetate, 2 ml acetic glacial acid), mobile phase 2, ethyl acetate/hexane/glacial acetic acid 70:30:2 (70 ml ethyl acetate, 30 ml hexane, 2 ml acetic glacial acid).

The chromatography plate has been cut and the area where the standard of bile acid and biotransformation samples have been placed with a capillary (2 cm from the lower edge of the chromatographic plate) and indicated with a pencil. The capillary has been taken and washed with ethyl acetate, and the standard solution has been filed two to three times. The capillary has been washed another time, and the sample of each biotransformation sample has been sown four times approximately. Samples have been dried on the plate, and the latter has been placed in the chromatographic chamber containing the mobile phase. After the chromatographic run, the mobile phase has been dried using heat, sprayed with a solution of phosphomolybdic acid ($H_3Mo_{12}O_{40}P$), sulfuric acid, and acetic acid in volume 2.5:2.5:5 and heated (approximately 120 °C) in order to visualize the reaction.

The same procedure used for bacteria has been applied for fungi, with the differences that in this case it has been used a Sabouraud Destroxe Broth (SDB) liquid medium and the growth time has been 96 h.

4.2.3 *Scaling of Cultures by Microorganisms*

The procedure applied has been previously published by the authors [9].

4.2.4 *Extraction of the Biotransformation Product*

In sterile conditions (laminar flow chamber), the culture broth containing bacteria or fungus has been transferred in sterile disposable tubes. It has been then centrifuged for 20 min at 3500 rpm to separate the microorganism from the supernatant. The supernatant has been transferred in a separation funnel, 1 ml of HCl (37%) has been added to adjust the pH (it must be maintained at 2), and it has been mixed until a homogeneous mixture has been obtained. Then, 100 ml of ethyl acetate have been added to the mixture and the phases (aqueous and organic) have been separated in the funnel. The process has been repeated with another 50 ml of ethyl acetate, and finally, 5 g of anhydrous sodium sulfate have been placed in the organic phase for remove water residues. It has been filtered, and the solvent has been evaporated. Nitrogen has been used to evaporate the organic solvent residue, and finally, the sample was ready for NMR analysis.

4.3 Results

The sampling has been done in two slaughtering houses in the cities of Latacunga and Cayambe. The sampling areas and the assigned code are shown in Table 4.1. The

Table 4.1 Bacteria isolated in Latacunga and Cayambe slaughtering houses

Sampling location	Code	Sampling place	Code
Latacunga slaughtering house			
Water channel inside the plant	MMB1	Animal processing area	MM B9
Container for collecting the intestines	MM B2	Area for collecting the intestines	MM B10
Storage area of the skin	MM B3	Intestine collection area	MM B11
Internal water channel for the system	MM B4	Container containing the rumen	MM B12
Vesicle collection vessel	MM B5	External water channel of the plant	MM B13
Area for collecting the intestines	MM B6	Area for storing the skin	MM B28
Animal processing area	MM B7	Vesicle collection vessel	MM B41
External water channel for the plant	MM B8	Area for collecting guts	MM B60
Cayambe slaughtering house			
Stomach contents of fed animals	MM B14	Processing surface of animals	MM B46
Container for collecting the intestines	MM B15	Viscera of the reception area	MM B47
Collection vessel of the intestines	MM B16	Viscera of the reception area	MM B48
Processing of the intestines	MM B23	Viscera of the reception area	MM B51
Area for collecting the heads	MM B24	Container for collecting the intestines	MM B53
Walls of the slaughtering room	MM B25	Walls of the slaughtering room	MM B54
Walls and floor for disemboweling	MM B26	Walls and floor for disemboweling	MM B55
Processing of the intestines	MM B27	Sidewalls of the intestine	MM B56
Walls of the slaughtering room	MM B29	Walls of the working area	MM B57
Udder collection area	MM B30	Viscera of the reception area	MM B59
Walls and floor for disemboweling	MM B31	Shedding surface area	MM B61
Sidewalls of the intestine	MM B32	Sidewalls of the intestine	MM B62
Walls of the slaughtering room	MM B33	Container for collecting intestines	MM B63
Area for collecting the intestines	MM B34	Viscera of the reception area	MM B64
Walls of the working area	MM B35	Walls of the working area	MM B65
Intestine container located in the slaughtering room	MM B36	Processing equipment of animals	MM B66
Floor and walls of the eviscerate area	MM B37	Stomach content of fed animals	MM B67
Viscera of table processing	MM B39	Viscera of the reception area	MM B68
Contents of the stomach of animals Fed	MM B40	Walls of the slaughtering room	MM B69

(continued)

Table 4.1 (continued)

Sampling location	Code	Sampling place	Code
Sink to stone	MM B42	C basin	MM B70
Walls the processing zone	MM B43	Surface	MM B71
Offal processing	MM B44	Walls of the slaughtering room	MM B72
Viscera of the table processing	MM B45	Table of the rumen	MM B73
		Bath	MM B74

biotransformation percentages by the microorganisms obtained for bile acids (AH and AD) are shown in Table 4.2.

Microorganisms that showed good substrate conversion yield from TLC analysis were retested by scaling up biotransformation. The reasons for their selection are described in Table 4.3.

4.3.1 Identification of Efficient Bacteria in the Biotransformation of Bile Acids

The identification of bacteria with the highest biotransformation of HA and DA acids has been performed by biochemical analysis, the correspondence percentages are shown in Fig. 4.2.

4.3.2 Nuclear Magnetic Resonance (NMR) Analysis

A product obtained from the biotransformation with bile acid salt (AD) by the MM B26 bacteria has been identified using ^{1}H-NMR analysis. Structure changes were observed in the steroid compound, with the disappearance of the lateral chain in C_{17} position, oxidation of the hydroxyl group in C_3 position, and the appearance of two double bonds in C_1–C_2 and C_4–C_5 positions (Fig. 4.3).

4.4 Discussion

The most representative results for the biotransformation achieved for bile acids were those whose biotransformation percentage yields equal to or higher than 50%. As can be seen in Table 4.2 for HA, a total of 11 strains were able to biotransform the substrate providing good yield products (50–75%) while 7 reached a biotransformation yields between 80 and 85%. On the other hand, fungi showed a very low biodegradation ability toward HA. In the case of DA, and according to the information shown in Table

Table 4.2 Biotransformation of bile acids

Code	Biotransformation yield (%)	Code	Biotransformation yield (%)
HA		DA	
Bacteria			
MM B25	80	MM B26	100
MM B26	(0)	MM B30	100
MM B29	50	MM B33	(0)
MM B30	65	MM B36	100
MM B35	80	MM B42	80
MM B36	75	MM B44	(0)
MM B37	50	MM B45	(0)
MM B38	85	MM B46	70
MM B39	50	MM B48	50
MM B40	60	MM B50	100
MM B41	70	MM B54	80
MM B44	80	MM B55	80
MM B48	50	MM B57	100
MM B50	(0)	MM B66	100
MM B54	75	MM B67	100
MM B64	50	MM B68	80
MM B66	80	MM B69	75
MM B67	80	–	–
MM B68	70	–	–
MM B69	80	–	–
Fungi			
MM F12	10%	–	–
MM F15	(0)	–	–

Note (0)—use all carbon of bile acid salt as an energy source until the production of carbon dioxide

4.3, a total of 3 bacteria provided a biotransformation yields between 50 and 75% and 11 between 80 and 100%, the latter being representative for the investigation.

Bacteria have been shown to have a greater biotransformation capacity than the tested fungi. The low biotransformation ability may be due to the difference in metabolic pathways between the two types of microorganisms. From a total of 47 bacteria evaluated in the biotransformation of bile acids, 30 bacteria (64%) were able to biotransform HA while 21 bacteria (45%) were able to biotransform DA. From 30 bacteria capable of biotransforming HA, 6.45% use all the carbon from bile acid salt as an energy source, to the production of carbon dioxide. For DA, 14.98% out of the 21 bacteria with biotransformation capacity, use all the carbon from bile acid salt as an energy source, to the production of carbon dioxide. From a total of 30 fungi

Table 4.3 Scaled microorganisms

Hyodeoxycholic acid	
Bacteria	Description
MM B38	This bacterium presented one biotransformation product at 24 and 48 h with the mobile phase one 50:2 (50 ml ethyl acetate and 2 ml acetic glacial acid) corresponding to the 3-oxo derivate standard of the bile acids administered (standard extracted from the biotransformation of *Pseudomona paucimobilis*)
MM B26	This bacterium has been monitored with mobile phase two 70:30:2 (70 ml ethyl acetate, 30 ml hexane, 2 ml acetic glacial acid). In this case, the biotransformation leads to the complete metabolization of the substrate without providing products after 24 h from the administration of the substrate
MM B35	This bacterium has been monitored with the mobile phase two 70:30:2. A biotransformation product has been observed after 24 h of reaction. The biotransformation yield increased at 48 h
MM F12	The fungus has been monitored with mobile phase two and with the standard corresponding to the 3-oxo derivate (standard extracted from the biotransformation of *Pseudomona paucimobilis*). The fungus did not undergo biotransformation in the first five days, but after 7 days of incubation with the substrate, from the TLC analysis a spot appears in correspondence of the 3-oxo derivative with a rather low yield
MM F15	The fungus has been monitored with mobile phase two and with the standard corresponding to the 3-oxo derivate, as for MM F12, the fungus begins to metabolize the substrate by oxidizing HA in the C_3 position after 7 days of reaction
Deoxycholic acid	
MM B66	This bacterium has been chosen for preparing semi-preparative culture since TLC showed a good yield of biotransformation of DA. The product obtained corresponds to the 3-oxo derivative of DA (standard extracted from the biotransformation of *Pseudomona paucimmobilis*)
MM B67	This strain, as evidenced by the TLC analysis, completely metabolizes the substrate providing the 3-oxo derivative of DA as the only biotransformation product with a yield of 100%
MM B26	As for MMB67, from the TLC analysis, only one spot has been present corresponding to 3-oxo derivative of DA
MM B42	An almost complete biotransformation is observed, i.e., the biotransformation product presented a significant change in its chemical structure, using chromatographic techniques

evaluated with bile acids, 2 (7%) were capable of biotransforming to HA and none acted on DA.

The product mainly obtained from biotransformation with HA is the 3-oxo derivative with the following bacteria: MM B25, MM B35, MM B36, MM B38, MM B39, MM B40, MM B44, MM B66, MM B67, MM B69. With a high correspondence percentage, four of the bacteria mentioned were identified by biochemical tests (Microgen GN-ID system), since the identification kit had four tests, thus, the ones that showed the greatest efficiency in the biotransformation capacity were *Vibrio*

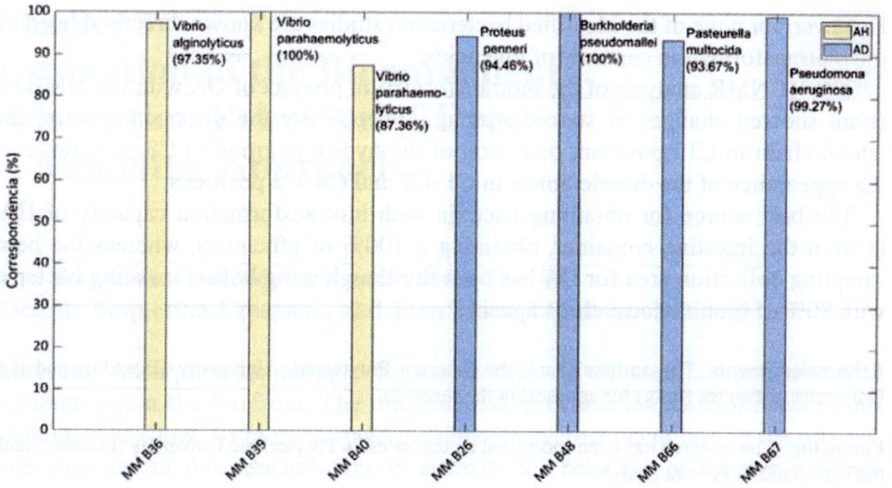

Fig. 4.2 Identified bacteria and their correspondence (%)

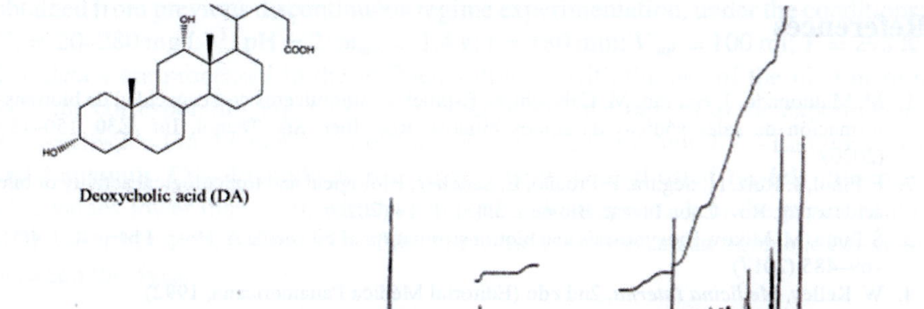

Fig. 4.3 1H-NMR spectrum of DA

alginolyticus (MM B36), Burkholderia pseudomallei (MM B40) and *Vibrio parahaemolyticus* (MM B40). The fourth test has been performed on MM B38; however, it has been not possible to identify it because this bacterium presented 85% match percentage. Hence, it is recommended to apply another test to identify its species.

For the biotransformations of DA, the 3-oxo derivative has been obtained as the main product, with the following bacteria: MM B26, MM B42, MM B35, MM B48, MM B54, MM B57, MM B66, MM B67, MM B68, MM B69. Four of the bacteria mentioned were identified with a high correspondence percentage by biochemical tests (Microgen GN-ID system): *Proteus penneri* (MM B26), *Burkholdeira pseudomallei* (MM B48), *Pasteurella multocida* (MM B66), and *Pseudomona aeruginosa* (MM B67). These microorganisms have pathogenic characteristics. Studies of *P. penneri* indicate its ability to intervene in the production of struvite acids [10].

5.1 Introduction

In recent years, the increase in colorants for industrial purposes has led to a higher concentration of these compounds in wastewater (WW). This leads to the deterioration of water sources as a consequence of the uncontrolled discharge of effluents, which threatens the vulnerability of the communities that use these water bodies as a source of supply [1]. Industrial liquid effluents present mixtures of dyes which generate interactions with each other, so it is of great importance to study the behavior of the mixtures. It is worth mentioning that the concentrations of the dyes in WW range from 10 mg L^{-1} to 200 mg L^{-1}, as a result of the great solubility of pigments in water [2]. For this reason, technology presents advances in the development of treatment of effluents contaminated by colorants [3]. The main technologies are: adsorption, advanced oxidative processes (AOPs), and microbiological digestion.

Adsorption, mainly with the use of biomass, is one of the most promising treatments in terms of water contaminated with dyes, due to its low cost, easy handling, and high efficiency [4]. However, obtaining good results with this technology requires the study of the most important phenomena that occur during the process: equilibrium, kinetics, and thermodynamic behavior [5]. The equilibrium of an adsorption system can be represented mathematically by conventional models such as the Langmuir and Freundlich isotherms. These models are the basis for describing the adsorption phenomenon [6]. These isotherms can be empirically modified with correction factors, which allow describing the behavior of the colorants in mixtures. According to Allaoui et al. [7], the modified forms of the Langmuir and Freundlich isotherms can better fit the adsorption data and represent the phenomenon more realistically. Empirical models, such as Freundlich's extended [8], pseudo–first order [9], pseudo–second order [10], and second-order polynomial [11], allow a good fit of the adsorption data in the mix. These models presented adequate adjustments in the batch system and for mixtures of contaminants such as heavy metals and colorants, mainly.

The study of the adsorption balance of dyes is of great importance because it allows determining the adsorbent's ability to remove them under certain conditions [12]. In addition, it contributes to the analysis of the phenomenon resulting from the mixture of dyes, by determining behaviors such as (i) antagonistic: the dye mixtures have interactions that reduce the adsorption capacity of at least one of the components; or (ii) synergistic: at least one component of the dye mixture increases its adsorption capacity [13]. The prediction of the behavior of liquid effluents in treatment systems allows the design, optimization, and control of treatment processes. For this reason, this work proposes to estimate the behavior of the adsorption isotherms of the dye mixtures Tartrazine Yellow (TY) and Brilliant Blue FCF (BB) in rice husk.

5.2 Materials and Methods

5.2.1 Data Selection

The experimental data were extracted from a two-component adsorption study with mixtures of TY and BB, with rice husk as adsorbent. The conditions of the experiment were: initial concentration $C_o = 20\text{–}280$ mg L^{-1}; pH = 2; adsorbent dose $m_{\text{ads}} = 1.4$ g; contact time $t = 180$ min; solution volume $V_{\text{sol}} = 100$ ml; temperature $T = 298$ K [14]. The design of the experiment evaluated the adsorption of one dye by maintaining a fixed initial concentration of the other dye. The initial concentrations of the dyes were from 20 to 120 mg L^{-1} with a step of 20, and from 120 to 280 mg L^{-1} with a step of 40. Therefore, the datasets were:

- Fixed initial concentrations for BB and variations of the initial concentrations of TY (Group 1).
- Fixed initial concentrations of TY and variations of initial concentrations of BB (Group 2).

Each group considers a total of 100 combinations of dyes, divided according to the fixed initial concentration. Each combination presents the values of the adsorption capacity (Q_e) of each dye.

5.2.2 Mathematical Modeling of the Data

The model used for the estimation of the isotherms of the bicomponent mixtures considers an additive behavior of the monocomponent adsorption capacities Eq. (5.1).

$$Q_{i,e} = X_{\text{TY}} Y_{\text{TY}} q_{i,\text{TY}} + X_{\text{BB}} Y_{\text{BB}} q_{i,\text{BB}} \tag{5.1}$$

where Q_e is the estimated adsorption capacity of the mixture, X_{BB} and X_{TY} are the equilibrium fractions of the BB and TY dyes, respectively. Y_{BB} and Y_{TY} are correction factors for model fit, $q_{i,\text{BB}}$ and $q_{i,\text{TY}}$ are the theoretical adsorption capacities estimated independently for TY and BB, respectively. The subindex i refers to the theoretical model selected to estimate the component behavior.

The theoretical models for estimating the monocomponent adsorption capacity were Langmuir Eq. (5.2) and Freundlich Eq. (5.3).

$$q_e = \frac{qmK_L C_e}{1 + K_L C_e} \tag{5.2}$$

$$q_e = K_F C_e^{1/n_F} \tag{5.3}$$

Table 5.1 Parameters of Langmuir and Freundlich adsorption isotherms for BB and TY at 298 K

Isotherms	Parameter	BB	TY
Langmuir	q_m (mg g^{-1})	9.98	6.02
	K_L (L mg^{-1})	0.164	0.255
Freundlich	K_F (mg g^{-1})	2.83	2.28
	$1/n_F$	0.275	0.213

where q_e is amount of material adsorbed per unit mass of adsorbent (mg g^{-1}), C_e—the equilibrium concentration (mg L^{-1}), q_m—the maximum absorption capacity (mg g^{-1}), K_L—the adsorption rate (L mg^{-1}), K_F—the constant absorption capacity (mg g^{-1}), and n_F—is the magnitude of the adsorption driving force (–).

The parameters of the monocomponent isotherms for BB and TY are results of the study referenced for this work, which are specified in Table 5.1.

Equation (5.1) adapts to the behaviors adopted by the dyes, so it considers each combination between the Langmuir and Freundlich isotherms. The development of Eq. (5.1) with the combinations of Eqs. (5.2) and (5.3) derives the expressions Eqs. (5.4)–(5.7) for each dataset.

$$Q_e = X_{\mathrm{TY}} Y_{\mathrm{TY}} q_{L_{\mathrm{TY}}} + X_{\mathrm{BB}} Y_{\mathrm{BB}} q_{L_{\mathrm{BB}}} \tag{5.4}$$

$$Q_e = X_{\mathrm{TY}} Y_{\mathrm{TY}} q_{L_{\mathrm{TY}}} + X_{\mathrm{BB}} Y_{\mathrm{BB}} q_{F_{\mathrm{BB}}} \tag{5.5}$$

$$Q_e = X_{\mathrm{TY}} Y_{\mathrm{TY}} q_{F_{\mathrm{TY}}} + X_{\mathrm{BB}} Y_{\mathrm{BB}} q_{L_{\mathrm{BB}}} \tag{5.6}$$

$$Q_e = X_{\mathrm{TY}} Y_{\mathrm{TY}} q_{F_{\mathrm{TY}}} + X_{\mathrm{BB}} Y_{\mathrm{BB}} q_{F_{\mathrm{BB}}} \tag{5.7}$$

Equations (5.4)–(5.7) codes were MOD11, MOD12, MOD21, and MOD22, respectively. The values of the statistical parameters were used to compare the fit of the data generated by the mathematical model with the experimental data. The determining statistical parameter to select the best-fit model was the standard residual error (RSE), obtained using the R software.

5.2.3 *Analysis of the Effect of Mixtures*

Equation (5.8) calculates the interaction coefficient for mixture (ICM). This coefficient uses the monocomponent adsorption capacities as well as the adsorption capacities in a mixture of each of the species, to determine the effect produced in the adsorption of binary mixtures of dyes in solution.

$$\mathrm{ICM}_i = \frac{(q_{i,\mathrm{TY}} - q_{\mathrm{TY}m}) + (q_{i,\mathrm{BB}} - q_{\mathrm{BB}m})}{q_{i,\mathrm{TY}} + q_{i,\mathrm{BB}}} \tag{5.8}$$

where ICM_i is the interaction coefficient for each mixture, $q_{\mathrm{TY}m}$ and $q_{\mathrm{BB}m}$ are the amount of TY and BB absorbed per unit mass of adsorbent in the mixture, respectively. This analysis considered the comparison of the results obtained from the experimental values and those estimated by the additive models presented in Eqs. (5.4)–(5.7). The comparison of the experimental value with the estimates considered the residual standard error (RSE) Eq. (5.9).

$$\mathrm{RSE} = \sqrt{\frac{1}{n-2}\sum(\mathrm{ICM}_{i\,\mathrm{exp}} - \mathrm{ICM}_{i\mathrm{cal}})^2} \tag{5.9}$$

where subindexes "exp" and "cal" refer to experimental and calculated values, and n is the number of evaluated points.

5.3 Results and Discussion

Figures 5.1 and 5.2 represent the models fitted for groups 1 and 2, respectively. Each group had a total of 10 cases evaluated, whose representation is shown with the letters from A to J. The curves represented show the behavior of the models in relation to the data of the interactions in mixture of the TY and BB dyes. The dependent variable (Q_e) represents the absorption capacity of the interactions and was plotted with respect to the equilibrium concentration of each of the components.

The four proposed models present curves that largely fit the experimental behaviors. In some cases, the models are superimposed on each other, which indicates that in those cases both the Langmuir and Freundlich models in their combinations can estimate the isotherms of the mixtures. Table 5.2 presents the values corresponding to the factors of Y_{TY} and Y_{BB} adjusted for each case.

The values of the adjusted constants refer to effects on the adsorption of the evaluated component. Thus, for group 1, in which BB is fixed, the adsorption isotherm for TY is evaluated. Also, for group 2, TY concentrations are fixed, and blue isotherms are evaluated. Thus, for the process at low concentrations of blue, in group 1, this refers to an interference in the adsorption of yellow, under the assumptions of the Langmuir-Langmuir and Langmuir–Freundlich behaviors. In the case of group 2, the yellow dye supposes an interference in most of the cases evaluated in the behaviors of the Langmuir–Langmuir and Freundlich–Langmuir type.

To establish the type of behavior, the graphical analysis required the evaluation of the fit. The adjustment of the model presented RSE values (Table 5.3) below 0.5, which indicates a great approximation between the proposed models and the experimental data.

For the case of group 1 (Fig. 5.1), the Langmuir–Langmuir and Langmuir–Freundlich behavior models overlap each other and have slightly lower RSE values

Table 5.2 Adjusted coefficients for the cases evaluated

	Fixed C_0 (mg L^{-1})	Y_{TY}				Y_{BB}			
	MOD	11	12	21	22	11	12	21	22
Group 1	20	0.501	0.499	0.484	0.484	−0.060	−0.063	0.005	0.006
	40	0.547	0.548	0.508	0.507	−0.009	−0.011	0.038	0.045
	60	0.549	0.549	0.496	0.494	−0.003	−0.004	0.032	0.037
	80	0.593	0.589	0.518	0.513	0.057	0.061	0.095	0.102
	100	0.575	0.568	0.498	0.490	0.084	0.087	0.116	0.120
	120	0.565	0.559	0.490	0.482	0.081	0.079	0.107	0.104
	160	0.760	0.758	0.650	0.645	0.033	0.029	0.060	0.054
	200	0.702	0.691	0.587	0.575	0.124	0.107	0.150	0.129
	240	0.908	0.900	0.757	0.749	0.080	0.065	0.109	0.088
	280	0.989	0.980	0.805	0.795	0.069	0.054	0.101	0.079
Group 2	20	−1.092	−0.383	−1.222	−0.439	0.837	0.711	0.832	0.711
	40	−0.538	−0.021	−0.639	−0.027	0.924	0.754	0.928	0.755
	60	−0.321	0.103	−0.378	0.120	0.950	0.760	0.956	0.759
	80	−0.270	0.001	−0.301	0.001	0.892	0.719	0.897	0.719
	100	−0.162	0.070	−0.172	0.074	0.857	0.671	0.860	0.670
	120	−0.024	0.187	−0.026	0.193	0.845	0.646	0.846	0.642
	160	−0.026	0.124	−0.026	0.120	0.816	0.627	0.816	0.624
	200	−0.019	0.105	−0.018	0.096	0.815	0.619	0.816	0.617
	240	0.024	0.142	0.021	0.125	0.848	0.631	0.848	0.628
	280	0.067	0.172	0.057	0.146	0.827	0.605	0.826	0.601

Table 5.3 Statistical analysis for the four models in a mixture at different concentrations

Fixed C_0 (mg L^{-1})	RSE							
	Group 1 fixed BB				Group 2 fixed TY			
	MOD11	MOD12	MOD21	MOD22	MOD11	MOD12	MOD21	MOD22
20	0.252	0.253	0.149	0.149	0.868	0.462	0.885	0.465
40	0.248	0.248	0.252	0.252	0.415	0.415	0.418	0.415
60	0.162	0.162	0.142	0.142	0.221	0.476	0.223	0.476
80	0.145	0.146	0.183	0.183	0.337	0.302	0.337	0.302
100	0.252	0.251	0.289	0.287	0.245	0.350	0.244	0.350
120	0.170	0.169	0.207	0.203	0.337	0.545	0.337	0.544
160	0.138	0.139	0.138	0.138	0.255	0.420	0.255	0.421
200	0.161	0.159	0.201	0.197	0.232	0.360	0.232	0.360
240	0.172	0.174	0.163	0.163	0.259	0.380	0.259	0.380
280	0.170	0.169	0.200	0.197	0.353	0.504	0.354	0.504

than those of the Freundlich–Langmuir and Freundlich–Freundlich combinations (which are overlapping mutually). This indicates that for the TY dye, the Langmuir model is the one that best interprets its behavior. For group 2, the Langmuir–Langmuir and Freundlich–Langmuir interactions presented the lowest RSE values, ensuring a better fit of the models. In addition, these models overlapped each other, which makes it possible to establish that the behavior of the BB dye was Langmuir type.

For each mixture, an interaction coefficient (ICM) was estimated, the observed values presented values between 0.502 and 0.825 for group 1 and between 0.363 and 0.841 for group 2. This refers to the fact that in the adsorption process, the dye mixtures competed each other, thus reducing the adsorption capacity of the material. The intervals for the ICMs calculated from the values estimated with the proposed models presented differences with respect to those proposed by the experimental data. The estimated values for group 1 (Fig. 5.3) expanded their lower and upper limits to the values of 0.421 and 0.998, respectively.

On the other hand, the values of group 2 (Fig. 5.4) presented a displacement of the interval, and this refers to an increase in the lower and upper values to 0.551 and 0.944, respectively. The predicted values in group 1 showed similarity with the observed values, and the RSE values for the four models were close to 0.1. In the case of group 2, the values showed similarity, but the estimates are less precise, the RSE values were between 0.203 and 0.210.

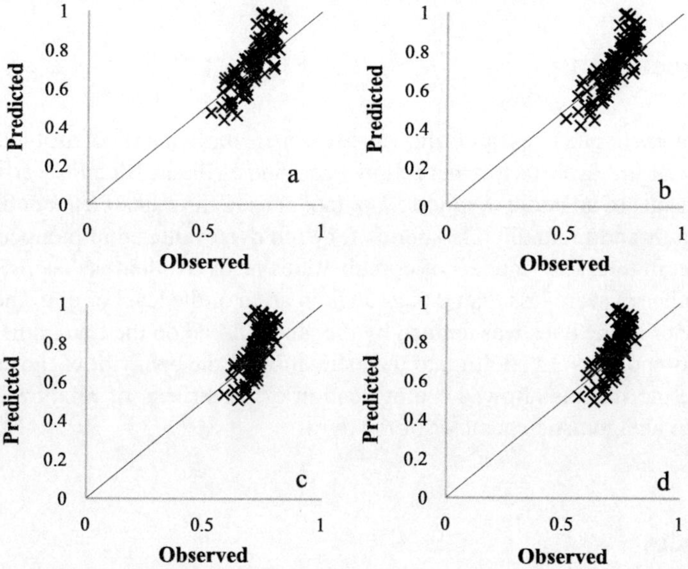

Fig. 5.3 Interaction coefficient predicted and observed for group 1. **a** Langmuir–Langmuir. **b** Langmuir–Freundlich. **c** Freundlich–Langmuir y **d** Freundlich–Freundlich

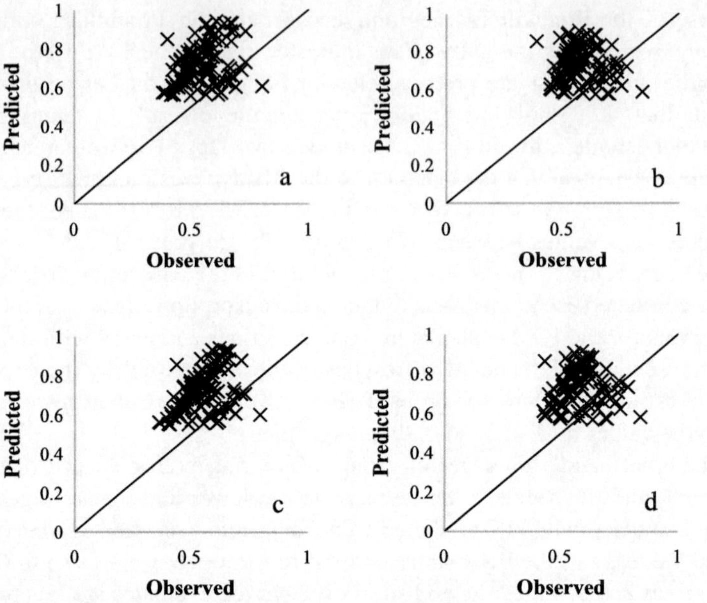

Fig. 5.4 Interaction coefficient predicted and observed for group 2. **a** Langmuir–Langmuir. **b** Langmuir–Freundlich. **c** Freundlich–Langmuir y **d** Freundlich–Freundlich

5.4 Conclusions

The development of the research allowed to estimate the behavior of the bicomponent adsorption of the dyes Tartrazine Yellow (TY) and Brilliant Blue FCF (BB) in rice husk by means of an additive model. The model used the adjustment coefficients of the Langmuir and Freundlich isotherms for each dye evaluated in monocomponent adsorption, in total four models of combinations of the isotherms. The RSE values for the isotherms were less than 0.6, as an indication of the level of fit of the models. The behavior of the dyes was defined by the model based on the Langmuir isotherm for both components. In addition to the estimation of the behavior of the adsorption isotherms, the models allowed the evaluation of the effects of mixtures, and this verified the antagonistic character of the dyes.

References

1. C. Macías Socha, M. García Colmenares, S.P. Chaparro, Determinación electroquímica de plomo y cadmio en aguas superficiales. Luna Azul **44**, 27–38 (2017). https://doi.org/10.17151/luaz.2017.44.3
2. S.J. Allen, G. Mckay, J.F. Porter, Adsorption isotherm models for basic dye adsorption by peat in single and binary component systems. J. Colloid Interface Sci. **280**, 322–333 (2004). https://

doi.org/10.1016/j.jcis.2004.08.078

3. J.M. Giler-Molina, L.A. Zambrano-Intriago, L.S. Quiroz-Fernández, D.C. Napoleão, J. dos Santos Vieira, N. Simões Oliveira, J.M. Rodríguez-Díaz, Degradation of oxytetracycline in aqueous solutions: Application of homogeneous and heterogeneous advanced oxidative processes. Sustain. **12**, 8807 (2020). https://doi.org/10.3390/su12218807
4. E.V. Daza-López, K.J. Fernández-Andrade, R. de Queiroz Nóbrega, L.Á. Zambrano-Intriago, G. Villanueva Ramos, L.S. Quiroz-Fernández, M.C.B.S.M. Montenegro, J.M. Rodríguez-Díaz, Modified or functionalized natural bioadsorbents: new perspectives as regards the elimination of environmental pollutants, in *Advances in the Domain of Environmental Biotechnology: Microbiological Developments in Industries, Wastewater Treatment and Agriculture*, ed. by N.R. Maddela, L.C. García Cruzatty, S. Chakraborty (Springer, Singapore, 2021), pp. 195–225
5. C.A. Andrade, L.A. Zambrano-Intriago, N.S. Oliveira, J.S. Vieira, L.S. Quiroz-Fernández, J.M. Rodríguez-Díaz (2020) Adsorption behavior and mechanism of oxytetracycline on rice husk ash: Kinetics, equilibrium, and thermodynamics of the process. Water. Air. Soil. Pollut. **231**:103. https://doi.org/10.1007/s11270-020-04473-6
6. R. Ezzati, Derivation of pseudo-first-order, pseudo-second-order and modified pseudo-first-order rate equations from Langmuir and Freundlich isotherms for adsorption. Chem. Eng. J. **392**, 123705 (2020). https://doi.org/10.1016/j.cej.2019.123705
7. M. Allaoui, M. Berradi, J. Bensalah, H. Es-sahbany, O. Dagdag, S. Ibn Ahmed, Study of the adsorption of nickel ions on the sea shells of Mehdia: kinetic and thermodynamic study and mathematical modelling of experimental data. Mater. Today Proc. (2021). https://doi.org/10.1016/j.matpr.2021.02.234
8. K. Sharma, R.K. Vyas, K. Singh, A.K. Dalai, Degradation of a synthetic binary dye mixture using reactive adsorption: experimental and modeling studies. J. Environ. Chem. Eng. **6**, 5732–5743 (2018). https://doi.org/10.1016/j.jece.2018.08.069
9. M.C. Silva, L. Spessato, T.L. Silva, G.K.P. Lopes, H.G. Zanella, J.T.C. Yokoyama, A.L. Cazetta, V.C. Almeida, H_3PO_4–activated carbon fibers of high surface area from banana tree pseudo-stem fibers: adsorption studies of methylene blue dye in batch and fixed bed systems. J. Mol. Liq. **324**, 114771 (2021). https://doi.org/10.1016/j.molliq.2020.114771
10. E. Khalilzadeh Shirazi, J.W. Metzger, K. Fischer, A.H. Hassani, Removal of textile dyes from single and binary component systems by Persian bentonite and a mixed adsorbent of bentonite/charred dolomite. Colloids Surf. Physicochem. Eng. Asp. **598**, 124807 (2020). https://doi.org/10.1016/j.colsurfa.2020.124807
11. C.E. Onu, J.T. Nwabanne, P.E. Ohale, C.O. Asadu, Comparative analysis of RSM, ANN and ANFIS and the mechanistic modeling in eriochrome black-T dye adsorption using modified clay. South Afr. J. Chem. Eng. **36**, 24–42 (2021). https://doi.org/10.1016/j.sajce.2020.12.003
12. J.P. Lima, G. Alvarenga, A.C.F. Goszczynski, G.R. Rosa, T.J. Lopes, Batch adsorption of methylene blue dye using Enterolobium contortisiliquum as bioadsorbent: Experimental, mathematical modeling and simulation. J. Ind. Eng. Chem. **91**, 362–371 (2020). https://doi.org/10.1016/j.jiec.2020.08.029
13. K. Xu, Y.-X. Liu, X.-F. Wang, S.-W. Li, J.-M. Cheng, Combined toxicity of functionalized nano-carbon black and cadmium on Eisenia fetida coelomocytes: the role of adsorption. J. Hazard Mater. **398**, 122815 (2020). https://doi.org/10.1016/j.jhazmat.2020.122815
14. L.A.Z. Mendoza, Remoción de colorantes sintéticos de las aguas residuales de la industria alimentaria usando como material adsorbente biomasa de arroz. Ph.D. thesis (2018)

Chapter 6
An Initial Approach About Data Preprocessing Techniques Applied to Polymer Electrolyte Fuel Cells: A Case Study

Ester Melo, James Peñafiel, Julio Barzola-Monteses, and Mayken Espinoza-Andaluz

Abstract Like other fields, a great amount of data is present when a polymer fuel cell is analyzed. Several variables are involved during fuel cells' operation, i.e., pressure, temperature, and chemical compounds in the anode and cathode side. A correct data preprocessing allows an appropriate quality of data to evaluate the fuel cell's behavior during the energy conversion process. It also helps to predict a fuel cell's future performance, considering techniques such as machine learning. This paper aims to provide the pathway to follow when a fuel cell's collected dataset is used to solve a supervised learning problem such as classification and regression. Therefore, data preprocessing techniques play an essential step before feeding it into any model to learn. It impacts on better learning, generalization, prediction, and computational time.

E. Melo · J. Peñafiel
Escuela Superior Politécnica del Litoral, ESPOL, Campus Gustavo Galindo Km. 30.5 Vía Perimetral, P.O. Box 09-01-5863, Guayaquil, Ecuador

J. Barzola-Monteses
Artificial Intelligence and Information Technology Research Group, University of Guayaquil, Av. Delta y Av. Kennedy, 090514 Guayaquil, Ecuador

Department of Computer Science and Artificial Intelligence, Escuela Técnica Superior de Ingenierías Informática y de Telecomunicación, Universidad de Granada, 18071 Granada, Spain

J. Barzola-Monteses
e-mail: julio.barzolam@ug.edu.ec

M. Espinoza-Andaluz (✉)
Escuela Superior Politécnica del Litoral, ESPOL, Facultad de Ingeniería Mecánica y Ciencias de La Producción, Centro de Energías Renovables y Alternativas, Campus Gustavo Galindo Km. 30.5 Vía Perimetral, P.O. Box 09-01-5863, Guayaquil, Ecuador
e-mail: masespin@espol.edu.ec

Á. Rocha et al. (eds.), *Communication, Smart Technologies and Innovation for Society*, Smart Innovation, Systems and Technologies 252,
https://doi.org/10.1007/978-981-16-4126-8_6

6.1 Introduction

A polymer electrolyte fuel cell (PEFC) represents a suitable response to the increasing demand for green energy technologies. It is an electrochemical device that converts the chemical energy from the reactant gases into electrical energy, heat, and water. The energy conversion process is clean and noiseless, producing electrical power with high efficiency [1]. However, some of the bottlenecks that have to be solved are related to durability and manufacturing costs [2]. The mentioned issues are a limitation to increase the presence of PEFCs in the global market. To overcome the limitations, most recent studies are focused on model development to diagnose and predict the aging parameters into the PEFCs [3–5].

Most of the models do not consider the previous exploratory analysis stage of the data [3–5]. Therefore, there is a lack of a detailed description of the different preprocessing techniques that can be used when PEFC's data will be analyzed. The obtained information from the preprocessing data will help to determine and select the most suitable algorithm for diagnosis and fault prediction in a PEFC.

This study proposes a review of the procedures and techniques that must be considered to treat the data obtained from a PEFC functional behavior before feeding it into a supervised learning problem of classification. The analyzed data comes from the fault diagnosis of a PEFC performed at the Department of Aeronautical and Automotive Engineering, Loughborough University [6].

The rest of the paper is organized as follows: Sect. 6.2 explains the methodology used in this work, and Sect. 6.3 shows the results. Section 6.4 shows the discussion of the results. Finally, some concluding remarks and directions for future work are given.

6.2 Methods

The data's preprocessing stage is one of the most critical steps to be performed before implementing any machine learning (ML) algorithm. As known, data in their natural state usually are incomplete, inconsistent, and erroneous. Besides, it has been demonstrated that the application of preprocessing techniques facilitates the focus on the interest variable and improves the precision or performance of the models [7].

Regarding the data preprocessing, Alasadi and Bhaya [8] determine that it consists of four well-defined steps: cleaning, integration or exploratory data analysis (EDA), transformation, and data reduction. Simultaneously, each step is constituted by techniques and procedures to be applied according to the bunch of data's characteristics. The different stages applied to the case study are shown in Fig. 6.1 and described in the following sections.

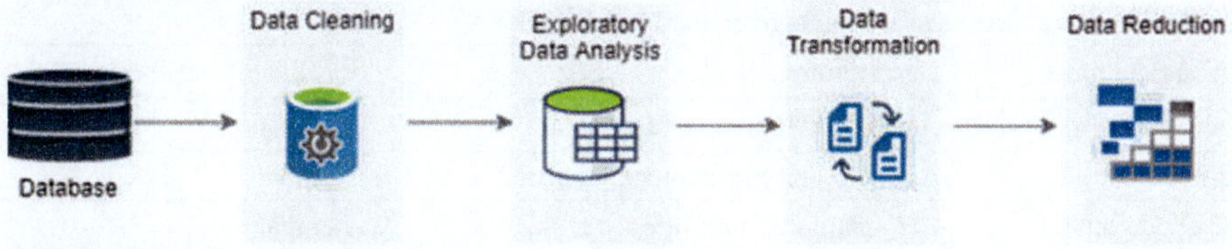

Fig. 6.1 Stages of data preprocessing applied to PEFC

6.2.1 Data Cleaning

According to García et al. [7], data cleaning includes or overlapped with other preprocessing techniques. This step aims to obtain a bunch of data free of incorrect, inconsistent, and unnecessary data. Several operations are carried out to solve issues related to missed, noisy, and inconsistent data in this stage. They are considered the most common in data mining processing [9].

Generally, missing values are usually filled manually or using a global constant or the average from the application of inferences, Bayesian formulas, decision tree, or the expectation–maximization (EM) algorithm [8, 10, 11].

On the other hand, for the treatment of noisy data, i.e., data with random errors, clustering, binning, and combined "human–machine" inspection are applied. Each of the mentioned methods matches the same objective: to detect, group, and delete atypical values [12, 13].

Inconsistent data are commonly related to incorrect values in the collecting process. They usually are suggested to be corrected manually. However, it is also possible to employ distance measures to identify similar or different data with entering errors [7].

In this case study, the data were stored in multiple Microsoft Excel files whose names indicated their status: normal, faulty, or unknown. For this study's purposes of classification, only the data corresponding to the normal and faulty states were considered. According to the researchers, the "unknown" state refers to conditions that were not good or bad enough to categorize it in that binary way.

Overall, there were about 25 files per state, each of which contained about 25,000 records. Moreover, each file had 20 features, which were assigned a name and described in Table 6.1.

Subsequently, missing, noisy, or inconsistent values were analyzed through the visualization and codes executed in Python.

Table 6.1 Recorded measurements from the PEFC database

Assigned name	Description	Unit	Item
Anode out p1	Anode outlet pressure from stack 1	mbar	1
Anode out p2	Anode outlet pressure from stack 2	mbar	2
Cathode out p1	Cathode outlet pressure from stack 1	mbar	3
Cathode out p2	Cathode outlet pressure from stack 2	mbar	4
Current	Current	A	5
Anode react	Anode reactant flow	SLPM	6
Anode in p1	Anode inlet pressure from stack 1	mbar	7
Anode in p2	Anode inlet pressure from stack 2	mbar	8
Cathode air in	Cathode air inlet flow	SLPM	9
Cathode in p1	Cathode inlet pressure from stack 1	mbar	10
Cathode stoi	Cathode stoichiometry	N/A	11
Cathode out t2	Cathode outlet temperature from stack 2	°C	12
Cathode in t2	Cathode inlet temperature from stack 2	°C	13
Cathode out t1	Cathode outlet temperature from stack 1	°C	14
Cathode in t1	Cathode inlet temperature from stack 1	°C	15
Water in p2	Primary water inlet pressure from stack 2	mbar	16
Water in p1	Primary water inlet pressure from stack 1	mbar	17
Water in f2	Primary water inlet flow from stack 2	SLPM	18
Water in f1	Primary water inlet flow from stack 1	SLPM	19
Water in tp	Water inlet temperature	°C	20

6.2.2 Exploratory Data Analysis

EDA is the first research of the data mainly based on visualization. EDA aims to understand the data type involved and the trends, to later apply the corresponding statistical tests and algorithms [14]. EDA is helpful for the following objectives [15]:

1. Improving the understanding of the involved variables in the collected data.
2. Proposing early hypotheses about the possible correlations between the variables.
3. Determining the tools and techniques to apply on a dataset in the preprocessing phase.

Moreover, EDA encompasses a broad set of methods. In the case of classification problems, the following is mainly considered [14]:

1. Adding, modification, and elimination of data from multiple sources.
2. A statistical overview of the variables.
3. Visualization of the variables.

In the current case study, two files with information about the cell's status were incorporated: normal and faulty. Besides, it was evaluated if there is any inconsistency in the integration. Visualization and extraction of the variables' statistics using Python code were performed like in the previous phase.

6.2.3 Data Transformation

Data transformation is the next step in EDA. It is vital to consolidate the data and facilitate the data mining process [10]. Like cleaning data, this step requires human audit since it includes the construction of features and attributes from the existing data. In this stage, normalization, smoothed, and cleaning procedures are carried out if needed [7].

In the current study, this step helps construct a new categorical variable named "status". It is composed of two types of status for the cell. Additionally, the presence of missing values, noisy, and inconsistent were again evaluated, i.e., the analysis considers the newly created variable, and later the data were normalized and smoothed.

6.2.4 Data Reduction

Data reduction is the process performed before implementing any ML algorithm, especially when the dataset has a large number of observations. The dataset has different features (variables), some of which may be redundant or not very significant. Data reduction should only consider the relevant features that will affect the model's learning, generalization, prediction, and computational time. If this process is not considered, it can lead to poor model performance due to overfitting characteristics or data that act as noise [16].

In general, all the data reduction algorithms are grouped in two techniques: variable selection and variable extraction. Variable selection determines the data's most relevant features to have a smaller dimensionality and the needed information. Variable extraction decreases the dataset without discard too many beginning features.

Data reduction techniques aim to reduce the data through linear combinations of the starting features [17]. In this case study, variable selection techniques were used since any ML algorithm is needed to identify relevant feature' set when classification models are built [16].

The filter, wrapper, and embedded are among the most widely variable selection method used. This study uses the filter method since the wrapper method needs an ML algorithm to select the dataset's relevant features. It is not required to use ML algorithms in the data preprocessing stage. On the other hand, the embedded method

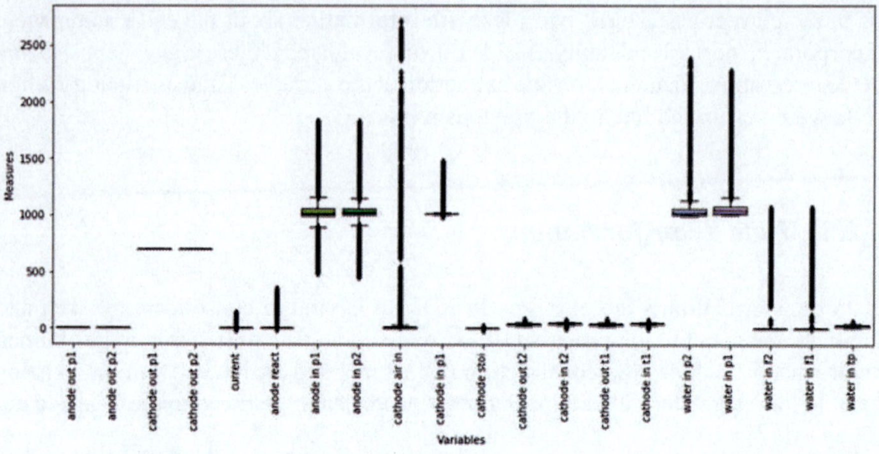

Fig. 6.2 Data cleaning process from the faulty state data

selects the relevant features during the training process where an ML algorithm is specified [18].

The filter method is a variable selection algorithm that, dislike the other methods, does not depend on an ML algorithm. It is based on direct computations of a metric without feedback, and its computational cost is low [19].

6.3 Results

6.3.1 Data Cleaning

The generated report from the evaluation of the lost and inconsistent values showed no presence of errors in both normal and faulty status. However, when the presence of noisy data is evaluated, the presence of outlier values was observed in the faulty status according to Fig. 6.2. It is due to the data that comes from the inadequate operating behavior of the cell.

As observed in Fig. 6.3, "normal" status showed outlier values. In this stage, the mentioned data were not analyzed but in the following phases.

6.3.2 Exploratory Data Analysis

In this stage, a visualization related to the variables is obtained. Anode reactant flow, cathode air inlet, and cathode in pressure were the variables with outlier values in normal status from stack 1. Outlier values from stack 2 were found in cathode

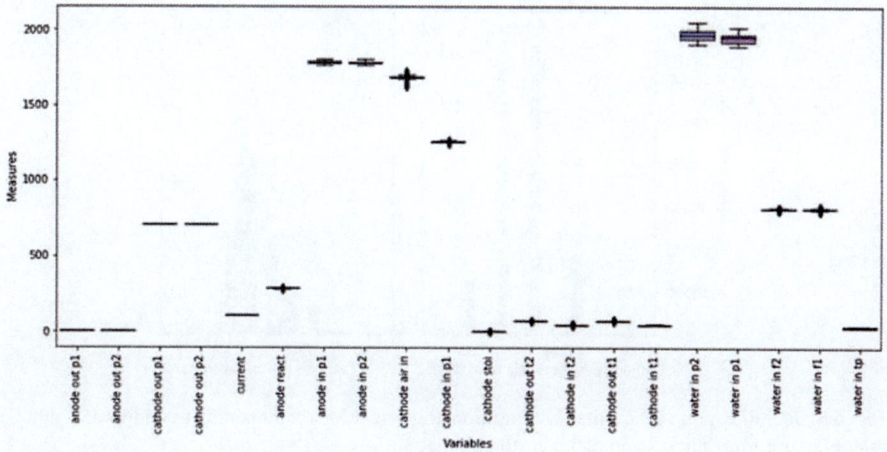

Fig. 6.3 Data cleaning process from the normal state data

stoichiometry and cathode inlet temperature. Water inlet temperature has outlier values from both stack 1 and stack 2.

Based on the obtained results, only the current and voltage behavior was considered to determine if the cell is normal or faulty. Moreover, it is important to note the cathode trend since 66% of the variables has outlier data, as shown in Figs. 6.4, 6.5, and 6.6.

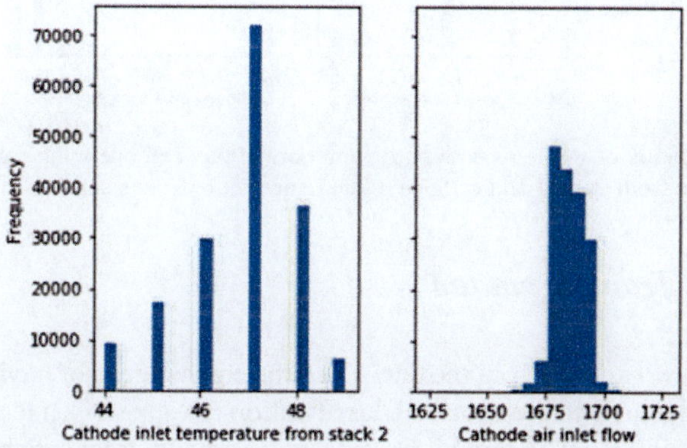

Fig. 6.4 Identification of variables with outliers in normal fuel cell operating conditions: cathode inlet temperature from stack 2 and cathode air inlet flow

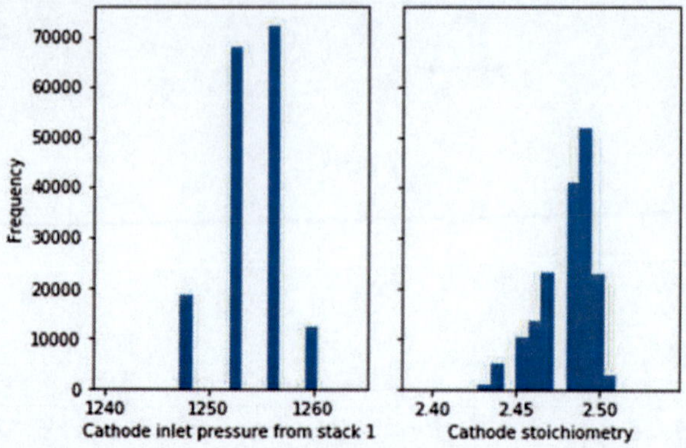

Fig. 6.5 Identification of variables with outliers in normal fuel cell operating conditions: cathode inlet pressure from stack 2 and cathode stoichiometry

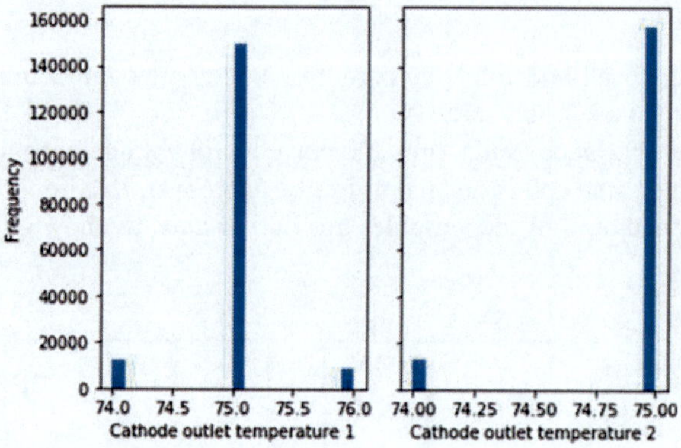

Fig. 6.6 Identification of variables with outliers in normal fuel cell operating conditions: cathode outlet temperature from stack 1 and cathode outlet temperature from stack 2

6.3.3 Data Transformation

It was built a new variable from the data since the convenience of having at least one categorical variable from the dataset. Classification is a supervised learning problem where the target is a label. In this study, this label is a categorical variable named "status" and precisely includes the description of the variable's status considering mainly the current and voltage. The mentioned variable helps to segment the analysis based on the status and compare the values. It was possible to find that the "normal" status has more atypical values than the "faulty" status. In this prefinal step, atypical values were eliminated, and the data were normalized. Figure 6.7 shows the normalization and distribution of significant variables mentioned in the previous section.

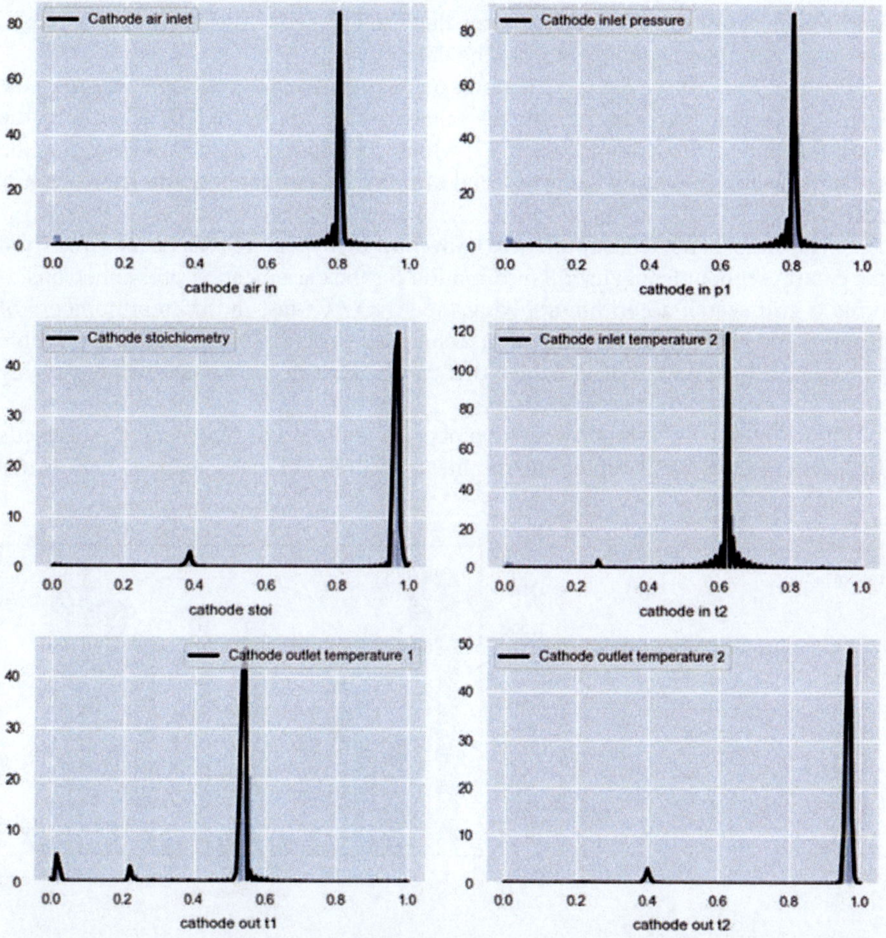

Fig. 6.7 Variable normalization and distribution

6.3.4 Data Reduction

The filter methods are recommended because they perform better in classification problems. These can be ANOVA f-test, chi-square (chi2), f_classif, and mutual_info_classif. These filter methods need to receive the value of the parameter k, which indicates the number of features to select. In order to find the optimal value of k, each of the filter selection methods should be evaluated, and the value k with the best performance should be identified [19].

In our case study, the dataset has two types of variables: numeric (features) and categorical (target variable), input and output, respectively. Due to these mentioned characteristics of the dataset, the ANOVA f-test reduction method is applied. This statistical test discriminates the features that are independent of the target variable; therefore, they can be removed from the dataset.

Figure 6.8 shows the dataset's features on the *x*-axis and the relevance result of the statistical test on the *y*-axis, i.e., higher values, greater relevance [20]. Therefore, the most relevant features are 4, 5, and 17, which correspond to the variables: cathode outlet pressure from stack 2, current and primary water inlet pressure from stack 1, respectively.

On the other hand, it is important to know how many features can be removed from the dataset without losing useful information for the classification problem. Considering a grid search algorithm applying the ANOVA f-test in different numbers of features and applying cross-validation techniques as detailed in [20], it was possible to establish the number of variables that can be kept in the dataset without losing useful information for predictions.

Figure 6.9 shows box-and-whisker plots, where the x-axis represents the dataset's features, and the *y*-axis represents the mean accuracy of classification. The results

Fig. 6.8 Bar chart of feature importance

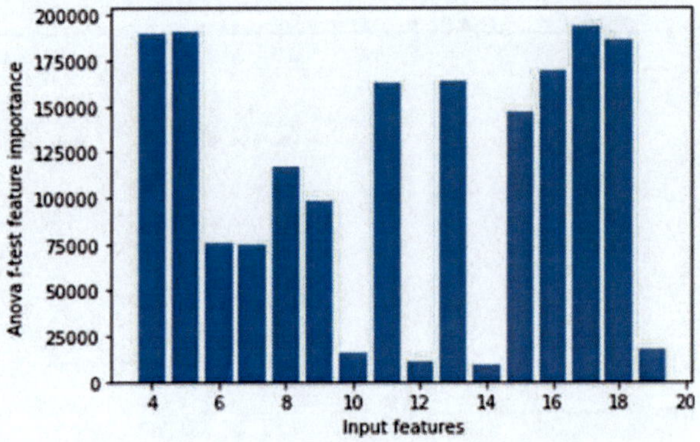

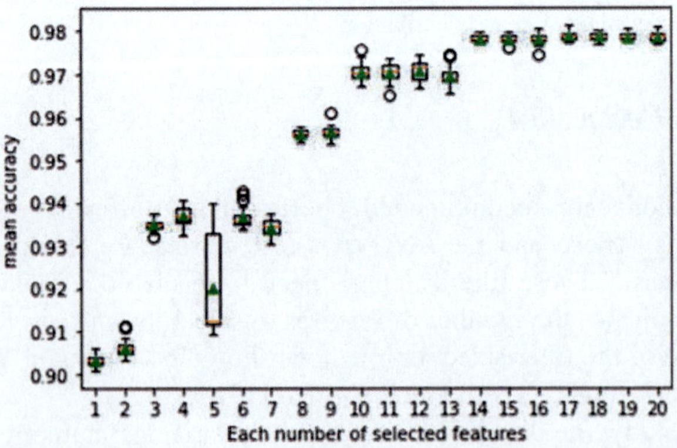

Fig. 6.9 Box and Whisker plots of classification accuracy for each number of selected features using ANOVA f-test

show that when selecting the best 17 features of the dataset, i.e., close to 1 of mean accuracy, the prediction problem is excellent. Additionally, it was noted, if up to 14 more relevant features were selected, the prediction results did not change significantly. Therefore, six features can be removed from the dataset and achieve an accuracy of over 95%. These six variables are the following: anode outlet pressure from stack 1, anode outlet pressure from stack 2, cathode outlet pressure from stack 1, cathode outlet temperature from stack 1, cathode outlet temperature from stack 2, and cathode inlet pressure from stack 1.

6.4 Discussion

The application of preprocessing techniques in the current study has allowed us to observe the trends and behaviors that have been omitted in previous studies. One of the trends is associated with the atypical pressure, air, and stoichiometry in the cathode. It is noted that the cathode is a sensible component to changes in the operating conditions. The last mentioned is confirmed when observing the same trend in faulty conditions.

Another fact to stand out is that 45% of the atypical values come from variables different to the current. It was considered in previous studies to determine the cell's status: normal or faulty [6]. However, applying a preprocessing technique shows that other factors should be considered to determine the cell's normality.

Finally, it is noted that ANOVA f-test, chi-square (chi2), f_classif, and mutual_info_classif are filter methods with better performance in classification problems. For this case study, the ANOVA f-test filter was applied, which allowed reducing the dimensions of the original dataset by 30%, i.e., in 14 features relevant to the prediction of the categorical variable. This procedure can be implemented through the scikit-learn library in Python.

After performing each phase of the data preprocessing specified in this article, we will have a high-quality dataset ready to be used in ML models.

6.5 Conclusions

The current study offers an initial view about the procedures to follow in applying preprocessing datasets' techniques originated in fuel cells. At the same time, it demonstrates the importance of applying the mentioned techniques to determine the features more relevant to estimate cell status and, therefore, help predict its useful life.

The results showed that the cell states (normal and failure) do not depend solely on current and voltage but also result from changes in various components of the energy transformation process. Specifically, the cathode's temperature, pressure, and stoichiometry are a factor little considered in detecting failures in fuel cells.

Further work is focused on the application of ML algorithms in polymer electrolyte fuel cell technologies.

References

1. L. Mao, L. Jackson, B. Davies, Effectiveness of a novel sensor selection algorithm in PEM fuel cell on-line diagnosis. IEEE Trans. Ind. Electron. **65**(9), 7301–7310 (2018)
2. B. Wang, B. Xie, J. Xuan, K. Jiao, AI-based optimization of PEM fuel cell catalyst layers for maximum power density via data-driven surrogate modeling. Energy Convers. Manag. **205**, 112460 (2020)
3. Z. Li, R. Outbib, S. Giurgea, D. Hissel, A. Giraud, P. Couderc, Fault diagnosis for fuel cell systems: a data-driven approach using high-precise voltage sensors. Renew. Energy **135**, 1435–1444 (2019)
4. L. Mao, L. Jackson, B. Davies, Investigation of PEMFC fault diagnosis with consideration of sensor reliability. Int. J. Hydrogen Energy **43**(35), 16941–16948 (2018)
5. D. Yu, Y. Wang, H. Liu, K. Jermsittiparsert, N. Razmjooy, System identification of PEM fuel cells using an improved Elman neural network and a new hybrid optimization algorithm. Energy Rep. **5**, 1365–1374 (2019)
6. L. Mao, L. Jackson, S. Dunnett, Fault diagnosis of practical polymer electrolyte membrane (PEM) fuel cell system with data-driven approaches. Fuel Cells **17**(2), 247–258 (2017)
7. S. García, J. Luengo, F. Herrera, *Data Preprocessing in Data Mining* (Springer, Canberra, 2015)
8. S. Alasadi, W. Bhaya, Review of data preprocessing techniques in data mining. J. Eng. Appl. Sci. **12**(16), 4102–4107 (2017)
9. D. Tomar, S. Agarwal, A survey on pre-processing and post-processing techniques in data mining. IJDTA **7**(4), 99–128 (2014)
10. S. García, S. Ramírez-Gallego, J. Luengo, J.M. Benítez, F. Herrera, Big data preprocessing: methods and prospects. Big Data Anal. **1**(1), 1–22 (2016)
11. J. Huang, Y.-F. Li, M. Xie, An empirical analysis of data preprocessing for machine learning-based software cost estimation. Inf. Softw. Technol. **67**, 108–127 (2015)
12. M. Gueraichi, A. Nacer, H. Moulai, Application of principal component analysis (PCA) and SVMs for discharges radiated fields discrimination, in *Intelligent Computing*. ed. by K. Arai, S. Kapoor, R. Bhatia (Springer, Cham, 2019), pp. 99–108
13. A. Khoshrou, E.J. Pauwels, Data-driven pattern identification and outlier detection in time series. arXiv:1807.03386 [stat] **858**, 471–484 (2019)
14. V. Cox, *Translating Statistics to Make Decisions* (Apress, 2017)
15. J. Vierheller, Exploratory data analysis, in *Process Design for Natural Scientists: An Agile Model-Driven Approach* (Springer, New York, 2015), pp. 110–126
16. S. Visalakshi, V. Radha, A literature review of feature selection techniques and applications: review of feature selection in data mining, in *2014 IEEE International Conference on Computational Intelligence and Computing Research* (2014), pp. 1–6
17. R. Zebari, A.M. Abdulazeez, D.Q. Zeebaree, D.A. Zebari, J.N. Saeed, A comprehensive review of dimensionality reduction techniques for feature selection and feature extraction. JASTT **1**(2), 56–70 (2020)
18. I. Guyon, A. Elisseeff, An introduction to variable and feature selection. J. Mach. Learn. Res. **3**, 1157–1182 (2003)
19. W. Duch, Filter Methods, in *Feature Extraction: Foundations and Applications*. ed. by I. Guyon, M. Nikravesh, S. Gunn, L.A. Zadeh (Springer, Berlin, Heidelberg, 2006), pp. 89–117
20. J. Brownlee, How to perform feature selection with Numerical Input Data Machine learning Mastery (2020)

Chapter 7
Simulation of a Condenser Component of a Thermal System Driven by Solar Energy for Seawater

William Quitiaquez, Elizabeth Suquillo-Gómez, Jorge Marcalla, and Fernando Toapanta-Ramos

Abstract Due to the population increase, the natural resources have been affected in a considerable percentage. This causes the scarcity of freshwater in several regions of the world, whereby different hybrid systems are analyzed that allow obtaining the highest efficiency in brackish water desalination. For this study, authors chose to use for the multistage flash distillation system (MSF) which makes up a large fragment of desalination capacity a worldwide. This one works powered by solar energy and takes full advantage of the hot steam that comes out of the turbine to be able to desalinate. In order to understand and know the condensation process, the ANSYS software is used to simulate a helical condenser with a sixteen-spiral coil, through which brine enters at a temperature of 26 °C, at 1 MPa of pressure, achieving an outlet temperature of 114.1 °C, which is why it is defined that under certain operating conditions the desalination of water can be much faster.

7.1 Introduction

Since ancient times, it has been considered that water is an inexhaustible resource, without considering that the Earth has approximately only 3% of freshwater, which is fundamental for sustainable development supplying much of the plant, animal needs, and human. Many freshwater systems have been threatened by human population growth, agriculture, industrialization, pollution, and anthropogenic climate change that often alter the patterns of water availability that have motivated new designs of the water system.

Kumar et al. [1] point out that, by 2030, population growth will have the greatest weight in the deterioration of water quality compared to other factors. This suggests that transdisciplinary research is required for the sustainable management of water resources. This suggests that transdisciplinary research is required for the sustainable management of water resources. For this reason, Okati et al. [2] point out that desalination industries can offer us alternative sources of available water, such as seawater

W. Quitiaquez (✉) · E. Suquillo-Gómez · J. Marcalla · F. Toapanta-Ramos
Universidad Politécnica Salesiana, Quito 170607, Ecuador
e-mail: wquitiaquez@ups.edu.ec

Á. Rocha et al. (eds.), *Communication, Smart Technologies and Innovation for Society*, Smart Innovation, Systems and Technologies 252,
https://doi.org/10.1007/978-981-16-4126-8_7

or brackish water. Mehmood and Ren [3] point out that the MSF technique currently represents 40% of the market due to its function with several stages/vacuum chambers as that the pressure in each stage decreases progressively compared to the last one. On the other hand, Al-Othman et al. [4] developed a simulation study of an MSF system, it was simulated with the Aspen HYSYS V8.8 software, having as results that this process produces freshwater at a speed of 1880 m^3/day for a seawater flow rate of around 4×10^4 m^3/day and a salinity estimated at 4.8%, the authors manifest that MSF's desalination can be completely powered by solar energy.

Lin and Yunran [5] mentioned that computer-aided simulation is economically more efficient, since detailed information in terms of flow pattern, temperature distribution, and concentration can be obtained, whereby Henghua et al. [6] performed a numerical simulation and optimization of the flash chamber for seawater desalination, and the CFD software was adopted to analyze the vapor–liquid flow nephogram of the velocity vector and the mass fraction within the flash chamber; the simulation results showed that the flash efficiency can be significantly increased by adding the baffle and changing the angle of the vertical baffle, concluding that the greater the pressure drop at the inlet and outlet of the flash chamber, the faster the flash speed.

Nigim and Eaton [7] developed a CFD prediction of the flashing processes in an MSF desalination chamber, using CFD software based on the outlet temperature and the vapor temperature of a real MSF plant of 105.6 and 104.7 °C, respectively, and the results give values of average temperatures of 106 and 102 °C, respectively, which are similar to those of the real plant; authors manifest that computational field modeling can be invaluable for a better understanding of heat transfer combined.

Dai and Li [8] performed a numerical investigation of the performance analysis of a heat pump water heater through a coupling model, the CFD software was used to simulate the transfer of the water tank, and the heat transfer coefficient of the variable diameter coil was 19.06% higher than the constant diameter coil. Also, compared to the constant diameter coil design, the average COP was 3.97% higher. The authors suggest that decreasing coil pitch and proper design would improve water temperature and reduce temperature stratification. The objective of the present work is to design and establish the efficiency of a helical condenser of an MSF desalination system using seawater to convert it into freshwater by simulation in CFD software.

7.2 Materials and Methods

Jones et al. [9] point out that the two main thermal technologies, MSF and multiple effects distillation (MED), produce most of the desalinated water; however, Hanshik et al. [10] mentioned that the MSF plant composes a big fragment of the ability of desalination worldwide, having a considerable market share.

There are numerous investigations on the different thermal desalination technologies, whose primary objective is the desalination of seawater or brackish water; therefore, comparative characteristics can be established between the systems as shown in Table 7.1.

Table 7.1 Characteristics of the different desalination systems

Parameter	Technologies		
	MSF [11, 12]	MED [13, 14]	Inverse osmosis [15–17]
Process	Evaporation	Evaporation–condensation	Membrane
Capacity (m^3/dia)	1285.2	0.76	1200
Water quality (ppm)	< 40	10–20	75
Energy consumption (kWh/m^3)	5	4.5–12.5	2.75
Maximum process temperature (°C)	115	55–75	85

7.2.1 System Description

Multistage flash distillation system technology makes up a large chunk of desalination capacity worldwide and is made up of flat plate solar collectors, heater, brackish water tank, condenser, and distilled water, as shown in Fig. 7.1.

In this system, as observed in Fig. 7.1, seawater enters for the condenser and passes through of the condensers tubes, causing the steam to condense and be collected in the distillate trays; the water is heated in a container called a heater in which the solar collectors provide the energy necessary to raise its temperature; the water vapor decreases its temperature when making the transfer and therefore it will condense and will have to be recirculated and heated [18].

The process is continuous since only a part of the salty water evaporates in a first stage, the remaining liquid phase is sent to the next stage, which will be at a lower pressure than the previous one, thus achieving an instantaneous boiling [10]. These systems consist of several stages (Fig. 7.2) where the pressure decreases progressively, succeeding in successively evaporating water. In the last stage, water remains that did not evaporate, called remaining fluid or brine [19].

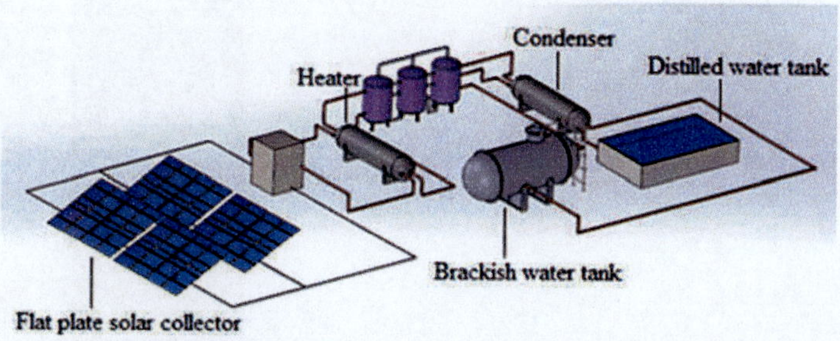

Fig. 7.1 Solar powered MSF system

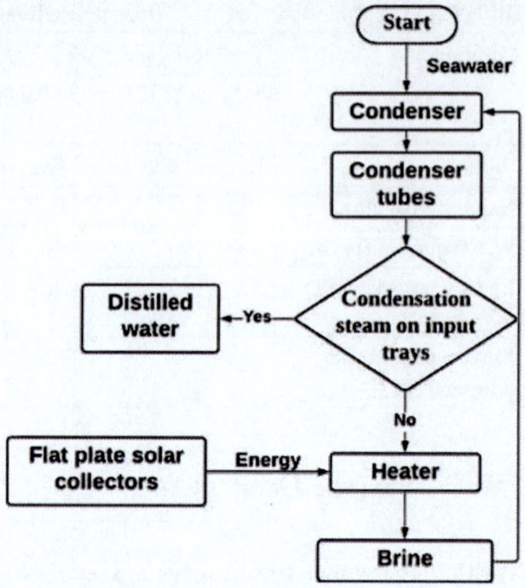

Fig. 7.2 MSF desalination process flowchart

7.2.2 *Condenser*

For this study, a helical condenser was used as shown in Fig. 7.3, due to that which said condenser is used for numerous continuous systems of small and medium powers [20]:

Condenser design. For the helical condenser to be used in this study, the design parameters shown in Table 7.2 were used.

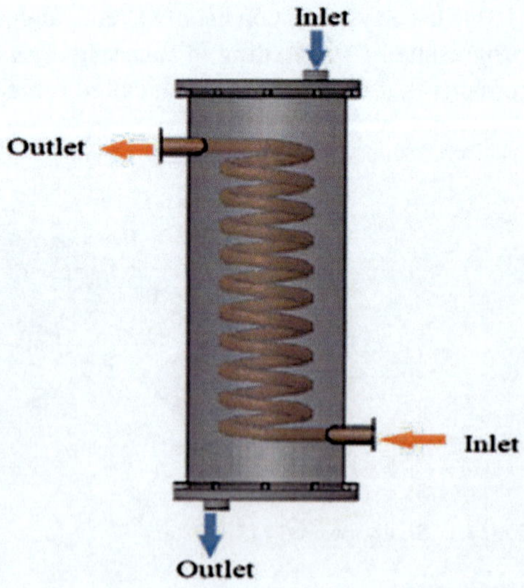

Fig. 7.3 Helical condenser

Table 7.2 Helical capacitor parameters [21]

Parameter	Value
Inside diameter, D_i	8 mm
External diameter, D_o	9.52 mm
Height, h	0.30 m
Thickness, e	1.52 mm
Material	Copper

7.2.3 *Heat Transfer*

To find the heat required for heat transfer during condensation, Eq. 7.1 was used to determine the mean logarithmic temperature difference LMTD [21].

$$\text{LMTD} = \frac{(T_{\text{Ri}} - T_{\text{Ao}}) - (T_{\text{Ro}} - T_{\text{Ai}})}{\ln\left[\frac{(T_{\text{Ri}} - T_{\text{Ao}})}{(T_{\text{Ro}} - T_{\text{Ai}})}\right]} \tag{7.1}$$

where T_{Ri} is the inlet temperature of the coolant, T_{Ro} is the outlet temperature of the coolant, T_{Ai} is the inlet temperature, and T_{Ao} is the outlet temperature of the water. Heat is linked to the first principle of thermodynamics, and according to this, two bodies in contact exchange heat between them until reaching thermal equilibrium; therefore, heat is proportional to mass, specific heat, and temperature variation, as shown in Eq. 7.2 [21].

$$Q = \dot{m} C_p \Delta T \tag{7.2}$$

where $\dot{m}$ is the mass flow, C_p is the specific heat of the fluid, and ΔT is the temperatures variation. The heat loss from one substance will be gained by the other until it reaches thermal equilibrium, in the way that is established in the law of the conservation of heat energy exchange as shown in Eq. 7.3 [21].

$$Q_{\text{gained}} = -Q_{\text{loss}} \tag{7.3}$$

Simultaneously, the heat transfer area is found using Eq. 7.4 [21], where A is the heat transfer area and U is the overall heat transfer coefficient.

$$A = \frac{Q}{U\ \text{LMTD}} \tag{7.4}$$

7.2.4 Numerical Model

The new simulators use complicated physical formulas that are implemented in CFD, in which the efficiency of the thermal resource is evaluated to know exact data in terms of phase, temperatures, pressures, liquid–vapor percentages, phase change times and behaviors [22]. CFD software can study the three-dimensional heat transfer process between the condenser coil and the water tank. The main seawater conditions in the CFD model description are presented in Table 7.3 [11].

The water tank adopts a two-dimensional axisymmetric model since the CFD model, which covers the conservation of energy, momentum, and energy is given by Eqs. (7.5)–(7.7) where it shows the constant temperature placement in different heating times [23].

Conservation of mass equation:

$$\frac{\partial u_x}{\partial x} + \frac{\partial u_x}{\partial y} = 0 \tag{7.5}$$

Equation of conservation of momentum:

$$\frac{\partial u_x}{\partial t} + u_x \frac{\partial u_x}{\partial x} + u_y \frac{\partial u_x}{\partial y} = \frac{1}{\rho_{w0}} \frac{\partial P}{\partial x} + v_w \left[\frac{\partial^2 u_x}{\partial x^2} \frac{\partial^2 u_x}{\partial y^2} \right] \tag{7.6}$$

$$\frac{\partial u_y}{\partial t} + u_x \frac{\partial u_y}{\partial x} + u_y \frac{\partial u_y}{\partial y} = \frac{1}{\rho_{w0}} \frac{\partial P}{\partial y} + v_w \left[\frac{\partial^2 u_y}{\partial x^2} \frac{\partial^2 u_y}{\partial y^2} \right] g\beta(T_w\, T_{w0}) \tag{7.7}$$

Energy conservation equation:

$$\frac{\partial T}{\partial t} + u_x \frac{\partial T}{\partial x} + u_y = \lambda_w \left[\frac{\partial^2 T}{\partial x^2} + \frac{\partial^2 T}{\partial y^2} \right] \tag{7.8}$$

Table 7.3 Design parameters

Parameter	Symbol	Value
Brine heater inlet temperature	T_{ft}	115 °C
Feed water inlet temperature	T_f	26 °C
Feed water salinity	S_f	36,000 rpm
Brine mass flow per stage width	V_b	180 kg/ms

(a)

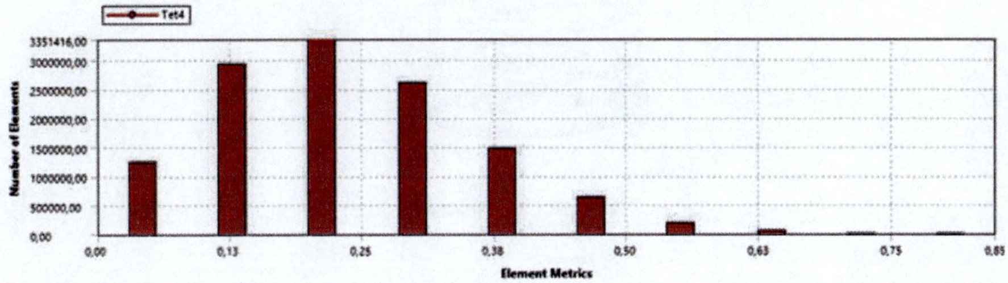

(b)

Fig. 7.4 **a** Helical condenser coil mesh. **b** Helical condenser coil mesh convergence

7.2.5 Mesh Generation

The mesh model is done on the coil of the helical condenser as shown in Fig. 7.4a, obtaining a mesh generation of 0.23 on average, which is found in the range of excellence, thus providing a possibility in obtaining accurate results at the time of simulation [24].

Thus, several 2,641,707 nodes and 12,468,715 total elements obtained in the meshing were also established. After the meshing process, the proportional results of the meshing convergence were reviewed, as shown in Fig. 7.4b.

7.3 Results and Discussion

For the simulation of the design made in this study, as shown in Fig. 7.5, the coil presents the ideal conditions for a helical copper condenser with the conditions detailed in Table 7.3 with the following inlet environments, brine´s inlet pressure corresponding to 1 MPa, inlet temperature 26 °C, flow rate of 180 kg/m s, and an inlet velocity of 6 m/s.

For the analysis of results, sixteen spirals have been considered for the coil as shown in Fig. 7.6, and the results obtained show that the brine presents a temperature change of 26 °C at an outlet temperature of 114.109 °C, achieving an adequate heat

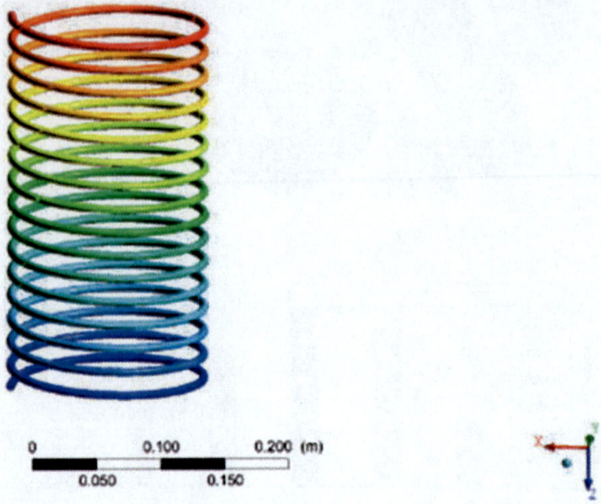

Fig. 7.5 Helical condenser in ideal condition

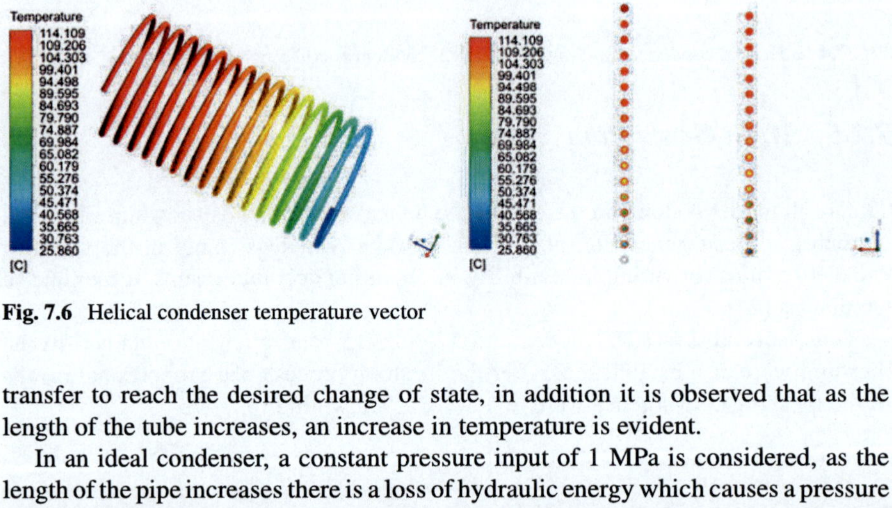

Fig. 7.6 Helical condenser temperature vector

transfer to reach the desired change of state, in addition it is observed that as the length of the tube increases, an increase in temperature is evident.

In an ideal condenser, a constant pressure input of 1 MPa is considered, as the length of the pipe increases there is a loss of hydraulic energy which causes a pressure drop; however, there are also some factors that will change this situation, such as mass transfer, surface tension, and turbulences that will cause oscillations and pressure losses as shown in Fig. 7.7.

7.4 Conclusions

For this work, the MSF technology was studied, which is essential for its application in a seawater desalination system, which makes the most of the hot steam that comes out of the turbine to be able to desalinate, this being an ecological system since it is

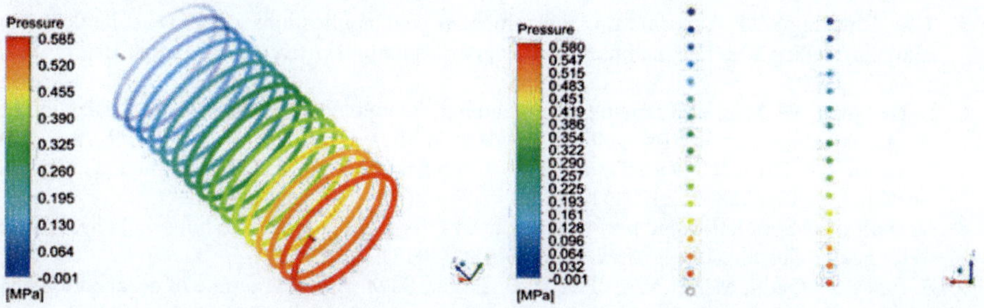

Fig. 7.7 Helical condenser pressure contour

drives with solar energy. For the simulation, the model of a helical condenser was established due to its thermal characteristics, having the following conclusions:

- Seawater condensation has been studied by CFD simulation, which successfully covered heat transfer in the fluid. Due to this, no assumptions were made about wall temperature, transfer coefficient, or the heat flux of the wall of the condensate coil, on the other hand, the ideal pressure variable of the condenser was worked with 1 MPa, showing that as the length of the pipe increases there is a loss of hydraulic energy; however, it was found that it is more efficient to do so in a vertical cylinder in which a computational study for turbulent fluids is provided.
- The design of the proposed model is efficient for the MSF technology condensation process, considering that the coil consists of sixteen spirals and the fluid (brine) enters at 6 m/s, with a flow of 180 kg/m s, reaching a temperature of 114.1 °C, which shows that, under specific operating conditions, the desalination of water can be much faster, this is due to the contributions of heat transfer by convection generated in the walls of the coil.

Acknowledgements Authors would like to thank to the Renewable Energies and Mechanical Implementation of SMEs Research Group (GIERIMP), Universidad Politécnica Salesiana.

References

1. P. Kumar, Y. Masago, B. Kumar Mishra, K. Fukushi, Evaluating future stress due to combined effect of climate change and rapid urbanization for Pasig-Marikina River, Manila. Groundw. Sustain. Dev. **6**, 227–234 (2018)
2. V. Okati, A. Ebrahimi-Moghadam, A. Behzadmehr, M. Farzaneh-Gord, Proposal and assessment of a novel hybrid system for water desalination using solar and geothermal energy sources. Desalination **467**, 229–244 (2019)
3. A. Mehmood, J. Ren, *Renewable Energy-Driven Desalination for More Water and Less Carbon* (Academic, China, 2021), pp. 333–372
4. A. Al-Othman, M. Tawalbeh, M. EL Haj Assad, T. Alkayyali, Novel multi-stage flash (MSF) desalination plant driven by parabolic trough collectors and a solar pond: a simulation study in UAE. Desalination **443**, 237–244 (2018)

5. T.W. Lin Lu, W.H. Yunran Min, Fundamentals and applications of CFD technology on analyzing falling film heat and mass exchangers: a comprehensive review. Appl. Energy **261**, 114473 (2020)
6. L. Henghua, W. Yan, W. Lianying, H. Yangdong, Numerical simulation and optimization of the flash chamber for multi-stage flash seawater desalination. Desalination **465**, 69–78 (2019)
7. T. Nigim, J.A. Eaton, CFD prediction of the flashing processes in a MSF desalination chamber. Desalination **420**, 258–272 (2017)
8. N. Dai, S. Li, Simulation and performance analysis on condenser coil in household heat pump water heater. Sustain. Urban Areas **36**, 176–184 (2018)
9. E. Jones, M. Qadir, M.T.H. van Vliet, V. Smakhtin, S.-M. Kang, The state of desalination and brine production: a global outlook. Sci. Total Environ. **657**, 1343–1356 (2019)
10. C. Hanshik, H. Jeong, K.-W. Jeong, S.-H. Choi, Improved productivity of the MSF (multi-stage flashing) desalination plant by increasing the TBT (top brine temperature). Energy **107**, 683–692 (2016)
11. H.D. Pedraza Corzo, D.M. Vesga Gómez, *Dimensionamiento y simulación de un sistema solar térmico* (Bucaramanga, 2019)
12. R.A. Rojas Opazo, *Propuesta de sistema de filtrado para agua de mar, planta de desalinización* (Valparaíso, 2020)
13. G. Chaouki, D. Kabakebji, I. Douba, A. Yassin, Performance analysis and optimization of hybrid multi-effect distillation adsorption desalination system powered with solar thermal energy for high salinity sea water. Energy **215** (2021)
14. A. Liponi, C. Wieland, A. Baccioli, Multi-effect distillation plants for small-scale seawater desalination: thermodynamic and economic improvement. Energy Convers. Manag. **205** (2020)
15. K. Harby, E.S. Ali, K. Almohammadi, A novel combined reverse osmosis and hybrid absorption desalination-cooling system to increase overall water recovery and energy efficiency. J. Clean. Prod. (2020)
16. S.-Y. Pan, A.Z. Haddad, A. Kumar, S.-W. Wang, Brackish water desalination using reverse osmosis and capacitive deionization at the water-energy nexus. Water Res. **183** (2020)
17. L.S.B. Lópeza, V.G. Chirino, D.C.L. Espitia, Estudio para la implementación de una planta desalinizadora y distribución del agua potable en la Alta Guajira (2017)
18. S. Ozlu, I. Dincer, Performance assessment of a new solar energy-based multigeneration system. Energy **112**, 164–178 (2016)
19. S.M. Alsadaie, I.M. Mujtaba, Generic model control (GMC) in multistage flash (MSF) desalination. J. Process Control **44**, 92–105 (2016)
20. E. Nadia, F. Ali, M. Christophe, R.J. Mohamed, Design of helical coil condenser of small-capacity Water/Lithium Bromide absorption cooling machine. Int. J. Hydrogen Energy **42**(13), 8888–8897 (2017)
21. F. Toapanta Ramos, C. Nieto Londoño, W. Quitiaquez, J.K. Toapanta, Ansys CFD analysis of the thermal behavior of coolant 134a in a condenser within a refrigeration cycle. Int. J. Eng. Trends Technol. **63** (2018)
22. A. Brief, *Thermal Management: A Critical CFD App You Have to Get Right* (2017), pp. 1–18
23. Q. Ye, S. Li, Investigation on the performance and optimization of heat pump water heater with wrap-around condenser coil. Int. J. Heat Mass Transfer **143**, 118556 (2019)
24. W. Quitiaquez, J. Estupiñán-Campos, C. Isaza-Roldán, F. Toapanta-Ramos, A. Lobato-Campoverde, Análisis numérico de un sistema de calentamiento de agua utilizando un colector solar de placa plana. Ingenius **24**, 97–106 (2020)

Chapter 8
Combined Methods for Seawater Desalination. Solar Active Desalination Process

Henry Alberto Salinas Freire, Osney Pérez Ones, and Susana Rodríguez Muñoz

Abstract Fresh water deficits have reached most of the world population in recent years, and factors like climate change, population growth and Covid pandemic have increased this problem. Over the nineteenth century, several studies have developed operative methods for seawater desalination, which resulted in the incorporation of desalination plants all around the world, but the most used methods are the following three: reverse osmosis, multi-stage distillation and multi-effect evaporation. The studies have explored promising processes, some of them are very auspicious from the energy analysis, but not many of them have surpassed the pilot plant stage. One of the simplest and oldest methods include the use of solar stills to obtain fresh water and the improvements have come about with the combination of solar stills and mechanisms to facilitate water evaporation, heat exchange, water steam cooling and condensation collecting. This work presents the results of an active solar still seawater desalination process research and compares it with the most commonly used methods.

8.1 Introduction

A fresh water supply is essential for all human life, and, while more than 75% of the planet surface is covered with water, just 0.014%, a small fraction of that is available for use in human activities like food preparation, sanitation, agriculture and industry [1, 2].

H. A. S. Freire (✉) · O. P. Ones
Grupo de Análisis de Procesos, Facultad de Ingeniería Química, Universidad Tecnológica de La Habana José Antonio Echeverría (CUJAE), Ave. 114 N° 11901 e/119 y 127, Marianao, La Habana, Cuba
e-mail: ha.salinas@quimica.cujae.edu.cu

S. R. Muñoz
Grupo de Ingeniería Ambiental, Facultad de Ingeniería Química, Universidad Tecnológica de La Habana José Antonio Echeverría (CUJAE), Ave. 114 N° 11901 e/119 y 127, Marianao, La Habana, Cuba

Á. Rocha et al. (eds.), *Communication, Smart Technologies and Innovation for Society*, Smart Innovation, Systems and Technologies 252,
https://doi.org/10.1007/978-981-16-4126-8_8

Over the past 100 years, water use had a six fold increase [3]. It has been estimated that in 2017, 785 million people worldwide still lacked access to basic services and 144 million still used superficial water they collected directly near the area where they lived [4]. Around 45% of gross domestic product, 52% of world population and 40% of global grain production will be at risk in the year 2050 [5] due to water shortages.

Water scarcity affects more than one billion people worldwide [5], and the deficit grows due to factors such as increased consumption in emerging countries, decreased availability due to pollution and glacial thawing. The consequences for the affected populations are deterioration of sanitary conditions, an increase in diseases and difficulty maintaining sustainable development [6] based on agriculture and industry.

Given the growing need for fresh water around the world, as Khawaji [7] points out, seawater has been considered an essential source for desalination, in order to supply the population of coastal areas for their basic activities such as drinking, hygiene, sanitation services and food preparation.

8.2 Desalination

8.2.1 Seawater Desalination

This process has been mentioned by ancient authors. In his studies, Delyannis [8] reports the first references in the Ancient Testament around year 1500 B.C., and that Aristotle (384–322 B.C.) describes a procedure that consists of boiling the seawater and condensing vapors, obtaining desalinated water [1, 6, 7].

During the Hellenistic Period, a Greek alchemist named Cleopatra the Wise developed several distillers, part of which were called *ambix*. Later, when Arabs wrote about these devices around VII century they wrote their name as *Al-ambiq*, which is where the word *alembic* comes from, and it is how they are currently known [8].

In the Middle Ages, Giovani Batista Della Porta (1535–1615) in his book Magiae Naturalis refers the use of large clay pots that were exposed to solar radiation to evaporate seawater and condense the steam in other containers where desalinated water was collected [8, 10]. Later, Mouchot (1869, 1879) describes in his books a procedure consisting of using polished mirrors from Damascus to concentrate solar radiation on glass containers and thus obtaining fresh water from sea water [8].

Wheeler and Evans in 1870 registered the first patent in the United States for a solar desalination plant and in 1872 a Swedish engineer named Carlos Wilson installed the first known large-capacity solar desalination plant in Las Salinas, Chile. It consisted of a sheet of glass on wooden frames. The covered area was 4450 m^2. The facility comprised a total area of 7896 m^2 and from a mining effluent with a salinity of 140,000 ppm produced 22.7 m^3 of fresh water per day [8]. This plant ran for 40 years.

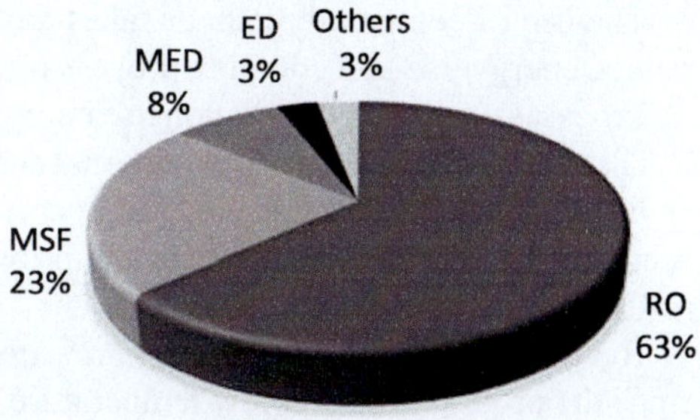

Fig. 8.1 Installed desalination capacity by technology [9]

In addition to the aforementioned procedures, many new desalination methods have been developed [9], but the plants currently installed in the world mainly use these three: reverse osmosis (RO), multi-stage flash evaporation (MSF) and multi-effect distillation (MED) [10]. They are shown in Fig. 8.1 along with electro dialysis (ED) and others.

8.2.2 *Desalination Methods*

These methods can be classified according to the principle they use: mainly, distillation, membranes and electricity and there are new methods based on freezing, ion exchange and others. Nuclear desalination ND [11–13], microbial desalination MBD [14–16], solar desalination SD [1, 17, 18] and others are combined methods. This is shown in Table 8.1 [19].

Table 8.1 Desalination methods based on their principles

Distillation	Membranes	Electricity	Others
MED	RO	ED	SD
MSF	DO	CDI	MBD
MVC	UF		IEX
ND	NF		AD
VD			FR
MD			GHYD
			LLE

Note MED Multi-effect distillation; *MSF* Multi-stage flash evaporation; *MVC* Mechanical vapor compression; *ND* Nuclear desalination; *VD* Vacuum distillation; *MD* Membrane distillation; *RO* Reverse osmosis; *DO* Direct osmosis; *UF* Ultra filtration; *NF* Nano filtration; *ED* Electro dialysis; *CDI* Capacitive de ionization; *SD* Solar desalination; *MBD* Microbial desalination; *IEX* Ion exchange; *AD* Adsorption; *FR* Freezing; *GHYD* Gas hydrate desalination; *LLE* Liquid–liquid extraction; internal face temperature

The use of combined methods takes advantage of specific conditions in order to reduce energy use and complex process operation.

Processes based on distillation use energy to heat the seawater and later evaporate it. These methods require a lot of thermal energy and must deal with energy loss to the environment and associated equipment corrosion. Nevertheless, they are still widely used and have achieved significant improvements in their efficiency and materials resistance to corrosion.

Reverse osmosis is the most economic desalination method using membranes and, since the process works at low temperature, no thermal energy is required, therefore there is no heat loss.

New methods and technical improvements are still in laboratory phases or pilot plants, which makes it difficult for an economic analysis in a commercial installation.

8.3 Methodology

8.3.1 Process Description

This process is based on air humidification–dehumidification and solar desalination methods. Additionally, it has a condensation heat recovery system. Forced air circulation is used in a closed loop and it does not re use the brine.

Sea water is pumped into a pre-heating heat exchanger and then sprayed into an evaporation glass chamber, exposed to solar radiation. Air is blown from above and evaporates part of the sprayed water, increasing air temperature and humidity. Humid air is taken to a cooling heat exchanger where water condenses, and air lowers its temperature. Brine is evacuated from the chamber and cool air is blown again over sprayed water, working in a closed loop. The water heating–air cooling system works with a refrigeration compressor. The scheme can be seen in Fig. 8.2, where the heat removed from the condensation is recovered and serves to preheat the incoming seawater before it passes into the solar evaporation chamber.

The process uses three motors: water pump, air blower and refrigeration system compressor. Evaporation chamber was made of 3-mm transparent glass cube of 1 m on each face with an impermeable plastic base. Technical specifications of these motors are detailed on Table 8.2.

Temperature sensors were located at in and out points for water and temperature-humidity sensors were used for air. Water and air flow were measured by flow sensors after the pump and compressor. Pressure gages were used for water, air and refrigerant gas systems. Solar irradiation and condensed collected were registered, with all sensor readings, every 5 min.

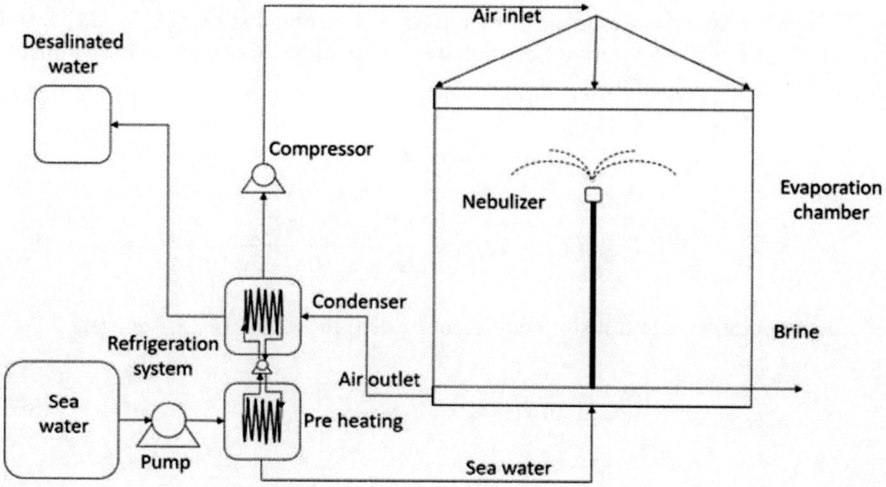

Fig. 8.2 Active solar desalination system

Table 8.2 Technical specification of the equipment used

Characteristics	Water peripherical pump	Refrigeration system compressor	Air compressor
Brand	Pedrollo	Embraco	Makita
Model	PKM60	EGS 100 HLC	UB1103
Rated power	0.5 HP	1/3 HP	0.5 HP
Flow rate/capacity	25 gpm	1050 BTU/h	4.1 m^3/min
Voltage	220 V	220 V	220 V

8.3.2 Experimental Tests

Tests were running with the variation of two factors, with three levels for each one: water flow L and air flow G, as shown in Table 8.3. In each case, the following data was recorded: temperature, humidity, solar irradiation and desalinated water obtained.

Table 8.3 Experimental design for tests

		Air flow G (m^3/h)		
		G1: 1.6	G2: 2.4	G3: 3.2
Water flow L	L1: 0,35	E11	E12	E13
(m^3/h)	L2: 0,70	E21	E22	E23
	L3: 1,00	E31	E32	E33

Theoretical analysis was made using equations for solar distillers [1, 20–22] based on convective and evaporative heat transfer calculations. Convective heat transfer coefficient h_c is defined by:

$$h_c = 0.884\left[\Delta T'\right]^{1/3} \tag{8.1}$$

$$\Delta T' = \left(T_w - T_g\right) + \frac{\left(P_w - P_{gi}\right)\left(T_w\right)}{\left(268.9 \times 10^3 - P_w\right)} \tag{8.2}$$

For evaporative heat transfer coefficient h_e determination, Eq. 8.3 is used:

$$h_e = 16.273 \times 10^{-3} h_c \frac{\left(P_w - P_{gi}\right)}{\left(T_w - T_{gi}\right)} \tag{8.3}$$

Evaporative heat q_e as defined by Cooper [23] is:

$$q_e = 16.273 \times 10^{-3} q_c \frac{\left(P_w - P_{gi}\right)}{\left(T_w - T_{gi}\right)} \tag{8.4}$$

Water evaporation rate $\dot{m}_{ew}$ [24], as a function of evaporative heat h_{fg} is defined by Eq. 8.5.

$$\dot{m}_{ew} = \frac{q_{ew}}{h_{fg}} \tag{8.5}$$

Theoretical total evaporation m based on convective and evaporative coefficients in t time T_{hc-e} is calculated using Eq. 8.6.

$$T_{hc-e} = m = \int_0^t \dot{m}_{ew} \mathrm{d}t \tag{8.6}$$

In another analysis, heat and mass transfer coefficients were calculated using Ranz and Marshall correlations [25]. These equations used Nusselt, Sherwood, Reynolds, Prandtl and Schmidt adimensional numbers [26, 27] as is shown in Eqs. 8.7 and 8.8.

$$\mathrm{Nu} = \frac{h_t(2R_{in})}{k_g} = 2 + 0.65\mathrm{Re}^{0.5}\,\mathrm{Pr}^{0.33} \tag{8.7}$$

$$\mathrm{Sh} = \frac{k_m(2R_{in})}{D_{wv}} = 2 + 0.65\mathrm{Re}^{0.5}\mathrm{Sc}^{0.33} \tag{8.8}$$

where Nu is the adimensional Nusselt number for air; Re is adimensional Reynolds number for air; Sh represents the adimensional Sherwood number for air; Pr stands for the adimensional Prandtl number for air; Sc is the adimensional Schmidt number for air; h_t represents the total heat transfer coefficient (J/s m^2 K); R_{in} is the particle equivalent radius (m); k_g is the air thermal conductivity (J/m °C); k_m is the mass transfer coefficient (m/s); and D_{wv} stands for the water diffusivity in air (m^2/s).

With h_t and using Eqs. 8.3–8.6, theoretical total evaporation T_{R-M} with Ranz-Marshall correlations is calculated. These two theoretical analyses were made with acquired data and compared with experimental desalted water E_m results.

8.4 Results

Experimental tests were performed in Lima, Peru between October 2019 and February 2020. The values obtained are shown in Table 8.4.

With experimental data obtained, mass and energy transfer coefficients and theoretical productivity T_{hc-e} and T_{R-M} were calculated, as explained in Sect. 8.3.2, and plotted with the experimental E_m obtained, as seen on Fig. 8.3.

Experimental productivity E_m is lower than both theoretically calculated. The difference can be explained because of non-homogenous temperature, humidity and pressure of fluids in evaporation chamber, solar irradiance and wind speed variation. It is important to notice that discrepancies are bigger at low-flow rates and tend

Table 8.4 Experimental desalinated water E_m obtained

Desalinated water, E_m (l/h)	G1	G2	G3
L1	1.2	1.3	1.6
L2	1.3	1.5	1.7
L3	1.7	2.0	2.5

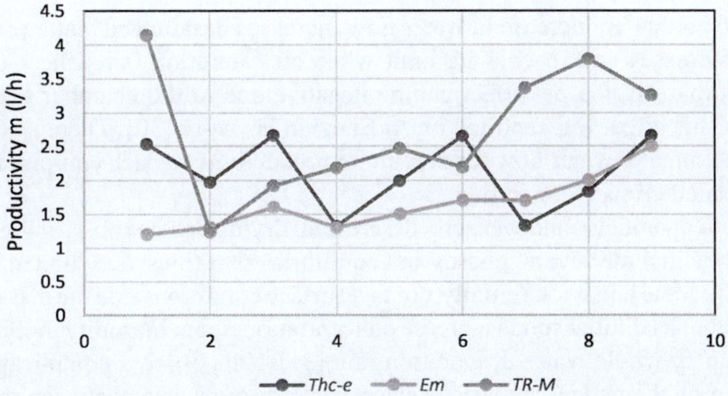

Fig. 8.3 Theoretical and experimental productivity

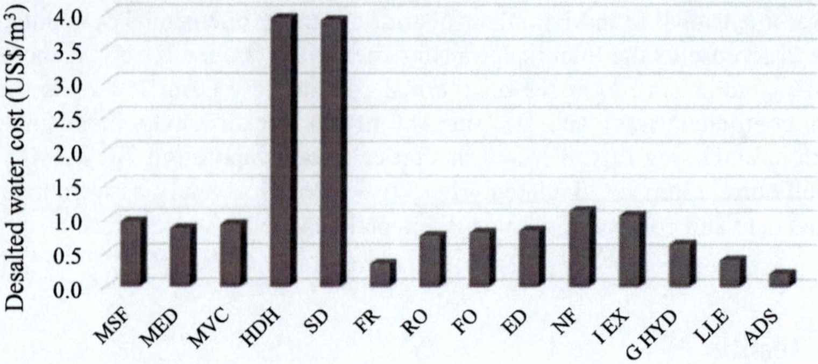

Fig. 8.4 Desalinated water costs comparison [28]

to reduce at high water and air flow rates. The experimental data and theoretical calculations allow us to compare energy use of the active solar desalination system with other methods costs presented by Youssef [28]. This is shown in Fig. 8.4.

Large capacity industrial desalination plants using MSF, MED, RO and ED correspond around 97% of desalination plants installed around the world [9, 29] and have been working for several decades, so those methods have been widely studied, improved and their costs reduced. Higher costs correspond to humidification–dehumidification and solar desalination stills, methods like freezing, adsorption, gas hydrates, etc. are still in laboratory research phase, and cost reported were obtained in small capacity equipment; as for solar stills, the high cost corresponds to their low productivity, and studies are being conducted to increase productivity and efficiency in simple technology desalinators to make them more competitive.

8.5 Discussion

It is observed that an increase in water flow increases desalinated water production, but this decreases and reaches its limit when air saturation is reached. Once this point is surpassed, the net effect can be negative due to the chamber temperature decrease. This effect was reported by Al Enezi in his work [30]. There is, therefore, an optimal value of water flow for maximum productivity in each equipment system and associated climate conditions.

Theoretical models cannot exactly describe all drying process on a droplet surface, because they include several phases and equilibria conditions, and diluted solids do not have the same behavior. Initially, contact surface conditions rule the mass and heat exchange, but as diluted solids increase concentration, inner humidity mobility starts to determinate whole water evaporation rate [31]. This process complicates calculations and mathematical analysis causing some discrepancies between theoretical and experimental results for spray drying and seawater evaporation.

Desalinated water costs depend on local installation, operations and energy costs. The main cost of this method is due to electrical energy, therefore water cost depends on local energy cost, mainly produced using fossil fuels [32].

An active solar desalinator uses solar irradiation as thermal energy without transforming it, as in photovoltaic systems for example, and does not have efficiency losses, although this limits its operation to maximum efficiency during daylight hours.

8.6 Conclusions

The conclusions of this research work are the following: (1) the air flow in the process directly affects the volume of distillate obtained, but at high-air speed ranges seawater particles are dragged; (2) combined desalination method of active solar still is addressed to produce water for human consumption in small towns; (3) the active solar still is focused to produce fresh water in areas and/or stations where there is a high incidence of solar radiation, as well as high-environmental temperature; (4) the active solar still can be used 24 h a day, although maximum efficiency will only be achieved when high-solar irradiation is available during daylight hours, and (5) the overall cost of desalinated water obtained using the active solar still, under the conditions studied, is higher than the other reported methods.

References

1. G. Tiwari, H. Singh, R. Tripathi, Present status of solar distillation. Sol. Energy **75**(5), 367–373 (2003)
2. S. Rashidi, N. Karimi, O. Mahian, J.A. Esfahani, A concise review on the role of nanoparticles upon the productivity of solar desalination systems. J. Therm. Anal. Calorim. **135**(2), 1145–1159 (2019)
3. UNESCO, United Nations World Water Development Report 2020: Water and Climate Change (UNESCO, Paris, 2020). Available at: https://unesdoc.unesco.org/ark:/48223/pf0000372985.locale=en. Accessed 02 March 2021
4. United Nations Children's Fund, World Health Organization, *Progress on Household Drinking Water, Sanitation and Hygiene 2000–2017. Special Focus on Inequalities* (New York, 2019)
5. United Nations, *Sustainable Development Goal 6 Synthesis Report 2018 on Water and Sanitation* (United Nations, New York, 2018). Available at: https://www.unwater.org/publication_categories/sdg-6-synthesis-report-2018-on-water-and-sanitation/
6. K.P. Lee, T.C. Arnot, D. Mattia, A review of reverse osmosis membrane materials for desalination—development to date and future potential. J. Membr. Sci. **370**(1–2), 1–22 (2011)
7. A.D. Khawaji, I.K. Kutubkhanah, J.-M. Wie, Advances in seawater desalination technologies. Desalination **221**(1–3), 47–69 (2008)
8. E. Delyannis, Historic background of desalination and renewable energies. Sol. Energy **75**(5), 357–366 (2003)
9. V.G. Gude, Desalination and sustainability—an appraisal and current perspective. Water Res. **89**, 87–106 (2016)

10. S.A. Kalogirou, Seawater desalination using renewable energy sources. Prog. Energy Combust. Sci. **31**(3), 242–281 (2005)
11. A. Adak, I. Rao, V. Srivastava, P. Tewari, Nuclear desalination by waste heat utilisation in an advanced heavy water reactor. Int. J. Nucl. Desalin. **2**(3), 234–243 (2007)
12. A. Sudi, S. Alimah, Economic aspect for nuclear desalination selection in Muria Peninsula, in *Proceedings of an International Conference on Opportunities and Challenges for Water Cooled Reactors in the 21st Century* (2011)
13. A. Al-Othman, N.N. Darwish, M. Qasim, M. Tawalbeh, N.A. Darwish, N. Hilal, Nuclear desalination: a state-of-the-art review. Desalination **457**, 39–61 (2019)
14. H.M. Saeed et al., Microbial desalination cell technology: a review and a case study. Desalination **359**, 1–13 (2015)
15. S. Sevda, H. Yuan, Z. He, I.M. Abu-Reesh, Microbial desalination cells as a versatile technology: functions, optimization and prospective. Desalination **371**, 9–17 (2015)
16. J.S. Sravan, A. Tharak, J.A. Modestra, I.S. Chang, S.V. Mohan, Emerging trends in microbial fuel cell diversification—critical analysis. Bioresource Technol. 124676 (2021)
17. R. Dunkle, Solar water distillation: the roof type still and the multiple effect diffusor. Int. Dev. Heat Transfer (1961)
18. N. Mathew, A.N. Jejurkar, P.S. Nair, *A Review on Existing Solar Distillation Systems* (2021)
19. R.K. AL-Dadah, P.G. Youssef, S.M. Mahmoud, Comparative analysis of desalination technologies. Energy Procedia **61**, 4 (2014)
20. K. Sampathkumar, T. Arjunan, P. Pitchandi, P. Senthilkumar, Active solar distillation—a detailed review. Renew. Sustain. Energy Rev. **14**(6), 1503–1526 (2010)
21. N. Setoodeh, R. Rahimi, A. Ameri, Modeling and determination of heat transfer coefficient in a basin solar still using CFD. Desalination **268**(1), 103–110 (2011)
22. S. Kumar, G. Tiwari, Estimation of convective mass transfer in solar distillation systems. Sol. Energy **57**(6), 459–464 (1996)
23. P. Cooper, The maximum efficiency of single-effect solar stills. Solar Energy **15**(3), 205–217 (1973)
24. D. Mowla, G. Karimi, Mathematical modelling of solar stills in Iran. Sol. Energy **55**(5), 389–393 (1995)
25. W. Ranz, W.R. Marshall, Evaporation from drops. Chem. Eng. Prog. **48**(3), 141–146 (1952)
26. R. Wisniewski, Spray drying technology review, in *45th International Conference on Environmental Systems* (2015)
27. K. Masters, *Spray Drying Handbook* (Longman Scientific and Technical, London, 1991)
28. P. Youssef, R. Al-Dadah, S. Mahmoud, Comparative analysis of desalination technologies. Energy Procedia **61**, 2604–2607 (2014)
29. S. Burn et al., Desalination techniques—a review of the opportunities for desalination in agriculture. Desalination **364**, 2–16 (2015)
30. G. Al-Enezi, H. Ettouney, N. Fawzy, Low temperature humidification dehumidification desalination process. Energy Convers. Manag. **47**(4), 470–484 (2006)
31. R. Vehring, W.R. Foss, D. Lechuga-Ballesteros, Particle formation in spray drying. J. Aerosol Sci. **38**(7), 728–746 (2007)
32. B.E. Logan, *The Global Challenge of Sustainable Seawater Desalination* (ACS Publications, 2017)

Chapter 9
Simulation of Centralized and Decentralized Photovoltaic Generation Systems for Isolated Sectors of the Electricity Grid

Jorge Tinajero, Wellington Mayorga, and Gary Ampuño

Abstract The path to reliable and safe electricity supply requires changes in generation systems, decentralized solutions is a promising concept for villagers living in isolated areas, these systems are characterized by the use of local energy, producing clean energy, however the intermittency of renewable sources leads to analyzing the best configuration of the type of system. Photovoltaic solar systems today are one of the most promising technologies for on-site generation and accessible to communities with little economic resources. Each configuration of photovoltaic systems has advantages and disadvantages compared to the other, which are reviewed so that the reader has a clear idea of the convenience of implementing this type of systems. The document reviews commercial software for the simulation, analysis, and sizing of renewable energy systems, selecting two software's for the sizing and comparison of centralized and decentralized systems. Simulation results indicate that the centralized system has a lower energy cost, however the software analyzed does not consider the laying costs of the micro-network distribution lines of the case or study or distribution losses. The cost of energy in the decentralized system is $0.66/kWh and 0.64 $/kWh for centralized, centralized installations require fewer batteries to store surplus energy produced by photovoltaic panels.

9.1 Introduction

Ensuring access to electricity is one of the sustainable development goals (SDGs) approved for the 2030 agenda by the United Nations general assembly, seeking to improve quality of life [1] and prosperity for approximately one million people who do not have access to electricity and live in remote areas. The implementation of renewable energy systems in isolated areas has been developed in recent years due

J. Tinajero · W. Mayorga · G. Ampuño (✉)
Universidad Politécnica Salesiana, Guayaquil, Ecuador
e-mail: gampuno@ups.edu.ec

J. Tinajero
e-mail: jtinajero@est.ups.edu.ec

Á. Rocha et al. (eds.), *Communication, Smart Technologies and Innovation for Society*, Smart Innovation, Systems and Technologies 252,
https://doi.org/10.1007/978-981-16-4126-8_9

to the technical and environmental advantages they offer, in addition to the modern technology they offer giving a high reliability of energy supply [1].

Renewable energy systems can be centrally installed for a community or in a decentralized manner for each of the homes, the second option arises alternatively, avoiding costs of laying down electricity networks that in many cases becomes unsustainable for the project due to the distance between homes and the limited energy produced by renewable generators, thus ensuring that all dwellings are fully supplied [2].

For reliable power supply, it is important to evaluate and compare the centralized generation (GC) and decentralized generation (GD) options of renewable systems, the lower cost of energy plays a key role in decision-making, the GD is applied on a small scale, offering flexibility and modularity, with a more efficient design built close to loads, even combining more than one renewable source. However, it is necessary to work with the right tools at the design and simulation stage of different configurations taking advantage of the availability of renewable resources and the load profile [3].

Bertheau et al. [4] compare the feasibility of implementing centralized and decentralized systems for 132 islands ensuring affordable, reliable and sustainable electricity supply. They apply a grid contour optimization tool where it calculates for each island the lowest cost hybrid system consisting of a diesel generator, photovoltaic panels, a battery bank, sensitivity analysis indicates that decentralized systems are a cost-effective and feasible option for most islands reaching the energy cost up to 0.21 USD/kWh [4].

Liu et al. [5] investigate the advantages of becoming independent of centralized supply even when it has no economic benefits, so researchers present a regional system model where they optimize the expansion of the capacity of photovoltaic systems and battery bank, the authors quantify and evaluate the degree of autonomy for 166 regions of southern Germany [5].

Several research and implementations of centralized and decentralized renewable energy systems have been conducted in communities in need of reliable and cost-effective electricity, in Ref. [6–10] they use commercial programs such as HOMER, iHOGA, TRNSYS, to design, optimize and simulate hundreds of configurations, using a number of commercial generators such as solar panels, wind turbines, diesel generators, biomass generators, biomass generators, batteries, hydrogen system, thermal sources etc. The tools allow to obtain the optimal system with the lowest cost of COE energy, in addition to the sensitivity analysis for any type of system configuration [6–10].

The document aims to compare the feasibility of centralized and decentralized photovoltaic systems for isolated sectors of the electricity grid, in the first part the state of photovoltaic systems in the country is presented, the second part compares the advantages and disadvantages of photovoltaic systems along with the software's used, the third part presents a case study in village San José del Carmen de Mata de Cacao Province Los Ríos, in the end the conclusions are presented.

9.2 Methods

In Ref. [11] reviewed the status of simulation, optimization and control technologies for hybrid solar-wind power systems without network connection, with storage in a battery bank, however by 2010 the need for continuous research and development effort in the area is detected to improve system performance, seek and standardize techniques to predict the energy production of systems and reliably integrate them with other sources renewable energy generation [11].

Currently you have several software tools for sizing, analysis, economic evaluation etc. Each with its own characteristics, advantages and user needs, this section introduces 13 software's used in hybrid systems and discusses 2. Sizing is used to determine the optimal size of each of the system elements and indicate component performance (HOMER), other simulation software with open architecture, the user modifies the algorithm that controls component interaction (TRNSYS).

According to [12] simulation programs can be classified into four categories, simulation, sizing, micro-network design and research, analyzed 37 energy system integration tools, including three hybrid renewable energy system simulation programs, another important point is the cost of the programs, Table 9.1 shows the selected programs that are commercially located, the developer, the type of analysis and the price along with availability [12].

9.2.1 *Case Study*

The approach of this work is the comparison of different types of systems and way of implementation to satisfy energy and reliably microgrid for the generation of the San José del Carmen de Mata de Cacao Campus belonging to the Province of the Rivers, the village is located at the coordinates – 1.976952, – 79,368,202 to 33 km from the city of Babahoyo, where 11 families with a total of 50 people live, the main activity is agriculture.

The energy consumption of families has been carried out according to field visits that have been made to nearby communities that have electricity service. The maximum power per family is 0.27 kW, while the minimum power is 0.02 kW, the load profile is shown in Fig. 9.1.

9.2.2 *Renewable Resources*

Through Pvsyst and Homer software, the weather base of the variables present at the study site has been obtained, Fig. 9.2 shows the overall radiation for the whole year by intervals of 1 h, this data is the input variable for the simulation, the place

Table 9.1 Characteristics of various software tools for hybrid systems simulation

Software	Developer	Latest version	Type of analysis
HOMER	NREL, USA (1993)	HOMER PRO 3.14	Technical analysis; economic analysis; emissions analysis
HYBRID2	University of Massachusetts, USA, and NREL (Hybrid 1 in 1994, Hybrid 2 in 1996)	Version 1.3	Technical analysis; economic analysis
RET Screen	Developed by the Ministry of Natural Resources, Canada in 1998	RETScreen EXPERT	Environmental financial analysis
IHOGA	University of Zaragoza, Spain	Version 3.0	Multi-lens or monobjective optimization using genetic algorithm
Island	German University of Oldenburg (1986–1991)	Version 8	Planning to track electrical and thermal power systems
TRNSYS	University of Wisconsin and University of Colorado (1975)	Version 17.1	simulate the transient behavior of the system
¡GRHYSO	University of Zaragoza, Spain	–	Technical analysis; economic analysis
YBRIDS	Solaris Houses	–	Technical analysis
RAPSIM	Australian University Energy Research Institute (1996)	Version 2 (1997)	Simulates the performance of a range of hybrid power systems
SOMES	University of Utrecht, Netherlands (1987)	–	Technical analysis; economic analysis
SOLSTOR	SNL (late 1970s and early 1980s)	–	Technical analysis
HySim	SNL (late 1980s)	–	Financial analysis
HybSim	SNL	–	Cost–benefit analysis

receives 127.9 kWh/m^2/month, but the month is June with 3.7 peak solar hours, data of importance for the sizing of the system.

Wind resource is limited in the locality, less than 4 m/s for this reason is not considered for sizing or analysis, commercial wind turbines for general the rated power need a power greater than 10 m/s [13].

9.3 System Components

Knowing the weather variables in the selected place, it is proposed that the centralized or decentralized system will consist of solar panels as a generator, a battery bank

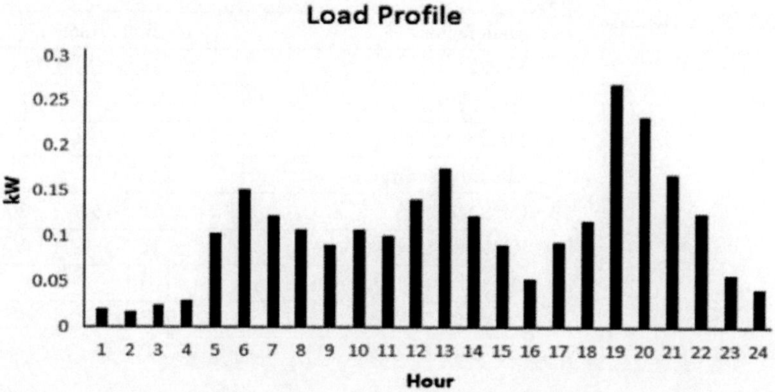

Fig. 9.1 Load profile of the Mata de Cacao

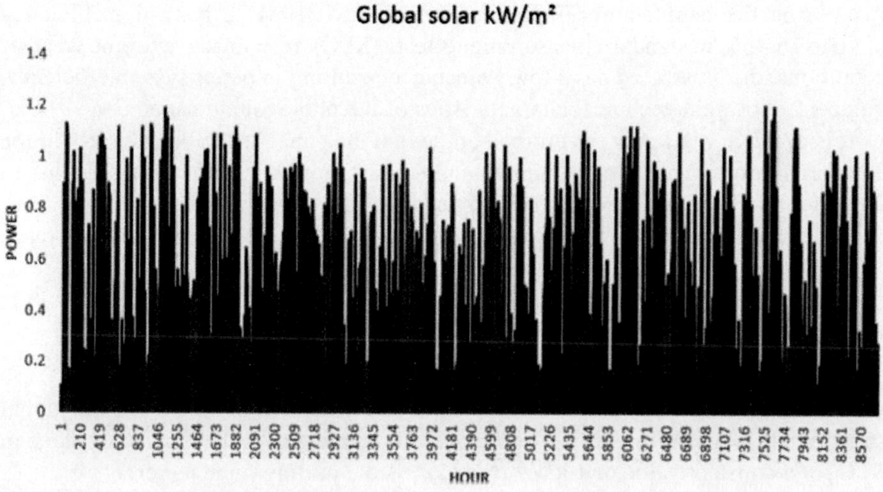

Fig. 9.2 Solar radiation at the simulation site

that does the function of storage and backup in case the photovoltaic panels do not generate the necessary amount of energy required by the load, an inverter that converts the DC generation into AC for the proper functioning of the loads of the house.

9.3.1 Photovoltaic Panels

Its function is to take advantage of solar energy from the sun in the form of electrons, which give up the energy to produce the photovoltaic effect. The union of several

Table 9.2 PV technical characteristics

Manufactory	Jinko solar
Model	JKM 210M-72
Power PV (Pmax)	210 W
Maximum voltage	37.4 VDC
Maximum current	5.6 A
Capital cost	US$ 160/210 Wp
Replacement cost	US$ 160/210 Wp
Operating and maintenance cost	$10
Lifetime	25 years

cells form the so called photovoltaic solar panel or module, commercially there are panels of 36, 60, and 72 cells.

Photovoltaic solar panels are polycrystalline type with a maximum power of 210 W from the manufacturer Jinko solar, model JKM 210M-72, has a high efficiency of up to 16.45% in standard measurements tests (STC), responds to extreme weather conditions, the glass used has a low reflectance resulting in better system efficiency. Table 9.2 shows the technical characteristics of the photovoltaic panel.

It is necessary to know the optimal angle that the panel has to be installed, in the equatorial zone it has an approximate angle of 5°, however, to avoid losses due to dirt, and the panels are placed at 15° so that dust and rain can roll.

Equation 9.1 is used to calculate the power delivered by the photovoltaic generator.

$$P_{\mathrm{PV}} = P_{\mathrm{PV,Rated}} \times d_{\mathrm{r}} \times \left(\frac{G_{\mathrm{T}}}{G_{\mathrm{T,STC}}} \right) \tag{9.1}$$

where, the power of the photovoltaic field P_{PV} is measured in kW, G_{T} is the global solar radiation in kW/m^2, $G_{\mathrm{T,STC}}$ is the overall radiation over the horizontal plane in STC for a constant value of 1 kW/m^2, and d_{r} is the performance factor.

9.3.2 *Battery Bank*

The battery bank is the only component that performs two functions that of storing energy when the generation is greater than demand, the excess energy is converted into chemical energy. The second function is to generate the energy contained in the bank to meet the needs of the load, especially in the hours of the night or when you have a peak consumption. Commercial acid lead batteries are used for simulation, despite how polluting they are, the competitive price for the number of operating cycles makes it suitable for insulated photovoltaic systems. Table 9.3 shows the characteristics of the batteries used.

Table 9.3 Battery technical characteristics

Model	Dural SC
Nominal voltage (V)	12
Nominal capacity (kWh)	1.5
Maximum capacity (Ah)	150
Roundtrip efficiency (%)	97
Capital cost ($)	290
Replacement cost ($)	290
Operating and maintenance ($/year)	6

Equation 9.2 can be used to calculate the number of components of the battery bank during project uptime (N_{bat}).

$$N_{\mathrm{bat}} = \mathrm{ceil}\left(\frac{\mathrm{Life}_{\mathrm{HS}} \times \mathrm{Life}_{\mathrm{bt}}^{\mathrm{pu,year}}}{T_{\mathrm{bat}}^{\mathrm{life}}}\right) \tag{9.2}$$

where $T_{\mathrm{bat}}^{\mathrm{life}}$ is the time from the start of the project to the last change of batteries; $\mathrm{Life}_{\mathrm{HS}}$ is the useful life of the photovoltaic system; $\mathrm{Life}_{\mathrm{bt}}^{\mathrm{pu,year}}$ represents the lifespan of batteries, and ceil brings the expression to the sum.

9.3.3 *Power Coupling*

The power coupling consists of the charge controller that performs multiple functions among which it is highlighted to regulate the voltage between the solar panel and the batteries, check the optimal charge of the batteries and protect against deep discharges, in this work MPPT controls, model ProStar 25M@12V, has an efficiency of 94%. While the inverter used is from the manufacturer Studer, model AJ 500-12, the component has the functionality of converting direct current to alternating current at a voltage of 110 or 220 V, the inverter supports an input voltage of up to 16 VDC, the yield is 93%, the output is sine wave.

9.4 Results

Decentralized systems have several advantages compared to centralized, economic resources have less difficulty in obtaining them, the project is reduced, maintenance work is much simpler and cheaper since complete equipment can be changed, while centralized systems They have the advantages of optimizing the size of the system, lower investment and costs compared to individualized systems, ease of operation and maintenance plans, additional rates can be set according to consumption.

The two systems have been simulated in two commercial software and widely used by researchers, and Table 9.4 shows the main results obtained in the simulations. The results are marked in difference in cost due to the sensitivity and entry of values in each software, however it has a similarity, the centralized system (SC) has a lower cost than the decentralized system (SD), in HOMER the leveled cost of energy (COE) for sc is 0.66 USD/kWh and for SD is 0.66 USD/kWh, in the second software PVsyst the difference is much greater with 0.58 USD/kWh and 0.41 USD/kWh, respectively.

To check the operation of the centralized system has been simulated in TRNSYS for 96 h, Fig. 9.3 shows the generation of the photovoltaic system which is higher than the charge and at that time the excess energy is stored in the battery bank, when the generation is less than the charge, the batteries deliver the energy needed to meet the energy needs.

The energy produced by the photovoltaic system is 23,327 kWh/yr, the energy consumed by the load is 9223 kWh/yr, the unmet load is 11.8 kWh/yr which corresponds to 0.12%, the results delivered by Homer, indicates that 90 batteries are needed, with a range of 51.3 h, the cost of storage is 0.419 $/kWh, the life expectancy of the bank according to the number of cycles is 10 years, the stored energies 6348 kWh/yr, while the delivered is 1272 kWh/yr with losses of 1272 kWh/yr.

9.5 Conclusions

In this work, two types of renewable energy systems are simulation to meet the needs of an isolated population, the results show that the two types of systems work correctly and efficiently, with respect to the size the photovoltaic generator in the centralized system is of greater power, while in this system the size of the battery bank is smaller if we multiply the battery needed for each house by the number of homes.

Economic analysis is evaluated using the total system cost (NPC) parameter where all costs incurring project lifetime such as the cost of capital are, component replacement costs, and operating and maintenance costs, the economic result indicates that centralized systems are more efficient than decentralized from the overall size of the photovoltaic system and economic analysis, the former costing $0.64/kWh and the second 0.66 $/kWh.

It remains as an open line of research to add costs for power lines, poles, isolators etc. and technical losses on distribution lines, these values increase the initial costs of the project and have not been considered in this study.

Table 9.4 Battery characteristics technical

Software	Decentralized system				Centralized system			
	PPV (kW)	Battery (One)	NPC (USD)	COE (USD/kWh)	PPV (kW)	Battery (One)	NPC (USD)	COE (USD/kWh)
HOMER	1	8	7261	0.66	18	85	77,163	0.64
Pvsyst	0.84	6	4490	0.58	8.4	24	41,710	0.41

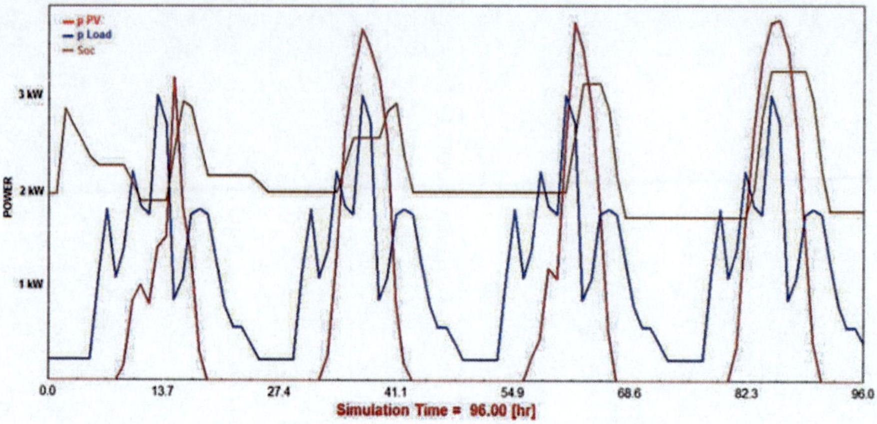

Fig. 9.3 Solar radiation at the simulation site

References

1. United Nations, United Nations Office of the High Representative for the Least Developed Countries, Landlocked Developing Countries, and Small Island Developing States (2nd SE4All Forum Commitments)—United Nations Partnerships for SDGs Platform (2011). https://sustainabledevelopment.un.org/partnership/?p=10125. Accessed 16 May 2017
2. M. Yaqoot, P. Diwan, T.C. Kandpal, Review of barriers to the dissemination of decentralized renewable energy systems. Renew. Sustain. Energy Rev. **58**, 477–490 (2016)
3. M. Yazdanie, M. Densing, A. Wokaun, The role of decentralized generation and storage technologies in future energy systems planning for a rural agglomeration in Switzerland. Energy Policy **96**, 432–445 (2016). https://doi.org/10.1016/j.enpol.2016.06.010
4. P. Bertheau, C. Cader, Electricity sector planning for the Philippine islands: considering centralized and decentralized supply options. Appl. Energy **251**, 113393 (2019). https://doi.org/10.1016/j.apenergy.2019.113393
5. W.H. Liu, W.S. Ho, M.Y. Lee et al., Development and optimization of an integrated energy network with centralized and decentralized energy systems using mathematical modelling approach. Energy **183**, 617–629 (2019). https://doi.org/10.1016/j.energy.2019.06.158
6. M.R. Elkadeem, S. Wang, S.W. Sharshir, E.G. Atia, Feasibility analysis and techno-economic design of grid-isolated hybrid renewable energy system for electrification of agriculture and irrigation area: a case study in Dongola, Sudan. Energy Convers. Manag. **196**, 1453–1478 (2019). https://doi.org/10.1016/j.enconman.2019.06.085
7. M. Baneshi, F. Hadianfard, Techno-economic feasibility of hybrid diesel/PV/wind/battery electricity generation systems for non-residential large electricity consumers under southern Iran climate conditions. Energy Convers. Manag. **127**, 233–244 (2016). https://doi.org/10.1016/J.ENCONMAN.2016.09.008
8. Y. Kalinci, A. Hepbasli, Techno-economic analysis of a stand-alone hybrid renewable energy system with hydrogen production and storage options. Int. J. Hydrogen Energy **40**, 7652–7664 (2015). https://doi.org/10.1016/J.IJHYDENE.2014.10.147
9. H. Shamachurn, Optimization of an off-grid domestic Hybrid Energy System in suburban Paris using iHOGA software. Renew. Energy Focus (2021). https://doi.org/10.1016/j.ref.2021.02.004
10. M.S. Behzadi, M. Niasati, Comparative performance analysis of a hybrid PV/FC/battery stand-alone system using different power management strategies and sizing approaches. Int. J. Hydrogen Energy **40**, 538–548 (2015). https://doi.org/10.1016/j.ijhydene.2014.10.097

11. W. Zhou, C. Lou, Z. Li et al., Current status of research on optimum sizing of stand-alone hybrid solar-wind power generation systems. Appl. Energy **87**, 380–389 (2010)
12. D. Connolly, H. Lund, B.V. Mathiesen, M. Leahy, A review of computer tools for analysing the integration of renewable energy into various energy systems. Appl. Energy **87**, 1059–1082 (2010)
13. R. Hosseinalizadeh, G.H. Shakouri, M.S. Amalnick, P. Taghipour, Economic sizing of a hybrid (PV-WT-FC) renewable energy system (HRES) for stand-alone usages by an optimization-simulation model. Renew. Sustain. Energy Rev. **54**, 139–150 (2016)

and challenge for users. A mobile device [7, 8] with access to the Internet has an important and notorious role in daily activities in today's new normal.

Although access to Web services remained a challenge for many users, mobile learning as a teaching methodology became especially useful for both private and public organizations. Its rapid acceptance created new ways to accompany students through a new teaching experience based on mobile technology for the formation and development of formative and collaborative learning areas for students and teachers [9–14]. Although the COVID-19 health emergency obliged educational institutions to adopt virtual platforms, the experience has shown an opportunity to use technologies through mobile access devices [15–17] facilitating virtual meetings [18–21] under free access through communication applications allowing teachers to virtually fulfil their role as educators [22–27]. Web services that offer virtual spaces such as those from Google, especially Google Classroom, are part of a set of platforms that generate mobile interaction creating a virtual classroom without the need for infrastructure [28]. These platforms allow access from anywhere and include tools and technology [29] that have been used to create an effective learning experience through mobile learning [30–32].

The Salesian University Association (ASU) of the Salesian Polytechnic University (UPS) is an association created to offer extracurricular activities for UPS students in the cities of Quito, Cuenca, and Guayaquil (Ecuador) (see Fig. 10.1). Each student group aims to educate students on different topics such as arts and culture, academics, sports, and religion. The groups share a common objective: to generate change in society starting from the university. The association is structured independently of the student political association and welcomes young people in each of the three cities where the university functions.

The students who participate in the ASU are from different years and majors at the university. The association seeks to provide activities that offer students additional experiences related to their fields of study and further promote their professional formation. This paper shows the results of using virtual classrooms provided by Web services to obtain said experiences.

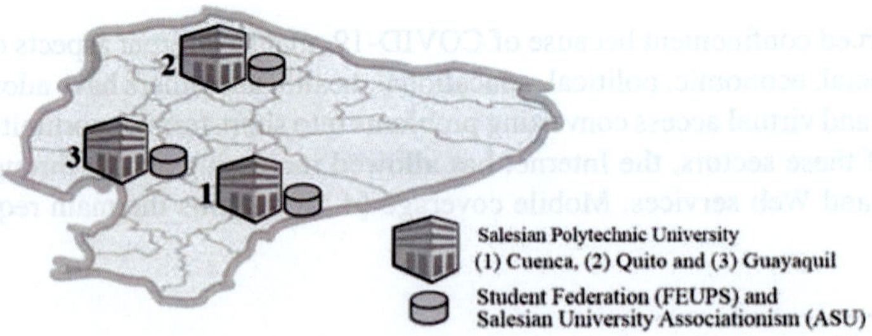

Fig. 10.1 Structure of student participation in each of the spaces proposed by the university

10.1.1 Background Literature

Online interaction spaces for a learning process require the use of resources and meaningful activities [8]. The realization of these activities depends to a large extent on the knowledge that participants can share with each other through a group methodology in the virtual world [10]. The processes involved in the construction of learning spaces are accompanied by technological tools and services that allow greater accessibility to participants from their computers or mobile devices with access to the Internet [13, 25]. The strategies that are evident in higher education institutions with a particular style of education focus on initiatives that allow for greater student participation in these spaces of expression during extracurricular times [32, 33].

The relationship that generates meaningful learning from a relational environment allows for a motivational teaching–learning process and positive manifestations among the participants [34–36], thus learning from a self-motivated [24, 37] and supported by the use of technologies [38] to enhance comprehension, attention, and retention skills in a conducive environment [39, 40].

It is evident that the improvements in digital skills by participants in online workspaces using Google Classroom technologies allow its applicability in different scenarios for the same objective, the improvement of its users [41, 42].

10.2 Methods

This is an empirical investigation using a quasi-experimental quantitative approach in which a survey is applied to a group of 19 participants from the academic and sports ASU associations. Participants are at least 19 years old. The association's objective is to use Google Classroom's mobile platform to generate virtual spaces through the creation of two classrooms (see Fig. 10.2) called "ASU ACADEMICO" (ASU Academic Group) and "ASU DEPORTIVO" (ASU Sports Group) creating

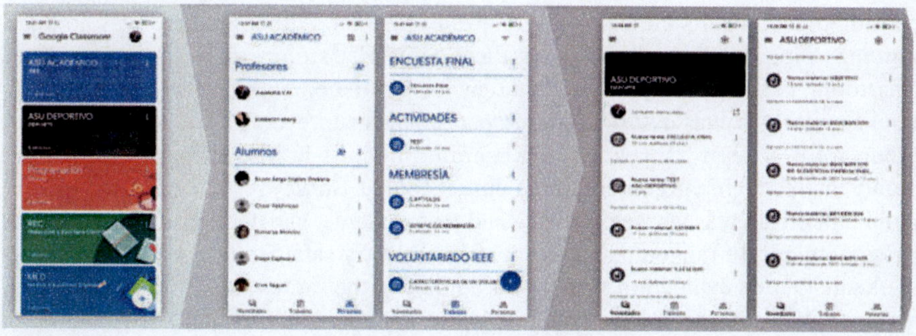

Fig. 10.2 Virtual spaces for communication, learning, and exchange

an asynchronous training space, providing opportunities for communication and exchange of activities.

Each group uploaded meeting materials, training activities, as well as interactive resources to the platform. The virtual ASU Academic Group was led by the current student board of the student branch of the Institute of Electrical and Electronic Engineers UPS (IEEE UPS) in the city of Guayaquil. They used the following resources to measure the knowledge acquired by students during the Web services mobile training process: documentation of its Objectives, Mission, Membership, IEEE Volunteering Manual, IEEE UPS-GYE Contribution, Activities, and Evaluations. The ASU Sports Group used course information, ASU Sports classification, physical recovery techniques, sports diet, and evaluation (see Fig. 10.2).

The groups themselves created the classrooms and sections and prepared supporting materials. The invitations to participate were sent as links through the Google Classroom virtual classroom and the Gmail account of each participant. Participants included students, tutors, and researchers from the GIEACI[1] group.

The ASU volunteer participants have several factors in common such as their year at the university, the fact that they have little free time during the day, and their participation in work–study activities. During the semester, individual and group training activities are completed under the planned supervision of a tutor. The mobile learning methodology encourages the tutor to rigorously prepare content published on the platform to promote the professional and personal growth of ASU participants.

10.3 Results and Discussion

Using Google Forms, a set of questions was developed for the application of a survey to the group of selected participants. Taking into consideration participant variation in each group, independent results were obtained from each virtual classroom (ASU Sports-ASU Academic).

Regarding ASU Sports respondents, it was determined that the participants' age range is between 19 and 21 years old. Participants' interest in course content was consulted, and results indicate a high level of interest. Although there was no question regarding the difficulty of using the Google Classroom application as a virtual classroom, participants indicated that they were satisfied with the use of the application. Similarly, it was determined that participants surveyed by the ASU Academic group were at least 21 years old. Regarding interest levels, all participants were highly interested in the content of the virtual course. Survey results indicated that the mobile platform was very easy to use and that students' satisfaction with the Google Classroom course ranged between satisfied and very satisfied (see Fig. 10.3).

A satisfaction evaluation survey was applied to the sample of 19 volunteer participants to measure the level of interest in the content used in each ASU course for the contribution of knowledge in each specific area. The results show that 78% of

[1] Available in: https://gieaci.blog.ups.edu.ec/.

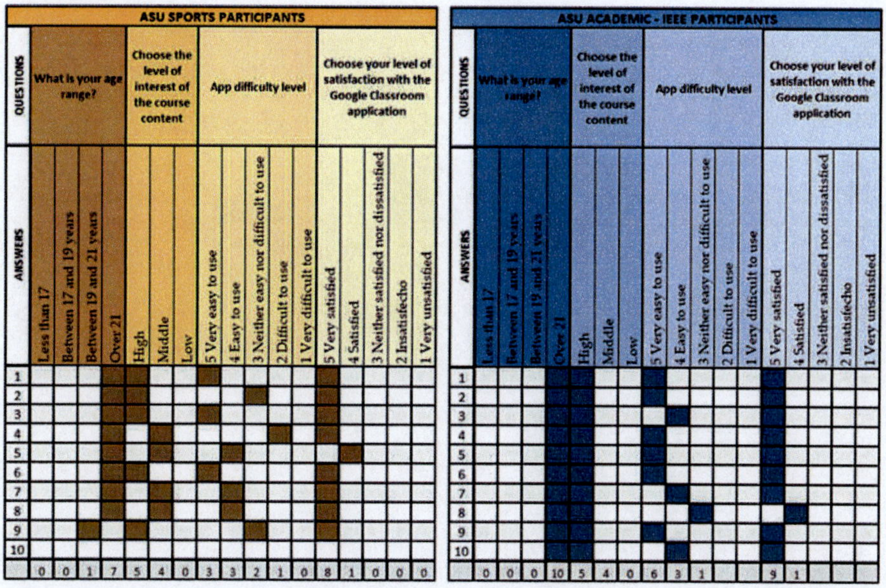

Fig. 10.3 Results of the questions asked to the participants of the virtual class in Google Classroom

participants indicate that the content was of high interest and the remaining 28% indicate that it was somewhat interesting (see Fig. 10.4).

Participants' difficulty level while using the Google Classroom application for their virtual meetings was consulted and varying results were obtained (see Fig. 10.5): 44% indicate that the platform is very easy to use, 33% indicate that it is easy to use, 17% indicate that it is neither easy nor difficult to use, and 6% indicate that it is difficult to use. It is concluded that more than 77% agree on its ease of use.

100% of surveyed students from the ASU groups (see Fig. 10.6) agreed that the Google Classroom application is recommended for training in the context of the

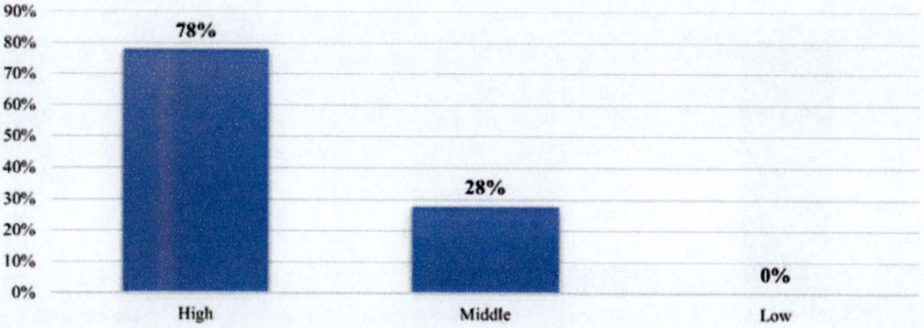

Fig. 10.4 Level of interest in the content of the course resources

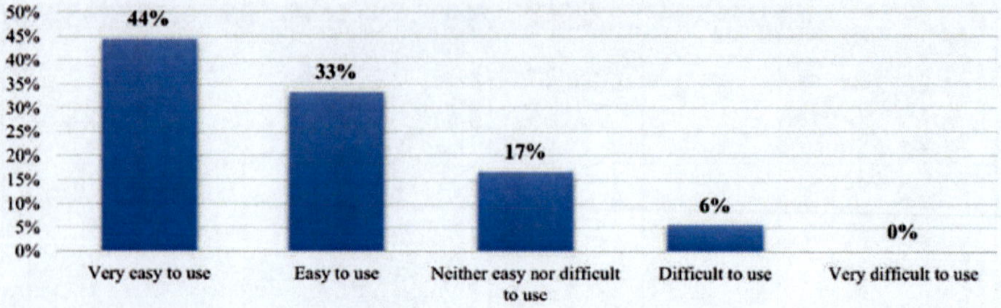

Fig. 10.5 Difficulty level using Google Classroom application

Fig. 10.6 Percentage of participants who would recommend using Google Classroom for training or regular classes

reality in which they are living and when appropriate content for intellectual growth is used.

The ASU training courses generated a considerable appreciation and interest for learning and participatively preparing class materials according to participants' area of interest. The online platform allowed participants to access course material in a comfortable way which promoted learning. 89% of the total participants indicate that they are very satisfied with the use of the mobile application, whereas 13% are satisfied (see Fig. 10.7).

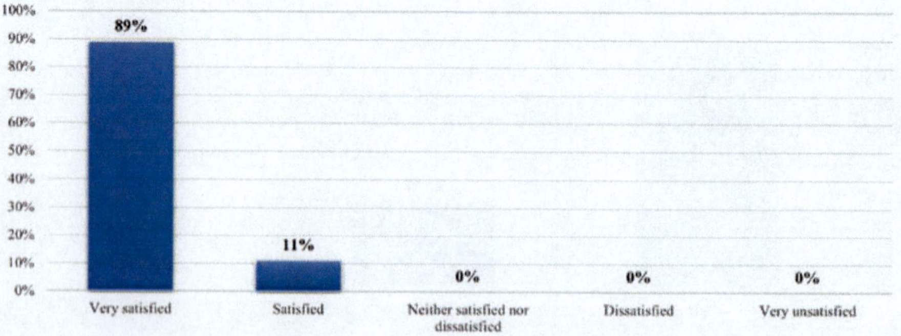

Fig. 10.7 Percentage of student satisfaction with the Google Classroom application

It is left as a discussion, that existing learning resource management systems are indispensable for educationally focused institutions or organizations [43]. The pandemic has unveiled the current state of online education [44], and the need for the use of technologies is an obvious strong support to all forms of access to knowledge, integration, and learning [45, 46]. This work provides a training opportunity for university students [47] in a context of associationism in the style and charisma of a Salesian university. A contribution to the student community is evidenced in the context of formative group activities, both during and before the pandemic [32]. The contribution is reflected in the use of Google Classroom by the student groups and their motivation and satisfaction with the results achieved in each of their online classrooms [48].

10.4 Conclusions

The use of free service platforms on the Web and the application of mobile learning methodologies used in virtual education have made it possible to minimize the physical resources required in traditional face-to-face education. With the COVID-19 pandemic and the ongoing changes it generates in society, the virtual environment educational and formative proposal adapted by the ASU has been acceptable. Thanks to the tutors' strategic design of course content, a great level of interest has been generated for each specific topic and unit.

A mobile learning strategy was designed for the use of virtual spaces provided by Google Classroom for the training and learning of participants of the ASU groups. The strategy improved the existing internal processes of each group aimed at communicating activities in person and was able to transform traditional club activities into a virtual environment. Finally, the integration of Web applications, Web-oriented services, and accessibility through mobile devices are highlighted as important factors to create new forms of interaction in the university environment for student groups and associations challenged by COVID-19 restrictions.

Acknowledgements Thanks are given to the GIEACI research group GIEACI (https://gieaci.blog.ups.edu.ec/) as well as to the authorities of the Universidad Politécnica Salesiana for accompanying this research, and to the participants and collaborators of the ASU Sports and Academic groups for the support provided.

References

1. R. Kanna, COVID-19 pandemic: tapping opportunities in the face of adversity. Indian Spine J. **4**, 2 (2021). https://doi.org/10.4103/isj.isj_1_21
2. D. Tashanova, A. Sekerbay, D. Chen, Y. Luo, S. Zhao, Q. Zhang, Investment opportunities and strategies in an era of coronavirus pandemic. SSRN Electron. J. (2020). https://doi.org/10.2139/ssrn.3567445

3. H. Al Zubayer Rony, S. Tahmina Awal, University teachers' training on online teaching–learning using online platform during COVID-19: a case study. Bangladesh Educ. J. **18** (2019)
4. V.A. Ballesteros-Ballesteros, Ó.I. Rodríguez-Cardoso, S. Lozano-Forero, J.L. Nisperuza-Toledo, El Aprendizaje Móvil en Educación Superior: Una Experiencia desde la Formación de Ingenieros. Rev. Científica. **38**, 243–257 (2020). https://doi.org/10.14483/23448350.15214
5. D. Anaby, M. Ryan, R.J. Palisano, M.F. Levin, J.W. Gorter, L. Avery, I. Cormier, R. Teplicky, J. Coulter, J. Hanes, Participation during a pandemic: forging new pathways. Phys. Occup. Ther. Pediatr. 1–4 (2021). https://doi.org/10.1080/01942638.2021.1875739
6. J.P. Chigwada, Opportunities and challenges offered by the effects of the COVID-19 pandemic on academic libraries, in *Handbook of Research on Library Response to the COVID-19 Pandemic* (IGI Global, 2021), pp. 342–357
7. O.V. Bondarenko, S.V. Mantulenko, A.V. Pikilnyak, Google classroom as a tool of support of blended learning for geography students. CEUR Workshop Proc. **2257**, 182–191 (2018). https://doi.org/10.31812/pedag.v51i0.3671
8. J. Llerena, R. Ayala-Carabajo, Significant learning activities (ASA) in the modality of face-to-face studies with integration of virtual educational media in engineering careers, in *Proceedings of the 15th Latin American Conference on Learning Technologies, LACLO 2020* (2020), pp. 1–9. https://doi.org/10.1109/LACLO50806.2020.9381134
9. M.A. Betancourt-Odio, A. Sartor-Harada, O. Ulloa-Guerra, J. Azevedo-Gomes, Self-perceptions on digital competences for M-learning and education sustainability: a study with teachers from different countries. Sustainability **13**, 1–12 (2021). https://doi.org/10.3390/su13010343
10. J. Llerena-Izquierdo, O. Barcia-Ayala, R. Ayala-Carabajo, Faculty training through crowdlearning for emerging online education, in *2020 IEEE ANDESCON* (2020), pp. 1–7. https://doi.org/10.1109/ANDESCON50619.2020.9272103
11. R.M. Tawafak, G. AlFarsi, J. Jabbar, S.I. Malik, R. Mathew, A. AlSidiri, M. Shakir, A. Romli, Impact of technologies during COVID-19 pandemic for improving behavior intention to use E-learning. Int. J. Interact. Mob. Technol. **15**, 184–198 (2021). https://doi.org/10.3991/IJIM.V15I01.17847
12. S. Arfan, M. Sundari Nasution, Responsivity of public services in Indonesia during the Covid-19 pandemic. Budapest Int. Res. Critics Inst. Humanit. Soc. Sci. **4**, 552–562 (2021). https://doi.org/10.33258/birci.v4i1.1638
13. J. Llerena-Izquierdo, R. Ayala-Carabajo, University teacher training during the COVID-19 emergency: the role of online teaching-learning tools, in *ICITS 2021* (2021), pp. 90–99. https://doi.org/10.1007/978-3-030-68418-1_10
14. R. Ayala, *Introducción a la concepción y aplicación del Método Fenomenológico-Hermenéutico de M. Van Manen* (2007)
15. J. Llerena-Izquierdo, F. Procel-Jupiter, A. Cunalema-Arana, Mobile application with cloud-based computer vision capability for university students' library services. Adv. Intell. Syst. Comput. **1277**, 3–15 (2021). https://doi.org/10.1007/978-3-030-60467-7_1
16. L. Budianto, Y. Arifani, Utilizing WhatsApp-driven learning during COVID-19 outbreak: EFL users' perceptions and practices. Comput. Lang. Learn. **22**, 264–281 (2021)
17. J.L. Izquierdo, M.R. Alfonso, M.A. Zambrano, J.G. Segovia, Aplicación móvil para fortalecer el aprendizaje de ajedrez en estudiantes de escuela utilizando realidad aumentada y m-learning. Rev. Ibérica Sist. Tecnol. Informação 120–133 (2019)
18. J. Llerena-Izquierdo, J. Idrovo-Llaguno, Introducing gamification to improve the evaluation process of programing courses at the Salesian Polytechnic University (Guayaquil, Ecuador). Adv. Intell. Syst. Comput. **1273**, 402–412 (2021). https://doi.org/10.1007/978-3-030-59194-6_33
19. J. Makira, E. Owino, The use of technology for learning during the Covid-19 pandemic season: a case of rural schools in Kenya. IJISRT (2021)
20. G. Kaisara, K.J. Bwalya, Investigating the E-learning challenges faced by students during Covid-19 in Namibia. Int. J. High. Educ. **10**, 308 (2020). https://doi.org/10.5430/ijhe.v10n1p308

21. O.M. Prince, Addressing teaching and learning challenges amidst the Covid 19 pandemic: implications for the Nigeria Primary Schools. Int. J. Integr. Educ. **4**, 1–7 (2021). https://doi.org/10.31149/ijie.v4i1.1094
22. J. Llerena-Izquierdo, L. Atiaja-Balseca, Gamification Within the Learning Evaluation Process Using Ardora at the Salesian Polytechnic University (Guayaquil, Ecuador) (2021), pp. 139–150. https://doi.org/10.1007/978-3-030-71503-8_11
23. M.D.H. Rahiem, Remaining motivated despite the limitations: university students' learning propensity during the COVID-19 pandemic. Child. Youth Serv. Rev. **120**, 105802 (2021). https://doi.org/10.1016/j.childyouth.2020.105802
24. R. Ayala Carabajo, Pedagogical relationship: Max van Manen's pedagogy in the sources of educational experience. Rev. Complut. Educ. **29**, 27–41 (2018). https://doi.org/10.5209/RCED.51925
25. R. López-Chila, J. Llerena-Izquierdo, N. Sumba-Nacipucha, Collaborative work in the development of assessments on a moodle learning platform with ExamView. Adv. Intell. Syst. Comput. **1277**, 131–141 (2021). https://doi.org/10.1007/978-3-030-60467-7_11
26. W.P. Flynn, N. Kumar, R. Donovan, M. Jones, P. Vickerton, Delivering online alternatives to the anatomy laboratory: early experience during the COVID-19 pandemic. Clin. Anat. **34** (2021). https://doi.org/10.1002/ca.23722
27. R. Ayala Carabajo, La esperanza pedagógica en la experiencia vivida de los educadores: un estudio fenomenológico-hermenéutico. Tomos I y II. 4ESIS Dr. INÁDITA. Dep. Pedagog. Sist. y Soc. Univ. Autónoma Barcelona-España (2009)
28. Ó. Rodríguez-Cardoso, V. Ballesteros-Ballesteros, S. Lozano-Forero, Tecnologías digitales para la innovación en educación: una revisión teórica de procesos de aprendizaje mediados por dispositivos móviles. Pensam. Acción 83–103 (2020)
29. J. Llerena-Izquierdo, J. Zamora-Galindo, Using H5P services to enhance the student evaluation process in programming courses at the Universidad Politécnica Salesiana (Guayaquil, Ecuador). Artif. Intell. Comput. Softw. Eng. Adv. **1326**, 216–227 (2021). https://doi.org/10.1007/978-3-030-68080-0_16
30. A. Agarwal, S. Sharma, V. Kumar, M. Kaur, Effect of E-learning on public health and environment during COVID-19 lockdown. Big Data Min. Anal. **4**, 104–115 (2021). https://doi.org/10.26599/bdma.2020.9020014
31. N. Aisyah, Rahmatullah, H. Sayfullah, Anggraini, Kusmiati, Sa'diya, Application of Google Classroom applications as media thematic learning in the pandemic period. J. Phys. Conf. Ser. **1779**, 012067 (2021). https://doi.org/10.1088/1742-6596/1779/1/012067
32. J. Llerena-Izquierdo, A. Valverde-Macias, Google Classroom as a blended learning and M-learning strategy for training representatives of the student federation of the Salesian Polytechnic University (Guayaquil, Ecuador). Adv. Intell. Syst. Comput. **1273**, 391–401 (2021). https://doi.org/10.1007/978-3-030-59194-6_32
33. R. Ayala-Carabajo, Retorno a lo esencial: Fenomenología hermenéutica aplicada desde el enfoque de Max Van Manen. Caligrama (2017)
34. J. Llerena Izquierdo, El Trabajo en Equipo Aligera la Carga. https://pure.ups.edu.ec/es/publications/el-trabajo-en-equipo-aligera-la-carga
35. J.J. Rocha Espinoza, Metodologías activas, la clave para el cambio de la escuela y su aplicación en épocas de pandemia. INNOVA Res. J. **5**, 33–46 (2020). https://doi.org/10.33890/innova.v5.n3.2.2020.1514
36. J.F. Flores Moran, La relación docente- alumno como variable mediadora del aprendizaje. Rev. San Gregor. **1** (2019). https://doi.org/10.36097/rsan.v1i35.957
37. C. Briones, E. Caballero, J. Flores, El aprendizaje autodirigido y la Flipped Classroom. Yachana **3**, 13–18 (2014). https://doi.org/10.1234/YCH.V3I1.7
38. L. Daniela, A. Visvizi, C. Gutiérrez-Braojos, M.D. Lytras, Sustainable higher education and Technology-Enhanced Learning (TEL). Sustainability **10** (2018). https://doi.org/10.3390/su10113883
39. L. Daniela, Smart pedagogy for technology-enhanced learning, in *Didactics of Smart Pedagogy* (Springer, Cham, 2019), pp. 3–21

40. R. Ayala Carabajo, Social science and humanities research training based on van Manen's hermeneutic phenomenological approach within the Hispanic American context. Educ. XX1 **19**, 359–381 (2016). https://doi.org/10.5944/educXX1.13945
41. J.A. Kumar, B. Bervell, S. Osman, Google classroom: insights from Malaysian higher education students' and instructors' experiences. Educ. Inf. Technol. **25**, 4175–4195 (2020). https://doi.org/10.1007/s10639-020-10163-x
42. A. Gupta, P. Pathania, To study the impact of Google Classroom as a platform of learning and collaboration at the teacher education level. Educ. Inf. Technol. **26**, 843–857 (2021). https://doi.org/10.1007/s10639-020-10294-1
43. N.W. Wulan Sari, T.T. Wulansari, W. Yanuartha, M.S. Sipayung, A.S. Pribadi, Factors which influence the success of Google Classroom based on age. J. Phys. Conf. Ser. **1807**, 012016 (2021). https://doi.org/10.1088/1742-6596/1807/1/012016
44. E. Susanto, R.N. Sasongko, M. Kristiawan, N. Nipriansyah, P. Purdiyanto, Constraints of online learning using Google Classroom during Covid-19. Educ. Q. Rev. **4** (2021). https://doi.org/10.31014/aior.1993.04.02.201
45. M. Zakaria, J. Hanim Ahmad, R. Bahari, S. Jeslyn Hasan, S. Zolkaflil, Benefits and challenges of adopting Google Classroom in Malaysian University: educators' perspectives. İlköğretim Online **20**, 1296–1304 (2021). https://doi.org/10.17051/ilkonline.2021.01.123
46. R. Ayala Carabajo, La metodología fenomenológico-hermeneútica de M. Van Manen en el campo de la investigación educativa. Posibilidades y primeras experiencias. Rev. Investig. Educ. RIE **26**, 409–430 (2008)
47. A. Henukh, I.M. Astra, The use of Google classroom as ICT literacy to improve physics students collaboration skill in industrial revolution 4.0. AIP Conf. Proc. **2331**, 030002 (2021). https://doi.org/10.1063/5.0041660
48. A.M. Guswara, W. Purwanto, The contribution of Google Classroom application and motivation to the learning outcomes of web programming. J. Educ. Technol. **1**, 424–432 (2020). https://doi.org/10.23887/JET.V4I4.29896

Chapter 11
Combining Escape Rooms and Google Forms to Reinforce Python Programming Learning

Joe Llerena-Izquierdo and Loqui-Larroza Sherry

Abstract This research presents a gamification proposal using *Escape Rooms* as a strategy to reinforce learning in classes covering programming in Python, for students in their first years at the Universidad Politécnica Salesiana. The participants of the study, a population of 80 students from two classes from the computer engineering major, ages 17–21, are challenged to complete four different scenarios developed with Google Forms and incorporating videos, QR codes, and questions related to the Python programing language, with the goal of completing a sentence at the end of the four Escape Rooms, that leaves a mark of teamwork. An empirical–analytical methodology was used, with a quasi-experimental quantitative approach using the survey technique, with a confidence level of 95% and a margin of error of 2%. The results showed high levels of interest, attention, and anxiety while participating in the gamification activity. It was determined that 99% of participants would recommend this methodology for use in other classes to stimulate academic reinforcement, and 78% of participants stated that they discover new knowledge while completing the activity.

11.1 Introduction

In recent years, teaching methods in higher education have evolved from conservative, teacher-centered approaches, such as the classical lesson, to more inductive, student-centered approaches. Some innovations such as game-based learning (GBL) are enjoying increasing popularity [1–3] as subject-specific improvements in education can be demonstrated by a variety of studies while using this technique [4]. With new education-oriented technologies [5], the emergence of new generations of young people with the skills to use them and a greater capacity to learn and work in groups

J. Llerena-Izquierdo (✉) · L.-L. Sherry
Universidad Politécnica Salesiana, Guayaquil, Ecuador
e-mail: jllerena@ups.edu.ec

L.-L. Sherry
e-mail: sloqui@est.ups.edu.ec

Á. Rocha et al. (eds.), *Communication, Smart Technologies and Innovation for Society*,
Smart Innovation, Systems and Technologies 252,
https://doi.org/10.1007/978-981-16-4126-8_11

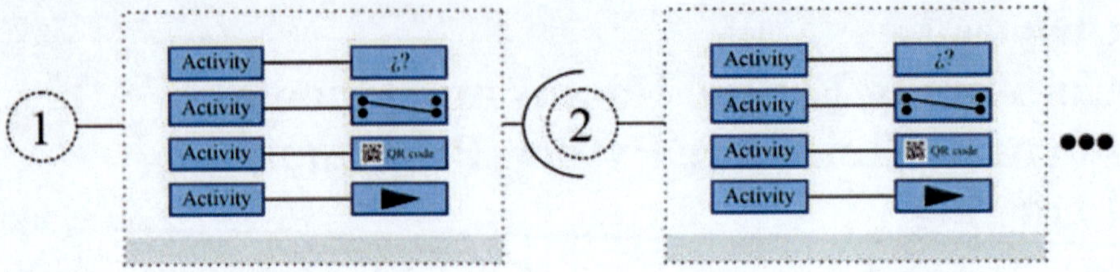

Fig. 11.1 *Escape Rooms* idea using gamification for learning

[6–8], the education field continues to incorporate new techniques and technologies such as gamification [9, 10]. Gamification is frequently used as it offers design flexibility and can be easily tailored to a specific educational area.

In an educational institution, gamification is defined as a set of rules relating to games that promote interactive learning. The application of gamification in a pedagogical context is an engaging alternative solution for many students who do not thrive with traditional learning methods [11, 12]. Students find that gamification can be more engaging than traditional learning because it encourages students through elements such as challenges, rewards, points, levels, options, and badges [13, 14]. It is up to teachers to incorporate techniques that allow gamification to be inserted pedagogically into their curriculum and to motivate students to participate in this new way of learning so that a subject such as programming languages can be taught in a very different way [15]. The term gamification is subject to game designs for non-context games [16]. One way to increase motivation, interest, and competence in educational settings is to play games. This methodology is called "gamification" and is associated with "serious games" that use recreational technologies such as video games that are oriented toward educational activities [17, 18]. The meeting room concept can also be transferred to computer-generated virtual environments (see Fig. 11.1). Virtual and augmented reality escape rooms produced in physical locations are an emerging trend in the escape room industry [19, 20]. These rooms have a high cost due to the devices and technology used.

Gamification is a technique that helps improve the effectiveness of teaching in the classroom; it is also versatile because it can be applied to any area. In education, gamification can create an attractive environment that, when combined with effective teaching techniques, can enhance student's learning experience [21, 22]. Based on these assumptions, various investigations have empirically affirmed that gamification can be used in education to motivate students in the learning process and improve the enjoyment of and student performance while completing educational tasks [22, 23]. One of the objectives within higher education is to use gamification as a tool to deepen knowledge and understanding of the material covered in class in a more recreational way. As gamification uses games, students are more excited to consciously work through problems, as they are trying to solve a puzzle or reach the end of a challenge [24–27].

Not only students but also teachers benefit from gamification, as they learn new platforms that can speed up and enhance learning activities, improve student grades, reduce repetition, and other factors that affect the performance of the group, in addition to being able to motivate their students more effectively learn their subject [28–31]. Gamification can be applied in non-recreational situations, which is why this methodology is being used more frequently in higher education. In this approach, students deepen their mastery of certain topics through personal and or group challenges; the reward is not necessarily a physical medal or a certain course grade, but rather feeling recognized and publicly demonstrating their progress in the subject of study, being able to stand out among and with others for their knowledge [32].

11.2 Methods

In this study, a quasi-experimental empirical–analytical methodology with a quantitative approach is used. A survey is applied to a group of 78 students from a sample of 80, from two classes from the computer engineering major at the Universidad Politécnica Salesiana in Guayaquil, Ecuador. The confidence level of the study is 95% with a 2% margin of error. This research studies the effects of incorporating gamification through the *Escape Rooms* game into the first semesters of programming where students need to master algorithms and programming in Python.

The technique uses virtual *Escape Rooms*, presenting scenarios including puzzle games or questions to be solved (see Fig. 11.2). It challenges participants to reason and remember what they have learned to continue advancing toward the final goal. In the prototype of this technique, during class time, the Google Forms platform was used to interact simultaneously with all students participating in the activity.

First-year students put their knowledge of programming in Python to the test. Teachers can adapt this tool, making it harder or easier, to keep students engaged and motivated during the activity. In Google Forms, rooms are created by teachers for a specific group of students who are given a determined amount of time for each task. In each room, the students must reach a part of a sentence that is at the end of

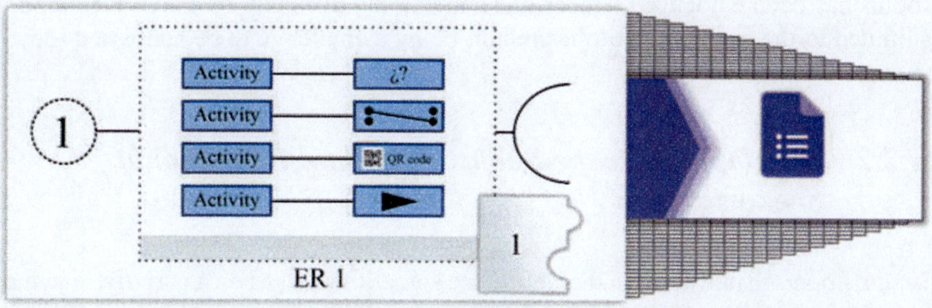

Fig. 11.2 Use of Google forms technology to create Escape Rooms

Fig. 11.3 Complete sentence formed by clues

the room. By completing each challenge, a sentence-like puzzle is assembled (see Fig. 11.3).

11.2.1 Escape Rooms as a Learning Technique

Researchers define gamification as game-based learning (GBL) [33, 34]. The Escape Rooms are related to the gamification methodology, from the approach of design and conformation of the game. From a context in which the participants are active and collaborative, in the Escape Rooms, student-centered animation strategies are established in the classroom for the development of group activities using a scenario and its components. These collaborative activities, in teams, allow an experience for solving problems from a game-based approach [35]. Meeting rooms are converted into virtual environments created by a computer. In a changing industry, virtual and augmented reality have begun to make possible virtual reproductions of traditionally expensive physical *Escape Rooms* [19]. Gamification is a tool adaptable to many learning fields. The teacher's abilities are a contributing factor in the creation of active learning activities for students [36]. In higher education, gamification is considered an important part of the change in the educational model. The benefits of its application are visible and measurable for those who are already utilizing this innovation. The goals of incorporating *Escape Rooms* into the educational process include the possibility of modifying behaviors, deepening student's engagement, and generating learning [23]. In education, the lack of design guides for the development of Escape Rooms has been evidenced. Under this context, the difficulty of a broader adoption is limited to the experience of the creator, being a challenge to be addressed [35].

11.2.2 Developing Escape Rooms as Python Educational Scenarios

Escape Rooms usually have a theme and are sometimes based on a narrative in which the players participate in role-play. Popular themes include the modern era, historic scenes, horror, fantasy, science, science fiction, and military action [19, 37].

This research uses an empirical–analytical methodology. To create the educational scenarios to be evaluated, it uses platforms offered by Google as they are online platforms permitting collaborative text editing, and they are easily accessible for those who have a Google account. This research aims to show how to create this application through Google Forms, with the hope that the detailed explanation will be useful to the entire educational community as there are very few studies showing the application of specific new gamification applications. The following is the process used to create an *Escape Room* using Google's platforms:

Using Google Forms, open a new form. This form will be used to create the *Escape Room*. A new form can be created, or a predetermined template can be used. Once the form is selected, the activities that will be inserted into the *Escape Room* must be selected according to class material. An *Escape Room* exercise allows students to apply their knowledge in a new way while compiling the data each student enters as he or she solves each exercise. The *Escape Room* exercise can be worked on by students from multiple classes. As it is the teacher who implements the game, he or she can create new alternatives and changes within the *Escape Room* to encourage students to work together across classes to reach the end and discover the hidden message.

The *Escape Room* prototype presented in this research is made for programing students in superior levels of their training. To create *Escape Rooms* in Google Forms, different scenarios are used to help connect different classes making the exercise more dynamic. When creating questions in the form, it is best to use a section for each question. This way, the student cannot advance without first answering the question correctly. Escape Rooms offer teachers a variety of alternative scenarios to choose a learning activity and elaborate questions using familiar technological tools.

11.3 Results and Discussion

The students access each of the four created *Escape Rooms* through the university's virtual learning platform using their institutional username and password. Once inside, the user is prompted to enter his or her name to enter the *Escape Room*. 69% of participants are between 17 and 19 years old, whereas 24% are between 20 and 21.

An 80% of participants are familiar with the concept of gamification, whereas 20% are unfamiliar with the concept. 95% of participants indicate that they know what an Escape Room is and how it functions (see Fig. 11.4). It is assumed that the group is conformed of youth considered part of the post-millennial, or centennial, generation, known for being digital natives with greater knowledge and capacity to use technology. 72% of participants show a high level of attention when participating in this type of gamification activity. 28% show a medium level of attention, and 0% are inattentive (see Fig. 11.5, Level of attention). 60% of the participants showed a high interest in the content proposed in the four Escape Rooms. 37% of the participants indicated they were somewhat interested, and 3% indicated that they were not interested (see Fig. 11.5, Content interest level). It is understood that 97% of the participants had an interest in the content, which is much higher than average.

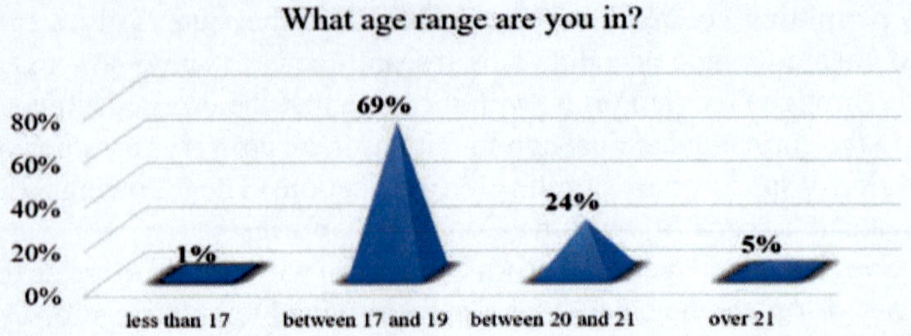

Fig. 11.4 Participant's age range in percentage

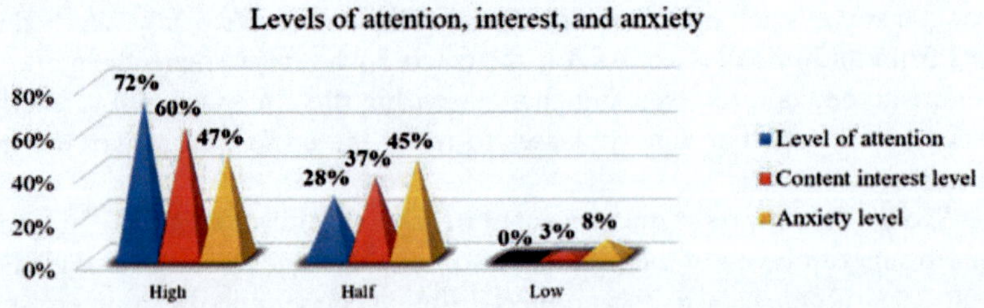

Fig. 11.5 Level of attention, level of interest in the content, and level of anxiety of the students as they advance in the development of the Escape Rooms, in percentages

Given this result, it is possible to continue making new learning scenarios using this technology to help students engage with and gain mastery over their academic subjects. Regarding anxiety while participating in the activity, 47% of participants experienced high levels of anxiety, 45% of participants experienced moderate levels of anxiety, and 8% experienced low levels of anxiety (see Fig. 11.5, Anxiety level).

A student's anxiousness to continue working through the *Escape Room* is considered a positive factor related to the student's interest in the activity. Anxiety in this case is a positive reaction to stress that indicates an emotional state generated in participants immersed in the learning context and the challenge of understanding content and achieving their goal of solving the problem. In this sense, it is a motivating factor for the student. When asked to define the level of difficulty of the activity on a scale from 1 to 5, with 1 being the least difficult and 5 being the most difficult, 17% of participants indicate that the difficulty level is 5, 40% indicate that the level is 4, 40% indicate that the level is 3, 4% indicate that the level is 2, and 0% indicate that the level is 1. The level of difficulty of the activity is concentrated between a level 3 and 4, and for 80% of participants, the activity was from medium to very difficult.

In addition, participants must not only complete an *Escape Room* but also manage to find part of a phrase at the end of each room of the four rooms and combine the four pieces at the end of the challenge. Being of medium difficulty, the results show that 62% manage to complete the four parts of the final phrase, 15% manage to complete

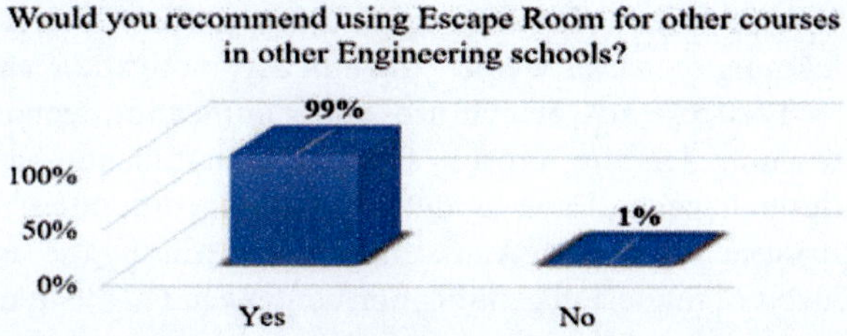

Fig. 11.6 Percent of students that would recommend using *Escape Room* activities in classes in other engineering majors

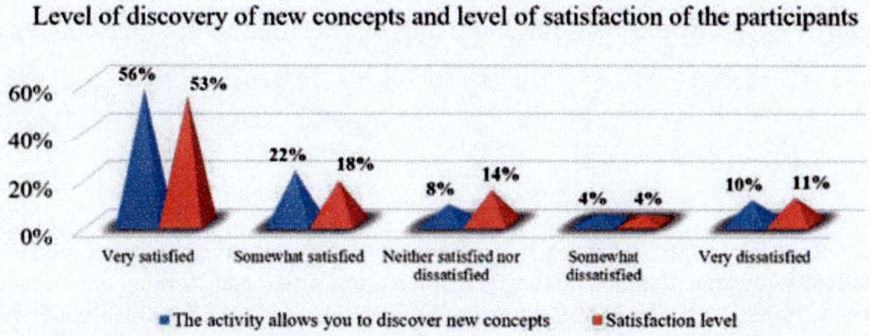

Fig. 11.7 Percentages of levels of discovery of new concepts in the activity and level of satisfaction of the participants

three parts of the final phrase, 10% manage to complete two parts of the final phrase and 13% achieve complete only one part of the sentence.

When asked if they would recommend the use of the *Escape Room* technique in other engineering classes, 99% of participants indicated that they would recommend it (see Fig. 11.6). Finally, the results show that regarding the discovery of new knowledge, 78% of participants are very satisfied or somewhat satisfied. Similarly, 71% of the participants indicate that they are very satisfied or somewhat satisfied with this activity. These results fulfill the objectives of the study. It is also evidenced that the percentage of dissatisfaction does not exceed 15% of participants (Fig. 11.7).

11.4 Conclusions

Gamification within the educational field is being increasingly used and adapted as an educational activity that allows students to engage with class materials in a dynamic way, promoting their active learning. Given this, the utility of enhancing teaching and learning methods with these technologies is obvious. This study shows that students' motivation to learn increases when participating in interactive games using existing technologies. When faced with a new challenge, students become anxious and adopt an investigative stance to find a solution and continue playing the game. The conformation and structuring of *Escape Rooms* challenge teachers to achieve learning outcomes. Escape Rooms are presented using the free service

Google Forms. Teachers acquire new skills while using this tool to develop new learning scenarios, evidencing efficacy, motivation, and interest.

This research, which uses the gamification technique as a complement to the teaching–learning process for first-year students, has motivated students to learn through games. Each *Escape Room* has incorporated interactive resources for participants, consolidating knowledge covered during the semester. The results show high levels of interest in gamification content and high attention levels while participating in gamification activities. Satisfaction levels to reinforce knowledge and acceptance of discovery of new concepts while using the *Escape Room* technique exceed 70%. 99% of participants would recommend using this technique in other classes.

References

1. C.Y. Chen, H.J. Huang, C.J. Lien, Y.L. Lu, Effects of multi-genre digital game-based instruction on student's conceptual understanding, argumentation skills, and learning experiences. IEEE Access **8**, 110643–110655 (2020). https://doi.org/10.1109/ACCESS.2020.3000659
2. C. Harteveld, N. Javvaji, T. Machado, Y.V. Zastavker, V. Bennett, T. Abdoun, Gaming4all: reflecting on diversity, equity, and inclusion for game-based engineering education, in *2020 IEEE Frontiers in Education Conference* (2020), pp. 1–9. http://doi.org/10.1109/FIE44824.2020.9274176
3. D.K.M.P.M. Dassanayake, S. Wijesinghe, T.L. Jayasiri, K.A.R. Keenawinna, W.H. Rankothge, N.D. Gamage, AwareME: public awareness through game-based learning, in 2020 *IEEE Region 10 Conference*, November 2020, pp. 662–666. http://doi.org/10.1109/TENCON50793.2020.9293720
4. J. Llerena-Izquierdo, A. Valverde-Macias, Google classroom as a blended learning and m-learning strategy for training representatives of the student federation of the Salesian Polytechnic University (Guayaquil, Ecuador). Adv. Intell. Syst. Comput. **1273 AISC**, 391–401 (2021). http://doi.org/10.1007/978-3-030-59194-6_32
5. J.L. Izquierdo, M.R. Alfonso, M.A. Zambrano, J.G. Segovia, Aplicación móvil para fortalecer el aprendizaje de ajedrez en estudiantes de escuela utilizando realidad aumentada y m-learning. Rev. Ibérica Sist. e Tecnol. Informação. 120–133 (2019)
6. Y. Zhan, P. Li, K. Wang, S. Guo, Y. Xia, Big data analytics by crowd learning: architecture and mechanism design. IEEE Netw. **34**, 143–147 (2020). https://doi.org/10.1109/MNET.001.1900286
7. J. Llerena-Izquierdo, O. Barcia-Ayala, R. Ayala-Carabajo, Faculty training through crowd learning for emerging online education, in *2020 IEEE ANDESCON* (2020), pp. 1–7. http://doi.org/10.1109/ANDESCON50619.2020.9272103
8. J. Llerena-Izquierdo, R. Ayala-Carabajo, University teacher training during the COVID-19 emergency: the role of online teaching-learning tools (2021), pp. 90–99. http://doi.org/10.1007/978-3-030-68418-1_10
9. R. Kazhamiakin, E. Loria, A. Marconi, M. Scanagatta, A gamification platform to analyze and influence citizen's daily transportation choices. IEEE Trans. Intell. Transp. Syst. 1–15 (2021). http://doi.org/10.1109/TITS.2021.3049792
10. J. Llerena-Izquierdo, J. Idrovo-Llaguno, Introducing gamification to improve the evaluation process of programing courses at the Salesian Polytechnic University (Guayaquil, Ecuador). Adv. Intell. Syst. Comput. **1273 AISC**, 402–412 (2021). http://doi.org/10.1007/978-3-030-59194-6_33
11. E.P. Do Carmo, A.C.T. Klock, E.H.T. De Oliveira, I. Gasparini, A study on the impact of gamification on student's behavior and performance through learning paths, in *Proceedings*

of IEEE 20th International Conference on Advanced Learning Technologies, ICALT 2020, pp. 84–86. http://doi.org/10.1109/ICALT49669.2020.00032
12. J.V. Paragulla, L. Andrade-Arenas, A.L.F. Barriales, Gamification in university teaching for systems engineering students at a Lima North University, in *EDUNINE 2020—4th IEEE World Engineering Education Conference. Challenges of Education in Engineering, Computing and Technology Without Exclusions: Innovation in the Era of the Industrial Revolution 4.0* (2020), pp. 20–23. http://doi.org/10.1109/EDUNINE48860.2020.9149524
13. A. Saggah, A.S. Atkins, R.J. Campion, A review of gamification design frameworks in education, in *4th International Conference on Intelligent Computing in Data Sciences, ICDS 2020.* (2020). http://doi.org/10.1109/ICDS50568.2020.9268748
14. J. Llerena-Izquierdo, F. Procel-Jupiter, A. Cunalema-Arana, Mobile application with cloud-based computer vision capability for university student's library services. Adv. Intell. Syst. Comput. **1277**, 3–15 (2021). https://doi.org/10.1007/978-3-030-60467-7_1
15. A.F. Barriales, J.V. Paragulla, L. Andrade-Arenas, Gamification as part of teaching and its influence on learning computational algorithms, in *EDUNINE 2020—4th IEEE World Engineering Education Conference. Challenges of Education in Engineering, Computing and Technology Without Exclusions: Innovation in the Era of the Industrial Revolution 4.0* (2020). http://doi.org/10.1109/EDUNINE48860.2020.9149510
16. M. Venter, Gamification in STEM programming courses: state of the art, in *IEEE Global Engineering Education Conference, EDUCON*, April 2020 (2020), pp. 859–866. http://doi.org/10.1109/EDUCON45650.2020.9125395
17. J.F. Llerena Izquierdo, R. Ayala Carabajo, El Uso De Grabaciones Por Video Como Recurso De Evaluación De Conocimientos De Aprendizajes. https://www.researchgate.net/publication/327426650_El_Uso_De_Grabaciones_Por_Video_Como_Recurso_De_Evaluacion_De_Conocimientos_De_Aprendizajes
18. J. Llerena-Izquierdo, R. Ayala-Carabajo, The use of video recordings as a resource for the evaluation of learning knowledge. https://www.academia.edu/44257612/The_use_of_video_recordings_as_a_resource_for_the_evaluation_of_learning_knowledge
19. S. Mystakidis, E. Cachafeiro, I. Hatzilygeroudis, Enter the serious e-scape room: a cost-effective serious game model for deep and meaningful e-learning, in *10th International Conference on Information, Intelligence, Systems and Applications, IISA 2019* (2019). http://doi.org/10.1109/IISA.2019.8900673
20. R. Ross, S. Bennett, Increasing engagement with engineering escape rooms. IEEE Trans. Games (2020). https://doi.org/10.1109/TG.2020.3025003
21. A. Makri, D. Vlachopoulos, R.A. Martina, Digital escape rooms as innovative pedagogical tools in education: a systematic literature review. Sustainability **13**, 4587 (2021). https://doi.org/10.3390/su13084587
22. H. Yin, What motivates Chinese undergraduates to engage in learning? Insights from a psychological approach to student engagement research. High. Educ. **76**, 827–847 (2018). https://doi.org/10.1007/s10734-018-0239-0
23. C. Dichev, D. Dicheva, Gamifying education: what is known, what is believed and what remains uncertain: a critical review. Int. J. Educ. Technol. High. Educ. **14**, (2017). http://doi.org/10.1186/s41239-017-0042-5
24. S. Seebauer, S. Jahn, J. Mottok, Learning from escape rooms? A study design concept measuring the effect of a cryptography educational escape room, in *IEEE Global Engineering Education Conference, EDUCON*, April 2020 (2020), pp. 1684–1685. http://doi.org/10.1109/EDUCON45650.2020.9125333
25. J. Llerena-Izquierdo, L. Atiaja-Balseca, Gamification within the learning evaluation process using Ardora at the Salesian Polytechnic University (Guayaquil, Ecuador). 139–150 (2021). http://doi.org/10.1007/978-3-030-71503-8_11
26. J. Llerena Izquierdo, El Trabajo en Equipo Aligera la Carga. https://pure.ups.edu.ec/es/publications/el-trabajo-en-equipo-aligera-la-carga
27. J. Llerena-Izquierdo, J. Zamora-Galindo, Using H5P services to enhance the student evaluation process in programming courses at the Universidad Politécnica Salesiana (Guayaquil, Ecuador) 216–227 (2021). http://doi.org/10.1007/978-3-030-68080-0_16

28. R. López-Chila, J. Llerena-Izquierdo, N. Sumba-Nacipucha, Collaborative work in the development of assessments on a moodle learning platform with ExamView. Adv. Intell. Syst. Comput. **1277**, 131–141 (2021). https://doi.org/10.1007/978-3-030-60467-7_11
29. J. Llerena Izquierdo, Codifica en Python. https://pure.ups.edu.ec/es/publications/codifica-en-python
30. R. Ayala Carabajo, La esperanza pedagógica en la experiencia vivida de los educadores: un estudio fenomenológico-hermenéutico. Tomos I y II. 4ESIS Dr. INÁDITA. Dep. Pedagog. Sist. y Soc. Univ. Autónoma Barcelona-España (2009)
31. R. Ayala, Introducción a la concepción y aplicación del Método Fenomenológico-Hermenéutico de M. Van Manen (2007)
32. J. Llerena, R. Ayala-Carabajo, Significant learning activities (ASA) in the modality of face-to-face studies with integration of virtual educational media in engineering careers, in *Proceedings of the 15th Latin American Conference on Learning Technologies, LACLO 2020* (2020), pp. 1–9. http://doi.org/10.1109/LACLO50806.2020.9381134
33. A. Manzano-León, P. Camacho-Lazarraga, M.A. Guerrero, L. Guerrero-Puerta, J.M. Aguilar-Parra, R. Trigueros, A. Alias, Between level up and game over: a systematic literature review of gamification in education. Sustainability **13**, 2247 (2021). https://doi.org/10.3390/su13042247
34. R.S. Alsawaier, The effect of gamification on motivation and engagement. Int. J. Inf. Learn. Technol. **35**, 56–79 (2018). https://doi.org/10.1108/IJILT-02-2017-0009
35. K. Kapp, The gamification of learning and instruction: game-based methods and strategies for training and education (2012)
36. M.H. Abdul Rahman, I. Ismail Yusuf Panessai, N.A.Z. Mohd Noor, N.S. Mat Salleh, Gamification elements and their impacts on teaching and learning a review. Int. J. Multimed. Appl. **10**, 37–46 (2018). http://doi.org/10.5121/ijma.2018.10604
37. S.Y. Huang, Y.H. Kuo, H.C. Chen, Applying digital escape rooms infused with science teaching in elementary school: learning performance, learning motivation, and problem-solving ability. Thinking Skills Creativity **37**, 100681 (2020). http://doi.org/10.1016/j.tsc.2020.100681

Chapter 12
Model of State Spaces for Estimating the Dynamic Origin–Destination Matrix for a Public Transport Network by Applying the Kalman Filter

Lina Patricia Zapata, Victor Manuel Larios, Francisco Castro Carrasco, and José Luis Aguayo

Abstract A model to estimate the passenger flow based on the dynamic origin–destination matrix applied at the South-Eastern Corridor from the terrestrial transport network in the city of Quito was the goal of this project. In the model, first: The state space was defined for the transport network, and later, was applied the Kalman Filter algorithm to get the dynamic origin–destination matrix. The state variables defined were the origin–destination pairs, the temporal variation, and the speed changes in the origin–destination pairs. Finally, the measure vector, the evolution matrix model, the process noise matrix, and the measurement error matrix were developed.

12.1 Introduction

An inefficient operation in public transport system causes excessive boarding times [1] that may be solved by knowing the supply of vehicles that meets passenger demand in a real-time, but to get the passenger demand value is very difficult, costly, and takes a lot of time [2]. A way to value passenger demand is through estimation origin–destination matrix (ODM), where each element in the array represents the number of people moving from a source node to destination node in the transport network [3].

L. P. Zapata (✉)
Grupo de Investigación en Desarrollo de Aplicaciones y Software basado en las TICs (DASTIC), Universidad Politécnica Salesiana, Quito, Ecuador
e-mail: lzapata@ups.edu.ec

V. M. Larios
Centro de Innovación de Ciudades Inteligentes (CICI), Universidad de Guadalajara, Guadalajara, Mexico

F. C. Carrasco
Escuela Politécnica Nacional, Quito, Ecuador

J. L. Aguayo
Grupo de investigación DASTIC, Universidad Politécnica Salesiana, Quito, Ecuador

Á. Rocha et al. (eds.), *Communication, Smart Technologies and Innovation for Society*, Smart Innovation, Systems and Technologies 252,
https://doi.org/10.1007/978-981-16-4126-8_12

Models for ODM estimation are categorized as static and dynamic. In static methods, the traffic flows are considered time-independent and the average demand can be determined over a period of time, as mentioned by Cascetta [4], Katharina Parry [5], Tebaldi [6], Hazelton [7], and Li [8], but the dynamic origin–destination matrix (DODM) is dependent on the time. Also, the project wants to support the planning of the operation of the bus fleet throughout the day, considering that the passenger demand changes in short periods of time; therefore, it is necessary pay attention to DODM.

Currently, there are several algorithms to estimate ODM, like of the Hazelton [9], that estimates a sequence of destination–origin matrices within a period a day, from the link flows in the transport network, it has been developed the Monte Carlo algorithm in Markov Chains (MCMC), to generate samples a posteriori, that propose a solution method by selecting a sequence of measures on the links, in a period of time between 10 and 15 min. However, Ying [10], estimates and predicts the DODM using the Kalman Filter (KF) considering a polynomial trend on the deviation of origin–destination (OD) flow within a period of time and concludes that the accuracy of sliding over the DODM estimated by this method, KF and polynomial tendency, is better than the self-recurrency used by Ashok [11]. Sherali [12], estimates the DODM considering the diary fluctuation known traffic flows in a sequence of days, using the MCMC technique that integrates Gibbs sampling, and proposes to use a DLM (dynamic linear model) to estimate the diary DODM. On the other hand, Pitombeira-Neto [13] in his work focuses on the application of the Kalman Filter in the DynaMIT-R Simulator for online traffic calibration. His methodology is to calibrate the measured input data over several days, estimate the OD(t) flow for five days: The process starts with a static matrix for the first day, then, the estimated OD flow of each interval is used as a historical estimate for the next interval. Bierlaire [14] mentions that when traffic conditions are normal and when time intervals are small, KF presents very accurate results from estimating the OD tables. Welch and Bishop [15], do a deep study on KF, as in continuous systems as in discrete systems, and conclude that KF is powerful, helping to estimate of past, present, and future state.

The purpose of this work is to get the DODM estimate optimal, through the approach of a model based on Kalman's algorithm to estimate passenger demand over time, in order to develop adequate planning and management of the bus fleet on the South-Eastern Corridor in the City of Quito (SECCQ). So, we started to build the state space model of the transport network. Later, we applied the KF algorithm, without historical information in the DODM, and it was generated the a priori information that enters the KF and estimated the static origin–destination matrix (EODM). The known links and destinations at various consecutive time intervals, besides, using Bayesian inference in Markov chains (MCMC) and following the Tebaldi and West methodology, were possible to perform the calculations and verify the validity of the model, that finally, it was developed in RStudio.

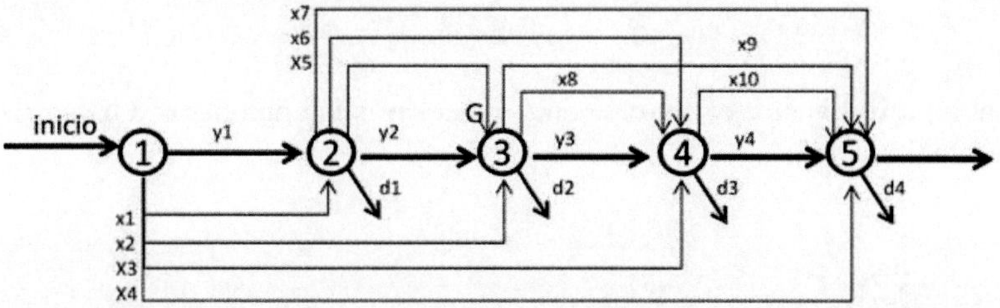

Fig. 12.1 Scheme of a simple hypothetical network

12.2 Formulation of the Model Applying Kalman Filter

12.2.1 State Spaces

In general, a system in the state space is defined by determining the state vector, X_t, the state evolution matrix, F_t, and the measurement vector, Y_t

To identify the various quantities, we use the hypothetical simple network shown in Fig. 12.1.

Where the nodes represent the passenger stations, and the arrows represent the people flows.

y_i, $i = 1 \ldots 4$, represent the passenger flows in the links between pairs of nodes.

x_j, $j = 1 \ldots 10$, represent the OD pairs.

d_k, $k = 1 \ldots 4$, represent the number of passengers whose destinations are stations 2, 3, 4, and 5, respectively.

12.2.2 Status Vector

It is formed by the state variables, we define as State Variables:

- The OD pairs.
- The variations in time of the OD pairs.
- The speed of variation of the OD pairs.

Where the variation of the OD pairs in time is given by:

$$v_{t,t-1} = (x_t - x_{t-1})/\Delta t \tag{12.1}$$

and the speed of this variation is given by:

$$a_{t+1,t} = \left(v_{t+1,t} - v_{t,t-1}\right)/2\Delta t \tag{12.2}$$

Then, we define the state vector as:

$$X = (x_{1,t}, v_{1,t}, a_{1,t}, x_{2,t}, v_{2,t}, a_{2,t}, \ldots, x_{n,t}, v_{n,t}, a_{n,t})^{\mathrm{T}} \tag{12.3}$$

That is, it is a vector of ***3n*** elements, where **n** is the number of OD pairs, and *t* represents time.

12.2.3 Measurements Vector

$\boldsymbol{Y_t}$ of dimension $\boldsymbol{y_{2m-1,1}}$is made up of the links and the destinations, which are the variables measured on site:

$$\mathbf{Y} = (y_{1,t}, y_{2,t}, \ldots, Y_{m,t}, d_{1,t}, d_{2,t}, \ldots, d_{m-1,t})^{\mathrm{T}}$$

where $y_{i,t;i=1,\ldots,m}$, are the measures of the links, *m* is the number of links, and $d_{i,t;i=1,\ldots,m-1}$, are the measurements of the destinations, except for the last station on the network.

In this way, the model of the network in the state space, for a discrete system, is determined by the equation of state:

$$x_t = F_t x_{t-1} + w_t \tag{12.4}$$

and the measurement equation:

$$y_t = H_t x_t + e_{yt} \tag{12.5}$$

where

- x_t is the current state vector,
- x_{t-1} is the state vector at the previous instant,
- F_t is the state evolution matrix.
- w_t is a vector that represents the noise of the process, with Gaussian distribution, of zero mean value, $E[w_{i,t}] = 0$, for $i = 1, 2, \ldots, n$ and covariance Q.
- y_t is a vector of observations or measurements.
- H_t is the matrix that relates the measurements to the state variables,
- e_{yt} is a vector of the errors in the measurements, with Gaussian distribution, with a mean value of zero, $E[e_{i,t}] = 0$ and variance R_t, para $i = 1, 2, \ldots, m$.

The vectors $w_1, e_2, \ldots, w_t, e_{y1}, e_{y2}, \ldots, e_{yt}$ are mutually independent.

12.2.4 Kalman Filter Algorithm

The Kalman filter [15] estimates the state $\boldsymbol{x_n} \in R^n$, $\boldsymbol{X_t} = (x_{1t}, x_{2t}, \ldots, x_{nt})$, a discrete linear system which can be represented by equations **¡Error! No se**

encuentra el origen de la referencia. and **¡Error! No se encuentra el origen de la referencia.** given a sequence of noisy observations $y_0, y_1, y_2, \ldots, y_{t-1}$ such that the estimation noise is minimized.

The Kalman filter algorithm is summarized in five equations, namely:

$$\hat{x}_{t|t-1} = F_t \hat{x}_{t-1|t-1} + w_t \tag{12.6}$$

$$P_{t|t-1} = F_t P_{t-1|t-1} F_t^{\mathrm{T}} + Q_t \tag{12.7}$$

$$K_t = P_{t|t-1} H_t^{\mathrm{T}} \left(H_t P_{t|t-1} H_t^{\mathrm{T}} + R_t \right)^{-1} \tag{12.8}$$

$$\hat{x}_{t|t} = \hat{x}_{t|t-1} + K_t \left(y_t - H_t \hat{x}_{t|t-1} \right) \tag{12.9}$$

$$P_{t|t} = (I - K_t H_t) P_{t|t-1} \tag{12.10}$$

where

$\hat{x}_{t-1|t-1}$ is the estimated state vector, at the previous instant.
$\hat{x}_{t|t-1}$ is the estimated state vector in the transition from the previous to the next instant.
$\hat{x}_{t|t}$ is the current estimated state vector.
$P_{t|t}$ is the noise covariance matrix $X_t - \hat{x}_{t|t}$ that is, between the actual value and current estimated value. The goal of the Kalman filter is to minimize $P_{t|t}$.
$P_{t-1|t-1}$ is the value of P at the previous instant.
$P_{t|t-1}$ is the estimated value of P in the transition from the instant before to the present.
K_t is the KF gain.

The optimization criterion used by the KF consists of obtaining the KF gain, K_t, which seeks to minimize the mean square error, that is, to minimize the sum of variances of the estimation noise $P_{t|t}$, in other words, find the value of K_t that minimizes $P_{t|t}$.

12.3 Formulation of Models on the Transport Network

To implement the Kalman algorithm, we should formulate the models of the different quantities for the transport network.

12.3.1 Evolution Matrix

The evolution matrix F_t has the following form for discrete systems:

$$\boldsymbol{F}_t = e^{\Delta t A} = I + \sum_{i=1}^{\propto} \frac{\Delta t^i A^i}{i!} \tag{12.11}$$

In practice, there is an $i_0 > 0$ for which, A^i is a matrix with all its terms equal to zero for all, $i >= i_0$, so the sum of $\boldsymbol{F}_t$ is finite.

Given that the state vector has terms of up to second order in terms of variation over time, then we select up to the second term of the series 12.11, so that the matrix $\boldsymbol{A}$ and matrix $\boldsymbol{F}$ take the form:

$$A_k = \begin{bmatrix} 0 & \Delta t & \frac{1}{2}\Delta t^2 \\ 0 & 0 & \Delta t \\ 0 & 0 & 0 \end{bmatrix} \quad F_k = \begin{bmatrix} 1 & \Delta t & \frac{1}{2}\Delta t^2 \\ 0 & 1 & \Delta t \\ 0 & 0 & 1 \end{bmatrix}$$

The matrix H depends on the network topology. To better illustrate, look at in Fig. 12.1 the following relationships:

$$y_1 = x_1 + x_2 + x_3 + x_4 + e_{y1}$$

$$y_2 = x_2 + x_3 + x_4 + x_5 + x_6 + x_7 + e_{y2}$$

$$y_3 = x_3 + x_4 + x_6 + x_7 + x_8 + x_9 + e_{y3}$$

$$y_4 = x_4 + x_7 + x_9 + x_{10} + e_{y4}$$

$$d_1 = x_1 + e_{y5}$$

$$d_2 = x_2 + x_5 + e_{y6}$$

$$d_3 = x_3 + x_6 + x_8 + e_{y7}$$

Then, the matrix H is determined as follows, where m is the number of links and n is the number of OD pairs:

$$H_{m,3n} = \begin{bmatrix} [1\,0\,0] & [1\,0\,0] & [1\,0\,0] & [1\,0\,0] & [0\,0\,0] & \cdots [0\,0\,0] \\ [0\,0\,0] & [1\,0\,0] & [1\,0\,0] & [1\,0\,0] & [0\,0\,0] & \cdots [0\,0\,0] \\ [0\,0\,0] & [0\,0\,0] & [1\,0\,0] & [1\,0\,0] & [0\,0\,0] & \cdots [0\,0\,0] \\ \vdots & \vdots & \vdots & \vdots & \vdots & \vdots \\ [0\,0\,0] & [0\,0\,0] & [0\,0\,0] & [0\,0\,0] & [0\,0\,0] & \cdots [1\,0\,0] \end{bmatrix}$$

12.3.2 Noise Model

Be σ_i the standard deviation of the OD x_i, then the variation in time over the standard deviation, σ_v-based pm σ_i, is $\sigma_v = \sigma_i/\Delta t$ and the speed of variation over the standard deviation, $\sigma_a = 2\sigma_i/\Delta t^2$. In short, the error model related to the OD x_i is:

$$w = \sigma_i \begin{bmatrix} 1 \\ 1/\Delta t \\ 2/\Delta t^2 \end{bmatrix}$$

For its part, the covariance matrix $Q_{3n,3n}$ is determined by $Q = w.w^{\mathrm{T}}$

$$Qt = \begin{bmatrix} \sigma_1\sigma_1 \begin{bmatrix} 1\,1\,2 \\ 1\,1\,2 \\ 2\,2\,4 \end{bmatrix} & \sigma_1\sigma_2 \begin{bmatrix} 1\,1\,2 \\ 1\,1\,2 \\ 2\,2\,4 \end{bmatrix} & \ldots\ \sigma_1\sigma_n \begin{bmatrix} 1\,1\,2 \\ 1\,1\,2 \\ 2\,2\,4 \end{bmatrix} \\ \sigma_2\sigma_1 \begin{bmatrix} 1\,1\,2 \\ 1\,1\,2 \\ 2\,2\,4 \end{bmatrix} & \sigma_2\sigma_2 \begin{bmatrix} 1\,1\,2 \\ 1\,1\,2 \\ 2\,2\,4 \end{bmatrix} & \ldots\ \sigma_2\sigma_n \begin{bmatrix} 1\,1\,2 \\ 1\,1\,2 \\ 2\,2\,4 \end{bmatrix} \\ \sigma_n\sigma_1 \begin{bmatrix} 1\,1\,2 \\ 1\,1\,2 \\ 2\,2\,4 \end{bmatrix} & \sigma_n\sigma_2 \begin{bmatrix} 1\,1\,2 \\ 1\,1\,2 \\ 2\,2\,4 \end{bmatrix} & \ldots\ \sigma_n\sigma_n \begin{bmatrix} 1\,1\,2 \\ 1\,1\,2 \\ 2\,2\,4 \end{bmatrix} \end{bmatrix}$$

And covariance matrix over the measurement error R is a diagonal matrix with values corresponding to the error, $\sigma_{y,}$ of the measurement of each link y_i.

12.3.3 Tests of the Formulated Model

To estimate the DODM by applying the FK algorithm, it is enough to have the sequential and simultaneous count in some time intervals. In our case, this information is generated by estimating the EODM, through the known data of the links and destinations. Next, using the methodology of TEBALDI and WEST [6], we calculated the most probable mean values of the OD flows of the EODM, in consecutive intervals of 10 min. These calculated mean values constitute the a priori values for entering the KF, in the first run, the resulting values are entered into the filter as a priori values for a second run, and this is repeated by several times until the results converge.

For the test network, Fig. 12.1, the state vector is made up of 30 state variables: ten OD pairs, ten temporal variations of the OD pairs, and ten speed of variation of the OD pairs, and is defined by:

$$X = \left(x_{1,t}, v_{1,t}, a_{1,t}, x_{2,t}, v_{2,t}, a_{2,t}, \ldots, x_{10,t}, v_{10,t}, a_{10,t}\right)^{\mathrm{T}}$$

In this example, we do it for ten time intervals, $t = 1, \ldots, 10$. The result is a matrix formed by those ten state vectors.

The vector of measurements is given by:

$$Y = \left(y_{1,t}, y_{2,t}, y_{3,t}, y_{4,t}, d_{1,t}, d_{2,t}, d_{3,t}\right)^{\mathrm{T}}$$

Each measurement vector has seven elements, four correspond to the measurements of the links and three to the measurements of the destinations.

The evolution matrix is expressed according to the indicated model of dimensions $F_{30,30}$ and is invariant in time. In a similar way, we determined the covariance matrices of the process noise $Q_{30,30}$ and the error of the measurements $R_{7,30}$, considering that the standard deviations are variable in time. To carry out the tests, we consider the standard deviations 10% of the a priori value in all cases.

The calculations made, by the algorithm that we have designed and implemented in RStudio, are shown in Fig. 12.2, where we can see the graphs as a function of time of the true values, a priori values, and those resulting from the KF of some OD pairs.

In Fig. 12.2, it can also be observed how the resulting KF values, in each new pass through the filter, follow more closely the true value curve.

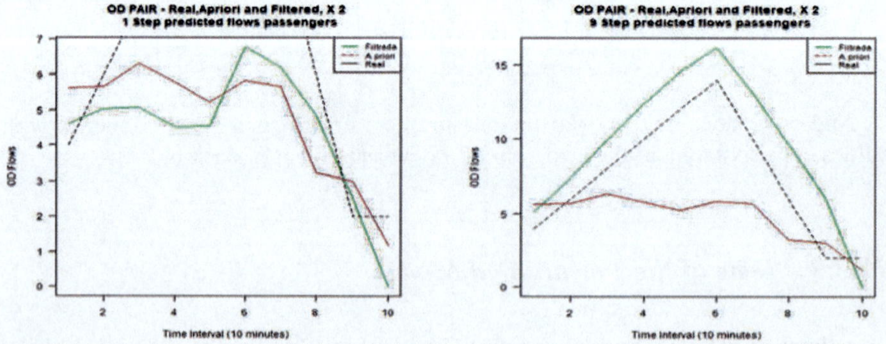

Fig. 12.2 Sequence of real, estimated, and resulting FK values for the OD x_1 pair of the test network

12.4 Case Study

As a case study, we estimate the DODM applying KF for the transport network South-Eastern Corridor in the City of Quito (SECCQ). In Fig. 12.3, the scheme of a network consisting of 19 bus stops or stations is shown. It has a single route to move from one node to another and has three transversal branches, the Quitumbe-Capulí, S. Martin, and La Lucha.

The SECCQ network consists of 19 nodes and 18 direct links in each direction of circulation. In our study, we only considered the direction of circulation from south to north. The direct links are detailed in Table 12.1.

The destination-origin matrix consists of 145 origin–destination pairs.

The state vector is defined by the variables: 145 OD pairs; 145 variables corresponding to the variation in time (velocity) of each OD pair and 145 variables representing the speed of variation in time (acceleration), and is defined by:

$$X = (x_{1,t}, v_{1,t}, a_{1,t}, x_{2,t}, v_{2,t}, a_{2,t} \ldots, x_{145,t}, v_{145,t}, a_{145,t})^{\mathrm{T}}$$

In this study case, we have measurements between 6:00 and 9:00 divided into 18 equal intervals of 10 min each, that is, $t = 1, \ldots, 18$. The result of the KF will yield 18 state vectors that form a matrix.

The vector of measurements is given by:

$$Y = \left(y_{1,t}, y_{2,t}, \ldots, y_{18,t}, d_{1,t}, d_{2,t}, \ldots, d_{14,t}\right)^{\mathrm{T}}$$

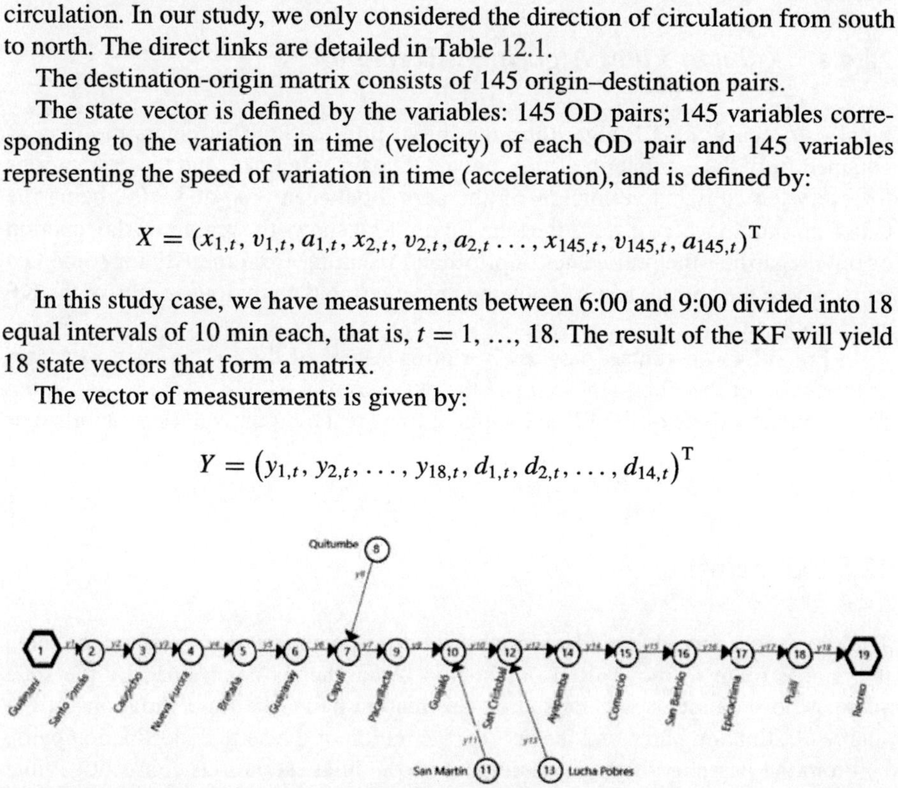

Fig. 12.3 Diagram of the SECCQ public transport network

Table 12.1 CSOQ network links

y_1	Guamaní—S. Tomás	y_7	Quitumbe—Capuli	y_{13}	S. Cristobal—Ayapamba
y_2	S. Tomás-Caupicho	Y_8	Capuli—Pacari	y_{14}	Ayapamba—Comercio
y_3	Caupicho—N. Aurora	y_9	Pacari—Guajaló	y_{15}	Comercio—S. Bartolo
y_4	N. Aurora—Bretaña	y_{10}	S. Martín Guajaló	y_{16}	S. Bartolo—Epiclachima
y_5	Bretaña—Guayanay	y_{11}	Guajaló—S. Cristobal	y_{17}	Eplicachima—Pujili
y_6	Guayanay—Capuli	y_{12}	La Lucha—S. Cristobal	y_{18}	Pujili—Recreo

Formed by 32 elements, 18 corresponding to the link measurements and 14 to the destination measurements.

The rest of the generated matrices, in the study case, have the following dimensions: evolution $F_{435,435}$, process noise $w_{435.1}$, process noise covariance $Q_{435,435}$, and covariance of the measurement error $R_{32,32}$. The $H_{32,435}$ matrix, as we already mentioned, is elaborated from the route matrix according to the physical topology of the network.

12.4.1 Kalman Filter Algorithm Execution

Finally, to apply the KF algorithm, the initial information that enters the filter is obtained from the estimation of the known OD pairs, the links, and the destinations for every ten consecutive minutes in the period between 6:00 and 9:00, being the Guamaní station where it starts the tour. Figure 12.4 shows the graphs on the function of time regarding the real values, a priori and resulting from the KF, for some OD pairs, where the current result is obtained by entering the previous results of the KF as a priori values.

In Fig. 12.4, we can see how the resulting values of the FK, in each new pass or execution of the filter, approximate the curve of true values. In the example, ten passes or executions of the FK are applied, thus, reaching the values to stabilize or converge.

12.5 Conclusions

The state space model formulated in this job for a transport network in which there is a single route to move from one station to another is determined by the state vector whose elements are the source–destination pairs, the time variations of the source–destination pairs, and the speed of variation of the source–destination pairs.

From the passenger flow measurement on the links, at various consecutive time intervals (in our case every 10 min), for a transport network in which there is a single route to move from one station to another, we can determine the dynamic origin destination matrix a priori by applying Bayesian inference following the Tebaldi and West methodology.

With the information contained in the dynamic origin–destination matrix a priori, considering the error in the measurements and defined the state space model in the algorithm, the Kalman filter is applied to estimate the dynamic origin–destination matrix to posteriori. The information obtained is useful for decision making, such as for planning the operation of the transport system.

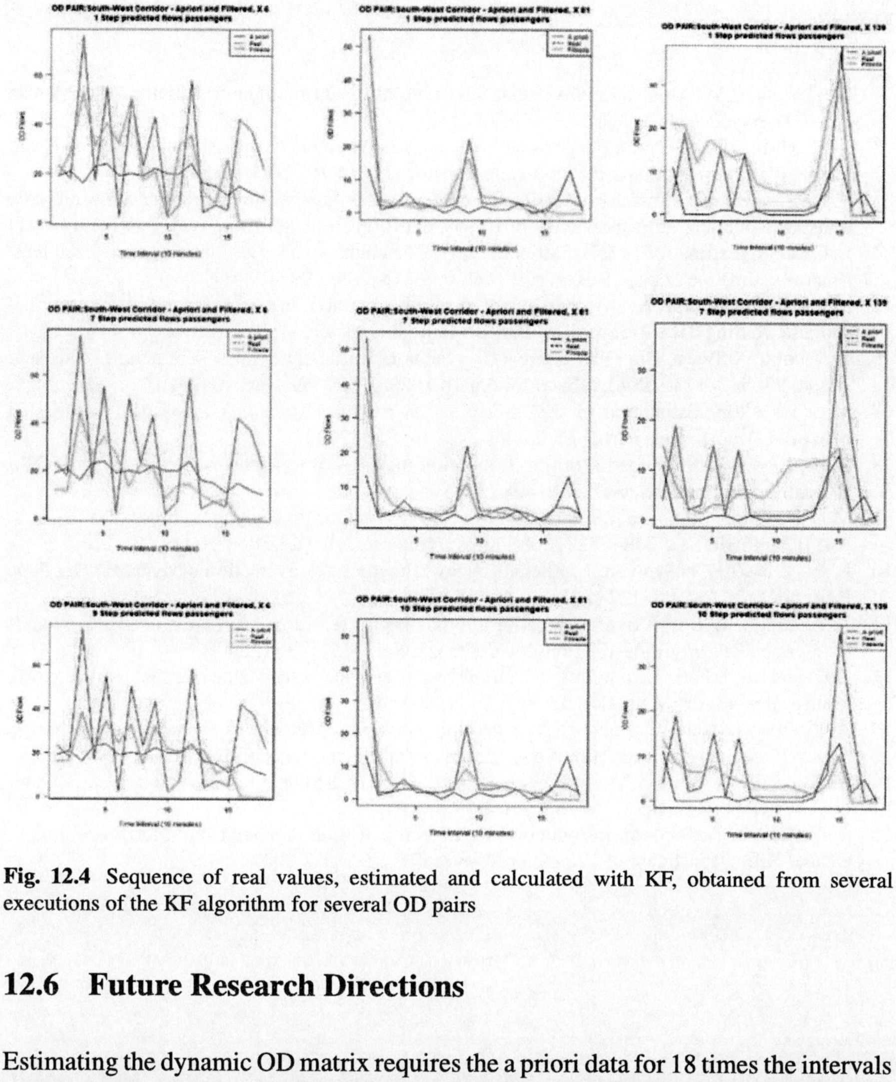

Fig. 12.4 Sequence of real values, estimated and calculated with KF, obtained from several executions of the KF algorithm for several OD pairs

12.6 Future Research Directions

Estimating the dynamic OD matrix requires the a priori data for 18 times the intervals obtained through the Tebaldi and West algorithm. By the large number of times intervals, the demand for processing times of the algorithm is quite significant. It is recommended to apply parallel programming in the Tebali and West algorithm to calculate static source–destination matrix concurrently or in parallel to reduce execution times and computational resource usage.

References

1. N.B. Aletà, C.M. Alonso, R.M.A. Ruiz, Smart mobility and smart environment in the Spanish cities. Transp. Res. Procedia **24**, 163–170 (2017)
2. D.M. Municipio de Quito, Diagnóstico de la movilidad en el Distrito Metropolitano de Quito para el Plan Metropolitano de Desarrollo Territorial (PMOT). Secr. Movil. (2014)
3. F.A.Z. González, Estimación y predicción de matrices dinámicas de viaje sobre un corredor de transporte público, utilizando datos históricos e información en tiempo real. Pontificia (2011)
4. E. Cascetta, Estimation of trip matrices from traffic counts and survey data: a generalized least squares estimator. Transp. Res. Part B Methodol. **18**, 289–299 (1984)
5. M.L.H. Katharina Parry, Estimation of origin-destination matrices from link counts and sporadic routing data. Transp. Res. Part B Methodol. **46**, 175–188 (2012)
6. C. Tebaldi, M. West, Bayesian inference on network traffic using link count data. J. Am. Stat. Assoc. **93**, 557–573 (1998). https://doi.org/10.1080/01621459.1998.10473707
7. M.L. Hazelton, Estimation of origin-destination matrices from link flows on uncongested networks. Transp. Res. Part B Methodol. **34**, 549–566 (2000)
8. B. Li, Bayesian inference for origin-destination matrices of transport networks using the EM algorithm. Technometrics **47**, 399–408 (2005)
9. M.L. Hazelton, Statistical inference for time varying origin-destination matrices. Transp. Res. Part B Methodol. **42**, 542–552 (2008). https://doi.org/10.1016/j.trb.2007.11.003
10. L. Ying, J. Zhu, W. Huiyan, L. Zhenyu, A novel method for estimation of dynamic OD flow. Procedia Eng. **137**, 94–102 (2016)
11. K. Ashok, M.E. Ben-Akiva, Alternative approaches for real-time estimation and prediction of time-dependent origin-destination flows. Transp. Sci. **34**, 21–36 (2000)
12. H.D. Sherali, T. Park, Estimation of dynamic origin-destination trip tables for a general network. Transp. Res. Part B Methodol. **35**, 217–235 (2001)
13. A. Pitombeira-Neto, C. Loureiro, L. Carvalho, Bayesian inference on dynamic linear models of day-to-day origin-destination flows in transportation networks. Urban Sci. **2**, 117 (2018)
14. M. Bierlaire, F. Crittin, An efficient algorithm for real-time estimation and prediction of dynamic OD tables. Oper. Res. **52**, 116–127 (2004)
15. G. Welch, G. Bishop, An introduction to the Kalman filter. University of North Carolina at Chapel Hill, Department of Computer Science. TR 95-041 (2004)

Chapter 13
Multi-objective Logistic Regression for Anomaly Detection in Water Distribution Systems

Gilberto Reynoso-Meza and Elizabeth Pauline Carreño-Alvarado

Abstract Providing clean and safe drinking water is a crucial task for any water supply company. In such an activity, automatic anomaly detection plays a critical role in drinking water quality monitoring. Recent anomaly detection techniques use tools from the machine learning area, where after a supervised learning process, a given machine learning representation can classify an event as normal or as an anomaly. Most of the times this learning process is performed via optimisation with a single objective cost function. Nevertheless, a trade-off between false positives and false negatives in the classifying task must be achieved. For this task, multi-objective optimisation could be an interesting tool. In this exploratory work, it is used the simplest classifier machine and logistic regression, and it is explored the advantages of using a multi-objective optimisation approach for its training.

13.1 Introduction

Providing clean and safe drinking water is a crucial task for any water supply company [14]. In order to guarantee such a supply, automatic anomaly detection plays a critical role in drinking water quality monitoring [1].

Recent anomaly detection techniques incorporate tools from the machine learning area [4]. After a supervised learning process, a given machine learning representation can identify anomaly events from normal events [18]. In such cases, they are referred as virtual or soft sensors: instead of having a physical device, information from other sources is mixed, in order to infer an anomaly event [21].

G. Reynoso-Meza (✉)
Industrial and Systems Engineering Graduate Program (PPGEPS), Pontifícia Universidade Católica do Paraná (PUCPR), Rua Imaculada Conceição 1155, Curitiba, Brazil
e-mail: g.reynosomeza@pucpr.br

E. P. Carreño-Alvarado
Departamento de Ingeniería Geomática e Hidráulica, División de Ingenierías, Campus Guanajuato, Universidad de Guanajuato, Av. Juárez 77, C.P. 36000 Guanajuato, Mexico
e-mail: ep.carreno@ugto.mx

Á. Rocha et al. (eds.), *Communication, Smart Technologies and Innovation for Society*, Smart Innovation, Systems and Technologies 252,
https://doi.org/10.1007/978-981-16-4126-8_13

Most of the times, such a learning process is performed via optimisation with a single objective cost function [5]. Nevertheless, for classification purposes, trade-off between false positives and false negatives is evident. That means, it is worthwhile to analyse the trade-off of a given classifier for anomaly detection between (1) triggering a false alarm and (2) letting pass a risk situation. Even if both instances are considered equivalent misclassifications, for practical purposes they are not. On the one hand, false positives trigger an alarm which needs to be attended; an excess of false alarms could overwhelm the technical staff. On the other hand, false negatives are undetected anomalies that could be hazardous for the plan operation. Therefore, there is a trade-off that a learner must achieve through its learning process.

In such instances, where there is a clear trade-off between conflicting objectives, multi-objective optimisation could be an interesting tool [8]. Multi-objective optimisation deals with conflicting objectives simultaneously. As consequence, a Pareto front is approximated: a set of Pareto optimal solutions. In such a set, the only difference between two solutions is the trade-off that they exhibit between conflicting objectives. Therefore, it is possible to ponder actively benefits versus drawbacks when the decision maker favours one objective over another. Such an idea could bring interesting solutions for machine learning [19].

In this exploratory work, we dealt with the simplest classification machine, logistic regression and explore the advantages of using a multi-objective optimisation approach for its training. It will be used as experimental setup the GECCO competition in anomaly detection [14].

The remainder of this paper is as follows: in Sect. 13.2, a brief backgrounds on multi-objective optimisation and supervised machine learning is given. In Sect. 13.3, the proposal of this exploratory work is presented, whilst in Sect. 13.4, the study design is explained. In Sect. 13.5, results are commented and discussed and finally some conclusion and future work is presented.

13.2 Background

Next, a brief background on machine learning, logistic regression and multi-objective optimisation is given.

13.2.1 Machine Learning and Supervised Learning

Machine learning refers to computer algorithms which improve themselves automatically through experience in the form of data. Such a learning process could be supervised (requiring inputs and targets), unsupervised (inputs required) or by reinforcement (via interaction with the surroundings) [10].

Supervised learning uses as inputs M instances (observations) with N features (explanatory variables) to train a given learner structure using reliable information of the targets for each one of the instances. The main goal is to construct a relationship to provide an output (target prediction) for any new instance (with its features). Such training could be oriented for classification or regression.

In both cases, several representations for learners exist as follows: artificial neural networks [12], support vector machines [3] or decision trees [7], among others. Nevertheless, the simplest one could be considered the logistic regression, which will be commented next.

13.2.2 Logistic Regression

Logistic regression is the simplest statistical model to predict a categorical value. It could be considered as the simplest classifier in the machine learning domain. The logistic regression uses the sigmoid function (Eq. 13.1, Fig. 13.1) to compute the probability of a given observation to be 1.

$$h_\theta(x) = \frac{1}{1 + e^{\beta_0 + \beta_1 x_1 + \beta_2 x_2 + \cdots + \beta_n x_n}} \tag{13.1}$$

where $x = [x_1, \ldots, x_N]$ are the N explanatory variables or features of the learner, $\beta = [\beta_1, \ldots, \beta_N]$ are the regression coefficients that are adjusted given M observations or instances. Usually, given this set of M instances, the parameter β is adjusted using the loss function of Eq. (13.2):

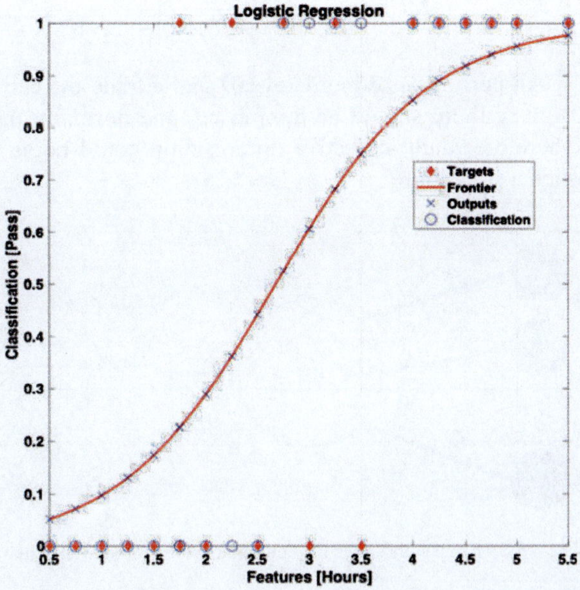

Fig. 13.1 Sigmoid function example. A classification task involving study hours versus grading is depicted

$$\text{Loss}\big(h_\beta(x), y\big) = \begin{cases} -\log\big(h_\beta(x)\big) & \text{if } y = 1 \\ -\log\big(1 - h_\beta(x)\big) & \text{if } y = 0 \end{cases} \tag{13.2}$$

and the cost function for optimisation of (13.3):

$$J(\beta) = -\big[y \log\big(h_\beta(x)\big) + (1 - y) \log\big(1 - h_\beta(x)\big)\big] \tag{13.3}$$

Finally, after such a learning process, a classifier is implemented as indicated in Fig. 13.2. As it can be noticed, cost function for optimisation or learning process is an aggregation function of classification and misclassifications. Furthermore, it is usual evaluating the final performance of such a model in the same way, with the true positives (TP), false negatives (FN), false positives (FP) and true negatives (TN), for example:

$$\text{Error rate} = \frac{\text{FP} + \text{FN}}{\text{TP} + \text{FP} + \text{TN} + \text{FN}} \tag{13.4}$$

$$\text{F1 score} = \frac{2 * \text{TPR} * \text{PPV}}{\text{TPR} + \text{PPV}} \tag{13.5}$$

$$\text{TPR} = \frac{\text{TP}}{\text{TP} + \text{FN}} \tag{13.6}$$

$$\text{PPV} = \frac{\text{TP}}{\text{TP} + \text{FP}} \tag{13.7}$$

$$\text{MCC} = \frac{\text{TP} * \text{TN} - \text{FP} * \text{FN}}{\sqrt{(\text{TP} + \text{FP})(\text{TP} + \text{FN})(\text{TN} + \text{FP})(\text{TN} + \text{FN})}} \tag{13.8}$$

All performance index reveals that a trade-off between FN and FP is pondered. Both of them should be minimised, and normally they are conflicting objectives. Therefore, multi-objective optimisation could be an interesting tool to deal with such a situation.

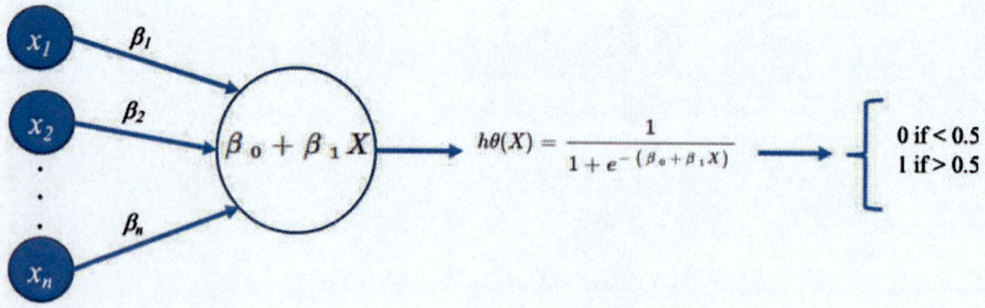

Fig. 13.2 Binary classifier implementation example after a learning process

13.2.3 Multi-objective Optimisation

As referred in [9], a MOP with m objectives can be stated as follows:

$$\min_{\theta} J(\theta) = [J_1(\theta), \ldots, J_m(\theta)] \tag{13.9}$$

subject to:

$$K(\theta) \leq 0 \tag{13.10}$$

$$L(\theta) \leq 0 \tag{13.11}$$

$$\underline{\theta_i} \leq \theta_i \leq \overline{\theta_i}, \quad i = [1, \ldots, n] \tag{13.12}$$

where $\theta = [\theta_1, \theta_2, \ldots, \theta_n]$ is defined as the decision vector with $\dim(\theta) = n$, $J(\theta)$ as the objective vector and $K(\theta)$, $L(\theta)$ as the inequality and equality constraint vectors, respectively, $\underline{\theta_i}, \overline{\theta_i}$, are the lower and the upper bounds in the decision space.

It has been noticed that there is not a single solution in MOPs, because there is not generally a better solution in all the objectives. Therefore, a set of solutions, the Pareto set Θ_P, is defined. Each solution in the Pareto set defines an objective vector in the Pareto front J_P (see Fig. 13.3). It is important to notice that most of the times we rely only in Pareto front and set approximations, J_P^*, Θ_P^*.

All the solutions in the Pareto front are a set of Pareto optimal and non-dominated solutions, where:

- Pareto optimality [9]: An objective vector $J(\theta^1)$ is Pareto optimal if there is not another objective vector $J(\theta^2)$ such that $J_i(\theta^2) \leq J_i(\theta^1)$ for all $i \in [1, 2, \ldots, m]$ and $J_j(\theta^2) < J_j(\theta^1)$ for at least one j, $j \in [1, 2, \ldots, m]$.
- Dominance [2]: An objective vector $J(\theta^1)$ is dominated by another objective vector $J(\theta^2)$ iff $J_i(\theta^2) \leq J_i(\theta^1)$ for all $i \in [1, 2, \ldots, m]$ and $J_j(\theta^1) < J_j(\theta^1)$ for at least one j, $j \in [1, 2, \ldots, m]$. This is denoted as $J(\theta^2) \preccurlyeq J(\theta^1)$.

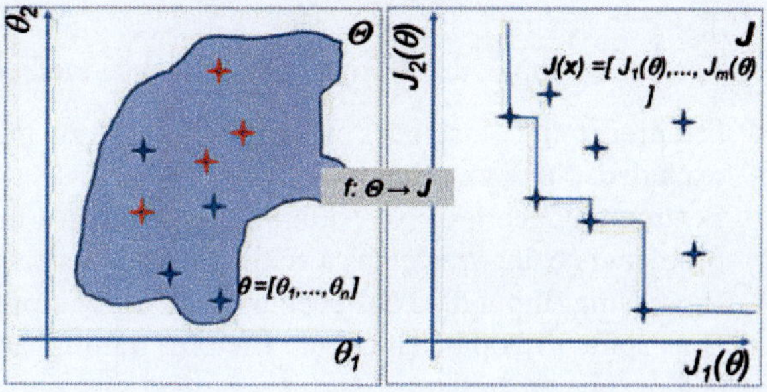

Fig. 13.3 Pareto optimality and dominance concepts for a min-min MOP. Dark solutions are non-dominated solutions which approximate (solid line) the Pareto front (dotted line) in the objective space J. Remainder solutions are dominated solutions

13.3 Tools and Methods

Instead using an aggregation function for FP and FN, such performance will be evaluated simultaneously via multi-objective optimisation. Therefore, multi-objective problem under consideration is as shown in Eq. (13.13):

$$\min_{\beta} J(\beta) = [\text{FP}, \text{FN}] \tag{13.13}$$

subject to:

$$-1 \leq \beta_i \leq 1, \quad i = [1, \ldots, n] \tag{13.14}$$

where FN is the number of false negatives in the training subset and *FP* the number of false positives. The following tools are used:

- Coding platform: MATLAB Online R2020b;
- Hardware: MacBook Air, Core i5, 1.8 GHz, 8 GB;
- Optimisation algorithm: spMODEx [15]. It is a multi-objective evolutionary algorithm based on Differential Evolution [13], using as diversity mechanism a spherical pruning [16]. This algorithm (or previous versions) has been used with success in the machine learning field [20]. Default values for hyper-parameters are used, and *3e4* function evaluations are used to approximate the Pareto front;
- Visualisation tool: 2D plots, parallel coordinates [6] and boxplots.

13.4 Study Design

This work uses a real-world dataset generated in a research project on drinking water. The dataset consists of data from Thüringer Fernwasserversorgung, a major German water supplier located in central Germany. This dataset has been used for different competitions about anomaly detection for drinking water in major international conferences [14]. The data and additional documentation are available for download [11]. In Table 13.1, the features used are depicted. More details are provided in [18].

For this example, the following data science methods are implemented:

- Pre-processing: instances with missing or not interpretable values have been excluded. No detrending has been performed.
- Feature engineering: No additional features are included. This means that this problem is being treated as a static problem instead of a dynamic time series.
- Processing: Input data has been normalised within the interval [0, 1].
- Data split: 13,900 instances are used for training and 566 instances for testing.

Table 13.1 Available parameters from the Thu ringer Fernwasserversorgung dataset

Parameter	Unit	Description
Time	Datetime	Timestamp
WT	C	Water temperature
ClO2-1, CLO2-2	mg/l	Chlorine dioxide (2 values)
pH	pH	pH value
Redox	mV	Redox potential
EC	micros/cm	Conductivity
TURB	NTU	Turbidity
FR-1, FR-2	M3/h	Water flow rate (2 values)
EVENT	Binary	Anomaly label

13.5 Multi-objective Learning Discussion

In Fig. 13.4, the Pareto front approximation is depicted. With blue squares are represented the Pareto front approximation in the learning process (that is, with the training subset) whilst the red diamonds represent the deformation of the approximation when using the subset for test. Such an analysis could reveal when an overfitting occurs: it is expected to keep the trade-off coherence, as well as similar trade-off or a reasonable mismatch between performance with the training set and the test set (see [17] for additional ideas on this topic).

In Fig. 13.5, the values of the weighting factor β of the approximate Pareto set are depicted using boxplots and parallel coordinates. With such an analysis, it is possible to appreciate the impact of a given feature in the prediction capabilities of the classifier. For example, small variation on β_9 (feature 9, FR2) indicates that this measure could be potentially omitted. In a practical sense, it could mean saving one sensor. On the opposite, β_3 (feature 3, pH) seems to be the one that the most impact has in the trade-off in the set.

Figure 13.4 together with Eqs. 13.4–13.8 could lead to a decision-making process, where it is possible to ponder practical aspects of the solution to be implemented.

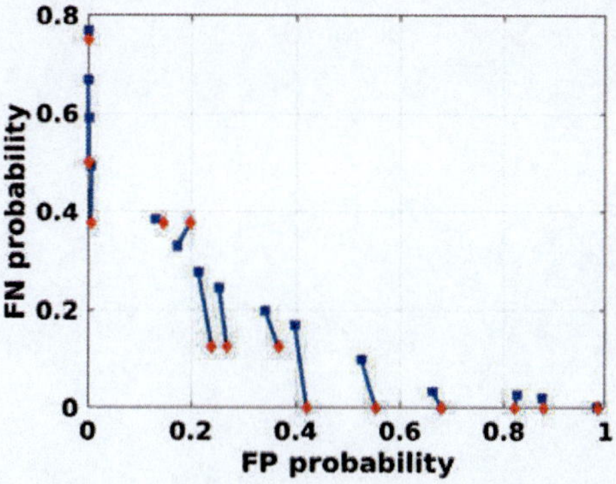

Fig. 13.4 Pareto front approximation of the logistic regression classifier. Blue squares indicate Pareto front approximation in the training phase, whilst red diamonds the deformation of the Pareto front with test data. Lines indicate translation of the Pareto optimal solution

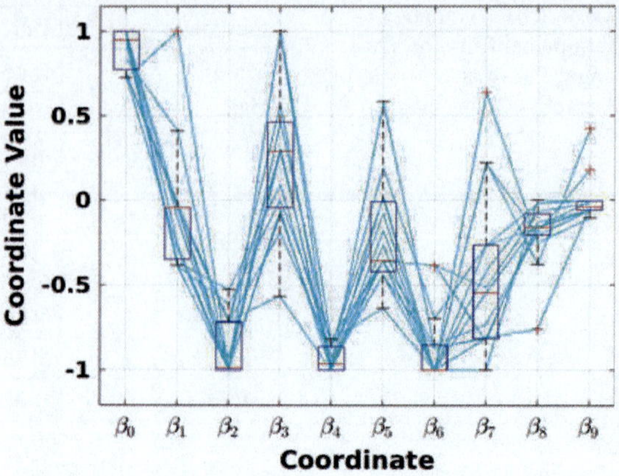

Fig. 13.5 Distribution of the weighting factors β after learning process via multi-objective optimisation

For example, according to F1 score criteria (5), learning vector 3 should be selected. Confusion matrix of such a solution is depicted in Fig. 13.6. Please notice that it is possible to interpret this matrix in the problem's context: on the one hand, 1017 false negatives mean that this automatic anomaly detection system won't trigger 1017 times the required alert; therefore, a potentially risk situation is generated. On the other hand, 229 false positives mean that a false alarm to the operators will be trigger 229 times; this usually will activate some safety (but unnecessary) protocol. This is why a trade-off should be achieved. Using this score and the Pareto front, decision makers can ponder this fact in the final selection.

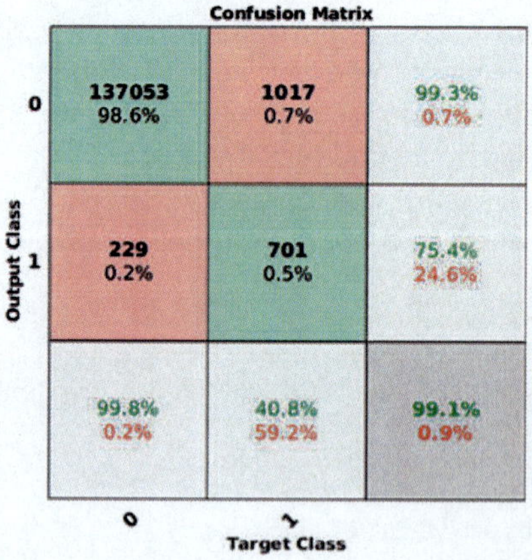

Fig. 13.6 Confusion matrix of the learning vector selected using the *F1 score* criteria

13.6 Conclusions

It has been presented an exploratory work, identifying the advantages of using a multi-objective optimisation approach for the learning process in machine learning. It has been selected the logistic regression, the simplest classifier. As case study, an anomaly detection for drinking water quality was selected. Such an analysis could bring some enlighten about the most important information that should be used and selecting a learner that fits the best the decision maker expectations. Future work should point towards the development of toolboxes to aid in such process.

References

1. M.H. Banna, S. Imran, A. Francisque, H. Najjaran, R. Sadiq, M. Rodriguez, M. Hoorfar, Online drinking water quality monitoring: review on available and emerging technologies. Crit. Rev. Environ. Sci. Technol. **44**(12), 1370–1421 (2014)
2. C.A.C. Coello, G.B. Lamont, *Applications of Multi-objective Evolutionary Algorithms*, vol. 1 (World Scientific, 2004)
3. P.C. Deka et al., Support vector machine applications in the field of hydrology: a review. Appl. Soft Comput. **19**, 372–386 (2014)
4. E.M. Dogo, N.I. Nwulu, B. Twala, C. Aigbavboa, A survey of machine learning methods applied to anomaly detection on drinking-water quality data. Urban Water J. **16**(3), 235–248 (2019)
5. P. Domingos, A few useful things to know about machine learning. Commun. ACM **55**(10), 78–87 (2012)
6. A. Inselberg, The plane with parallel coordinates. Vis. Comput. **1**(2), 69–91 (1985)
7. S.B. Kotsiantis, Decision trees: a recent overview. Artif. Intell. Rev. **39**(4), 261–283 (2013)
8. G.R. Meza, X.B. Ferragud, J.S. Saez, J.M.H. Durá, *Controller Tuning with Evolutionary Multiobjective Optimization* (2017)
9. K. Miettinen, Concepts, in *Nonlinear Multiobjective Optimization* (Springer, 1998), pp. 5–36
10. T.M. Mitchell, et al., *Machine Learning* (McGraw-Hill, New York, 1997)
11. S. Moritz, F. Rehbach, S. Chandrasekaran, M. Rebolledo, T. Bartz-Beielstein, GECCO Industrial Challenge 2018 Dataset: a water quality dataset for the "Internet of Things: Online Anomaly Detection for Drinking Water Quality" competition at the Genetic and Evolutionary Computation Conference 2018, Kyoto, Japan (2018). https://doi.org/10.5281/zenodo.3884398
12. A. Najah, A. El-Shafie, O.A. Karim, A.H. El-Shafie, Application of artificial neural networks for water quality prediction. Neural Comput. Appl. **22**(1), 187–201 (2013)
13. M. Pant, H. Zaheer, L. Garcia-Hernandez, A. Abraham, et al., Differential evolution: a review of more than two decades of research. Eng. Appl. Artif. Intell. **90**, 103479 (2020)
14. F. Rehbach, S. Chandrasekaran, M. Rebolledo, S. Moritz, T. Bartz-Beielstein, GECCO Challenge 2018: Online Anomaly Detection for Drinking Water Quality (2018), http://www.spotseven.de/gecco/gecco-challenge/gecco-challenge-2018/
15. G. Reynoso-Meza, J. Sanchis, X. Blasco, S. García-Nieto, Physical programming for preference driven evolutionary multi-objective optimization. Appl. Soft Comput. **24**, 341–362 (2014)
16. G. Reynoso-Meza, J. Sanchis, X. Blasco, M. Martínez, Design of continuous controllers using a multiobjective differential evolution algorithm with spherical pruning, in *European Conference on the Applications of Evolutionary Computation* (Springer, 2010), pp. 532–541
17. G. Reynoso-Meza, A. Vignoni, Y. Boada, J. Picó, E. Picó, Model mis-match in multi-objective optimisation and preservation of trade-off order. IFAC-PapersOnLine **52**(26), 249–254 (2019)

18. V.H.A. Ribeiro, S. Moritz, F. Rehbach, G. Reynoso-Meza, A novel dynamic multi-criteria ensemble selection mechanism applied to drinking water quality anomaly detection. Sci. Total Environ. **749**, 142368 (2020)
19. V.H.A. Ribeiro, G. Reynoso-Meza, Ensemble learning by means of a multiobjective optimization design approach for dealing with imbalanced data sets. Expert Syst. Appl. **147**, 113232 (2020)
20. V.H.A. Ribeiro, G. Reynoso-Meza, Feature selection and regularization of interpretable soft sensors using evolutionary multi-objective optimization design procedures. Chemom. Intell. Lab. Syst. **212**, 104278 (2021)
21. F.A. Souza, R. Araújo, J. Mendes, Review of soft sensor methods for regression applications. Chemom. Intell. Lab. Syst. **152**, 69–79 (2016)

Chapter 14
What Affects Household Electricity Demand in Ecuador: Using Analytical Hierarchy Process to Read Homeowners' Perception

J. Hidalgo-Crespo, J. L. Amaya, and M. Soto

Abstract Analytical hierarchy process was applied to empirically prioritize the relevance of different electrical consumption indicators separated into criterions and subcriterions, according to the point of view of 543 homeowners' validated information from the city of Guayaquil. This research offers a reasonable perspective of the major factors that drive electricity usage at home from the point of view of the users. Evaluation of perceptions by social status was also studied dividing the sample in socioeconomic groups. The findings are important, as they differ from most existing literature that normally illustrates the correlation among indicators and the dependent variable. On the criteria level, homeowners believe that the criterions regarding electrical equipment (EE), structural factors (SF) and habits and behaviours (HB) are of more significant importance. Criterion inhabitants' characteristics (IC) is the least important for all socioeconomic groups with only 11% (5th place), being the lowest value for the high socioeconomic group with less than 9%.

14.1 Introduction

Energy has been acknowledged as the most essential and crucial input for social and economic development of a country [1]. It is ubicated in the centre of a triangle between nature, society and economy, and it is treated as an essential element of global progress [2]. Along the various energy resources available, electricity is a key secondary energy source for developing countries and a future primary energy resource for the developed world that looks to use it to power up their engines for transportation applications mostly. Globally almost 70% of the estimated electricity

J. Hidalgo-Crespo (✉)
Universidad Politécnica Salesiana, Guayaquil 090101, Ecuador
e-mail: jhidalgo@ups.edu.ec

J. L. Amaya
Escuela Superior Politécnica del Litoral, ESPOL, Guayaquil 090112, Ecuador

M. Soto
Universidade da Coruña, 15008 A Coruña, Spain

Á. Rocha et al. (eds.), *Communication, Smart Technologies and Innovation for Society*, Smart Innovation, Systems and Technologies 252,
https://doi.org/10.1007/978-981-16-4126-8_14

is produced using fossil fuels [3] and represents a major environmental problem because of the emission of toxic chemical pollutants and greenhouse gases like carbon dioxide and others [4].

Worldwide, residential electricity usage accounts for 30 per cent of total electricity generation [5]. Households are major drivers of an economy, triggering economic activities along all the supply chain [6]. They represent a wide group of consumers of energy resources such as electricity [7]. In Ecuador, household electricity consumption represented 37% of total electricity consumed (20,000 GWh) in the year 2018 [8], being the most consuming sector of the country. In this context, identifying the determinants and factors that have a meaningful impact to predict electricity consumption is of great importance and a key step towards the design and implementation of energy saving strategies and policies. The authors have already tried to correlate many factors with the residential electricity consumption for the country, such as behavioural, socioeconomic, demographic, using tools as ordinary least square regressions or sorting algorithms with the help of general surveys [9–13]. However, after all this research, the authors have explained as far as ten per cent of the variability and found as a general conclusion that Ecuadorian households do not behave as other third world countries and that there is a lack of environmental consciousness from all the members of the households.

To investigate this problem, a survey was taken to homeowners from the city of Guayaquil, following multi-criteria decision making format to evaluate the principal criterions and sub-criterions that influence the most in the electricity usage of households. The remainder of this paper is organized as follows: the Literature Review introduces the theoretical foundations of determinants selection and hierarchy construction, following that Methodology is presented to collect the information and data evaluation, after that Results are summarized and the ranking of criterions and sub-criterions is shown by the weights of each of them, and finally, the Conclusions and Discussion are given.

14.2 Literature Review

The amelioration of living standards for the Ecuadorian community calls for the necessity to pay attention to the promotion of sustainable electricity consumption and to seek new energy-related consumption practices. Relevant social impacts evaluation studies are required for sharpening detailed plans and connecting and improving public acceptance of government policies. However, making policies without considering homeowners' voices may lead to public dissatisfaction and distrust [14]. To establish a comprehensive set of household electricity indicators for the efficient assessment of homeowners' opinions towards the energy consumption of their households, a combination of reviewing existing methods, energy codes and academic research papers need to be conducted [15].

14.2.1 Household Electricity Indicators

Critical indicators typically include energy behaviours, personality trait, demographic information, building features and weather conditions [16]. For instance, one author [17] discovered a relationship between residential electricity consumption per capita and the ratio of the registered unemployed but in working-age population in Lithuania. Ye et al. [18] found that income and private tenure are the most effective determinants in vast majority of the neighbourhoods. Also, higher incomes consume more electricity due to larger households and bigger possession of appliances [19]. Other research [20] established a set of physical and behavioural factors to determine an energy demand load profile such as: climate, household composition, family income, culture background and human factors, building design and number of appliances. One paper [21] found that the employment rate, residential space and installation of electrical appliances can lead to an increase in residential electricity consumption. Other authors found that annual electricity expenditure depends on number of rooms, years of education completed by household head, employment, and tenancy status. Other research [22] indicated that ownership of electrical appliances and changes in socioeconomic and building factors such as energy efficiency awareness and practices, income, household size and floor space showed high statistical significance and explained 57% of total electricity consumption of Ghanaian households.

Different authors [23, 24] found a relative importance of income, dwelling size, family composition and routines, appliance age and region in determining household electricity consumption in Montenegro. Household electricity consumption is driven by household income [25], fuel price, demographics, building attributes and lifestyles. One researcher [26] revealed that the floor area, household size, disposable household income and gender are significant in explaining the variation of electricity consumption in Perth Social Housing in Western Virginia. Other information [27] showed that family size, household living area, age, education level and gender of the household gender influence household electricity demand. A survey to households in Indonesia showed [28] that family size, time spent at home, education level, home appliances and lightning had a significant positive effect on the monthly electricity bill. Household or building envelope characteristics: wall thermal resistance, window-to-wall ratio, wall outside painting and plastering, height, age [28, 29], as well as the relationship with the exteriors, such as shading, sunlight access [30] can affect electric consumption. The use of day lighting by consumers to reduce electrical power is a sustainable approach for energy optimization, considering that 30% of global electricity consumption is attributed to lighting.

14.2.2 Multicriteria Decision Making: Analytical Hierarchy Process

Among the various MCDM techniques available, the analytical hierarchy process (AHP) proposed by Saaty in 1980 is the most popular and has been used in several areas including planning, selection of the best choice, resource allocation and conflict resolution [31]. It displays a flexible way to assist the decision makers in formulating their sustainable problems in a rational and logical path [32, 33]. Although it is simple to understand and have good software support, it can be cognitively draining for the decision makers and in some cases produce a weak sustainability perspective when not used appropriately [34]. In many industrial engineering applications, the final decision is based on the evaluation of several alternatives in terms of several criteria and AHP is an effective approach in dealing with these kind of decision problems. In this context, group decision making tends to give better results because of the broader knowledge available and the possibilities of debate.

14.3 Methodology

14.3.1 Criterions and Sub-Criterions List

In step 1, based on reviewing the existing references listed in Sect. 14.2, energy codes for the country and researchers' experience, a total of 44 indicators were identified. After, they were refined and formed a draft indicator list according to the following three rules: feasibility, applicability and effectiveness applied by Ye et al. [18].

Feasibility: This rule measures the chances that an indicator could be assessed by the homeowner, keeping in mind it should be as understandable as possible. In the light of this rule, the following indicators "Orientation of the building", "Shape of the building", "Air tightness of envelope", "HV&AC facilities", "GDP", "Social class", "Wall thermal resistance", "Window-wall ratio", and "Household orientation" were no longer considered.

Applicability: This rule requires that the identified indicators cover the main aspects of residential home electricity usage. After, this rule the following indicators were discarded: "Insulation of envelope", "Outdoor and indoor shadows", "Lightning facilities" and "HoH gender".

Effectiveness: This rule requires that the identified indicator list should ignore repeatability and group similar indicators into one indicator that holds them all. Therefore, all the indicators written in the literature review section have been grouped into one as shown in Table 14.1. For instance: "Unemployed working age" and "Employment age" form "Household working people", and also, "Lights", "Indoor

Table 14.1 Criterions and sub-criterions

Code	Criterions	Code	Sub-criterions
EF	Economic factors	EF01	Household working people
		EF02	Income
		EF03	Disposable Income
		EF04	Electricity price
		EF05	Tenure status of household
SF	Structural factors	SF01	Height of household
		SF02	Wall plastering
		SF03	Natural light access
		SF04	Building characteristics
IC	Inhabitants' characteristics	IC01	Householders´ characteristics
		IC02	Householders routines
		IC03	Time spent at home
		IC04	Education level
EE	Electrical equipment	EE01	Energy efficiency knowledge
		EE02	Appliance ownership and age
		EE03	Power supply
		EE04	Use of electrical appliance
HB	Habits and behaviours	HB01	Energy consumption habits
		HB02	Behaviour regarding new equipment buying
		HB03	Environmental consciousness

lightning" and "Sunlight access" unite to form "Natural light access". Additionally, "Residential space", "Building attributes", "Number of rooms" and "Building age" form "Building characteristics". Finally "Age of householders", "Household size", "Family composition" and "HoH age" form "Householders' characteristics" and "Daily use of electrical appliances" and "Time of use" form "Use of electrical appliances".

Finally, the draft of indicators has been grouped in criterions, following the literature and researchers' expertise. The criterions selected for the present paper were as follows: economic factors, structural factors, inhabitants' characteristics, electrical equipment, habits and behaviours. All 20 indicators were separated after in the five criterions and became sub-criterions. Codes were assigned to each criterion and sub-criterion for better after management, as shown in Table 14.1.

14.3.2 Data Collection

In step 2, a survey has been conducted to read homeowners' perceptions on the importance of different indicators in the consumption of households' electricity. The data was collected from Guayaquil, capital of Guayas Province in Ecuador. The city was chosen because it is the largest and most populated in the country with around 2.70 million people.

With several barriers during data collection such as time constraints and financial limitations, sampling was done with the help of almost 200 students from a local university in the city to simplify data collection. Each student took the survey to three homeowners including their own household in the month of June 2020. For this study, it was considered a population of 430,000 households for the city of Guayaquil. The total sample size was rounded to 600 respondents.

Next, a pairwise comparison questionnaire was designed to prioritize the criterions and sub-criterions. Homeowners were asked to compare the importance of two indicators at the time assigning a weight in the form of a numeric value, between 1 and 9 according to their preference that measured the relative importance of one criterion over the other. The 1–9 scale was explained to the participants at the beginning of the questionnaire.

In step 3, the analytical hierarchy process (AHP) method is used to deliver the relative importance of the criterions and sub-criterions with the final goal of understanding how householders perceive the causes of electricity consumption. Following the steps of Lin [35], the relative importance between the ith indicator and jth indicator is given by $a_{ij} = w_i / w_j$.

14.4 Results

Analyses were made following the completion of pairwise comparisons. The first stage of the analysis was to check the consistency of judgments. Data from 590 households were received, from which, only 543 were validated with consistency ratios equal to or lower than 0.1. The second stage was to form the respective matrixes using the geometrical average of the validated data. Finally, the calculation of the relative weights of both criterions and sub-criterions are done and the ranking by importance developed. Figure 14.1 shows the social class stratification for the sample.

14.4.1 Weights and Rankings of Criterions

Following this stratification information, five different scenarios were evaluated to calculate the relative weights of both criterions and sub-criterions. One for all the data and one for each of the social-economic groups to know if there exist changes in

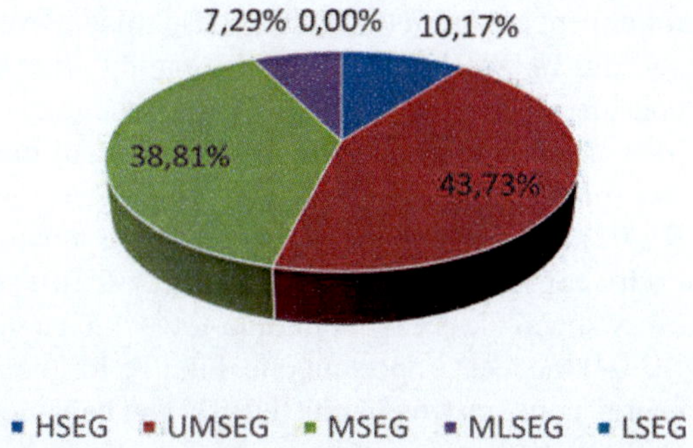

Fig. 14.1 Household stratification for the sample

opinions depending on the social class level. At the criterions level, electrical equipment (EE), structural factors (SF) and habits and behaviours (HB) are of the greatest importance for homeowners. The summation of the weights of these three criterions accounts for over 71% of the total weight of the five criterions. Electrical equipment (EE), economic factors (EE) and inhabitants' characteristics (IC) occupy the first, fourth and fifth positions, respectively, in the ranking for all the socioeconomic groups. Structural factors (SF) and habits and behaviours (HB) exchange the second and third place of the list as we move from higher to lower socioeconomic groups. Oddly enough, economic factors (EF) is not considered the most important criterion for lower social classes, although access to efficient electrical equipment depends on the economic situation of each household. This can be explained by the facilities of credit given to buyers by electrical equipment stores that even though they charge high taxation rates, they offer a long-term payment structure (up to 24 months or more), which give the buyer the sensation of low monthly fees.

14.4.2 Weights and Rankings of Sub-criterions

It is asserted by the author that the consistency of these criterions between terms are determined by the role that each sub-criterion plays in the electricity consumption process for a household. The weight change of each sub-criterion will reveal how the homeowners' thoughts vary among socioeconomic groups. For the economic factor sub-criterions, the electricity price (EF04) and tenure status of household (EF05) occupy the first and second positions in the ranking. Household working people (EF01), income (EF02) and disposable income (EF03) have similar weights that vary from 0.167 to 0.198. For the structural factor sub-criterions, the building characteristics (SF04), height of household (SF01) and wall plastering (SF02) occupy the first, second and third places, respectively, summing a total of 81% of weight. Natural light (SF03) access is considered the less important indicator according to homeowners and occupies the last position for every socioeconomic group except the middle-low socioeconomic group that considers wall plastering (SF02) the least important. For the inhabitant's characteristic sub-criterions, householders' characteristics (IC01), time spent at home (IC03) and householders' routines (IC02) occupy the first, second and third places, respectively, summing almost 84% of weight. Education level is considered the least important for all socioeconomic groups, which has given us an idea of the Ecuadorian way of thinking and the lack of confidence in the education system. The electrical equipment sub-criterions locate appliance ownership and age (EE02) and power supply (EE03) as the most important indicators that affect electricity consumption. Energy efficiency (EE01) knowledge is the least important for every social class expect middle-low which considers the use of electrical appliance (EE04) the least important one. Finally, for the habits and behaviours sub-criterions, energy consumption habits (HB01) and behaviours regarding new equipment buying (HB02) are the two most important, adding up to almost 76% of weight. Their values are very close to each other, and they exchange first and second positions along the different socioeconomic groups. However, environmental consciousness (HB03) is considered the least important for all socioeconomic groups.

14.5 Conclusions and Discussion

Attempting to diminish the gap of the disaccord between government perceptions and research assumptions, this paper identified the factors of significant influence to the electricity consumption of a household for the city of Guayaquil based on the perceptions of the people that consume the energy, the homeowners. Factors were collected through literature review. Questionnaires distributed to over 590 institutes were used to evaluate the influences of all the factors quantitatively and the result was generated from the AHP process. Homeowners´ attitudes towards various factors regarding electricity consumption presented in this paper can help policymakers, decision makers and researchers better understand the point of view of home users when asked about indicators that affect electricity consumption and establish sensibilization campaigns and household energy efficiency courses. On the criteria level, homeowners believe that the criterions regarding electrical equipment (EE), structural factors (SF) and habits and behaviours (HB) are of more significant importance among the five. Regarding the judgements on the electrical equipment (EE), the most important one, sub-criterions that hold stronger importance are appliance ownership and age (EE02) and power supply (EE03). Energy efficiency knowledge (EE01) is the least important according to homeowners for every socioeconomic group. On the other hand, building characteristics (SF04) and height of the household (SF01) occupy the first and second ranking for structural factors (SF), while natural light access (SF03) stands on the last position with almost 19% of weight, except for the middle-low socioeconomic, where it's wall plastering (SF02). This could happen, because of the climate of the city which is hot and humid, giving cooling and refrigeration more importance than natural light access. Also, for habits and behaviours (HB), sub-criterion energy consumption habits (HB01) is the most important with 38.5% of weight and sub-criterion environmental consciousness (HB03) the least important with 24.2% of weight. Criterion inhabitants' characteristics (IC) is the least important for all socioeconomic groups with only 11% (5th place), being the lowest value for the high socioeconomic group with less than 9%. One thing to further analyse is the fourth position of the economic factors (EF) criterion with a weight of 17% when research finds it as one of the most important. Additionally, oddly enough, homeowners consider the sub-criterions electricity price (EF04) and income (EF02) as the first and last one in the ranking of economic factors (EF), when Ecuador has one of the lowest tariffs in Latin America.

The authors also realize that city´s homeowners' distance themselves from the problem of electricity usage. This can be seen in the last position locations of the sub-criterions education level (IC04), energy efficiency knowledge (EE01) and environmental consciousness (HB03), which analyse the responsibility of the homeowner for their own consumption. This study offers a reasonable perspective of the major factors that drive electricity usage at home from the point of view of the users. This finding is important, as it differs from most existing literature that normally illustrate the correlation among indicators and the dependent variable, and hence, further emphasizing on sensibilization campaigns to cut energy use in both Guayaquil and other affluent cities around the country. We hope future researchers can pick this up and add on to the analysis in their work, as it will certainly help contribute towards a better understanding of Guayaquil's electricity driving factors.

References

1. J. Ren, L. Dong, Evaluation of electricity supply sustainability and security: multi-criteria decision analysis approach. J. Clean. Prod. **172**, 438–453 (2018)
2. S.H. Zolfani, J. Saparauskas, New application of SWARA method in prioritizing sustainability assessment indicators of energy system. Eng. Econ. **24**, 408–414 (2013)
3. M.N. Hussain, M. Uqaili, K. Harijan, M. Mustafa, M. Rahman, M. Khan, Multi-criteria analysis of electricity generation scenarios for sustainable energy planning in Pakistan. Energies **11**(4), 757
4. J.-J. Wang, Y.-Y. Jing, C.-F. Zhang, J.-H. Zhao, Review on multi criteria decision analysis aid in sustainable energy decision-making. Renew. Sustain. Energy Rev. **13**(9), 2263–2278
5. F. Taale, C. Kyeremeh, Drivers of households' electricity expenditure in Ghana. Energy Build. (2019)
6. Y. Long, Y. Yoshida, J. Meng, D. Guan, L. Yao, H. Zhang, Unequal age-based household emission and its monthly variation embodied in energy consumption—a case study of Tokyo, Japan. Appl. Energy **247**, 350–362 (2019)
7. International Energy Agency (IEA), *Electricity Information* (Paris, France, 2015). Available from: https://www.iea.org/bookshop/666-Electricity_Information_2015
8. J. Hidalgo-Crespo, Y.G. Cañizalez, S.C. Pisco, J.B. Monteses, L.M. Alonso Águila, T. Crespo-Vaca, Household electricity demand in marginal Ecuador: estimation and impact of explanatory variables. RISTI—Revista Iberica de Sistemas e Tecnologias de Informacao **E23**, 378–391 (2019)
9. C.Y. González, J. Hidalgo-Crespo, P.S. Coello, M.J. Barzola, A.L. Alonso, Sorting urban-marginal areas in Ecuador by weighting energy consumption indicators. RISTI—Revista Iberica de Sistemas e Tecnologias de Informacao **E23**, 444–457 (2019)
10. J. Hidalgo, T. Crespo, S. Coello, Y. González, Household sustainable behaviour evaluation and its relationship with socioeconomic indicators in the city of Guayaquil, in *Proceedings of*

the LACCEI International Multi-conference for Engineering, Education and Technology, July 2019
11. J. Hidalgo, S. Coello, Y. González, The determinants of household electricity demand in marginal Ecuador: "a case study at Monte Sinai", in *Proceedings of the LACCEI International Multi-conference for Engineering, Education and Technology*, July 2018
12. L.M. Alonso, J. Barzola-Monteses, Y. González-Cañizalez, J. Hidalgo-Crespo, S. Coello-Pisco, Sorting algorithm applied in geographical regions using residential electricity consumption indicators. Int. J. Adv. Sci. Eng. Inf. Technol. **10**(2), 562–566 (2020)
13. G. Montibeller, D. Von Winterfeldt, Cognitive and motivational biases in decision and risk analysis. Risk Anal. **35**(7), 1230–1251 (2015)
14. Y. Yang, B. Li, R. Yao, A method of identifying and weighting indicators of energy efficiency assessment in Chinese residential buildings. Energy Policy **38**(12), 7687–7697 (2010)
15. M. Shen, Y. Lu, K.H. Wei, Q. Cui, Prediction of household electricity consumption and effectiveness of concerted intervention strategies based on occupant behaviour and personality traits. Renew. Sustain. Energy Rev. **127** (2020)
16. S. Vojtovic, A. Stundziene, R. Kontautiene, The impact of socio-economic indicators on sustainable consumption of domestic electricity in Lithuania. Sustainability **10**(2), 162 (2018)
17. B. Mashhoodi, A. van Timmeren, Local determinants of household gas and electricity consumption in Randstad region, Netherlands: application of geographically weighted regression. Spat. Inf. Res. **26**(6), 607–618 (2018)
18. Y. Ye, S.F. Koch, J. Zhang, Determinants of household electricity consumption in South Africa. Energy Econ. (2018)
19. R. Yao, K. Steemers, A method of formulating energy load profile for domestic buildings in the UK. Energy Build. **37**(6), 663–671 (2005)
20. Y.-T. Chen, The factors affecting electricity consumption and the consumption characteristics in the residential sector—a case example of Taiwan. Sustainability **9**(8), 484 (2017)
21. M. Sakah, S. de la Rue du Can, F.A. Diawuo, M.D. Sedzro, C. Kuhn, A study of appliance ownership and electricity consumption determinants in urban Ghanaian households. Sustain. Cities Soc. (2018)
22. V. Đurišić, S. Rogić, J.C. Smolović, M. Radonjić, Determinants of household electrical energy consumption: evidences and suggestions with application to Montenegro, in *Proceedings of TMREES, EURACA*, Athens, Greece, 4–6 Sept 2019
23. M.E. Biresselioglu, M.H. Demir, A. Rashid, B. Solak, E. Ozyorulmaz, What are the preferences of household energy use in Pakistan?: findings from a national survey. Energy Build. (2019)
24. J. Zhang, F. Teng, S. Zhou, The structural changes and determinants of household energy choices and energy consumption in urban China: addressing the role of building type. Energy Policy **139** (2020)
25. P. Esmaeilimoakher, T. Urmee, T. Pryor, G. Baverstock, Identifying the determinants of residential electricity consumption for social housing in Perth, Western Australia. Energy Build. **133**, 403–413 (2016)
26. X. He, D. Reiner, Electricity demand and basic needs: empirical evidence from China's households. Energy Policy **90**, 212–221 (2016)
27. M.E. Wijaya, T. Tezuka, A comparative study of households' electricity consumption characteristics in Indonesia: a techno-socioeconomic analysis. Energy Sustain. Dev. **17**(6), 596–604 (2013)
28. M. Cinelli, S.R. Coles, K. Kirwan, Analysis of the potentials of multi criteria decision analysis methods to conduct sustainability assessment. Ecol. Ind. **46**, 138–148 (2014)
29. E. Triantaphyllou, S.H. Mann, Using the analytic hierarchy process for decision making in engineering applications: some challenges. Int. J. Ind. Eng.: Theory Appl. Pract. **2**(1), 35–44 (1995)
30. M. Bedir, E.C. Kara, Behavioral patterns and profiles of electricity consumption in Dutch dwellings. Energy Build. **150**, 339–352 (2017)
31. D. Brounen, N. Kok, J.M. Quigley, Residential energy use and conservation: economics and demographics. Eur. Econ. Rev. **56**(5), 931–945 (2012)

32. Y. Lin, W. Yang, K.S. Gabriel, A study on the impact of household occupants' behavior on energy consumption using an integrated computer model. Front. Built Environ. (2015)
33. D. Godoy-Shimizu, P. Steadman, I. Hamilton, M. Donn, S. Evans, G. Moreno, H. Shayesteh, Energy use and height in office buildings. Build. Res. Inf. 1–19 (2018)
34. E. Resch, R.A. Bohne, T. Kvamsdal, J. Lohne, Impact of urban density and building height on energy use in cities. Energy Procedia **96**, 800–814 (2016)
35. C. Xu, Y. Li, X. Jin, L. Yuan, H. Cheng, A real-time energy consumption simulation and comparison of buildings in different construction years in the Olympic Central Area in Beijing. Sustainability **9**(12), 22–45 (2017)

Chapter 15
Preliminary Studies of Performance and Lipid Profiles of Ecuadorian *P. volubilis L.* as Contribution to Agricultural Innovation

Tamara Fukalova-Fukalova, Johanna Castillo, Klever Parreño, Marco Gaibor, and Pablo Londoño-Larrea

Abstract Sacha inchi, *Plukenetia volubilis* L., is an oilseed cultivated in different localities in Ecuador and has not been widely recognized due to a lack of information about its nutritional benefits and its high content of essential fatty acids. Here, the oil obtained by cold pressing of Sacha inchi seeds from four different regions of Ecuador, El Oro, Cumandá, Chontal, and Napo, has been characterized. The oil extraction yield was determined, and the lipid profile was obtained by GC-FID gas chromatography, which allowed determining the composition of polyunsaturated (oleic, linoleic, linolenic) and saturated (palmitic and stearic) fatty acids. With the results, a one-factor analysis of variance ANOVA, a variability analysis, and a principal component analysis PCA was performed to determine the significant differences between the oils from the different localities. It was concluded that the concentration of essential fatty acids varies for omega 3 between 58.75 and 60.50%; for omega 6 between 31.56 and 32.66%; for omega 9 between 1.33 and 5.18%.

15.1 Introduction

Currently, there is great interest in recovering ancestral knowledge, as established by the United Nations through Article 8 literal j of the Convention on Biological Diversity [1]. This knowledge is related to the use of biodiversity, which includes the relationship between the properties of plants and their uses. The diversity of the flora in Ecuador is a contribution of raw material to obtain a range of products for the

T. Fukalova-Fukalova (✉) · J. Castillo · K. Parreño
Facultad de Ciencias Químicas, Universidad Central del Ecuador, Francisco Viteri y Gilberto Gatto Sobral, CP: 170521 Quito, Ecuador

P. Londoño-Larrea
Facultad de Ingeniería Química, Universidad Central del Ecuador, Ritter S/N y Bolivia, CP: 170521 Quito, Ecuador

M. Gaibor
AgroG2Ec S.A., Quito, Ecuador

Á. Rocha et al. (eds.), *Communication, Smart Technologies and Innovation for Society*, Smart Innovation, Systems and Technologies 252,
https://doi.org/10.1007/978-981-16-4126-8_15

industrial sector and that can generate an important contribution to its development. Among the most relevant raw materials, recent interest is given to the Sacha inchi plant native to tropical forests [2] and which has been successfully cultivated in other countries. It is an oilseed plant and is found mainly in Colombia, Ecuador, Brazil, and Peru; in the latter, there are records that show that this plant was part of the diet and of pre-Inca cultures [3]. Its production begins 6.5 months after transplantation and reaches ages up to 10 years.

Sacha inchi is a common name used to designate a series of different species (*Plukenetia ssp*) consisting of 19 species, whose taxonomic limits are not fully known [4]. *Plukenetia volubilis* L. is the most widely used species. The oil extracted from its seeds is the main product of interest since it has a high content of antioxidants and is rich in unsaturated fatty acids, among which the essential fatty acids of the omega group stand out, highly appreciated in the medicinal, nutritional, and cosmetic fields for their health benefits [5–7]. These functional compounds even when found in low concentrations, they provide positive effects on the health of consumers and help to prevent non-communicable diseases. Likewise, according to a study carried out by the World Health Organization [8] in 2003, 0.1% of total world morbidity is directly related to diseases caused by prolonged exposure to UV radiation. Sacha inchi oil is a promising sunscreen that can be an alternative to highly efficient sunscreens such as zinc oxide and titanium dioxide, whose prolonged use generate adverse skin effects [9]. Long-chain polyunsaturated fatty acids, such as omega 3 (ω-3), followed by omega 6 (ω-6), are especially appreciated by humans because they are not biosynthesized in the body. These functional fatty acids possess pharmacological properties, among which are antioxidant, antirhythmogenic, antitumor, anti-inflammatory, and anticonvulsant activity. Its natural sources are mainly of plant origin, especially oils and nuts.

At the time of the COVID-19 pandemic, demand has increased to transform more resilient dietary habits and drive more sustainable production [10] to promote food and nutrition security. The cultivation of Sacha inchi is emerging as a sustainable, short-cycle, and very profitable green trade [11], and its oil has a healthy and nutritious food. Sacha inchi is cultivated in the three continental regions of Ecuador, which are the Highlands, the Coast, and the Amazon region, with different geoclimatic conditions that inevitably lead to differences between the extraction yield and characteristics of the extracted oils. Given the serious economic situation, a key strategy is to develop tools to respond the commercial requirements through research and new knowledge to generate value and strengthen entrepreneurship that involves derivative products.

Despite being a country rich in agriculture, in Ecuador the Sacha inchi seed is sold as a raw material and no added value has been given to its commercially derived by-products. However, its commercialization has had a favorable evolution since the first plantings began in the country in 2012 to the present. In Ecuador, the production of Sacha inchi is carried out in 12 provinces, with 16 Producers Associations, of which five have Sacha inchi collection centers. Under the SumaSacha Ecosystem, 70% of this production is exported to Peru as raw material, and 30% is industrialized and manufactured finished products (food and cosmetics) in the specialized processing plant owned by the company Agroindustrias Ecuador G2.

The objective of this study was to determine the lipid profile of Sacha inchi oils, by gas chromatography, which allows differentiating the oils obtained from seeds from four areas in Ecuador, as well as comparing the cold extraction yields to evaluate the effect of the geoclimatic conditions of its origin.

15.2 Material and Methods

15.2.1 Raw Material

Sacha inchi (*P. volubilis* L.) seeds were used from four regions of Ecuador: Chontal, Napo, Cumandá y El Oro. Each locality is characterized by different geoclimatic conditions summarized in Fig. 15.1.

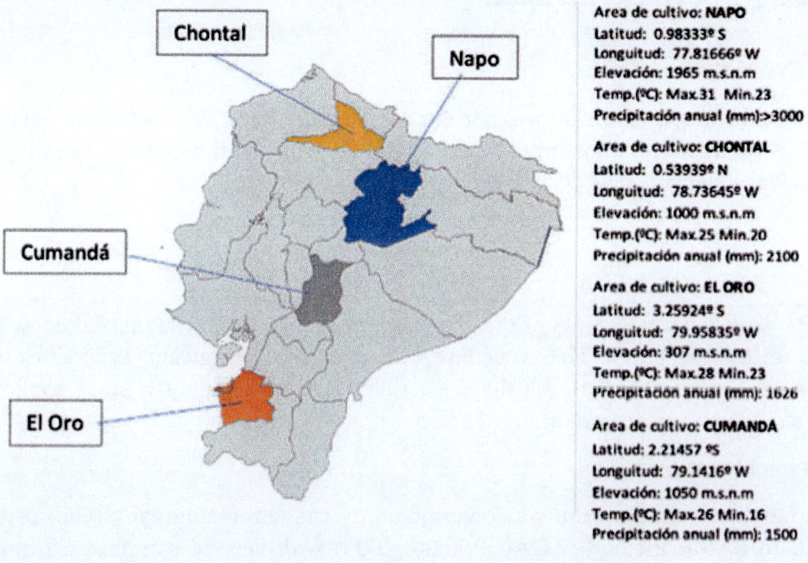

Fig. 15.1 Location of the producing regions of the Sacha inchi samples

15.2.2 Sample Preparation

Seeds for Extraction

Only healthy seeds were handpicked randomly from each zone, then the seed husks were carefully removed by hand, and nuts were stored at 4 °C until processed. The extraction process was carried out following a previously developed and tested protocol. A bench size stainless steel pneumatic press was used for oil extraction with a maximum pressure of 1000 psi.

Samples for Lipid Profile

The derivatization of fatty acids was carried out according to the official method (AOAC, 2005) that included two stages:

(1) saponification.
(2) preparation of methyl esters.

15.2.3 Analytical Methods

Oil Extraction

After cold extraction, the oil was decanted from the sides of the vessel, filtered using a cellulose filter and transferred to dark glass GC vials stored at 4 °C. The oil yield (OY) was determined as percentage using the equation:

$$OY = (wo/wn) * 100 \tag{15.1}$$

where wo is the total weight of the extracted oil and wn is the weight of the raw nuts used for the extraction. The procedure was repeated by triplicate, and values were reported in percent (w/w). All the extraction process and analysis steps applied in this study are represented on Fig. 15.2.

Fatty acid composition

The fatty acid composition was determined by gas chromatography (GC) using a standard FAME method (AOAC 966.06, 2005) with derived samples. Chromatographic analyses were performed using an Agilent Technologies 7890A GC System Model, equipped with flame ionization detector (FID), split-splitless injector. The process conditions are summarized in Table 15.1.

The values were reported in w/w, and the analysis was carried out by triplicate.

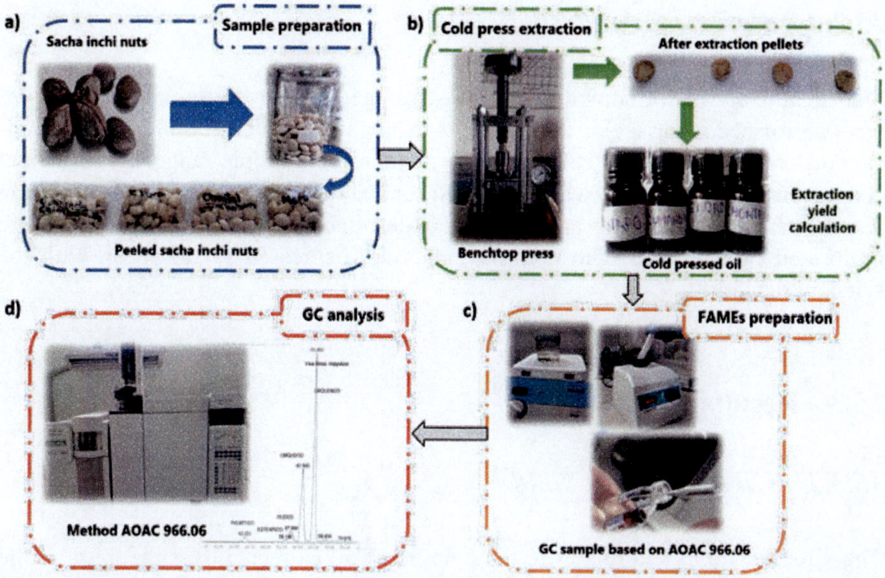

Fig. 15.2 Extraction process and analysis scheme for oil from Sacha inchi nuts **a** sample preparation, **b** cold press extraction, **c** FAMEs preparation, **d** GC analysis

Table 15.1 Chromatographic analysis conditions

Parameters	Conditions
GC-detector	FID
Column	100 m, 250 μm, 0.20 μm Agilent 112-88A7/HP-88
Injector temperature, °C	240
Injector temperature, °C	250
Split rate	0.1:1
Injection volume, μL	2
Carrier gas	Carrier Gas Helium
Linear speed, cm/s	21.402
Oven temperature sequence	
Initial	60 °C, 2 min
Ramp	8 °C/min up to 145 °C, 26 min
Ramp	2 °C/min up to 220 °C
Post run	240 °C/1 min

15.2.4 *Statistical Analysis*

Statistical analyses for comparison of the average results of different nuts samples were performed using a one-way analysis of variance (ANOVA). The least significant difference (LSD) was calculated from Duncan's multiple range test. Principal component analysis (PCA) was performed for standardized composition data using R statistical software 4.0.4 and Excel software package 2016. All analyses were performed by triplicate. The presented data are expressed as the mean with the standard deviation.

15.3 Results

15.3.1 *Oil Extraction Yield*

The oil extraction yield relates to the amount of oil extracted over the mass of the seeds, expressed as a percentage (g/100 g). The variability between the means of each yield is shown in Fig. 15.3.

Seeds from the Chontal locality show a higher average with 31.05%, followed by the average of Cumandá locality with 30.96%. Cumandá stood out among them with the highest variability (min. 27% and max. 33). On the other hand, Napo has more homogeneous performance data (min. 28.2% and max. 28.7%). Both the diagram (a) and the mean with the confidence intervals (b) confirm the latter, even though its performance is lower among all the localities, and its seeds produce a more uniform quantity of oil.

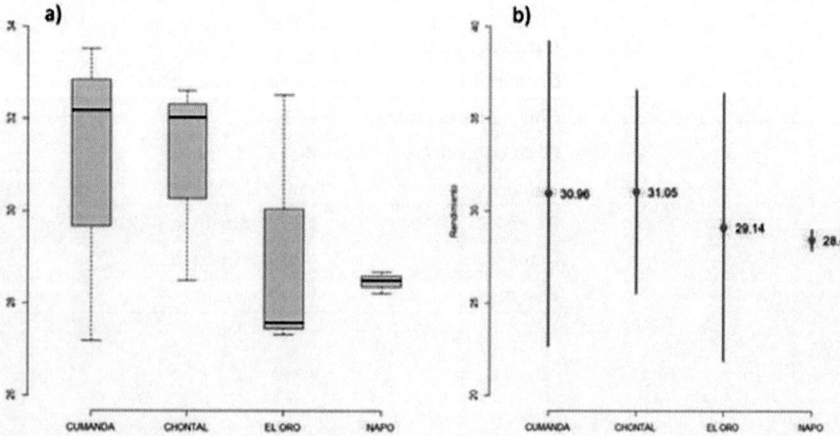

Fig. 15.3 Variability analysis for Sacha Inchi oil extraction performance: **a** box-and-whisker plots; **b** the mean and confidence intervals

15.3.2 Lipid Profile Analysis

The lipid profile analysis was carried out for each oil obtained to establish differences due to cultivation in different geoclimatic zones. The analysis is detailed in Table 15.2.

When comparing the mean values of each component, it was evidenced for the case of Napo region, the percentage of omega 3 (60.50%) is greater than the rest of the regions. In the case of omega 6, a higher percentage was evidenced in the Cumandá region (32.66%). For omega 9, the region that gave the highest percentage was El Oro (5.18%). Palmitic acid had a higher concentration in the case of El Oro (3.36%), while in the case of stearic acid there is a higher concentration in the oil of Napo region (0.69%).

The range of coefficients of variation for omega 3 was presented in the range between 0.68 and 4.73%; for omega 6 between 0.75 and 6.61%; for omega 9 between 1.33 and 13.01%. The fatty acids with the lowest dispersion correspond to the Napo region since it has the lowest standard deviation in all parameters, except stearic acid. In the case of stearic acid, the highest dispersion was presented in the lipid profile of El Oro oil with the coefficient variation of 49.36%.

15.3.3 Discriminant and Principal Component Analysis (PCA)

From experimental data, the principal components analysis was carried out to transform the original variables into a smaller number of variables (principal components) that are expressed as a linear function (PC) and their interrelationships, represented in Fig. 4a. In addition, the discriminant analysis was applied to verify the differences between the oils of the four studied localities, represented in Fig. 4b.

The fatty acids are presented in decreasing order according to the percentage of their abundance in each of the samples and were obtained by GC-FID analysis.

The balance of the relationships reveals correlations between the components of the fatty acid composition obtained. The omega 3 and 6 components have a strong negative correlation with each other because they are at opposite extremes. The variables omega 9, stearic and palmitic acids have weaker relationships, as does yield. All parameters are essentially linked to the seed variable.

Confidence ellipses for the regions allowed discriminant differentiation for the seeds by regions. The Napo and El Oro regions, being very far apart, are different from the others, while the Cumandá and Chontal regions are similar due to the proximity of their ellipses.

Table 15.2 Lipid profile of Sacha inchi oil from each locality. Average values and standard deviation $N = 3$/coefficient of variation CV

Locality	Omega 3 (%)/CV	Omega 6 (%)/CV	Omega 9 (%)/CV	Palmitic acid (%)/CV	Stearic acid (%)/CV
Cumandá	59.36 ± 0.98/1.65	32.66 ± 1.13/3.47	4.35 ± 0.20/4.69	3.01 ± 0.21/7.10	0.62 ± 0.20/32.94
Chontal	59.37 ± 2.81/4.73	32.30 ± 2.14/6.61	4.56 ± 0.59/13.01	3.09 ± 0.24/7.78	0.68 ± 0.18/26.23
El Oro	58.75 ± 1.57/2.68	32.11 ± 0.91/2.83	5.18 ± 0.42/8.07	3.36 ± 0.13/3.89	0.60 ± 0.30/49.36
Napo	60.50 ± 0.41/0.68	31.56 ± 0.24/0.75	4.39 ± 0.06/1.33	2.85 ± 0.11/3.89	0.69 ± 0.06/8.81

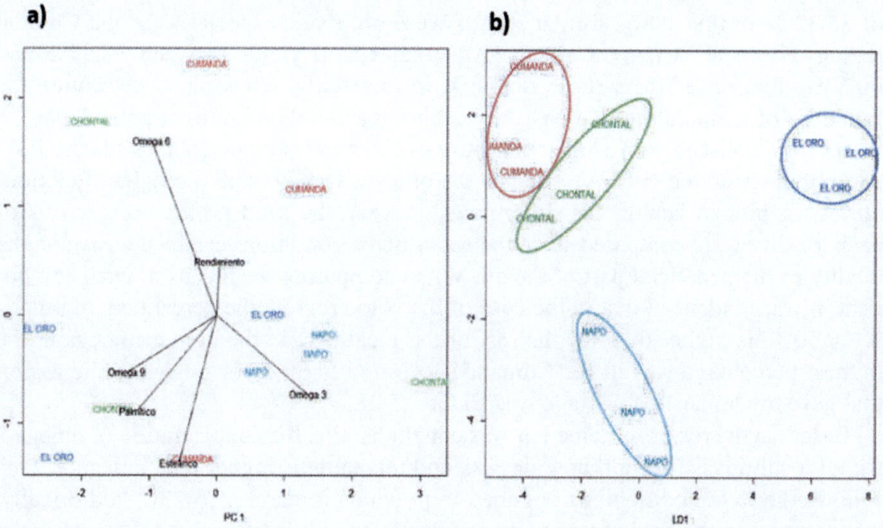

Fig. 15.4 Statistical analysis of localities differences.a Principal component analysis, b discriminant analysis

15.4 Discussion

Vegetable oils are widely used in human food, due to their high content of polyunsaturated fatty acids, which provide and satisfy the required daily energy requirements. On the other hand, they are used as raw material for the pharmaceutical and cosmetic industries, among others. Currently, the Sacha inchi seed is regaining prominence worldwide due to the nutritional value represented by its oil with a high content of fatty acids. In this sense, the studies of Sacha inchi of local origin require the identification of the yield and the characterization of the lipid profiles of this raw material according to the area of its production. Although Sacha inchi oil has been studied in some producing countries such as Brazil, Colombia [12–14] and, especially, Peru [15–18], the results obtained in this study are the first to examine the oil extraction yield of the seed by cold pneumatic pressure and its lipid composition depending on the growing area in Ecuador.

In the analysis of variance at 95% confidence, it is observed that the origin of the seeds according to the localities does not have a significant effect on the oil extraction yield. The analysis of variability indicated that, for Chontal, El Oro and Cumandá localities this parameter is heterogeneous, highlighting the Cumandá locality in the highest variability. However, the seeds from the Napo locality had the lowest yield and the most homogeneous variability.

The results were compared with the results of the research by Romero et al. [19] with seeds from the city of San Vicente, Manabí-Ecuador with a yield of 26.92%. Also, the results obtained by Chasquibol et al. [16] with seeds from the province of Rodríguez Mendoza, department of Amazonas-Peru, reported an average of 35.32%. In this study, similar results were observed, especially for the Chontal region. However, Muangrat et al. [20] obtained oil yields between 37.97% and 40.63%, that differ from those obtained in this study, allowing to determine the influence of temperature as a possible technological aid on improving yield.

As it is an oilseed with a high contribution of monounsaturated fatty acids (MUFA) and polyunsaturated (PUFA), such as the omegas family, with biological functionalities for human health, the differences between the lipid profiles of the oils of each locality were compared since the geoclimatic conditions can be decisive on the quality of the beneficial part of the oil. When comparing the means of each component, it was evidenced that in the case of the Napo region, the percentage of omega 3 (60.50%) is higher than for the rest of the localities; in the case of omega 6, the highest percentage was in the Cumandá region (32.66%); for omega 9, the region that gave the highest percentage was El Oro (5.18%).

Based on the results obtained, it was concluded that the concentration of omega 3 for all localities is higher than omega 6, and this latter is higher than omega 9. This follows the expected trend as obtained in previous research [21–23]. Additionally, when applying the analysis of variance, it was concluded that for polyunsaturated fatty acids there are no significant differences between localities. On the contrary, for two major saturated fatty acids, which are stearic and palmitic, the concentration of palmitic acid implies a statistically significant difference between the localities, being able to establish a differentiation criterion in the raw material.

Palmitic acid is the most abundant saturated fatty acid in the diet. Its concentrations found in this research exceed the range between 1.58 and 2.25% obtained in the study carried out by Singanuson and Jimyangyuen [24]. Another frequently present fatty acid is stearic acid, and it was found below the reference to the range reported between 1.20 and 1.70% in the same study. Being saturated acids with an atherogenic characteristic, its low levels in Sacha inchi oil make it healthier.

The high amount of mono- and polyunsaturated fatty acids found, in total around 96.0% for each locality, exceeds that found in the studies carried out by Cisneros et al. [25] with 91.2%; by Fanali et al. [26] with 92%, and by Liu et al. [27] with 93%. Due to its predominant presence, linolenic acid (ω-3) stands out, whose consumption provides benefits in reducing the risks of cardiovascular diseases. Linoleic acid (ω-6) is the second with the highest percentage within the lipid composition of oils from four locations and contributes to the prevention of inflammatory diseases [28–30]. When comparing the content of oleic acid (ω-9) whose quantity is accentuated in olive oil (71.22%) studied by Romero Aroca [31], in the present study this component is low, being characteristic in Sacha inchi oils. However, in the work carried out by Ramos Escudero [32] with Sacha inchi oils from different regions of Peru, a range between 10.54 and 14.30% of oleic acid was obtained, being a relatively high concentration.

In the principal component and discriminant analysis, it was possible to accentuate the differences between the oils from four regions. The oil of the seeds from the Cumandá and Chontal regions have similar characteristics in all the parameters studied. Napo and El Oro seed oils differ from other oils due to their omega 3 and omega 6 components. Components such as omega 9, stearic and palmitic acids, have similar characteristics in all regions. Omega 6 is not characteristic of Chontal seed oil. The extraction yield is not characteristic for any locality. On the other hand, omega 3 is the most characteristic of Napo seed oil.

Although high proportions of monounsaturated and polyunsaturated fatty acids are good for human health, they are also highly susceptible to oxidation. The study carried out by Rodríguez et al. [33] on the oxidative stability of Sacha inchi oil showed the decay of monounsaturated and polyunsaturated fatty acids, observing that cis isomers are more easily oxidized than their trans counterparts. The cold extraction process increases the useful life of the oil against oxidation [34].

15.5 Conclusions

The high yields, easy handling, and resistance to pests, and diseases have made Sacha inchi crop a very attractive alternative for small and medium producers in Ecuador. This study was helpful as an informative starting point to relate the geoclimatic conditions in which Sacha inchi is cultivated with the oil extraction yield and fatty acid composition. This in turn will provide a decision tool for future agro-industrial investors.

After cold-pressed extraction, the statistical analysis confirmed the oil yield was similar among the four regions analyzed. The lipid profile showed the presence of fatty acids (in descending concentration): omega 3, omega 6, omega 9, palmitic acid, and stearic acid in all samples, being omega 3 the higher in content in the oil extracted from Napo seeds. In general, the content of essential fatty acids found in this study is higher than the reported by previous literature.

References

1. PNUMA, Convenio sobre la Diversidad Biológica 2020–2021: viviendo en armonía con la naturaleza. Homepage: https://www.cbd.int/undb/media/factsheets/undb-factsheets-es-web.pdf. Accessed 2 Mar 2021
2. R. Chirinos, et al., Sacha inchi (*Plukenetia volubilis*): a seed source of polyunsaturated fatty acids, tocopherols, phytosterols, phenolic compounds and antioxidant capacity. Food Chem. **141**(3), 1732–1739 (2013)
3. D. Flores, O. Lock de Urgaz, Revalorizando el uso milenario del sacha inchi (*Plukenetia volubilis* L.) para la nutrición, la salud y la cosmética. Rev. Fitoterapia **13**(1), 23–30 (2012)
4. M. Corazon Guivin, et al., Caracterización genética de accesiones Sanmartinenses del Banco Nacional de Germplasma de Sacha Inchi *Plukenetia volubilis* L. (E.E. El Porvenir-INIA). Folia Amazonica **18**(1–2), 23–31 (2009). https://doi.org/10.24841/fa.v18i1-2.276
5. F. Garmendia, G.P. Ronceros, Efecto del aceite de sacha inchi (*Plukenetia volúbilis* L.) sobre el prefil lipídico en pacientes con hiperlipoproteinemia. Rev. Peru Med. Exp. Salud Publica **28**(4), 628–632 (2011)
6. L.M. Fernández y Carrillo, J.D. Serra, J.R.M. Álvarez, R.S. Alberich, F.P. Jiménez, Grasas de la dieta y salud cardiovascular. An. Pediatr. **74**(3), 192.e1–192.e16 (2011). https://doi.org/10.1016/j.anpedi.2010.12.005

7. H. Kahiluoto, Food systems for resilient futures. Food Security **12**, 853–857 (2020). https://doi.org/10.1007/s12571-020-01070-7
8. WHO/SDE/OEH/02.2, Índice UV solar mundial: guía practica. Homepage: https://www.who.int/uv/publications/globalindex/es/. Accessed 25 Apr 2021
9. M. Sáez-de-Ocariz, M. Orozco-Covarrubias, Protección solar en el paciente pediátrico. Acta Pediatr. Mex. **36**, 364–368 (2015)
10. S. Wang, F. Zhu, Y. Kakuda, Sacha inchi (*Plukenetia volubilis* L.): nutritional composition, biological activity, and uses. Food Chem. **265**(1), 316–328 (2018). https://doi.org/10.1016/j.foodchem.2018.05.055
11. Estudio de viabilidad económica del cultivo de Plukenetia volubilis Linneo, sacha inchi, en el departamento de San Martín. Avances Económicos Nº3 (Instituto de Investigaciones de la Amazonía Peruana (IIAP), 2009)
12. Castaño T, D. L., Valencia G, M. del P., Murillo P, E., Mendez A, J. J. and Eras Joli, J. composición de ácidos grasos de sacha inchi (Plukenetia volúbilis Linneo) y su relación con la bioactividad del vegetal. Rev. Chil. Nutr. **39**(1), 45–52(2012)
13. Ayala, G. A. análisis de crecimiento y producción de 3 variedades de sacha inchi (Plukenetia volubilis L.), en el municipio de Tena Cundinamarca. (Universidad de Ciencias Aplicadas y Ambientales, 2016)
14. G. Sanchez, Caracterización y cuantificación de los ácidos grasos omega 3 y omega 6 presentes en el aceite de Sacha inchi (*Plukenetia volubilis* L.) (Universidad Nacional de Colombia-Medellín, 2012)
15. E.F. Ramos, L. Ponce, E. Barnett, L. Celi, M. Ramos, Perfil de ácidos grasos de aceite de Sacha Inchi (*Plukinetia volubilis*) en comparación con otros aceites comestibles. Campus **XXI**(21), 101–108 (2016)
16. N. Chasquibol, A. Guinda, C. Del Águila, J.C. Yácono, M.C. Pérez, Estudioas preliminares sobre la caracterización de aceites de semillas de sacha inchi (*Plukinetia Huayllabambana*), cultivadas en la provincia de Rodríguez de Mendoza, departamento de Amazonas-Perú (2013)
17. D. Sihuayro Larico, Evaluación del rendimiento en la extracción del aceite de sacha inchi (*Plukenetia volubilis* L.) del ecotipo predominante en el valle del Río Apurímac (Ayacucho) y su caracterización físico-química y sensorial (Universidad Nacional Jorge Basadre Grohmann, Tacna, 2013)
18. N. Adrianzén, C. Rojas, G. Linares, Efecto de la temperatura y tiempo de tratamiento térmico de las almendras trituradas de Sacha Inchi (*Plukinetia volubilis* L.) sobre el rendimiento y las características físico-químicas del aceite obtenido por prensado mecánico en frio. Agroind. Sci. **2**, 45–55 (2011)
19. L. Romero Hidalgo, C. Valdivieso Rogel, S. Bonilla Bermeo, Caracterización del aceite de la semilla de Sacha Inchi (Plikinetia Volubilis) del cantón San Vicente, Manabi, Ecuador, obtenida mediante procesos no térmicos de extrusión. La Granja: Revista de Ciencia de la Vida **30**(2), 70–79 (2019). https://doi.org/10.17163/lgr.n30.2019.07
20. R. Muangrat, P. Veeraphong, N. Chantee, Screw press extraction of Sacha inchi seeds: oil yield and its chemical composition and antioxidant properties. J. Food Process Preserv. **42**(1), e13635 (2018). https://doi.org/10.1111/jfpp.13635
21. N.R. Gutiérrez Suquillo, I.A. Saá Arévalo, A.F. Vinueza Lozada, Diseño y construcción de un prototipo para la extracción continúa de aceite de la semilla Sacha Inchi con un proceso de prensado en frío. Enfoque UTE **8**(2), 15–32 (2017)
22. P. Bondioli, I. Della Bella, P. Rettke, Alpha linolenic acid rich oils. Composition of *Plukenetia volubilis* (Sacha Inchi) oil from Peru. La Rivista Italiana Delle Sostanze Grasse **LXXXIII**, 120–123 (2006)
23. L.A. Follegatti-Romero, C.R. Piantino, R. Grimaldi, F.A. Cabral, Supercritical CO_2 extraction of omega-3 rich oil from Sacha inchi (*Plukenetia volubilis* L.) seeds. J. Supercrit. Fluids **49**, 323–329 (2009). https://doi.org/10.1016/j.supflu.2009.03.010
24. R. Singanusong, S. Jiamyangyuen, Effects of maturity on chemical composition and antioxidant activity of Sacha Inchi (*Plukenetia volubilis* L.) cultivated in Northern Thailand. Walailak J. Sci. Technol. **17**(9), 998–1009 (2018)

25. F.H. Cisneros, D. Paredes, A. Arana, L. Cisneros-Zevallos, Chemical composition, oxidative stability and antioxidant capacity of oil extracted from roasted seeds of Sacha-inchi (*Plukenetia volubilis* L.). J. Agric. Food Chem. **62**(62), 5191–5197 (2014). https://doi.org/10.1021/jf500936j
26. C. Fanali, L. Duga, F. Cacciola, M. Beccaria, S. Grasso, M. Dachá, P. Dugo, L. Mondello, Chemical characterization of Sacha Inchi (*Plukenetia volubilis* L.) oil. J. Agric. Food Chem. **59**(24), 13043–13049 (2011). https://doi.org/10.1021/jf203184y
27. Q. Liu, Y.K. Xu, P. Zhang, Z. Na, T. Tang, Y.X. Shi, Chemical composition and oxidative evolution of Sacha Inchi (*Plukentia volubilis* L.) oil from Xishuangbanna (China). Grasas Aceites **65**(1), e012 (2014). https://doi.org/10.3989/gya.075713
28. L. Carrillo Fernández, J. Dalmau Serra, J.R. Martínez Álvarez, R. Solà Alberich, F. Pérez Jiménez, Grasas de la dieta y salud cardiovascular. Aten Primaria **43**(3), 157.e1–157.e16 (2010). https://doi.org/10.1016/j.aprim.2010.12.003
29. J.J. Carrero, L. Martín-Bautista, L. Baró, J. Fonollá, J. Jiménez, J.J. Boza, E. López-Huertas, Efectos cardiovasculares de los ácidos grasos omega-3 y alternativas para incrementar su ingesta. Nutr. Hosp. **XX**(1), 63–69 (2005)
30. D. Aires, N. Capdevila, M.J. Segundo, Ácidos grasos esenciales. Offarm **24**(4), 96–102 (2005)
31. A.J. Romero Aroca, Caracterización y diferenciación de los aceites vírgenes de oliva de la comarca Priorat (Tarragona) dentro del mercado global de aceites de la variedad 'Arbequina' (Universitat de Lleida, 2011)
32. F. Ramos Escudero, Caracterización y trazabilidad del aceite de sacha inchi (*Plukinetia volubilis* Linneo) (Universidad de Sevilla, 2013)
33. G. Rodríguez, E. Villanueva, P. Glorio, M. Baquerizo, Estabilidad oxidativa y estimación de la vida útil del aceite de sacha inchi (*Plukenetia volubilis* L.). Scientia Agropecuaria **6**(3), 155–163 (2015). https://doi.org/10.17268/sci.agropecu.2015.03.02
34. H. Nusselder, P. Cloesen, El cultivo de semillas nobles: ¿Sacha inchi de la Amazonía a la cuenca caribeña? Homepage: https://docplayer.es/12364385-El-cultivo-de-semillas-nobles-sacha-inchi-de-la-amazonia-a-la-cuenca-caribena.html. Accessed 2 Mar 2021

Chapter 16
Toxicological Evaluation of Essential Oil and Ethanolic Extracts of *Ruta graveolens* in *Artemia salina*

Mónica Espadero, Sandy Gavilanes, Tatiana Mosquera, and Gabriela Alvarez

Abstract Traditional medicine is part of humanity's cultural heritage and has been used for thousands of years to the present day. Different plants have extensive pharmacological activity, and it is used against various diseases, but the safe use of plants with pharmacological effects is still uncertain. *R. graveolens* is a specie known as medicinal plant since ancient times and currently used for treatment of various diseases, but at the same time, there are certain toxic components in its chemical composition. The toxicity of essential oil and ethanolic extracts at 50 and 96% of *R. graveolens* in *Artemia salina* was evaluated in this research. In addition, a qualitative comparison of the chemical components present in the essential oil and in the extracts was made by infrared spectroscopy (FT-IR). For the preparation of the extracts and the essential oil, flowers, leaves, and stems of the species previously dried were used, which were subjected to a process of maceration and distillation by steam drag, respectively, obtaining 0.13% yield of the essential oil. The FT-IR comparison of the essential oil with the spectra of 2-nonanone and 2-undecanone indicates the presence of C=O groups, C–H, C–C bonds, allowing the identification of ketones. For the toxicity test, the essential oil and the extracts at doses 0.1, 1, 10, and 100 ppm were used in *A. salina* for 24 h. Essential oil at concentrations > 10 ppm and ethanolic extract at concentrations > 25 ppm caused the death of more than 50% of individuals.

16.1 Introduction

Ruta graveolens is a specie commonly known as rue, is a perennial plant in the form of a small shrub, strongly aromatic, belonging to the *Rutaceae* family, native to southern Europe and Asia, and its distribution is cosmopolitan [1]. This plant has been introduced in Ecuador, especially in the sierra region in the provinces of Bolívar, Chimborazo, Pichincha, Tungurahua, and Azuay, located at an altitude of 1500–3000 m above sea level [2, 3]. In addition, it is used in medicinal treatments by

M. Espadero (✉) · S. Gavilanes · T. Mosquera · G. Alvarez
Universidad Politécnica Salesiana, Quito, Ecuador
e-mail: mespadero@ups.edu.ec

Á. Rocha et al. (eds.), *Communication, Smart Technologies and Innovation for Society*, Smart Innovation, Systems and Technologies 252,
https://doi.org/10.1007/978-981-16-4126-8_16

their as antiseptic, abortion, anthelmintic, anti-parasitic, anti-diarrheic, anti-febrile, antirheumatic, and antimicrobial properties [4–6].

The main compounds of *R. graveolens* are alkaloids such as quinolone, coumarins, and flavonoids; the essential oil that contains volatile substances such as terpenoids and oxygenated compounds such as ketones, where 2-undecanone and 2-nonanone, stand out [7, 8], which represent more than 50% of its total composition. Some of the active ingredients may have antifungal and antimicrobial activities [9].

Although there is a belief that medicinal plants are innocuous, previous research has shown the toxic effects of some species used in traditional medicine [10, 11]. Pharmacological trials have confirmed the adverse effects of *R. graveolens*, such as phototoxicity, genotoxicity, and mutagenicity; even in studies evaluating the cytotoxicity of hydroalcoholic extracts of *R. graveolans*, it has been shown to cause maximum mortality at 48 h of exposure in larvae of *A. aegypti* [12]. In addition, the nematicidal effect of *R. graveolens* extracts on nematodes that attack roots of plants has been analyzed, observing mortality in the larvae [13]. This motivated the present investigation to determine the toxicity of the essential oil and ethanolic extracts at 50 and 96% of *R. graveolens* in *A. salina*, as well as an analysis of the extracts and essential oil by infrared spectroscopy (FTIR).

16.2 Materials and Methods

16.2.1 Essential Oil Extraction

Leaves, stems, and flowers of *R. graveolens* were obtained at the Cuenca city, province of Azuay. The samples were collected in July 2020 and, subsequently, were taxonomically identified and deposited at the Herbarium of the University of Azuay.

The collected material of *R. graveolens* (3064 g) was selected manually and cut with a pruning shears. After drying, the *R. graveolens* parts were placed in a steam distillation equipment. The oil obtained by condensation of the volatile components of the *R. graveolens* leaves, stems, and flowers was centrifuged for 15 min at 1500 rpm to remove impurities, and finally the yield of the essential oil was calculated.

To evaluate the toxicity effect of *R. graveolens* essential oil, a stock solution was prepared with 0.5 ml of dimethylsufoxide (DMSO), 20 mg of essential oil, and 1.5 ml of distilled water, and it was diluted to 100, 10, 1, and 0.1 ppm.

16.2.2 Preparation of Ethanolic Extracts

R. graveolens leaves, steams, and flowers were dried in an oven at a temperature lower than 40 °C for approximately 5 days and powdered mechanically to obtain particles of 1 mm [14]. The powder (65 g) was subjected to prepare the ethanolic extracts at 50 and 96% and macerated with solvent such as ethanol and water in the ratio of 1:10 (w/v). Subsequently, the solvent was evaporated with a rotary evaporator at the boiling temperature of ethanol. The extracts were diluted to concentrations of 100, 10, 1, and 0.1 ppm with distilled water.

16.2.3 Chemical Identification by FTIR

Fourier transform infrared spectroscopy, also known as FTIR analysis, was done in a Thermo Scientific equipment (Nicolet iS10 model) with ATR and diamond crystal, and the spectra were measured in a range of 400–4000 cm^{-1} with 32 repetitions each one, the essential oil was compared with spectra proper from the HR Aldrich library.

16.2.4 Acute Toxicity Study

The toxicological activity of the essential oil and the ethanolic extracts at 50 and 96% of *R. graveolens* was evaluated according to the method described [15], which the specie *A. salina* is used as biological agent. A saline water solution (20%) was prepared to stimulate the hatching of the eggs. Subsequently, 0.5 g of cysts were placed in the water and incubated at a temperature of approximately 25 °C, with a source lighting and aeration system for 48 h to guarantee hatching. Ten larvae or nauplii were introduced in the extracts at concentrations of 0.1, 1, 10, and 100 ppm. In addition, a control group was included and introduced in salt water without extracts, and another in DMSO. After 24 h, the living and dead nauplii were counted to determine the lethal concentration (50%). The mortality rate of each treatment was calculated with the following Abbott's formula [16].

$$\%M = \frac{m_e - m_b}{1 - m_b} * 100 \quad (16.1)$$

$\%M$ Mortality rate.
m_e Mortality in essential oil and extracts.
m_b Mortality in the control group.

16.2.5 Statistical Analyses

A completely randomized design was applied with a factorial arrangement of (3 × 4) + 2 additional controls, with three repetitions. The data were analyzed by one-way analysis of variance (ANOVA) with Tukey's test ($\alpha = 0.05$). In addition, a logistic regression with Probit adjustment was carried out using InfoStat software.

16.3 Results and Discussion

16.3.1 Essential Oil Extraction

The yield of the essential oil of *R. graveolens* was 0.13%, a normal value in comparison with yields obtained in similar investigations, because the average is 0.23%, thus indicating that the species presents very low yields [14]. In addition, a slightly yellow, aromatic, penetrating, and highly volatile liquid was observed in the preliminary analysis of the essential oil.

16.3.2 Chemical Identification by FTIR

The frequencies emitted by the FTIR spectrum are assigned to the stretching and bending of vibrations that each functional group has. In the spectrum obtained from the essential oil of *R. graveolens*, bands were observed where the characteristic absorptions of 2-undecanone and 2-nonanone overlap, these being the major components of essential oil [15]. The principal component found was ketones (> 75%), data that are ratified with previously research [15], where the presence of 2-undecanone (43%) and of 2-nonanone (23%) was demonstrated. Therefore, these components are the principal in essential oils of the *Rutaceae* family [5, 7].

The FTIR spectra of the essential oil showed regions at 1100–1500 cm^{-1}, and this region is considered the "fingerprint" since there are vibrations that generate interactions of the C–C, C–N, and C–O bonds [16]. The formation of C=O groups is observed in the vibrations corresponding to the region of 1650–1800 cm^{-1} (aldehydes, ketones, esters) and finally, the bands of C–H bonds appear at 2800–3000 cm^{-1} [17].

The comparison of the essential oil spectra with 2-nonanone and 2-undecanone spectra shows similar frequencies, with 1718, 2854, and 2923 cm^{-1} being the peaks with the highest coincidence between the spectra (Fig. 16.1), decreasing the intensity in those corresponding to C=O vibrations (ketones).

The chemical composition of essential oil of *R. graveolans* is important in the activity biological and the identification by FTIR of essential oil and ethanolic extracts is a relevant analysis to corroborate their chemical composition. The frequencies corresponding to the C=O bonds, characteristic of ketones, reaffirm the spectra obtained from the essential oil of *R. graveolens* [15, 18].

The spectra obtained from the ethanolic extracts at 50% (Fig. 16.2) show regions defined by peaks at 1640 cm^{-1} corresponding to the N–H bending, at 2980 cm^{-1}

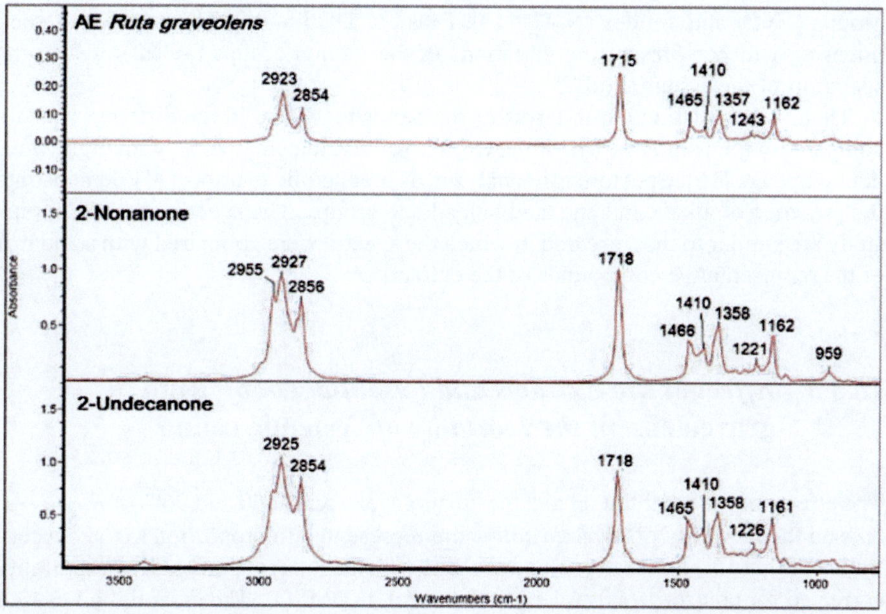

Fig. 16.1 FTIR spectra of essential oil, 2-undecanone and 2-nonanone for the chemical identification of the essential oil *R. graveolens*

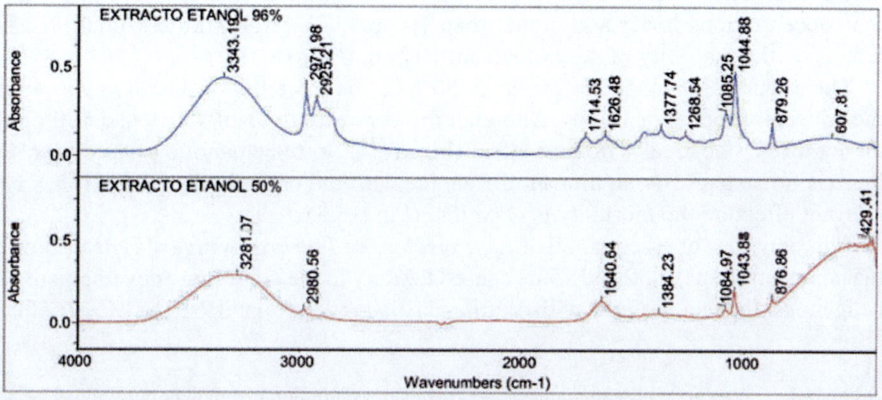

Fig. 16.2 FTIR spectra of ethanolic extracts at 50 and 96% of *R. graveolans*

appear bands linked to the C–H stretching bond and at 3281 cm^{-1} appear the O–H bond of ethanol [16]. The spectrum corresponding to the ethanolic extract at 96% evidence bands at 607 cm^{-1} with a low intensity that represents the C–Cl bonds; at 879 cm^{-1} with an average intensity that represents to the alkenes (=C–H). In addition, in the spectrum appear peaks in the area of 1000–1400 cm^{-1} assigned to the alkyl halides and regions from 1600 to 1760 cm^{-1} that represent the aromatic groups (C=C) and amides (N–H) [17]. Peaks at 2925 and 2971 cm^{-1} with strong intensity, which represent the vibrations of the methyl groups (–CH3) present the spectrum of the essential oil.

There was a distinction in terms of the bands identified in the different spectra, compared with a study in which the extract was evaluated at 70%, with eight well-defined peaks [16]. Spectometric analyzes of methanolic extracts [17] demonstrate the presence of alkane, alkene, and haloalkane groups. The spectra obtained in this study are similar to that research in which the spectra were compared with standards of the representative compounds of the extract.

16.3.3 *Effect of the Extracts and Essential Oil of* Ruta graveolens *on the Mortality of* Artemia salina

The effects of the essential oil and the ethanolic extracts at 50 and 96% of *R. graveolens* on the mortality of *Artemia salina* are represented in proportion to the concentration (Table 16.1). A scatter plot was made with the data for percentage of mortality corrected for control 1 (salt water) and control 2 (DMSO). The essential oil caused 73% mortality of *A. salina* at 100 ppm considering control 1 as a reference. However, comparing with the control 2, it caused 66.6% mortality at 10 ppm.

The results obtained for the mortality of *A. salina* after the application of the essential oil of *R. graveolens* evidence a statistically significant difference between the concentrations lower and higher than 10 ppm. Therefore, the essential oil has effects on the mortality of *A. salina* from 10 ppm (Fig. 16.3).

The effect of the ethanolic extract at 50% on the mortality of *A. salina* was with the concentration from 1 ppm. Although the concentrations of 1, 10, and 100 ppm do not differ statistically in their effect (Fig. 16.4). In the ethanolic extract at 96%, there is no statistically significant difference with the control. Therefore, it has no relevant effect on the mortality of *A. salina* (Fig. 16.5).

A toxicity test of essential oil of *R. graveolens* in *Tetrapygus niger* [7] determined that concentrations of 50 and 75 μl caused toxicity in the embryonic development of hedgehogs. In other research with *A. salina* [19], it is determined that the LC_{50} (Lethal

Table 16.1 Efecto de los tratamientos de *R. graveolens* sobre el porcentaje de mortalidad de *A. salina*

Concentration (ppm)	Formulation—% mortality		
	Essential oil	Ethanolic extract 50%	Ethanolic extract 96%
0.1	0	0	0
1	6.67	16.67	10.00
10	73.33	26.67	16.67
100	73.33	73.33	13.33
Control DMSO	20% Artemias death		

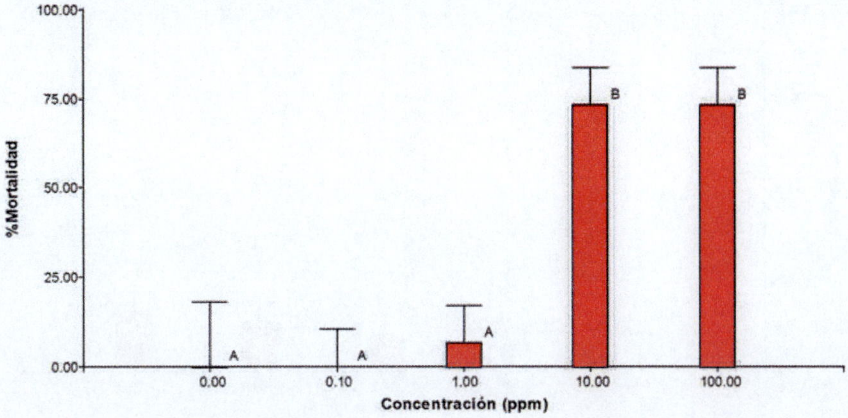

Fig. 16.3 Treatment with essential oil of *R. graveolens* at 10 ppm. *Note* Two treatments without a common letter are statistically significant ($p \leq 0.05$)

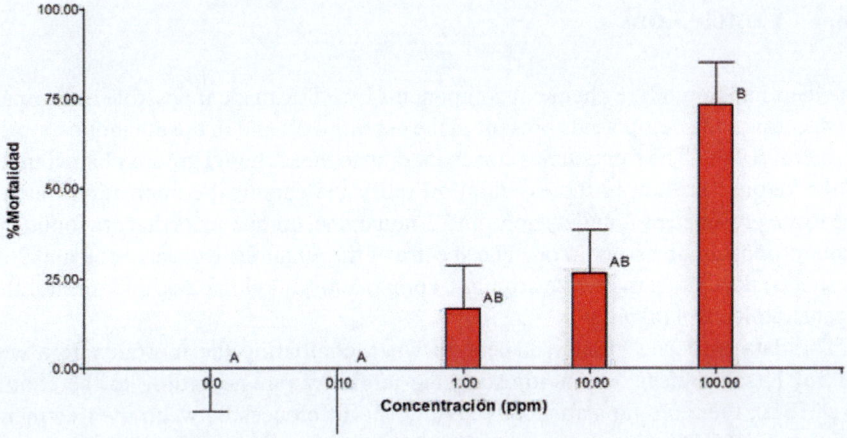

Fig. 16.4 Treatment with ethanolic extracts of *R. graveolens* at 50% in a concentration of 100 ppm. *Note* Two treatments without a common letter are statistically significant ($p \leq 0.05$)

concentration) of the methanolic extract of *R. graveolens* is 0.18 ppm. There is not research about the toxicity of ethanolic extracts of *R. graveolans*. However, this study is a beginning for further research, taking into account the great pharmacological potential of the species. Considering that the extract presents toxicity > 25 ppm and the essential oil > 10 ppm, in research with cell lines, the essential oil evidences toxic effects at 79 and 97 μg/ml [20].

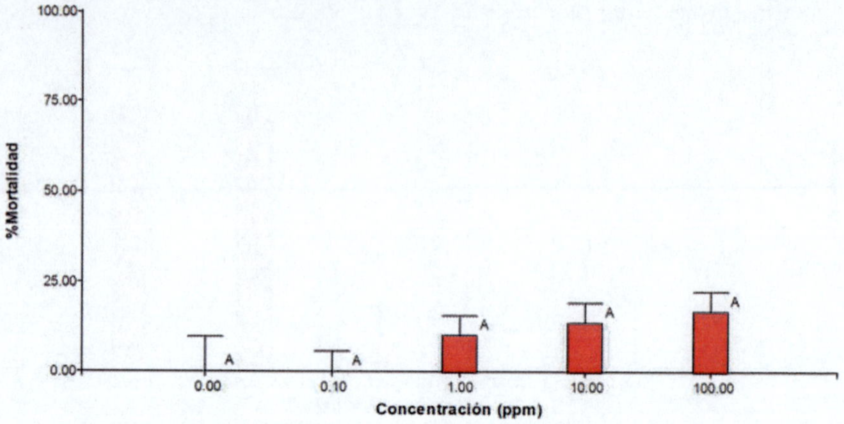

Fig. 16.5 Treatment with ethanolic extracts of *R. graveolens* at 96% in a concentration of 100 ppm. *Note* Two treatments without a common letter are statistically significant ($p \leq 0.05$)

16.4 Conclusions

The identification of the chemical components by FTIR made it possible to compare the spectra of the compounds present in the essential oil and in the ethanolic extracts of *R. graveolens.* The frequencies corresponding to the carbonyl groups characteristic of the ketones present in the essential oil ratify the chemical composition, mainly due to the presence of 2-undecanone and 2-nonanone, components that are found in a major proportion in essential oil. The spectra of the ethanolic extracts at 50 and 96% present peaks that characterize aromatic components, alkyl halides, and unsaturated organic molecules (alkenes).

The data evidence greater dispersion when comparing the mortality rate with control 1 (salt water). When adjusting the mortality rate according to the control 2 (DMSO), there are no statistically significant differences between the treatments and the control. Therefore, salt water should be considered as unique control for this research.

The dose used significantly influences in the mortality of *A. salina*, with concentration from 10 ppm. However, the essential oil caused the death of *A. salina* at concentrations > 10 ppm. The ethanolic extract at 50% caused significant mortality from concentrations > 25 ppm and finally, the ethanolic extract at 96% did not produce statistically significant effects. The logistic regression analysis with Probit adjustment corroborated the results.

This preliminary research seeks to determine if an ethanolic extract can be used in the formulation of different cosmetic and pharmaceutical products, in order to take advantage of the diversity of compounds present in *R. graveolens.*

References

1. L. Serrano Gallardo, A. Soto Domínguez, P. Ruiz Flores, M. Nava Hernández, J. Morán Martínez, R. García Garza, E. Martínez Pérez.: Efecto Tóxico del Extracto Acuoso de *Ruta graveolens* del Norte de México sobre el Hígado de Rata Wistar. Morphol **31** (3), 1041–1048 (2013)
2. Missouri Botanical Garden: Tropicos|Name—*Ruta graveolens* L. http://legacy.tropicos.org/name/28100014?tab=specimens. Last accessed 05 Feb 2021
3. L. De La Torre, H. Navarrete, M. Marcia, H. Balslev, Plantas útiles del Ecuador, Quito: Herbario QCA de la Escuela de Ciencias biológicas de la Pontificia Universidad Católica del Ecuador (2008), pp. 89–100
4. J. Gonzales, V. Benavides, R. Rojas, J. Pino.: Efecto embriotóxico y teratogénico de *Ruta chalepensis* L. "ruda", en ratón (*Mus musculus*). Revista Peruana de Biología **13** (3), 223–226 (2007)
5. A. Arwa, O. Sawsan A y B. Yasser K.: Phytochemical analysis, in vitro assessment of antioxidant properties and cytotoxic potential of *Ruta chalepensis* L. essential oil. J. Essent. Oil Bearing Plants **23** (6), 1409–1421 (2021)
6. D. González.: La legislación vigente en Ecuador para la fabricación, uso, comercialización de plantas medicinales y fitomedicamentos. Boletín Latinoamericano y del Caribe de Plantas Medicinales 8 (1), 51–57 (2009)
7. A. Castro, J. Juárez, N. Ramos, S. Suárez.: Structural elucidation of essential oil of *Ruta graveolens* L. *Ruda*, antioxidant activity and cytotoxicity bio-assay. Ciencia e Investigación **14** (1), 25–28 (2011)
8. A. Ivanova, B. Mikhova, H. Najdenski, I. Tsvetkova, I. Kostova: Antimicrobial and cytotoxic activity of *Ruta graveolens*. Fitoterapia **76** (1), 344–347 (2005)
9. J. Asgarpanah, R. Khoshkman: Phytochemistry and pharmacological properties of *Ruta graveolens* L. J. Med. Plants Res. **6** (23), 3942–3949 (2012)
10. H. Inamul: Safety of medicinal plants. Pak. J. Med. Res. **43** (4), 85–89 (2004)
11. International Federation of Pharmaceutical Manufacturers & Associations (IFPMA), de The Pharmaceutical Industry and Global Health-Facts and Figures (2014)
12. A.V. Velásquez Cueva: Efecto de tres concentraciones del extracto hidroalcohólico de *Rosmarinus officinalis* L. "romero" y *Ruta graveolens* L. "ruda" en la mortalidad de larvas III de *Aedes aegypti* L. (2018)
13. E.C. Castro, L. Vargas, A.R. Bautista: Efecto toxico del extracto acuoso de *Ruta graveolens* L. (*Rutaceae*) sobre larvas de *Anopheles albimanus Wiedemann*, 1820 y *Culex quinquefasciatus Say*, 1823 (Diptera: *Culicidae*), en condiciones experimentales. Entomotropica **25** (1), 11–18 (2011)
14. A.V. Carrión, C.R. García: Preparación de extractos vegetales: determinación de eficiencia de metódica. Repositorio Ucuenca, Cuenca (2010)
15. A. Miranda, D. Vinueza, K. Acosta, C. Toaquiza, E. Mendez, F. Arias: Estudio fitoquímico preliminar y evaluación de la actividad citotóxica del látex de *Euphorbia laurifolia* Juss. Ex. lam sobre *Artemia salina*, http://dspace.espoch.edu.ec/handle/123456789/9383. Last accessed 05 Feb 2021
16. W.S. Abbott.: A method of computing the effectiveness of an insecticide. J. Econ. Entomol. **18** (2), 265−267 (1925)
17. J.A. Ramón Alvarado.: Extracción y caracterización de aceites esenciales de la ruda *Ruta graveolens* L. y el marco *Ambrosia chamissonis* para su potencial uso como plaguicida, http://repositorio.ug.edu.ec/handle/redug/50266. Last accessed 29 Feb 2021
18. O. Pino, Y. Spanchez, M. Rojas, Y. Abreu, Correa y Teresa: Composición química y actividad antibacteriana del aceite esencial de *Pimpinella anisum* L.. Available: http://scielo.sld.cu/scielo.php?pid=S1010-27522012000300007&script=sci_arttext&tlng=pt. Last accessed 17 May 2021

19. C. Grande Tovar, J. Castro, C. Valencia Llano, D. Navia Porras, J. Delgado Ospina, M. Valencia Zapata, H. Mina Hernandez, M. Chaur: Synthesis, characterization, and histological evaluation of chitosan-*Ruta Graveolens* essential oil films. Molecules, **25** (7), 1688 (2020)
20. R. Bañuelos Valenzuela, L. Delgadillo Ruiz, F. Echavarría Cháirez, O. Delgadillo Ruiz, C. Meza López.: Composición química y ftir de extractos etanólicos de *Larrea tridentata, Origanum vulgare, Artemisa ludoviciana* y *Ruta graveolens*. Agrociencia **52** (1), 309–321 (2018)
21. E. Al-Shareef, A.H. Abdal Sahib, I. Hameed.: Phytochemical screening by FTIR spectroscopic analysis and anti-fungal activity of fruit extract of selected medicinal plant of *Ruta graveolens*. Indian J. Public Health Res. Dev. **10** (1), 994–999 (2019)

Chapter 17
Physicochemical Characteristics and Antioxidant Capacity of Yellow and Orange Andean Floral Species

Elena Coyago-Cruz, Manuel Coronel, Manuel Baldeón, and Edwin Vera

Abstract Flowers are used for ornamental purposes and to enhance the appearance, taste, and value of foods. Ecuador is a country with great biodiversity including flowers. There is limited information on the physicochemical characteristics of these plants, particularly those related with human health. In this context, thirteen floral species in the Andean region of the Province of Pichincha, Ecuador were characterized. The following parameters were evaluated: pH, soluble solids, ash, moisture, and color; the content of carotenoids and total phenolic compounds by spectrophotometry; and the antioxidant capacity, which was determined by the Trolox equivalent antioxidant capacity. These results demonstrated that the *Cassia fistula* and *Tecoma stans* flowers had high pH; *Pyrostegia venusta* showed high soluble solids content; *Tecoma stans* had high total titratable acidity; *Abutilon pictum* had high ash; *Gazaniax hybrida* had high moisture. The highest contents of total carotenoids and total phenolics were presented by *Gazania x hybrida* (69.28 ug/g) and *Glaucium flavum* (355.46 mg eq. Gallic acid/g), respectively. The highest values of antioxidant capacity were presented by *Oxalis tetraphylla* and *Glaucium flavum* (166.90 and 164.00 μMeq trolox/g, respectively). These results indicated that Andean floral species could have potential beneficial properties that can be applied for human health.

E. Coyago-Cruz (✉)
Grupo de Investigación y Desarrollo en Ciencias Aplicadas a los Recursos Biológicos, Universidad Politécnica Salesiana, Carrera de Ingeniería en Biotecnología de los Recursos Naturales, Av. 12 de octubre, Quito, Ecuador
e-mail: ecoyagoc@ups.edu.ec

M. Coronel
Centro de Investigación de Alimentos (CIAL), Facultad de Ciencias de la Ingeniería e Industrias, Universidad UTE, Ingeniería de Alimentos, Quito, Ecuador

M. Baldeón
Facultad de Ciencias Médicas, Universidad Internacional del Ecuador, de La Salud y la Vida, Quito, Ecuador

E. Vera
Departamento de Ciencia de los Alimentos y Biotecnología, Facultad de Ingeniería Química, Escuela Politécnica Nacional, Quito, Ecuador

Á. Rocha et al. (eds.), *Communication, Smart Technologies and Innovation for Society*, Smart Innovation, Systems and Technologies 252,
https://doi.org/10.1007/978-981-16-4126-8_17

17.1 Introduction

Flowers have typically been used for ornamental purposes, although in various parts of the world they are used to enhance the appearance, taste, and value of foods [1]. For example, in ancient Rome and Greece, various species of roses were used in the preparation of purees and omelets [2]. The flowers are rich in a wide variety of natural antioxidants, including flavonoids, anthocyanins, and other phenolic compounds. In this regard, various studies have shown that a diet rich in antioxidants could prevent chronic diseases such as type 2 diabetes, cancer, cardiovascular, and neurodegenerative diseases [3, 4].

Since flowers are composed mainly of water, it is recommended to consume them in large quantities, similar to the consumption of other vegetables, to obtain nutritional benefits [5]. From a nutritional point of view, flowers can be divided into three main parts: (1) pollen, that is a source of protein, amino acids, and carbohydrates, although in very small quantities [6]; (2) nectar, which contains a mixture of glucose, fructose, sucrose, amino acids, and proteins [7]; and (3) the petals, which are a source of vitamins and antioxidants [8].

In flowers, there are a great variety of compounds that have antioxidant properties such as phenolic compounds and carotenoids, which could indicate that their consumption, even in small amounts, could have beneficial health effects [8]. Flowers also possess compounds that have anti-inflammatory [9], antimicrobial [10], and antiviral [11] properties. Overall, flowers are an excellent source of numerous phytochemical compounds whose benefits for health and their mechanisms of action have been confirmed [12]. In this sense, it is important to point out that there are limited studies on the potential beneficial effects of flowers, particularly in Ecuador a megabiodiverse country. Consequently, it is important to study the potential bioactive compounds present in the Ecuadorian flora.

Several factors can affect the acceptability of edible flowers including: species of the flowers and their characteristics, type of social group, personal characteristics of the consumers, and their processing [5]. The study of the flowers will determine those with the greatest potential for commercialization and those that could be toxic. In Ecuador, there is a great variety of floral species of which there are no nutritional studies and their possible health benefits are not known. Thus, the main purpose of this study was to determine the physicochemical characteristics and antioxidant capacity of yellow and orange Andean floral species.

17.2 Methods

The collection of thirteen randomly selected flowers from different places in Pichincha-Ecuador was carried out under the framework contract MAE-DNB-CM-2017-0080-UTE, Project MAE-DNB-2019-0911-O which contemplates the collection of species in Ecuador and the respective deposit in the herbarium of the Pontifical

Catholic University of Ecuador. The techniques and procedures for the collection of the plants were in accordance with the Department of Plant Genetic Resources of INIAP, the International Code of Conduct for the collection and transfer of plant germplasm of FAO [13], and the publications of Biodiversity International [14].

Thus, on fresh flower petals, physical–chemical analyzes (weight; equatorial diameter and longitudinal diameter [15]; soluble solids [16]; titratable acidity [17]; pH [18]; moisture; and ash [19, 20]) and color of the flowers were evaluated using a Minolta brand tristimulus colorimeter (Spectrocrom, Ecuador) in CIELAB color space (L*, a* and b*)) were carried out. After their characterization, samples were lyophilized in a Martin Christ equipment (Osterode, Germany).

17.2.1 Total Carotenoids, Total Phenolic Compounds, and Antioxidant Activity Analysis

The lyophilized flower powder was subjected to a carotenoid extraction process using a micro-extraction in 2 mL ppendorf tubes with approximately 10 mg of lyophilized material and 300 μL of a solution of acetone: methanol: dichloromethane (1:1:2), it was stirred for 1 min in ultrasound model VWR (VWR International, EE.UU.) and the supernatant was recovered after centrifuging the solution at 14,000 rpm for 4 min in a microcentrifuge model MiniSpin (Eppendorf, EE.UU.). The extract was evaporated to dryness in a rotary evaporator, redissolved in ethanol HPLC, and quantified in a spectrophotometer Thermo Scientific Evolution 60S BioTek (Agilent Scientific Instruments, EE.UU.), at a wavelength of 450 nm to then be expressed as total carotenoids (μg equivalent of β-carotene/g of lyophilized sample) [21, 22]. Results were expressed as the mean + standard deviation; all experiments were performed in triplicate.

For phenolic compounds, approximately 10 mg of lyophilized sample were extracted with 500 μL of an 80% solution of methanol acidified with HCl and subjected to ultrasound shaking for 2 min. The supernatant was collected and the Folin Ciocalteu protocol was applied using a BIOTEK SYNERGY H1 spectrophotometer (BioTek, EE.UU.) [23]. Gallic acid was used for the calibration curve and the results were reported as mg of gallic acid equivalent/g of dry weight. Results were expressed as the mean ± standard deviation; all experiments were performed in triplicate.

The antioxidant activity was evaluated using approximately 0.1 g of lyophilized powder and a solution of methanol: water (50:50), the mixture was stirred and sonicated for 10 min, then it was extracted with an acetone: water solution (70:30). The calibration curve was made with a Trolox standard and quantified by the microplate technique using a BIOTEK SYNERGY H1 (Winooski, VT, EE.UU.) [24]. Results were expressed as the mean ± standard deviation; all experiments were performed in triplicate.

17.3 Results

The collection was carried out between the months of February and July 2020 in different cantons of the Province of Pichincha (Table 17.1), as shown by the coordinates and the altitude in meters above sea level (masl) of the collection point.

17.3.1 Physicochemical Characteristics

Table 17.2 shows the physical properties (dimensions) of analyzed flowers.

Solandra grandiflora has the least number of petals, but they are the widest and longest in relation to the other flowers, while *Gazania x hybrida* has the largest number of petals that are small and thin compared to the other flowers.

Table 17.1 Sites of sampling collection of studies plants

Scientific name	Province	Canton	Coordinates	Altitude (masl)[a]
Abutilon pictum	Pichincha	Quito	S0° 12′ 9.888″ W78° 22′ 59.999″	2339
Arnica montana	Pichincha	Quito	S0° 19′ 0.01″ W78° 22′ 59.995″	2644
Cassia fistula	Pichincha	Quito	S0° 0′ 59.691″ W78° 28′ 59.922″	2775
Crocosmia x crocosmiiflora	Pichincha	Mejia	S0° 26′ 14.831″ W08° 31′ 55,056″	2750
Cytisus x praecox	Pichincha	Quito	S0° 12′ 9.888″ W78° 22′ 59.999″	2339
Gazania x hybrida	Pichincha	Quito	S0° 11′ 19.000″ W78° 23′ 46.000″	2360
Glaucium flavum	Pichincha	Quito	S0° 12′ 9.888″ W78° 22′ 59.999″	2359
Jacobaea maritima	Pichincha	Rumiñahui	S0° 25′ 10.000″ W78° 24′ 28.000″	2860
Leptocarpha rivularis	Pichincha	Quito	S0° 19′ 0.01″ W78° 22′ 59.995″	2644
Oxalis tetraphylla	Pichincha	Mejia	S0° 26′ 14.831″ W78° 31′ 55.000″	2750
Pyrostegia venusta	Pichincha	Quito	S0° 11′ 34.000″ W78° 23′ 45.000″	2360
Solandra grandiflora	Pichincha	Quito	S0° 19′ 59.000″ W78° 22′ 57.000″	2380
Tecoma stans	Pichincha	Quito	S0° 11′ 19.000″ W78° 23′ 46.000″	2360

[a]*masl* Meters above sea level

Table 17.2 Physical properties of studies plants

Scientific name	Weight (g)	#petals	Length (cm)	Width (cm)
A. pictum	0.70 ± 0.15	4.80 ± 0.63	3.81 ± 0.16	1.75 ± 0.28
A. montana	4.59 ± 1.04	12.40 ± 1.07	2.20 ± 0.38	9.14 ± 1.10
C. fistula	0.55 ± 0.01	10.00 ± 0.00	5.88 ± 0.77	4.95 ± 0.09
C. x crocosmiiflora	0.14 ± 0.03	6.00 ± 0.00	3.30 ± 0.28	2.24 ± 0.78
C. x praecox	0.19 ± 0.03	3.20 ± 0.79	2.09 ± 0.23	0.39 ± 0.06
G. x hybrida	2.70 ± 0.02	23.00 ± 2.11	1.50 ± 0.11	9.70 ± 0.53
G. flavum	0.64 ± 0.02	4.00 ± 0.00	3.83 ± 0.75	0.26 ± 0.02
J. maritima	0.11 ± 0.01	21.00 ± 4.00	1.52 ± 0.20	0.60 ± 0.19
L. rivularis	1.06 ± 0.12	10.50 ± 2.76	1.37 ± 0.20	5.33 ± 0.79
O. tetraphylla	0.02 ± 0.00	5.00 ± 0.00	0.97 ± 0.10	0.28 ± 0.04
P. venusta	0.39 ± 0.08	5.00 ± 0.00	7.01 ± 1.37	0.86 ± 0.10
S. grandiflora	0.16 ± 0.03	1.20 ± 0.42	47.37 ± 4.88	29.11 ± 7.28
T. stans	0.14 ± 0.02	5.00 ± 0.00	4.13 ± 0.32	3.67 ± 0.32

Table 17.3 shows the chemical properties (pH, soluble solids, total acidity, and ash) of the analyzed flowers.

The pH of the flowers under study varied between 6.30 (*T. stans*) to 2.45 (*O. tetraphylla*); the soluble solids from 15.03 (*P. venusta*) to 0.13°Brix (*G. flavum*); the titratable acidity from 0.22 (*T. stans*) to 0.02% (*C. x crocosmiiflora*, *J. maritima* and *S. grandiflora*); ash from 2.82 (*A. pictum*) to 0.36% (*C. x praecox*), and moisture from 95.30 (*G. x hybrida*) to 74.52% (*A. pictum* and *C. x crocosmiiflora*).

Table 17.3 Chemical properties (pH, soluble solids, total acidity, ash, and moisture)

Scientific name	pH	SS (Brix)	TA (%)	Ash (%)	Moisture (%)
A. pictum	6.00 ± 0.00	6.00 ± 0.00	0.07 ± 0.00	2.82 ± 0.07	74.52 ± 1.00
A. montana	5.15 ± 0.24	7.58 ± 0.18	0.03 ± 0.00	0.87 ± 0.09	82.72 ± 1.10
C. fistula	6.30 ± 0.63	9.33 ± 1.53	0.03 ± 0.00	0.38 ± 0.08	87.28 ± 0.55
C. x crocosmiiflora	3.00 ± 0.00	9.30 ± 0.48	0.02 ± 0.00	1.58 ± 0.18	74.76 ± 0.17
C. x praecox	5.45 ± 0.44	2.74 ± 0.22	0.03 ± 0.00	0.36 ± 0.00	84.36 ± 0.07
G. x hybrida	3.00 ± 0.00	2.00 ± 0.00	0.10 ± 0.03	0.98 ± 0.00	95.30 ± 2.42
G. flavum	5.00 ± 0.00	0.13 ± 0.04	0.13 ± 0.02	0.89 ± 0.11	86.62 ± 3.20
J. maritima	6.00 ± 0.00	5.59 ± 0.19	0.02 ± 0.00	1.39 ± 0.03	78.78 ± 0.28
L. rivularis	4.65 ± 0.24	5.34 ± 0.10	0.03 ± 0.01	0.57 ± 0.01	84.87 ± 2.14
O. tetraphylla	2.45 ± 0.44	4.55 ± 0.44	0.04 ± 0.01	1.17 ± 0.09	90.25 ± 0.60
P. venusta	4.15 ± 0.24	15.03 ± 0.13	0.07 ± 0.01	0.62 ± 0.02	78.31 ± 0.28
S. grandiflora	5.30 ± 0.48	8.00 ± 0.00	0.02 ± 0.01	1.26 ± 0.15	81.24 ± 0.86
T. stans	6.30 ± 0.48	1.00 ± 0.00	0.22 ± 0.02	1.99 ± 0.19	84.87 ± 0.29

Table 17.4 Color analysis of studied flowers

Scientific name	L	a*	b*	Color zone
A. pictum	68.71 ± 1.74	11.69 ± 1.69	45.02 ± 1.59	Orange
A. montana	73.22 ± 0.55	9.31 ± 0.17	103.52 ± 0.52	Yellow
C. fistula	71.04 ± 3.11	−0.31 ± 2.48	64.90 ± 4.84	Yellow
C. x crocosmiiflora	56.58 ± 7.29	26.44 ± 9.06	53.14 ± 9.08	Orange
C. x praecox	72.01 ± 1.54	−5.88 ± 1.44	67.57 ± 4.17	Yellow
G. x hybrida	Na	Na	Na	Orange
G. flavum	67.69 ± 1.88	−2.69 ± 1.78	50.12 ± 1.54	Yellow
J. maritima	66.36 ± 5.52	6.71 ± 0.44	64.36 ± 11.72	Yellow
L. rivularis	72.61 ± 1.30	−1.16 ± 1.47	89.52 ± 3.65	Yellow
O. tetraphylla	Na	Na	Na	Yellow
P. venusta	57.53 ± 3.55	27.28 ± 0.74	−1.97 ± 0.34	Orange
S. grandiflora	77.50 ± 0.64	1.84 ± 0.47	52.93 ± 3.68	Yellow
T. stans	69.48 ± 1.79	−4.36 ± 1.09	62.70 ± 1.83	Yellow

Table 17.4 shows the color of the flowers analyzed.

All the flowers analyzed have an orange or yellow color according to the CIELAB system; four of the thirteen flowers are found in the orange zone (*A. pictum*, *Crocosmia x crocosmiiflora*, *Gazania x hybrida*, and *P. venusta*) and the others correspond to the yellow zone; these results correspond to the color observed for the collection of the flowers.

17.3.2 Total Carotenoids, Total Phenolic Compounds, and Antioxidant Activity

The concentration of total carotenoids, total phenolic compounds, and the antoxidant activity of the flowers analyzed is shown in Table 17.5.

Total carotenoids ranged from 69.28 (*G. x hybrida*) to 0.69 μg of β-carotene/g of dry weight (DW) (*S. grandiflora*), total phenols from 355.46 (*G. flavum*) to 66.51 mg of gallic acid equivalent/g DW (*J. maritima*), and antioxidant activity from 166.90 (*G. flavum and O. tetraphylla*) to 12.20 μmol Eq trolox/g DW (*L. rivularis*).

Table 17.5 Total carotenoids, total phenolic compounds, and antioxidant activity

Scientific name	Total carotenoids µg of β-carotene/g of dry weight	Total phenolic compounds mg of gallic acid equivalent/g of dry weight	Antioxidant activity µmol Eq trolox/g of dry weight
A. pictum	2.73 ± 0.07	263.89 ± 9.57	106.20 ± 17.3
A. montana	22.31 ± 0.07	83.29 ± 0.28	63.10 ± 7.7
C. fistula	16.13 ± 0.29	202.86 ± 15.51	90.70 ± 4.6
C. x crocosmiiflora	4.18 ± 0.02	345.92 ± 12.27	142.20 ± 2.9
C.s x praecox	4.87 ± 0.47	144.08 ± 14.72	32.70 ± 3.4
G. x hybrida	69.28 ± 0.04	323.60 ± 8.46	88.70 ± 1.7
G. flavum	3.43 ± 0.09	355.46 ± 8.83	164.00 ± 4.6
J. maritima	14.18 ± 6.32	66.51 ± 12.77	54.80 ± 5.2
L. rivularis	21.57 ± 5.10	224.84 ± 13.61	12.20 ± 0.6
O. tetraphylla	13.62 ± 0.20	211.92 ± 7.24	166.90 ± 0.3
P. venusta	3.35 ± 0.03	208.99 ± 9.54	82.00 ± 12.3
S. grandiflora	0.69 ± 0.11	128.40 ± 1.08	53.40 ± 5.5
T. stans	29.92 ± 0.09	305.25 ± 14.63	102.20 ± 4.6

17.4 Discussion

17.4.1 Physicochemical Properties (pH, Soluble Solids, Total Titratable Acidity, Ash, and Humidity) and Color

These results demonstrated that the *C. fistula* and *T. stans* flowers had high pH; *P. venusta* showed high soluble solids content; *T. stans* had high total titratable acidity; *A. pictum* had high ash; *G. x ybrid* had high moisture; and four of the thirteen flowers were orange and the others were yellow. This new physical–chemical analysis of the vegetal matter could be the basis for future biology and nutritional studies.

These results agree with previous studies that have characterized similar flowers. For instance, the floral dimensions for *S. grandiflora* are in accordance with that presented in literature that reported corolla length ranges of 13.5–37.0 cm [25]. The length of the petals of *A. pictum* showed comparable results with other studies [26], further the length of the flowers of *C. x praecox* within the expected range for the genus *Cytisus* [27]. The number of petals, width, and length of the *G. x ybrid* flowers are within the values established for *G. rigens* [28]; *G. flavum* [29] and *L. rivularis* [30] are within the expected range; the maritime *Jacobaea* flowers are within the values established for *Senecio Jacobaea* [31].

In addition, these species showed a lower pH than 4.6, like that reported for lemon, raspberry, mandarin, orange, among others. The soluble solids content of *G. flavum* (0.13°Brix), *G. x ybrid* (2.0°Brix), *O. tetraphylla* (4.55°Brix), *L. rivularis* (5.36°Brix)

A. pictum (6.00°Brix). *A. montana* (7.58°Brix) is equal to the value reported for carrot, lemon, red beet, orange, sour cherry, tangerine, and plum, respectively, [32], while the other flowers are found in a low acidity range such as pumpkin, carrot, and beet [32]. The lowest titratable acidity obtained for the flowers corresponds to *Solandra grandiflora* with a value of 0.01%, equal to that reported for plum, while *T. stans* presents the highest value with 0.22% lower than that reported for apple (0.31%) [32]. The moisture showed a range between 74.52 and 95.30%, these values are similar to those reported in other studies of edible flower petals [33]. The flowers studied that have high acidity were *G. flavum* (0.13%) and *G. x ybrid* (0.10%).

The color coordinates of *A. pictum* that are located in the orange color zone agree with those reported in literature [26]. The color coordinates of the petals of *A. montana* are located in the area of the yellow-orange color, as suggested by other authors [34] and *C. x praecox* showed a marked yellow color [35].

The results, based on the nutritional characteristics, suggest that *O. tetraphylla* would contribute an acid pH to the diet; *P. venusta* high content of soluble solids which are mainly related to the sugar content; *T. stans* higher content of organic acids such as citric acid; *A. pictum* high ash content which is related to mineral content; and *G. x ybrid* high water supply. However, these properties must be analyzed together with a toxicity analysis for xenobiotics or secondary metabolites specific to each species.

17.4.2 Total Carotenoids, Total Phenolic Compounds, and Antioxidant Activity

Flowers are an important source of bioactive compounds, including carotenoids and phenolic compounds, the latter giving these species important antioxidant properties and providing benefits for human health. Thus, the present results showed that the *G. x ybrid* flowers had high total carotenoids content; *G. flavum* high total phenolic compounds content; and *G. flavum* and *O. tetraphylla* high antioxidant activity.

Regarding the content of total carotenoids *C. x praecox* presents a concentration of total carotenoids very similar to that of *C. scoparius* and that could be represented mainly by chrysanthemaxanthine, xanthophylos, and xanthophillepoxid [36]. *G. ybrid* has a carotenoid content lower to that of *G-. rigens* that is 64.2% [37]. The concentration of total carotenoids in *G. flavum* is higher than that found in other study, 1.6 μg equivalent of β-carotene/g of DW, which may be related to the color intensity of the flower and its flowering status [38]. *L. rivularis* in this study has a higher carotenoid content than that presented in other studies which is 5.6 μg/g [39]. *G. x ybrid* has the highest content of total carotenoids compared to the other flowers studied.

On the other hand, *A. montana* presents a content of total phenolic compounds total of 83.29 mg gallic acid equivalents, that indicates that *A. montana* contains phenolics like flavonoids: atragaloside, isoquercitroside, and quercetrol-3-glucogalacturonic as suggested by other authors [34]. The content of phenolic compounds of *C. fistula* is 202.86 mg gallic acid equivalents/g of dry weight (DW). These data vary with those reported in other studies with 32 mg gallic acid equivalents/g of DW, while 453 μmol/g DW this may be due to the environmental conditions of plant growth [40]. Some studies determined that phenolic compounds for *Cytisus scoparius* were found in 134.67 mg/g of gallic acid equivalents which closely resembles that found in *C.* x *praecox* [41]. In terms of antioxidant compounds, *G. rigens* has a higher concentration of phenolic compounds than in relation to *G. krebsiana* which is 251.40 mg/g of gallic acid equivalents [42]. The concentration of phenolic compounds of *G. flavum* is much higher than those presented in other studies which are 158.3 mg/g of gallic acid equivalents [38], which may be mainly due to the extraction method used. For *L. rivularis* DC the concentration of phenolic compounds presents similar results to those in other investigation which are 267.43 mg/g of gallic acid equivalents [43]. *O. tetraphylla* traditionally called "Trebol de la Buena Suerte." This plant has been used for human consumption; it is mostly used for cattle grazing. Hydroalcoholic extract of *O. tetraphylla* contains compounds that belong to the flavonol group (flavonoid class) with anthelmintic activity [44]. In literature is found that in an hydroetanolic extract for *P. venusta*, a higher level of polyphenolics compounds (998 mg/g extract) where flavonoids represents 543 mg of quercetin equivalent/g extract [45]. The content of polyphenolic compounds for *T. stans* is higher than that reported in literature on methanolic extract (216 mg gallic acid (GAE)/g extract) [46].

Finally, the antioxidant capacity of *C. x praecox* is higher than that found in *C. scoparius* which is 12.06 μmol Eq trolox/g of DW [36]. *C. fistula* presents an antioxidant capacity of 90.70 μmol Eq trolox/g of DW, different with those reported in other studies [40]. *G. rigens* has a higher antioxidant capacity than *G. krebsiana* that is a 0.83 μmol Eq trolox/g of DW [42]. *L. rivularis* DC has lower values of antioxidant capacity than the values presented in another study; this may be mainly due to the method of extraction of samples [43]. *P. venusta* has a significant inhibitory effect of extracts of flowers on ABTS free radicals and the orange and yellow petal colors were due to carotenoids [47]. For *T. alata* in chemical systematic studies cataloged the species of the genus *Thunbergia* with antioxidant, antibacterial, antifungal capacity [48].

17.5 Conclusions

Ecuador, due to its great biodiversity, hosts a great variety of floral species with potential medicinal and nutritional properties. Here we showed that different species had high values of pH [*C. fistula* (6.30)], soluble solids [*P. venusta* (15.03°Brix)], titratable acidity [*G. flavum* (0.13%)], ash (*A. pictum* 2.82%), moisture (*G.* x *hybrida* 95.30%), total carotenoids (*G.* x *hybrida* 69.28 μg of β-carotene/g of dry weight), total phenolic compounds (*G. flavum* 355.46 mg of gallic acid equivalent/g of dry weight) and antioxidant activity [*G. flavum* (164.00 μmol Eq trolox/g of dry weight) and *O. tetraphylla* (166.90 μmol Eq trolox/g of dry weight)]. Present results could contribute to develop novel functional foods with unique nutritional composition. It will be important to carry out studies to evaluate the nutritional value of the studied flowers.

Acknowledgements This research was carried out under the framework contract MAE-DNB-CM-2017-0080-UTE. Project MAE-DNB-2019-0911-O and was financed by the Ecuadorian Corporation for the Development of Research and the Academy (CEDIA) within the CEPRA-XII-2019-Flores Andinas Project "Physical-chemical characterization and bioactivity tests of Andean floral species with nutritional potential and preventive effect of certain human diseases."

References

1. A. Tanji, F. Nassif, Edible weeds in Morocco. Weed Technol. **9**(3), 617–620 (1995)
2. L. Melillo, Diuretic plants in the paintings of Pompeii. Am. J. Nephrol. **14**(4–6), 423–425 (1994)
3. K. Kopec, J. Balik, "Kvalitologie zahradnickych produktu," *Brno MZLU*, pp. 34–40 (2008)
4. Z. Youwei, Z. Jinlian, P. Yonghong, A comparative study on the free radical scavenging activities of some fresh flowers in southern China. LWT Food Sci. Technol. **41**(9), 1586–1591 (2008)
5. L. Fernandes, S. Casal, J.A. Pereira, J.A. Saraiva, E. Ramalhosa, *Edible flowers: A review of the nutritional, antioxidant, antimicrobial properties and effects on human health* (2017)
6. M. Weber, The existence of a special exine coating in *Geranium robertianum* pollen. Int. J. Plant Sci. **157**(2), 195–202 (1996)
7. S.W. Nicolson, M. Nepi, E. Pacini, *Nectaries and Nectar* (Springer, Netherlands, 2007)
8. J. Mlcek, O. Rop, Fresh edible flowers of ornamental plants—A new source of nutraceutical foods. Trends Food Sci. Technol. **22**(10), 561–569 (2011 Oct)
9. E.M. Choi, J.K. Hwang, Investigations of anti-inflammatory and antinociceptive activities of *Piper cubeba, Physalis angulata* and *Rosa hybrida.* J. Ethnopharmacol. **89**(1), 171–175 (2003)
10. C. Pérez, C. Anesini, In vitro antibacterial activity of Argentine folk medicinal plants against *Salmonella typhi.* J. Ethnopharmacol. **44**(1), 41–46 (1994)
11. N. Mahmood, S. Piacente, C. Pizza, A. Burke, A.I. Khan, A.J. Hay, The anti-HIV activity and mechanisms of action of pure compounds isolated from *Rosa damascena.* Biochem. Biophys. Res. Commun. **229**(1), 73–79 (1996)
12. B. Lu, M. Li, R. Yin, Phytochemical content, health benefits, and toxicology of common edible flowers: a review (2000–2015). Crit. Rev. Food Sci. Nutr. **56**, S130–S148 (2016)
13. Organización de las Naciones Unidas para la Alimentación y la Agricultura, "Texto concertado del proyecto de código internacional de conducta para la recolección y transferencia de germoplasma vegetal" (1993)
14. L. Guarino, V. Ramanatha Rao, E. Goldberg, *Collecting plant genetic diversity: technical guidelines. 2011 update* (2011)
15. E. Coyago-Cruz, M. Corell, C.M. Stinco, D. Hernanz, A. Moriana, A.J. Meléndez-Martínez, Effect of regulated deficit irrigation on quality parameters, carotenoids and phenolics of diverse tomato varieties (*Solanum lycopersicum* L.) (2017)
16. E. Coyago-Cruz et al., Antioxidants (carotenoids and phenolics) profile of cherry tomatoes as influenced by deficit irrigation, ripening and cluster. Food Chem. **240** (2017 Aug), 870–884 (2018)
17. L.P. Esther González, G.A. Giraldo Giraldo, P. Titular, A.L. Duque, P. Asistente, Periodo de cosecha y método de conservación del hongo comestible Pheurotus ostreatus (jacq.) p. kumm. Harvest and conservation method Pleurotus ostreatus edible mushroom (jacq.) p. kum
18. H. Egan, *Pearson's Chemical Analysis of Foods*, 8th edn. (Churchill Livingstone, Edinburgh, New York, 1981)

19. S.S. Nielsen, *"Food Analysis"* (Springer, New York, 2010)
20. R. Lane, "Association of Official Analytical Chemistry, 925.10-32.1.03." AOAC (2000)
21. E. Coyago-Cruz, M. Corell, C.M. Stinco, D. Hernanz, A. Moriana, A.J. Meléndez-Martínez, Effect of regulated deficit irrigation on quality parameters, carotenoids and phenolics of diverse tomato varieties (*Solanum lycopersicum* L.). Food Res. Int. **96**, 72–83 (2017)
22. C.M. Stinco, A.M. Benítez-González, D. Hernanz, I.M. Vicario, A.J. Meléndez-Martínez, Development and validation of a rapid resolution liquid chromatography method for the screening of dietary plant isoprenoids: Carotenoids, tocopherols and chlorophylls. J. Chromatogr. A **1370**, 162–170 (2014)
23. E.A. Ainsworth, K.M. Gillespie, Estimation of total phenolic content and other oxidation substrates in plant tissues using Folin-Ciocalteu reagent. Nat. Protoc. **2**(4), 875–877 (2007)
24. G. Bobo-García, G. Davidov-Pardo, C. Arroqui, P. Vírseda, M.R. Marín-Arroyo, M. Navarro, Intra-laboratory validation of microplate methods for total phenolic content and antioxidant activity on polyphenolic extracts, and comparison with conventional spectrophotometric methods. J. Sci. Food Agric. **95**(1), 204–209 (2015)
25. L.M. Bernardello, A.T. Hunziker, A synoptical revision of Solandra (Solanaceae). Nord. J. Bot. **7**(6), 639–652 (1987)
26. N.G. Odenwald, J.R. Turner, *Identification, selection, and use of southern plants: for landscape design.* Claitor's Law Books and Publishing (2006)
27. F. González-Andrés, J.M. Ortiz, Morphometrical characterization of cytisus and allies (Genisteae: Leguminosae) as an aid in taxonomic discrimination. Isr. J. Plant Sci. **44**(2–3), 95–114 (1996)
28. F. Zulfiqar, A. Younis, Z. Abideen, A. Francini, A. Ferrante, Bioregulators can improve biomass production, photosynthetic efficiency, and ornamental quality of *Gazania rigens* L. *Agronomy* **9**(11), 773 (2019, Nov)
29. J. Dickson, Studies in floral anatomy—II. The floral anatomy of *Glaucium flavum* with reference to other members of the Papaveraceae. J. Linn. Soc. London, Bot. **50**(333), 175–224 (1935, Sept)
30. O. Urban, Estudios sobre la biología de algunas plantas endémicas (1927)
31. S. Jacobaea, E. Cameron, A study of the natural control of Ragwort (1935)
32. N.K. Sinha, J.S. Sidhu, J. Barta, J.S.B. Wu, M.P. Cano, *Handbook of Fruits and Fruit Processing* (Wiley-Blackwell, Oxford, UK, 2012)
33. L. Fernandes, S. Casal, J.A. Pereira, J.A. Saraiva, E. Ramalhosa, Edible flowers: a review of the nutritional, antioxidant, antimicrobial properties and effects on human health. J. Food Compos. Anal. **60**, 38–50 (2017)
34. M.C.A. Mejía, *Plantas medicinales: botanica de interest medico.* María Cristina Arango Mejía (2006)
35. S.S. Gavini, A. Farji-Brener, La importancia del color: Morfos florales, tasas de visita y éxito reproductivo en el arbusto sarothamnus scoparius. Ecol. Austral **25**(3), 204–211 (2015)
36. N. González et al., Potential use of *Cytisus scoparius* extracts in topical applications for skin protection against oxidative damage (2013)
37. L.R.G. Valadon, R.S. Mummery, Carotenoids of compositae flowers. Phytochemistry **10**(10), 2349–2353 (1971)
38. M. Boulaaba et al., Antioxidant, antiproliferative and anti-inflammatory effects of *Glaucium flavum* fractions enriched in phenolic compounds. Med. Chem. Res. **28**(11), 1995–2001 (2019)
39. E. Uquiche, F. Garcés, Recovery and antioxidant activity of extracts from *Leptocarpha rivularis* by supercritical carbon dioxide extraction. J. Supercrit. Fluids **110**, 257–264 (2016)
40. A. Luximon-Ramma, T. Bahorun, M.A. Soobrattee, O.I. Aruoma, Antioxidant activities of phenolic, proanthocyanidin, and flavonoid components in extracts of *Cassia fistula.* J. Agric. Food Chem. **50**(18), 5042–5047 (2002)
41. C. Gil, A.P. Duarte, Antioxidant activity of extracts of Portuguese shrubs: *Pterospartum tridentatum, Cytisus scoparius* and Erica spp. (2009)
42. F.O. Balogun, A.O.T. Ashafa, Antioxidant, hepatoprotective and ameliorative potentials of aqueous leaf extract of *Gazania krebsiana* (Less.) against carbon tetrachloride (CCl4)-induced liver injury in Wistar rats. Trans. R. Soc. South Africa **71**(2), 145–156 (2016)

43. A. Jiménez-González et al., UHPLC-ESI-ORBITRAP-MS analysis of the native Mapuche medicinal plant palo negro (*Leptocarpha rivularis* DC.—Asteraceae) and evaluation of its antioxidant and cholinesterase inhibitory properties. J. Enzym. Inhib. Med. Chem. (2018)
44. B.J. González-Cruz et al., In vitro larvicidal and in vivo anthelmintic effects of *Oxalis tetraphylla* (Oxalidaceae) hydroalcoholic extract against *Haemonchus contortus* in lambs. J. Helminthol. **92**(3), 309–316 (2018)
45. C.C. Veloso et al., Anti-inflammatory and antinociceptive effects of the hydroethanolic extract of the flowers of *Pyrostegia venusta* in mice. Braz. J. Pharmacogn. **22**(1), 162–168 (2011)
46. M. Govindappa, S. Ts, R. Channabasava, J. Mk, P. Ks, V.B. Raghavendra, Antioxidant activity and phytochemical screening of *Tecoma stans* (L.) Juss. ex Kunth. J. Phytol. **3**(3), 68–76 (2011)
47. N.M. Mostafa, *Pyrostegia venusta* (Ker Gawl.) Miers: a botanical, pharmacological and phytochemical review. Med. Aromat. Plants **02**(03), 123 (2013)
48. A. Charles, V.A. Ramani, Chemosystematics of genus Thunbergia (A-mini review)

Chapter 18
Dynamics of Nitrates and Phosphates in Chone Multipurpose Dam, Ecuador

David Carrera-Villacrés and Michael Vega

Abstract Eutrophication is a phenomenon caused by excess nutrients in a body of water, which results in accelerated growth of aquatic plants and a decrease in aerobic processes. The reservoir of the chone multipurpose dam (CMD) is undergoing eutrophication processes, which is contaminating the water and restricting its use for domestic consumption and agricultural irrigation. The objectives of this article were to determine the growth history of phosphate (PO_4^{3-}) and nitrate (NO_3^-) nutrients in the years from 2014 to 2020 to determine the acceleration of the eutrophication of the reservoir and the characterization of the quality of water for human consumption and agricultural irrigation. The study showed a growth of nutrients since 2014, with growths of more than 41% in (NO_3^-) and (PO_4^{3-}) this allowed to conclude the presence of accelerated eutrophication in the water due to phosphates exceeding 0.1 mg/L every year. Also, in the year 2020 water samples were not suitable for human consumption because the chemical oxygen demand (COD) and biochemical oxygen demand (BOD_5) exceeded the maximum permissible limits of current Ecuador regulations, but the water is acceptable for agricultural irrigation.

18.1 Introduction

The city of Chone has extreme climatic conditions, with annual rainfall greater than 3500 mm and less than 500 mm, and an average temperature of 25 °C [1]. The chone multipurpose dam (CMD) was built to minimize the damage caused by the variation in rainfall, to supply the population with water for irrigation and human consumption in the dry season and to prevent flooding during the rainy season [2]. Despite the advantages of the dam construction, it must also be considered that it

D. Carrera-Villacrés (✉)
Universidad de las Fuerzas Armadas ESPE. Departamento de Ciencias de la Tierra y la Construcción, Carrera de Ingeniería Civil, Av. Gral. Rumiñahui, Sangolquí, Ecuador
e-mail: dvcarrera@espe.edu.ec

D. Carrera-Villacrés · M. Vega
Universidad Central del Ecuador, Facultad de Ingeniería en Geología, Minas, Petróleos y Ambiental (FIGEMPA), Carrera de Ingeniería Ambiental, Av. Universitaria, Quito, Ecuador

Á. Rocha et al. (eds.), *Communication, Smart Technologies and Innovation for Society*, Smart Innovation, Systems and Technologies 252,
https://doi.org/10.1007/978-981-16-4126-8_18

causes irreversible environmental impacts such as changes in water quality, quantity band flow [3]. Decomposition of organic-matter and the use of fertilizers upstream add nutrients to the water that accumulate, causing eutrophication [4]. Eutrophication is one of the most frequent problems, caused by high-phosphorus concentrations that accelerate the growth of unwanted aquatic plants, and therefore, reduce the quality of the water [5].

The rivers of the Chone, such as the Grande River, present the progressive problem of eutrophication of their waters, where agricultural activities have an impact, causing high concentrations of phosphorus in the soil and infiltrating into the water [6]. The objectives of the present work were to compare nitrate and phosphate concentrations of data obtained from the year 2014 to 2020 to determine if there is an acceleration in eutrophication, in addition to analyze the water quality for irrigation and human consumption through measure physicochemical parameters such as pH, chemical oxygen demand (COD), biochemical oxygen demand (BOD_5), electrical conductivity (EC), dry residue (DR), calcined residue (CR), organic concentration (OC) and inorganic concentration (IC).

18.2 Methods

18.2.1 Study Area

The city of Chone is located in the western part of Ecuador, South America, in the province of Manabí, at coordinates 601,512 E, 9,923,462 N zone 17 south [7]. Chone is a subtropical city whose predominant climate is dry in the rainy season (from June to November) and warm rainy (from December to May), its temperature in summer fluctuates between 23 and 28 °C, while in the dry season it can reach 34 °C, therefore, the climate is considered one of the most unstable at regional level [8].

The city has been affected by floods (caused by the El Niño phenomenon), which harms its economic activities based on livestock and agriculture [1]. The solution to this problem was the construction of the CMD, this has been in operation since November 2015 and whose geographic location can be seen on Fig. 18.1. The sampling points P0, P1, P2, P3 and P4 were taken in the CMD which covers an area of approximately 60 km^2 and the CMD is located east of the city of Chone, province of Manabí and runs approximately 30 km to join the Mosquito River [9].

18.2.2 Data Collection

For the present study, nitrates, and phosphates data for the years from 2014 to 2017, were taken from the soil and water characterization study of the upper Grande River Basin [10]. The authors determined 26 sampling points for the dry and rainy seasons

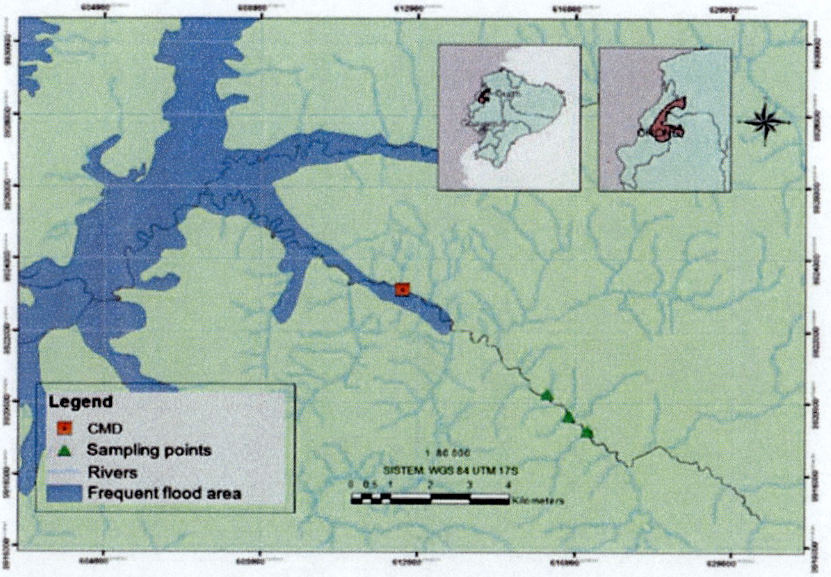

Fig. 18.1 Location map of the CMD, Grande River, Chone canton

and the sampling program were based on INEN 2169:98 on water quality, sampling, handling and conservation of samples and INEN 2179:98 on water quality, sampling and sampling techniques [11]. From the total of 26 samples, four were considered, and shown in Table 18.1: one corresponding to the CMD reservoir (P0) and the remaining three (P1, P2, P3) were upstream sampling points near the CMD reservoir because we wanted to analyze the behavior of nitrates and phosphates before reaching the reservoir.

The values of these 4 points correspond to the rainy season. This period was chosen because the hydrographic system of the Grande River shows an accelerated increase in eutrophication [8]. In [12] is shown that there is a higher concentration of nitrates and phosphates in times of rainy in Chone city.

To obtain data from the same sampling points P0, P1, P2, P3 and P4 for the years 2018 and 2019, was necessarily looking for a relationship between time and phosphate and nitrate concentrations in order to obtain an equation and be able to estimate the values for the years without data [13]. It was observed that for the period 2014–2017 there was growth in these nutrients so an attempt was made to fit the data available (from 2013 to 2017) to growth rates of order 0, 1 and 2 [7]. The linear regressions that were performed did not faithfully represent the behavior of the data for each sampling point, obtaining R^2 values between 0.3 and 0.7 (see Fig. 18.2).

Thus, it was not possible to construct an equation to fill in the data, we proceeded to interpolate the existing data [14], and these data were added for the year 2018.

Table 18.1 Phosphate and nitrate concentrations (mg/L) per sampling point (2014–2017)

Pt.	Description	UTM(WGS 84)		2014		2015		2016		2017	
		E	N	NO_3^-	PO_4^{3-}	NO_3^-	PO_4^{3-}	NO_3^-	PO_4^{3-}	NO_3^-	PO_4^{3-}
P0	CMD	611,624	9,923,151	0.98	1.04	1.04	1.35	1.1	1.28	1.1	1.28
P1	Grande River	615,292	9,920,250	0.85	1.02	1.02	0.63	1.2	1.13	1.2	1.13
P2	Grande River	615,846	9,919,650	0.14	0.72	0.72	0.43	1.3	0.72	1.3	0.72
P3	Grande River	616,320	9,919,240	0.14	0.77	0.77	0.95	1.4	1.75	1.4	1.75

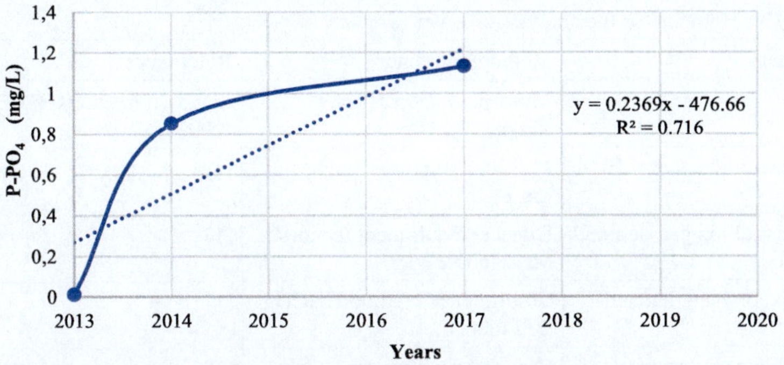

Fig. 18.2 Linear regression for velocity of order 0 of point M1

Table 18.2 Results of the laboratory analysis of the three water point samples (January, 2020)

Pt.	Chemical parameters of CMD samples							
	COD (mg/L)	BOD_5 (mg/L)	DR (ppm)	CR (ppm)	OC (ppm)	IC (ppm)	PO_4^{3-} (mg/L)	NO_3^- (mg/L)
P1	4	< 7	1400	280	1120	280	0.202	0.8
P2	8		1160	240	920	240	0.207	0.8
P3	–		1080	440	640	440	0.4	0.7
Mean P4	6	< 7	1213	320	893.3	320	0.27	0.77

To obtain the 2020 data, an in situ sampling was carried out in the CDM reservoir and simple samples were taken from 3 different sites. Physicochemical analyses were performed in the Environmental Engineering laboratory of the Faculty of Geology, Mines, Petroleum and Environmental Engineering (FIGEMPA) of Universidad Central del Ecuador, their averages were taken as sampling point P4 (Table 18.2).

To obtain the results of the analysis of the samples, the different standardized methods presented in Table 18.3.

18.3 Results and Discussion

Data for sampling points P0, P1, P2 and P3 were found in the literature for 2013, 2014, 2017 [15]. As one of the objectives was to analyze the behavior of nitrates and phosphates over the period from 2014 to 2020. To complete this work, it was necessary to search for data from the same points in 2016 and 2017 and these data is in [16]. After analyzing the data from all sampling points and plotting the behavior of phosphates and nitrates from 2014 to 2017, it was possible to observe that from 2014

Table 18.3 Methods for the analysis of water samples

Parameter	Method/equipment	References
pH	Multi-parameter HACH Ión pH	Standard methods, HACH (2000) [17]
Electrical conductivity (EC)	Multi-parameter HACH Ión EC	[18]
Biochemical oxygen demand (BOD_5)	Digester Equipment tecno. Modelo DR 4500	[17]
Chemical oxygen demand (COD)	Colorimetric method HACH DR/4000 V	[17]
Dry Residue (DR) and Calcined Residue (CR)	Gravimetric method	[19]
Phosphates	Spectrophotometer 8190 method HACH DR 5000	[17]
Nitrates	Spectrophotometer 8039 method HACH DR 5000	[17]

to 2016 there was an almost linear growth of the two parameters, so it was possible to interpolate data between these two years and obtain an approximate value for 2015.

18.3.1 Analysis of Nitrate and Phosphate Behavior for the Rainy Season in the CMD Reservoir and Grande River (2014–2020)

Dynamic of nitrates. Table 18.4 presents nitrate concentrations in the reservoir (P0), three sampling points in the Rio Grande surveyed from 2014 to 2017 and the 2020 reservoir sampling point (P4). Data for the year 2018 were obtained from other

Table 18.4 N–NO_3 concentrations in the Grande River reservoir (2014–2020)

Year	Nitrates (mg/L)—sampling points				
	P0	P1	P2	P3	P4
2014	0.98	0.13	1.06	1.06	0.8
2015	1.04	1.02	0.72	0.77	0.85
2016	1.1	1.2	1.3	1.4	0.9
2017	1.1	1.2	1.3	1.4	0.9
2018	2.1	1.3	1.4	1.5	1.6
2019	1.43	1.275	1.5	1.52	1.18
2020	0.77	1.25	1.6	1.55	0.77

studies [20, 21]. The data for the year 2019 and the remaining data for point P4 were obtained thanks to an interpolation.

In general, it can be observed that at each sampling point nitrates increased from 2014 to 2017. The increase in this nutrient may be since the reservoir generates a greater water impoundment, decreasing turbulence in the rivers, which reduces their oxygenation because the exchange of oxygen on the water surface is restricted [22]. The little oxygen available together with the continuous input of nutrients causes the aerobic population to begin to die off developing anaerobic processes together with eutrophication [23]. Sampling point P4 presents lower nitrate concentration for the year 2020 with respect to the year 2017, it is likely that there is low-fertilizer discharge at this location, the geology of the surrounding rocks is not rich in nitrates, and the soil profile does not favor nutrient percolation [24].

According to the Table in Annex 1 of Ministerial Agreement 097-A from Ecuador, which stipulates the "Quality criteria for water sources for human and domestic consumption," the maximum allowable limit (MPL) of nitrates for Ecuador is 50 mg/L [25]. It means that the 5 samples are under the statute in terms of nitrates, however the water is not necessarily suitable for human consumption as it should comply with the rest of the parameters.

The values of Table 18.4 comply with the Ecuadorian environmental standard for water use for agricultural irrigation. This is detailed in Table 18.3 of Annex 1 of Ministerial Agreement 097-A [25].

Dynamic Phosphates. Table 18.5 presents the phosphate concentration for the sampling points from 2014 to 2017. A growth of the nutrient is observed from 2014 to 2017 except for point P0 that decreased. The increase in phosphate concentration may be since the reservoir generates greater water impoundment, decreasing turbulence in the rivers, which reduces their oxygenation due to restricted oxygen exchange at the water surface [22]. The available dissolved oxygen (DO) was low, the continuous discharge of phosphorus-based pesticides and the thermal stratification in the water room could cause the aerobic population to begin to die off as all the DO is consumed and anaerobic processes develop along with eutrophication [26].

Table 18.5 P–PO_4 concentrations in the reservoir and Grande River (2014–2020)

Year	P–PO_4 sampling points				
	P0	P1	P2	P3	P4
2014	1.42	0.85	0.14	0.14	1.1
2015	1.35	0.63	0.43	0.95	1.05
2016	1.28	1.13	0.72	1.75	1.03
2017	1.28	1.13	0.72	1.75	1.03
2018	0.33	0.63	0.77	0.64	0.56
2019	0.29	0.84	0.78	0.92	0.41
2020	0.27	1.05	0.8	1.2	0.27

Point P4 presents a lower value of phosphates 0.27 mg/L in 2020 with respect to the values of the 4 sampling points in 2017; this may be due to many factors such as the site where P4 is located, this is a place where there is not much discharge of pesticides, there is no data from previous years of phosphates to estimate their dynamic, the surrounding rock strata are not rich in phosphates and orthophosphates [27].

It can be noticed that phosphates exceed the value of 0.1 mg/L, phosphate values above 0.1 indicate that eutrophication processes are accelerated [28]. Table 18.5 shows that there are occasions when the limit has been exceeded by 17 times more than the starting values, this indicates that there is a very accelerated eutrophication from 2017 onward.

According to Ecuadorian environmental regulations, Table 18.1 of Annex 1 of Ministerial Agreement 097-A, which stipulates the "Quality criteria for water sources for human and domestic consumption," does not present the maximum permissible limit (MPL) for phosphates [25]. It can be said that between 2014 and 2020 the sampling points are under the standard for phosphates, this does not imply that the water is fit for human consumption as it should comply with the rest of the parameters of the national environmental standard.

The values in Table 18.5 also comply with the Ecuadorian national environmental standard for the use of water for agricultural irrigation, this is detailed in Table 18.3 of Annex 1 of Ministerial Agreement 097-A [25].

When analyzing the percentage growth of the concentration of nitrates and phosphates from 2014 to 2017 for the sampling points, exorbitant results can be found, as shown in Table 18.6. This demonstrates the growth of nutrients in the CMD, and consequently the acceleration of the eutrophication and degrading of the quality of water for human consumption and agricultural irrigation [29].

Growth is not compared until 2020, because the data for sampling points P0, P1, P2 and P3 are interpolated and not from laboratory analysis.

Table 18.6 Percentage growth of nitrate and phosphate in the CMD reservoir and at 3 sampling points on the Grande River

Sample		Nitrates (mg/L)			Phosphates (mg/L)		
		2014	2017	Growth (%)	2014	2017	Growth (%)
P0	CDM	0.98	1.1	12.24	1.42	1.28	−9.86
P1	Grande River	0.85	1.2	41.18	0.13	1.13	769.23
P2	Grande River	0.14	1.3	828.57	0.14	0.72	414.29
P3	Grande River	0.14	1.4	900.00	0.14	1.75	1150.00

18.3.2 Determination of Water Quality for Human Consumption and Agricultural Use in the CDM and Grande River

Due to the lack of data for the years from 2014 to 2019 for COD, BOD_5, DR, CR, OC, and IC, these were not possible to know how they have varied over time at each sampling point. However, in this article we present the results obtained in the FIGEMPA laboratory for 3 sampling points that were done in the reservoir in situ in CMD, January 2020 because the water quality for domestic use and agricultural irrigation can be known at the time of sampling. This information can be seen in Table 18.2.

The analysis of COD and BOD_5 showed that they are above the maximum permissible limits (MPL) of Table 18.2 "Quality criteria for water sources for human and domestic consumption" of Annex 1, Ministerial Agreement 097-A [25]. According to FAO, the MPL for DR and CR is 1500 ppm if the water is to be used for human consumption. Under this premise, Table 18.2 determined that the average of the three samples for DR was 1213 ppm and for CR was 320 ppm, this indicates that the water samples taken in the reservoir in January 2020 are within the standard. The international standard was considered because no limits for this parameter were found in the national regulations from Ecuador [30].

18.4 Conclusions

The existence of an increase in nutrients (expressed as nitrates and phosphates) in the different sampling points from 2014 to 2020 was demonstrated and this may be due to direct discharges of organic fertilizers (cattle manure, chemical fertilizers), flooding caused by water damming and low aerobic microbial activity in the water body due to the decrease in available dissolved oxygen. If the concentration of nutrients in the chone multipurpose dam (CDM) continues to increase, there will be greater eutrophication of the waters and this will affect the economic and social aspects of the surrounding communities.

The point samples taken at CMD in January 2020 can be concluded that the water was contaminated by organic load if its use for human consumption is required, because its COD and BOD_5 values (6 mg/L and 7 mg/L, respectively) exceed the maximum permissible limits of the national environmental regulations from Ecuador (< 4 and < 2 mg/L). Although the nitrates and phosphates values are within the maximum permissible limits of the national regulations for human consumption and agricultural use, it does not mean that the water can be used for these purposes because it must comply with the rest of the parameters of the standard and these analyses have not been performed for this study.

The instant the samples were taken in January 2020, it could be concluded that the water was below the maximum allowable limit for DR and DC of the FAO that is 1500 mg/L.

References

1. Gobierno Autonomo Descentralizado, *Plan de Desarrollo y Ordenamiento Territorial. Del Cantón Chone* (2011), pp 8–13
2. Secretaría de Agua, *Multipropósito chone: Mitigará los impactos del Fenómeno del Niño* (2015)
3. S. Utrera, "Libro: La incidencia ambiental de las obras hidráulicas—9788481559132—Utrera Cano, Sebastián Félix—Marcial Pons Librero" (2002)
4. J. Escobar, *La contaminación de los ríos y sus efectos en las áreas costeras y el mar*, vol. 50, no. 1680–9017 (2002)
5. P. Torres, C. Cruz, P. Patiño, Índices de Calidad de Agua en Fuentes Superficiales utilizadas en la producción de agua para consumo humano, una revisión crítica. Rev Ing Univ Medellín 8 (2008) https://doi.org/10.1039/9781788012669-00386
6. H.P. Jarvie et al., Phosphorus and nitrogen limitation and impairment of headwater streams relative to rivers in Great Britain: a national perspective on eutrophication. Sci Total Environ **621**, 849–862 (2018). https://doi.org/10.1016/j.scitotenv.2017.11.128
7. FAO, Estudio Hidrológico de inundaciones en la cuenca alta del río Chone (Subcuencas: Garrapata, Mosquito y Río Grande) (2012)
8. D. Carrera-Villacrés, E. Guerrón Varela, L. Cajas Morales, T. González Farinango, É. Guamán Pineda, P.A. Velarde Salazar, Relación de temperatura, pH y CE en la variación de concentración de fosfatos en el Río Grande, Cantón Chone. Congr. Cienc. y Tecnol. ESPE **13**(1) (2018). https://doi.org/10.24133/cctespe.v13i1.807
9. A. Anchundia, *Análisis de violación de derechos Población Río Grande frente al proyecto de Propósito Múltiple Chone*. INREDH (2012)
10. D. Carrera-Villacrés, M. Pazmiño-Montero, R. Durán, E. Mayorga, Soil characterization from the Rio Grande high basin for preparing a technosol that retains phosphorus to prevent eu-Trophication in the dam water of the Presa Propósito Multiple Chone. IOP Conf. Ser. Earth Environ. Sci. **344**(1). https://doi.org/10.1088/1755-1315/344/1/012038
11. INEN 2176, *Agua. Calidad del agua. Muestreo. Técnicas de Muestreo* (2013), p. 15
12. D. Carrera-Vihacrés, M. Pazmiño-Montero, R. Durán, E. Mayorga, Soil characterization from the Rio Grande high basin for preparing a technosol that retains phosphorus to prevent eu-Trophication in the dam water of the Presa Propósito Multiple Chone, in *IOP Conference Series: Earth and Environmental Science*, vol. 344 (no. 1, 2019) https://doi.org/10.1088/1755-1315/344/1/012038
13. OMM, *Guía de prácticas climatológicas* (2018)
14. O. Borja-Goyes, D. Carrera-Villacrés, D. González-Riera, E. Guerrón-Varela, P. Montalvo-Alvarado, A. Moreno-Chauca, *Affectations in soil fertility in the direct influence area downstream of the Grande River: Chone multiple purpose dam in Ecuador*, vol. 893 (2019)
15. K. Calvopiña, A. Vilela, Diseño de tecnosoles para la retención de fosfatos en el agua, de la presa propósito múltiple Chone (PPMCH), a partir de muestras de suelos del cantón Chone, Manabí, Ecuador (Universidad de las Fuerzas Armadas ESPE, 2017)
16. D. Carrera-Vihacrés, M. Pazmiño-Montero, R. Durán, E. Mayorga, Soil characterization from the Rio Grande high basin for preparing a technosol that retains phosphorus to prevent eu-Trophication in the dam water of the Presa Propósito Multiple Chone. IOP Conf. Ser. Earth Environ. Sci. **344**(1) (2019). https://doi.org/10.1088/1755-1315/344/1/012038
17. APHA/AWWA/WPCF, *Standard methods for the examination of water and wastewater*, 19th ed. (1995)

18. NMX-AA-093-SCFI, *Análisis De Agua—Determinación De La Conductividad Electrolítica* (2000), p. 27
19. DOF, NMX-AA-034-SCIFI-2015 Análisis De Agua—Medición De Sólidos Y Sales Disueltas En Aguas Naturales, Residuales Y Residuales Tratadas—Método De Prueba. *D. la Fed.* (2015), p. 16
20. R. Arias-Carrera et al., Metamorphosis and dynamics of inorganic species in the waters of the Chone multiple purpose dam, Ecuador. Adv. Intell. Syst. Comput. **893** (2019). https://doi.org/10.1007/978-3-030-04447-3_15
21. O. Borja-Goyes, D. Carrera-Villacrés, A. Moreno-Chauca, P. Montalvo-Alvarado, Affectation in soil fertility in the direct influence area downstream of the Grande River: chone multiple purpose Dam in Ecuador. Adv. Intell. Syst. Comput. **893**, 73–88 (2019). https://doi.org/10.1007/978-3-030-04447-3_15
22. S. Daneshgar, A. Callegari, A. G. Capodaglio, D. Vaccari, The potential phosphorus crisis: resource conservation and possible escape technologies: a review. Resources **7**(2) (2018). https://doi.org/10.3390/resources7020037
23. I. Urdalen, *Phosphorus recovery from municipal wastewater* (2014) [Online]. Available: http://ivarurdalen.com/wp-content/uploads/2014/03/Phosphorus-recovery-technologies_19.12.2013.pdf
24. A.J.P. Smolders, L.P.M. Lamers, E.C.H.E.T. Lucassen, G. Van Der Velde, J.G.M. Roelofs, Internal eutrophication: how it works and what to do about it—a review. Chem. Ecol. **22**(2), 93–111 (2006). https://doi.org/10.1080/02757540600579730
25. MAE, *TULSMA, Libro VI, Anexo I*, vol. tabla 3 (2015), pp. 12–60
26. S. Fernandez, Interferencia de los embalses, y su régimen de explotación, con algunos procesos y especies relacionados con la fauna. Dam Maint. Rehabil. 175–182 (2018). https://doi.org/10.1201/9781315140728-29
27. M. Stutter, A. Wyness, H. Watson, N. Dodd, Coupled macronutrient cycling in stream biofilms: effects of stoichiometry, light and temperature. Sci. Total Environ. **703** (Feb 2020). https://doi.org/10.1016/j.scitotenv.2019.134880
28. H.P. Jarvie, A.N. Sharpley, P.J.A. Withers, J.T. Scott, B.E. Haggard, C. Neal, Phosphorus mitigation to control River eutrophication: Murky waters, inconvenient truths, and 'Postnormal' science. J. Environ. Qual. **42**(2), 295–304 (2013). https://doi.org/10.2134/jeq2012.0085
29. D.V. Carrera-Villacrés, H.M. Ortega-Escobar, J. Ramírez-García, C. Ramírez-Ayala, C. Trejo-López, A.E. Becerril-Román, Boron, phosphates, and salinity indexes from the Zahuapan-Atoyac hydrographic system, Mexico | Boro, fosfatos e índices de salinidad del sistema hidrográfico Zahuapan-Atoyac, México. Tecnol. y Ciencias del Agua **2**(2), 5–21 (2011)
30. NMX-AA-034-SCFI-2001, *NMX-AA034-SCFI-2001 Análisis de Agua—Determinación de Sólidos y Sales Disueltas en Aguas Naturales, Residuales y Residuales Tratadas—Método de Prueba (Cancela a las NMX-AA-020-1980 Y NMX-AA-034-1981)* (2001), p. 18

Chapter 19
The Effect of Ascorbic and Citric Acids in the in Vitro Establishment of *Solanum betaceum* to Prevent Phenolic Oxidation

Katherine Sofía Salazar-Vega, Sebastián Andrés Pillalaza Montalvo, and Ivonne Vaca-Suquillo

Abstract Tamarillo (*Solanum betaceum*) is an Andean woody plant. Due to the chemical composition and high-nutritional value of its fruits, it is considered a viable option for research and mass production through plant tissue culture. This species is susceptible to phenolic oxidation caused by mechanical damage such as cuts made to obtain axillary buds as explants. In order to solve this problem, explants were cut and added into a solution of citric acid and ascorbic acid at a concentration of 100 ppm, respectively; subsequently, the disinfection protocol was conducted to avoid contamination in the phase of in vitro establishment. This research aims to analyze the response of axillary buds and seeds. Phenolic oxidation, plantlet regeneration, number and coloration of leaves, as well as callus induction, were evaluated. Axillary buds showed 8% of phenolic oxidation, while seeds did not present any. The highest plantlet regeneration (68.42%), leaf coloration (3/5) and number of leaves (3.26) were obtained by axillary buds. The phenolic oxidation showed by buds (8%) was very low compared to previous literature.

19.1 Introduction

Tamarillo (*S. betaceum*) is a tropical fruit native to the Andes, with high-economic importance due to its various physiological characteristics that make it optimal for cultivation, in addition to having excellent chemical compounds for its consumption, such as soluble sugars, organic acids, minerals, carotenoids and phenolic compounds [1, 2]. As a consequence of the characteristics shown by this species, different culture techniques have been implemented, among which in vitro tissue culture stands out. This technique is based on the induction of the totipotency of plant cells, which when found in a medium with adequate levels of growth regulators will result in morphogenesis and, subsequently, in the regeneration of the whole plant [3, 4]. When cultivating the tissue of woody plants, browning of the explants occurs in response to

K. S. Salazar-Vega · S. A. Pillalaza Montalvo · I. Vaca-Suquillo (✉)
Group of Research and Development BIOARN, Universidad Politécnica Salesiana, Avenida 12 de Octubre N°2422 y Wilson, Quito 170109, Ecuador
e-mail: ivaca@ups.edu.ec

Á. Rocha et al. (eds.), *Communication, Smart Technologies and Innovation for Society*, Smart Innovation, Systems and Technologies 252,
https://doi.org/10.1007/978-981-16-4126-8_19

mechanical damage produced by the cuts, because the phenolic compounds accumulated in the vacuoles react with catechol oxidase giving rise to browning of the tissue and subsequently its necrosis [5]. This problem can be counteracted through the action of various antioxidants which inhibit the action of catechol oxidase, ensuring the survival of the tissue in the medium and allowing plant cells to react to the various compounds in the culture media, thus obtaining a morphological response induced by totipotency [6, 7]. This study compares the in vitro performance of two *S. betaceum* explants (seed and axillary bud), and stands out the antioxidant effect of a citric acid and ascorbic acid solution during the disinfection protocol, while making cuts to obtain explants, as a measure to avoid the effects of browning produced in axillary tamarillo buds, and ensure the survival of the explants in the establishment stage and their subsequent morphogenesis.

19.2 Methods

This research project was developed in the Plant Biotechnology laboratory of the Universidad Politécnica Salesiana, in Quito, Ecuador. The growth chamber was regulated at a photoperiod of 16/8 and a temperature of 24 ± 2 °C. The procedure used for sterilizing the seeds and preparing culture medium for both seeds and buds is described by Salazar and Vaca, on a previous study on *S. betaceum* seeds [8].

19.2.1 Axillary Buds and Antioxidant Solution

Regarding axillary buds, plant material was collected from 9-month-old *S. betaceum* plants, kept under greenhouse conditions in the city of Quito. Twigs of 30 cm in length were cut, whose leaves and apical buds were removed; these were wrapped in paper and moistened with running water. In order to reduce the browning of the tissue, the plant material was collected 4 h prior to sowing in culture medium. The rods were divided into halves and rinsed under running water, placed in a 2 L capacity container with 1% (w/v) powdered detergent solution, stirring for 10 min constantly. The length of the explants was reduced to 2 cm and they were immersed in a solution with antioxidant properties composed of citric acid (100 ppm) and ascorbic acid (100 ppm); solution that was also used in each rinse. Next, the explants were immersed in alcohol (EtOH) at 70% (v/v) for half a minute, followed by a rinse; then were added in iodine tincture at 1.3% (v/v) for 30 s and rinsed again. The explants were kept for 15 s in a 1 mL/L solution of pentahydrate copper sulfate and were then rinsed for the last time.

19.3 Results

For the variables related to morphogenetic response, axillary buds showed better plantlet regeneration with an average of 68.42%, number of leaves with 3.26%, and color of leaves with a 3/5 average, indicating a medium green coloration and 86% callus induction. Meanwhile seed showed 50% of plantlet regeneration and an average of 1/5 for the variable leaf color, meaning weak coloration. The oxidation shown by browning of the explant and/or culture medium was analyzed in both seed and bud; however, this problem is produced due to the cuts made in the tissue [9]. Despite, the fact that seeds presented 0% phenolization, axillary buds reached 8%, which represents a very low percentage of tissue oxidation, ensuring their survival and a subsequent cell regeneration in all the stages of in vitro culture (see Figs. 19.1

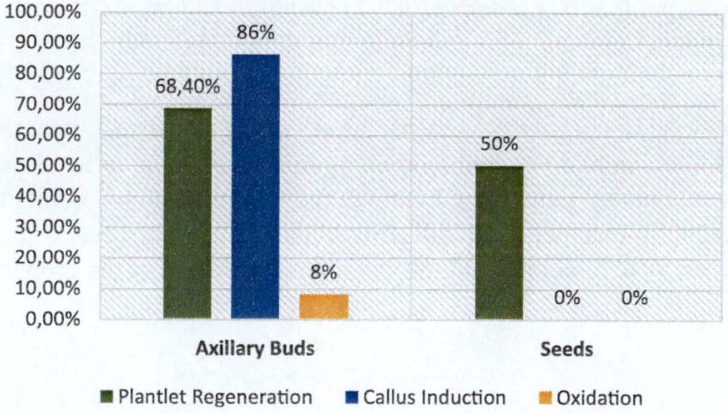

Fig. 19.1 Plantlet regeneration, number of leaves, callus induction and oxidation showed by two explants (axillary bud and seed) of *S. betaceum* in vitro culture

and 19.2).

19.4 Discussion

19.4.1 Seed Versus Axillary Bud

When cultivating *S. betaceum*, seedlings are the most common chosen explants [10, 11]. The literature that describes axillary buds is not as extensive, since researchers may have found easier to use somatic embryogenesis [12, 13]. Although the use of axillary buds as explants has been reported [14], the explants were obtained from plants that had already been established in vitro, and in this case the oxidation is not a problem as when obtaining explants from *ex vitro* plants. The development of callus tissue is attributed to wounds, as an accumulation of cytokinin occurs before

Fig. 19.2 Development of *S. betaceum* plantlets obtained by axillary bud explants: **a** 5-day-old plantlet, **b** 15-day-old plantlet and **c** Callus tissue in a 15-day-old plantlet

callus formation [15–17], agreeing with the results of this investigation. In the case of seedlings there were no wounds, resulting in the absent of callus tissue. The fact that there were not any growth regulators in the culture medium, coincides with the previous study [18] that mentions the formation of callus at the base of the explants of nodal segments with *S. betaceum* buds occurred regardless of the growth regulators dose.

19.4.2 Antioxidant Effect of Citric and Ascorbic Acids

There had been investigations in which the application of antioxidants was performed after the disinfection protocol to reduce the browning caused by phenolic oxidation. When working with *Polylepis tomentella*, the explants were rinsed in a sterile solution of 300 ppm of citric acid and 300 ppm of ascorbic acid, significantly reducing the browning of the explants, and ensuring their survival during the stage of introduction to the culture medium [19]. On the other hand, the application of antioxidants added to the culture medium has had high efficiency in the control of browning in the tissue culture of orchids. The addition of antioxidants such as citric acid, ascorbic acid and cysteine-HCl as treatments of the explants in the culture media managed to reduce phenolic oxidation by 80–90%, ensuring the survival of the explants and their subsequent regeneration [20]. Likewise, when using explants of *Musa paradisiaca*, adequate concentrations of citric acid and ascorbic acid of 100 ppm, respectively, in the M&S culture media were determined as effective antioxidant agents against the browning of the developing plant [21]; the same yields occurred in the present investigation. *Acrocomia aculeata* embryos were cultured in a semisolid M&S culture medium with salts and vitamins, supplemented with citric acid, cysteine, gibberellic acid and sucrose, increasing the viability of the embryos by reducing the necrosis generated by oxidation [22].

By evaluating the interaction of antioxidants in different concentrations, its potential against the browning of explants has been taken into account. It has been shown that the application of citric acid or ascorbic acid in conjunction with Polyvinylpyrrolidone (PVP) in the culture medium, managed to reduce the oxidation of the explants at different stages of plant tissue culture with overall positive results. In *Cedrela odorata*, the interaction of the antioxidants ensured the survival of young explants [23]. While in *Psidium guajava*, the interaction of citric acid and ascorbic acid has proven to be inefficient in controlling the browning of apical explants; therefore, PVP was established as a viable antioxidant agent to reduce oxidation [24]. In addition, the application of PVP without the interaction of any other antioxidant ensured the survival of *Fragaria chiloensis* explants by significantly reducing the oxidation caused by cuts [25]. However, PVP as an antioxidant can be inefficient in species like *Premna serratifolia*, in which activated charcoal was more effective in reducing browning than PVP [26].

Nevertheless, there had been many success cases of the use of ascorbic acid as an antioxidant in plant tissue culture [27, 28]. Ascorbic acid has also been reported as an inefficient antioxidant when applied in a *Strelitzia reginae* in vitro culture [29]. Citric acid has not only proven to be an excellent antioxidant when applied exogenously, but it has also been used to help as a phytostabilizing agent during the accumulation of heavy metals as arsenic [30].

19.5 Conclusions

S. betaceum axillary buds showed the best results for plantlet regeneration. Even though there is still a low percentage of oxidation, the browning of axillary buds is not significantly important when compared to all the benefits, being one of them the reduced chance of genetic variability, caused by using buds instead of seeds. Phenolic oxidation is still a major issue when using nodal explants in plant tissue culture, reason for which research in the field of antioxidants should be encouraged to find new strategies to battle it.

Acknowledgements The authors would like to thank the Universidad Politécnica Salesiana for supporting this research.

References

1. P. Acosta-Quezada, M. Raigón, T. Riofrío-Cuenca et al., Diversity for chemical composition in a collection of different varietal types of tree tomato (*Solanum betaceum* Cav.), an Andean exotic fruit. Food Chem. **169**, 327–335 (2015)

2. S. Espin, S. Gonzalez-Manzano, V. Taco et al., Phenolic composition and antioxidant capacity of yellow and purple-red Ecuadorian cultivars of tree tomato (*Solanum betaceum* Cav.). Food Chem. **194**, 1073–1080 (2016)
3. B. Buchanan, W. Gruissem, R. Jones, Biochemistry and molecular biology of plants, 2nd edn. Wiley (2015)
4. J. Concha, C. Roman, R. Velasco, Evaluación del efecto de diferentes concentraciones de fitohormonas en la micropropagación del plátano dominico hartón (*Musa* aab Simmonds). Biotecnología en el Sector Agropecuario y Agroindustrial **6**, 99–104 (2008)
5. S. EL-Gioushy, R. Liu, H. Fan, A complete protocol to reduce browning during coconut (*Cocos nucifera* L.) tissue culture through shoot tips and inflorescence explants. Plant Arch. **20**(2), 2196–2204 (2020)
6. Á. Azofeifa-Delgado, Problemas de oxidación y oscurecimiento de explantes cultivados in vitro. Agrononomía Mesoamericana **20**(1), 153–175 (2009)
7. E. Criollo, K. Insuasti, W. Degaldo, Regeneración *in vitro* de plántulas de tomate de árbol (*Solanum betaceum* Cav. Sendt.). Rev Colomb Ciencias Hortícolas **10**(2), 252–261 (2016)
8. K. Salazar-Vega, I. Vaca-Suquillo, *Solanum betaceum* in vitro seed germination and seedling development in response to pregerminative treatments. Bionatura **6**(1), 1520–1523 (2021)
9. J. Gao, P. Zhang, J. Xue et al., Advances in phenolic substances and their effects on browning in woody plant tissue culture. Acta Horticulturae Sinica **46**(9), 1645–1654 (2019)
10. J. Prohens, F. Nuez, The tamarillo (*Cyphomandra betacea*): a review of a promising small fruit crop. Small Fruits Rev **1**(2), 43–68 (2001)
11. P. Acosta-Quezada, T. Riofrío-Cuenca, J. Rojas et al., Phenological growth stages of tree tomato (*Solanum betaceum* Cav.), an emerging fruit crop, according to the basic and extended BBCH scales. Sci. Hortic. **199**, 216–223 (2016)
12. S. Correia, A. Alhinho, B. Casimiro et al (2019) NEP-TC a rRNA methyltransferase involved on somatic embryogenesis of tamarillo (*Solanum betaceum* cav.). Front. Plant Sci. **10**, 438 (2019)
13. D. Graça, S. Correia, E. Ozudogru et al., Cryopreservation of Tamarillo (*Solanum betaceum* Cav.) embryogenic cultures. In: Step wise protocols for somatic embryogenesis of important woody plants, pp 95–101. Springer, Cham 2 (2018)
14. A. Copatti, F. Loy, J. Cruz, C. Schwartz, Reguladores vegetais na propagação *in vitro* de Tamarileiro (*Solanum betaceum*). Rev. da 13ª Jorn. pós Grad. e Pesqui. Congrega (2016)
15. W. Ling, F. Liew, W. Lim et al., Shoot induction from axillary shoot tip explants of fig (*Ficus carica*) cv. Japanese BTM 6. Trop. Life Sci. Res. **29**(2), 165–174 (2018)
16. K. Dharmayanti, E. Sulistyaningsih, R. Wulandari, Callus induction on true shallot seed explant using a combination of BA and 2,4-D. Ilmu Pertan (Agricultural Science) **2**(3), 137–143 (2017)
17. S. Bhatia, K. Sharma, R. Dahiya, T. Bera, *Modern applications of plant biotechnology in pharmaceutical sciences*, 1st edn. Elsevier Inc. (2015)
18. A. Copatti, F. Loy, J. Cruz, C. Schwartz, Reguladores vegetais na propagação *in vitro* de Tamarileiro (*Solanum betaceum*). Rev. da 13ª Jorn. pós Grad. e Pesqui. Congrega **13**, 10 (2016)
19. C. Vega-Krustulovic, J. Bermejo-Franco, G. Villegas-Alvarado et al., Propagación masiva de "*Polylepis tomentella*" Weddell ssp. "*nana*" mediante técnicas de cultivo "*in vitro*." Ecol en Boliv **42**(2), 102–120 (2007)
20. S. Nahuat, J. Giorgana, N. Santana, Efecto de diferentes antioxidantes en el control de la fenolización en el cultivo *in vitro* de orquídeas. Rev. Sociedad Mexicana de Biotecnología y Bioingeniería, XVII Congreso Nacional de Biotecnología y Bioingeniería (1996)
21. R. Ancasi-Espejo, J. Montero-Tonconi, N. Ferreira-Castedo, I. Muñoz-Guzmán, Determinación un mejor medio de cultivo en la fase de establecimiento para la propagación in vitro de plátano (*Musa paradisiaca* L.). J. Selva Andin Res. Soc. **7**(2), 104–111 (2016)
22. K. Sánchez-Zúñiga, E. Arnáez-Serrano, I. Moreira-González, G. Vargas-Hernández, Optimización de un protocolo de cultivo *in vitro* de embriones de Coyol (*Acrocomia aculeata*). Rev. Tecnología en Marcha **32**, 30–35 (2019)

23. S. Sampayo-Maldonado, J. Jasso-Mata, M. Jiménez-Casas et al., Efecto del medio de cultivo en la propagación *in vitro* de genotipos de *Cedrela odorata* L. (Eff Cultiv Mediu Vitr Propag Cedrela odorata L. genotypes) **9**, 62–69 (2016)
24. O. Concepción, L. Nápoles, A. Pérez et al., Efecto de tres antioxidantes en el cultivo *in vitro* de ápices de guayaba (*Psidium guajava* L.). Relación entre el origen del explante y el contenido de compuestos fenólicos. Cultiv. Trop. **26**, 33–39 (2005)
25. K. Quiroz, M. Berríos, B. Carrasco et al., Meristem culture and subsequent micropropagation of Chilean strawberry (*Fragaria chiloensis* (L.) Duch.). Biol. Res. **50**, 20 (2017)
26. R. Chinnappan, N. Ruthar, S. Sethu, Rapid *in vitro* propagation of *Premna serratifolia*, a medicinally important declining shrub, India. Conserv. Evid. **8**, 66–73 (2011)
27. W.H. Ko, C.C. Su, C.L. Chen, C.P. Chao, Control of lethal browning of tissue culture plantlets of *Cavendish banana* cv. Formosana with ascorbic acid. Plant Cell Tissue Organ Cult. **96**, 137–141 (2009)
28. C.F. Ndakidemi, E. Mneney, P.A. Ndakidemi, Effects of ascorbic acid in controlling lethal browning in *in vitro* culture of *Brahylaena huillensis*; using nodal segments. Am. J. Plant Sci. **5**(1), 187–191 (2014)
29. J.J. North, Effects of antioxidants, plant growth regulators and wounding on phenolic compound excretion during micropropagation of *Strelitzia reginae*. Int. J. Phys. Sci. **7**(4), 638–646 (2012)
30. H. González, D. Fernández-Fuego, A. Bertrand, A. González, Effect of pH and citric acid on the growth, arsenic accumulation, and phytochelatin synthesis in *Eupatorium cannabinum* L., a promising plant for phytostabilization. Environ. Sci. Pollut. Res. **26**, 26242–26253 (2019)

Chapter 20
The Technification of Irrigation as a Strategy of Community Resilience. Case Study: Pisque River Basin, Ecuador

Charles Cachipuendo

Abstract The technification of irrigation improves the productivity of food and raw matter for the industry, so in the face of environmental and economic changes, civilizations have generated technologies that allow the access, conduction, distribution and application of water in production units, however, the technification has concentrated only on improving the infrastructure, while irrigation for the Andean communities in Ecuador is a system that is interrelated with nature, community and infrastructure. Thus, the objective was to determine the influence of technification of irrigation on the change of production and efficient use of water in agriculture. The data were collected using interviews and dialogs with focus groups was found to since the agrarian reform change the methods of irrigation, from gravity to sprinkler and drip, where investments are made by the State, non-governmental organizations, and water users, it was also found that rose cultivation uses drip and presented the highest physical and economic efficiency with a productivity of 7.67 USD/m^3, different from the production of pasture that uses sprinkler reaching 0.12 USD/m^3. The technification of community irrigation concentrated in the pressurization of water using the spray and drip method causing changes in the production and increasing the efficiency of water use.

20.1 Introduction

In the face of constant environmental and socio-economic changes, the population of rural territories implements actions that allow them to change, learn, adapt, and sustain over time [1], so that in relation to food production, expensive goods and raw materials, they have managed to generate and adapt technology that leads to being more efficient in the use of water from conservation, access, control, distribution, and

C. Cachipuendo (✉)
Universidad Politécnica Salesiana, Av. Isabel La Católica N. 23-52 y Madrid, Quito, Ecuador
e-mail: ccachipuendo@ups.edu.ec

Á. Rocha et al. (eds.), *Communication, Smart Technologies and Innovation for Society*, Smart Innovation, Systems and Technologies 252,
https://doi.org/10.1007/978-981-16-4126-8_20

application in the production unit [2]. Water, which is taken to the agricultural production units (PAUs) by a series of physical infrastructure components, generate socio-cultural, socio-environmental, and socio-economic relationships typical of community irrigation systems [3]. The Ecuadorian Andes by individual, collective and community actions of users and with the support of the State and international cooperation through local non-governmental organizations (NGOs), has technified irrigation to claim rights of access, control and use of water.

One of the differential factors of communities is the organization capacity to implement actions within their territories in accordance with the socio-cultural, environmental and economic dynamics, and not only the irrigation canal unites them, but also they maintain common objectives of good living. Thus, irrigation technology can turn them into resilient communities through efficient use of water, according to Boelens [4]. Including these concepts in community management processes can jeopardize the rights of communities by ignoring local knowledge to enter into global dynamics. Concepts of social, technical efficiency, environmental and economic management can be applied without jeopardizing the social and cultural dynamics of community organizations and strengthening community irrigation management in the territories by generating changes in the relationships of water distribution systems and production systems.

Community organizations that manage irrigation water with their ability to change, learn, and adapt implement actions that allow them to optimize the use of water in agriculture, thus, becoming resilient [5]. Environmentally, resilience is focused on the conservation of water supplies, avoiding waste and pollution. Socially, they implement actions that change their organizational structures according to state regulations, i.e., they are dynamic. They change their production systems by considering socio-economic dynamics, thus seeking economic profitability. They also manage to learn and adapt technology by articulating ancestral and contemporary knowledge.

In this context, indigenous communities consider irrigation as a system coupled between community, infrastructure, and nature [6]. Then, the irrigation technology, also called modernization, must be focused on improving physical infrastructure, generating spaces to share knowledge, and maintaining a natural environment respecting other living beings; hence, its purpose is the efficient use of water in agriculture and development in irrigated territories [7]. Irrigation efficiency is commonly measured by technical or physical efficiency and water productivity or economic efficiency [8]. Technical efficiency can be performed at two levels: the irrigation system or the production unit; thus, at the system level, the efficiency is measured in the different components, such as conduction and distribution [9], while at the production unit it depends not only on the irrigation method but also on climatic, agronomic conditions and mainly on the capacities of human action in the management of the system. The main variables to be considered are irrigation frequency and time, water requirements of the crop [10] and performance of the application device: sprinkler or dropper. Finally, an indicator that demonstrate the efficient use of water in agriculture is the applied and consumed flow [11]. Economic efficiency will depend on the amount of water applied and the profitability of the crop [12]. In the Pisque River basin, there are three production systems: roses, bovine milk, and food products, the technification processes show changes in agricultural production systems, targeting those that are more profitable, which are generally export products [13]. These changes generate intra- and inter-community conflicts over water use, with agro-exporting systems generally more technically efficient [14], but these production systems jeopardize the food sovereignty and economy of small producers.

In this article, with the case study of several irrigation systems in the Pisque River basin, the aim is to analyze how the technification of irrigation influences these production systems and the efficient use of water. Hence, the historical review of the irrigation methods used according to the production system, and the definition of indicators of technical and economic efficiency were carried out.

20.2 Methods

The study was carried out in the basin of the Pisque river located 75 km north-east of Quito, Ecuador, with a temperature of 14 °C, annual precipitation of 800 mm, no rain between June to September and winds up to 17 m/s. The topography presents slopes that reach up to 45 degrees of inclination. The soils are volcanic and vary from sandy, loamy, and clayey, according to altitudinal floors ranging from 2680 to 3600 masl. The main production in the area corresponds to bovine milk, flowers, and subsistence production (vegetables, breeding of minor animals).

Sixteen community irrigation systems, 3 inter-community irrigation systems and 3 collective irrigation systems were selected. The information collected refers to: (i) technification processes and changes in production systems obtained through interviews and group dialogues with leaders; (ii) efficiency assessment of water application at the unit level in 30 flower production units, 20 milk and 10 subsistence production units; and (iii) production and productivity of crops estimated on the basis of information provided by 90 farmers, 30 flower producers and 64 subsistence producers. The analysis was carried out according to the size of the production unit, and for the cattle raising it is divided into 4 strata proposed by Requelme [15], $P_1 <$ 1 ha, $P_2 > 1 < 3$ ha, P_3 $3 > 5$ ha and $P_4 > 5$ ha. In the case of flower producers, it was divided into three strata $F_1 < 0.5$ ha, $F_2 > 0.5 < 2$ ha, and $F_3 > 2$ ha.

Percentage assessment by time periods was used to demonstrate the participation of the actors in irrigation technology, changes in production systems and irrigation methods used.

For the technical efficiency of the irrigation system, the proposal by Playan [9] was applied: To calculate the relationship between the water needs of the crop and water applied at the conduction, distribution and at the plot level, Eq. 20.1.

$$UEA = \frac{Wr}{RDW} \times 100 \tag{20.1}$$

Where UEA means the efficient use of water at the plot level expressed as a percentage; Wr is the amount of water required by the crop and was calculated with CROPWAT software; and RDW is the real amount of water applied in m^3/ha.

For the economic analysis, Clemmens [12] proposed the relationship between crop productivity and water used, Eq. 20.2.

Where UEA means the efficient water used at the plot and expressed as a percentage; WR is the amount of water required by the crop and was calculated with CROPWAT software; and RDW is the amount of water applied in m^3/ha.

For the economic analysis, Clemmens [12] proposed the relationship between crop productivity and water used, Eq. 20.2.

$$\mathrm{WUI} = \frac{\mathrm{NR}}{\mathrm{RDW}} \tag{20.2}$$

Where WUI is the index of water consumed from an economic point of view; NR is the net return for the crop ($); and RDW is the amount of water applied in m^3.

The gross margin analysis to determine the performance of each production system was calculated using Eq. 20.3.

$$R = Y.P_y \tag{20.3}$$

Where R is the revenue ($/ha); Y is the performance; P_y is the product price per unit ($/u).

Net returns were estimated from the following expression:

$$\mathrm{NR} = R - \mathrm{TVC} \tag{20.4}$$

Where NR is the net return ($/ha); R is the income ($/ha); and TVC represents the total costs ($/ha).

20.3 Results

20.3.1 Actors in the Technification of Community Irrigation

Technification processes using new technologies that generate economic, environmental, and social benefits originate from the communities, which are involved from the conception to the execution of the project. The financing of the technification comes from the State, NGGs, and irrigators, but the management of irrigation systems over time is a role of the users (see Fig. 20.1).

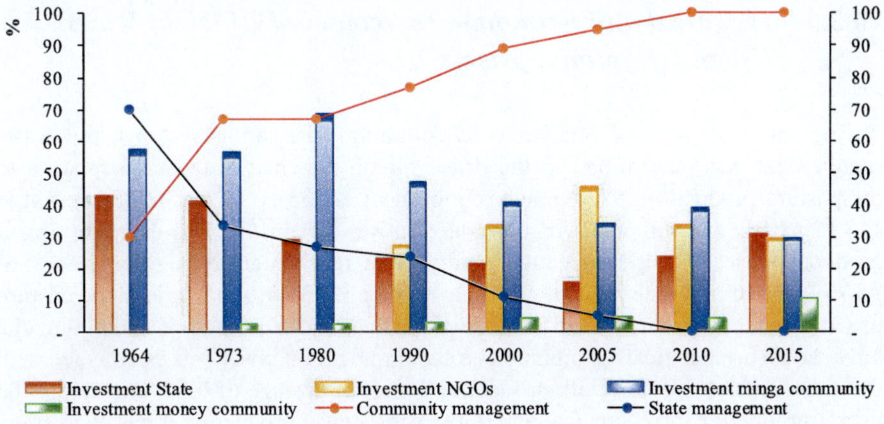

Fig. 20.1 Participation of the community, the State and NGOs in the historical process of technification of irrigation and management

20.3.2 Changes in the Agricultural Production System

The effect of technified irrigation is observed in the change in agricultural production systems due to their profitability. Thus, producers before the agrarian reform cultivated cereals such as barley, wheat, quinoa and maize, and vegetables such as potatoes, onion, beans, and after the agrarian reform they looked for alternatives that would generate higher economic income, and there was a twist to milk production thanks to spraying irrigation. The production of flowers under the greenhouse began in 1988 in the Canan-valle community, and they used drip irrigation (see Fig. 20.2).

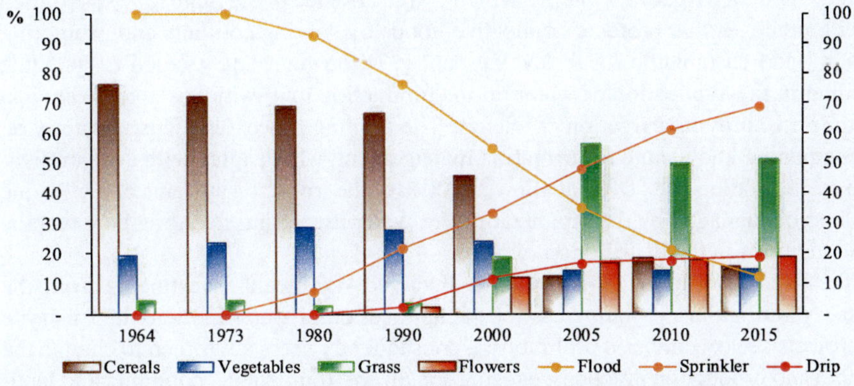

Fig. 20.2 Production systems and irrigation methods in the technification process

20.3.3 Technical and Economic Indicators of Efficient Water Use in Irrigation Systems

To measure the physical efficiency of community irrigation systems, the water requirement was determined in the driest month which is August. In relation to the pasture production, the average requirement is 536.10 m^3/ha; for flowers it is 1132.2 m^3/ha; and for subsistence production it is 521 m^3/ha. However, the flower production uses the drip irrigation method that reaches an average efficiency of 96.75%, avoiding waste by applying an average 1170.31 m^3/ha, while in the pasture production, sprinkler irrigation is used with an average efficiency of 1106.99 m^3/ha have shown no statistical difference. The economic productivity was established with the water data applied, in milk production it has an average of 0.12 \$/m^3/ha, in the flower production 7.67 \$/m^3/ha, and in the subsistence production it has an average of 0.29 \$/m^3/ha (Table 20.1).

20.4 Discussion

The Andean population of Ecuador maintains a relationship with nature, so they built an infrastructure to use water and avoid its waste, and they applied their knowledge to various crops, improving their production and productivity, maintaining the economy and food sovereignty of the territory. In the Pisque basin, the communities began technification in the years 60–80, when they had access to the land [16] due to the agrarian reform, mainly improving the infrastructure for water management under State programs. During the technification of local NGOs with the financing of international cooperation to improve infrastructure, they also trained users in the management of irrigation systems, which by the absence of the State they substituted it with practices that protected collective goods by solving conflicts and promoting integral and sustainable local development [17]; however, they failed to generate training in the application of water on the production unit, which is a key technical factor for improving irrigation efficiency. The training processes focused mainly on strengthening knowledge for irrigation management, which agree with decentralization stages of the State [18] that allowed 100% of the irrigation systems of the Pisque basin to be managed by the organizations of water users, thus avoiding bureaucracy and inequities in the distribution of water.

Technification of irrigation systems involved significant investments from the State, NGOs and users in devices for the application of water in the unit. This type of projects seeks economic profitability, consequently users are forced to change the system and production practices, causing social, environmental, economic, and landscape problems in the territories [19], in addition to jeopardizing food sovereignty and nature. However, these investments in infrastructure allow for the achievement of some objectives that are normally set in modernization processes of irrigation

Table 20.1 Physical and economic efficiency indicators of water use in three agricultural production systems

Variables	Pasture (Milk) production					Flower production				Subsistence agriculture
	P1	P2	P3	P4	$\overline{P}$	F1	F2	F3	$\overline{F}$	
Physical efficiency indicators										
Calculated water requirements to year (m^3/ha/year)	3209.50	3209.50	3209.50	3209.50	***3209.50***	10,669.2	10,669.2	10,669.2	***10,669.2***	3126.00
Calculated water requirements for dry season (m^3/ha/August)	536.1	536.1	536.1	536.1	***536.1***	1132.2	1132.2	1132.2	***1132.2***	521
Amount of water applied in the year (m^3/ha)	13,819.8	7568.7	4905.1	4963.5	***7814.3***	12,668.6	12,721.3	12,528.0	***12,639.3***	4760.0
Amount of water applied in August (m^3/ha)	1961.20	1071.40	692.9	702.5	***1107.00***	1173.0	1177.90	1160.0	***1170.3***	680.0
Water application efficiency on the plot %	27.3	50	77.4	76.3	***57.8***	96.5	96.1	97.6	**96.8**	**76.6**
Economic efficiency indicators										
Productivity (l/cow; plot/month; kg/m^2)	650.89	1578.43	2919.14	6702.76	2692.80	0.8	0.78	0.9	0.83	2
Net return ($/ha/month)	117.3	111.6	101.9	123.6	113.6	7184.2	8615.1	11,108.5	8969.95	197.6
Economic water productivity ($/$m^3$/ha/August)	0.06	0.10	0.15	0.18	**0.12**	6.12	7.32	9.58	**7.67**	**0.29**

(continued)

Table 20.1 (continued)

Variables	Pasture (Milk) production					Flower production				Subsistence agriculture
	P1	P2	P3	P4	$\overline{P}$	F1	F2	F3	$\overline{F}$	
Economic water productivity ($/m^3/ha/year)	0.10	0.18	0.25	0.30	**0.21**	6.81	8.13	10.64	**8.52**	**0.50**

Values in bold and italics in the table mean they are averages

P1 ≤ 1; P2 ≥ 1 < 3; P3 ≥ 3 < 5; P4 ≥ 5 (ha). F1 ≤ 1; F2 ≥ 1 < 3; F3 ≥ 3 (ha)

systems: to increase water yields and productivity [20] that allows communities to be resilient.

Technification aims to avoid the waste of water, which in the Andean mountain range is achieved by pressurizing water and using more efficient irrigation methods, such as aspersion and drip. For the pasture production, the spraying method is used and for the flower production the drip irrigation is used, because it is more efficient by the automation, i.e., in a technification process, the production, social, and environmental relations change [21].

It is evident that users have to implement crops that generate greater profits, such as flowers for export, thus the economic productivity of water increases versus crops destined for local or national food.

The economic and physical efficiency of water use in agriculture does not depend solely on the technological level of the infrastructure, but instead other factors are involved such as the type of crop, soil, production area, social water distribution and the knowledge users have about its application. Thus, the application efficiency of water in the pasture production is less in surfaces whose size is smaller than a hectare, because the water distribution is carried out by irrigation times and not by irrigated area. As for the flower production, it is less efficient in areas whose size is smaller than one hectare because its technological drip irrigation level is manual.

20.5 Conclusions

The technification or modernization of the community irrigation systems was promoted by users after the agrarian reform, which is currently fully managed by the users.

Irrigation technification is an irrigation strategy of communities to be resilient, allowing them to be more efficient in water use and to be more profitable production systems, covering the investments made. The economic water production is higher in flower crops, followed by subsistence crop and pasture production.

Community irrigation systems pressurize water by gravity and use it in the pasture production with the spraying method; however, efficiency will also depend on the cultivation area and the knowledge of the irrigators; meanwhile as regards the flower production, the application efficiency will depend on the automation irrigation level.

References

1. C. Folke, Carpenter, Ss., Walker, B., Scheffer, M., Chapin, T., Rockström, J.: Resilience thinking: integrating resilience, adaptability and transformability. Ecol. Soc. 15, 20 (2010). https://doi.org/10.1038/nnano.2011.191
2. L.S. Pereira, I. Cordery, I. Iacovides, Improved indicators of water use performance and productivity for sustainable water conservation and saving. Agric. Water Manag. **108**, 39–51 (2012). https://doi.org/10.1016/j.agwat.2011.08.022

3. A. Sandoval-Moreno, M.G. Günther, La gestión comunitaria del agua en México y Ecuador: otros acercamientos a la sustentabilidad. Ra Ximhai. **9**, 165–179 (2013)
4. R. Boelens, J. Vos, The danger of naturalizing water policy concepts: water productivity and efficiency discourses from field irrigation to virtual water trade. Agric. Water Manag. **108**, 16–26 (2012). https://doi.org/10.1016/j.agwat.2011.06.013
5. M. Hurlbert, P. Mussetta, Creating resilient water governance for irrigated producers in Mendoza. Argentina. Environ. Sci. Policy. **58**, 83–94 (2016). https://doi.org/10.1016/j.envsci.2016.01.004
6. C. Cachipuendo, Modelo dinâmico para a gestão e manejo sustentável de sistemas de irrigação comunitários, no marco do Bom Viver : estudo do caso na bacia do Rio Pisque (2017)
7. J. Bebel, C. Gutiérrez-Martín, Efectos de la modernización de regadíos en España (2017)
8. M.E. Jensen, Beyond irrigation efficiency. Irrig. Sci. **25**, 233–245 (2007). https://doi.org/10.1007/s00271-007-0060-5
9. E. Playán, L. Mateos, Modernization and optimization of irrigation systems to increase water productivity. Agric. Water Manag. **80**, 100–116 (2006). https://doi.org/10.1016/j.agwat.2005.07.007
10. L. Zhuo, A.Y. Hoekstra, The effect of different agricultural management practices on irrigation efficiency, water use efficiency and green and blue water footprint. Front. Agric. Sci. Eng. (2017). https://doi.org/10.15302/J-FASE-2017149
11. J.M. Tarjuelo, J.A. Rodriguez-Diaz, R. Abadía, E. Camacho, C. Rocamora, M.A. Moreno, Efficient water and energy use in irrigation modernization: lessons from Spanish case studies. Agric. Water Manag. **162**, 67–77 (2015)
12. A.J. Clemmens, D.J. Molden, Water uses and productivity of irrigation systems. Irrig. Sci. **25**, 247–261 (2007). https://doi.org/10.1007/s00271-007-0067-y
13. J. Vos, R. Boelens, Sustainability standards and the water question. Dev. Change. **45**, 205–230 (2014). https://doi.org/10.1111/dech.12083
14. B.E. Bravo-Ureta, R. Jara-Rojas, M. Lachaud, V. Moreira, S.M. Scheierling, Water and farm efficiency: insights from the frontier literature. Agric. Appl. Econ. Assoc. (2015). https://doi.org/10.1017/CBO9781107415324.004
15. N. Requelme, A.A. Gallegos, J. Herrán, C. Cachipuendo, Prácticas tecnológicas y productividad en fincas lecheras de agricultura familiar en el norte de Ecuador. In: 21th International congress on project management and engineering. Cadiz (2017)
16. M. Becker, S. Tutillo, Historia agraria y social de Cayambe. Yala, Abya, Quito (2009)
17. I. Gentes, Políticas sustentables de agua, gestión comunitaria de propiedad hídrica y sistemas alternativas de solucionar disputas. Algunas lecciones para los países Andinos. CEPAL 1–33 (2004)
18. J. Palerm-viqueira, Gobierno y administración de sistemas de riego. Región y Soc. XVII, 31 (2005)
19. J.M. Gómez-Espín, Modernización de regadíos en España: experiencias de control, ahorro y eficacia en el uso del agua para riego. Agua y Territ. 69–76 (2019). https://doi.org/10.17561/at.13.3972
20. G. Angella, S. Prieto, J. López, G. Barraza, R. Salgado, P. Tomsic, … E. Fereres, La modernización de los sistemas de riego y la importancia de los aspectos no estructurales. Congr. Nac. del Agua. **25**. Conagua (2015)
21. S. Sampedro, Cambios en el patrón de cultivo en los regadíos de la cuenca del Guadalquivir ¿modernización de regadío o reforma intermedia de la PAC? Treinta años Política Agrar. Común en España Agric. y multifuncionalidad en el Context. la nueva Rural. 2016, págs. 172–183 (2016)

Chapter 21
Limitations and Potentialities of Organic or Non-conventional Milk Production in Ecuador

Narcisa Requelme and Ana Afonso

Abstract Hunger, poverty, and production are at the core of worldwide problems; this makes it imperative to promote sustainable agro-ecosystems. There is evidence that dairy production under the conventional model is unsustainable, but there is an organic alternative. The aim of this research was to define the limiting and enhancing factors of organic milk production to guide the sustainability of livestock activity in Ecuador. Considering limits, inputs, outputs, and interactions, as well as production factors in the social, environmental, and economic dimensions, we describe the alternative practices of three productive units are described through participant observation, interviews, and laboratory analysis of forage, soil, milk, and water. The results show that there are social factors, such as productive orientation about available resources and the associativity, that are a determinant driver for eliminating dependencies in the establishment of inputs and the commercialization of milk. Environmental factors include the recovery of the biodiversity of forage species in silvopastoral systems that restore biological cycles and soil fertility. Economic factors relate to the availability of resources for acquiring or breeding animals under the non-conventional system and for certification processes. We conclude that organic milk production can be implemented as in the case of the Marullacta farm.

21.1 Introduction

The world's population grows at an average annual rate of 1.1%, having reached 7.8 billion people in 2020 [1]; it is estimated that 88 to 115 million people suffer from extreme poverty, hunger, or food insecurity [2]. Countries that experience famine coincide with those with less economic capacity (in Asia 11% and in Africa 20% of the population is undernourished, and 75% of malnourished people live in rural areas).

The conventional model created by the green revolution, which became massified and thus increased production as if natural resources were infinite, is being exhausted

N. Requelme (✉) · A. Afonso
c/José Gutiérrez Abascal 2, Universidad Politécnica de Madrid, 28006 Madrid, España
e-mail: narcisa.requelme@alumnos.upm.es

Á. Rocha et al. (eds.), *Communication, Smart Technologies and Innovation for Society*, Smart Innovation, Systems and Technologies 252,
https://doi.org/10.1007/978-981-16-4126-8_21

because inputs, mechanization, and irrigation depend on increasingly expensive and scarce fossil fuels [3]. The objective of producing enough food to prevent hunger has not necessarily been met and has chosen a techno-economic emphasis, thus leading to the advance of the agricultural frontier, monocultures, and the indiscriminate use of highly toxic pesticides with environmental and public health consequences.

Faced with this multidimensional problem of hunger, poverty, and production, the UN Agenda for 2030 [4] aimed at zero hunger (SDG 2) and sustainable production and consumption (SDG 12), which imply a change in the production model. The bet is on a sustainability that Gallopín [5] sees as a shift that is not directed to an indefinite growth of energy and materials, but to a process of improving the socio-ecological system. There are direct relationships between social, economic, and environmental sustainability dimensions, and their analysis cannot fail to address this complexity. Economic development has been based on obsolete technologies that have not responded to agricultural problems and have rather generated devastating climatic effects, evidencing the limits of sustaining life on the planet and the collapse of ecosystems [6, 7].

According to Naredo [8], the sustainability of agricultural systems will depend on their ability to supply resources and eliminate waste without losing their productive capacity, depending on the configuration and harmony of the social systems [9]. This must be assessed in terms of efficiency, stability, use of organic techniques, food security, self-sufficiency, biodiversity, preservation of values, traditional knowledge [10, 11], family farming, and stakeholder participation in decision-making processes [12, 13].

To develop production systems as complex and dynamic agro-ecosystems, where productivity is not the exclusive goal but also stability, equity, autonomy of production, cultural acceptance of technologies and resilience, the incorporation is needed of an agro-ecological perspective [14]. Therefore, changes are urgent to initiate transition processes at different scales, levels, and dimensions.

Different alternatives, such as precision farming, have been developed in comparison with the conventional model, which optimizes crop management by controlling variables and using techniques to compensate deviations in the production processes [15]. Ecological agriculture is based on diversified and self-sufficient systems [16], highlighting the presence of multiple benefits such as high biodiversity, diversification of activities, organic crops, and self-sufficiency; they point to inadequate soil management and average productivity as key aspects in conventional systems. Another alternative is organic agriculture, defined as a global production management system that promotes and enhances agro-ecosystem health, soil diversity, cycles, and biological activity. Organic livestock husbandry, as an alternative, comprises a system based on the harmonious relationship between land, plants, and animals with respect to the physical and behavioral needs of livestock, including organic fodder under the principles of health, ecology, equity, and caution [17]. Ecological livestock husbandry is another alternative oriented to a sustainability where the quality and safety of products, respect for public health and animal welfare, as well as the efficiency of organic and inorganic matter cycles are prevalent, thus boosting energy flows, soil biodynamics, and the reduction of erosion [18].

Organic production is a model established through transition processes, restoring ecosystem functions to a multifunctional state capable of providing diverse ecosystem services [19]. Since 1990, consumers have been demanding organic dairy products due to the introduction of hormones, the use of soy, maize, and other GMOs crops produced with synthetic inputs and pesticides, as well as cases of "mad cow" disease and the excessive use of medicines, antibiotics, and steroids that leave residues in food [20]. They are considered organic if they come from livestock companies that meet environmental protection and animal welfare requirements, without the use of chemicals or GMOs [21].

The various positions on the possibility of changing the production model represent a wake-up call for academia to investigate the foundations of these alternatives. In addition, it is necessary to review the way of doing research, since classic agricultural sciences develop under an approach characterized by the dominance over nature, a short-term productivity approach, atomist and reductionist visions of reality, evaluation done without considering social and environmental costs, supremacy of scientific knowledge over local knowledge, prevalence of modernization, excessive confidence in technology, and belief in the neutrality of science [22].

The aim of this study is to define the limiting and enhancing factors of organic milk production by describing the alternative practices as implemented in three production units that aim at reorienting dairy farming toward sustainability.

21.2 Methods

This research is descriptive. It contributes to the generation of knowledge under a new paradigm based on sustainability, complexity, and uncertainty [22]. Regarding sustainability, it seeks to overcome the short-term approach, incorporating sustainability and solidarity criteria. As for complexity, it seeks a vision of multidimensional agro-ecosystems in differentiated contexts. In terms of uncertainty, it is expressed in principles such as that of caution applied for the making of decisions. Systematization is proposed under this new paradigm as a methodology [23] to collect knowledge from the experience of producers based on what Sousa [24] sees as the "ecology of knowledge".

The systematization was carried out having as its axis the systems approach, considering the limits, inputs, outputs, and interactions [25] as well as the factors of production in the social, environmental, and economic spheres. The study was developed in three provinces of the Ecuadorian highlands at family productive units in Marullacta, in Pichincha and El Carmen in Carchi, and one in the Atapo Palmira community irrigation system in Chimborazo.

Participant observation and semi-structured interviews were applied to actors to collect data related to means of production, animal load, feeding strategies, health and reproduction, infrastructure, records management, and labor. The parameters of productivity and nutritional content of forages and the quality of soil, water, and milk were determined in laboratories.

Based on the information collected, the limiting and enhancing factors of organic milk production were defined in accordance with the regulations that promote and control organic, ecological, and biological production in Ecuador [26].

21.3 Results

21.3.1 Limits and Components of the Systems

The surface area of the systems under study, as well as the quality and fertility of the soils, vary (Table 21.1); it is imperative to apply mechanisms to recover biological cycles by reducing the use of fertilizers [27] and look for organic fertilization such as bio-coal, which acts as a long-lasting reservoir that delays its return to the atmosphere as CO2 [28], that has improved yield at a concentration of 2.25 mg/ha + conventional fertilization in forage oat production [29]. Similarly, the use of organic fertilizers in forage oat production increased the quality of fodder compared to the reference material [30]. The use of suitable forage mixtures depends on the adaptability of the species to ecosystems, their management, and other sustainability criteria within the system [31].

For the selection of breeds, the standards state that organic livestock must be born and grown in organic production units considering adaptability and vitality and resistance to diseases, and giving preference to creole breeds [26]. In Marullacta and El Carmen, cattle are born inside farms, not so in Atapo Palmira, where they are bought in fairs. In addition, there is genetically improved cattle, with few specimens of creole breeds such as Pizán (El Carmen) and black (Atapo Palmira).

21.3.2 Inputs and Outputs of the Systems

Continuous grazing is practiced in all three localities. The availability of forage in Atapo Palmira is constant, but due to the surface (<1 ha), it is insufficient to raise more than 2–3 animals. Windbreak barriers have been implemented with trees and shrubs in agrosilvopastoral systems. Several studies point to them as more sustainable than conventional systems, with greater generation of environmental services and more efficient use of renewable local resources [32]. These are also a strategy against climate change [33] and allow the development of feeding regimes that reduce methane production while tannins intake improves meat production and milk quality [34, 35].

Women in Atapo Palmira make mineral salts using grain salt (50 kg), ash (30 kg), and bone meal (20 kg); 10 g/day are administered to all animals every two days. Productively, there are significant differences as the highest production is obtained in Marullacta (30 l/cow/day) while in Atapo Palmira it is only 3.5 L/cow/day. In terms

Table 21.1 Limits, inputs, outputs, and interactions of the systems

Criteria	Marullacta	Atapo Palmira	El Carmen
Limits			
Total surface (ha)	90	305	150
Type of soil	Sandy loamy	Sandy	Sandy loamy
Soil pH	5.9	ND	6.43
OM content	13.89	ND	11.2
P content	5.57	ND	9.57
K content	0.89	ND	1.02
Ca content	10.75	ND	18.35
Grassland surface (ha)	15	1 ha/family	70
Paddock rotation	60–80	In the field	30–45
Crop rotation	No	No	Potato–fodder
Fodder composition			
Raw protein	7.3	ND	19.6
Ether extract	1.6	ND	3
Raw fiber	18	ND	17.90
Ash	8.575	ND	14
Nitrogen-free extract	39.525	ND	45.5
Milking cows	20	1 to 2	192
Cows	15	1	77
Input			
Fodder availability (kgMS/ha/day)	50–70	ND	ND
Mineral salts (g/day)	20	10/2 days	180
Output			
Cow production/day (L)	30	3.1	18–20
Milk quality			
Fat (% g/100 mL)	3,6	ND	3.24
Total protein (%g/100 mL)	2,9	ND	3.01
Lactose (% g/100 mL)	4,7	ND	4.35
Total solids (%g/100 mL)	12,0	ND	11.38

(continued)

Table 21.1 (continued)

Criteria	Marullacta	Atapo Palmira	El Carmen
CCS (x 1000 mL)	474	ND	489
CBT (x 1000 mL)	237	ND	6.846
Main interactions			
Animal load (UBA/ha)	2.5	3	2.2
	100% fodder, free grazing	100% fodder	100% fodder, forage beet
Food strategy	5 kg of kernel food for 8 days to milking cows	No	Prepared balanced food
Reproduction strategy	Natural mating	Artificial insemination	Artificial insemination
Mandatory vaccination	Brucellosis, tuberculosis, aphthous fever	Aphthous fever	Brucellosis, tuberculosis, aphthous fever

Note ND = not determined

of quality, studies in Indonesia found differences in the composition of milk, with a higher amount of omega-3 fatty acid content reported in milk from the organic model compared to the conventional model [36]. Fresh cheese, yogurt, ghee, and milk are elaborated in Marullacta and have been certified as kosher. The trade motto "From happy cows" has been adopted, and the farm is certified as organic.

The soils are fertilized by conducting the cleaning water from the stables to the paddocks; the feces are dispersed with water to avoid soil contamination. In El Carmen, fertilizing costs were reduced by collecting manure that is mixed with stubble; it is flipped every three days and kept for three months until compost and *oil* fertilizer are obtained to incorporate into the paddocks. Under the agro-ecology approach, manure-based *oil* in Atapo Palmira is produced in 18.1 kg/200 L of water, and alfalfa, oats and nettle are added; after one month the product obtained is applied in 1 l/20 L of water every 8 days when forage is growing, after which the ratio is lowered to ½ L, and then 1–2 l/20 L of water is applied again.

21.3.3 Main Interactions in the Systems

The standard establishes an animal load of 1 AUE/ha to prevent overgrazing and erosion. In order to ensure the availability of animals, calves in Marullacta are raised with mother's milk for six months. Feed of cows is complemented with 5 kg/animal of palm kernel meal every 8 days. In El Carmen, calves receive mother's milk up to four months, as well as supplementation food. Fodder beet (Beta vulgaris) reduced production costs as 104 T/ha can be produced. In Atapo Palmira, the animal load has an average of 3 AUE/ha; there producers prepare supplementation food with corn, wheat, quinoa or oats (2.04 kg), beans or peas (0.22 kg), bone meal (10 g) and eggshell (10 g), which is administered to cows. Studies for the reduction of methane emissions refer to the use of monensin to increase fermentative efficiency [37], in addition to positive effects for controlling meteorism and ketosis. Croissant [38] suggests a forage-based diet, with small amounts of balanced food only as a complement, to balance the food ration and fulfill physiological requirements that ensure good health [39].

Organic milk production also includes drug-free anti-parasitic strategies. Homeopathic treatments and the use of medicinal or ethnoveterinary plants are alternative tools to reduce anti-parasitic and antibiotic use. Some medicinal plants modulate the immune system and reduce inflammation and can be used to control diseases [40]. However, treatments can take longer or may be difficult to get. Norms establish the immediate treatment of animals via chemically synthesized allopathic medicines or antibiotics. In Marullacta, Oxalis spiralis is used to treat inflammation of the udder and fever. In El Carmen, autovaccines are applied to treat papilloma, and *Minthostachys mollis* is used to treat diarrhea. In Atapo Palmira, women prepare deworming tinctures, expectorant syrups, healing ointments, a powder for tympanism and a bruise ointment, among others. The use of hydroalcoholic *Leucaena leucocephala* pod extract on ovine gastrointestinal nematodes at a concentration of 50 mg/ml inhibits the hatching of eggs or nematodes in 20%, and at 6.25 mg/mL, it kills 22% of the larvae [41]. Other research shows that deworming is not always necessary if the nematode load can be maintained at moderate levels [42], which could reduce production cost and the waste in the milk.

A frequent disease in dairy farms, the incidence of mastitis in Marullacta, is low because animals are in good health and get adequate nutrition. Homeopathic treatments are also an alternative for mastitis [43–46]. Blanco [43] notes that the use of iodine and selenium reduces the risk of mastitis. Almeida and Coutinho [44] report that homeopathic treatments based on minerals and medicinal plants can be used as alternatives to antibiotics. In El Carmen, mastitis is controlled with conventional treatments, and in Atapo Palmira, 100 g/L of boiled water from white mallow (*Sphaeralcea bonariensis*) is used, the dose being 1 L for three days to animals; then, another 100 g are cooked, the udder is washed with the resulting water, and leaves are placed as a plaster for three days.

Marullacta is the first certified organic farm. The process begins by applying to AGROCALIDAD through the certification body; the applicant signs a service contract and presents all the requisites. Verified documentation should be kept for five years. To maintain certification, laboratory analyses of both inputs and processed products are carried out. The certificate of "organic producer" lasts 1 year. The instruction manual also provides for the issuance of group certifications.

21.4 Discussion

The information collected and analyzed made it possible to determine interrelated economic, environmental, and social factors that limit or enhance the transition processes from conventional production to organic milk production. One of the most important ones is the producers' change of vision that defines the orientation of the land in terms of knowledge, profitability, and relationship with the market. Efficient food and health measures should be implemented, and many of them based on ancestral, traditional, or local knowledge may not have a scientific-experimental test but that have shown effectiveness. Training and the socialization of experiences are essential to widely divulge knowledge of organic, agro-ecological, ecological, or other practices. There are also innovations that go hand in hand not only to improve the production process, but also enhance the machinery for soil preparation, planting, harvesting, processing of organic inputs, infrastructure, etc.

Organic and conventional production systems have a substantial condition for sustainability that they are based on social and solidarity economy for the supply of inputs and the marketing of the produce, which entails establishing direct short circuits with consumers based on associativity schemes at the local and regional levels. This is aimed at breaking the dependencies of suppliers, as well as of industries that condition volumes, price, and quality under specific market standards [45].

One of the main environmental factors is the enhancement of biodiversity, which incorporates grasses and legumes into silvopastoral systems to recover flora and fauna, as well as to provide shade for livestock. Restoration of biological cycles requires the use of organic fertilization mechanisms that increase soil biological activity. Sustainable systems cannot be established as monocultures; diversification and crop rotation are efficient practices in the use and recycling of the applied nutrients. Resilience to climate change is essential for sustainability since ecosystems have changed and the availability of resources such as water is decreasing. Irrigation efficiency is another indicator of resilience. Methane emissions from animals are losses of untapped food energy and are one of the causes of global warming. The incorporation of legumes in diets is a strategy to reduce up to 50% of methane production in relation to grass-only diets.

In terms of safety, dairy products with residues of antibiotics, anti-parasitic, or other supplies cause public health problems that reduce their consumption. Production under agro-ecological systems can increase levels of safe food consumption for the entire population.

Sustainability of the systems is also related to access to productive resources. The Atapo Palmira system shows the limitations of access to land and water. There are 305 ha in the whole system, and each family has <1 ha, 57% corresponding to small dairy farmers. With irrigation water and the incorporation of organic matter (1 m^3 for 1/2 ha), soil fertility has been improved. The money required for investments is a constraint to get cattle from non-conventional farms (as they are scarce and raising them under alternative systems could take two or more years) and to implement agrosilvopastoral systems. In Europe, the states subsidize producers who change their production model.

Concerning profitability, there is a reference price for conventional milk production ($0.42), but not for organic milk, which should be differentiated because it requires certification, inspection, auditing, and direct marketing costs, and because it guarantees environmentally friendly practices and seeks animal welfare. Economic analyses of organic production systems show positive net income results, meaning that organic milk production is a viable economic alternative [46]. Certification can be other limiting factor because of the established deadlines; follow-up and control visits are made every six months, and recertification must be requested after two years. The process involves expenses for certifications and various other procedures. Producers in a community or territory could be certified in order to optimize processes and avoid individual certification costs.

In productivity and profitability terms, the conventional production model is superior to non-conventional production due to the decrease in production costs. Organic systems with a difference from $0.5 to $0.7 higher than conventional systems are reported in the U.S. [47]. Overall, studies in France indicate that the vulnerability of dairy farms decreases when drastic changes occur, since productivity is improved through self-sufficiency, profitability, economic efficiency, and even independence of subsidies from agricultural policy [48].

21.5 Conclusions

Organic or non-conventional milk production is implementable. However, it does not solve all the problems, especially in economic terms, because the market is not large enough and the population is not yet able to pay for products with high quality standards and whose cost includes environmental and social externalities.

There is potential for developing sustainable production models. Alternative treatments for diseases can be applied, fodder produced with agrosilvopastoral systems and cattle adapted to local climatic conditions can be rescued; however, the technical-scientific basis and correct use of alternative practices must be further investigated.

The results obtained provide elements for farmers to decide to change the conventional production model taking advantage of local knowledge and the suitable conditions available for organic milk production. It is important for the state to propose public policies for sustainable production together with programs and projects that help producers make the transition to a sustainable model and obtain organic production certifications.

References

1. UNFPA: World Population Dashboard, https://www.unfpa.org/data/world-population-dashboard
2. The World Bank: Poverty, https://www.worldbank.org/en/topic/poverty/overview
3. M. Altieri, C. Nicholls, Agroecología: única esperanza para la soberanía alimentaria y la resiliencia socioecológica. Agroecología. **7**, 65–83 (2012)
4. ONU: Transformar nuestro mundo: la agenda 2030 para el desarrollo sostenible. in Resolución Aprobada Por La Asamblea General Del 25 De Septiembre Del 2015. p. 40 (2015)
5. G. Gallopín, Sostenibilidad y desarrollo sostenible: un enfoque sistémico, División de desarrollo sostenible y asentamientos humanos (2003)
6. P. Zarta Ávila, La 1sustentabilidad o sostenibilidad: un concepto poderoso para la humanidad. Tabula Rasa. 409–423 (2018). https://doi.org/10.25058/20112742.n28.18
7. R. Laserna, J. Cortés, C. Ledo, A. Ramírez, R. Valdivieso, W. Hurtado, J. Casazola, F. Calderón, Sostenibilidad y desarrollo humano. La calidad de vida en Cochabamba. Los amigos del libro/CERES, Cochabamba (1995)
8. J.M. Naredo, Sobre el origen, el uso y el contenido del término "Sostenible." Doc. Soc. **102**, 48–57 (1996)
9. M. Altieri, C. Nicholls, Agroecología: teoría y práctica para una agricultura sustentable, México (2000)
10. H. Mancini, L. van den Berg, I. Cardoso, A. Vermue, F. Bianchi, M. Peña-Claros, P. Tittonell, Understanding farm diversity to promote agroecological transitions. Sustainability. **10**, 1–20 (2018). https://doi.org/10.3390/su10124337
11. E. Sevilla, Agroecología y desarrollo rural sustentable: una propuesta desde Latino América, in Sarandon (ed.) Agroecología; el camino para la agricultura. Argentina (2000), pp. 1–28
12. A. Smith, S. Snapp, R. Chikowo, P. Thorne, M. Bekunda, J. Glover, Measuring sustainable intensification in smallholder agroecosystems: a review. Glob. Food Sec. **12**, 127–138 (2017). https://doi.org/10.1016/j.gfs.2016.11.002
13. C. Bockstaller, L. Guichard, D. Makowski, A. Aveline, P. Girardin, S. Plantureux, Agri-Environmental Indicators to Assess Cropping and Farming Systems: a review BT–sustainable agriculture. Presented at the (2009)
14. R. Martínez, Agroecología: atributos de sustentabilidad. InterSedes Rev. las Sedes Reg. III, 25–45 (2002)
15. D. Silva, E. Vasques, L. Nogueira, D. De Oliveira, A sustainability rereading of agrarian production systems. Interações (Campo Gd. 18, 43 (2017). https://doi.org/10.20435/inter.v18i4.1527
16. P. Vasquez, M. Vignolles, Establecimiento agroproductivo ecologico vs. agricultura convencional: Partido de Tandil, provincia de Buenos Aires. Soc. Nat. **27**, 267–280 (2015). https://doi.org/10.1590/1982-451320150206
17. IFOAM: Normas para la producción y el procesamiento orgánicos. (2005)
18. Á. Noguera, F. Salmerón, N. Reyes, Bases teórico-metodológicas para el diseño de sistemas agroecológicos. Rev FCA UNCUYO. **51**, 273–293 (2018)
19. P. Tittonell, Las transiciones agroecológicas: múltiples escalas, niveles y desafíos. Rev. FCA UNCUYO. **51**, 231–246 (2019)
20. H. Hamadani, A. Khan, Organic dairy farming—an overview. J. Livest. Sci. **6**, 4–9 (2015)
21. S. Olarte, La producción de leche orgánica en la región Puno: una alternativa de desarrollo sostenible. Mundo Agrar. 13 (2013)
22. S.J. Sarandón, Potentialities, challenges and limitations of agroecological research as a new paradigm in agricultural sciences [Potencialidades, desafíos y limitaciones de la investigación agroecológica como un nuevo paradigma en las ciencias agrarias]. Rev. la Fac. Ciencias Agrar. **51**, 383–394 (2019)
23. P. Castañeda, Sistematización y generación de conocimientos en trabajo social. Aportes metodológicos a la formación profesional. Altern. Cuad. Trab. Soc. **23** (2015). https://doi.org/10.14198/altern2015.22.02

24. B. de Sousa, Descolonizar el saber, reinventar el poder. Ediciones Trilce, Montevideo–Uruguay (2010)
25. L. Bertalanffy, Teoría general de sistemas. (1976)
26. AGROCALIDAD: Instructivo de la normativa general para promover y regular la producción orgánica-ecológica-biológica en el Ecuador (2013)
27. G. Pardo, J. Aibar, P. Ciria, C. Lacasta, J. Lezaún, C. Zaragoza, Organic versus conventional methods of fertilization and weed control in a long term rotation of cereals in semiarid Spain. Chil. J. Agric. Res. **71**, 187–194 (2011). https://doi.org/10.4067/S0718-58392011000200002
28. A. Escalante, G. Pérez, C. Hidalgo, J. López, J. Campo, E. Valtierra, J. Etchevers, Biocarbón (biochar) I: naturaleza, historia, fabricación y uso en el suelo. Terra Latinoam. **34**, 367–382 (2016)
29. A. Concilco, M. Moreno, M. García, H. Quiroga, O. García, Influence of biochar applied to soil on yield and quality attributes of fodder oats (2018)
30. A. Moreno, J. Cantú, J. Reyes, V. Contreras, Forage maize nutritional quality according to organic and inorganic. Sci. Agropecu. **8**, 127–135 (2017). https://doi.org/10.17268/sci.agropecu.2017.02.05
31. J. Iglesias, F. Funes, O. Toral, S. Milera, M. Milera, Diseños agrosilvopastoriles ene l contexto de desarrollo de una ganadería sustentable. Apuntes para el conocimiento. Pastos y Forrajes. 34, 241–258 (2011)
32. P. Del Pozo, V. de Melo, E. Ortega, El análisis emergético como herramienta para evaluar la sustentabilidad en dos sistemas productivos. Rev. Ciencias Técnicas Agropecu. **23**, 59–63 (2014)
33. C. Mbow, M. van Noordwijk, R. Prabhu, T. Simons, Knowledge gaps and research needs concerning agroforestry's contribution to Sustainable Development Goals in Africa. Curr. Opin. Environ. Sustain. **6**, 162–170 (2014). https://doi.org/10.1016/j.cosust.2013.11.030
34. A. Piñeiro, J. Canul, J. Alayón, A. Chay, A. Ayala, C. Aguilar, F. Solorio, J. Ku, Potential of condensed tannins for the reduction of emissions of enteric methane and their effect on ruminant productivity. Arch Med Vet. **47**, 263–272 (2015)
35. C. Lascano, E. Cárdenas, Alternativas para mitigação de emissão de metano em sistemas de criação de animais domésticos. Rev. Bras. Zootec. **39**, 175–182 (2010). https://doi.org/10.1590/S1516-35982010001300020
36. V. Wanniatie, M. Sudarwanto, T. Purnawarman, D. Jayanegara, Kualitas susu dari peternakan organik. Wartazoa. **27**, 125–134 (2017). https://doi.org/10.14334/wartazoa.v27i3.1597
37. J. Carmona, D. Bolívar, L. Giraldo, El gas metano en la producción ganadera y alternativas para medir sus emisiones y aminorar su impacto a nivel ambiental y productivo. Rev. Colomb. Ciencias Pecu. **18**, 49–63 (2005)
38. A. Croissant, S. Washburn, L. Dean, M. Drake, Chemical properties and consumer perception of fluid milk from conventional and pasture-based production systems. J. Dairy Sci. 90, 4992–4953 (2007). https://doi.org/10.3168/jds.2007-0456
39. G. Haas, Evaluacion del balance de nutrientes de las granjas de produccion lacteos organicos en diferentes niveles de intensidad en Alemania. Agric. Renov. y Sist. Aliment. 223–232 (2007)
40. H. Ayrle, M. Mevissen, M. Kaske, H. Nathues, N. Gruetzner, M. Melzig, M. Walkenhorst, Medicinal plants—prophylactic and therapeutic options for gastrointestinal and respiratory diseases in calves and piglets? A systematic review. BMC Vet. Res. **12**, (2016). https://doi.org/10.1186/S12917-016-0714-8
41. N. Rivero, A. Jaramillo, M. Rivas, G. Ballesteros, A. Zaragoza, Actividad antihelmintica de la vaina de Leucaena leucocephala sobre nematodos gastrointestinales de ovinos (in vitro). Abanico Vet. 9, 1–9 (2018). https://doi.org/10.21929/abavet2019.95
42. J. Da Silva, C. Passos, A. Da Fonseca, J. Guimarães, Gastrointestinal helminths in calves and cows in an organic milk production system. Rev. Bras. Parasitol. Veterinária. **21**, 87–91 (2012). https://doi.org/10.1590/s1984-29612012000200003
43. I. Blanco, T. Lundh, K. Holtenius, N. Fall, U. Emanuelson, The status of essential elements and associations with milk yield and the occurrence of mastitis in organic and conventional dairy herds. Livest. Sci. **168**, 120–127 (2014). https://doi.org/10.1016/j.livsci.2014.07.016

44. R. Almeida, C. Coutinho, Uso de medicamentos homeopáticos para o tratamento da mastite bovina : Revisão. pubvet. **12**, 1–10 (2018)
45. A. Villa, S. Bracamonte, Procesos de aprendizaje y modernización productiva en el agro del noroeste de México: los casos de la agricultura comercial de la Costa de Hermosillo, Sonora y la agricultura orgánica de la zona sur de Baja California Sur. Estud. Front. **14**, 217–254 (2013). https://doi.org/10.21670/ref.2013.27.a08
46. A. Alves, A. Quintão, L. Takao, L. Magalhães, Análise de desempenho econômico da produção orgânica de leite: estudo de caso no Distrito Federal. Ciência e Agrotecnologia. **33**, 567–573 (2009). https://doi.org/10.1590/s1413-70542009000200032
47. W. McBride, C. Greene, Costs of organic milk production on U.S. dairy farms. Rev. Agric. Econ. **31**, 793–813 (2009). https://doi.org/10.1111/j.1467-9353.2009.01467.x
48. M. Bottes, Conversion to organic farming decreases the vulnerability of dairy farms. Agron. Sustain. Dev. (2019)

Chapter 22
I + PD Temperature Controller Tuned for Ant Colony Optimization (ACO) on an ARM Platform

William Montalvo, Deisy Guanochanga, and John Chapaca

Abstract In recent years, the development of new industrial processes in engineering areas has allowed the implementation of new control strategies to ensure the effective operation of the control process within the industry. This paper presents the development of an I + PD controller for a temperature process applied to a control plant trainer (CPT); using the bio-heuristic method of ant colony optimization (ACO) to obtain the appropriate parameters for the tuning of the controller in the plant. On the STM32F board with advanced RISC machine (ARM) technology, the controller is implemented, allowing the sending and receiving of data in real time; for the design of the I + PD control, MATLAB/SIMULINK software is used, working together with the Wainjung library, which allows data collection. To verify the efficiency of the I + PD control, the Wilcoxon test is used, which analyzes the Integral of the time-weighted absolute error (ITAE) of the I + PD control versus the ITAE of a conventional PID control.

22.1 Introduction

In industrial process control systems, PID family controllers, such as PI and PD, have been used in industry to meet the desired productivity, adjusting each of its control parameters appropriately [1]. Thus, PID control is the most applied composition in industrial processes, thanks to the simplicity of its design and the strong performance obtained in linear control systems. However, once the control process presents some complication, as in the case of nonlinear systems, parameters that change over time, disturbance models and variables with slow responses as in the case of temperature, level and pressure; the performance of the response of the PID control system becomes unpleasant [2].

Due to the structural restrictions of PID controllers and in order to obtain more efficient control processes, alternative control structures have been developed, for

W. Montalvo (✉) · D. Guanochanga · J. Chapaca
Universidad Politécnica Salesiana, UPS, 170146 Quito, Ecuador
e-mail: wmontalvo@ups.edu.ec

Á. Rocha et al. (eds.), *Communication, Smart Technologies and Innovation for Society*, Smart Innovation, Systems and Technologies 252,
https://doi.org/10.1007/978-981-16-4126-8_22

example, the I + PD controller, which makes it possible to control unstable processes while minimizing the error signal [3].

In the I + PD controller, the integral term acts on the error, while the proportional and derivative terms work on the process variable; therefore, the overpulse and settling time is greatly decreased compared to conventional PID [4]. For verifying the changes obtained in the control process, the I + PD control is implemented in the temperature variable of a control plant trainer (CPT). The STM32F407-Discovery electronic card is used as controller, which effectively simplifies the hardware design and the development environment, since it enables data validation in real time [5]; since it has an ARM architecture, which provides high performance in the processing of analog and digital signals.

For optimizing the I + PD controller on the plant, it is tuned by the bio-heuristic method of ant colony optimization (ACO), based on the simulation of the behavior of an ant in its population, where it transfers or guides other ants by means of the control of the number of information that remains in the path (pheromones) [6].

In this work, an I + PD temperature control optimized by ACO was developed, analyzing it comparatively by means of a performance index with a conventional PID control also optimized by ACO, showing the efficiency obtained in the control process of the temperature variable in a didactic plant. The control of this variable is important in different industrial processes, in the food, pharmaceutical, gas, oil and chemical industries, due to its help in food preservation, reduction of labor accidents, resource savings and improvements in product quality. Verifying the main advantage of the I + PD control, in the attenuation of the overpulse that accompanies a change of reference value [7].

22.2 Methodology

22.2.1 Identification of the System

The black box method allows to identify the mathematical model with which the plant works (CPT), so that data is collected during a sampling period of 0.01 s, during 22 min of plant operation, where 7 min are used for the plant to reach its maximum temperature and 15 min for the plant to cool down to room temperature.

By means of the temperature sensor integrated in the CPT, to digitize the behavior of the high brightness temperature LED, data corresponding to a PWM with a duty cycle of 100% was collected. For data acquisition, the STM32F407G discovery card with serial communication to the computer is used, which through a UART module, it allows the data to be sent to the computer for storage and graphing, which are processed by MATLAB software with the help of the system identification toolbox (IDENT), establishing the mathematical model of the plant. Figure 22.1 shows the response obtained from the system both in simulation and in its real form, obtaining a 78.59% of similarity.

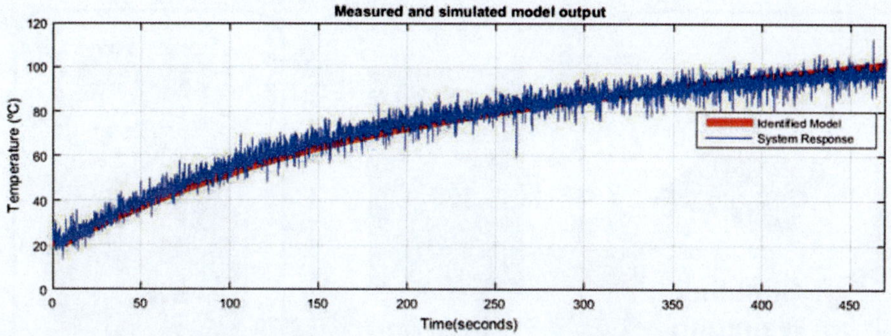

Fig. 22.1 System characteristic curve

The continuous-time transfer function results in Eq. (22.1)

$$G(s) = \frac{0.004861s + 0.000008245}{s^2 + 0.007813s + 0.000003447} \tag{22.1}$$

The transfer function obtained in discrete time results in Eq. (22.2)

$$G\left(z^{-1}\right) = \frac{0.00004898z + 0.004898}{z^2 - 1.999z + 0.9992} \tag{22.2}$$

22.2.2 Commissioning of the I + PD Control in the Control Plant Trainer (CPT)

The structure shown in Fig. 22.2 is for both an I + PD control and a conventional PID, which is implemented with the help of MATLAB software, using Simulink and the Wainjung library, involving serial communication for sending and receiving data, which are obtained in real time.

The control is designed in Simulink, together with the Waijung library, since it offers the construction of block diagrams, to experiment control schemes; in this case, the conventional I + PD and PID control, where the high-brightness LED together with the temperature sensor, works to obtain temperature values and perform a favorable control of the high brightness LED. Figure 22.3 shows the implementation of this control with blocks of the Waijung library in Simulink, through serial communication with the MATLAB software and the STM32F407G card, using a UART module, which allows sending and receiving data that the CPT provides; so that, according to the programming done in blocks, and the data obtained, the software proceeds to perform the I + PD control or PID control of the high-brightness LED.

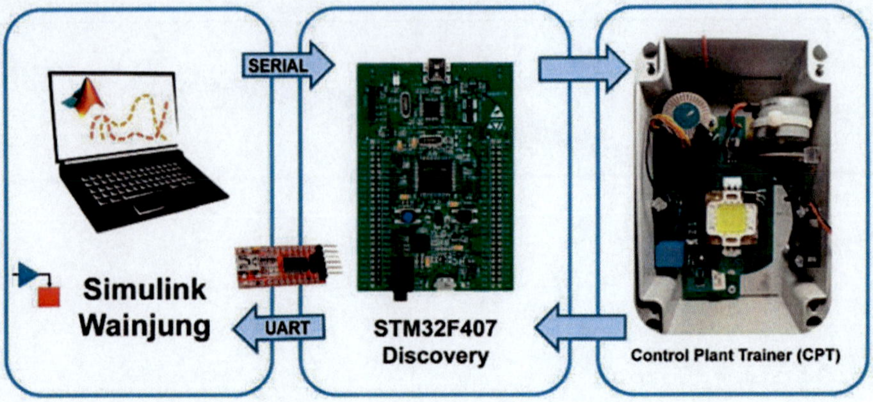

Fig. 22.2 Structure for I + PD and PID control operation

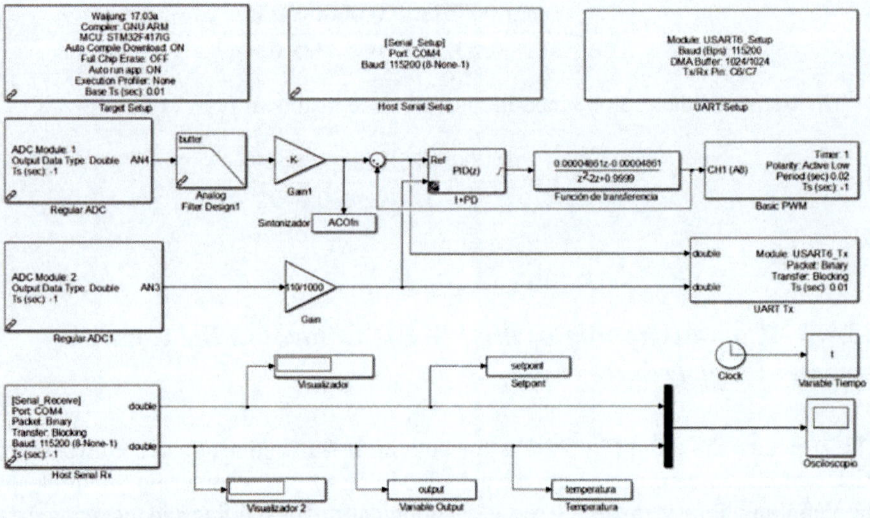

Fig. 22.3 Program block for I + PD control implementation

22.2.3 *I + PD Control Loop*

There are several modifications of the PID control, one of them is called I + PD, which establishes its operation in the integration of the error, and in turn, the sum of the output is subtracted, that is, the proportional and derivative part of the control; Fig. 22.4 shows the block diagram of the control that was implemented.

Where $u(s)$ represents the input; $y(s)$ the control output of the plant; $e(s)$ is the error corresponding to the reference signal $r(s)$.

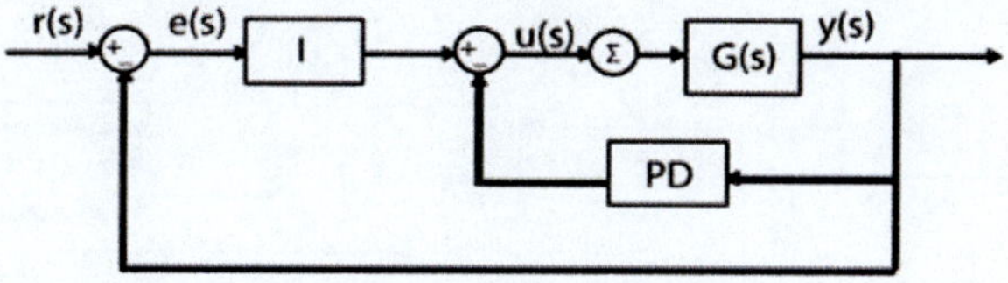

Fig. 22.4 Block diagram for an I + PD control loop

This type of control is used to improve the industrial process, i.e., avoid sudden changes in the variable being controlled, either by external or internal disturbances that cause error signals in the control.

The control law representing the signal [1] of the I + PD controller is expressed by Eq. (22.3).

$$u_{\mathrm{I+PD}}(t) = K_i \int_0^t e(t)\mathrm{d}t - K_{\mathrm{p}} y(t) - K_{\mathrm{d}} \frac{\mathrm{d}y(t)}{\mathrm{d}t} \tag{22.3}$$

where

K_i Proportional gain
$e(t)$ Error signal
K_{p} Proportional gain
$y(t)$ Control output signal
K_{d} Derivative gain

The transfer function of a conventional PID control system is expressed by Eq. (22.4)

$$G(s) = \frac{Y(s)}{R(s)} = K_{\mathrm{p}} + \frac{K_i}{s} + K_{\mathrm{d}} s \tag{22.4}$$

In which, Y(s) is the output, and R(s) is the input of the control system.

The transfer function that is applied for the I + PD control adds parameters that serve for the stabilization of the control, which are e and L, obtaining the expression of Eq. (22.5)

$$G(s) = \frac{Y(s)}{R(s)} = \frac{k}{s} e^{-Ls} \tag{22.5}$$

Thus, the I + PD block in Simulink provides values of the parameters $I = 6.8728$, $P = 8.6862$, $D = 0.1229$ and other data that can be observed in the MATLAB command window, and these values are useful for the control of the ant colony tuned plant (ACO). Figure 22.5 shows the block diagram in Simulink for an I + PD control.

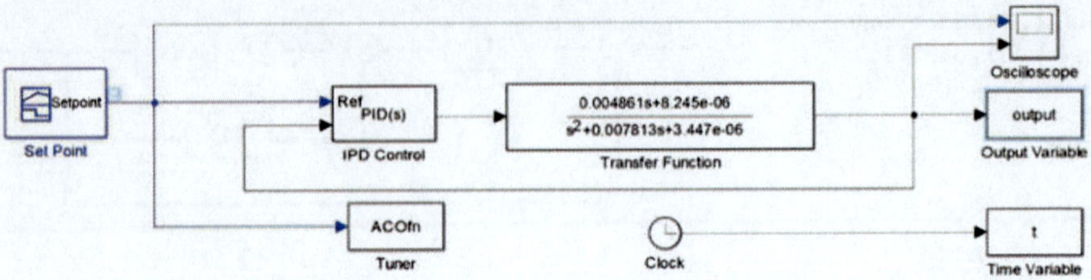

Fig. 22.5 I + PD control block diagram

22.2.4 *Tuning by Bio-heuristic Algorithm by Ant Colony (ACO)*

The source of inspiration for the ant colony algorithm is the real behavior of ants, looks for the path that contains the highest concentration of pheromones. The higher the concentration of pheromones along the route, the more likely it is that the path will be chosen, since they evaluate its quantity and quality, so this method leads to optimal routes for efficient control within the process.

In [8], an ACO analysis is performed on the behavior of ants when searching for the best route to travel from one point to another, and this is expressed by Eq. (22.6).

$$p_{i,j}^{k} = \left\{ \frac{[\tau_{i,j}(t)]^{\alpha}[\eta_{i,j}]^{\beta}}{\sum t \in \text{allowed}[\tau_{i,t}(t)]^{\alpha}[\eta_{i,t}]^{\beta}} \right. \tag{22.6}$$

where

$p_{i,j}^{k}$	is the probability that ant K chooses to move from point i to point j.
$\tau_{i,j}$	is the pheromone
$\eta_{i,j} = \frac{1}{d_{i,j}}$	is the inverse of the distance
β	is a parameter that determines the relative importance of the pheromone versus distance ($\beta > 0$)

The result of Eq. 22.6 leads to the selection of shorter edges with more pheromones.

The experimental indicators obtained for the efficient development of the I + PD and conventional PID control process within the EPC are shown in Table 22.1.

This algorithm associates heuristic information which measures the preference of the shortest path that the ant has to reach its nest; it also gives us the learned desirability at the time of depositing the pheromone on the road. While the ant has the property of finding valid solutions with the lowest computational cost, it also has a memory to store, evaluate and reconstruct the path; allows updating of associated pheromone traces. The application of ant colony algorithms to combinatorial optimization problems and the development of better variants allows to improve the speed of the processes and solutions obtained.

Table 22.1 ACO bio-heuristic method indicators

Ant colony optimization indicators	
Iterations	2
Number of ants	25
Number of parameters	3
α	0.8
β	0.2
Pheromone evaporation rate	0.7
Number of parameters	10,000

Table 22.2 P, I, D parameters for controllers

I + PD parameters	PID parameters
P = 6.8728 I = 8.6862 D = 0.1229	P = 2.7465 I = 7.8260 D = 0.0709

22.3 Results

22.3.1 *Development of I + PD Control Tuned by ACO Method and Analysis of the Data Obtained*

The block diagram that represents the I + PD control is shown in Fig. 22.5, and the control simulation is performed; in the PID block, modifications are made to the parameters b and c, where $b = 1$ and $c = 0$. This modification is used to represent an I + PD control.

Table 22.2 shows the values of the PID parameters for both I + PD and PID control, which are useful for tuning by the ant colony bio-heuristic (ACO) method.

Figure 22.6 shows the control process in the real part of the plant, where it is shown that the rise time slows down due to the temperature being a slow variable, and the tuning process takes time to find the shortest route to reach the set point of 50° C being this time $t_s = 175$ s, and its settling time is immediate. Therefore it is corroborated the elimination of the overpulse in both the simulated and real part of the control process.

22.3.2 *Development of the Conventional PID Control Tuned by the ACO Method and Analysis of the Obtained Data*

Figure 22.7 shows the behavior of the conventional PID temperature control in real time, where it is observed that the control signal has a rise time $t_s = 300$ s, until reaching the established set point of 50 °C; in addition, it has an overpulse of 5 °C despite the fact that the same controller is used for the same variable; this is due to the implementation of the ant colony optimization (ACO), as can be seen in Figs. 22.7 and 9, where the system response is almost ideal.

In addition, in [3], a time-delayed I + PD control is performed, where time-weighted performance criteria such as ISTE and $[IST]_2$ E are used, so that the over-pulse peak generated when the variable control process is initiated which is reduced; however, by taking the ant colony bio-heuristic (ACO) method as the tuning method, the overpulse peak can be reduced by a value very close to 0%, thus giving a more efficient transient state response.

22.5 Conclusions

Figures 22.6 and 22.7 show the behavior of the I + PD control as PID, where it can be seen that the results obtained in rise time (t_s), settling time (t_e) and overpulse have a large margin of difference, thus giving a significant reduction in settling time (t_e) of approximately 60%; in addition, a 100% reduction in overpulse is denoted.

The use of ARM technology such as the STM32F407 board allows a fast control of variables such as temperature, due to its easy acquisition of data in real time, which can be linked to MATLAB software and its SIMULINK utility. The bio-heuristic tuning by ACO allows to improve the control strategies based on its iterations, number of population, evaporation rate of the pheromone; however, these variables must be parameterized carefully because its computational cost is high becoming a problem in real control processes.

The Wilcoxon test is an ideal method to evaluate real plants, because it requires more than 25 experimental tests, to perform a statistical analysis of the performance index values of both the I + PD and PID control, demonstrating the acceptance of the alternative hypothesis (H_a), given that, $Z > Z_a$, showing a confidence level of 95% for the I + PD control; it is also emphasized that in the I + PD control, there is less energy consumption.

References

1. I.J. Conference, W. Chatrattanawuth, N. Suksariwattanagul, T. Benjanarasuth, J. Ngamwiwit, Fi Fe (T) Dt, no. 3, pp. 5649–5652 (2006)
2. D. Sandhya, B.A. Reddy, K.A.G. Rao, in *Fuzzy I-PD and fuzzy PID control of non-linear systems*, in 2009 International Conference on Control Automation Communication and Energy Conservation. INCACEC 2009, 1–6 June 2009
3. I. Kaya, Optimum analytical tuning of I-PD controllers for unstable time delay systems, 2018 6th Int. Conf. Control Eng. Inf. Technol. CEIT **2018**(October), 25–27 (2018). https://doi.org/10.1109/CEIT.2018.8751752
4. S. Vinodhini, S.J. Suji Prasad, B. Meenakshipriya, B. Venkatesan, Design of fuzzy I-PD controller for level control of spherical tank system, in *Proceedings of 2017 Internationa*

Conference on Innovation and Information, Embedded Communication System. ICIIECS 2017, vol. 2018, 1–5 Jan 2018. doi: https://doi.org/10.1109/ICIIECS.2017.8276046.
5. T. Li, F. Luan, M. Wang, Q. Song, Z. Shi, Design of Remote Monitoring System Based on STM32F407 Microcontroller, *2019 IEEE International Conference Power, Intelligence and Computing System. ICPICS 2019*, pp. 304–307 (2019). doi: https://doi.org/10.1109/ICPICS47731.2019.8942548
6. R. Jangra, R. Kait, Analysis and comparison among ant system; ant colony system and max-min ant system with different parameters setting, in *3rd IEEE International Conference*, 17–20 (2017). doi: https://doi.org/10.1109/CIACT.2017.7977376
7. M. Generalized, P. Control, A design method of multirate. 17–22 Dec 2004
8. K. Loubna, B. Bachir, Z. Izeddine, in Optimal digital IIR filter design using ant colony optimization. *Proceedings of 2018 International Conference Optimal Application. ICOA 2018*, pp. 1–5 (2018). doi: https://doi.org/10.1109/ICOA.2018.8370500.
9. H.K. Wu, J.G. Hsieh, Y.L. Lin, Study on kernel-based wilcoxon classifiers, *Proceedings of 2010 IEEE International Conference of Intelligence System Knowledge Engineering. ISKE 2010*, pp. 249–253 (2010). doi: https://doi.org/10.1109/ISKE.2010.5680870
10. S.J.S. Prasad, P.A. Balakrishnan, PSO based I-PD controller for barrel temperature control in plastic injection molding system. Eur. J. Sci. Res. **80**(3), 351–357 (2012)
11. W. Montalvo, M. Ortega, E. Avilés, Performance comparison of two electronic controllers on an ARM platform. Adv. Intell. Syst. Comput. **1277**, 185–197 (2021). https://doi.org/10.1007/978-3-030-60467-7_16

There are several tools available for text mining analysis designed to find patterns and trends from text data. Voyant, WordCounter, TagCrowd, and Sober are some of them.

Voyant tools (http://voyant-tools.org/) were launched in 2003 and is a free, open-source text analysis software package. This web-based set of tools has an intuitive and friendly interface that is easy to manage even by beginners. Its use and does not require creating an account. Voyant contains 28 different tools to visualize and analyze texts [5].

WordCounter (https://wordcounter.net/) is an online word count website that identifies the relevant words (excluding stop words like articles, pronouns, and prepositions) that repeat the most in the text. WordCounter is a simple user interface that features a prominent word and character count, a large text area with various controls, and a sidebar that allows detailed text analysis.

TagCrowd (https://tagcrowd.com) is a web application for visualizing word frequencies in any text by creating a representation known as word cloud, text cloud, or tag cloud. TagCrowd specializes in making word clouds easy to read, analyze and compare. This tool has options such as language to use, limit number of words, exclude words based on frequency, or exclude unwanted words.

Sobek (http://sobek.ufrgs.br) is a text mining tool that has been developed to support educational applications. It is a tool easy to use that allows visualization of a concise diagram containing the main terms and relationships of a text.

The Commission Brundtland [6] defined sustainable development as a paradigm that proposes to satisfy the actual needs without compromising the possibility of future generations to accomplish their own needs. This definition has been widely discussed and accepted by several authors and organizations that include in their research matters as sustainability [e.g., 7–11].

To identify development as sustainable, it has to evidence a balanced integration among environmental, socio-cultural, and economic dimensions [7, 10–13], which have been identified as essential components of sustainability, and their requirements are cited by Harris [14] as follows:

Economic sustainability: An efficient system should produce goods and services, maintaining a good level of governance and a limited external debt, to avoid extreme economic inequalities.

Environmental sustainability: A system based on environmental sustainability should avoid overexploitation of renewable resources systems, and consuming responsibly nonrenewable resources at the time that it is investing to find suitable substitutes. Environmental sustainability includes the care of all ecosystem services associated with biodiversity.

Social sustainability: This sustainability should achieve equality in the provision of social services as health and education, generating gender equity, and adequate opportunities and participation.

Smart tourism has been defined as a tourism platform integrated with information and communication technologies (ICT). This system integrates tourism sources and information technologies such as the internet of things, artificial intelligence, and cloud computing to provide explicit information and efficient services to tourists in certain territories based on the development of innovative mobile technology [15–17].

Tourism destinations are constituted as complex systems [18] that are composed of ensembles including touristic products and services [19], and their success is based on the combination of factors as innovation, human resources, and collaboration structures [20, 21]. In that context, smart tourism allows integration and managing of information in a centralized way, providing fast access to tourist that requires information and services. Additionally, the data collected by the elements considered in s smart tourism system are an effective source to analyze the tourist demands.

Both terms have been associated, and there are several approaches to investigate "sustainable development" and "smart tourism" as topics included in research that includes methods, identification of indicators, and the interactions with society, environment, and economy.

This study aims to contribute to the identification of trends in academic papers related to the terms: "sustainable development" and "smart tourism" using text mining tools.

23.2 Methodology

Our study aims to analyze trends in a corpus (total structure) compound by ten academic papers related to the terms "sustainable development" and "smart tourism using free online text mining tools.

To obtain the collection of papers used in this research, we insert the terms: "sustainable development" and "smart tourism" in Google Scholar web tool, allowing that the terms appear in any (all) fields of the paper. We select papers published in indexed journals in a period from 2014 to 2020 aiming to get recent publications and English as a language to broaden the scope of the research.

To clean the data included in the total content compiled, we manually removed "stop words," which include adverbs, conjunctions, and repetitive words that have no direct relationship with the content of the publications. Also were removed headings, footers, informative data of authors and their academic linkages, acknowledgments, and bibliographical references. The texts were saved in ".txt" format, widely accepted by text mining tools.

Then, the contribution of each paper to the total structure (corpus) of content was calculated, and it was determined the incidence percentage of the selected terms: "smart tourism" and "sustainable development," as well as the five terms with higher frequency. This analysis was performed using WordCounter tool, available in https://wordcounter.net.

The tendency of prevailing terms on the corpus and the incidence variation were identified with the use of Voyant, an online free tool available in https://voyant-tools.org which also generated a cloud of words based on the terms with the highest incidence on the corpus.

The terms of greater incidence in each paper were determined with the online free tool WordCounter, available in https://wordcounter.net, and the most incident terms were extracted using TagCrowd, an online free tool available in http://tagcrowd.com.

Finally, the most relevant terms of the corpus were graphically represented using Sobek, an educational free tool, available in http://sobek.ufrgs.br.

23.3 Results and Discussion

The first 10 papers provided by the search string were selected for the analysis and their information is summarized in Table 23.1 which include an assigned identification number (ID) attributed to each paper, title, year of publication, country of origin, journal or event where it was published and the number of words.

Of the 10 papers selected (Table 23.1), 3 are from Australia, and 2 are from the United Kingdom, representing 50%. It becomes an indicator that the terms "sustainable development" and "smart tourism" used in this research are studied and discussed in those countries. There is only one journal that presents two publications, showing that the topic has been approached from different areas and disciplines in several journals.

Relating to the contribution of each paper to the total structure of content analyzed (Fig. 23.1), the highest percentage was reported by the paper ID 2 (14%), followed by the papers ID 2 and ID 3 (13% each) and papers ID 3, ID 10, and ID 1 (11% each). The rest was contributed by the sum of the remaining papers (ID8, ID9, ID5, and ID4). This factor difficult to generalize our results due to the predominance of some papers regarding the others.

Table 23.2 shows the incidence percentage of the selected terms: "smart tourism" and "sustainable development" determined with WordCounter tool. The highest incidence of "smart tourism" was found in paper ID 6 (22%), while the highest incidence of the term "sustainable development" was identified in paper ID 1 (3%). The incidence percentage of the term "smart tourism" varied from 1 to 22% while in the term "sustainable development," we found values varying from less than 1 to 3%. In all selected papers, the term "smart tourism" has higher values of incidence percentage than "sustainable development."

The five terms with higher frequency (Table 23.3) including the number of times and the percentage in which they appeared were identified using WordCounter tool. When the analysis was performed for a single word, the term "tourism" (9%) followed by "smart" (8%) appeared as the most incidents while when the analysis was performed for a couple of words, the terms "smart tourism" (17%) and "smart city" (4%) were the words with highest percentages of incidence.

Our findings evidence that that studies about smart tourism are recurrent and presented the highest percentages in occurrence either considering one of these words or their combinations. Therefore, the research grouped in this study shows a trend in the use of "smart tourism" as a substantial part of its content. This trend is consistent with the findings presented by [22] and [23].

Table 23.1 Papers selected for the text mining analysis

ID	Paper title	Year	Country	Journal/Event	Number of words (%)
ID 1	Territorial capital, smart tourism specialization, and sustainable regional development: experiences from Europe	2017	New Zealand	Habitat international	6.177 (11.65%)
ID 2	Developing a model for sustainable smart tourism destinations: a systematic review	2019	Iran	Tourism management perspectives	7.646 (14.42%)
ID 3	Conceptualizing smart tourism destination dimensions	2015	Australia	Institute of information management	5.940 (11.21%)
ID 4	Conceptualizing smart tourism destination dimensions	2015	United Kingdom	Springer international publishing Switzerland	4.021 (7.59%)
ID 5	Conceptual foundations for understanding smart tourism ecosystems	2015	Australia	Computers in human behavior	3.623 (6.83%)
ID 6	The concept of smart tourism in the context of tourism information services	2016	China	Tourism management	7.160 (13.51%)
ID 7	Smart tourism destinations: ecosystems for tourism destination competitiveness	2018	United Kingdom	International journal of tourism cities	7.206 (13.59%)
ID 8	Smart tourism challenges	2015	Korea	Journal of tourism	2.145 (4.05%)
ID 9	Smart tourism as a pillar for sustainable urban development: an alternate smart city strategy from Mauritius	2019	Australia	Smart cities	3.146 (5.93%)

(continued)

Table 23.1 (continued)

ID	Paper title	Year	Country	Journal/Event	Number of words (%)
ID 10	How smart is your tourist attraction? Measuring tourist preferences of smart tourism attractions via a FCEM-AHP and IPA approach	2016	China	Tourism management	5.947 (11.22%)

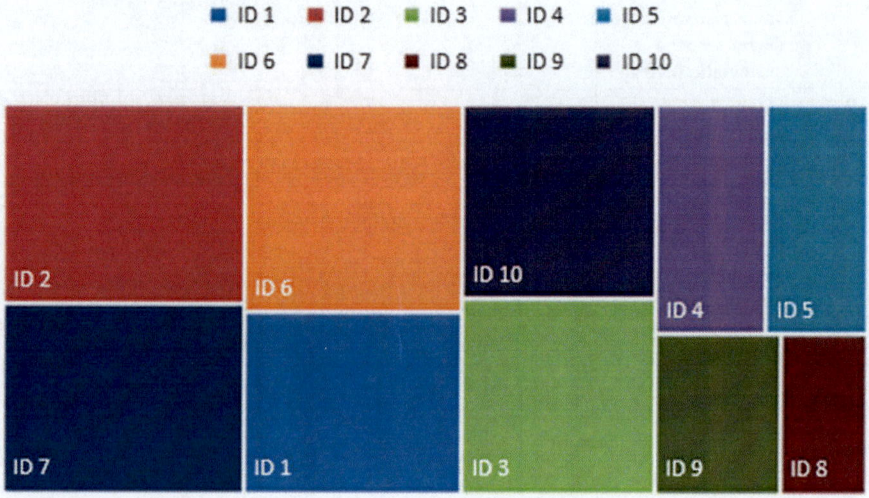

Fig. 23.1 Structure of content analyzed

Table 23.2 Incidence percentage of selected terms

ID	Selected terms	
	"Smart tourism" (%)	"Sustainable development" (%)
ID 1	1	3
ID 2	12	less than 1
ID 3	16	less than 1
ID 4	4	less than 1
ID 5	3	less than 1
ID 6	22	less than 1
ID 7	6	less than 1
ID 8	15	less than 1
ID 9	5	1
ID 10	5	less than 1

Table 23.3 Terms with higher frequency

Terms with higher frequency				
ID	Single word		Couple of words	
ID 1	Tourism	1357 (9%)	Smart tourism	557 (17%)
ID 2	Smart	1190 (8%)	Smart city	123 (4%)
ID 3	Information	301 (2%)	Tourism destination	75 (2%)
ID 4	Development	280 (2%)	Smart cities	74 (2%)
ID 5	Destinations	205 (1%)	Tourism industry	36 (1%)

On the other hand, the term "sustainable development" was not present on the highest positions of occurrence, only appearing "development" on the fourth position (2%) in the analysis considering one word, which evidences a weak linking between the term "sustainable" and the terms included by the scientific community responsible for the research part of our analysis. This could be explained by the fact that "sustainable development" is a term with a wide definition and scope, which on some occasions could lead to the researchers to be more specific in the use of terms related to their studies and even by the difficulties around using sustainable development goal for improving territories [24]. However, it is also possible to find studies where the importance of considering "sustainable development" in future research has been highlighted [25].

The analysis generated through Voyant tools shows the tendency of prevailing terms on the corpus under study (Fig. 23.2), divided into 10 ten segments of equal size. This evidences the variation in the incidence of the terms: "smart," "tourism," information," "development," and "destinations", which is consistent with the results obtained with WordCounter tool (Table 23.3).

The five terms identified keep on a prevalence throughout all segments, while the highest prevalence is shown by the terms "smart" and "tourism." This is consistent with the incidence variation Fig. 23.3 generated by TermRadio option in Voyant tools

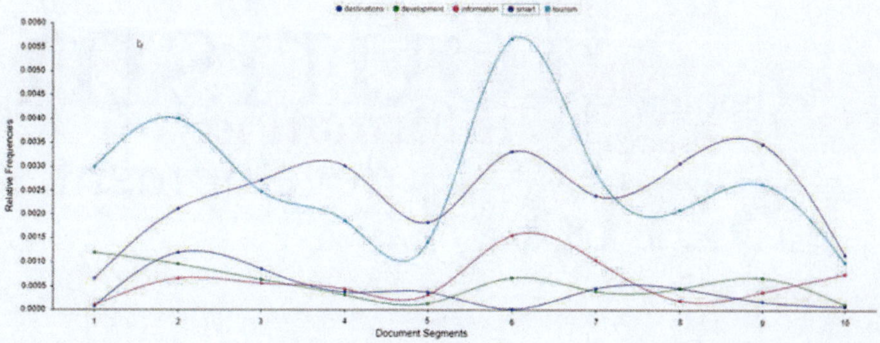

Fig. 23.2 Tendency of prevailing terms on the corpus

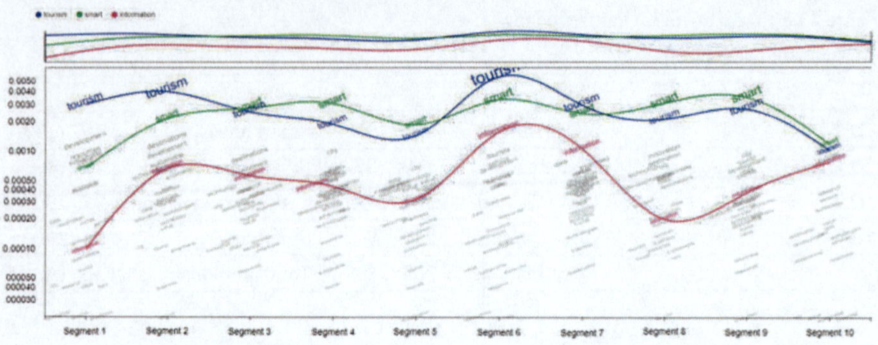

Fig. 23.3 Incidence variation on corpus

which allows to select any word generated on the chart and highlight their variance by color. Figure 23.2 also shows the variation of the three terms with the highest incidence on the corpus.

A cloud of words based on the terms with the highest incidence on the corpus (Fig. 23.4) was structured allowing the visualization and identification of the focus and subjects included on the corpus as an integral view. This analysis also was displayed using Voyant tools and reinforced our previous findings where the terms "smart," "tourism," "development," "information," and "development" occupied a preponderant position in the corpus panorama.

The terms "smart" and "tourism" are frequent and recurrent during all the corpus demonstrating that the information generated by researchers continuously links both terms.

In the analysis performed with TagCrowd Tool, we identified the terms of greater incidence in each paper (Table 23.4) and classified them in three positions according to the incidence order. The predominant terms present in most of the papers were

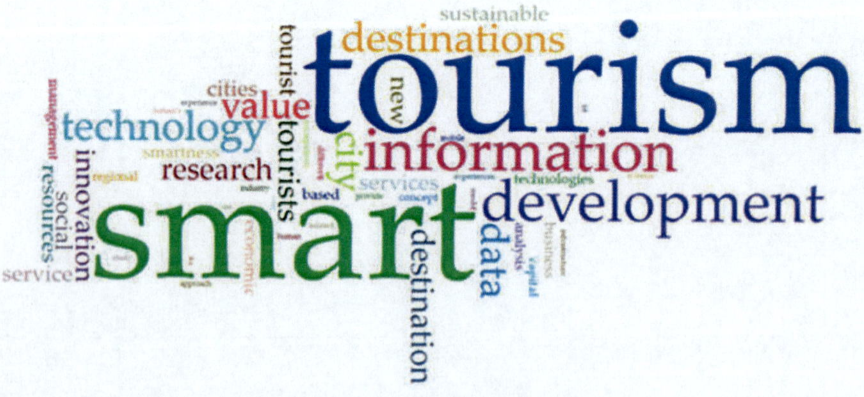

Fig. 23.4 Cloud of words with the more incident terms on the corpus

Table 23.4 Terms of greater incidence analyzed per paper

ID	Terms of greatest incidence					
	1st place term		2nd place term		3rd place term	
ID 1	Tourism	182**	Regional	146	Development	87
ID 2	Tourism	231**	Smart	181*	Destinations	127
ID 3	Tourism	129**	Smart	148*	Technologies	47
ID 4	Smart	135*	City	109	Destinations	51
ID 5	Smart	59*	Tourism	49**	Ecosystem	46
ID 6	Tourism	390**	Smart	231*	Information	127
ID 7	Smart	229*	Tourism	110**	City	98
ID 8	Smart	105*	Tourism	82**	Technology	35
ID 9	Smart	90*	City	72	Tourism	55**
ID 10	Tourist	136	Smart	122*	Evaluation	87

"tourism" and "smart" and were highlighted with 1 or 2 asterisks, respectively. These results demonstrate consistency between the analysis performed with the entire corpus and the analysis performed per paper.

Finally, the most relevant terms of the corpus were graphically represented (Fig. 23.5) using Sobek tool, and this representation shows separately the terms

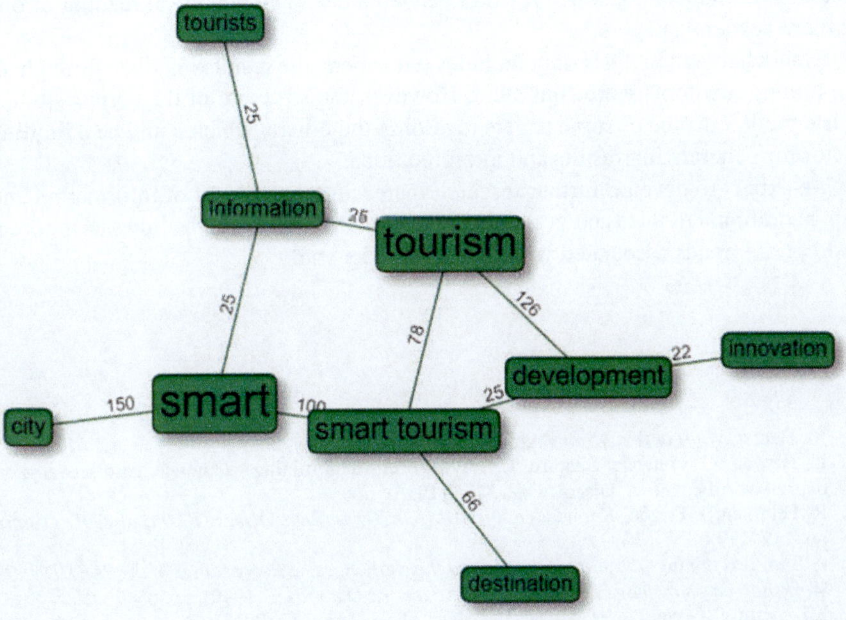

Fig. 23.5 Graphical representation of the most relevant terms on the corpus

"smart" and "tourism" as the words with the highest incidence on the corpus, but also appeared in the graph the term "smart tourism" as a relevant in the analysis, which evidenced the strong linking between both terms. The connections between the terms with the greatest incidence showed an association of the term "smart "with "city," information," and "tourism"; while the term "tourism" was directly connected with "innovation," "development," and "smart." Therefore, the core of the term network was "smart tourism."

23.4 Conclusions

Text mining techniques have a great potential to contribute to academic research through a wide range of analyses that generate many possibilities to explore textual corpus. This research evidence a constant trend in the use of the term "smart tourism" associated with "development" and "innovation" which clearly indicates that researchers are exploring the interactions between them.

This study included the application of diverse free online text mining tools and even when not have contemplated all the options available, our work aimed to support and promote the use of open access tools in research.

In academic work and research, time is a highly valued resource and the mining text tools used in this study proved to be efficient, with an intuitive user interface and demonstrated good levels of velocity, becoming an excellent option to analyze textual data in a context when it is not needed a deep and analytical reading of one or more academic papers.

Even when our analysis only includes ten papers, the results obtained through all the mining text tools were consistent. However, the structure of the corpus studied exists predominance of some papers regarding the others, which could be a limiting factor for generalizing results and interpretations.

We expect to develop further research increasing the volume of information and using additional tools to achieve higher precision in results, which allows us to reveal appropriate trends associated with the topic under study.

References

1. M. Hearst, *What is Text Mining*, vol. 5 (SIMS, UC Berkeley, 2003)
2. U. Fayyad, G. Piatetsky-Shapiro, P. Smyth, From data mining to knowledge discovery: an overview. Adv. Knowl. Discov. Data Min. (1996)
3. R. Feldman, I. Dagan, *Knowledge Discovery in Textual Databases (KDT)*, in *KDD* (1995), pp. 112–117
4. A. Tan, *Text mining: the state of the art and the challenges*, in *Proceedings of the PAKDD 1999 Workshop on Knowledge Discovery from Advanced Databases* (1999), pp. 65–70
5. L.J. Sampsel, Voyant tools. Music. Ref. Serv. Q. **21**, 153–157 (2018)
6. G. Bruntland, *World Commission on Environment and development, Our Common Future.* Presented at the 1 Janury 1987

7. D. Chhabra, Sustainable marketing of cultural and heritage tourism. Routledge (2010)
8. E.D. Larson, M.H. Ross, R.H. Williams, Beyond the era of materials. Sci. Am. **254**, 34–41 (1986)
9. L. Quevedo, M. Cejas, P. Vinueza Naranjo, W. Samaniego, Promoting sustainable development through touristic MTB (Mountain Bike) routes. The experience of an Andean region of Ecuador. Test Eng. Manage. **83**, 25968–25977 (2020)
10. D. Sitarz (ed), *Agenda 21: The Earth Summit Strategy to Save Our Planet* (1993)
11. UNWTO (ed) *UNWTO Annual Report 2013. World Tourism Organization (UNWTO)* (2014). https://doi.org/10.18111/9789284416110
12. P. Knox, H. Mayer, *Small Town Sustainability: Economic, Social, and Environmental Innovation.* Walter de Gruyter (2013)
13. H. Richins, J.S. Hull, *Mountain Tourism: Experiences, Communities, Environments and Sustainable Futures* (CABI, 2016)
14. J.M. Harris, Sustainability and sustainable development. Int. Soc. Ecol. Econ. **1**, 1–12 (2003)
15. Z. Ling-Yun, L. Nao, L. Min, On the basic concept of smarter tourism and its theoretical system. Tourism Tribune/Lvyou Xuekan **27** (2012)
16. D. Wang, X.R. Li, Y. Li, China's "smart tourism destination" initiative: a taste of the service-dominant logic. J. Destin. Mark. Manage. **2**, 59–61 (2013)
17. W. Zhu, L. Zhang, N. Li, *Challenges, Function Changing of Government and Enterprises in Chinese Smart Tourism*, vol. 4, (2014)
18. A. Fyall, Destination management: challenges and opportunities, in *Destination Marketing and Management: Theories and Applications* (CABI) (2011), pp. 340–357
19. D. Buhalis, Marketing the competitive destination of the future. Tour. Manage. **21**, 97–116 (2000)
20. L. Prats, J. Guia, F.-X. Molina, How tourism destinations evolve: the notion of tourism local innovation system. Tour. Hosp. Res. **8**, 178–191 (2008)
21. J.R. Ritchie, G.I. Crouch, *The Competitive Destination: A Sustainable Tourism Perspective* (Oxfordshire Cambridge MA, 2005)
22. F. Femenia-Serra, B. Neuhofer, Smart tourism experiences: conceptualisation, key dimensions and research agenda. J. Reg. Res. **42**, 22 (2018)
23. A. Kontogianni, E. Alepis, Smart tourism: state of the art and literature review for the last six years. Array **6**, 100020 (2020)
24. J.M. Klopp, D.L. Petretta, The urban sustainable development goal: indicators, complexity and the politics of measuring cities. Cities **63**, 92–97 (2017)
25. A. Kemmler, D. Spreng, Energy indicators for tracking sustainability in developing countries. Energy Policy **35**, 2466–2480 (2007)

Chapter 24
Design and Evaluation of a Distributed Architecture for GeoSparkSim

Verónica Alba Valencia and Julio Proaño Orellana

Abstract The human being throughout history has developed ways to optimize his time and resources. Nowadays, mobility is one of the biggest problems. Traffic simulation tools are frequently used to study mobility problems. GeoSparkSim is a vehicular traffic simulation tool that extends Apache Spark to generate datasets of road network traffic significantly with the simulation of microscopic traffic that includes traffic lights, change of track, and the pursuit of vehicles. This work presents an architecture to implement and evaluate the GeoSparkSim simulation tool. The architecture was developed on different containers in Docker, and two stages of the study were created for its evaluation. Each stage contains 26 simulations in different geographic zones from the Quito city. Results show the potentials of GeoSparkSim in order to study traffic mobility and compare its implementation on different Cloud services such as Google, Amazon, and Azure. In order to have an acceptable simulation, we found a machine that has at least 146 GB in storage and 8 GB in RAM is recommended.

24.1 Introduction

Simulation models are effective tools for analyzing a wide variety of dynamic problems. These models integrate the behavior of different entities and their interactions; currently, there are several models of vehicle traffic simulation, and one of the ways to classify them is according to the abstraction level with which they represent the system studied, among which are the following: The microscopic model describes with a low level of abstraction the entities and their interactions, and it is a detailed representation and close to reality; the mesoscopic model represents many of its

V. Alba Valencia · J. Proaño Orellana (✉)
Universidad Politécnica Salesiana, Quito, Ecuador
e-mail: jproanoo@ups.edu.ec

V. Alba Valencia
e-mail: malba@est.ups.edu.ec

Á. Rocha et al. (eds.), *Communication, Smart Technologies and Innovation for Society*, Smart Innovation, Systems and Technologies 252,
https://doi.org/10.1007/978-981-16-4126-8_24

entities with a higher level of abstraction, but can describe its activities and interactions at a similar level of abstraction to the microscopic, and the macroscopic model describes the entities and their activities and interactions at a maximum level of abstraction, presents a study not close to reality, in which global variables are of interest [1].

Within simulation applications for microscopic vehicle traffic scenarios, there are several simulators like VISSIM [2], SUMO [3], OLSIMv4 [4], GeoSparksim [5], and among others. Most existing traffic simulators are designed on ineffective or inconvenient distributed programming models. Many of them still use message transfer services and do not use advanced calculation models or job planners [6].

A new proposal to simulate traffic with a distributed system and under a scalable microscopic model in Apache Spark is GeoSparkSim, which works in conjunction with Geospark Spatial RDD. Spatial RDDs are collections of objects partitioned into a group of machines. Each RDD is built using parallelized transformations (filter, join or groupBy) that can be tracked to retrieve RDD and Spark data to provide a holistic approach that allows data scientists to simulate, analyze, and visualize large-scale urban traffic data [6]. The authors' assessments of the capabilities of the GeoSparkSim tool in the experimental analysis show that GeoSparkSim can simulate the movements of 200,000 vehicles on a very large road network (250,000 road crossings and 300,000 road sections) [6].

The GeoSparkSim tool is an open-source software. It is built in a distributed environment that optimizes data processing and its code can be adapted to different vehicle traffic scenarios [5]. However, its implementation is not trivial because there are many elements that should be considered such as database, analysis of results tool, etc. On the other hand, GeoSparkSim generates a large amount of information as output. So, the study carried out in this article is an architecture based on Docker containers that includes the GeoSparkSim simulator, a database, an environment for the analysis of the output data. All of them are connected to each other. The prosed architecture is to facilitate the analysis of the data obtained and to provide a guideline for the hardware and software capabilities needed by the GeoSparkSim tool. Additionally, a brief Cloud implementation evaluation is also described. This article is distributed as follows: In the current section, there is a tour of the relevance of the simulators, their classification according to their degree of abstraction, and the design proposal and evaluation of the GeoSparkSim application. Section 24.2 describes the hardware and software features that were used to test containers with GeoSparkSim, MongoDB, and Jupyter-Notebook. Section 24.3 presents the results of the architecture design in Docker for GeoSparkSim according to the two stages analyzed. Section 24.4 discusses the results obtained, and finally, Sect. 24.5 concludes the article.

24.2 Methods

The city of Quito is located in the middle of two green strips, which has allowed its longitudinal expansion [7], and the structure of the selected scenarios is due to the variability of constants in the parameters chosen for the analysis as shown in Table 24.3 as well as the distribution of the geographical areas shown in Table 24.5. For the study of the GeoSparkSim tool, two stages were created, and each contains 26 simulations divided into four scenarios that are show in Table 24.1; each stage has a virtual machine with different features in the hardware, and the architecture proposed in the software is maintained.

To perform the different simulations, the hardware capacity of each virtual machine was considered to define the input data of each simulation. The characteristics of each virtual machine can be seen in Table 24.2.

24.2.1 Description of the Proposed Architecture

In each virtual machine is installed: (1) Docker Engine[8] to manage the containers, GeoSparkSim, Jupyter-Notebook, and MongoDB container, (2) GUI for MongoDB: Compass [9] and Studio 3 T [10], (3) Ganglia [11] to monitor the performance of the virtual machine, (4) Bashtop [12] monitors and displays the behavior of the processor, memory, disks, network, and internal processes in a short interval of time, (5) Folders are created to share documents with each of the containers.

The proposed architecture communicates to the containers through a shared Docker network, which facilitates communication between applications, and data can be observed through shared folders between the containers and the virtual machine, and the generated data is stored in the database for further analysis; all these interactions are monitored with Ganglia and Bashtop, as can be seen in Fig. 24.1.

Table 24.1 Distribution of the study stages of the GeoSparkSim tool

Stages I—Virtual machine 1	Stages II—Virtual machine 2
Scenario 1–5 simulations—parameters A1	Scenario 1–5 simulations—parameters B1
Scenario 2–9 simulations—parameters A2	Scenario 2–9 simulations—parameters B2
Scenario 3–9 simulations—parameters A3	Scenario 3–9 simulations—parameters B3
Scenario 4–3 simulations—parameters A4	Scenario 4–3 simulations—parameters B4

Table 24.2 Features of virtual machines in hardware and software

Hardware				Software					
Stage	CPU	RAM (GB)	Hard disk (GB)	Operative system	Container—Docker	Support and monitoring applications			
1	3	6	80	Ubuntu 20.04 LTS	1 GeoSparkSim 2 Jupyter-Notebook 3 MongoDB	Ganglia	Bashtop	Compass	Studio 3 T
2	8	8	146						

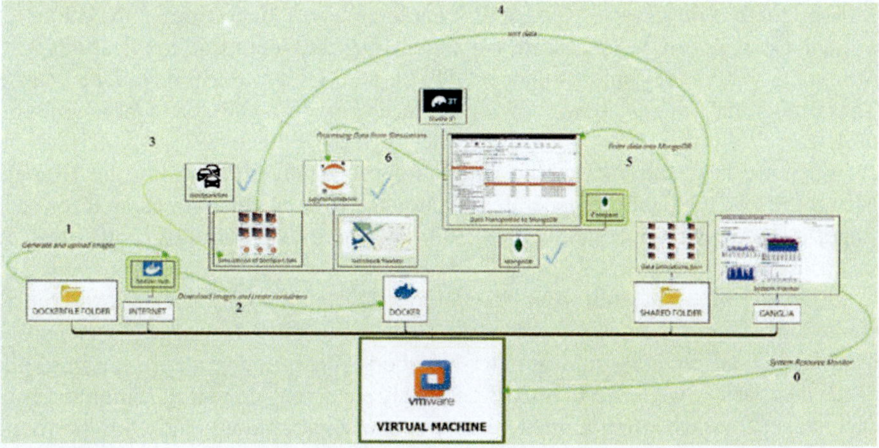

Fig. 24.1 Interaction of the proposed architecture to evaluate the GeoSparkSim tool

24.2.2 *Configuration of the Architecture*

The configuration of each image was made with DockerFile documents which are configured with libraries according to the needs of each of the GeoSparkSim [5] and Jupyter-Notebook [13] applications. Docker-hub has several ready-to-use images including the MongoDB image [14]. Once the operation of each container has been tested, they are loaded to the public repository in Docker-hub to be able to use them in virtual machines for Stage I and II, after installing Docker Engine.

When assembling the architecture, the images are downloaded from the Docker-hub repository, and these images become containers when running on the Docker Engine [8], once created the containers are started and you can work with the ID of each container created in the virtual machine. To perform the simulations, data scripts are configured according to each scenario proposed for Stages I and II. These scripts start the simulation by calling the GeoSparkSim container and entering the data for the simulation by lines of code.

GeoSparkSim container. The GeoSparkSim container performs each simulation through two geographical points given by their latitude and longitude, then delimits this area with a rectangular shape, and the starting point of these coordinates can be given by the data space-oriented approach (DSO) or the network-based approach (NB) and according to the selected parameters distributed by the area vehicles with different times which are determined according to each scenario of each stage. When the GeoSparkSim container performs a simulation, it creates different folders with JSON files containing the data from the simulations, such as the position of each vehicle, intersections, signals, reports, etc. To collect all the JSON files in each folder, the scripts are supported by a code written in Python that allows you to group all the JSON documents in a folder into a single file, so that the data transport is optimal between the containers.

MongoDB container. MongoDB container saves data from simulations in a database that can be accessed via two MongoDB compass applications [14] MongoDB native GUI and Studio 3 T [15] a paid GUI with 30 days free license. These tools make it easy to view, browse, insert, modify, and delete large amounts of data.

When we have the data of the simulations in a single file, using the GUI for MongoDB, we can add the generated documents to each database created for each stage, and each of them contains 26 collections where each of the 26 simulations are saved.

Jupyter-Notebook container. The Jupyter-Notebook container allows you to create and share documents that contain code in different languages, and the one used for this article is Python. Jupyter-Notebook makes a direct connection to one of the databases hosted in MongoDB and analyzes one by one all the simulations of each stage; for each simulation, a vehicle is chosen at random, and with the found vehicle data, the position of the vehicle is observed at each step of the vehicle when driving on the tracks; with this information, the vehicle speed can be calculated in each time.

Monitors of the System. Each time simulations are performed, uploading JSON documents into MongoDB, analyzing data in Jupyter-Notebook, Ganglia and Bashtop monitor the virtual machine resources according to the behavior of each container that uses them.

24.2.3 Evaluation

The experiment consists of two stages each with four scenarios as described in Table 24.4 for Stage I and Stage II. For each scenario of the experiment, we have a variable parameter and the remaining three constants of the four parameters that are entered for the simulations with GeoSparkSim, according to the variations that can be seen in Tables 24.3 and 24.4.

There are five geographical areas of the Quito city that were used to carry out the simulations, for scenario one was employed all five geographical areas described in Table 24.5, while scenarios two, three, and four are used geographical areas one, three, and five.

Table 24.3 Conditions of scenarios to evaluate with GeoSparkSim

Scenario	Geographical area	The number of vehicles	The simulation steps	Vehicle generation type (DSO or NB)
1	Variable	Constant	Constant	Constant
2	Constant	Variable	Constant	Constant
3	Constant	Constant	Variable	Constant
4	Constant	Constant	Constant	Variable

Table 24.4 Scenarios for stage I and stage II simulations

Scenario	Input for stage I and II		Input for stage I			Input for stage II		
	Number of geographical areas	Vehicle generation type	Number of vehicles	Time		Number of vehicles	Time	
				Hours	Seconds		Hours	Seconds
1	1, 2, 3, 4, 5	DSO	1000	0.5	1800	3000	0.5	1800
2	1, 3, 5	DSO	2000	0.5	1800	4000	1	3600
	1, 3, 5	DSO	3000	0.5	1800	6000	1	3600
	1, 3, 5	DSO	4000	0.5	1800	8000	1	3600
3	1, 3, 5	DSO	1000	0.9	3000	3000	1.7	6000
	1, 3, 5	DSO	1000	1	3600	3000	2	7200
	1, 3, 5	DSO	1000	1.5	5400	3000	3	10,800
4	1, 3, 5	NB	1000	0.5	1800	3000	1	3600

Table 24.5 Geographical areas description of studio in Quito

N°	Geographical area	Coordinates			
		Lat1	Lat2	Long1	Long2
1	North	−0.0621414062	−0.1505468544	−78.52666855	−78.42367172
2	North Central	−0.1320074821	−0.2204126644	−78.52151.871	−78.41852188
3	Central	−0.1901146261	−0.2343171264	−78.52804184	−78.47654343
4	South Central	−0.2217859452	−0.2659883482	−78.533535	−78.48203659
5	South	−0.2561179215	−0.3548217982	−78.58812332	−78.4828949

24.2.4 Metric

The first phase analyzes in a fast way the behavior of the simulations carried out, with respect to the execution time, taking as reference a vehicle chosen at random from each of the simulations of the four scenarios proposed for Stage I and II. The second phase studies the hardware resources that each of the two stages occupies when performing the simulations, such as the memory consumed in each process. The amount of JSON documents generated and the amount of space each simulation occupies. The third phase examines the cost benefit of using virtual payment machines in the Cloud to perform simulations.

24.3 Results

In the first stage when performing the simulations with GeoSparksim, we have 44,000 simulated vehicles in a time of 741 min, and in the second stage when performing

Table 24.6 Simulation time results in stage I and II

	Scenarios	Stage I			Stage II		
		Number of simulated vehicles	Run time (min)	Vehicles/min	Number of simulated vehicles	Run time (min)	Vehicles/min
Results	1	5000	83	60.24	15,000	450	33.33
	2	27,000	255	105.88	54,000	1765	30.59
	3	9000	363	24.79	27,000	3000	9.00
	4	3000	40	75	9000	285	31.58
	5	44,000	741	66.48	105,000	5500	26.13

Table 24.7 Results of disk and RAM usage of GeoSparkSim simulations in stage I and II

	Scenarios	Stage I			Stage II		
		Documents generated	Disk space employed (GB)	Memory (GB)	Documents generated	Disk space employed (GB)	Memory (GB)
Results	1	9,679,252	1.50	3.10	15,250,371	2.40	3.5
	2	44,616,713	6.94	3.30	166,547,027	26.00	5.60
	3	42,497,056	6.52	3.56	200,411,391	31.00	6.56
	4	5,485,777	0.86	3.06	29,365,464	4.50	3.63
	5	102,278,798	15.82	3.25	411,574,253	63.90	4.82

the simulations with GeoSparksim, we have 105,000 simulated vehicles in a time of 5500 min, and the number of vehicles generated per minute can also be observed, giving an overall average of 66.48 and 26.13 simulated vehicles for Stage I and II, respectively, as can be seen in Table 24.6.

When performing the simulations with GeoSparksim, the resources used by the storage application were 15.82 GB, and the use of RAM was on average 3.25 GB in Stage I, while for Stage II the resources used by the storage application was 63.90 GB, and the use of RAM on average was 4.82 GB according to Table 24.7.

According to the results obtained in Phase II of Stage I and II, the costs of a virtual machine in the Cloud to perform simulations of similar characteristics to those of this article are shown in Tables 24.8 and 24.9, respectively.

24.4 Discussion

The first phase focuses on the time that the application takes to perform the simulations from the container in Docker, and the application responds in an efficient way in Scenarios 1, 2, and 4 for Stages I and II; in these scenarios, the variables that

Table 24.8 Features and prices of a virtual machine for the phase I

Item	Features virtual machine 1				730 h (1 month)		
	Operative system	RAM (GB)	Disk (GB)	CPU	Company	Product	Price
1	Ubuntu 20.04 LTS	6	80	3	AWS	Amazon EC2	$54.94
2		8	120	4			$58.94
3		8	200	8			$113.81
1		15	375	4	GOOGLE CLOUD	Compute engine n1-standard-4	$114.88
2		16	375	4		Compute engine n2-standard-4	$127.02
3		30	375	8		Compute engine n1-standard-8	$204.04
1		7	128	4	AZURE	Virtual machine	$185.59
2		14	128	8			$356.79
3		14	256	8			$362.23

Table 24.9 Features and prices of a virtual machine for the phase II

Item	Features virtual machine 2				730 h (1 month)		
	Operative system	RAM (GB)	Disk (GB)	CPU	Company	Product	Price
1	Ubuntu 20.04 LTS	16	140	8	AWS	Amazon EC2	$107.81
2		16	170	8		Amazon EC3	$113.81
3		32	200	16		Amazon EC4	$207.81
1		15	375	4	GOOGLE CLOUD	Compute Engine n1-standard-4	$114.88
2		16	375	4		Compute Engine n2-standard-4	$127.02
3		30	375	8		Compute Engine n1-standard-8	$204.04
1		32	200	8	AZURE	Virtual Machine D8a v4	$327.09
2		56	340	6		Virtual Machine NC6 Promo	$346.75
3		14	256	8		Virtual Machine A4	$350.45

change were the geographical area, the number of vehicles, and the type of simulation for vehicles. On the other hand, in Scenario 3, where the simulation time is variable, the number of simulated vehicles in one minute is smaller than the others. At the beginning of the experiment, there was a lower storage capacity to perform the simulations, this led both in stage I and in stage II to a breaking point in the experiment when simulations of scenario 3 were performed in each of them, this is because the program took up 98% of the hardware resources in the simulations of scenarios 1 and 2, and this tells us that the more simulations are performed the more resources in RAM and storage are required to improve the simulation time.

The second phase refers to the hardware resources that the application occupies to perform simulations from the container in Docker, the space employed on the hard disk by the generated files corresponding to Stage I, it is 15.82 GB, and on average in each simulation, 3.25 GB was performed in RAM, which tells us that the simulator uses 54.17% of the total RAM and 19.78% of storage, while the space used in the hard disk by the files corresponding to Stage II is 63.90 GB and on average in each simulation used 4.82 GB in RAM, which tells us that the simulator occupies 60.25% of the total RAM and 43.77% of storage. The use of hardware resources in Stage II was greater in storage and RAM than in Stage I, which indicates that by increasing the values to be simulated the hardware resources grow exponentially, despite maintaining the geographical areas where the simulations are conducted.

With the virtual machines 1 and 2 defined for this experiment, simulations of stages 1 and 2 were performed. AWS, Google Cloud, and Azure are companies that offer the service of virtual machines to run containers and perform tests in a certain time. Based on the experience with the tests of this experiment in Stage I, it is recommended to use a Google Cloud virtual machine the Compute Engine n1 product as it doubles hardware expectations for a cost like AWS's Amazon EC2, while for Stage II it is recommended to use an AWS virtual machine the EC2 or Google Cloud product, as they have simulated prices and features.

24.5 Conclusions

The architecture made in Docker to test the performance of the GeoSparkSim tool offers a safe working environment, because it has an exclusive internal network for containers, which allows direct and fast communication between applications. GeoSparkSim performs simulations in working environments with low hardware features such as virtual machine 1, and simulations must be according to the capabilities of the machine for optimal performance. For simulations with more than 105,000 vehicles and more than 3 days of simulation, we recommend virtual machines with features greater than 8 GB in RAM and 146 or 200 GB in storage. Consider not only the space of the files generated by GeoSparkSim, but also the space that those files will need to be processed in both MongoDB and Jupyter-Notebook. The use of containerized applications optimizes those of machine resources, in the case of the GeoSparkSim container, the simulation time takes up 54.17% of RAM and 19.78%

of storage in Stage I; while in Stage II, it takes up 60.25% of the total RAM and 43.77% of storage, which allows us to obtain data in a shorter time, according to the resources of each machine; once the process is completed by the container, the resources are released for used by another container or other application. To use the Docker-hub images, it is recommended to search for official images and keep in mind the recommendations or specifications of the author. As future work, this architecture can be used to performed simulations with a greater amount of data in a HPC cluster or in a Cloud environment.

References

1. R. Pérez Prieto, J. Canca Ortiz, Módulo de simulación microscópica para la herramienta TRAMOS Integración de modelos y animación gráfica (2005)
2. K. Chen, L. Yu, Microscopic traffic-emission simulation and case study for evaluation of traffic control strategies. J. Transp. Syst. Eng. Inf. Technol. **7**, 93–99 (2007). https://doi.org/10.1016/S1570-6672(07)60011-7
3. M. Behrisch, L. Bieker, J. Erdmann, D. Krajzewicz, SUMO—simulation of urban mobility: an overview. Proc. SIMUL 2011, Third Int. Conf. Adv. Syst. Simul. (2011)
4. J. Brugmann, M. Schreckenberg, W. Luther, Real-time traffic information system using microscopic traffic simulation. Proc. 8th EUROSIM Congr. Model. Simulation, EUROSIM 2013, 448–453 (2015). https://doi.org/10.1109/EUROSIM.2013.83
5. Z. Fu, J. Yu, M. Sarwat, Building a large-scale microscopic road network traffic simulator in apache spark. Proc. IEEE Int. Conf. Mob. Data Manag. 2019–June (2019), pp. 320–328. https://doi.org/10.1109/MDM.2019.00-42
6. J. Yu, Z. Fu, M. Sarwat, Dissecting GeoSparkSim: a scalable microscopic road network traffic simulator in Apache Spark. Distrib. Parallel Databases. (2020). https://doi.org/10.1007/s10619-020-07306-x
7. T.I. Daniel, S. Pesantes, E. Bilbao, Contrastes topográficos: centro de artes escénicas en el teleférico, Universidad San Francisco de Quito, Colegio de Arquitectura y Diseño Interior, architecture thesis (2015)
8. Docs Docker: Docker Engine overview, https://docs.docker.com/engine. Last accessed 16 Jan 2021
9. Compass MongoDB: Compass, https://www.mongodb.com/products/compass. Last accessed 16 Jan 2021
10. Studio 3T: The Professional Client, IDE & GUI for MongoDB, https://studio3t.com. Last accessed 16 Jan 2021
11. Ganglia: Ganglia Monitoring System, http://ganglia.sourceforge.net. Last accessed 16 Jan 2021
12. Aristocratos: GitHub - aristocratos/bashtop: Linux/OSX/FreeBSD resource monitor, https://github.com/aristocratos/bashtop. Last accessed 16 Jan 2021
13. Jupyter: Project Jupyter, https://jupyter.org. Last accessed 16 Jan 2021
14. MongoDB: Introduction to MongoDB—MongoDB Manual 2.6.4. https://docs.mongodb.com/manual. Last accessed 16 Jan 2021

Chapter 25
Toward a Social Media Emergency Events Detection Based on Supervised Learning

Yahir Mendoza, Jorge Santillan, Roberth Alcivar-Cevallos, and Jorge Parraga-Alava

Abstract Urban emergency events affect many lives and property of people every day. They should be tackled in a fast, precise, and with an early detection goal. In recent years, social media has become popular, turning its users into proper "citizen sensors," which means a person who, through devices, records incidents on these platforms. Therefore, they become real-time communicators of emergency events. This paper presents a preliminary study based on supervised learning toward social media emergency events detection. This approach has four phases: the first two correspond to data acquisition and preprocessing, where tweets available in a public repository about emergency and non-emergency were taken, filtered, and converted through a process of lemma feature extraction. In the third phase, four most commonly used supervised algorithms were implemented to detect tweets containing emergency events. Finally, using validation metrics and algorithms to detect emergency and non-emergency reported in the tweets was verified. Computational results showed that all evaluated algorithms achieved average rates of 85.38% of accuracy meanwhile that in execution time about 23 seconds, to detect a real emergency event. The best algorithm when trade-offs between execution time and emergency detection capability are considered was the SVM.

25.1 Introduction

Technology will undoubtedly be a key tool in the management of future disasters [1]. The next generation of emergency managers will have purposefully selected this career. They will be better prepared than their predecessors to use the tools, technologies, and strategies needed for the future. The only thing that has not changed is the fundamental aspect of emergency management since it must respond effectively by demonstrating the control of communications and the coordination mechanisms that it has; the recovery process in an emergency must be understood, and most

Y. Mendoza · J. Santillan · R. Alcivar-Cevallos · J. Parraga-Alava (✉)
Facultad de Ciencias Informáticas, Universidad Técnica de Manabí, Avenida José María Urbina, Portoviejo 130105, Ecuador
e-mail: jorge.parraga@usach.cl

Á. Rocha et al. (eds.), *Communication, Smart Technologies and Innovation for Society*, Smart Innovation, Systems and Technologies 252,
https://doi.org/10.1007/978-981-16-4126-8_25

importantly, it must be innovated to prepare our community for different types of situations [2]. When an emergency occurs, people tend to post what is happening on social media, either to inform or to express their concern about the issue that has occurred. In this sense, social media becomes a valuable source of information for the knowledge of an emergency, for example, traffic accidents, fires, robberies, murders, and some natural disasters.

In recent years, as social networks develop rapidly and the number of active users increases, many crowdsourcing researchers for emergency management [3] have seen an opportunity to enrich themselves with new types of data. Social media can improve our communication systems by substantially increasing our ability to prepare for, respond to, and recover from events that threaten public health. Studies are needed to evaluate the reliability and validity of the information related to public health communicated through social networks. Some relevant metrics already exist and are used by the business community; however, few published scientific studies have applied these tools to assess the capabilities or effectiveness of social media in public health emergencies [4].

There are many data on social media, although a few are helpful; in most social media messages, data is sometimes overwhelmed by unnecessary data. Some previous work focused on the extraction of relevant messages using text classification models. The work detailed in [5] proposes a system for locating and analyzing events on Twitter to detect car traffic accidents. Another work, as [6] shows an approach to detect potentially suicidal messages on social networks such as Facebook and Twitter, but mainly in English. In [7], also addressing suicide as an emergency but with an approach to automatically increasing the number of texts in messages in the dataset. In [8], it has utilized machine learning models to perform classification and aggregation of tweets about the 2010 Haiti earthquake. In this same thematic, in [9], authors implemented a platform for the automatic classification of microblog communications related to the crisis that occurred and was felt with the 2013 Pakistan earthquake.

Although the previously mentioned works are fascinating and present a certain degree of classification accuracy in a type of emergency, they do not show significant results for an emergency in general or a disease. Also, most of them focus on the English language and not on Spanish. In this research, we intend to fill the gap related to detecting emergent events in social media but using Spanish. Our goal is to classify texts in this language into a category: emergency or non-emergency. It would be a start point toward a social media emergency event detection as we expect that new texts in social media be labeled as emergency or non-emergency. It will facilitate the monitoring of emergencies in Ecuador.

The paper is organized as follows. Section 25.2 presents the methodology carried out to develop the work. Section 25.3 presents the dataset and configurations used in the computational experiments and the results achieved in our research. The discussions of the study are given in Sect. 25.4, and, finally, Sect. 25.5 contains the conclusions of this work.

25.2 Methodology

25.2.1 *Data Acquisition*

The dataset in [10] called UrbangEnCy was used, which contains 25,547 tweets in Spanish about possible urban emergency events reported on Twitter within the Ecuadorian territory. Here, each tweet can belong to one of two classes: emergency and non-emergency. As the dataset has a class imbalance, it was necessary to perform a balancing procedure, choosing the same number of tweets with the "emergency" and "non-emergency" labels in random order. This version has 2982 labeled tweets for each of the two classes. Note that in this dataset, labels of the tweet were validated by human experts.

25.2.2 *Preprocessing*

A cleaning of tweets was carried out, which consisted of the elimination of words that did not provide relevant information for the text, such as unisons, vowels, connectors, numbers, special characters such as # or @, and also symbols such as brackets, parentheses or quotation marks or in the case of punctuation marks, exclamation marks or question marks and also the elimination of URLs. After that, each tweet is used as input of a featuring extraction process called lemmatization. This process guarantees that the number of words to be analyzed in each text is reduced by maintaining the base word from which each of the expressions expressed in each tweet is derived. Thereby, we obtain the context of the messages written on the tweet no matter how a user expressed it. In Fig. 25.1, a schematic representation of the first two phases carried out is observed.

25.2.3 *Supervised Learning Algorithms*

Supervised classification is the process where the existence of "emergency" or "non-emergency" is determined in each message (tweet). We considered the most known classification algorithms to be used in our computational experiments. These are the following: Naive Bayes (NB) [11], artificial neural networks (ANN) [12], decision trees (DT) [13], random forest (RF) [13], and support vector machines (SVM) [14]. The parameters used in those algorithms are shown in Table 25.1.

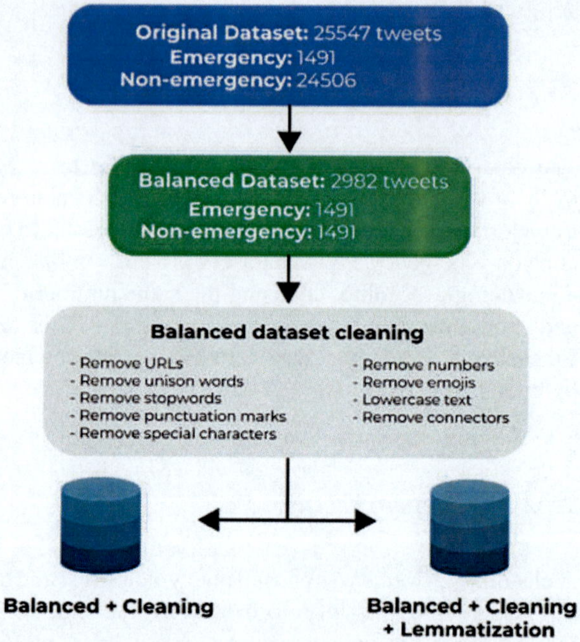

Fig. 25.1 Acquisition, preprocessing, and featuring extraction process of the dataset

Table 25.1 Parametrization used in the supervised learning models

Model	Parameters
NB	– Laplace 0—Na.action Na.pass—Usekernel F—Usepoisson F—prior null
DT	– Na.action Na.rpart—Model F—x F—y F
SVM	– Kernel linear—Scale T—Na.action Na.omit—Type null—Tolerance 0.0001—epsilon 0.1
RF	– Ntree 200—Keep.forest T -Replace T -nPerm 1
ANN	– Hidden C (20,15)—Threshold 0.05—Algorithm Rprop + -Act.fct Logistic -Linear.output F

25.2.4 Validation Metrics

To evaluate the performance of the algorithms, we used accuracy, precision, sensitivity, specificity, error, and execution time. They are computed as follows:

$$\text{Accuracy} = \frac{\text{VP} + \text{VN}}{n} \tag{25.1}$$

$$\text{Precision} = \frac{\text{VP}}{\text{VP} + \text{FP}} \tag{25.2}$$

$$\text{Sensitivity} = \frac{\text{VP}}{\text{VP} + \text{FN}} \tag{25.3}$$

$$\text{Specificity} = \frac{\text{VN}}{\text{FP} + \text{VN}} \tag{25.4}$$

$$\text{Error} = \frac{\text{FP} + \text{FN}}{n} \tag{25.5}$$

$$\text{Execution time (in seconds)} = \frac{\text{Tm}}{n} \tag{25.6}$$

where

True positive (TP)	The tweet is positive, and it is classified as positive.
True negative (TN)	The tweet is negative, and it is classified as negative.
False positive (FP)	The tweet is negative, and it is classified as positive.
False negative (FN)	The tweet is positive, and it is classified as negative.
Total time (Tm)	The time (in seconds) during the n executions.
n	number of executions of algorithms.

We considered as positive an emergency and as negative a non-emergency tweet.

25.3 Results

To measure the performance of the classifiers regarding their ability to identify emergency and non-emergency tweets, experiments were performed using the dataset with and without lemmatization. In each case, the data were divided into training and test subsets with 70% and 30% of the instances, respectively. The five supervised learning models were implemented in RStudio using the Google Colaboratory environment and running with 30 repetitions each one. From there, the average is obtained for each of the metrics described in the previous section. These results are observed in Tables 25.2 and 25.3, for the dataset with and without lemmas.

Table 25.2 Mean of validation metrics values obtained by classifiers over 30 runs using balanced dataset without lemmas

Supervised model	Accuracy	Sensitivity	Specificity	Error	Precision	Time
NB	68.63	77.12	60.37	31.26	66.22	31.14
DT	79.14	69.21	89.05	20.86	86.40	70.61
SVM	83.86	80.03	87.89	16.04	86.83	44.47
RF	84.91	76.95	92.87	15.09	91.55	788.4
ANN	83.68	84.36	83.01	16.32	83.23	178.80

Table 25.3 Mean of validation metrics values obtained by classifiers over 30 runs using balanced dataset with lemmas

Supervised model	Accuracy	Sensitivity	Specificity	Error	Precision	Time
NB	62.76	68.21	57.37	37.24	61.59	18.82
DT	79.06	70.89	87.26	20.94	84.86	40.65
SVM	83.22	80.80	85.64	16.78	84.93	23.34
RF	85.38	78.52	92.23	14.62	91.01	401.4
ANN	83.70	84.47	82.98	16.30	83.28	69.77

Tables 25.2 and 25.3 show that the performance of the supervised models is acceptable. To improve the visualization, it is better to observe Fig. 25.2. In the accuracy metric, the one with the best average value is the RF algorithm with 84.91% and 85.38%, when the dataset without and with lemmatization was used, respectively. The accuracy measures the relationship between the correctly predicted observations and the total observations. Thus, on average, with random forest, 8 out of 10 tweets are correctly classified by assigning the label "emergency" or "non-emergency" as appropriate. The second-best performance according to the accuracy corresponds to the ANN with a percentage ranging from 83.68 to 83.7%, with similar values we have the third place, which is the SVM algorithm, with values ranging from 83.22 to 83.96%, when the dataset without and with lemma were used, respectively. Then we have the fourth place where the DT algorithm is located, and its values go from 79.06 to 79.14% of accuracy. Finally, we have the NB algorithm, which gives us the values of 62.76% to 68.63% of accuracy when the dataset without and with lemma were used, respectively. Although random forests and ANN have good accuracy values compared to their competitors, it is necessary to analyze the execution times, also shown in Fig. 25.2.

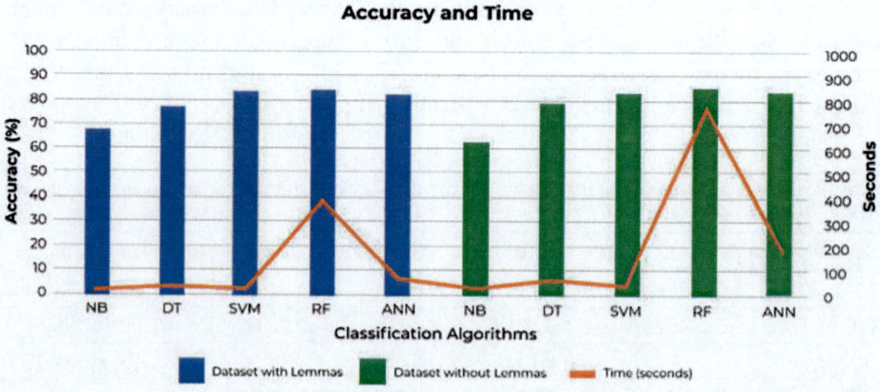

Fig. 25.2 Average values for accuracy and execution time during 30 runs of the classification models for the dataset with lemma (blue bar) and without lemma (green bar)

In Fig. 25.2, it is observed that the supervised model that has the highest average execution time is the RF algorithm with 788.4 s and 401.4 s, when the dataset without and with lemmatization was used, respectively. The time is obtained as the relationship between the sum of all the execution times and the total executions. In the second place, the ANN with a time that ranges from 178.8 to 69.77 s was found. The DT algorithm, with values that range from 70.61 to 40.65 s of the execution time, was obtained in third place. Then in the fourth place, the SVM algorithm is located, and its values go from 44.47 to 23.34 s. Finally, the NB algorithm gives a time from 31.14 s until 18.82 s, when the dataset with and without and lemmatization was used. Thereby, when execution time is considered a performance indicator, the best models are NB and SVM. While for the metric, the accuracy is SVM.

Another metric evaluated in Tables 25.1 and 25.2 was precision. The precision measures the ratio of the positive observations predicted correctly to the total of positive observations predicted. The highest average value belongs to the RF algorithm with 91.55%, which means that 9 out of 10 tweets classified as "emergency" really are emergencies. The sensitivity measures the relationship between correctly predicted observations and incorrectly predicted negative observations. The best classifier according to this metric was the ANN algorithm with 84.47%, which means that 8 out of 10 tweets classified as "emergency" are correctly predicted. Specificity measures the relationship between correctly predicted negative observations and wrongly predicted positive observations. For this metric, the RF algorithm highlights achieving 92.87%. Thus, 9 out of 10 tweets classified as "non-emergency" are correctly labeled by the algorithm. Finally, we evaluated the error. It measures incorrectly predicted observations and total observations. The RF algorithm obtained the slightest error with 14.62%; i.e., almost two out of ten tweets are misclassified by this algorithm.

25.4 Discussion

Because the results showed that the classification models had very homogeneous performance, we conducted statistical tests. In this sense, to test whether the classification algorithms work similarly or not, the Kruskal–Wallis test has been performed. This decision was made once it verified that the data did not meet the parametric criteria. Once the test is performed, p-values $< 2.2 \times 10^{-16}$ are obtained for all cases considered, both with and without lemmatization dataset. This value indicates a significant difference in the algorithm performance in one of the five metrics considered. As there is a significant difference, a pots-hoc test of Nemeyi [1] is carried out. In this test, as a null hypothesis, it is assumed that there are no significant differences between the pairs of algorithms, while the alternative hypothesis is that if there is a significant difference between them. As a threshold, a significance level of 5% was considered. The results in terms of the p-values are shown in Tables 25.4 and 25.5. Here, all pairs (except SVM-ANN) of algorithms have p-values that are lower than 0.005 (0.05/10) inaccuracy.

Table 25.4 *P*-values obtained by post-hoc tests comparing the algorithm performance regarding five quality metrics using the dataset of tweets without lemmatization

Pairs	Accuracy	Sensitivity	Specificity	Error	Precision	Time
RF-DT	3.0×10^{-10}	2.5×10^{-10}	7.0×10^{-09}	3.0×10^{-10}	9.4×10^{-13}	1.7×10^{-13}
NB-DT	3.0×10^{-10}	2.5×10^{-10}	2.7×10^{-10}	3.0×10^{-10}	2×10^{-16}	1.7×10^{-13}
NB-RF	3.0×10^{-10}	0.48710	2.7×10^{-10}	3.0×10^{-10}	2×10^{-16}	1.7×10^{-13}
ANN-DT	3.0×10^{-10}	2×10^{-16}	6.0×10^{-12}	3.0×10^{-10}	1.1×10^{-06}	1.7×10^{-13}
ANN-RF	2.1×10^{-07}	1.8×10^{-10}	2.7×10^{-10}	2.1×10^{-07}	2×10^{-16}	1.7×10^{-13}
ANN-NB	3.0×10^{-10}	6.4×10^{-12}	2.7×10^{-10}	3.0×10^{-10}	2×10^{-16}	1.7×10^{-13}
SVM-DT	3.0×10^{-10}	2×10^{-16}	0.072	3.0×10^{-10}	0.1	1.7×10^{-13}
SVM-RF	1.3×10^{-06}	1.4×10^{-08}	2.7×10^{-10}	1.3×10^{-06}	1.5×10^{-10}	1.7×10^{-13}
SVM-NB	3.0×10^{-10}	0.00096	2.7×10^{-10}	3.0×10^{-10}	1.5×10^{-10}	1.7×10^{-13}
SVM-ANN	0.1	5.9×10^{-12}	2.2×10^{-09}	0.1	2.7×10^{-07}	1.7×10^{-13}

[a]The values with an asterisk are cases with no significant difference in performance

Table 25.5 *P*-values obtained by post-hoc tests comparing the algorithm performance regarding five quality metrics using the dataset of tweets with lemmatization

Pairs	Accuracy	Sensitivity	Specificity	Error	Precision	Time
RF-DT	3×10^{-10}	1.8×10^{-10}	5.8×10^{-10}	3×10^{-10}	[a]2×10^{-16}	1.7×10^{-13}
NB-DT	3×10^{-10}	0.032	[a]2×10^{-16}	3×10^{-10}	2.4×10^{-10}	1.7×10^{-13}
NB-RF	3×10^{-10}	8.8×10^{-10}	2.1×10^{-10}	3×10^{-10}	2.4×10^{-10}	1.7×10^{-13}
ANN-DT	3×10^{-10}	[a]2×10^{-16}	1.5×10^{-05}	3×10^{-10}	0.032	1.7×10^{-13}
ANN-RF	4×10^{-09}	1.1×10^{-09}	2.1×10^{-10}	4×10^{-09}	[a]2×10^{-16}	1.7×10^{-13}
ANN-NB	3×10^{-10}	[a]2×10^{-16}	[a]2×10^{-16}	3×10^{-10}	2.4×10^{-10}	1.7×10^{-13}
SVM-DT	3×10^{-10}	[a]2×10^{-16}	0.00109	3×10^{-10}	0.271	1.7×10^{-13}
SVM-RF	3×10^{-10}	1.0×10^{-06}	2.1×10^{-10}	3×10^{-10}	2.4×10^{-10}	1.7×10^{-13}
SVM-NB	3×10^{-10}	4.7×10^{-16}	[a]2×10^{-16}	3×10^{-10}	2.4×10^{-10}	1.7×10^{-13}
SVM-ANN	0.034	8.2×10^{-08}	0.00031	0.034	0.011	1.7×10^{-13}

[a]The values with an asterisk are cases with no significant difference in performance

Results are strong evidence that RF obtains mean values of this metric better than other algorithms considered in the evaluation. A similar situation occurs with the other metrics; although for the case of sensitivity, RF is not able to achieve a significant difference. Of all the algorithms evaluated, the one with the fewest cases of significant difference is the SVM, which in four out of five metrics, presents at least a similar performance with the other algorithms. It is evident that when using the dataset without lemmas, the RF algorithm gives the best average results for accuracy but the worst in execution time. For a dataset with lemmas, the behavior is similar, highlighting the significant difference of RF against the other algorithms concerning its level of accuracy.

Although the SVM did not have significant differences in all cases, it is with this algorithm that a better tradeoff is achieved between accuracy and time, having average values greater than 80% and less than 100 s, respectively.

25.5 Conclusions

In this work, a first approach has been proposed to detect emergency events on social media. One of the crucial aspects is that our results show that it is possible to extract posts from social media and classify them into the "emergency" or "non-emergency" classes with up to 85.38% accuracy. The evaluation of five classification algorithms showed that they obtain similar performances in predictive capacity and in execution time, except for RF, which was the one that required the most computational time. The application of lemmatization to the dataset does not seem to positively influence the algorithms' performance since the results of the validation metrics when compared without lemmatized dataset, tend to be very similar.

As further work, we would like to improve the levels of emergency detection. After that, we can also classify types of emergencies that can be reported on social networks.

Acknowledgements The authors thank the project "Geospatial Patterns Recognition in Urban Emergency Service Events Data". Project reference code YTAUTO1889-2018-IINV0002, II-UTM.

References

1. B. Van De Walle, M. Turoff, S.R. Hiltz, *Information Systems for Emergency Management* (Routledge, New York, 2014)
2. G.D. Haddow, J.A. Bullock, D.P. Coppola, *Introduction to Emergency Management* (Butterworth-Heinemann, 2017)
3. H. Gao, G. Barbier, R. Goolsby, Harnessing the crowdsourcing power of social media for disaster relief. IEEE Intell. Syst. **26** (3), 10–14 (2011)
4. R.M. Merchant, S. Elmer, N. Lurie, Integrating social media into emergency-preparedness efforts. N. Engl. J. Med. **365** (4), 289–291 (2011)
5. R. Li, K. Hou, R. Khandiwala, K. Chen, TEDAS: a twitter-based event detection and analysis system, in *2012 IEEE 28th International Conference on Data Engineering* (IEEE, 2012), pp. 1273–1276
6. R.W. Acuña Caicedo, J.M. Gómez Soriano, H.A. Melgar Sasieta, Assessment of supervised classifiers for the task of detecting messages with suicidal ideation. Heliyon **6** (8) (2020)
7. J. Parraga-Alava, R.A. Caicedo, J.M. Gómez, M. Inostroza-Ponta, An unsupervised learning approach for automatically to categorize potential suicide messages in social media, in *2019 38th International Conference of the Chilean Computer Science Society (SCCC)*, Concepcion, Chile (2019), pp. 1–8
8. C. Caragea, N. McNeese, A. Jaiswal, G. Traylor, H.-W. Kim, P. Mitra, D. Wu, A.H. Tapia, L. Giles, B.J. Jansen, J. Yen, Classifying text messages for the Haiti earthquake. ISCRAM (2011)

9. M. Imran, C. Castillo, J. Lucas, P. Meier, S. Vieweg, AIDR: artificial intelligence for disaster response, in *Proceedings of the 23rd international conference on world wide web* (2014), pp. 159–162
10. J. Parraga-Alava, R. Alcivar-Cevallos, L. Vaca-Cardenas, J. Meza, UrbangEnCy: An emergency events dataset based on citizen sensors for monitoring urban scenarios in Ecuador. Data Brief **34** (2021)
11. I. Rish, An empirical study of the naive Bayes classifier, in *IJCAI 2001 Workshop on Empirical Methods in Artificial Intelligence* (2001), pp. 41–46
12. A. Delgado, Aplicación de las Redes Neuronales en Medicina. Revista de la Facultad de Medicina **47**(4), 221–223 (1999)
13. J.M. Carrillo, J. Parraga-Alava, How predicting the academic success of students of the ESPAM MFL? A preliminary decision trees based study, in *2018 IEEE Third Ecuador Technical Chapters Meeting (ETCM)*, Cuenca, Ecuador (2018), pp. 1–6
14. S.V.M. Vishwanathan, M. Narasimha, SSVM: a simple SVM algorithm, in *Proceedings of the 2002 International Joint Conference on Neural Networks. IJCNN'02* (IEEE, 2002), pp. 2393–2398

Chapter 26
Exploratory Study of Physic Informed Deep Learning Applied to a Step-Pool for Different Flow Magnitudes

Sebastián Cedillo, Esteban Sánchez-Cordero, Esteban Samaniego, and Andrés Alvarado

Abstract Physical laws governing a certain phenomenon can be included in a deep-learning model within a new paradigm: the so-called physical informed deep learning (PIDL). Physical laws in hydraulics consist of partial differential equations (PDEs) resulting from balance laws. The potential use of PIDL in a step-pool reach having a complex flow and geometric characteristics is tested in this article. The studied morphology belongs to a hydraulic observatory in a mountain river in Ecuador where flow and geometric data are available. The water level profile of PIDL was compared to a stationary one-dimensional HEC-RAS model and water levels measured at three staff gauges in the reach. Saint–Venant equations, geometry data, and boundary conditions were used to implement a PIDL-based model. The chosen PIDL architecture is based on the one with the lowest value for the loss function. The resulting water level profile of the PIDL model does not have instabilities, and according to dimensionless RMSE is slightly less efficient in its predictions than the HEC RAS model. Moreover, the difference between HEC-RAS and PIDL water profile decreases as flow increases.

26.1 Introduction

Deep learning is a technique able to approximate the behaviour of a system based on data input [1, 2]. In some physical systems, the availability of data is limited, so the introduction of the governing physics as additional information in deep learning has resulted in the so-called physics informed deep learning (PIDL) [1]. The inclusion

S. Cedillo (✉) · E. Samaniego · A. Alvarado
Departamento de Recursos Hídricos y Ciencias Ambientales, Universidad de Cuenca, Av. 12 de Abril s/n, 010107 Cuenca, Ecuador
e-mail: sebastian.cedillog@ucuenca.edu.ec

E. Sánchez-Cordero
Departamento de Ingeniería Civil, Universidad de Cuenca, Av. 12 de Abril s/n, 010107 Cuenca, Ecuador

E. Sánchez-Cordero · E. Samaniego · A. Alvarado
Facultad de Ingeniería, Universidad de Cuenca, Av. 12 de Abril s/n, 010107 Cuenca, Ecuador

Á. Rocha et al. (eds.), *Communication, Smart Technologies and Innovation for Society*, Smart Innovation, Systems and Technologies 252,
https://doi.org/10.1007/978-981-16-4126-8_26

of physical laws in the deep-learning technique reduces dependency on large, high-quality data sets and provides a robust model with physical significance [3, 4]. In river hydraulics, the fluid flow pattern is governed by partial differential equations based on balance laws [4]. The conservation laws comprise the mass and momentum conservation laws leading to the so-called Saint–Venant equations for the one-dimensional case. The computations of the derivatives in deep learning is through automatic differentiation (AD) [2]. AD refers to a family of techniques where the derivative calculation is divided into elementary operations where symbolic differentiation is used, and then the values are stored and accumulated to generate the overall derivative value [5].

PIDL can solve forward and inverse problems. In a forward problem, the training points are obtained from boundary and initial conditions where the conservative equations are enforced to obtain a stable solution [6]. An inverse problem consists on deducing the hidden quantities such as parameters using scattered and noisy observations [7].The solution of balance laws in fluid flows can develop discontinuities, however PIDL has been proposed as an alternative methodology to solve those equations [6].

Physics informed deep learning has been successfully used to solve forward and inverse hydraulic benchmark cases. Raissi et al. [5] used concentration data as training data in an incompressible Newtonian flow. Wang et al. [4] developed a deep-learning methodology based on multi-scale decomposition for turbulent flows. Mao et al. [7] worked with compressible flow in the inviscid limit with smooth and discontinuous solutions, whilst Rao et al. [8] successfully developed a mixed-variable scheme to study steady and unsteady viscous incompressible laminar flow. Nevertheless, the manuscript' authors have not found an application with a complex geometry in a mountain river reach.

Step-pool is a common morphology in mountain rivers with a staircase shape where risers are steps and pools are treads [9]. Steps are generally formed of different materials, such as alluvial boulders, large wood debris, or bedrock [10, 11], whilst pools are made of finer granulometry materials [12]. In steps, water jets are produced where the flow is supercritical. When the water falls into the pools, a hydraulic jump is produced, forming air pockets. [13]. This morphology plays an important environmental and hydraulic role in steep rivers [11], but due to the process described, its hydraulic modelling presents an important challenge.

As mentioned previously, PIDL has shown its potential for hydrodynamic modelling. Thus, exploring its capabilities in this regard appears as a worthwhile task. In this study, the prediction capacity of PIDL is explored in a step-pool, a complex morphology in a mountain river, which entails a challenging task in terms of flow modelling. The PIDL results are compared against a widely used hydrodynamic model called HEC RAS. The flow was modelled as steady and one-dimensional for both PIDL and HEC-RAS because of the validation data, consisting of water depth measurements at three staff gauges for three flow magnitudes. Even though PIDL uses limited flow and geometric data, it has shown that it is a viable option to model a complex morphology such as a step-pool.

26.2 Methods

26.2.1 Study Site and Field Data

The geometric data for this study comes from a hydraulic observatory in the Quinoas river located in Azuay-Ecuador (see Fig. 26.1). The observatory is 1500 m long and consists of three cascades, two step-pool, and one plane-bed. Each studied morphology has three staff gauges except for Cascade 3 where five staff gauges were installed. The sub reach used in this study was step-pool 1 having a length of 12 m and an average slope of 6.1%.

The available data is divided into three categories. Firstly, detailed geometry data comprising cross sections and profile data. Secondly, roughness values, water depth, wetted width, mean velocity for thirty flows ranged from 0.035 to 0.878 m^3 s^{-1}. Thirdly, bed material size. The data previously mentioned was used in HEC RAS models as well as PIDL model. Further information is provided in the next subsections.

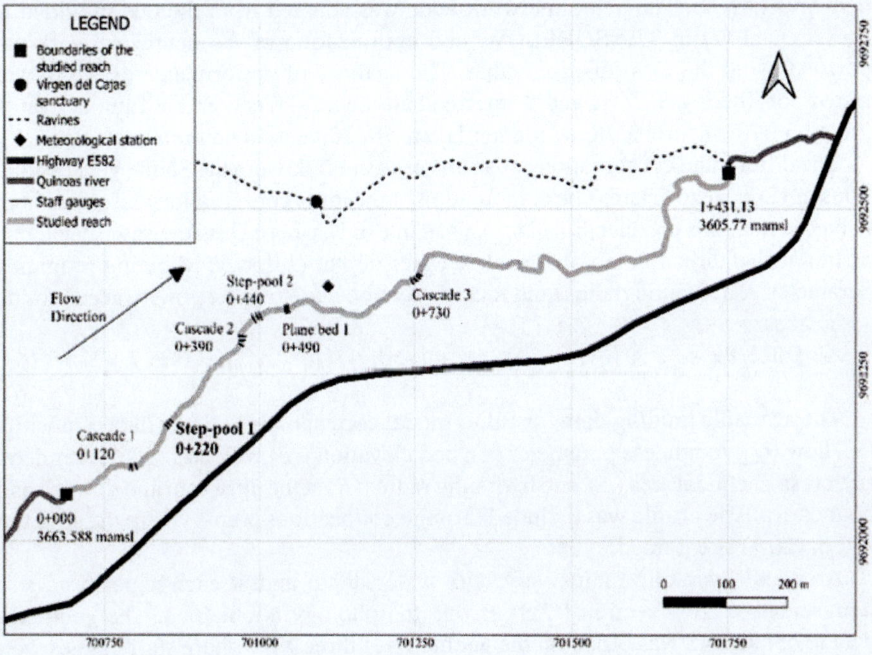

Fig. 26.1 Plan view of the studied 1500 m river reach showing the location of the sub reach Step-pool 1 which is used in this study

26.2.2 HEC-RAS Model

Three HEC RAS models were implemented one for low flow (0.035 m^3s^{-1}), middle flow (0.443 m^3s^{-1}), and high flow (0.878 m^3s^{-1}). The detailed available topographic data was used for the geometry in the hydrodynamic models. A mixed flow regime was taken as mixed based on the tumbling flow explained in the introduction section, so boundary conditions upstream and downstream were required. Normal depth was assumed at the boundaries. The model ran under steady conditions with some additional cross sections (XS) added to improve numerical stability. HEC RAS for steady flow solves iteratively the energy equation with a mean energy slope between two XS [14]. Nevertheless, a momentum equation is used when there is transition between supercritical to subcritical flow or vice versa.

26.2.3 Physic Informed Neural Network

The deep-learning methodology used in this study is based on artificial neural networks (ANN). The neural network code was adapted from the one provided by Raissi et. al [1] in TENSORFLOW, and optimization of the neural network was pursued using the loss function value. The analysis of performance of ANNs was tested considering 3, 5, 7, and 9 intermediate neural layers, and a Tanh activation function in the neurons. In the hidden layers 10, 20, and 40 neurons were used.

The differential conservation equations used in PIDL were the Saint–Venant equations in 1D at steady state. The combination of the mass conservation and momentum balance equations produce Eq. (26.1), used in PIDL, where U is streamwise velocity, x streamwise direction, zo bed level, h water depth, Q flow, n Manning roughness parameter, R hydraulic radius, and A cross-sectional area, and g gravity acceleration.

$$Q\partial U/\partial x + \; g\,A\,\left(\partial(zo+h)/\partial x + \left(Q^2n^2\right)/\left(R^{4/3}A^2\right)\right) = 0 = F \tag{26.1}$$

The available training data for PIDL model corresponds to thirty data of position (x), flow (Q), roughness parameter (n), bed elevation (zo), velocity (U), water depth (h), cross-sectional area (A), and hydraulic radius (R) at the upstream and downstream boundaries. The profile was included through collocations points where the structure of Eq. (26.1) is enforced.

Area and hydraulic radius vary with water depth and at each x position, so a representative cross section (rXS) of this morphology is obtained. The geometric data to obtain rXS was based on the geometry of three XSs where staff gauges were installed. All the XS points were translated so that its lowest points coincide in an axis origin. The rXS points are obtain with the geometric mean (GM) of the translated three XSs points. GM is used due to the high variability of XS, so some points act as outliers. This mean is less sensible to outliers allowing to capture the mean tendency

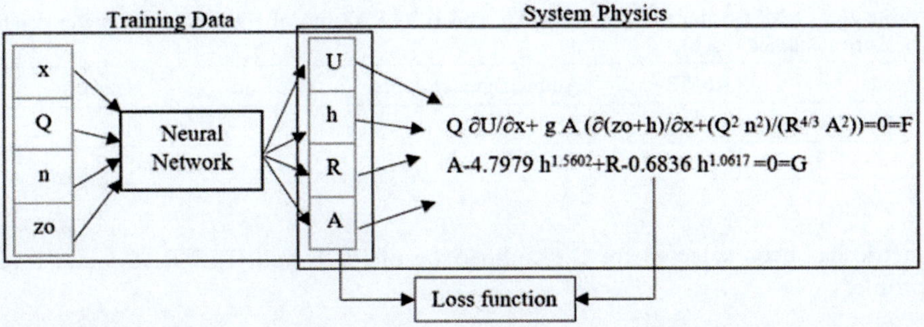

Fig. 26.2 Physics informed deep learning scheme where the contribution of training points and physics influence the loss function

of the measured XSs. From rXS points was possible to find potential equations of area and hydraulic ratios as functions of water depth (Eqs. (26.2) and (26.3)).

The loss function used in PIDL was the mean squared error with two components: (1) the enforcement of training data and (2) the enforcement of system physics: the conservation equations (F), and the rXS (G) (see Fig. 26.2). The predictions of water depth from PIDL model at the three flow values (0.035 m^3s^{-1}, 0.443 m^3s^{-1}, and 0.878 m^3s^{-1}) were used for comparison purposes.

$$A = 4.7979\, h^{1.5602} \tag{26.2}$$

$$R = 0.6836\, h^{1.0617} \tag{26.3}$$

26.2.4 Metrics

The root mean square error (RMSE) (see Eq. (26.4)) is a widely metric used in meteorological and environmental studies such as: air quality or climate research [15–18]. This metric is suitable to compare model outputs with field measurements because larger errors has more weight than smaller ones [15, 16]:

$$\text{RMSE} = \left(\sum (S_i - O_i)^2 \big/ N\right)^{0.5} \tag{26.4}$$

where S_i are the simulated values of a variable, O_i are observed values of a variable, and N is the number of data.

A dimensionless version of this metrics is obtained when it is divided by the mean of the observations.

The mean average error (MAE) (see Eq. (26.5)) metric is widely used during model evaluations [15]. In this metric, all errors have the same weight. Thus, this

Table 26.1 RMSE values of the best PIDL and HEC-RAS model with respect to water depths measure at studied reach

Model	RMSE	Adimensional RMSE	Layers	Neurons
PIDL	11.30	31.01	7	40
HEC-RAS	9.62	26.39	–	–

metric has been selected for the comparison of PIDL and HEC RAS water level profile:

$$\mathrm{MAE} = \left(\sum \left| h_{textPIDL} - h_{\mathrm{HEC}} \right| \Big/ N\right) \tag{26.5}$$

where h_{PIDL} is the water depth predicted by PIDL model, h_{HEC} is the water depth predicted by HEC RAS model, and N is the number of data.

26.3 Results

26.3.1 Experiments Results

The RMSE value of the best performing PIDL model and the HEC-RAS model, with respect to the water level values measured in the river reach, are summarized in Table 26.1. The best PIDL model with 7 hidden layer 40 neurons were used, which was deter-mined by trial and error. The HEC-RAS model presents better performances than the PIDL model with a difference $\approx 4\%$.

26.3.2 Difference Between PIDL and HEC-RAS Profile

Figure 26.3 shows the water level profile of both models, PIDL and HEC-RAS, for the three flow magnitudes. It is clearly depicted that there are no instabilities of the water profiles for any of the models. A detailed inspection of the water level profile of the HEC-RAS model shows better performances with respect to the measured values, corroborated with the values in Table 26.1.

Table 26.2 shows the MAE values calculated between the PIDL and HEC-RAS water level profiles. A decrease in the MAE value is observed when the value of the flow rate in analysis increases.

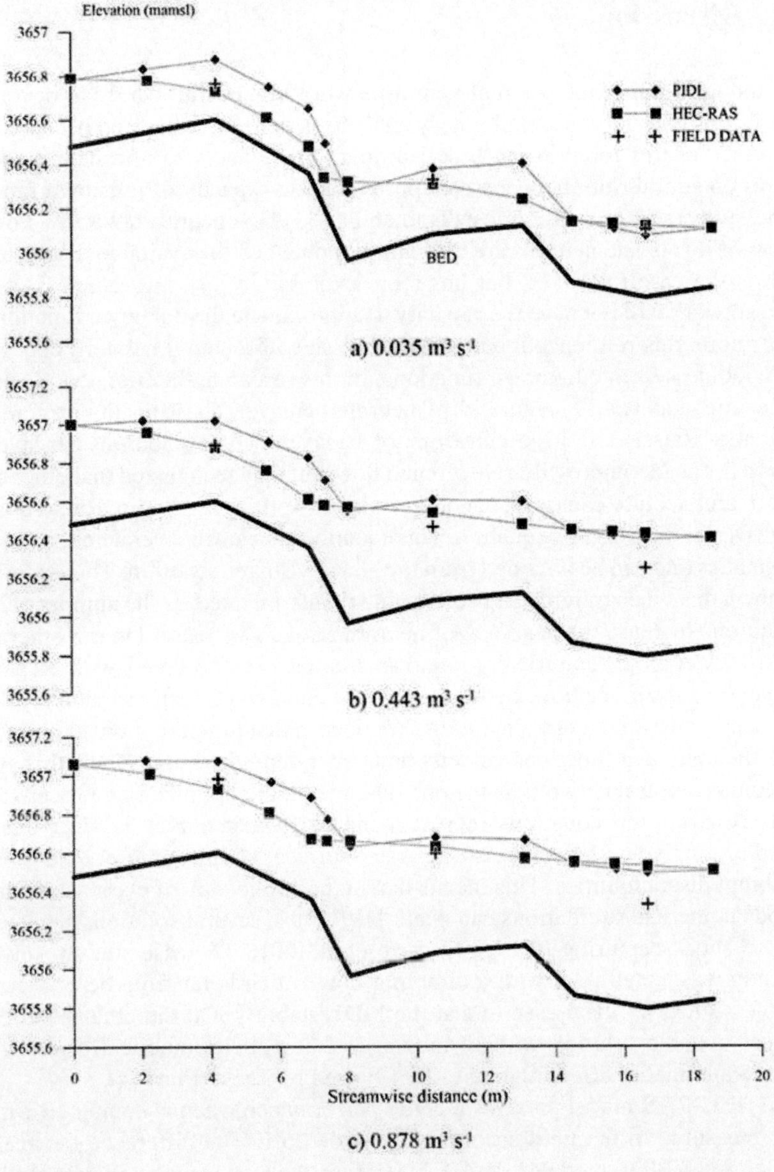

Fig. 26.3 Comparison of water surface profiles resulting from PIDL and HEC-RAS and the available field data

Table 26.2 MAE between PIDL and HEC-RAS profile

Flow ($m^3\ s^{-1}$)	MAE
0.035	10.20
0.443	7.99
0.878	7.88

26.4 Discussion

Even though finding the optimal neural network architecture study is beyond the scope of this research, this exploratory analysis is done as a first step because of the importance of this topic in machine learning [4, 6]. The size of a neural network requires an equilibrium between the approximation capacity of unknown functions (Expressivity) and the training cost (Trainability) [1, 2]. A neural networks with a high number of layers and neurons are difficult to train when the available data is limited which is the case for PIDL, but has more expressivity [4]. Nevertheless, a small neural network will not have the capacity to approximate the unknown functions [2]. Furthermore, this reference mentions that the size of a neural network depends on the smoothness of the unknown functions. In this research, the best neural network architecture model had 7 layers with 40 neurons per layer. These results are consistent with that of Raissi et al. [4] architecture of 10 layers with 50 neurons per layer in a research for an incompressible Newtonian flow, but they recognized that other neural network architecture can provide good results as well. Furthermore, Raissi et al. [1] used PIDL for the Burger's equation. This equation appears in several fields of applied mathematics and can be obtained from the Navier–Stokes equation. This experiment has shown that when training and collocation points are fixed, as the number of layers and neurons increase the accuracy of neural networks increase. On the other hand, Tartakovsky et al. [2] empirically found an architecture of 3 layers with 50 neurons per layer for subsurface flow. In this system, the unknown functions include velocity, water depth, area, and hydraulic radius. We believe that tumbling flow in this system makes the unknown functions velocity and water depth irregular. Thus, this system will require a neural network as the one obtained in the optimization process.

The conservation equations represent the backwater curves in river channels applied to long smooth profiles of the water surface, so they are not able to represent abrupt discontinuities. This means that in the application of classical numerical methods numerical oscillations can occur [19]. Thus, several solutions schemes are focus on shock-capturing (García-Navarro et al. [20]). The case study resolved in this study was a step-pool with a tumbling flow which is far from being a smooth solution, so it is an ideal case to test the PIDL stability. On the other hand, PIDL produces a stable solution without the need of a explicit numerical discretization thanks to automatic differentiation (AD) [2] used by TensorFlow [4].

The HEC-RAS model presents a better performance when is compared with the values measured in the field according to Table 26.1. The difference between the HEC-RAS and PIDL models is $\approx 4\%$. Therefore, the results obtained with the PIDL model can be considered as a satisfactory approximation for this complex case study in spite of the limited training data available. Note that the PIDL model presented in this study can be improved by changing neural network architecture, including additional physical restrictions, or by adding flow and geometric data at staff gauges points.

There has been a decreasing difference between PIDL and HEC-RAS profile as flow increases. In the step-pool under study, the steps are formed by boulders and cobbles so the lowest the flow the highest is the interaction of water with these particles. At low flows, there is flow division at steps. This phenomenon cannot be represented in a 1D model. Indeed, there is not a single-jetting flow plunging into pools, but several jetting flows falling into pools at several points. As flow increases the flow division disappears and one single-jetting flow falls into the pools. Thus, the behaviour is closer to the one expected in 1D.

26.5 Conclusions

In this study, a first attempt to use physical informed deep learning (PIDL) in a mountain river has been implemented. The river reach studied is a step-pool located in a hydraulic observatory with limited data on both the flow and the geometry. The deep learning methodology used is based on artificial neural networks. Artificial neural networks were implemented in TensorFlow, with an optimization process in the network architecture.

There were several promising findings in this study. Even though the neural network input was limited, it has an acceptable performance comparable to HEC-RAS. The best PIDL structure agrees with some structures used in the literature, but this research requires further studies to be conclusive. Both models, HEC-RAS and PIDL, as flow increases, get closer to each other according to MAE. This difference at low flows is believed to be produced by a flow division at steps.

Acknowledgements This research was developed within the framework of the project "Towards a sound mathematical description of the dispersion of pollutants in mountain rivers" funded by Research Directorate of the University of Cuenca (DIUC) under the XVIII competition of research projects. This manuscript is an outcome of the Doctoral Program in Water Resources by the first author, offered by Universidad de Cuenca, Escuela Politécnica Nacional, and Universidad Técnica Particular de Loja.

References

1. M. Raissi, P. Perdikaris, G.E. Karniadakis, Physics Informed Deep Learning (Part I): Data-driven Solutions of Nonlinear Partial Differential Equations. arXiv preprint arXiv:1711.10561 (2017). http://arxiv.org/abs/1711.10561. Accessed 21 Aug 2020
2. A.M. Tartakovsky, C.O. Marrero, P. Perdikaris, G.D. Tartakovsky, D. Barajas-Solano, Physics-informed deep neural networks for learning parameters and constitutive relationships in subsurface flow problems. Water Resour. Res. **56**(5) (2020). https://doi.org/10.1029/2019WR026731
3. R. Wang, K. Kashinath, M. Mustafa, A. Albert, R. Yu, Towards physics-informed deep learning for turbulent flow prediction, in *Proceedings of the 26th ACM SIGKDD International Conference on Knowledge Discovery & Data Mining*, pp. 1457–1466 (2020). https://doi.org/10.1145/3394486.3403198.

4. M. Raissi, A. Yazdani, G.E. Karniadakis, Hidden Fluid Mechanics: A Navier-Stokes Informed Deep Learning Framework for Assimilating Flow Visualization Data. arXiv preprint arXiv: 1808.04327 (2018). http://arxiv.org/abs/1808.04327. Accessed 14 Sept 2020
5. A. Güneş Baydin, B.A. Pearlmutter, A. Andreyevich Radul, J. Mark Siskind, Automatic differentiation in machine learning: a survey. J. Machine Learn. Res. **18** (2018)
6. Z. Mao, A.D. Jagtap, G.E. Karniadakis, Physics-informed neural networks for high-speed flows. Comput. Methods Appl. Mech. Eng. **360** (2020). https://doi.org/10.1016/j.cma.2019.112789
7. M. Raissi, P. Perdikaris, G.E. Karniadakis, Physics informed deep learning (Part II): data-driven discovery of nonlinear partial differential equations. arXiv preprint arXiv:1711.10566 (2017). http://arxiv.org/abs/1711.10566. Accessed 1 Sept 2020
8. C. Rao, H. Sun, Y. Liu, Physics-informed deep learning for incompressible laminar flows. Theor. Appl. Mech. Lett. **10**(3), 207–212 (2020). https://doi.org/10.1016/j.taml.2020.01.039
9. A.R. Maxwell, A.N. Papanicolaou, Step-pool morphology in high-gradient streams. Int. J. Sedim. Res. **16**, 380–390 (2001)
10. J.H. Curran, E.E. Wohl, Large woody debris and flow resistance in step-pool channels, Cascade Range. Washington. Geomorphology **51**(1–3), 141–157 (2003). https://doi.org/10.1016/S0169-555X(02)00333-1
11. W.A. MacFarlane, E. Wohl, Influence of step composition on step geometry and flow resistance in step-pool streams of the Washington Cascades. Water Resour. Res. **39**(2) (2003). https://doi.org/10.1029/2001WR001238
12. A.J. Lee, R.I. Ferguson, Velocity and flow resistance in step-pool streams. Geomorphology **46**(1–2), 59–71 (2002). https://doi.org/10.1016/S0169-555X(02)00054-5
13. A. Chin, E. Wohl, Toward a theory for step pools in stream channels. Prog. Phys. Geogr. **29**(3), 275–296 (2005). https://doi.org/10.1191/0309133305pp449ra
14. G. Brunner, HEC RAS, River Analysis System Hydraulic Reference Manual. Version 5.0, USACE CEC: Davis, CA, USA (2016)
15. T. Chai, R.R. Draxler, Root mean square error (RMSE) or mean absolute error (MAE)? Arguments against avoiding RMSE in the literature. Geosci. Model Dev. **7**(3), 1247–1250 (2014). https://doi.org/10.5194/gmd-7-1247-2014
16. C.J. Willmott, K. Matsuura, Advantages of the mean absolute error (MAE) over the root mean square error (RMSE) in assessing average model performance. Climate Res. **30**(1), 79–82 (2005). https://doi.org/10.3354/cr030079
17. P.C. Nayak, B. Venkatesh, B. Krishna, S.K. Jain, Rainfall-runoff modeling using conceptual, data driven, and wavelet based computing approach. J. Hydrol. **493**, 57–67 (2013). https://doi.org/10.1016/j.jhydrol.2013.04.016
18. A. Ritter, R. Muñoz-Carpena, Performance evaluation of hydrological models: Statistical significance for reducing subjectivity in goodness-of-fit assessments. J. Hydrol. **480**, 33–45 (2013). https://doi.org/10.1016/j.jhydrol.2012.12.004
19. R.C. Berger, R.L. Stockstill, Finite-element model for high-velocity channels. J. Hydraul. Eng. **121**(10), 710–716 (1995)
20. P. García-Navarro, F. Alcrudo, J.M. Savirón, 1-D Open-channel flow simulation using TVD-McCormack Scheme. J. Hydraul. Eng. **118**(10), 1359–1372 (1992). https://doi.org/10.1061/(ASCE)0733-9429(1992)118:10(1359)

Chapter 27
Effects of Data Augmentation on the Identification of Cough Sound Using Convolutional Neural Networks

David Naranjo, Juan Chica, and Christian Salamea Palacios

Abstract Cough identification employing deep learning models has generated significant interest in fields such as telemedicine. However, it is challenging to have a considerable number of cough sounds to train such models, and consequently, it is necessary to apply techniques to generate more data. In this work, the effect of data augmentation on the identification of cough was studied using convolutional neural networks (CNN). For this, a system composed of Front-End and Back-End was implemented. In the Front-End, the original recordings and those generated by data augmentation by frequency shift and noise injection were processed and transformed into Mel spectrograms. On the Back-End, the spectrograms were analyzed and classified using a CNN. The results show that the transformations proposed to increase the amount of data help reducing the overfitting by 42.5% while bettering accuracy by a relative improvement of 1%. Therefore, applying data augmentation techniques to train CNNs for differentiating coughs from other sound events results in a simple but effective strategy.

27.1 Introduction

A cough is a sound and convulsive movement that appears as an inherent reflex to the protective mechanisms of the respiratory system. The cough presents with a characteristic sound and an associated body movement and conveys information about the state of the respiratory system. Thus, through cough analysis, it is possible

D. Naranjo (✉)
Escuela Politécnica Nacional, Quito, Ecuador
e-mail: david.naranjo@epn.edu.ec

J. Chica · C. Salamea Palacios
Interaction, Robotics and Automation Research Group, Universidad Politécnica Salesiana, Cuenca, Ecuador
e-mail: csalamea@ups.edu.ec

C. Salamea Palacios
Speech Technology Group, Information and Telecommunication Center, Universidad Politécnica de Madrid, Madrid, Spain

Á. Rocha et al. (eds.), *Communication, Smart Technologies and Innovation for Society*, Smart Innovation, Systems and Technologies 252,
https://doi.org/10.1007/978-981-16-4126-8_27

to inquire about different health conditions that can range from an allergic reaction to more severe clinical conditions such as chronic obstructive pulmonary disease, lung cancer, or other lung diseases that, according to Agustí and Hogg [1], were the cause of one-sixth of the deaths in the world.

Traditionally, cough detection has been based on auscultation with the use of a stethoscope. The responsibility for this procedure to effectively diagnose various respiratory diseases rests mainly with the medical staff, so the right diagnosis will depend on how well trained the professional is. Besides, this procedure mostly requires the patient to go to a hospital or schedule a home medical visit, which does not always allow information to be collected when coughing episodes occur in real time [2].

On the other hand, the gradual emergence of computer-based devices for monitoring cough has been encouraged due to the development of digital technologies. These devices work on pattern recognition based mainly on characteristics extracted from cough sounds and complementary signals such as electromyography of chest movement.

However, most of these systems have only been tested in controlled environments where patients did not engage in any action or physical activity and were not urged to use complex recording systems. Therefore, a limitation of the audio processing methods implemented in these monitors is that they cannot cope with noisy environments [2, 3].

Conversely, a portable cough monitor would allow continuous real-time monitoring of patients with respiratory diseases. This way of monitoring cough will enable capturing reliable information for professionals, which will promote the potential of telemedicine in respiratory diseases. Also, from the point of view of the patient, this monitoring system would be more comfortable and cause minimal disruption to their daily activities. Nevertheless, the development of portable monitors implies implementing systems that can identify and discriminate the cough from any other sound event recorded in potentially noisy environments.

The problem of differentiating cough from other sound events has been approached from different perspectives. Thus, for example, in the field of machine learning, Khunarsal et al. [4] implemented an algorithm for the classification of ambient sound using spectrograms and K-nearest neighbors classifiers. Mitra and Wang [5] proposed an audio classification technique based on the analysis of audio content using multilayer perceptrons. Also, Costa et al. [6] addressed a similar problem using a support vector machine.

On the other hand, Matos et al. [7]implemented hidden Markov models to detect cough signals from continuous audio recordings. However, the most promising results have been achieved by implementing deep learning algorithms, especially using convolutional neural networks (CNN), as is the work carried out by Amoh and Odame [8] who effectively implemented a CNN to identify cough sounds.

When working with CNN, the input data must be images. In sound classification, authors such as Wang et al. [9] or Barata et al. [10] have suggested techniques such as Relative Spectral Transform—Perceptual Linear Prediction or Mel-Frequency Cepstral Coefficients to transform sound into spectrograms. In this way, the audio recognition task becomes a visual recognition task. Applying these techniques, they have modeled CNN with classification accuracies over 90%, which have been successfully implemented on portable cough monitors.

CNNs are used to process images simulating the behavior of neurons in the primary visual cortex of a biological brain. They can learn input–output relationships and they are based on convolution operations. Thus, conventionally, a CNN is made up of convolution layers, pooling layers, and fully connected layers [11, 12].

This structure involves a large number of internal parameters that are adjusted during learning. This large number of parameters makes CNNs prone to overfitting, which occurs when the model learns patterns underlying the training dataset but cannot generalize to recognize new data and therefore does not perform an efficient classification [8, 11, 12].

To avoid this problem, techniques such as dropout are applied, consisting of disconnecting a group of neurons in each training iteration [13]. However, one of the most efficient ways to deal with overfitting is to increase the number of samples in the training set as much as possible. Nevertheless, the data used to identify cough is relatively difficult to obtain because it involves recording a reflex action, that is, an unplanned event. Therefore, the available databases are often not large enough to train deep neural networks [14].

A technique employed to face this problem is data augmentation, which consists of creating new data through transformations on the original data.

When using CNN, these data augmentation transformations are performed on the input images. Transformations of rotation, translation, scaling, clipping, among others can be achieved. In this way, when augmenting data, it is possible to prevent the model from learning irrelevant patterns, optimizing the resolution of classification problems as in the case of the AlexNet [15] model that implements two data augmentation techniques to improve its performance [14, 16].

The present study focuses on implementing a CNN model to address the problem of identifying cough as a visual recognition task, and the effect on the performance of the model is studied when applying data augmentation techniques on the original audio recordings.

27.2 Methodology

27.2.1 Database

The database used in this study consists of 266 recordings of coughs and 69 recordings of other sound events such as conversations, laughter, or ambient noise. The recordings are in WAV format and last between 1 and 112 s. The participants made the recordings telematically and anonymously after accepting informed consent.

27.2.2 Data Augmentation Techniques

The augmented data was obtained by modifying original recordings. This was achieved through a function that took as inputs the audio files, the sampling rate, and the variation factor, either for noise injection or frequency shift.

Noise injection adds random amplitude values to the original audio sample. Whereas by frequency shift, the pitch of the original audio sample was randomly varied.

The variation factor, both for noise injection and frequency shift, was adjusted to a maximum of 20% to maintain the information on the problem domain invariants, which are used to classify the data [14, 17].

Applying data augmentation techniques made it possible to build a database five times the size of the original database.

27.2.3 System Architecture

The system architecture consisted of two components: Front-End and Back-End. In the Front-End component, the audio signal is preprocessed. Consequently, the detection of cough is transformed into a visual recognition task by generating the corresponding Mel spectrograms. In this context, first, the duration of the recordings was standardized to the modal value of 3 s. Thus, the recordings with longer duration were cut into 3 s segments, while the recordings with a shorter duration were replicated and subsequently concatenated until reaching 3 s. The resulting samples were reanalyzed to discard all those with silent intervals. This procedure also allowed balancing the data with a ratio of cough sounds to other sound events of around 1.08.

Subsequently, the processed audio samples were transformed into Mel spectrograms (MFCC). This graphic representation was chosen since it has been widely used in the fields of speech recognition and the identification of speaker or environmental sound events [18, 19]. The Mel spectrograms were generated with a window size of 30 ms, a sample frequency of 48000 Hz, and a hop length of 550.

On the other hand, in the Back-End component, the Mel spectrograms are analyzed through a convolutional neural network which enables to determine whether an audio signal came from a cough sound or another sound event. The convolutional neural network architecture used in this study is based on the LeNet-5 architecture, which works well on small datasets. Additionally, the number of neurons in each layer was adapted according to Amoh and Odame [8] who worked with a reduced size dataset.

Analogous to LeNet-5 [8, 20], the CNN consisted of five layers, divided into two convolutional layers, two fully connected layers, and a binary classification layer. The two convolutional layers have 16 rectified linear units (ReLU) each. The first convolutional layer has a 9×3 size filter and takes the 67×300 spectral segments of Mel as inputs. Next, a 2×1 layer of max-pooling is implemented, followed by the second convolutional layer with a filter size of 5×3 and another layer of $2 \times$

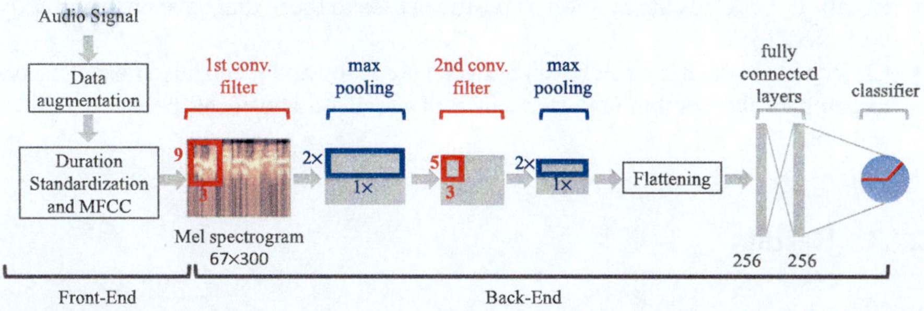

Fig. 27.1 System architecture

1 of max-pooling. Two fully connected layers follow the convolutional layers with 256 units for each one.

In each layer, in addition, batch regularization was implemented, and dropout regularization was used with a rate of 0.5 to reduce the probabilities of overfitting. Finally, the last layer implemented the ReLU function to determine if the analyzed audio signal is a cough sound or another sound event. The system and CNN architecture are shown in Fig. 27.1.

Four models were trained: the first on the original data set, the second on the original data set and the augmented data by frequency shift, the third with on original data set and the augmented data by noise injection, and the fourth on the original data set and the augmented data by noise injection and frequency shift.

The loss function used was the binary cross-entropy. The ratio between the size of the training dataset and the batch for all models was 8. Likewise, an early stopping regularization with a monitor in the validation dataset loss was implemented to reduce overfitting. All models were developed in a Python environment with the Keras library.

27.2.4 Performance Metrics

The evaluation of the model performance was based on the following metrics:

- Binary cross-entropy loss: It describes the loss between two probability distributions, using a logarithmic penalty that generates a large score for large differences and a small score for small differences. The binary cross-entropy loss is used when adjusting the model weights during training. The goal is to minimize the loss; that is, the lower the loss, the better the model [21].
- Accuracy: The overall success rate of the classifier or the fraction of predictions that the model made correctly [22, 23].
- Precision: It measures the relationship between the true positives and the instances classified as positive[23].

- Recall: It measures the relation of positive cases to those that have been correctly classified[23].
- F1-score: It is the harmonic mean between precision and recall, also makes it easy to compare the combined performance of precision and recall between various solutions [23, 24].

27.3 Results

Figures 27.2, 27.3, 27.4, and 27.5 show the graphs of the performance metrics of the four trained models. In all models, the binary cross-entropy loss at the time of early stopping was lower for the training data set than for the validation data set. Nevertheless, in all cases, except the one using data augmentation by noise injection, the binary cross-entropy loss for the validation data converged to values less than 0.5.

On the other hand, it is observed that both accuracy and f1-score of the training data set were higher compared to the values corresponding to the validation data set. In all cases, accuracy and f1-score for the validation data set reached values higher than 0.9.

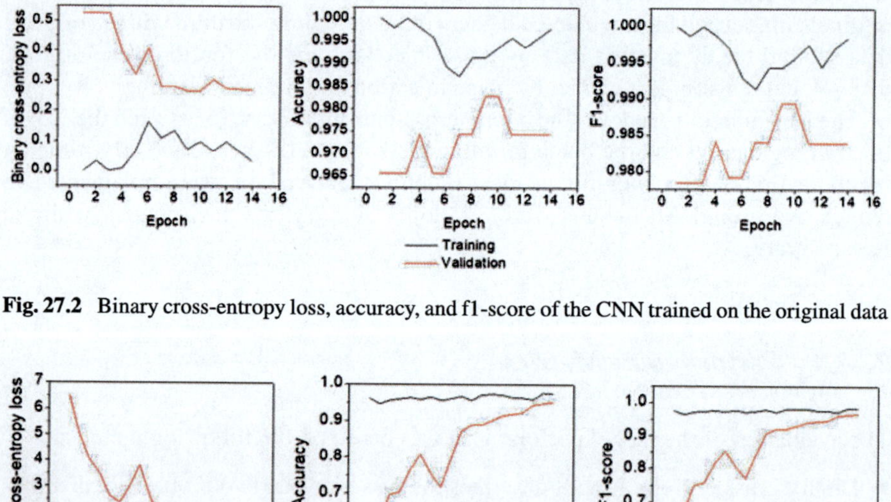

Fig. 27.2 Binary cross-entropy loss, accuracy, and f1-score of the CNN trained on the original data

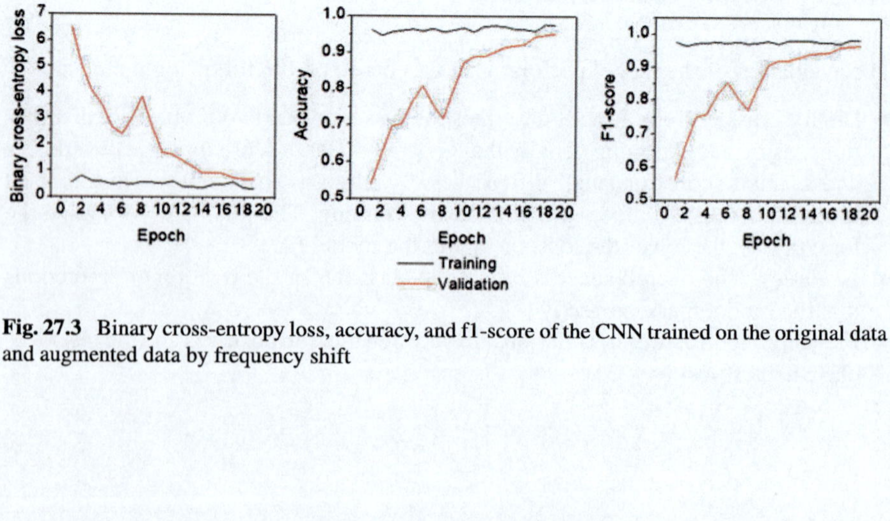

Fig. 27.3 Binary cross-entropy loss, accuracy, and f1-score of the CNN trained on the original data and augmented data by frequency shift

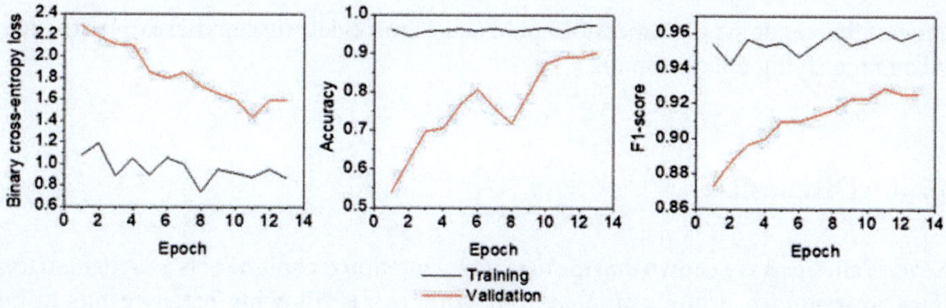

Fig. 27.4 Binary cross-entropy loss, accuracy, and f1-score of the CNN trained on the original data and augmented data by noise injection

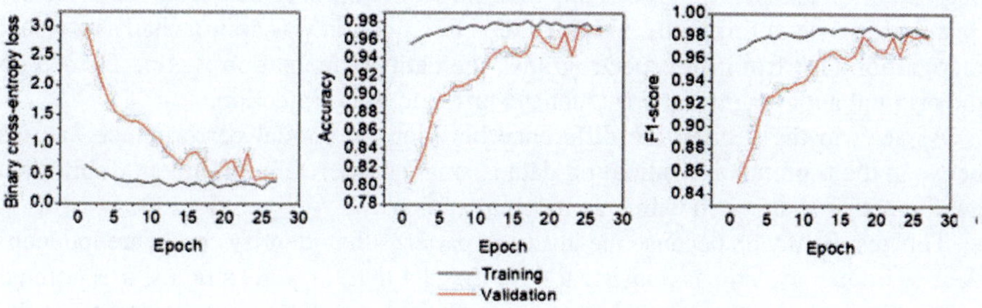

Fig. 27.5 Binary cross-entropy loss, accuracy, and f1-score of the CNN trained on the original data and augmented data by frequency shift and noise injection

Table 27.1 shows the differences between the performance metrics on the training dataset and those of the validation dataset at the time of the early stopping.

In all cases, it is observed that the differences for all performance metrics are smaller in the model in which data augmentation was used both by frequency shift and noise injection.

Furthermore, it is observed that when the data augmentation only by frequency shift is used, smaller differences are also obtained between the performance metrics than when using only the original data. On the contrary, when data augmentation is applied only by noise injection, smaller differences are obtained between the performance metrics of the model trained with the original data. This behavior could

Table 27.1 Absolute difference between training and validation performance metrics of the four trained models

Model	Binary cross-entropy loss	Accuracy	F1-score
Or	0.27040	0.02560	0.01600
Or + FS	0.34020	0.02030	0.01400
Or + NI	0.58380	0.03720	0.03090
Or + FS + NI	0.16740	0.01320	0.00920

Or Original data set, *FS* Frequency shift, *NI* Noise injection)

suggest that random noise injection could make the models susceptible to overtraining when identifying cough sounds.

27.4 Discussion

Several studies have shown that the data augmentation technique acts as a regularizer, thus preventing overfitting in neural networks while allowing better results to be obtained in unbalanced class situations [14–17].

In fact, the fewer data available, the fewer data to train, and the less chance of obtaining accurate predictions on data that the model has not yet seen. Thus, data augmentation takes the approach of generating more training data from the available data. In the case of this study, although the classification was approached as a visual recognition task from Mel spectrograms, the data augmentation was performed on the original audio signals, by frequency shift, and noise injection.

As seen in the results, the difference between the model performance metrics between the training and validation data is smaller when using frequency shift data augmentation than when using the original audios.

This result may be because the invariant patterns that identify cough are independent of frequency. This is consistent with the fact that, in some studies, it is pointed out that the sounds that give a characteristic pitch to coughing have their origin in the larynx. However, not all people produce these sounds at the same intensity. There are even people who do not produce them at all due to genetic factors or as a consequence of surgical interventions [25, 26]. Likewise, when there is a frequency shift in other sound events such as laughter or conversations, although the changes are notable, the timbral nature is not altered to a greater extent. Therefore, the model could perceive it as a completely different recording but belonging to the same class. Consequently, the frequency shift on the recordings, by a factor of no more than 20%, allowed the model to focus on patterns that are independent of the cough pitch.

On the other hand, it was observed that when applying data augmentation by noise injection, the difference between the performance metrics of the model between the training and validation sets is more significant than when the original recordings are used. In this context, according to Van Hirtum and Berckmans [27] and Korpáš et al. [25], sounds that could be characterized as noise can be indicators that a person smokes that they suffer from certain respiratory diseases, or the level of mucus in the trachea. Consequently, randomly adding noise could interfere with characteristic patterns that allow identification of the cough.

However, when using both data augmentation techniques, the model performance metrics difference between the training and validation data is smaller than when using the original recordings. Therefore, it is observed that in this study, the application of this technique allows the classification task to be generalized in a better way.

In this study, by training CNN with augmented data, a classification accuracy of 97.50% was achieved. This result is equipable with those reported in the literature, such as Amoh et al. [8], Wang et al. [9], or Barata et al. [10], who achieved classification accuracies greater than 90% on identifying cough sounds by implementing CNN. This fact corroborates that CNNs allow tackling the task of recognizing cough efficiently [28]. However, applying data augmentation increased accuracy on the validation set by less than 1% compared to the original recordings.

This slight improvement could be because, since the original recordings are relatively short and were recorded under different conditions, the modifications applied had to be very small. Therefore, it is likely that they will not have a significant effect on model learning. Also, the performance of the model without using data augmentation is quite good (96.7% accuracy), which makes the task of improving it more difficult.

27.5 Conclusions

Having a significant number of cough recordings is not always possible, making the task of developing automatic cough identification systems difficult. Therefore, this work studied the effects of data augmentation in identifying coughs from other sound events through the use of CNN.

The experiment results show that the combined use of the data generated by frequency shifting and noise injection in the original recordings decreased the difference in the model performance measure between the training and validation sets. At the same time, an absolute improvement of the accuracy of the model is achieved. Consequently, it is concluded that the transformations proposed to increase the amount of data help to reduce the overfitting of the model, which makes it possible to use these procedures to train deep convolutional neural networks oriented to the differentiation of coughs among other sound events.

Although the results of this study are encouraging, it is essential to emphasize that the new inputs received by the neural network are still highly correlated which is a limitation. This is because augmented data come from a small amount of original data, which could be triggered by the fact that no new information is being produced in reality, but only current information has been mixed. As such, this might not be enough to get rid of overfitting in some cases completely. For this reason, other techniques such as regularization or dropout were applied. Sometimes, even so, not having enough data does not enable to train good models. For these cases, future work should explore some techniques such as transfer learning or feature extraction.

Acknowledgements To the Corporación Ecuatoriana para el Desarrollo de la Investigación y Academia, CEDIA, for the financing provided to research, development, and innovation through the CEPRA projects, especially the project CEPRA-XV-2021-011: Caracterización de la tos provocada por el COVID-19 en pacientes con diagnóstico positivo.

The authors thank to Escuela Politécnica Nacional, Universidad Politécnica Salesiana, and Pontificia Universidad Católica del Ecuador.

References

1. A. Agustí, J.C. Hogg, Update on the pathogenesis of chronic obstructive pulmonary disease. J. New Engl. Med. Rev. **381**, 1248–1256 (2019). https://doi.org/10.1056/NEJMra1900475
2. A. Spinou, S.S. Birring, An update on measurement and monitoring of cough: what are the important study endpoints? J. Thorac. Dis. **6**(7), 728–734 (2014). https://doi.org/10.3978/j.issn.2072-1439.2014.10.08
3. C. Domingo, A. Sogo Sagardía, Tos crónica: viejos problemas, nuevas perspectivas. Rev. Asma **1**(3), 98–105 (2016)
4. P. Khunarsal, C. Lursinsap, T. Raicharoen, Very short time environmental sound classification based on spectrogram pattern matching. Inf. Sci. (Ny) **243**, 57–74 (2013). https://doi.org/10.1016/j.ins.2013.04.014
5. V. Mitra, C. Wang, Content based audio classification: a neural network approach. Soft Comput. **12**, 639–646 (2008). https://doi.org/10.1007/s00500-007-0241-4
6. Y.M.G. Costa, L.S. Oliveira, A.L. Koerich, F. Gouyon, J.G. Martins, Music genre classification using LBP textural features. Signal Process. **92**(11), 2723–2737 (2012). https://doi.org/10.1016/j.sigpro.2012.04.023
7. S. Matos, S.S. Birring, I.D. Pavord, D.H. Evans, Detection of cough signals in continuous audio recordings using hidden Markov models. IEEE Trans. Biomed. Eng. **53**(6), 1078–1083 (2006). https://doi.org/10.1109/TBME.2006.873548
8. J. Amoh, K. Odame, Deep neural networks for identifying cough sounds. IEEE Trans. Biomed. Circuits Syst. **10**(5), 1003–1011 (2016). https://doi.org/10.1109/TBCAS.2016.2598794
9. H.H. Wang, J.M. Liu, M. You, G.Z. Li, Audio signals encoding for cough classification using convolutional neural networks: a comparative study, in *Proceedings of the 2015 IEEE International Conference on Bioinformatics and Biomedicine (BIBM 2015)*, pp. 442–445 (2015). https://doi.org/10.1109/BIBM.2015.7359724
10. F. Barata, K. Kipfer, M. Weber, P. Tinschert, E. Fleisch, T. Kowatsch, Towards device-agnostic mobile cough detection with convolutional neural networks, in *2019 IEEE International Conference on Healthcare Informatics, ICHI 2019*, pp. 1–11 (2019). https://doi.org/10.1109/ICHI.2019.8904554
11. N. Aloysius, A review on deep convolutional neural networks, in *International Conference on Communication and Signal Processing*, pp. 588–592 (2017). https://doi.org/10.1109/ICCSP.2017.8286426
12. W. Rawat, Z. Wang, Deep convolutional neural networks for image classification : a comprehensive review. Neural Comput. **29**, 2352–2449 (2017). https://doi.org/10.1162/NECO
13. N. Srivastava, G. Hinton, A. Krizhevsky, I. Sutskever, R. Salakhutdinov, Dropout : a simple way to prevent neural networks from overfitting. J. Mach. Learn. Res. **15**, 1929–1958 (2014)
14. C. Shorten, T.M. Khoshgoftaar, A survey on image data augmentation for deep learning. J. Big Data **6**(60), 1–48 (2019). https://doi.org/10.1186/s40537-019-0197-0
15. A. Krizhevsky, G.E. Hinton, ImageNet classification with deep convolutional neural networks. Adv. Neural Inf. Process. Syst, 1–9 (2012)
16. M.A. García, Data Augmentation para la Clasificación Automática de la Calidad Vocal Data Augmentation in Automatic Classification of Voice Quality (2020)
17. X. Cui, S. Member, V. Goel, S. Member, B. Kingsbury, S. Member, Data Augmentation for deep neural network acoustic modeling. IEEE/ACM Trans. Audio, Speech, Lang. Process. **23**(9), 1469–1477 (2015). https://doi.org/10.1109/TASLP.2015.2438544
18. G. Arturo, M. Mascorro, A. Torres, Reconocimiento de voz basado en MFCC, SBC y Espectrogramas. Ingenius, pp. 12–20 (2013)
19. M.A. García, E.A. Destéfanis, Power cepstrum calculation with convolutional neural networks. J. Comput. Sci. Technol. **19**(2), 132–142 (2019). https://doi.org/10.24215/16666038.19.e13
20. Y. Lecun, L. Bottou, Y. Bengio, P. Hafner, Gradient-based learning applied to document recognition. IEEE Proc. **86**(11), 2278–2324 (1998). https://doi.org/10.1016/j.bbrc.2005.03.111

21. D. Ramos, J. Franco-Pedroso, A. Lozano-Diez, J. Gonzalez-Rodriguez, Deconstructing cross-entropy for probabilistic binary classifiers. Entropy **20**(208), 1–20 (2018). https://doi.org/10.3390/e20030208
22. X. Deng, Q. Liu, Y. Deng, S. Mahadevan, An improved method to construct basic probability assignment based on the confusion matrix for classification problem. Inf. Sci. (Ny) (2016). https://doi.org/10.1016/j.ins.2016.01.033
23. B. Juba, H.S. Le, Precision-recall versus accuracy and the role of large data sets. Proc. AAAI Conf. Artif. Intell. **33**, 4039–4048 (2019). https://doi.org/10.1609/aaai.v33i01.33014039
24. D. Chicco, G. Jurman, The advantages of the Matthews correlation coefficient (MCC) over F1 score and accuracy in binary classification evaluation. BMC Genomics **21**(6), 1–13 (2020). https://doi.org/10.1186/s12864-019-6413-7
25. J. Korpáš, J. Sadloňová, M. Vrabec, Analysis of the cough sound: an overview. Pulm. Pharmacol. **9**(5–6), 261–268 (1996). https://doi.org/10.1006/pulp.1996.0034
26. P.G. Gibson, A.E. Vertigan, Speech pathology for chronic cough: a new approach. Pulm. Pharmacol. Ther. **22**(2), 159–162 (2009). https://doi.org/10.1016/j.pupt.2008.11.005
27. A. Van Hirtum, D. Berckmans, Assessing the sound of cough towards vocality. Med. Eng. Phys. **24**(7–8), 535–540 (2002). https://doi.org/10.1016/S1350-4533(02)00055-3
28. J. Monge, C. Hoyos-Barcelo, P. Lesso, P. Casaseca, Robust detection of audio-cough events using local Hu moments. IEEE J. Biomed. Heal. Informatics **23**(1), 184–196 (2019). https://doi.org/10.1109/JBHI.2018.2800741

Chapter 28
Analysis of Energy Efficient in Traction and Breaking for Electric Vehicles Through Laboratory Experiments

Juan D. Valladolid, John S. Abril, and Alexis D. Valdiviezo

Abstract The following article performs an energy efficiency analysis in the traction and braking electric vehicle systems which delivers a maximum torque of 285 Nm, with a 90 kW battery, and 145 km/h as maximum speed, being the main objective to find the optimal vehicle performance points at the driving moment. In order to achieve the objective, "EMOLAB" acquisition software is used, and this one is linked to the vehicle's OBDII port via Bluetooth from which the obtained data were torque, angular speed, acceleration, current and an inclinometer to determine slopes. These were obtained through on board route tests to three different destinations in the city of Cuenca-Ecuador: Tres Cruces (4147 m.a.s.l), Nulti (2577 m.a,s.l) and Historic Center (2549 m.a.s.l,), the starting point being the Salesian Polytechnic University (2550 m.a.s.l). Data were used for efficiency calculations, input and output power, so in this way generate maps to find points where the vehicle wastes considerably the delivered energy by its battery to recommend driving strategies where this energy is used to increase the efficiency and vehicle autonomy.

28.1 Introduction

Nowadays, in terrestrial mobility, the vehicle is one of the main causes of environmental pollution in urban areas. This is the main reason that many countries including the USA, China, Japan, and European countries have chosen to make a transition within the automotive fleet and give priority to electric vehicles (EV), in this way reduce the air pollution caused by combustion vehicles emissions [1], considering that EVs do not present emissions and have high energy efficiency [2].

J. D. Valladolid (✉) · J. S. Abril · A. D. Valdiviezo
Department of Automotive Engineering, Universidad Politécnica Salesiana, Cuenca 101007, Ecuador
e-mail: jvalladolid@ups.edu.ec

J. D. Valladolid
Department of Electronics Engineering, Pontificia Universidad Javeriana, 110321 Bogotá, Colombia

Á. Rocha et al. (eds.), *Communication, Smart Technologies and Innovation for Society*, Smart Innovation, Systems and Technologies 252,
https://doi.org/10.1007/978-981-16-4126-8_28

However, electric vehicles have not been fully accepted in many countries, due to the reduced autonomy of their battery, power losses in the inverter, transmission and friction [3], its main disadvantage being limited travel driving [4].

For this reason, it is essential to generate battery optimization strategies for these vehicles, since they promise a more efficient travel path in terms of energy consumption [5], so it is proposed to carry out an energy efficiency analysis in the traction and braking systems of the electric vehicle.

Therefore, as a starting point, an article that raises the need to study the components of the electric vehicle (EV) is analyzed [3], where they generate considerable power losses, and thus find the optimal operating points of the EV, which presents opportunities and possibilities of proposing energy optimization strategies, thus seeking to obtain greater efficiency and autonomy.

In this way, this article proposes an energy efficiency analysis based on driving cycles and route tests on board on roads with different inclination angles, because the EV consumes its energy differently according to the route on which it circulates [6].

From this perspective, it is proposed to generate energy efficiency maps for the electric vehicle (EV) traction system, based on a mathematical model presented in [7]. To obtain speed, torque, current, voltage, and lean angle data, real driving tests will be carried out on the three routes.

For the on board route tests, three different routes have been proposed, starting from the Salesian Polytechnic University (2550 m.a.s.l) (Lat.: −2.8869248°, Long.: −78.9892568°) 2: Route One toward Tres Cruces (4147 m.a.s.l) (Lat.: −2.7775089°, Long.: −79.240712°) a distance of 40 km; Route Two toward Nulti (2577 m.a.s.l) (Lat.: −2.8641804°, Long.: −78.8942409°) a distance of 32 km, on these routes due to altitude variation the vehicle suffers an excessive energy consumption depending on the cycle of driving [14], and Route Three toward Cuenca Historic Center (2549 m.a.s.l,) (Lat.: −2.8974172°, Long.: −79.0044893°) a distance of 3 km. To establish optimization strategies, it must be considered that energy efficiency is very important for electric vehicles to improve their autonomy [15].

28.2 Methods

28.2.1 Energy Efficiency

The efficiency term in electric motors is directly related to the difference of the output power and the input power, considering the power losses, losses generated between the difference of the input power and the output power. Power losses are generated by different factors such as iron losses, copper losses, inverter losses, friction losses, and mechanical losses [8].

Thus, efficiency is expressed by the following equation:

$$\eta = \frac{P_{\text{Out}}}{P_{\text{In}}} \tag{28.1}$$

where P_{out} and P_{in} are the output power and the input power, respectively, and are represented as follows:

$$P_{\text{In}} = V_{\text{Bat}} I \tag{28.2}$$

where V_{Bat} is battery voltage and I is battery current.

$$P_{\text{Out}} = R_T V \tag{28.3}$$

where R_T is the drag and V is the speed of the vehicle in m/s.

Drag—This can be expressed by means of the following equation.

$$R_T = P + F_{xa} + mg\sin(\beta) + u_f m\cos(\beta) \tag{28.4}$$

where P is a weigh, F_{xa} is the aerodynamic force, m is the mass, g is the gravity, β is the inclination angle, and u_f is the aerodynamic drag coefficient.

Figure 28.1 shows the diagram of external forces acting on the vehicle. The motion dynamics model of electric car describes interaction between car, tires, roadway, and driving system of car [10, 11].

Aerodynamic Force—This force is given by the following equation:

$$F_{xa} = \frac{1}{2}\rho C_x A_f V^2 \tag{28.5}$$

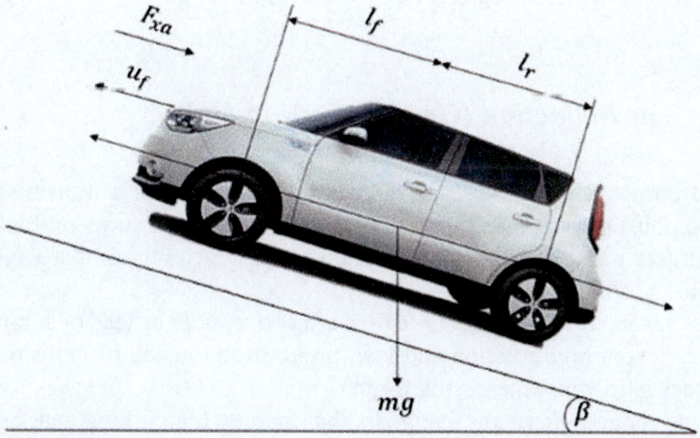

Fig. 28.1 Diagram of external forces acting on the vehicle

Table 28.1 Parameters for the dynamic model on the Kia Soul EV [10]

Parameter	Value	Unit
Total mass of the vehicle	1670	kg
Aerodynamic drag coefficient	0.35	–
Frontal area vehicle	2.3	m^2
Air density	0.96	kg/m^3
Tire radius	0.325	m
Rolling resistance coefficient	0.015	–
Distance from gravity center to front axle	1.2	m
Distance from gravity center to rear axle	1.4	m
Track width	1.576	m
Max. torque of electric motor	285	Nm
Ratio in the transmission system	8.2	–

where ρ represents the air density, C_x represents the aerodynamic coefficient, A_f represents the vehicle frontal area, and V represents the vehicle speed.

It is essential to mention that for our study, an altitude of 2550 m above sea level has been considered. For this reason, the air density of 0.96 kg/m^3 has also been considered [12]. The rolling resistance coefficient is calculated in relation to the different types of road. Taking into consideration the real slopes of the routes, a value of 0.015 has been considered. The aerodynamic coefficient has been considered by the value established by the Kia Soul EV manufacturer, thus obtaining a value of 0.35. Finally, the front area value of the vehicle has likewise been a value obtained by the data established by the manufacturer, thus obtaining a value of 2.3 m^2.

In Table 28.1, the data required for the calculation of the aerodynamic force have been represented, which are specified by the manufacturer.

28.2.2 *Gear Reduction (Gear Reduction Unit)*

The speed bumps for an EV serve the same role as that of a transmission in an internal combustion vehicle. Internal combustion vehicles have multiple speeds, and the reducer has a constant transmission ratio, depending on the engine power delivered to the shaft [13].

For this reason, this component is called a speed reducer instead of a transmission. These receive very high rotation and low torque from the engine and slow down to a proper gear ratio while increasing torque.

Inside the reducer there are five gears that include the parking gear and manual transmission oil that is not meant to be replaced. The main function of the reduction gears is to reduce the speed, that is, for a greater driving power of the motor, which has a transmission ratio of 8,206. In order to establish optimization strategies, it must be taken into account that energy efficiency is extremely important for electric vehicles to improve their autonomy [15].

28.3 Results and Discussion

For the experimentation, the real driving routes were carried out on the three routes and in this way proceeded to acquire data through the established equipment, for which certain considerations were established:

- Keep the tire inflation pressure recommended by the manufacturer (30 PSI).
- Correctly instrument the equipment before starting the route.
- Store and analyze the obtained data.

Different tests were carried out in each one of the established routes in order to obtain a greater amount of information and in this way to be able to select the best quality of the data to work on. The data acquisition system used in this experiment provides vehicle-specific variables such as torque (Nm), road inclination (Degrees °), battery current amperes (A), and vehicle speed (km/h) with a sampling rate of $T_s = 1$ s [16].

Before obtaining the surfaces and the efficiency maps of the EV in real driving routes, it is important to present curves respect to the variables obtained in the different routes with which the efficiency maps will be generated by means of surfaces, and the variables obtained can be seen in Fig. 28.2.

It is important to mention that presented variables in Fig. 28.2 are variables that have been obtained in the three different routes that have been analyzed, taking into consideration that the two routes are very similar in terms of the geographical situation of the city in different places, and in a specific route, a relatively flat surface has been considered when making the route in the Historic Center of the Cuenca city.

28.3.1 Route One Results

Figure 28.3 shows the efficiency maps of the aforementioned variables being the yellow area of the map where the best performance of the vehicle is shown. Depending on the speed, this generates an efficiency of 90.6% at 53.57 km/h with a current consumption of 70.7 A. On the other hand, depending on the torque, the EV generates an efficiency of 90% at 64.94 Nm with a current consumption of 90 A.

Figure 28.4 shows the efficiency maps of the regenerative processes of the EV, produced by the vehicle's inertia movement and by the decelerations caused by the brake use. The maximum peaks have an efficiency of 89.2% with a consumption of 7.12 A at 62.7 km/h and 82.1% with a consumption of 17.5 A at 12.1 Nm.

Table 28.2 shows the results obtained from the three planned routes, taking into account the surfaces presented in Fig. 28.3, surfaces that have been prepared taking

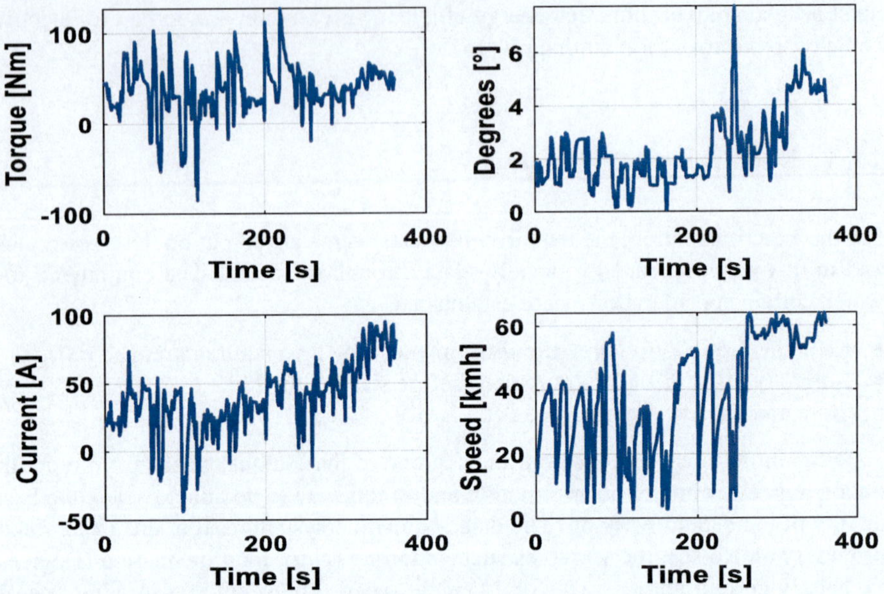

Fig. 28.2 Obtained variables from the EV

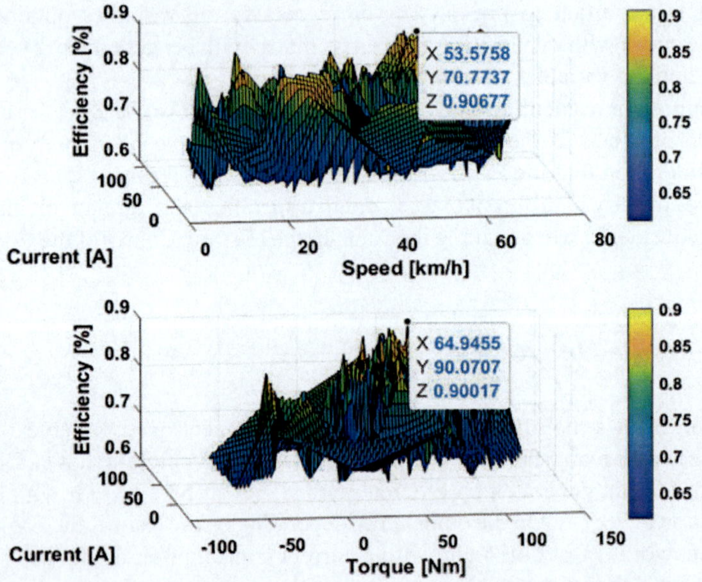

Fig. 28.3 Efficiency maps on Route One

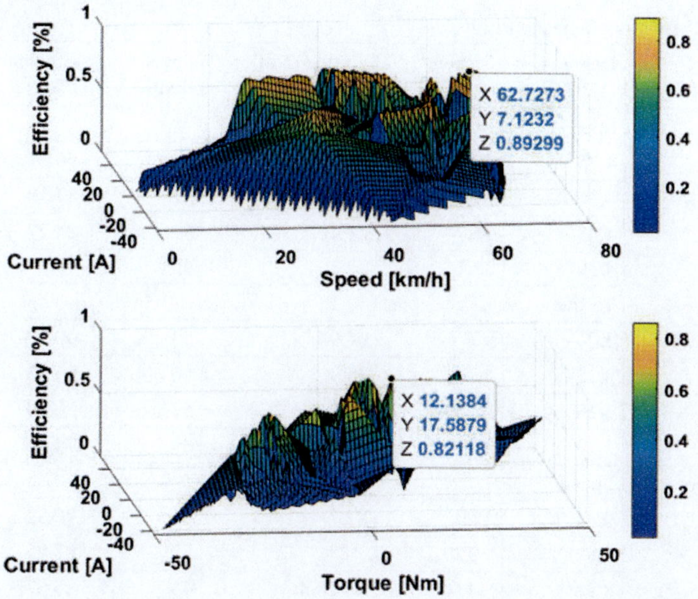

Fig. 28.4 Efficiency maps considering regenerative processes in Route One

into account the variables obtained from the EV, corresponding to torque, current, speed, inclination, and efficiency.

As the surfaces are presented in Fig. 28.3, the efficiency and current consumption are represented as a function of speed and torque, respectively, it is so that in Table 28.2 the most significant efficiency values have been represented in each one of the analyzed routes, thus obtaining the data shown, taking into consideration that the shaded data for both speed and torque represent the highest efficiencies in each of the routes.

It is also important to mention that the road inclination does not generate major repercussions in terms of vehicle efficiency, unlike speed and torque as already mentioned.

28.4 Conclusions

In this article, an energy efficiency analysis has been developed under real driving cycles based on current consumption on different routes in the city of Cuenca with different road inclinations. On Route One, the optimum efficiency point is 90.6% at a speed of 53.57 km/h, however, the EV generates efficiencies greater than 80% at speeds over 51 km/h and torques greater than 20.2 Nm.

Table 28.2 Energy efficiency results

	Degree [°]	Current [A]	Torque [Nm]	Speed [km/h]	Efficiency [%]
Route One	Speed				
	4.75	70.7	51.4	53.7	90.6
	4.3	67.7	50.5	52	87.18
	4.53	22.7	21.4	51	84.27
	4.39	22.1	20.2	40	75.3
	Torque				
	3.24	90	64.94	52	90
	3.23	46	35.9	50	85.93
	3.28	28.6	20.2	51	80.59
	3.17	21.1	19.7	39	74.96
Route Two	Speed				
	9.31	17.3	82.7	61	90.32
	8.82	−129.1	−87	59	87.95
	8.78	−116.6	−118.8	45	85.6
	9.26	31.1	44.5	42	80.69
	Torque				
	3.96	41.7	26.6	64	89.35
	3.90	34.1	27	52	83.55
	3.90	38.3	41.2	38	76.95
	3.88	25.7	62.7	16	69.92
Route Three	Speed				
	0.4	−6.1	−14	25	90.2
	0.5	−3.1	8.4	23	89.25
	0.4	3.2	16.1	21	87.9
	0.4	19.5	20.5	20	84.71
	Torque				
	0.2	10.5	22.7	19	89.45
	0.5	18.1	40	17	80.36
	0.3	15.8	42.2	15	73.76
	0.2	15.4	45.9	13	65.58

On Route Two, the optimum efficiency point is 90.32% at a speed of 61 km/h; however, the vehicle generates efficiencies greater than 80% at speeds over 42 km/h and torques less than 27 Nm.

In the case of Route Three, the most optimal point of efficiency is 90.2% at a speed of 25 km/h, considering that the surface where the vehicle travels, the slopes tend to zero, due to the characteristics of the geographical space of this route; therefore, the greatest efficiencies are generated with speeds greater than 17 km/h and torques less than 40 Nm.

Regarding the braking system, the efficiency is high, thus having values around the 90% at speeds between 20 and 62.7 km/h, and this is due to the fact that regenerative processes of the EV are considered during the journey, generally in negative slopes or descents due to the effect of the vehicle's inertia.

References

1. K. Kiddee, T. Sukcharoensup, Efficiency analysis of suspension energy harvesting based on supercapacitor for electric vehicles, in *2020 8th International Electrical Engineering Congress (iEECON)*, Chiang Mai, Thailand, pp. 1–4 (2020)
2. B. Liu, L. Li, X. Wang, S. Cheng, Hybrid electric vehicle downshifting strategy based on stochastic dynamic programming during regenerative braking process. IEEE Trans. Veh. Technol. **67**(6), 4716–4727 (2018)
3. J.D. Valladolid, R. Albarado, D. Mallahuari, D. Patiño, Experimental performance evaluation of electric vehicles (EV) based on analysis of power and torque losses, in *2020 IEEE International Conference on Industrial Technology (ICIT)*, Buenos Aires, Argentina, pp. 933–938 (2020)
4. H. Jung, G. Park, D. Kim, S. Jung, Optimal design and validation of IPMSM for maximum efficiency distribution compatible to energy consumption areas of HD-EV. IEEE Trans. Magnetics **53**(6), 1–4, Art no. 8201904 (2017)
5. R. Ristiana et al., Energy efficient speed control using IBEV model in an EV testbed platform, in *2018 IEEE 8th International Conference on System Engineering and Technology (ICSET)*, Bandung, pp. 46–50 (2018)
6. C.T. Krasopoulos, M.E. Beniakar, A.G. Kladas, Multicriteria PM motor design based on ANFIS evaluation of EV driving cycle efficiency. IEEE Trans. Transp. Electrification **4**(2), 525–535 (2018)
7. X. Zhang, D. Göhlich, J. Li, Energy-efficient toque allocation design of traction and regenerative braking for distributed drive electric vehicles. IEEE Trans. Veh. Technol. **67**(1), 285–295 (2018)
8. J. Gu, M. Ouyang, D. Lu, J. Li, L. Lu, Energy efficiency optimization of electric vehicle driven by in-wheel motors (2013)
9. M. Thompson, C. Hoff, J. Gover, A Model to Estimate the Effect of DC Bus Voltage on HEV Powertrain Efficiency (2010)
10. J.D. Valladolid, H.D. Montesdeoca, M.B. Ortiz, J.P. Ortiz, Optimal adjustment of parameters for the powertrain model of an electric vehicle using experimental data, in *2020 IEEE International Conference on Industrial Technology (ICIT)*, Buenos Aires, Argentina, pp. 939–944 (2020)
11. V. Ferencey, M. Omachelová, Mathematical modelling of the motion dynamics in the longitudinal direction of the electric vehicle, in *2016 Cybernetics & Informatics (K&I)*, Levoca, Slovakia, pp. 1–6 (2016)
12. Y. Aguilar, I. Caldas, A. Rivera, E. Tapia, Estudio de la influencia de la apertura de las ventanas y la velocidad de circulación en la aerodinámica de un vehículo de turismo. Ingenius. Revista de Ciencia y Tecnología **18**, 7–13 (2017)
13. Kia Motors Corp.: Guía del vehículo eléctrico (2014)
14. B. Zhao, N. Xu, H. Chen, K. Guo, Y. Huang, Design and Experimental Evaluations on Energy-Efficient Control for 4WIMD-EVs Considering Tire Slip Energy (2020)
15. J. Valladolid, D. Patiño, J. Ortiz, I. Minchala, G. Gruosso, Proposal for Modeling Electric Vehicle Battery Using Experimental Data and Considering Temperature Effects (2019)
16. D. Valladolid, J.P. Ortiz, F.A. Berrezueta, G.P. Novillo, Lithium-ion SOC optimizer consumption using accelerated particle swarm optimization and temperature criterion, in *2019 AEIT International Conference of Electrical and Electronic Technologies for Automotive (AEIT AUTOMOTIVE)*, Turin, Italy, pp. 1–6 (2019)

Chapter 29
Analysis of Energy Demand of Passenger Buses in the City of Cuenca

Néstor Diego Rivera, Paúl Andrés Molina, Gina Pamela Novillo, Byron David Llivicura, and Andrés Leonardo Peñaloza

Abstract This work presents a methodology capable of determining the energy demand of 5 bus lines in the city of Cuenca-Ecuador. The selected lines were chosen for being the ones with the highest number of passengers. The study was carried out by acquiring data from the OBD II port of the transport units through an open-source data logger device, which stores variables such as speed, acceleration, slope of the road, GPS location, etc. The acquired information was post-processed in MATLAB. By studying the analyzed variables, a model that allows identifying the relationship between the vehicle's specific power (VSP) and the energy consumption during the drives was determined. To calculate this value, the vehicle's mass, acceleration and its trajectory were taken into consideration. To avoid biases on the acquired information during the complete bus routes, statistical tools such as the analysis of variance and correlation were used to select schedules according to the city vehicular traffic. With use of machine learning techniques, the existing relationship between speed, acceleration, road slope and energy demand was evidenced, a factor that is directly related to emission of polluting gases from the units into the atmosphere. As a result of the investigation, a disparity was determined in the bus line routes design, with lines with maximum and minimum dispatch rates of 405 and 178, respectively. Several lines have extensive routes with consumption of 330.44 kW and slopes of 24.85%, for which a modification in the design of the routes could be suggested.

29.1 Introduction

During 2016, transport sector generated 36% of greenhouse gases in Latin America and the Caribbean [1]. According to the energy balance carried out by the MICSE (Coordinating Ministry of Strategic Sectors), transport sector is the largest fuel consumer in Ecuador [2]. In the world, there are more than 10 million buses, and this number increases especially in densely populated countries [3].

N. D. Rivera (✉) · P. A. Molina · G. P. Novillo · B. D. Llivicura · A. L. Peñaloza
Universidad Politécnica Salesiana, Cuenca, Ecuador
e-mail: nrivera@ups.edu.ec

Á. Rocha et al. (eds.), *Communication, Smart Technologies and Innovation for Society*, Smart Innovation, Systems and Technologies 252,
https://doi.org/10.1007/978-981-16-4126-8_29

In Ecuador during 2018, there were 159 buses that ran on gasoline and 8450 on diesel [4]. Even though public transportation is a clean and efficient mobility form compared to the use of private vehicles [5], the large number of units generate repercussions in air quality in cities. An urbanization rate of 80% causes traffic congestion problems [6] and energy consumption (fuel) is a parameter affected by: mode of transport, number of passengers, tons transported, distance traveled and type of vehicle [7].

Energy and polluting emissions related to the transport sector have aspects close to urban planning [8]. The city of Cuenca, the place used for this study, is located at 2550 m above sea level, south of Ecuador; it has a topography with slopes ranging between 0 and 15%, typical of the inter-Andean region [9]. Population growth has consequences on mobility, such as an increase in the demand for urban transport by 31% [10], which represents a high index compared to other mobility systems. This generates changes within urban mobility, causing saturation road network and significantly contributing to air pollution and energy consumption. Most of the countries have developed studies to know the energy demand of the transport units, meanwhile in Ecuador, there is a lack of methodologies that characterize this factor, reason for this research.

Several authors propose different solutions for this problem. Sierra et al. [11] develop a methodology to estimate fuel consumption of road transport in Ecuador and empirically determines the VKT and FE indices. The results were obtained from a non-representative sample, which is why it does not represent the national reality. Giraldo et al. [12] use a PEMS device to determine driving patterns, fuel consumption and polluting emissions (CO, CO_2, NO, NO_2) of a fleet of 15 buses driven at an altitude of 2000 masl. They estimate that CO_2 results are 290% higher than USEPA's recommendations and gases such as NOx are 44% lower compared to the IPCC.

Castagliola et al. [13] determined the impact of pollutant emissions measured under real driving cycles in Euro 5 diesel vehicles. It infers that the specific power of the vehicle (VSP is the instantaneous power per unit mass of the vehicle, necessary by the engine to overcome the opposing forces to movement) depends mainly on the road gradient. Aggarwal et al. [14] present a methodology for developing a work plan and achieving carbon-efficient mobility for the city of Deli.

Rawalpindi et al. [15] use LEAP an energy planning system that allows monitoring of pollution and energy demand of urban transport. For information processing, Molina [16] applies K-means algorithm to establish a model based on specific fuel consumption and engine performance to identify a driving mode. Rivera et al. [17] develop an algorithm capable of classifying engine start, gear change and brake maneuvers acquired through a data logger device. Molina et al. [18] uses supervised learning techniques to identify braking and clutch maneuvers through the study of PID 'signals acquired through the vehicle's OBD II port.

29.2 Material and Methods

Characteristics associated with driving style can be acquired through the vehicle's OBD II port [19] and through the control unit obtain sensors information (MAP, IAT, TPS, etc.) [20]. Data logger devices allow to acquire this information without attaching additional mass to the vehicle such as the portable emissions measurement system (PEM's) [21]. For this study, the Parameter Identification (PIDs) are acquired through the OBD II port through a free interface called Freematics ONE.

Data logger devices were placed on the five bus lines with the highest passenger demand in the city of Cuenca. With the collected information, it is possible to determine operational costs of urban transport and strategies that allow fuel savings. The following sections detail routes analysis (number of passengers, IPK index, etc.), characteristics of the units and how the specific power (VSP) and energy demand values were calculated.

29.2.1 Urban Transport Routes Analysis

This section analyzes different urban transport routes of the city, according to the demand of users, presented in Table 29.1, which correspond to the mobility plan of the city of Cuenca in 2015. The number of the offered places on a working day is a function of the number of expeditions per day on each bus route [9]. The index of passengers per kilometer of public transport (IPK) is an index that is a function of the supply and demand of the route.

As can be seen in Fig. 29.1, energy demand is affected by traffic of vehicles acceleration and speed; for this reason, it is essential to analyze traffic hourly intensity, resulting in peak and off-peak hours.

Table 29.1 Passenger demand

Line	Denomination	Passengers (Year)	Expeditions	IPK
27	Huizhil-Sinincay-Huizhil	11,292,005	82	5
16	Hospital del IESS-San Pedro–Hospital del IESS	10,374,760	120	4.5
28	Feria Libre-Llacao-Feria Libre	8,477,855	199	4
100	Ricaurte-Baños-Ricaurte	7,967,950	196	5.8
13	Hospital del IESS-Tejar-Hospital del IESS	7,479,580	125	3.3

Fig. 29.1 Hourly variation

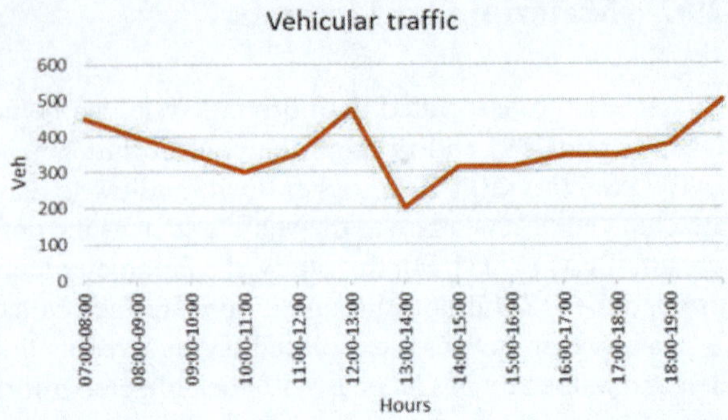

29.2.2 Bus Units' Characteristics

Table 29.2 presents buses characteristics used during this investigation. With information shown in Table 29.3, the necessary analysis can be built to estimate the different variables.

Table 29.2 Models of bus units

Model	Hino	International	Agrale	Volkswagen
Chassis	AK 8JRSA	4700 FE	MA 17	17–260
No. cylinders	6	6	6	6
Cylinder	7,790 cc	7,200 cc	7.118 cc	6.900 cc
Power	240 Hp @ 2350 RPM	250 HP @ 2200 RPM	280 HP @ 2200 RPM	253 HP @ 2300 RPM
Torque	706 kg-m @ 1450 RPM	660 lb-ft @ 1400 RPM	700 lb-ft @ 1200RPM	664 lb-ft @ 1200 RPM
Injection system	Common rail	Common rail	Common rail	Common rail
Fuel	Diesel	Diesel	Diesel	Diesel
Type approval stage	Euro III	EPA'98	EURO III	EURO V
Rim tipe	11R22.5	11R22.5 or 295/80R22.5	295/80R22.5	275/80R22.5

Table 29.3 Characteristics of the units

Lines	Chassis	Model	Front area (m^2)	PBV (Kg)
27	HINO	AK 8JRSA	7,580	14,200
16	INTERNATIONAL	4700 FE	7,611	15,876
28	AGRALE	MA 17	7,548	17,500
100	VOLKSWAGEN	17–260	6,936	17,300
13	HINO	AK 8JRSA	7,580	14,200

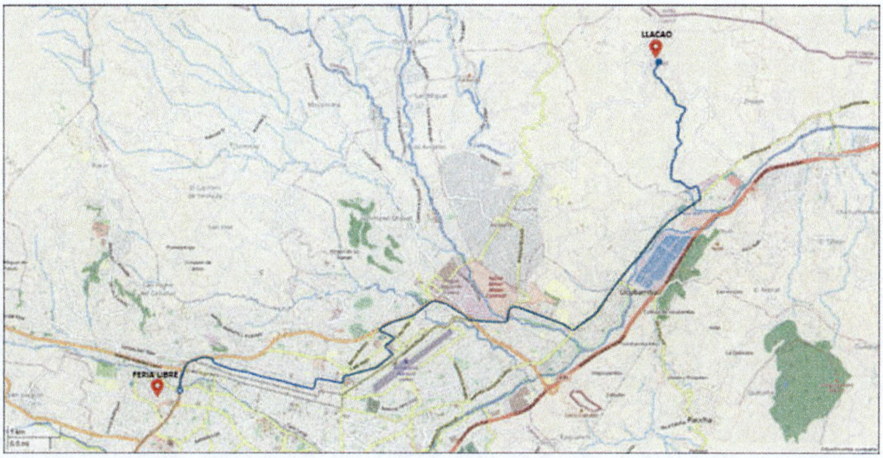

Fig. 29.2 Rout: Feria Libre—Llacao

29.3 Results and Discussion

29.3.1 Driving Cycle and Altitude Profile

To determine the driving profile and altitude, information obtained from the "data logger" device was used. With the collected data, different routes journey can be traced, using the GPS incorporated in the interface. In Fig. 29.2, line 28 route is shown, which starts from the sector called "Feria Libre" and ends in "Llacao," returning to "Feria Libre" and thus fulfilling its expedition cycle. This line covers a 16 km distance in a time of 52 min, with a total of 75 stops. The average speed of this route is 15 km/h and with a maximum of 46 km/h.

The altitude profile, presented in Fig. 29.3, corresponds to bus line number 16, obtaining altitude results as a function of time. The route begins at a point called "IESS Hospital," which is located at 2455 m above sea level, and ends in "San Pedro" sector located at 2891 m above sea level.

29.3.2 Obtaining Parameters

The five routes under study present different driving characteristics, with average speeds ranging from 16 to 19 km/h. The maximum speed reached was 69 km/h, as shown in Table 29.4. The buses must overcome slopes that go from 8.82% at least and maximum slopes of 24.85%; these routes take place through the city center, having as peak hours from 07:00 to 08:00 in the morning, from 12:00 to 13:00 in the afternoon and at night with the highest traffic flow from 6:00 p.m., which represents

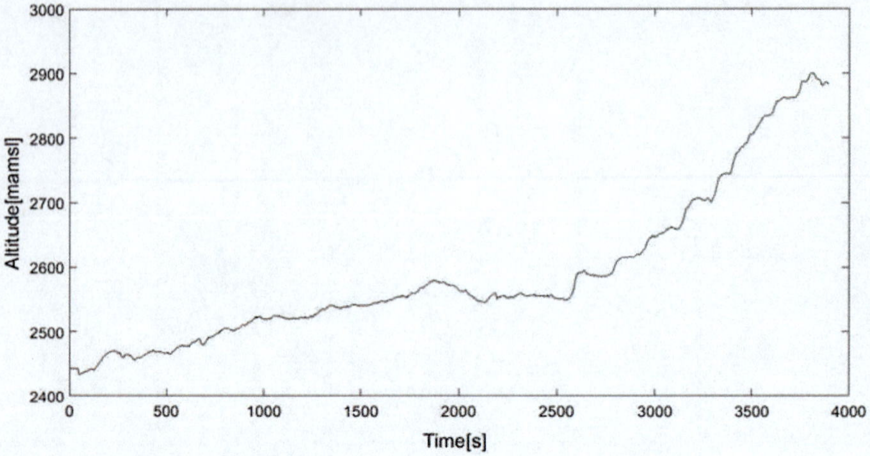

Fig. 29.3 Altitude profile of line 16 "Hospital del IESS-San Pedro"

Table 29.4 Bus driving cycles

Bus lines	16	28	13	27	100
Max speed. [km/h]	69	46	56	48	64
Average speed [km/h]	19	19	16	17	17
Max acceleration [m/s^2]	0.0833	0.1014	0.0747	0.0828	0.0775
Travel distance [km]	21	16	13	21	17
Travel time [min]	64	52	51	76	62
Max pending [%]	17.11	24.85	8.82	10.41	12.59

a high demand for passengers in the urban transport service, as these are the start and end times of working days, which produces a high level of traffic, resulting in accelerations and decelerations due to driving.

Urban transport system is directly affected by the entry and exit of users of transport units. This factor represents an additional load that the bus must overcome to travel. As can be seen in Table 29.5, users with the greatest use of public transport units are young adults between 24 and 45 years old, with 49.2%, while 37.8% of the

Table 29.5 Passenger demand and weight by age range

Age range	Years	%	Mass (men) [kg]	Mass (women)[kg]
Teenagers	16–24	37.8	60,44	54,28
Young adult	24–45	49.2	58,88	50,02
Adult	45–60	13	56,78	48,35
Total	–	100	176,098	152,65

Table 29.6 Passenger demand by expedition

Line	Demand by expedition
Line 27	405
Line 16	257
Line 28	127
Line 100	115
Line 13	178

Table 29.7 Weight of men and women by age range

Line	Teenagers (Men and women) [kg]	Young adult (Men and women) [kg]	Mass by expedition Men and women [kg]
Line 27	8712.60	10,724.07	22,172.51
Line16	5536.71	6814.96	14,090.25
Line 28	2728.27	3358.15	6943.12
Line 100	2481.30	3054.16	6314.62
Line 13	3831.97	4716.65	9751.90

demand corresponds to young people under 24 years. Therefore, 87% of users are under 45 years of age. The body mass of passenger transport users is obtained from the average age presented by the INEC in the journal of statistics and methodologies of 2017 [22], resulting in a body mass of 60.44 kg for young men and for women young people of 54.28 kg.

Passengers demand per expedition is presented in Table 29.6, taking line 27 as the route with the highest number of passengers, with a total of 405 people, which reflects a significant rate of passengers per kilometer (IPK) of 5.

Weight of users by age range is presented in Table 29.7. To evaluate the demand, it has been considered that male users correspond to 43% and women correspond to 57%. This represents a significant variation in weight for estimating the energy calculation of passenger transport.

Lines 27 and 16 are the public transport units that transport the most people per expedition, so the buses must support loads of 22,172.51 kg and 14,090.25 kg, respectively.

29.3.3 Power Calculation

Necessary power (kW) for the movement of the vehicle is presented in Table 29.8. Parameters with the greatest influence are weight of the passengers and gradient of the road. Line 16 has as its starting point "Hospital del IESS" and as the end of the route "San Pedro," presenting a PBV power of 181.84 kW, and on the return from "San Pedro" to the "Hospital del IESS" a PBV power of 30.32 kW, as presented in

Table 29.8 Necessary vehicle power

Lines	Necessary power H [kW]		
	Power (PBV)	Power (50% load)	Power (100% load)
Line 16	30.32	41.27	53.21
	181.84	255.62	330.44
Line 28	164.37	196.51	228.65
	89.26	104.53	120.22
Line 13	119.95	160.08	200.22
	33	42.11	51.21
Line 27	118.11	207.06	296.02
	80.28	141.95	203.62
Line 100	138.32	160.76	183.20
	121.34	142.47	163.60

Fig. 29.4. Weight of the occupants represents a significant power variation, requiring 330.44 kW because of a 100% load, as observed in Table 29.8.

Power required to move the vehicle is lower on negative slopes, having a maximum PBV power of 30.32 kW on the road, showing that the vehicle increases its power when traveling on roads with steep slopes.

As can be seen in Table 29.9, by calculating the specific vehicle power (VSP) [kW/ton], the speed, acceleration and slope of the road are considered as a result of a variation of the power respect to the road degree inclination, without involving the weight of the vehicle for its movement. Similarly, it is considered line 16, having as a route result "Hospital del IESS—San Pedro" a PBV power of 14.41 kW/ton and

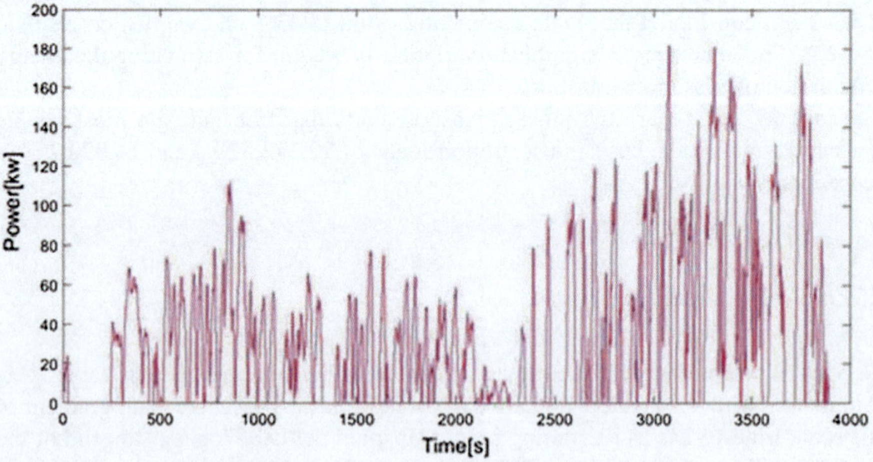

Fig. 29.4 Altitude profile of line 16

Table 29.9 Specific vehicle power (VSP)

Lines	Specific vehicle power (VSP) [kW/ton]		
	Power (PBV)	Power (50% load)	Power (100% load)
Line 16	14.41	14.41	14.41
	435.93	435.93	435.93
Line 28	436.77	436.77	436.77
	163.59	163.59	163.59
Line 13	368.72	368.72	368.72
	0	0	0
Line 27	344.26	344.26	344.26
	242.54	242.54	242.54
Line 100	281.88	281.88	281.88
	251.67	251.67	251.67

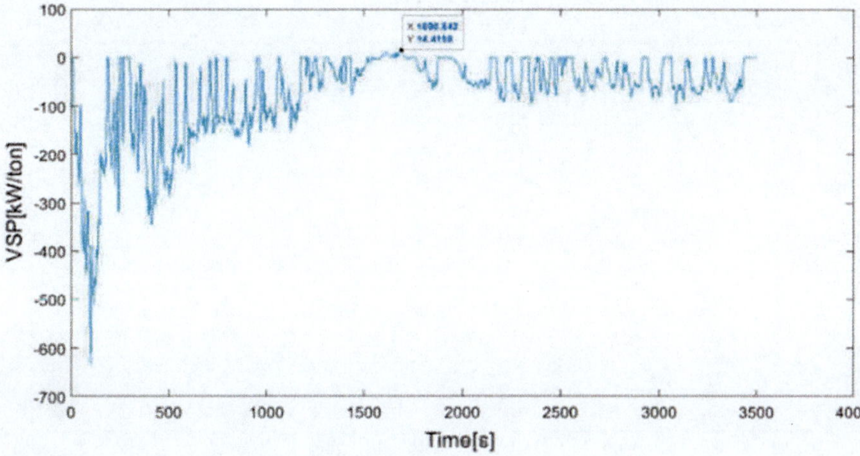

Fig. 29.5 VSP: Route “San Pedro–Hospital del IESS”

the route “San Pedro—Hospital del IESS” of 435.93 kW/ton, presented in Figs. 29.5 and 29.6 respectively.

29.3.4 Energy Demand

Required energy for the vehicle move forward during the journey will increase according to the power required by the engine to overcome the slopes of the road. As can be seen in Fig. 29.7, the “IESS Hospital–San Pedro” route corresponding to

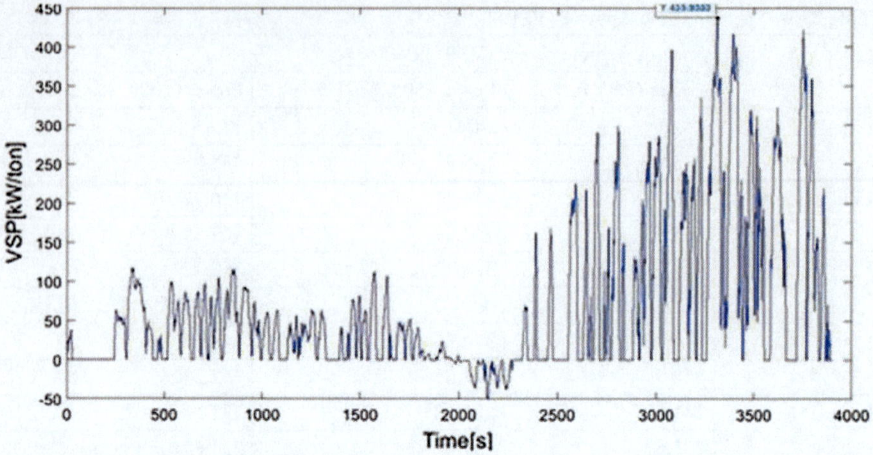

Fig. 29.6 Route: "Hospital del IESS–San Pedro"

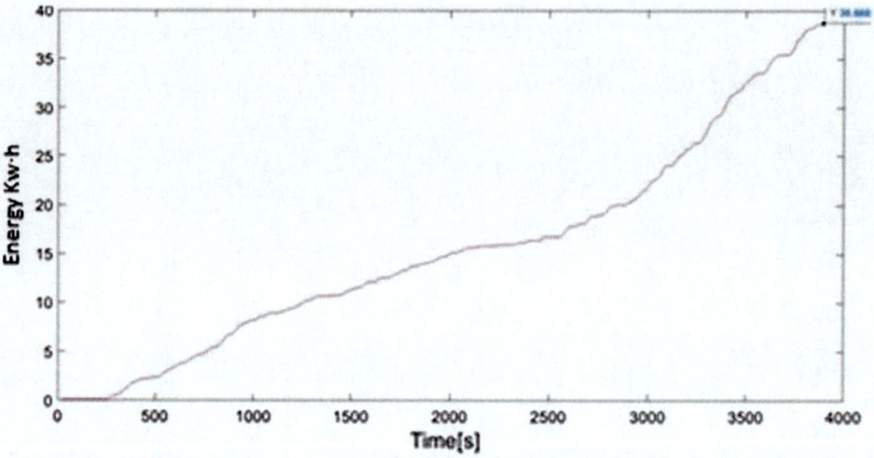

Fig. 29.7 Energy demand

line 16 has a higher energy demand due to the city's slopes, obtaining a maximum energy demand of 38.66 kWh in the highest part of the route.

For the energy demand, the additional weights or loads of the passengers during the journey were considered. The routes "San Pedro—Hospital del IESS" and "Tejar—Hospital del IESS" represent values of −7.66 and −4.75 kW/h, respectively; due to the fact that the passenger vehicle travels from the highest points located in the rural area to the lowest points of its routes located in the urban area in the "IESS Hospital" sector; so, energy consumption is affected by the negative slope of these. On the other hand, line 16 route presents a higher energy demand than the other bus lines analyzed by requiring 71.32 kW/h in its route (Table 29.10).

Table 29.10 Energy demand

Lines	Energy demand [KW/h]		
	Energy demand (PBV)	Energy demand (50% load)	Energy demand (100% load)
Line 16	−3.49	−5.57	−7.66
	38.66	54.99	71.32
Line 28	17.96	21.34	24.72
	12.55	14.87	17.19
Line13	22.88	30.48	38.08
	−2.53	−3.64	−4.75
Line 27	19.84	34.41	48.97
	12.93	22.08	31.24
Line 100	21.84	25.64	29.42
	10.97	12.76	14.54

29.4 Conclusions and Future Works

The result of this research is a real demand for the energy required by the routes for their movement, with data monitored under real driving conditions, in rural and urban routes of the city of Cuenca, during peak and off-peak hours. Type of driving affects energy consumption of the car, according to the results of the calculations made, the acceleration and the speed of the car are directly related. Considering peak hours as a preponderant factor, due to the high influx of vehicles. Drivers are constantly accelerating and decelerating having maximum accelerations of 0.1014 m/s^2 as observed in line 28.

Specific power of the vehicle (VSP) is the instantaneous power per unit mass and presents approximate values because the mathematical expression obtained does not consider the frontal area and the additional masses that intervene in the resistance of the vehicle. For the vehicle necessary power [kW], real data involved in its operation were considered, resulting in a load power per expedition of 330. 44 kW corresponding to the public transport unit of the route "Hospital del IESS—San Pedro," which reflects that this unit requires a greater amount of energy, because it travels a 21 km distance with a maximum slope of 17.11%.

Another factor that intervenes in the energy demand is the frontal area, for the bus on line 16 there is a frontal area of 7,611 m^2, being the bus that has the largest area of all the analyzed buses. The influx of passengers represented an additional burden to the forces that the bus must overcome, so it is necessary to consider the number of users of urban transport. The passenger per kilometer (IPK) index for line 16 is 4.5 which reflects a high demand from users of this route.

Another factor to consider is the air density since this varies respect to the altimetry of the bus. It presents maximum and minimum values of 0.9630 kg/m^3 and 0.9192 kg/m^3, respectively. The air density decreases when traveling along the "IESS-San Pedro Hospital" route, because the "San Pedro" sector is located at 2,914 m above sea level. With the energy demand estimation, a modification in the routes of the bus lines can be proposed, as there are transport lines that travel long distances and with high slope values, giving rise to excessive fuel consumption. Future work would aim to determine the appropriate routes for these transport lines, optimizing fuel consumption and collecting data to improve the model.

References

1. H. Martinez Salgado, El desafío del sector transporte en el contexto del cumplimiento determinadas a nivel nacional de América Latina. CEPAL, pp. 7–9 (2018)
2. MICSE, Balance energético nacional 2013, año base 2012. Ministerio Coordinador de Sectores Estratégicos, Quito Ecuador (2013)
3. T. Wei, H.C. Frey, Factors affecting variability in fossil-fueled transit bus emission rates. Atmos. Environ. **233**(December 2019), 117613 (2020). https://doi.org/10.1016/j.atmosenv.2020.117613
4. INEC, El trasporte terrestre de pasajeros en ecuador y quito: Perspectiva histórica y Situación actual. https://www.ecuadorencifras.gob.ec/documentos/web-inec/Bibliotecas/Estudios/Estudios_Economicos/Transporte_Quito.pdf. Accesed 12 Nov 2020
5. G. Correa, P.M. Muñoz, C.R. Rodriguez, A comparative energy and environmental analysis of a diesel, hybrid, hydrogen and electric urban bus. Energy **187** (2019). https://doi.org/10.1016/j.energy.2019.115906
6. L. Viscidi, R. O'Connor, La energía del transporte: Un enfoque en el transporte urbano en América Latina. EL DIÁLOGO, pp. 1–8 (2017)
7. P. Carretero Peña, G. Sánchez, Gestión de la eficiencia energética, cálculo del consumo, indicadores y mejora. AENOR ediciones, pp. 18–21 (2012)
8. N. Giacobbe, J. Frediani, L. Aon, O. Ravella, El consumo energético del sector transporte analizado desde la demanda. Aspectos metodológicos. Avances en Energías Renovables y Medio Ambiente **11** (2007)
9. Municipalidad de Cuenca, Una diagnosis en 9 puntos para el centro histórico, Plan de movilidad y espacios públicos (2015)
10. Municipalidad de cuenca, Plan de movilidad y espacios públicos. Ilustre municipalidad de cuenca (2015)
11. J.C. Sierra, Estimating road transport fuel consumption in Ecuador. Energy Policy **92**, 359–368 (2016). https://doi.org/10.1016/j.enpol.2016.02.008
12. M. Giraldo, J.I. Huertas, Real emissions, driving patterns and fuel consumption of in-use diesel buses operating at high altitude. Transp. Res. Part D Transp. Environ. **77**(October), 21–36 (2019). https://doi.org/10.1016/j.trd.2019.10.004
13. M. Castagliola, M. Costabile, M. Prati, Impact of road grade on read driving emision from two Euro diesel vehicles. Appl. Energy, 586–593 (2018). https://doi.org/10.1016/j.apenergy.2018.09.108
14. P. Aggarwal, S. Jain, Energy demand and CO_2 emissions from urban on-road transport in Delhi: current and future projections under various policy measures. J. Clean. Prod. **128**, 48–61 (2016). https://doi.org/10.1016/j.jclepro.2014.12.012
15. Shabbir, S. Ahmad, Monitoring urban transport air pollution and energy demand in Rawalpindi and Islamabad using leap model. Energy, 2323–2332 (2010). https://doi.org/10.1016/j.energy.2010.02.025
16. J.J. Molina Campoverde, Driving mode estimation model based in machine learning through PID's signals analysis obtained from OBD II, in *Applied Technologies*, Cham, pp. 80–91 (2020). https://doi.org/10.1007/978-3-030-42520-3_7

17. R.C. Néstor Diego, P.A. Molina Campoverde, G.P. Quirola Novillo, A.K. Naula Bermeo, Development of an algorithm capable of classifying the starting, gear change and engine brake variables of a vehicle by analyzing OBD II signals. In: M. Botto-Tobar, W. Zamora, J. Larrea Plúa, J. Bazurto Roldan, A. Santamaría Philco (eds) *Systems and Information Sciences. ICCIS 2020. Advances in Intelligent Systems and Computing*, vol. 1273. Springer, Cham (2021). https://doi.org/10.1007/978-3-030-59194-6_11
18. P.A. Molina Campoverde, N.D. Rivera Campoverde, G.P. Novillo Quirola, A.K. Bermeo Naula, Characterization of braking and clutching events of a vehicle through OBD II signals, in M. Botto-Tobar, W. Zamora, J. Larrea Plúa, J. Bazurto Roldan, A. Santamaría Philco (eds) *Systems and Information Sciences. ICCIS 2020. Advances in Intelligent Systems and Computing*, vol. 1273. Springer, Cham (2021). https://doi.org/10.1007/978-3-030-59194-6_12
19. E. Erikcsson, Independent driving pattern factors and their influence on fuel-use and exhaust emission factors. Transp. Res. Part D: Transport, 325–345 (2001)
20. J.E. Meseguer, C.K. Toh, C.T. Calafate, J.C. Cano, P. Manzoni, Drivingstyles: a mobile platform for driving styles and fuel consumption characterization. J. Commun. Netw. **19**(2), 162–168, abr. (2017). https://doi.org/10.1109/JCN.2017.000025
21. Y. Huanga, E.C.Y. Ng, J.L. Zhoua, N.C. Surawskia, E.F.C. Chan, G. Hong, Eco-driving technology for sustainable road transport: a review. Renew. Sustain. Energy Rev. (93), 596–609 (2018)
22. A. Peña, D. Barco, C. Patiño, Propuesta metodológica para la construcción de una. Revista de estadísticas y metodologías **3**, 49–73 (2017)

Chapter 30
Driving Style Analysis by Studying PID's Signals for Determination of Its Influence on Pollutant Emissions

Néstor Diego Rivera, Paúl Andrés Molina, Andrea Karina Bermeo, Oscar Enmanuel Bermeo, and José Luis Figueroa

Abstract This work analyzed the driving style influence on total polluting emissions emitted by an internal combustion vehicle. In this research, the most sold sedan vehicle in Ecuador was used. Parameters used to define the driving style were speed and acceleration; with these information, two styles were classified: normal and aggressive. There are no formal studies in the media about the relationship between pollutant emissions and driving style. Output variables used were vehicle fuel consumption and CO_2, CO, HC and NO_X pollutant emissions, and as input variables, driving parameters: intake manifold absolute pressure, throttle position, engine speed, speed and the acceleration of the vehicle. It was identified the most important variables such as MAP, TPS, VSS, and RPM with a determination index of 0.97519. Information was acquired by a data logger device, and post-processed using automatic learning techniques was verified a direct relationship between driving style and the polluting emissions, as well as fuel consumption. Therefore, it was verified that a normal driving style can reduce pollutant emissions by up to 22%, for which it is recommended that drivers should avoid sudden acceleration and sudden braking.

30.1 Introduction

Constant increase in vehicle number causes a deterioration in air quality, due to the polluting emissions emitted by internal combustion engines [1]. To regulate this situation, international standards have been implemented (e.g., Euro 6/VI standards) that limit fuel consumption and polluting emissions [2]. Electronic injection systems addition allows regulate fuel–air mixture that enters to the engine, thereby reducing the generation of polluting emissions [3]. Subsequently, gases can be treated with post-combustion systems such as three-way catalytic converter (TWC), common for light vehicles [4]. Many systems can be incorporated to control emissions and fuel consumption, but one of the factors that go unnoticed in vehicle performance is the

N. D. Rivera (✉) · P. A. Molina · A. K. Bermeo · O. E. Bermeo · J. L. Figueroa
Universidad Politécnica Salesiana, Cuenca, Ecuador
e-mail: nrivera@ups.edu.ec

Á. Rocha et al. (eds.), *Communication, Smart Technologies and Innovation for Society*, Smart Innovation, Systems and Technologies 252,
https://doi.org/10.1007/978-981-16-4126-8_30

driving style of the driver [5]. Different driving styles result in different levels of fuel consumption and polluting emissions [6].

Eco-driving is a term used to describe the energy-efficient used in vehicles, based on decisions and behaviors made by the driver [7], regardless of the vehicle's manufacture year [8], which goes from the choice, maintenance of the vehicle, selection of routes and driving style (e.g., aggressive, conservative, and shy) [9]. These types of techniques can reduce fuel consumption and polluting emissions such as CO_2 by up to 10% [10].

Common strategies include avoiding sudden accelerations and high speeds [11]; hence, the importance of studying driving style and its influence on polluting emissions for being a low-cost strategy that significantly reduces fuel consumption and greenhouse gases effect.

30.2 Literature Review

Characteristics associated with driving style can be acquired through the vehicle's OBD II port [11] and obtain log information from the sensors can be obtained (e.g., RPM (engine speed), vehicle speed sensor (VSS), manifold absolute pressure (MAP) and mass airflow (MAF)) via the control unit [12]. Data logger devices can acquire information under normal operating conditions without attaching additional mass to the vehicle, such as the portable emissions measurement system (PEM's) [5]. With the obtained data, several authors propose different methodologies for its study.

Corcoba et al. [13] present a driving assistant which allows the user to know information about his driving style, (acceleration, braking, fuel consumption, amount of CO_2 emitted) avoiding bad driving habits. Meseguer et al. [12] analyze and generate a classification of driving styles and fuel consumption, for which it applies an algorithm that allows characterizing the aggressiveness degree that each driver has and calculates fuel consumption and polluting emissions in real time.

Karginova et al. [14] classify bus drivers' driving styles (normal, hard) and provide real-time feedback on their driving. Pereira et al. [10] develop a mobile application that uses fuzzy logic with data obtained from a Bluetooth OBD II device and estimates fuel consumption. Dörr et al. [15] design a system that uses fuzzy logic to classify driving style, achieving a correct classification 68% of the time.

Dong et al. [16] propose a study of driving behavior based on data from vehicle sensors and GPS equipment, using statistical and deep learning techniques. Molina [17] applies K-means algorithm to establish a model based on specific fuel consumption and engine performance to identify a driving mode. Rivera et al. [18] develop an algorithm capable of classifying engine start, gear change and brake maneuvers acquired through a data logger device. Molina et al. [19] use supervised learning techniques to identify braking and clutching maneuvers through the study of PID signals.

30.3 Materials and Methods

In order to identify the different driving styles and know their influence on polluting emissions, Chevrolet Aveo Family 1.6 sedan vehicle was selected for being among the best-selling in Ecuador during the period of 2017–2018. [20]. One way to properly record driving style data was by acquiring information from the OBD II port with non-intrusive use of a data logger device. This information was saved and post-processed as shown in Fig. 30.1.

30.3.1 OBD Data Acquisition and Signal Filtering

In the laboratory where the test conditions can be controlled, the vehicle was placed on a chassis dynamometer (MAHA LP 3000). Synchronously, data acquisition was performed through the OBD II port by means of the data logger device called "Freematics ONE + ", free to use, which allows the sensors data extraction processed by the electronic control unit (ECU). Information of the concentration of pollutant emissions in percentages was recorded by the Brain Bee AGS-688 gas analyzer. To estimate the behavior of the output variables, different tests were carried out with different driving styles (normal and aggressive). A normal driving style is defined as a vehicle that is driven at low speeds and accelerations, with moderate use of the brake, in addition to adequate gear synchronization. Whereas, in a vehicle with an aggressive driving style it is characterized by driving at high speeds, sudden accelerations, and use of the braking system in an unexpected way. In each performed test for normal driving style, speed was in a range of 0 to 50 km/h, while in aggressive driving, speed was in a range of 0 to 100 km/h as can be seen in Fig. 30.2.

In Fig. 30.3, the distribution of the acceleration data for each of the different driving styles is shown. In a normal driving style, a distribution of the data closer to zero

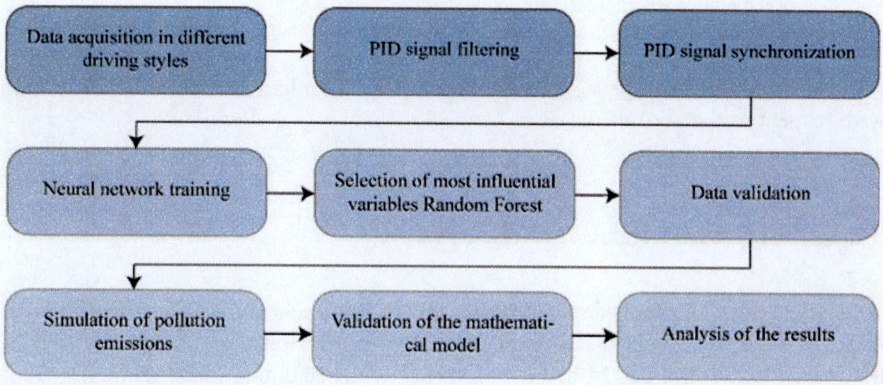

Fig. 30.1 Methodology

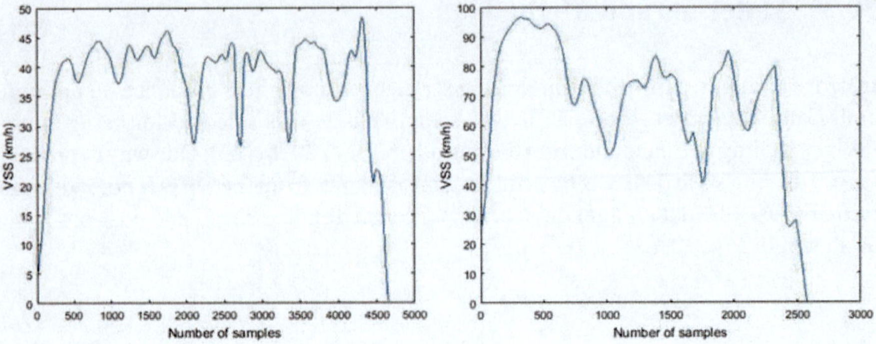

Fig. 30.2 Sensor VSS: normal and aggressive driving

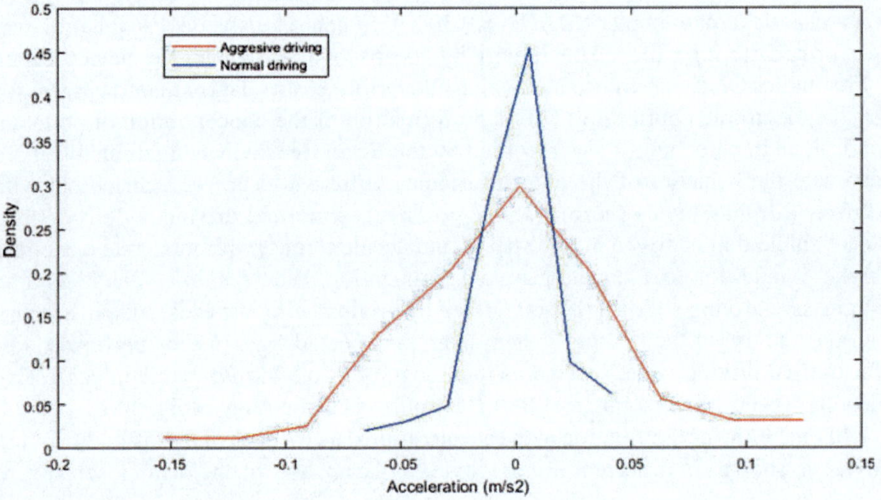

Fig. 30.3 Acceleration distribution (normal and aggressive driving)

was observed, while in an aggressive driving style we had a dispersed distribution with respect to the zero of the data.

30.3.2 Relevant Variables Estimation

To analyze pollutant emissions behavior, an artificial neural network (ANN) model was used, the input vector is the recorded data of the sensors (VSS, TPS, RPMS, MAP, IAT, ECT) obtained from the monitoring device. Acquisition "Freematics ONE + " and for the output vector (pollutant gases) the driving style was defined between

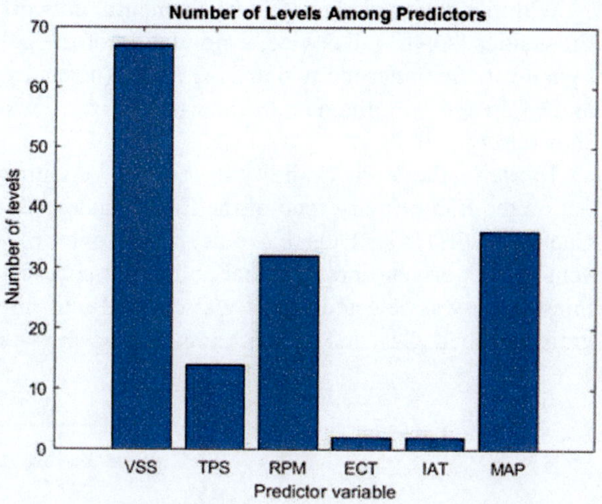

Fig. 30.4 Influential variables in polluting gases emission

normal and aggressive. During artificial neural network training used to estimate the emitted polluting gases, reliable determination indices were obtained, close to 1.

Using the random forest algorithm, variables most influential in emission of polluting gases were analyzed for tests carried out with normal and aggressive driving styles. According to the used algorithm, the most influential variables in the emission of polluting gases are the TPS, VSS, RPMS, and MAP sensors, while the least influential is the IAT and ECT sensors, as shown in Fig. 30.4.

30.4 Results

30.4.1 Artificial Neural Network Validation (RNA)

With the use of random forest algorithm and the neural network training, it was identified that polluting emissions depend directly on the variables of certain sensors such as VSS, TPS, RPM, and MAP with determination indices of 0.97519 for the carbon monoxide case as shown in table 30.1.

Table 30.1 Training of ANR

Polluting gas	R (determination index)
CO	0.97519
CO_2	0.95,483
HC	0.97132
NO_x	0.96717
O_2	0.95532

With the obtained algorithm by the neural network and the collected data by the Freematics "ONE + " device, a simulation of the polluting gases emission emitted by the vehicle under the two driving styles (normal and aggressive) was performed and compared with the data of the gas analyzer device to verify its functionality as shown in Fig. 30.5.

To verify the model validity in real road conditions, several tests were carried out on the E35 primary road of the Ecuadorian state network, located in the city of Cuenca at 2560 m.a.s.l. On this route, which covers 6.7 km (see Fig. 30.6), the selected vehicle was driven under normal and aggressive driving modes. In this interval, a microcycle was selected and it was divided into three sections: acceleration (S1), cruising speed (S2), and deceleration (S3), as shown in Fig. 30.7.

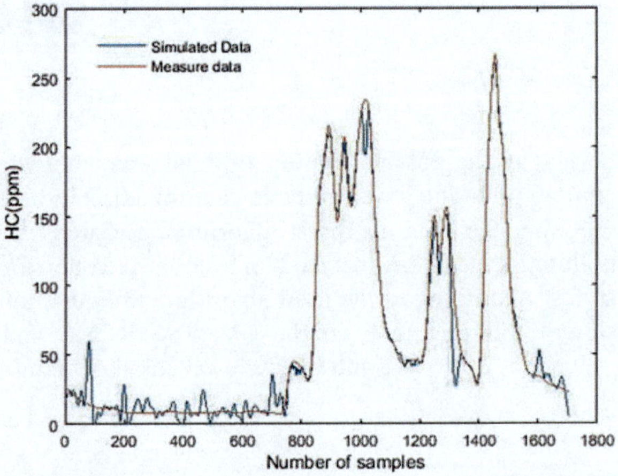

Fig. 30.5 HC Simulation using an RNA

Fig. 30.6 Road test route

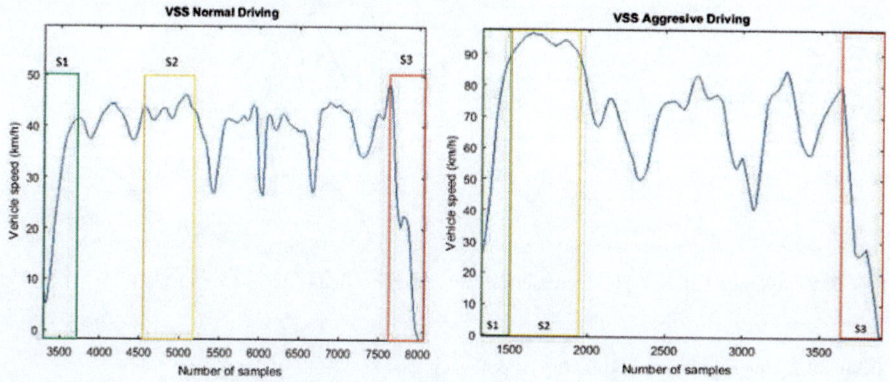

Fig. 30.7 Microcycles normal and aggressive driving

30.5 Discussion

In the data obtained from the simulation data, it can be seen that the HC (see Fig. 30.8), in a normal driving style had a maximum peak value of 437.2 ppm, while with an aggressive driving style it had a maximum peak value of 1376 ppm; therefore, there was an increase of 31.75% in peak values for the emission of HC in aggressive driving.

In an aggressive driving style, the average amount of polluting gases was high compared to a normal driving style in gases such as CO, HC, and NO_x (see Fig. 30.9) while with an efficient driving style the average amount of polluting gases was higher compared to a normal driving style for CO_2 and O_2 gases (see Fig. 30.9). In addition, the amount of polluting gases will be higher if the vehicle is in the acceleration section regardless of the driving style that it will present. Table 30.2 shows mean values of the amount of pollutant emission during the road test.

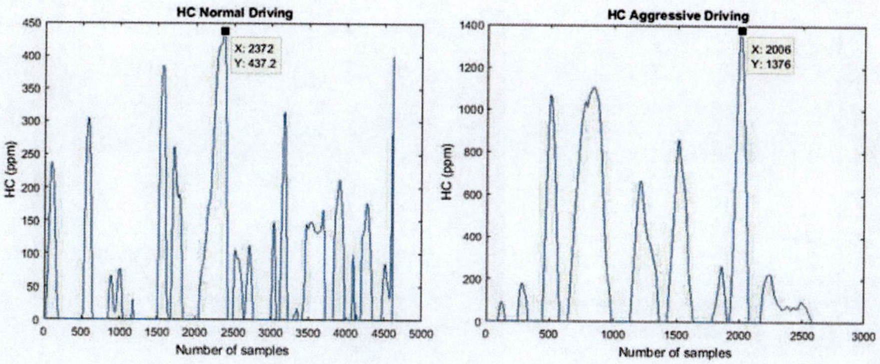

Fig. 30.8 HC simulated normal and aggressive driving

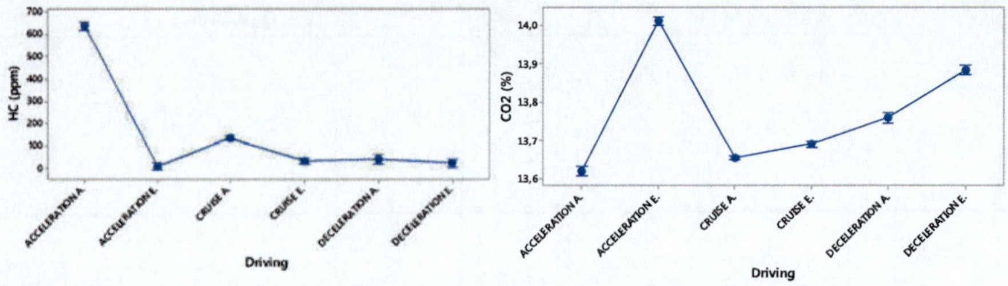

Fig. 30.9 Average value of HC-Conduction and average value of CO_2-Conduction

Table 30.2 Average value of quantity of polluting gases

Polluting gas	Normal acceleration	Aggressive acceleration	Normal cruise	Aggressive cruise	Normal deceleration	Aggressive deceleration
CO (%)	0.318	0.867	0.446	0.951	0.233	0.332
CO_2 (%)	14.011	13.618	13.689	13.652	13.884	13.759
HC (ppm)	7.457	639.8	34.17	135.85	23.152	42.03
NO_x (ppm)	40.09	323.3	26.33	99.17	154.99	37.32
O_2 (%)	2.399	0.0301	1.405	1.035	3.447	2.435

For a normal driving style, by analyzing the emission of polluting gases such as HC, respect to speed and acceleration (see Fig. 30.10) the amount of concentration was 250 ppm. When the acceleration increases, it had a value of 130 ppm and when there was no acceleration in the deceleration section, there was a 48% reduction in emissions with respect to the acceleration section.

For an aggressive driving style, analyzing the emission of the polluting gas NO_x, respect to speed and acceleration (See Fig. 30.11), the amount of concentration of this gas was 2500 ppm when acceleration increases and had a value of 750 ppm

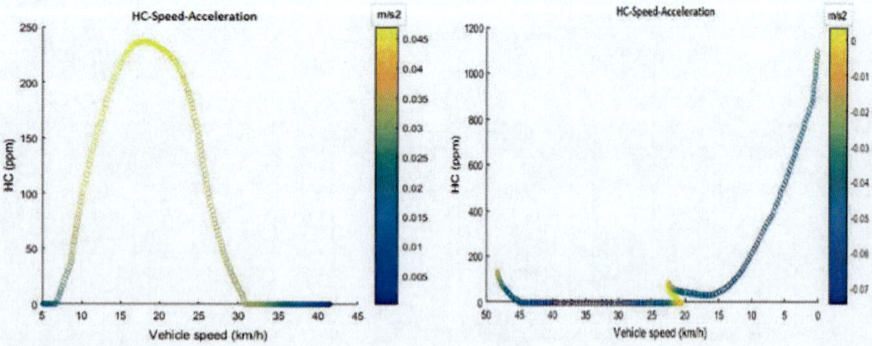

Fig. 30.10 Pollutant gas HC–speed–acceleration

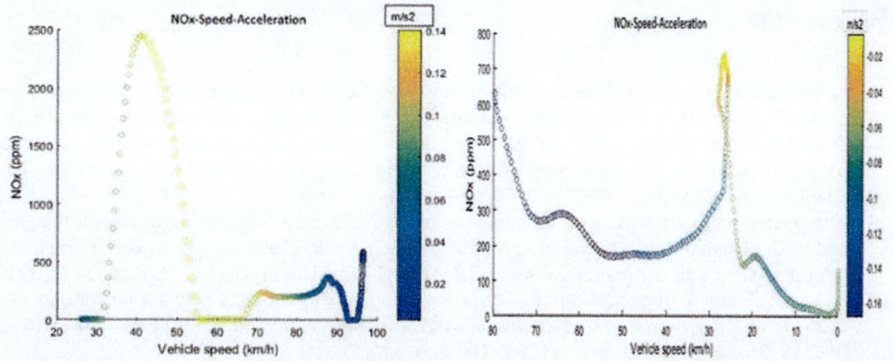

Fig. 30.11 Pollutant gas NO_x–speed–acceleration

when there was no acceleration; therefore in the deceleration section, there was a 70% emission reduction respect to the acceleration section.

30.6 Conclusions

In this research, the influence of normal or aggressive driving style on polluting emissions was verified through parameter identification data PID signals. By means of the random forest algorithm, it was shown that the PID signals corresponding to the MAP, TPS, VSS, and RPM sensors directly influence the emission of polluting gases, and by using the neural networks algorithm (RNA), a determination index of 0. 97,519 was obtained, which indicates a good fit of the data. In the analysis carried out in the road test, it can be observed that when driving a vehicle with an aggressive driving style, there was a greater increase in pollutant emissions compared to a normal driving style. By means of a statistical analysis, it can be observed that, having an aggressive driving style, due to the greater presence of fuel in the mixture, gases such as CO, HC, and NO_x have higher emissions compared to CO_2 and O_2. Likewise, normal driving style was analyzed in which CO_2 and O_2 gases had a higher concentration percentage with respect to CO, HC, and NO_x due to the lack of high speeds and accelerations. In the relationship between vehicle speed and acceleration analysis, it was observed that acceleration was the most influential variable in polluting emissions regardless of the speed at which the vehicle circulates, for this reason the importance of drivers avoiding sudden accelerations and sudden stops. It can be seen that a normal driving style results in a 22% reduction in pollution emissions compared to the aggressive driving style.

References

1. C. Sancan, D. Gregorio, Vehículos híbridos, una solución interina para bajar los niveles de contaminación del medio ambiente causados por las emisiones provenientes de los motores de combustión interna. Hybrid vehicles, an interim solution to lower levels of environmental pollution caused by emissions from internal combustion engines, dic. 2017. https://doi.org/10.33890/innova.v2.n12.2017.527
2. European Commission Amending regulation (EC) No 7152007 of the European Parliament and of the Council and Commission regulation (EC) No 6922008 as regards emissions from light passenger and commercial vehicles (Euro 6) O J Eur Union, vol 142 , pp. 16–24 (2012)
3. H. Dia, S. Panwai, Impact of driving behaviour on emissions and road network performance», in *2015 IEEE International Conference on Data Science and Data Intensive Systems*, dic. 2015, pp. 355–361. https://doi.org/10.1109/DSDIS.2015.68
4. X. Zheng, Y. Wu, S. Zhang, L. He, J. Hao, Evaluating real-world emissions of light-duty gasoline vehicles with deactivated three-way catalyst converters. Atmos. Pollut. Res. **9**(1), 126–132 (2018). https://doi.org/10.1016/j.apr.2017.08.001
5. Y. Huanga, E.C.Y. Ng, J.L. Zhoua, N.C. Surawskia, E.F.C. Chan, G. Hong, Eco-driving technology for sustainable road transport: a review. Renew. Sustain. Energy Rev. (93), 596–609 (2018)
6. C.F. Yeh, L.T. Lin, P.J. Wu., C.C.Huang, Using on-board diagnostics data to analyze driving behavior and fuel consumption, in Y. Zhao, T.Y. Wu, T.H. Chang, J.S. Pan., L. Jain (eds) *Advances in Smart Vehicular Technology, Transportation, Communication and Applications*. VTCA 2018. Smart Innovation, Systems and Technologies, vol. 128. Springer, Cham (2019)
7. Y. Xu, H. Li, H. Liu, M.O. Rodgers, R.L.Guenslerk, Eco-driving for transit: An e_ective strategy to conserve fuel and emissions. Appl. Energy (2016). https://doi.org/10.1016/j.apenergy.2016.09.101
8. A. Rionda, X.G. Paneda, R. Garcia, G. Diaz, D. Martinez, M. Mitre, D. Arbesu, I. Marin, *Blended learning system for efficient professional driving. Computers & Education*, vol 78, pp. 124-139 (2014)
9. M. Sivak, B. Schoettle, Eco-driving: strategic, tactical, and operational decisions of the driver that inuence vehicle fuel economy. Transp. Policy **22**, 96–99 (2012)
10. A. Pereira, M. Alves, H. Macedo, Vehicle driving analysis in regards to fuel consumption using Fuzzy Logic and OBD-II devices», en *2016 8th Euro American Conference on Telematics and Information Systems (EATIS)*, abr. 2016, pp. 1–4. https://doi.org/10.1109/EATIS.2016.7520160
11. E. Erikcsson, Independent driving pattern factors and their influence on fuel-use and exhaust emission factors. Transp. Res. Part D: Transport, 325–345J (2001)
12. J.E. Meseguer, C.K. Toh, C.T. Calafate, J.C. Cano, P. Manzoni, Drivingstyles: a mobile platform for driving styles and fuel consumption characterization. J. Commun. Netw. **19**(2), 162–168, abr. 2017. https://doi.org/10.1109/JCN.2017.000025
13. V. Corcoba Magaña, M. Muñoz Organero, *Eco-driving: Energy Saving Based on Driver Behavior*. Juan Antonio Ortega Ramírez, Alejandro Fernández-Montes, Juan Antonio Álvarez (2015)
14. N. Karginova, S. Byttner, M. Svensson, Data-driven methods for classification of driving styles in buses. *SAE International*, Warrendale, PA, SAE Technical Paper 2012–01–0744, abr. 2012. https://doi.org/10.4271/2012-01-0744
15. D. Dörr, D. Grabengiesser, F. Gauterin, Online driving style recognition using fuzzy logic, en *17th International IEEE Conference on Intelligent Transportation Systems (ITSC)*, oct. 2014, pp. 1021–1026. https://doi.org/10.1109/ITSC.2014.6957822
16. W. Dong, J. Li, R. Yao, C. Li, T. Yuan, Y.L. Wang, Characterizing driving styles with deep learning, *ArXiv160703611 Cs*, Oct. 2016, Accedido: oct. 29, 2020. [En línea]. Disponible en: http://arxiv.org/abs/1607.03611
17. J.J. Molina Campoverde, Driving mode estimation model based in machine learning through PID's signals analysis obtained from OBD II, in *Applied Technologies*, pp. 80–91. Cham (2020). https://doi.org/10.1007/978-3-030-42520-3_7

18. R.C. Néstor Diego, P.A. Molina Campoverde, G.P. Quirola Novillo, A.K. Naula Bermeo, Development of an Algorithm Capable of Classifying the Starting, Gear Change and Engine Brake Variables of a Vehicle by Analyzing OBD II Signals, in M. Botto-Tobar, W. Zamora, J. Larrea Plúa, J. Bazurto Roldan, A. Santamaría Philco (eds)*Systems and Information Sciences. ICCIS 2020. Advances in Intelligent Systems and Computing*, vol. 1273. Springer, Cham (2021). https://doi.org/10.1007/978-3-030-59194-6_11
19. P.A. Molina Campoverde, N.D. Rivera Campoverde, G.P. Novillo Quirola, N.A.K. Bermeo, Characterization of braking and clutching events of a vehicle through OBD II signals, in M. Botto-Tobar, W. Zamora, J. Larrea Plúa, J. Bazurto Roldan, A. Santamaría Philco (eds.) *Systems and Information Sciences. ICCIS 2020. Advances in Intelligent Systems and Computing*, vol. 1273. Springer, Cham (2021). https://doi.org/10.1007/978-3-030-59194-6_12
20. AEADE: Anuario 2018. http://www.aeade.net/wp-content/uploads/2019/03/Anuario%202018.pdf. Accesed 1 Nov 2020

Chapter 31
PWM Methods in SiC-Current Source Inverter in Motor Drives

E. Fernández, S. Zeas, and D. Rojas

Abstract This document presents a study of PWM methods of modulation in high frequency in current source inverter (CSI) in the motor drive for electric traction systems in order of evaluated total harmonic distortion (THD) and efficiency. The modulations used for the analysis are the phase-shifted carrier (SPWM), online PWM technique, and trapezoidal PWM technique (TPWM). For the use of silicon carbide device (SiC) in CSI topology is necessary to work with techniques of modulation a higher frequency to increase the commutation. The analysis proposed includes the study and comparison of techniques proposed to different frequencies of operation, the testing and simulations are analyzed using MATLAB software. Power losses and total harmonic distortion (THD) are evaluated and compared.

31.1 Introduction

The implementation of CSI topology in applications for the electric traction system is increased. In [1–4] authors mention that CSI has a good behavior at high frequency and is possible the decrease the size of the inductor and the capacitor of output. This topology has certain advantages over VSI that are presented in [5, 6].

The use of sinusoidal pulse modulation technique (SPWM) in CSI's limits the carrier frequency and shows a high harmonic distortion in the currents [7]. The use of different techniques, like harmonics elimination PWM technique (SHE), helps to solve this problem but is complex for the implementation. Therefore, each technique has different advantages and disadvantages in terms of harmonic content, power losses, and efficiency.

An important factor to obtain high efficiency in electric traction is a good design of the inverter system. For this, it is necessary to reduce power losses in semiconductor devices. The losses that are generated in these devices are conduction that is produced by the current that passes through the element and the commutation losses that are related to the commutation frequency.

E. Fernández (✉) · S. Zeas · D. Rojas
ERGON Center Research, University of Azuay, Cuenca, Ecuador
e-mail: efernandez@uazuay.edu.ec

Á. Rocha et al. (eds.), *Communication, Smart Technologies and Innovation for Society*, Smart Innovation, Systems and Technologies 252,
https://doi.org/10.1007/978-981-16-4126-8_31

The use of wide-bandgap devices in electric traction systems takes more interest in the last years, especially the use of devices of silicon carbide in power converters [7–10]. Several studies [11–14] show that the use of SiC devices allows improving the switching frequency and the operation in voltage and temperature. These devices work better than conventional silicon devices and improve efficiency at low power loss levels.

Several studies [15, 16] show the implementation of modulation techniques in CSI, and most are designed in low frequency and with silicon devices. The results show high harmonic content in the output currents and high-power losses.

The aim of this document is to show the study and the best method of PWM technique in a CSI with SiC devices in the motor driver to improve the efficiency and reduce the THD in the system. The methodology used consists of the study of different techniques of PWM used for CSI, the PWM techniques are implemented and compared to a high frequency of operation. For the analysis of the converter, a system is developed in MATLAB software. This paper contributes to the search for topologies with current sources and modulation techniques with greater efficiency using SiC devices. The document presents the following order, and the second section presents the operation of CSI power converter. In the third section, the PWM techniques proposed for the study (SPWM, online PWM, and TPWM) are presented. The fourth section includes the testing of techniques proposed on SiC devices to a high frequency of switching in motor drives complemented with an analysis of power losses, and total harmonic distortion to different frequencies comparing it with devices of conventional Si and show the results in simulation. Finally, in section fifth, the discussion and conclusion are presented.

31.2 CSI-SiC Power Converter in Motor Drive

The topology of the converter for the study is shown in Fig. 31.1. The topology is *V–I* converter is used to regulate the DC-link current input for the three-phase

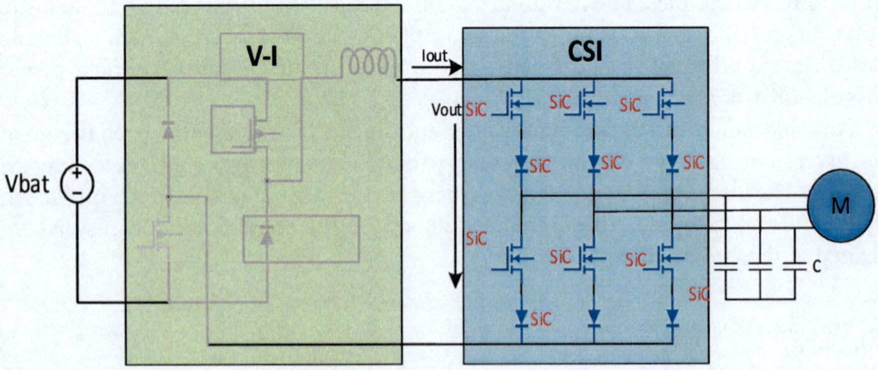

Fig. 31.1 CSI-SiC power converter

Table 31.1 Parameters of design of CSI inverter

Parameter	Features
Voltage of battery	300 V
Inductance	100 μH
Switching frequency	100 kHz
Capacitor	60 μF
Mosfet SiC	SCT30N120 1200 V, 45 A
Schkotty diode SiC	SCS220 KC VRRM = 1200 V, I = 50 A
Inverter power	3.5 kW
Motor stator resistor	0.294 Ω
Motor stator inductor	0.00139 H

CSI inverter [17]. The *V–I* has its own control for regulating the input current. This converter does not enter in the study. The CSI inverter is composed of six Mosfets with six diodes in series of SiC materials. Requirements of design of the converter being considered in Table 31.1.

31.3 Methods

For the use of PWM techniques in CSI, it is very important to consider the following rules described in [18, 19]: First, the transistors always operate to reduce the possibility of generating an open circuit in the DC link and avoid damage to devices and output capacitors; second, only two switches operate at any time. If this is not the form of the PWM current, meets are distorted. To meet these conditions, the use of digital systems is necessary for the generation of PWM pulses.

The simple phase-shifted carrier (SPWM) is implemented in [20–22]. This technique consists of the comparison between three sine signals called modulation signals and one triangle signal called carrier signal (Fig. 31.2), and then the pulses are treated with combinatorial logic to meet the operating conditions of the CSI.

The modulation index *m*, is calculated as:

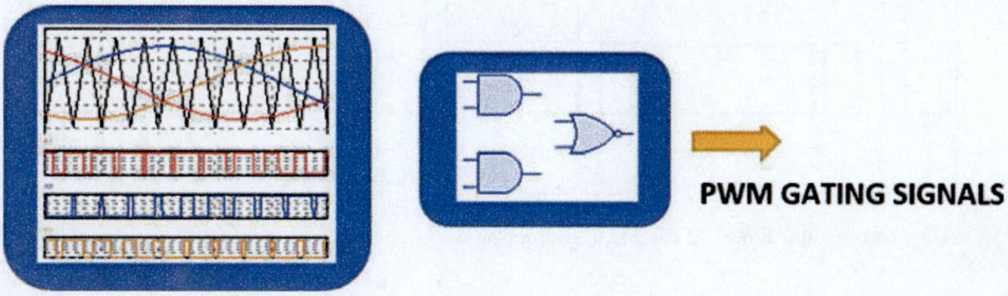

Fig. 31.2 Simple phase-shifted carrier (SPWM) implementation

$$m = \frac{A}{A_{\mathrm{m}}} \tag{31.1}$$

where A and A_{m} represent the amplitude of modulating signal and amplitude of carrier signal, respectively.

The trapezoidal PWM method (TPWM) consists of the comparison between the triangular carrier signal and trapezoidal signal to generate the sequence shown in Fig. 31.3. This method is discussed in [23–26].

The modulation index m, is expressed as:

$$m = \frac{\overline{V_{\mathrm{m}}}}{\overline{V_{\mathrm{cr}}}} \tag{31.2}$$

The online PMW technique proposed in [19, 27, 28] consists of an analog/digital combined circuit (Fig. 31.4) for obtaining the gating signals of control of the devices of commutation [19]. This method has four blocks of operation. First block is called

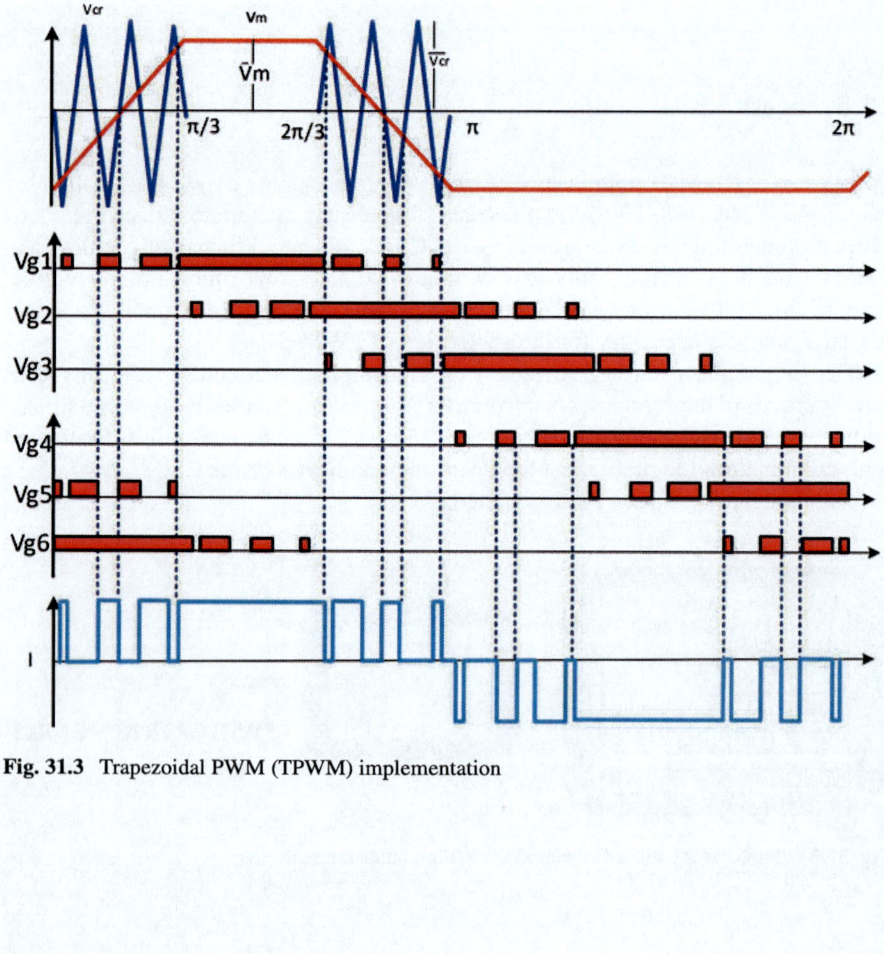

Fig. 31.3 Trapezoidal PWM (TPWM) implementation

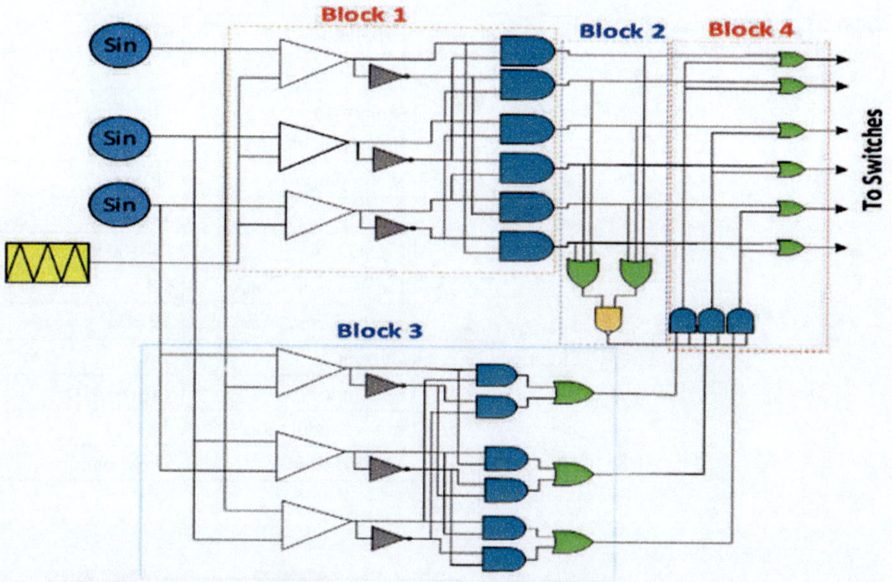

Fig. 31.4 Circuit and schematic of pulse in PWM online technique

the basic pulse generator, and the second block denominated generated complementary pulses. The third block is the distributor of short-circuit pulse. The fourth block is the combiner of pulses. Finally, the pulse of the gating signal.

31.4 Results and Discussions

This section presents an analysis of the techniques of PWM proposed for CSI with SiC devices to 100 kHz of switching in motor drives. This analysis includes a comparison of power losses, THD harmonic distortion, and efficiency of different frequencies of switching in each method of PWM previously described.

The model of Mosfet SiC used is developed in Simscape toolbox of Simulink. To check the model in the topology CSI-SiC and PWM technical, different simulations are developed with the parameters shown in Table 31.2.

The results in simulation for situation A are shown in Fig. 31.5; the results in simulation for situation B are presented in Fig. 31.6; and the results in simulation for situation C are presented in Fig. 31.7.

Table 31.2 Parameters of simulation for PWM technique

	Parameter	Features
Situation A	Current	5 A
	Frequency	15–100 kHz
	Index modulation	0.8
	Method used	SPWM
Situation B	Current	5 A
	Frequency	15–100 kHz
	Index modulation	0.8
	Method used	TPWM
Situation C	Current	5 A
	Frequency	15–100 kHz
	Index modulation	0.8
	Method used	PWM online

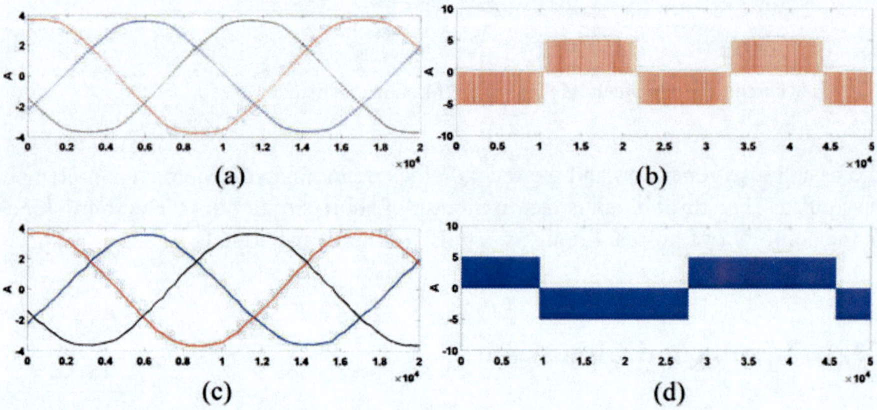

Fig. 31.5 Result of the model in CSI-SiC topology with SPWM technique. **a** Output currents in PMSM to 5 kHz. **b** SPWM currents to 5 kHz. **c** Output currents in PMSM to 100 kHz. **d** SPWM currents to 100 kHz

31.4.1 Power Losses

Conduction Losses. The power losses for conduction in the Mosfet SiC and Schottky diodes SiC are calculated by the Eqs. 31.3 and 31.4 [9, 29].

$$P_{\text{C\_ Mosfet}} = R_{\text{on}} I_{\text{rms}}^2 \quad (31.3)$$

$$P_{\text{C\_ Diode}} = I_{\text{rms}}^2 R_{\text{D}} + I_{\text{DC}} V_{\text{D}} \quad (31.4)$$

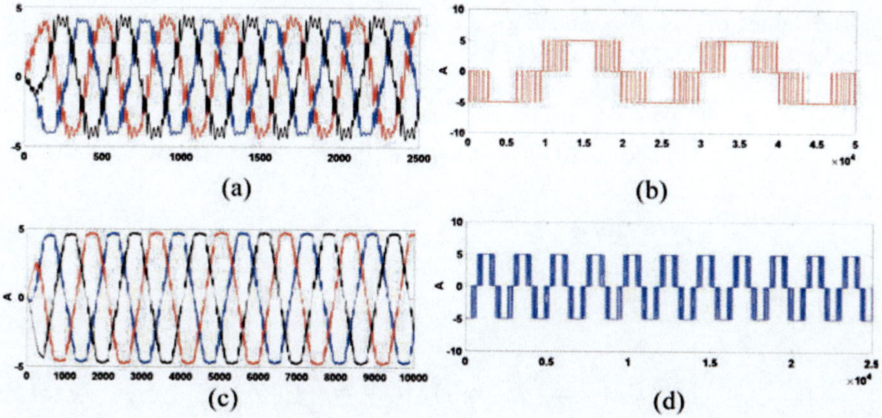

Fig. 31.6 Result of model in CSI-SiC topology with TPWM technique. **a** Output currents in PMSM to 5 kHz. **b** TPWM currents to 5 kHz. **c** Output currents in PMSM to 100 kHz. **d** SPWM currents to 100 kHz

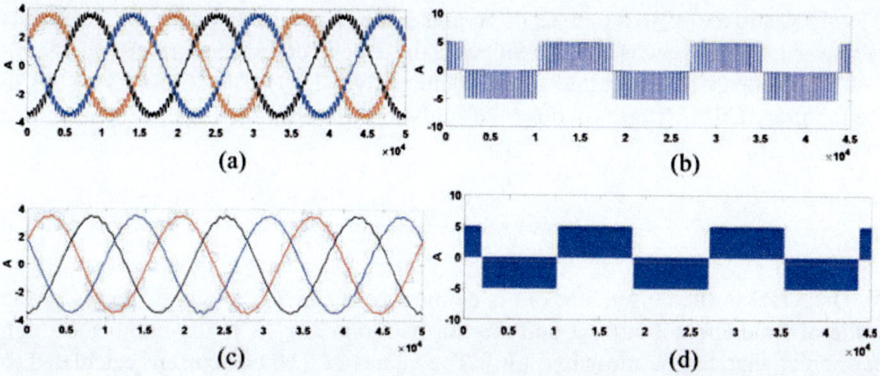

Fig. 31.7 Result of model in CSI-SiC topology with PWM online technique. **a** Output currents in PMSM to 5 kHz. **b** PWM online currents to 5 kHz. **c** Output currents in PMSM to 100 kHz. **d** PWM online currents to 100 kHz

R_{ds} is the drain-source resistor of the device, I_{rms} is the current flowing, I_{Dc} is the current of the diode, and V_D is the voltage of the diode [9, 28].

Switching Losses. The switching losses in the Mosfet SiC and SiC Schottky diode are expressed in Eqs. 31.5 and 31.6. E_{ON} and E_{OFF} represent the turn-on and turn-off switching energy in the Mosfet SiC, and E_{swD} is the energy of switching in the SiC Schottky diode [30].

$$P_{sw\_Mosfet} = f_{SW(E_{ON}+E_{OFF})} \tag{31.5}$$

$$P_{sw\_Diode} = f_{SW(E_{SWD})} \tag{31.6}$$

The values of E_{on}, E_{off}, and E_{swD} are considered the times of intervals and are calculated by (31.7), (31.8), and (31.9).

$$E_{\mathrm{ON}} = \int_{0}^{\mathrm{tri}+\mathrm{tfu}} V_{\mathrm{ds}}(t) I_{\mathrm{D}}(t)\mathrm{d}t = V_{\mathrm{dc}} I_{\mathrm{on\_rms}} \left(\frac{\mathrm{tri}+\mathrm{tfu}}{2}\right) + Q_{\mathrm{rr}} V_{\mathrm{dc}} \tag{31.7}$$

$$E_{\mathrm{OFF}} = \int_{0}^{\mathrm{tri}+\mathrm{tfi}} V_{\mathrm{ds}}(t) I_{\mathrm{D}}(t)\mathrm{d}t = V_{\mathrm{dc}} I_{\mathrm{off\_rms}} \left(\frac{\mathrm{tri}+\mathrm{tfi}}{2}\right) \tag{31.8}$$

$$E_{\mathrm{SWD}} = \int_{0}^{\mathrm{tri}+\mathrm{tfu}} V_{\mathrm{d}}(t) I_{\mathrm{f}}(t)\mathrm{d}t = \frac{1}{4} Q_{\mathrm{rr}} V_{\mathrm{df}} \tag{31.9}$$

The power losses in the CSI-SiC converter analyzed with the different modulation methods are presented in Fig. 31.9. In addition to this study, a comparative analysis of losses with conventional silicon devices is presented. The Mosfet device used is the STW45NM60 650 V, Rds 0.09 Ω, 45 A, and diode RURG3060 600 V 30. The results of power losses in power inverter SiC with these devices are shown in Fig. 31.9.

The efficiency of power converter is defined by (31.10). The efficiency results in the topology CSI-SiC proposed for the study to different PMM methods are shown in Fig. 31.10.

$$\eta(\%) = \frac{P_{\mathrm{Total}} - P_{\mathrm{Tlosses}}}{P_{\mathrm{Total}}} \tag{31.10}$$

The THD in the currents of out is defined in Eq. 31.11. I_1 and I_n represent the value of fundamental current and the nth harmonic, $n_{(\mathrm{max})}$ is the number of high harmonics that can be measured [30]. The values of THD of current calculated to high and low frequency for each method of modulation are shown in Fig. 11.

$$I_{\mathrm{THD}} = \frac{100\sqrt{\sum_{n=2}^{\max} I_n^2}}{I_{1\mathrm{rms}}} \tag{31.11}$$

31.5 Conclusions

This document presents a study and analysis of different modulation methods in CSI based on SiC devices for applications in motor drives. The results of the simulation demonstrate that the CSI converter implemented with the SPMW method has power losses compared with the other methods.

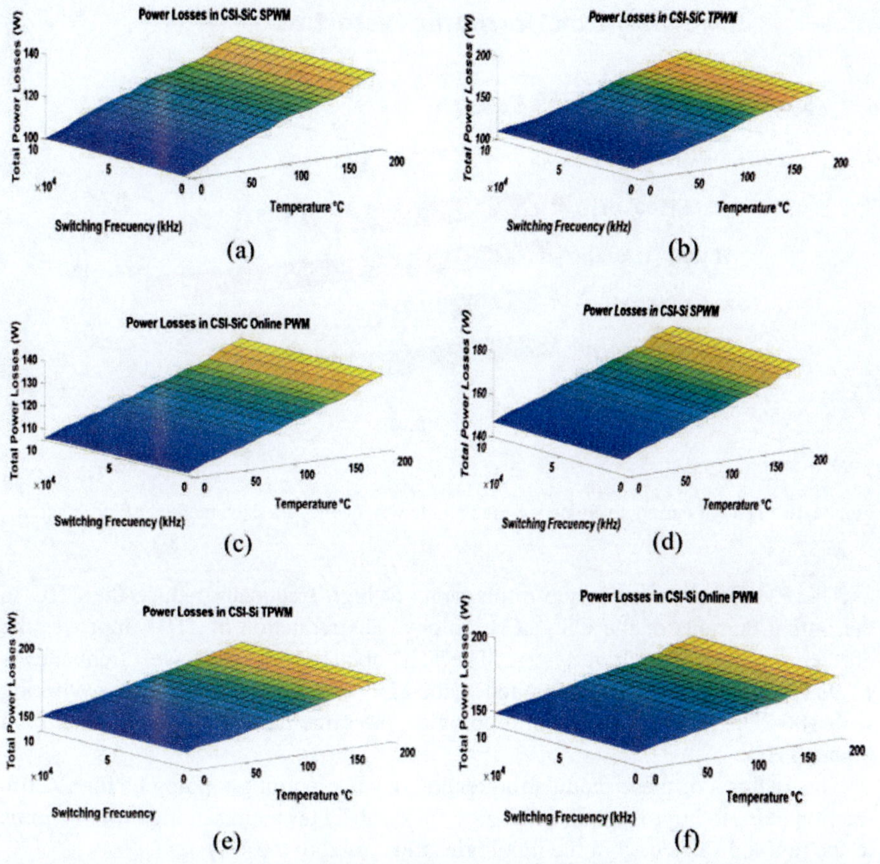

Fig. 31.8 Study of power losses in CSI-SiC and CSI-Si, **a** Power losses in CSI-SiC with SPWM, **b** Power losses in CSI-SiC with TPWM, **c** Power losses in CSI-SiC with online PWM, **d** Power losses in CSI-Si with SPWM, **e** Power losses in CSI-Si with TPWM, **f** Power losses in CSI-Si with online PWM

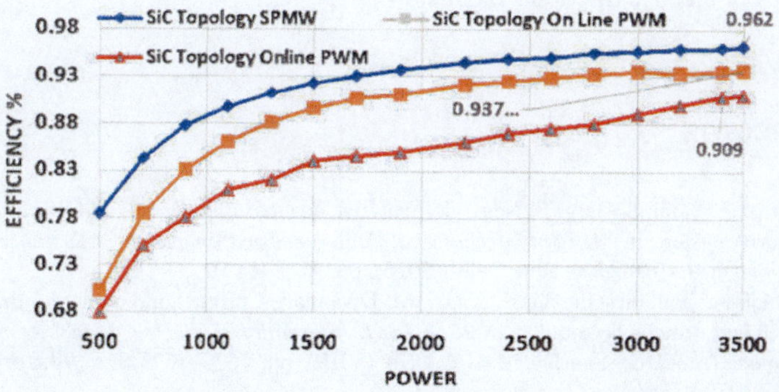

Fig. 31.9 Efficiency CIS- SiC topology in different methods of PWM

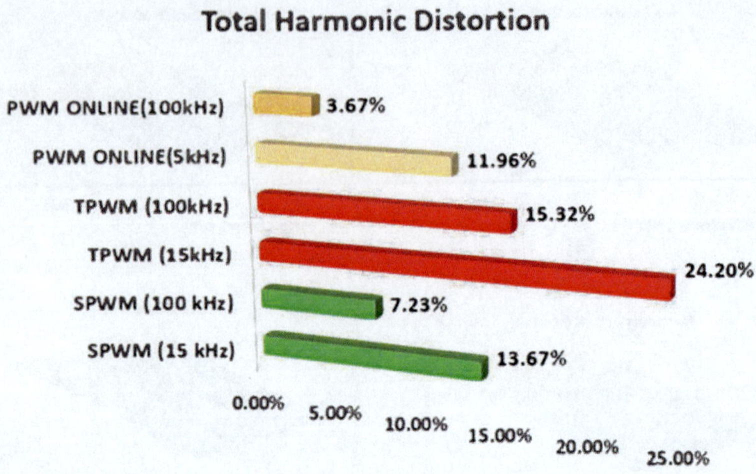

Fig. 31.10 THD of current evaluated to high and low frequency for each method of modulation

The PMW online technique implements to high frequency reduce the THD in the output currents of the CSI-SiC topology. The reduction of THD improves the efficiency of the motor drive system. The THD obtained with the lowest frequency is 11.96%. While increasing the frequency, the THD is reduced to 3.67% when working with 100 kHz. While in the other topologies, the same trend is observed but with a higher THD.

The analysis of these modulation techniques in emerging topologies for electric traction helps to improve their efficiency and contributes to presenting new solutions to the problems detected in the topologies most used in these systems.

The SiC devices in motor drive allow increasing the frequency of switching and improve the efficiency of power converter compared to inverters developed with conventional silicon devices. This shows that the use of this technology of wide gap band in the motor drive system helps to improve efficiency and allows obtaining more efficient electric traction systems.

References

1. V. Sharma, P. Saini, S. Garg, B. Negi, Comparative analysis of VSI, CSI and ZSI fed induction motor drive system, in *2016 3rd International Conference on Computing for Sustainable Global Development (INDIACom)*, New Delhi (2016), pp. 1188–1191
2. M. Dibaji, A. Damaki Aliabad, E. Amiri, M. Divandari, Current source inverter drive system for switched reluctance motors, in *2019 IEEE International Electric Machines and Drives Conference (IEMDC)*, San Diego, CA, USA (2019), pp. 1670–1673. http://doi.org/10.1109/IEMDC.2019.8785405

3. M.P. Maharajan, P. Muthu, M. Palpandian, S. Kannadasan, Analysis of low harmonics and high efficient BLDC motor drive system for automotive application, in *2011 International Conference on Recent Advancements in Electrical Electronics and Control Engineering*, Sivakasi (2011), pp. 526–531. http://doi.org/10.1109/ICONRAEeCE.2011.6129788
4. G.J. Su, P. Ning, Loss modeling and comparison of VSI and RB-IGBT based CSI in traction drive applications, in *2013 IEEE Transportation Electrification Conference and Expo (ITEC)*, Detroit, MI (2013), pp. 1–7
5. J. Martin, A. Bier, S. Catellani, L.G. Alves-Rodrigues, F. Barruel, A high efficiency 5.3 kW current source inverter (CSI) prototype using 1.2 kV silicon carbide (SiC) bi-directional voltage switches in hard switching, in *PCIM Europe 2016*; *International Exhibition and Conference For Power Electronics, Intelligent Motion, Renewable Energy and Energy Management*, Nuremberg, Germany (2016), pp. 1–8
6. H. Ye, Y. Yang, A. Emadi, Traction inverters in hybrid electric vehicles, in *2012 IEEE Transportation Electrification Conference and Expo (ITEC)*, Dearborn, MI (2012), pp. 1–6. http://doi.org/10.1109/ITEC.2012.6243442. K. Elissa, Title of paper if known (Unpublished)
7. F. Acosta-Cambranis, J. Zaragoza, L. Romeral, A comprehensive analysis of SVPWM for a five-phase VSI based on SiC devices applied to motor drives, in *IECON 2019—45th Annual Conference of the IEEE Industrial Electronics Society*, Lisbon, Portugal (2019), pp. 1490–1495. http://doi.org/10.1109/IECON.2019.8927394
8. A. Bindra, Wide-bandgap power devices: adoption gathers momentum. IEEE Power Electron. Mag. **5**(1), 22–27 (2018)
9. F. Shang, A.P. Arribas, M. Krishnamurthy, A comprehensive evaluation of SiC devices in traction applications, in *2014 IEEE Transportation Electrification Conference and Expo (ITEC)*, Dearborn, MI (2014), pp. 1–5
10. A. Merkert, J. Müller, A. Mertens, Component design and implementation of a 60 kW full SiC traction inverter with boost converter, in *2016 IEEE Energy Conversion Congress and Exposition (ECCE)*, Milwaukee, WI (2016), pp. 1–8
11. D. Han, S. Li, Y. Wu, W. Choi, B. Sarlioglu, Comparative analysis on conducted CM EMI emission of motor drives: WBG versus Si devices. IEEE Trans. Industr. Electron. **64**(10), 8353–8363 (2017)
12. X. Ding, J. Cheng, F. Chen, Impact of silicon carbide devices on the powertrain systems in electric vehicles. Energies **10**, 533 (2017)
13. D. Han, Y. Li, B. Sarlioglu, Analysis of SiC based power electronic inverters for high speed machines, in *2015 IEEE Applied Power Electronics Conference and Exposition (APEC)*, Charlotte, NC (2015), pp. 304–310
14. D. Han, J. Noppakunkajorn, B. Sarlioglu, Comprehensive efficiency, weight, and volume comparison of SiC and Si-based bidirectional DC–DC converters for hybrid electric vehicles. IEEE Trans. Veh. Technol. **63**(7), 3001–3010 (2014)
15. K. Gnanasambandam, A.K. Rathore, A. Edpuganti, D. Srinivasan, J. Rodriguez, Current-fed multilevel converters: an overview of circuit topologies, modulation techniques, and applications. IEEE Trans. Power Electron. **32**(5), 3382–3401 (2017)
16. R. Guedouani, B. Fiala, E.M. Berkouk, M.S. Boucherit, Modeling and control of multilevel three-phase PWM current source inverter, in *International Aegean Conference on Electrical Machines and Power Electronics and Electromotion, Joint Conference*, Istanbul (2011), pp. 455–460. http://doi.org/10.1109/ACEMP.2011.6490642
17. G. Su, L. Tang, Current source inverter based traction drive for EV battery charging applications, in *2011 IEEE Vehicle Power and Propulsion Conference*, Chicago, IL (2011), pp. 1–6. http://doi.org/10.1109/VPPC.2011.6043143
18. S. von Malottki, K. Hameyer, Extended base speed range by using a current-source-inverter-fed IPMSM for automotive application, in *2014 16th European Conference on Power Electronics and Applications*, Lappeenranta (2014), pp. 1–8. http://doi.org/10.1109/EPE.2014.6910968
19. E. Fernández, A. Paredes, V. Sala, L. Romeral, A simple method for reducing THD and improving the efficiency in CSI topology based on SiC power devices. Energies **11**(10), 2798 (2018). https://doi.org/10.3390/en11102798

20. J. Bao, W. Bao, Z. Zhang, W. Fang, A simple current-balancing method for a three-phase 5-level current-source inverter, in *2009 35th Annual Conference of IEEE Industrial Electronics*, Porto (2009), pp. 104–108. http://doi.org/10.1109/IECON.2009.5414790
21. Z. Liu, Z. Zheng, S.D. Sudhoff, C. Gu, Y. Li, Reduction of common-mode voltage in multiphase two-level inverters using SPWM With phase-shifted carriers. IEEE Trans. Power Electron. **31**(9), 6631–6645 (2016). https://doi.org/10.1109/TPEL.2015.2499380
22. E.R.C. da Silva, E.C. dos Santos, B. Jacobina, Pulsewidth modulation strategies. IEEE Ind. Electron. Mag. **5**(2), 37–45 (2011). https://doi.org/10.1109/MIE.2011.941120
23. K. Taniguchi, H. Irie, Trapezoidal modulating signal for three-phase PWM inverter. IEEE Trans. Industr. Electron. **33**(2), 193–200 (1986)
24. A. Agarwal, V. Agarwal, FPGA realization of trapezoidal PWM for generalized frequency converter. IEEE Trans. Industr. Inf. **8**(3), 501–510 (2012)
25. K. Thakre, K.B. Mohanty, Performance improvement of multilevel inverter through trapezoidal triangular carrier based PWM, in *2015 International Conference on Energy, Power and Environment: Towards Sustainable Growth (ICEPE)*, Shillong (2015), pp. 1–6. http://doi.org/10.1109/EPETSG.2015.7510170
26. E.E. Fernández Palomeque, Optimization of a CSI inverter and DC/DC elevator with silicon carbide devices, for applications in electric traction systems. Tesi doctoral, UPC, Departament d'Enginyeria Electrònica (2019)
27. A.K. Abdelsalam, A. Massoud, A. Darwish, S. Ahmed, Simplified generic on-line PWM technique for single phase grid connected current source inverters, in *2012 Twenty-Seventh Annual IEEE Applied Power Electronics Conference and Exposition (APEC)*, Orlando, FL (2012), pp. 1398–1403
28. J.R. Espinoza, G. Joos, Current-source converter on-line pattern generator switching frequency minimization. IEEE Trans. Industr. Electron. **44**(2), 198–206 (1997). https://doi.org/10.1109/41.564158
29. M.H. Bierhoff, F.W. Fuchs, Semiconductor losses in voltage source and current source IGBT converters based on analytical derivation, in *2004 IEEE 35th Annual Power Electronics Specialists Conference* (IEEE Cat. No.04CH37551), vol 4 (2004), pp. 2836–2842
30. E. Fernández, M. Coello, R. Rockwood, Optimization of weight-volume in impedance network SiC converter for electric motorcycle, in *2019 IEEE 4th Ecuador Technical Chapters Meeting (ETCM)*, Guayaquil, Ecuador (2019), pp. 1–6. http://doi.org/10.1109/ETCM48019.2019.9014910

Chapter 32
A Proposed Model for Electricity Demand Forecasting in Ecuador Considering Akaike Criterion

Manuel Jaramillo and Stalin Llamuca

Abstract The objective of electrical load forecasting is to satisfactorily and accurately predict the electrical demand that could increase or decrease in the future. A large number of engineering applications have accurate and reliable electricity demand forecasting models. Accurate load forecasting helps to plan the capacity and operation of different utilities to reliably supply power to consumers. The present work establishes the collection of annual and monthly historical data, obtained from the Ecuadorian electricity regulation and control agency. This study is focused on the comparison of the SARIMA regression model that considers the seasonality of electrical demand data, providing an evaluation of the model based on the Akaike information criterion. The input data have been divided into two data sets, one annual and one monthly, to build the forecasting model. By simulating the actual data, removing the last 5 years of data in the annual case and 2 years in the monthly case, the simulation results are checked against the actual data. The accuracy of the prediction models has been evaluated using different error matrices. R software was used for the prediction analysis having results with an error of less than 5% when comparing actual and estimated electrical demand.

32.1 Introduction

In any country, electrical planning for future energy demand is a key management tool that allows countries to strategically assign resources for investments in generation, transmission, and distribution.

In Ecuador, the electricity regulation and control agency ARCONEL is the government entity in charge of all electricity public services and public lighting. Furthermore, this government entity manages all data related to demand and production energy in Ecuador, which is available for everyone through its website. Although ARCONEL provides quite the amount of historical data for electrical demand in Ecuador, it does not have the same level of information or tools for energy demand

M. Jaramillo (✉) · S. Llamuca
Universidad Politécnica Salesiana, Cuenca, Ecuador
e-mail: mjaramillo@ups.edu.ec

Á. Rocha et al. (eds.), *Communication, Smart Technologies and Innovation for Society*, Smart Innovation, Systems and Technologies 252,
https://doi.org/10.1007/978-981-16-4126-8_32

forecasting which greatly limits the capability of the government and research for and Ecuadorian electrical planning for the next years.

Forecasting energy is relevant for the improvement of energy efficiency in the country, as well as reducing the potential environmental impact. Furthermore, short forecast horizons can be very relevant in certain application areas such as electrical energy, daily, monthly, or annual energy consumption [1]. Long-term demand forecasting is used to avoid buying too much or too little power from generators, while short-term forecasting is used to avoid distributing more power than is purchased. This type of forecasting requires the prediction error to be below the contract tolerances, which is usually around 5% [2, 3].

In addition, electric power demand forecasting presents the initial step in the development of future generation, transmission, and distribution facilities; however, the accuracy of electric power load forecasting often cannot meet the desired outcome because it is influenced by several uncertain and uncontrollable factors such as economic development, human social activities, country policies, and climate change [4].

Due to the critical importance of forecasting electric demand and the lack of proper models or tools for this purpose in Ecuador, this research proposes an algorithm to create a SARIMA model, which will allow forecasting of electric demand with an error of less than 5%. This model is by the Akaike criterion.

The Akaike information criterion, widely used in the field of biology and econometrics, is an objective tool that allows quantifying the fitness of a particular model relative to a finite set of models [5]. Finally, regression algorithms, such as the SARIMA models, are designed to capture seasonal patterns and are perfect for modeling electric demand time series [6], using a same-month approach as is the case with monthly data.

32.2 Methodology

32.2.1 Seasonal ARIMA (SARIMA)

SARIMA models are used when time series shows a seasonal variation, and to formalize the resulting model, the backward shift operator $\left(B^k\right)$ is used which is the observation of the time series backwards in time by a period k [7]. The backward shift operator presents a general seasonality transformation, where the time series is stationary only if the statistical properties (mean and variance) are constant over time [8]. The general seasonality transformation equation is presented below:

$$z_t = \nabla_s^D \nabla^d y_t = (1 - B^s)^D (1 - B)^d y_t \quad (32.1)$$

where

z Is the differentiation of time series

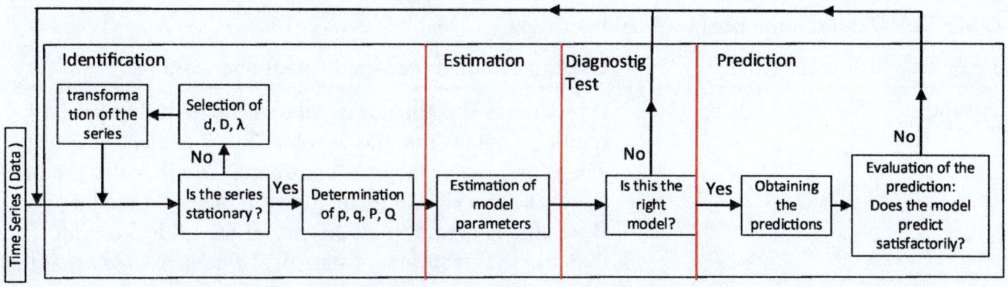

Fig. 32.1 Box Jenkins methodology pseudocode

d It is the degree of non-seasonal differentiation used
D Is the degree of seasonal differentiation used
s Seasonal period
B Regression coefficient.

Then, the general SARIMA form of the SARIMA (*p*, *P*, *q*, *Q*) model is:

$$\phi_p(B)\phi_p(B^s)z_t = \delta + \theta_q(B)\theta_q(B^s)a_t \tag{32.2}$$

where

ϕ Autoregressive process
θ Moving average process
δ Constant value
a_t White noise process.

32.2.2 *Box Jenkins Methodology*

The methodology used in this research for the development of estimation models and forecasting is based on the Box Jenkins as it is shown in Fig. 32.1. In addition, the parameters that have been considered to evaluate the best estimation model are the error sum of squares (SSE), *p*-value, and Akaike information criterion (AIC), and these parameters are shown in Table 32.1.

32.2.3 *Forecasting Considerations*

This research considers certain parameters for forecasting analysis and prediction verification for one-period electrical demand signals [9], and these parameters are:

- For any statistical model, the minimum data required for fitting depends on the number of parameters to be estimated and the number of random variations in the data.

Table 32.1 Prediction model evaluation criteria

Error sum of squares (SSE)	Used to evaluate predicted values and residuals
P-value	If *p*-value is less than or equal to the size of the critical region, it means that this *p*-value is in the rejection region. Therefore, the x_i factor is considered significantly greater than zero and should be included in the fitted model If *p*-value is greater than the size of the critical region, then, there is insufficient statistical evidence to say that the evaluated term is significant, and therefore, it should not be included in the adjusted model
AIC (Akaike Information Criterion)	AIC estimates the minimum value of the reference model; it is calculated by using the maximized value of the likelihood function. Therefore, the best model is selected according to the minimum AIC value obtained

- From a statistical perspective, it is important to have a larger number of observations than parameters; although, in practice, the data often contain a high number of randomness, which causes the size of the sample requirements to increase accordingly.
- Even if having the minimum amount of data required, it does not guarantee adequate estimates of seasonality.
- When forecasting using seasonal models, the number of data needed depends on the type of model and the amount of randomness in the data.

32.2.4 Electricity Demand in Ecuador

The Ecuadorian electricity sector needs to consider several parameters for the coming years; among them, it is necessary to carry out a study of electricity demand, which will allow the government to place itself in the possible scenarios to be triggered in the future, in order to ensure the supply of energy through investments in generation, transmission, and distribution.

Planning for the expansion of the Ecuadorian electrical system in a prospectus of the future evolution of the electrical market creates a plan for investments in the electric sector. If electrical demand does not increase, it would not be necessary to carry out expansions in the system; since it is assumed that in the starting point of the planning process, all the electrical demand is supplied [10].

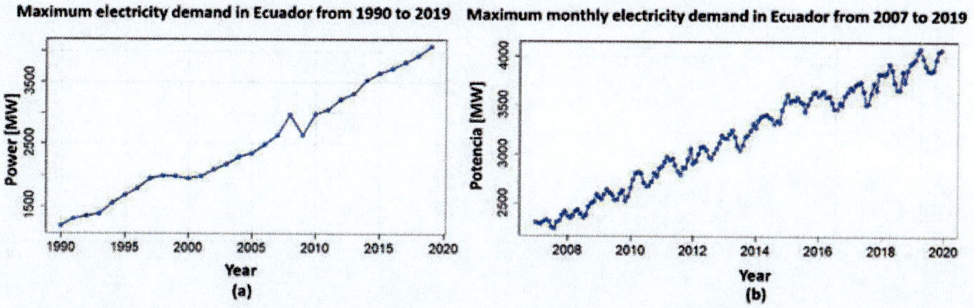

Fig. 32.2 **a** Maximum annual electricity demand in Ecuador from 1990 to 2019, **b** maximum monthly electricity demand in Ecuador from 2007 to 2019

32.2.5 Ecuadorian Electricity Regulation and Control Agency (ARCONEL)

ARCONEL manages updates, validates, and processes statistical and geographic information of the different entities of the electrical sector, by promoting free access to complete, timely and quality information on activities of generation, transmission, distribution, and public lighting. This information is published on the web portal www.regulacionelectrica.gob.ec for public access [11], and it is shown in Fig. 32.2.

32.2.6 Problem Formulation

For this research, seasonal ARIMA (SARIMA) models are used to predict the maximum demand of the Ecuadorian electricity sector for future years.

For data process and analysis, the software RStudio is used, in which the data obtained from the documents of the Statistics of the Ecuadorian Electricity Sector found in the ARCONEL web page are processed and through programming, forecasting maximum demand values are calculated. This research shows two study cases by considering yearly and monthly demand data.

Case Study 1: Maximum Annual Power Demand

With Ecuadorian annual maximum demand from 1990 to 2019, SARIMA models are programmed, followed by data differentiation in order to obtain the autocorrelation function (ACF) and partial autocorrelation function (PACF), both shown in Fig. 32.3.

Next, the proposed algorithm considers non-stationary and seasonal differentiation (d, D) values equals to 1. These values were considered because it is not necessary to differentiate the data more than once.

Following, the algorithm calculates the values of (p, q) and (P, Q), which represent the non-stationary autoregressive and moving average coefficients, and the stationary

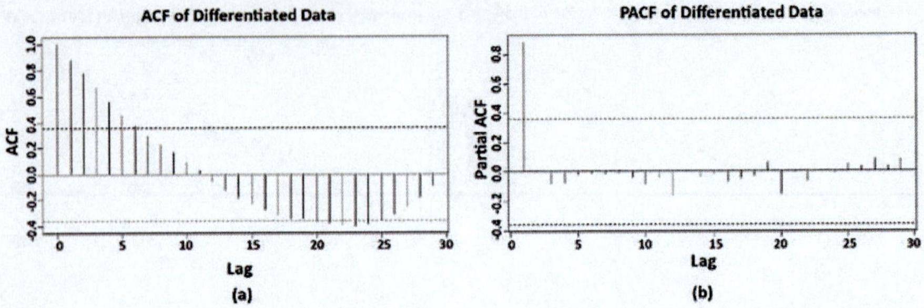

Fig. 32.3 **a** ACF of differentiated data, **b** PACF of differentiated data

Table 32.2 Annual SARIMA models, with their respective evaluation criteria

Model	(*p*, *d*, *q*)	(*P*, *D*, *Q*)	*s*	AIC	SSE	*p*-value
1	(0, 1, 0)	(1, 1, 1)	1	−85.6806	0.05622203	0.9205082
2	(0, 1, 0)	(0, 1, 1)	1	−85.45176	0.05966706	0.2249689
3	(0, 1, 1)	(0, 1, 1)	1	−83.76094	0.06300745	0.7833637
4	(1, 1, 0)	(1, 1, 1)	1	−83.54841	0.0565212	0.8839251
5	(0, 1, 1)	(1, 1, 1)	1	−82.40829	0.05854784	0.4971028

autoregressive and moving average coefficients. The seasonal period value takes a value of one, since the data set has one power demand value per year.

Finally, it is necessary to determine which one is the best SARIMA model in order to later calculate the best annual prediction. Table 32.2 shows the AIC, SSE, and p-values of the five models generated, and based on the evaluation criteria, the best SARIMA model is determined.

This research considers that the generated models must not exceed a certain level of complexity. This complexity is evaluated by considering the sum of the model coefficients (p, d, q, P, D, Q), and this value must be less than 10. Finally, forecasting is performed to predict the electricity demand values for the next 5 years. The results are shown in Fig. 32.4 and Table 32.3.

Case Study 2: Maximum Monthly Electricity Demand

The same analysis as in case study one is performed in this section. The main difference with section one is that monthly data is analyzed. ACF and PACF represented in Fig. 32.5 are calculated in the same way. The results of the best models are shown below in Table 32.4, where the best SARIMA model is determined by the same evaluation criteria.

After finding the best model, forecasting is carried out for the maximum monthly demand for 2 years (24 months). These results are shown in Fig. 32.6 and Table 32.5.

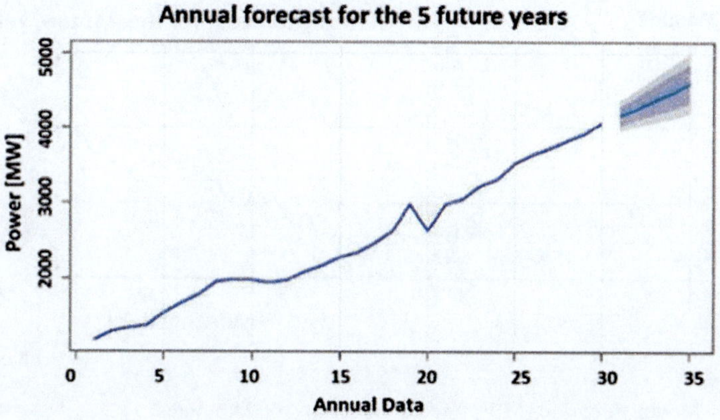

Fig. 32.4 Annual maximum demand forecasting for 5 years

Table 32.3 Annual maximum demand forecasting values results for the 5 years

Year	1	2	3	4	5
Average power [MW]	4164.88	4279.21	4381.90	4489.70	4595.26

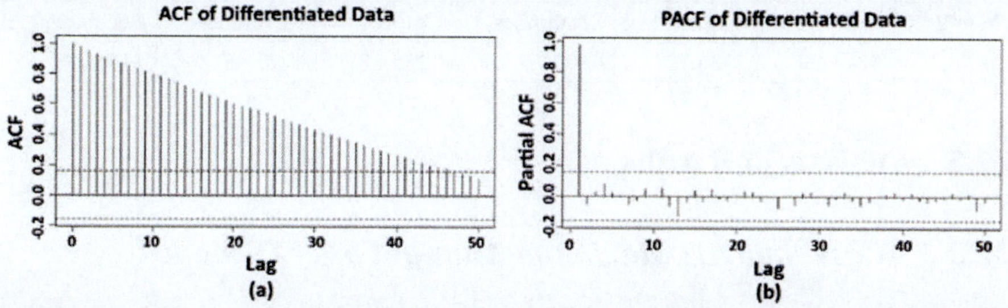

Fig. 32.5 **a** ACF of the differentiated data, **b** PACF of the differentiated data

Table 32.4 SARIMA models with monthly data

Model	(*p*, *d*, *q*)	(*P*, *D*, *Q*)	*s*	AIC	SSE	*p*-value
1	(2, 1, 1)	(0, 1, 1)	12	−792.1848	0.02545203	0.6738887
2	(1, 1, 1)	(0, 1, 1)	12	−792.1415	0.02579888	0.4475178
3	(0, 1, 2)	(0, 1, 1)	12	−791.6785	0.02589336	0.3713313
4	(1, 1, 2)	(0, 1, 1)	12	−791.3815	0.02558976	0.6126908
5	(1, 1, 1)	(0, 1, 1)	12	−790.2299	0.02588931	0.4732326

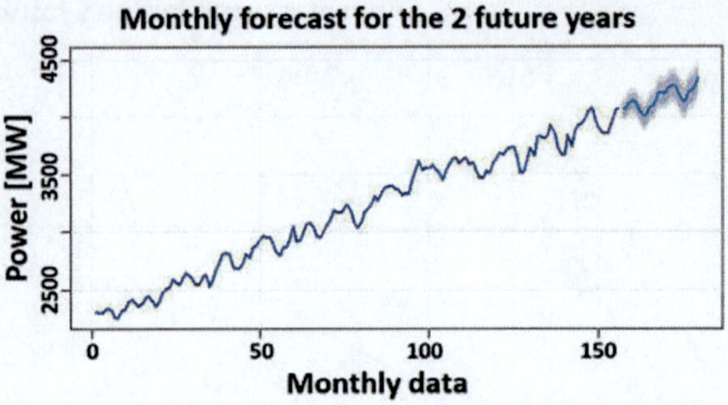

Fig. 32.6 Monthly maximum demand forecasting for 24 months

Table 32.5 Monthly demand [MW] forecasting results for 24 months

Year 1	January	February	March	April	May	June
	4061.52	4072.01	4114.09	4137.22	4124.90	4063.90
Year 1	July	August	September	October	November	December
	4007.98	4000.04	4073.02	4099.20	4140.75	4207.36
Year 2	January	February	March	April	May	June
	4202.55	4210.67	4251.46	4274.11	4261.64	4200.60
Year 2	July	August	September	October	November	December
	4144.66	4136.72	4209.70	4235.88	4277.43	4344.04

32.3 Analysis of Results

32.3.1 Case Study 1: Maximum Annual Power Demand

The results of the implemented models are validated by generating two test data sets. The first data set is generated by only consider the first 25 annual demand values, and the second data set has the complete 30 annual demand values. Then, forecasting is performed with the first data set, and the predicted values are compared with the real ones (second data set). This comparison is shown in Table 32.6 and Fig. 32.7.

Table 32.6 Error calculation between actual and estimated annual demand values

Year	Actual power [MW]	Estimated power [MW]	Error [%]
1	3643.36	3628.37	0.41
2	3723.95	3726.08	0.06
3	3825.82	3823.78	0.05
4	3928.93	3921.48	0.19
5	4077.03	4019.19	1.42

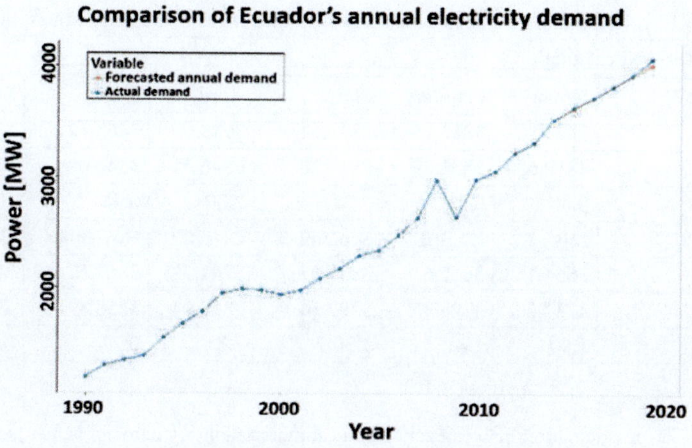

Fig. 32.7 Comparison between predicted and actual annual demand

32.3.2 Case Study 2: Maximum Monthly Electricity Demand

A similar analysis as in the previous section is performed for the monthly demand model. The first data set is generated by only consider 132 values (first 11 years, 12 demand values per year), and the second data set has the complete 156 values (13 years, 12 demand values per year). Then, forecasting is performed with the first data set, and the predicted values are compared with the real ones (second data set). This comparison is shown in Fig. 32.8, Tables 32.7, and 32.8.

Fig. 32.8 Comparison between predicted and actual monthly demand

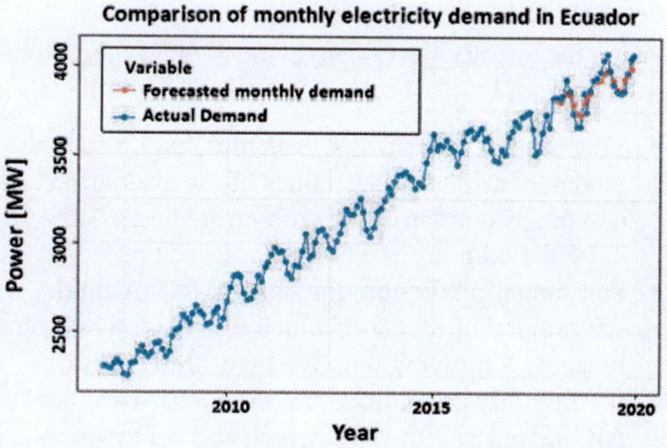

Table 32.7 Error calculation between actual and estimated monthly demand values, first year

First year forecast calculation						
Power [MW]	January	February	March	April	May	June
Actual	3825.86	3813.75	3838.64	3928.93	3863.12	3738.52
Estimated	3810.63	3801.79	3833.20	3846.65	3846.02	3792.53
Error [%]	0.40	0.31	0.14	2.09	0.44	1.44
Power [MW]	July	August	September	October	November	December
Actual	3664.15	3662.6	3847.37	3741.78	3867.57	3927.3
Estimated	3737.20	3728.95	3797.34	3828.04	3858.49	3930.24
Error [%]	1.99	1.81	1.30	2.31	0.23	0.07

Table 32.8 Error calculation between actual and estimated monthly demand values, second year

Second year forecast calculation						
Power [MW]	January	February	March	April	May	June
Actual	3943.99	3974.038	4045.88	4077.03	3984.73	3905.52
Estimated	3922.17	3928.54	3967.18	3982.68	3982.28	3928.66
Error [%]	0.55	1.14	1.95	2.31	0.06	0.59
Power [MW]	July	August	September	October	November	December
Actual	3862.35	3851.71	3863.33	3971.2	4050.62	4077.3
Estimated	3873.25	3864.96	3933.34	3964.05	3994.50	4066.25
Error [%]	0.28	0.34	1.81	0.18	1.39	0.27

32.4 Conclusions

From the results described in Sect. 32.3, the following conclusions have been summarized:

- This research performs, executes, and simulates the most optimal SARIMA model, considering the values of the evaluation criteria (SSE, AIC, and *p*-value), and this is demonstrated with errors under 1.5% for annual prediction and under 2.4% for monthly prediction.
- For annual prediction, the best SARIMA model was selected by considering an AIC valued of 85.6806, which was the lowest one among the different generated models (complying with the algorithm criteria).
- For monthly prediction, the best SARIMA model was selected by considering an AIC valued of 792.1848, which was the lowest one among the different generated models (complying with the algorithm criteria).
- In Table 32.2, it is shown that the best model for annual prediction has a *p*-value of 0.9205082, which is the closest to 1 (complying with the algorithm criteria).
- In Table 32.4, it is shown that the best model for monthly prediction has a *p*-value of 0.6738887, which is the closest to 1 (complying with the algorithm criteria).

- By analyzing results for monthly forecasting in Table 32.4, the model accuracy for seasonality is validated as it can be observed that for the last month of each year, and the maximum predicted demand 3930.24 and 4066.25, respectively, are the highest in a 12-month span, which complies with the seasonality concept for electrical demand.

References

1. A. Hošovský, J. Piteľ, M. Adámek, J. Mižáková, K. Židek, Comparative study of week-ahead forecasting of daily gas consumption in buildings using regression ARMA/SARMA and genetic-algorithm-optimized regression wavelet neural network models. J. Build. Eng. 1–3 (2020)
2. I.S. Msiza, F.V. Nelwamondo, T. Marwala, Electrical Energy demand prediction using artificial neural networks and support vector regression. J. Comput. **3**, 1–8 (2008)
3. A.A. Mir, M. Alghassab, K. Ullah, Z.A. Khan, Y. Lu, M. Imran, A review of electricity demand forecasting in low and middle income countries: the demand determinants and horizons. Sustainability **12**, 2–14 (2020)
4. M.A. Hammad, B. Jereb, B. Rosi, D. Dragan, Methods and models for electric load forecasting: a comprehensive review. Logist. Sustain. Transp. **11**, 51–76 (2020)
5. D.R. Martínez, J.L. Albín, J.C. Cabaleiro, T.F. Pena, F.F. Rivera, V. Blanco, El criterio de información de Akaike en la obtención de modelos estadísticos de Rendimiento. XX Jornadas Paralelismo 439–444 (2009)
6. A. Zubair, Short-term electric demand forecasting for power, pp. 122–126 (2019)
7. B. Fanoodi, B. Malmir, F.F. Jahantigh, Reducing demand uncertainty in the platelet supply chain through artificial neural networks and ARIMA models. Comput. Biol. Med. **113** (2019)
8. A.E. Permanasari, I. Hidayah, I.A. Bustoni, SARIMA (Seasonal ARIMA) implementation on time series to forecast the number of Malaria incidence, in *Proceedings of 2013 International Conference on Information Technology and Electrical Engineering, Intelligent and Green Technologies for Sustainable Development, ICITEE 2013*, pp. 203–207 (2013)
9. K. Goswami, A.B. Kandali, Electricity demand prediction using data driven forecasting scheme: ARIMA and SARIMA for real-time load data of Assam, in *2020 International Conference on Computational Performance Evaluation, ComPE 2020*, pp. 570–574 (2020)
10. J. Suntaxi, Y. Salazar, R. Loor, Pronóstico de la demanda en energía y potencia del Sistema Eléctrico Quito. Rev. Técnica Energía **15**, 12–21 (2019)
11. ARCONEL, Estadisticas Anuales y Multianual Del Sector Eléctrico Ecuatoriano 2018, p. 214 (2019)

Chapter 33
Control Strategy Applied to Smart Photovoltaic Inverters for Reactive Power Exchange Through Volt-Var Control to Improve Voltage Quality in Electrical Distribution Networks

Marcelo Montufar, Wilson Pavón, Manuel Jaramillo, and Silvio Simani

Abstract Photovoltaic (PV) generation systems have become part of the electrical distribution networks due to their clean energy and adaptability. The increasing energy consumption and the several load types in the distribution network could produce voltage deviation. Therefore, there are some strategies, which are used to solve this problem, for example, capacitor banks, voltage regulators, transformer tap variation, static Var compensators, and reactive power exchange employing inverters. This article presents a smart inverter able to deliver and absorb reactive power through Volt-Var control to solve voltage deviations. The inverter compensates for voltage deviation in the primary transformer, whether inductive or capacitive loads are connected in the point of common coupling (PCC). In consequence, the inverter improves the voltage quality in the entire network. Finally, total harmonics percentage (THD) and frequency deviation have an acceptable performance during the operation, due to the inverter hardware configuration, and the phase-locked loop (PLL) implementation.

33.1 Introduction

PV energy is growing all over the world, because this technology is effortless to install, such as rooftop; in this way, PV generation projects increase in the city [1–3]. Therefore, the inverters amount in distribution network is constantly increasing; it

M. Montufar · W. Pavón (✉) · M. Jaramillo
Universidad Politécnica Salesiana, Quito, Ecuador
e-mail: wpavon@ups.edu.ec

M. Montufar
e-mail: mmontufarc@est.ups.edu.ec

S. Simani
Università degli Studi Di Ferrara, Ferrara, Italy

Á. Rocha et al. (eds.), *Communication, Smart Technologies and Innovation for Society*,
Smart Innovation, Systems and Technologies 252,
https://doi.org/10.1007/978-981-16-4126-8_33

brings the presence of THD problems and, thus, the quality of voltage reduces in entire the system.

It is shown that the three-level inverter could improve the waveform [4]; due to this, device reduces the THD percentage [5, 6]. On the other hand, the inverter could also mitigate the voltage deviations produced by the connection of inductive or capacitive loads in the distribution network [7, 8]. Thus, to attenuate them, some strategies are required, like capacitor banks, voltage regulators, transformer tap changers, and static Var compensators.

Any possible solution represents device installation and maintenance costs, and their performance does not satisfy the system requirements [9]. The Volt-Var control strategy implemented on inverters is an excellent option for solving all those problems [10, 11].

This article analyzes the behavior of the Volt-Var control applied to PV inverters to mitigate voltage deviations. The strategy is called "smart inverter", and it detects the inductive or capacitive load to improve the voltage quality in distribution networks through the exchange of reactive power. The inverters must have the capacity to absorb and deliver reactive power; using these characteristics, the distribution voltage quality improves [12].

33.2 Volt-Var Control Methodology

The inverter receives active power from the DC converter and the PV system. The voltage and the current from the PV are the inputs for the maximum power point tracking (MPPT) block, while the PWM signal is the MPPT output. Moreover, the DC converter is driven by a PWM signal as shown in Fig. 33.1.

The converter relation among input and output voltage is given by Eq. (33.1) [13]. V_{pv} is the PV voltage, V_{dc} is the DC converter voltage output, and D is the duty cycle.

$$V_{pv} = V_{dc}[1 - D] \tag{33.1}$$

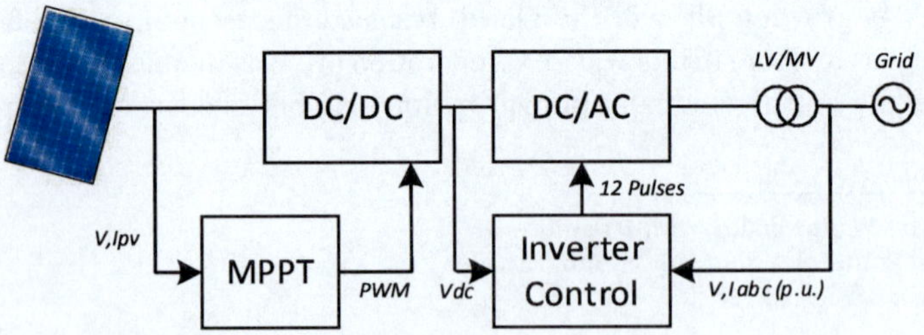

Fig. 33.1 General photovoltaic inverter system

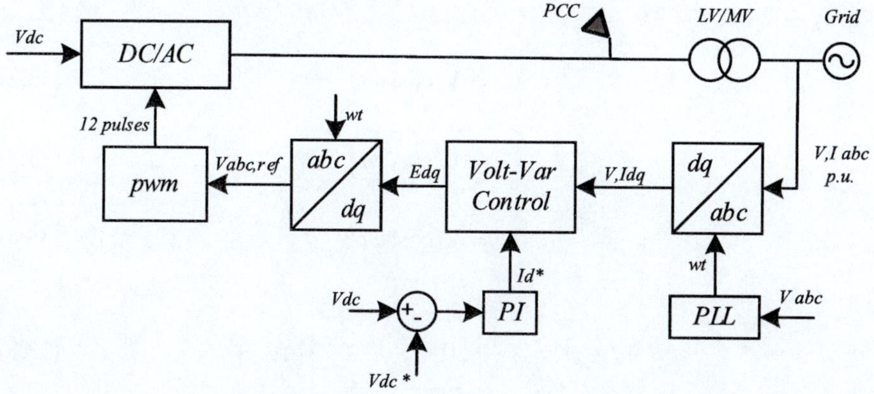

Fig. 33.2 Inverter control strategy

Manipulating reactive power in the distribution network is handled by inverter control. In Fig. 33.1, VI_{abc} and V_{dc} are inverter inputs, both expressed in per unit quantities, and a 12 pulses PWM is an output of the inverter control. The inverter control strategy is described in Fig. 33.2 and shows the Volt-Var control stage. The dq rotating reference frame variables are shown in Fig. 33.3, to obtain VI_{dq} is used the three-phase Clarke and Park transform from VI_{abc}.

The Volt-Var control is shown in Fig. 33.3. The reference V_{abc} is calculated by the three-phase inverse Clarke and Park transforms from E_{dq} components, shown in Fig. 33.2. Finally, reference V_{abc} is transformed into 12 pulses PWM signal to apply it at the three-level inverter [14].

The variables of a three-phase system are given by Eqs. (33.2), (33.3), (33.4), and then, the transformation is expressed by Eqs. (33.5) and (33.6). The inverse process of the transforms can be obtained by the inverse transformation, which are given by

Fig. 33.3 Volt-Var control

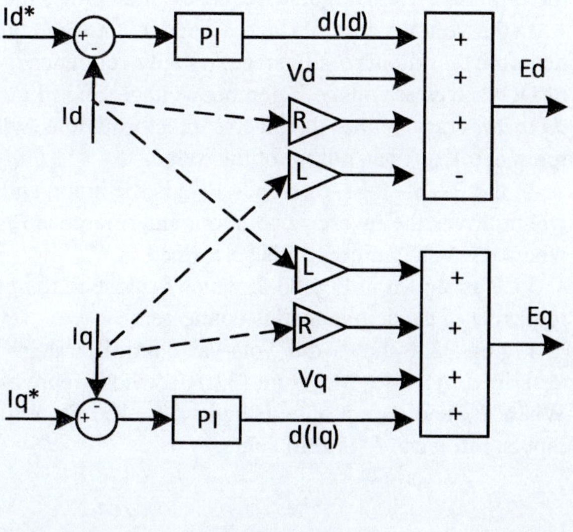

Eqs. (33.7) and (33.8).

$$V_a = V_p \cos(wt) \tag{33.2}$$

$$V_b = V_p \cos(wt) \tag{33.3}$$

$$V_c = V_p \cos(wt) \tag{33.4}$$

$$[f_{V_{\alpha\beta0}}] = \sqrt{\frac{2}{3}} \begin{bmatrix} 1 & -\frac{1}{2} & -\frac{1}{2} \\ 0 & \frac{\sqrt{3}}{2} & -\frac{\sqrt{3}}{2} \\ \frac{1}{\sqrt{2}} & \frac{1}{\sqrt{2}} & \frac{1}{\sqrt{2}} \end{bmatrix} [f_{V_{abc}}] \tag{33.5}$$

$$[f_{V_{dq0}}] = \sqrt{\frac{2}{3}} \begin{bmatrix} \cos\theta_d & \cos(\theta_d - \frac{2\pi}{3}) & \cos(\theta_d + \frac{2\pi}{3}) \\ -\sin\theta_d & -\sin(\theta_d - \frac{2\pi}{3}) & -\sin(\theta_d + \frac{2\pi}{3}) \\ \frac{1}{\sqrt{2}} & \frac{1}{\sqrt{2}} & \frac{1}{\sqrt{2}} \end{bmatrix} [f_{V_{\alpha\beta0}}] \tag{33.6}$$

$$[f_{V_{\alpha\beta0}}] = \sqrt{\frac{2}{3}} \begin{bmatrix} 1 & 0 & \frac{1}{\sqrt{2}} \\ -\frac{1}{2} & \frac{\sqrt{3}}{2} & \frac{1}{\sqrt{2}} \\ -\frac{1}{2} & -\frac{\sqrt{3}}{2} & \frac{1}{\sqrt{2}} \end{bmatrix} [f_{V_{dq0}}] \tag{33.7}$$

$$[f_{V_{abc}}] = \sqrt{\frac{2}{3}} \begin{bmatrix} \cos\theta_d & -\sin\theta_d & \frac{1}{\sqrt{2}} \\ \cos(\theta_d - \frac{2\pi}{3}) & -\sin(\theta_d - \frac{2\pi}{3}) & \frac{1}{\sqrt{2}} \\ \cos(\theta_d + \frac{2\pi}{3}) & -\sin(\theta_d + \frac{2\pi}{3}) & \frac{1}{\sqrt{2}} \end{bmatrix} [f_{V_{\alpha\beta0}}] \tag{33.8}$$

Figure 33.2 shows the methodology for inverter control, *abc* to *dq* stage defines the used transformations, where VI_{abc} and wt are inputs, and VI_{dq} is the output of this stage. In PLL, V_{abc} is the input, and wt is the output, and this phase used the three-phase system angular frequency and defines the inverter frequency [15].

In the Volt-Var control stage, *VI* and reference I_d are the inputs, while E_{dq} components are the outputs of this stage. Usually, reference I_d is obtained from the operation of DC converter voltage, reference voltage, and PI control, shown in Fig. 33.2. The *dq* to *abc* stage defines the inverse transformations, where E_{dq} and wt are inputs, and reference V_{abc} is an output of this stage.

In the PWM stage, reference V_{abc} is the input, and 12 pulses PWM is an output, which drives the inverter operation, and reference I_d is also presented in Fig. 33.3 where the Volt-Var control is performed.

PCC is shown in Fig. 33.2, which is close to the primary transformer, where the inductive or capacitive load is connected.

Figure 33.3 shows the Volt-Var control strategy of inverter control, which is described by Eqs. (33.9) and (33.10), whose representation is shown in Fig. 33.3. Where E_d and E_q are outputs and I_d, I_q, V_d, V_q, reference I_d, and reference I_q are inputs, reference I_q is zero value.

$$E_d = V_d + I_d R - I_q L + dI_d \tag{33.9}$$

$$E_q = V_q + I_d L + I_q R + dI_q \tag{33.10}$$

Active and reactive power control from Volt-Var methodology is described in Eqs. (33.11) and (33.12).

$$p = E_d I_d + E_q I_q \tag{33.11}$$

$$q = -E_d I_q + E_q I_d \tag{33.12}$$

At the PCC level, shown in Fig. 33.2, the inverter absorbs and delivers reactive power as the load requires. The inductive and capacitive effect in the inverter implies a phase difference between the voltage-current, and it is described in Eq. (33.13).

$$\varphi = \cos^{-1}\left(\frac{P^2}{\sqrt{P^2 + Q^2}}\right) \tag{33.13}$$

33.3 Results

The present section shows the more outstanding results of the proposed research. In order to evaluate the Volt-Var control, it was used two different types of loads and was evaluated their effects in voltage-current drag, THD, and frequency variation.

Table 33.1 shows the different theoretical load connected in PCC and their connection time.

Figure 33.4a shows the experimental inductive and capacitive load connected in PCC. In the first-time interval, according to Table 33.1, the experimental load absorbs 8.5 KVAr, and the second time interval shows the load delivers 10.2 KVAr. Therefore, there is a difference between the theoretical, see Table 33.1, and the experimental reactive power.

Figure 33.4b shows the inverter response due to the load, in the first-time interval for the reactive power axis, the inverter absorbs 10 KVAr. In the second time interval for the reactive power axis, the inverter delivers 8.5 KVAr, the voltage axis of Fig. 33.4b shows that in both time intervals, when the inverter is absorbing and

Table 33.1 Time intervals connection for different load types

Time (seg)	Load type	Active power (KW)	Reactive power (KVAr)
0.25–0.45	Inductive	50	10
0.45–0.65	Capacitive	50	10

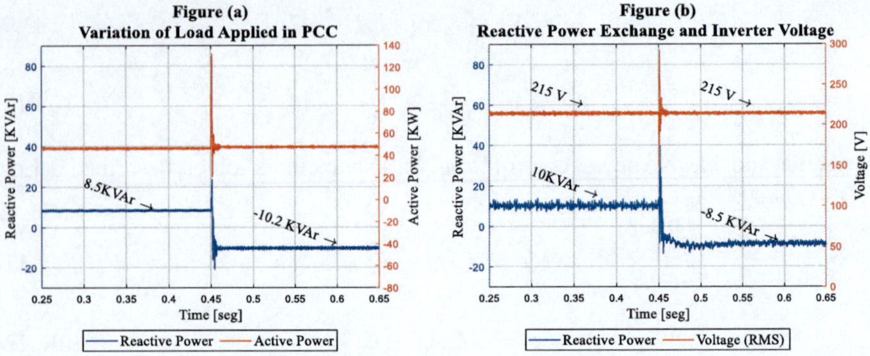

Fig. 33.4 Load applied in PCC and reactive power exchange by the inverter

delivering, the inverter has the same numeric voltage value of 215 V. In conclusion, there are not voltage deviations.

Figure 33.5a shows the system development with the inclusion of Volt-Var control in the reactive power axis case, for both time intervals and both load types, see Table 33.1, the primary transformer absorbs 1.5 KVAr. In the voltage axis case, for both time intervals, the voltage is 215 V; therefore, there are not voltage deviations for different load types.

Figure 33.5b shows the system response, without the Volt-Var control, in the reactive power axis case, the first-time interval with an inductive load, see Table 33.1, shows the primary transformer delivering 8.37 KVAr, and the second-time interval with a capacitive load shows the primary transformer absorbing 10.4 KVAr, while the voltage axis case, in the first-time interval with an inductive load, see Table 33.1, shows the primary transformer voltage is 214.4 V and the second-time interval with a capacitive load, see Table 33.1, the primary transformer voltage is 217 V; therefore, without the Volt-Var control, the primary transformer presents voltage deviations.

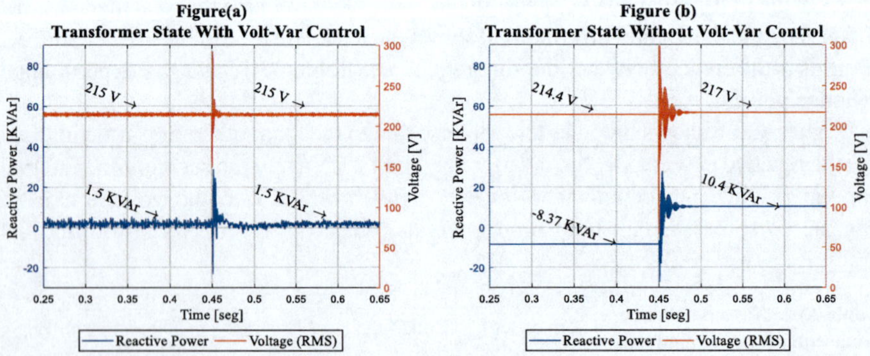

Fig. 33.5 State of transformer (with/without) Volt-Var control

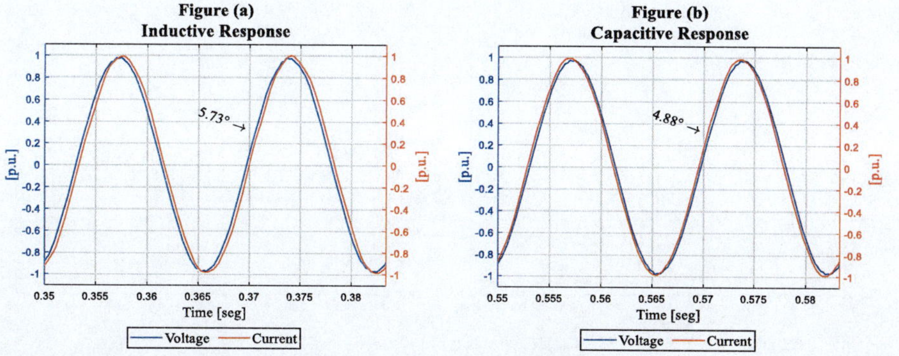

Fig. 33.6 Waveform and phase difference between voltage and current

Table 33.2 Active power, reactive power, and angle of the inverter

Type	Time (seg)	Active power (KW)	Reactive power (KVAr)	Angle (°)
Inductive	0.25–0.45	99.5	10	5.73
Capacitive	0.45–0.65	99.5	−8.5	4.88

Figure 33.6a shows the inverter inductive effect, where the current is delayed for the voltage, the angle is 5.73°, see Table 33.2, typical of an inductive load.

Figure 33.6b shows the inverter capacitive effect where the current is ahead for the voltage, the angle is 4.88°, see Table 33.2, typical of a capacitive load is shown in per unit to show the angle between the voltage and current.

Figure 33.7 shows the available active power of the PV generation and the reactive power that the inverter absorbs and delivers due to the load applied, see Table 33.1.

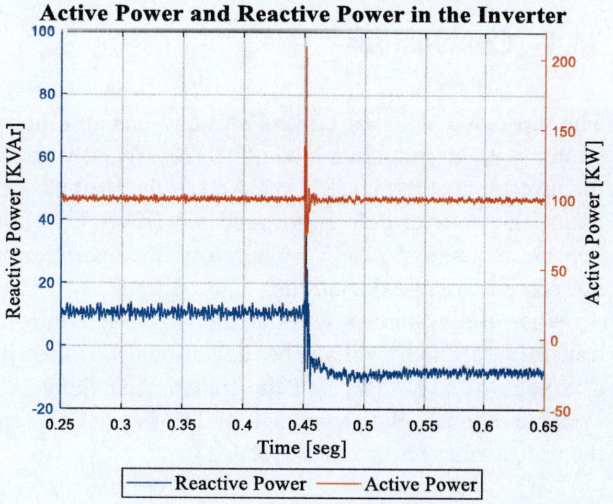

Fig. 33.7 Active power and reactive power in the inverter

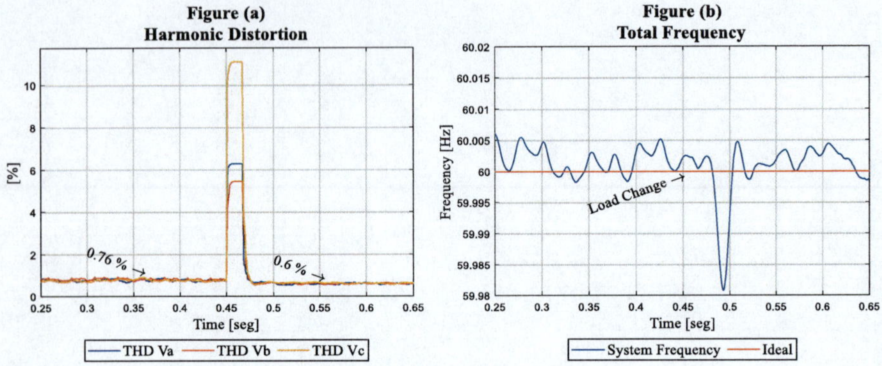

Fig. 33.8 THD and frequency for both load types

Table 33.2 defines the available active power, reactive power, and the inverter angle, and the angle is the function of active and reactive power available in the inverter, according to Fig. 33.7 and the time intervals.

Figure 33.8a shows the THD of the system, in the first-time interval corresponds to an inductive load, the harmonic distortion present mean values of 0.76%, and the second time interval corresponds to a capacitive load, the harmonic distortion present mean values of 0.6%.

Figure 33.8b shows the frequency for both load types. The first-time and second-time interval corresponds to an inductive and capacitive load, respectively (see Table 33.1). Frequency oscillates in the ideal reference with deviation peaks of 60.005 Hz. At the time 0.45 seg, the load change is presented, but frequency does not deviate immediately. Between 0.475 and 0.5 seg, the frequency has a deviation peak of 59.98 Hz.

33.4 Conclusions

The presented Volt-Var control strategy has an efficient performance. It is demonstrated with an inductive load of 10 [KVAr], which is connected in the system, and the inverter absorbs 10 [KVAr]. And, if the load has capacitance 10 [KVAr] components, the inverter delivers around 8.5 [KVAr], in both cases; the voltage keeps a constant voltage of 215 [V]. Therefore, the operation with the Volt-Var control does not present voltage deviations.

When the system is without the Volt-Var control strategy, it is found that the transformer primary voltage has deviations. An inductive load of 10 [KVAr] produces a voltage of 214.4 [V] and the transformer delivers 8.37 [KVAr]. Moreover, the system responses with a voltage of 217 [V] with a capacitive load of 10 KVAr, and the transformer absorbs 10.4 [KVAr].

The presented methodology demonstrated in terms of %THD, with an inductive load presents values close to 0.79% and 0.6% with capacitive load. Finally, both load types present maximum deviations of 0.5% in frequency of 60.005 [Hz] and a maximum deviation peak of 59.982 [Hz], corresponding to a load change. In conclusion, the performance of the proposed controller improved the overall system development.

References

1. R. Singh, Approximate rooftop solar PV potential of Indian cities for high-level renewable power scenario planning. Sustain. Energy Technol. Assessments **42**, 100850 (2020). http://doi.org/10.1016/j.seta.2020.100850
2. W. Pavon, E. Inga, S. Simani, Optimal distribution network planning applying heuristic algorithms considering allocation of PV rooftop generation, in *2020 IEEE Andescon, Andescon 2020* (2020). http://doi.org/10.1109/ANDESCON50619.2020.9272062
3. W.D.P. Vallejos, Standalone photovoltaic system, using a single stage boost DC/AC power inverter controlled by a double loop control, in *2017 IEEE PES Innovative Smart Grid Technologies Conference—Latin America ISGT Latin America 2017*, Janua, vol. 2017, pp. 1–6 (2017). http://doi.org/10.1109/ISGT-LA.2017.8126705
4. G. Yang, J. Yin, Z. Huang, Y. Zhang, A method to improve the reliability of three-level inverter based on equivalent input disturbance and repetitive control combinations. Microelectron. Reliab. **114**, 113837 (2020). http://doi.org/10.1016/j.microrel.2020.113837
5. S.S. Rangarajan, E.R. Collins, J.C. Fox, Efficacy of a smart photovoltaic inverter as a virtual detuner for mitigating network harmonic resonance in distribution systems. Electr. Power Syst. Res. **171**, 175–184 (2019). https://doi.org/10.1016/j.epsr.2019.02.001
6. I. Cagri, E. Karatepe, M. Boztepe, Electrical power and energy systems impact of harmonic limits on PV penetration levels in unbalanced distribution networks considering load and irradiance uncertainty. Electr. Power Energy Syst. **118**, 105780 (2020). http://doi.org/10.1016/j.ijepes.2019.105780
7. I. Kim, R.G. Harley, The transient-state effect of the reactive power control of photovoltaic systems on a distribution network. Int. J. Electr. Power Energy Syst. **99**, 630–637 (2018). https://doi.org/10.1016/j.ijepes.2018.01.015
8. I. Abdelmotteleb, T. Gomez, J.P. Chaves-Avila, Benefits of PV inverter volt-var control on distribution network operation, in *2017 IEEE Manchester PowerTech, Powertech 2017* (2017). http://doi.org/10.1109/PTC.2017.7981098
9. C. Li, V.R. Disfani, H.V. Haghi, J. Kleissl, Coordination of OLTC and smart inverters for optimal voltage regulation of unbalanced distribution networks. Electr. Power Syst. Res. **187**, 106498 (2020). http://doi.org/10.1016/j.epsr.2020.106498
10. L. Jiang, K.M. Smedley, Volt/VAR control with DERs, LTCs, and switched capacitors: case study with a commercial distribution system, in *2019 IEEE PES Innovative Smart Grid Technologies Asia, ISGT 2019*, pp. 1330–1334 (2019). http://doi.org/10.1109/ISGT-Asia.2019.8881145
11. Q. Long, J. Wang, D. Lubkeman, N. Lu, P. Chen, Volt-var optimization of distribution systems for coordinating utility voltage control with smart inverters, in *2019 IEEE Power Energy Society Innovative Smart Grid Technologies Conference, ISGT 2019*, pp. 0–4 (2019). http://doi.org/10.1109/ISGT.2019.8791600
12. A.M. Howlader, S. Sadoyama, L.R. Roose, S. Sepasi, Distributed voltage regulation using volt-var controls of a smart PV inverter in a smart grid: an experimental study. Renew. Energy **127**, 145–157 (2018). https://doi.org/10.1016/j.renene.2018.04.058

13. A.I.M. Ali, M.A. Sayed, E.E.M. Mohamed, Modified efficient perturb and observe maximum power point tracking technique for grid-tied PV system. Int. J. Electr. Power Energy Syst. **99**, 192–202 (2018). http://doi.org/10.1016/j.ijepes.2017.12.029
14. A.S. Pabbewar, M. Kowsalya, Three level neutral point clamped inverter using space vector modulation with proportional resonant controller. Energy Procedia **103**, 286–291 (2016). https://doi.org/10.1016/j.egypro.2016.11.287
15. S. Pinzón, W. Pavón, Diseño de Sistemas de Control Basados en el Análisis del Dominio en Frecuencia. Rev. Técnica "Energía" **15**(2), 76–82 (2019). http://doi.org/10.37116/revistaenergia.v15.n2.2019.380

Chapter 34
K-Factor Analysis to Increase the Actual Capacity of Electrical Distribution Transformers

Daniel Contreras Ramírez and Lata-García Juan

Abstract Currently, the different forms of use of electricity involves the massive use of electronic equipment, which are connected to distribution transformers, the same ones that have been operating for a few years under different operating conditions for which they have been designed. Electronic equipment constitutes nonlinear loads that are impedances that vary with the applied voltage and that when connected to the network introduce non-sine current waves known as harmonic currents, which cause several problems such as: the increase in losses in the conductive elements and the increase in temperature in the processing equipment. The K factor lets us know how much a transformer can have to power nonlinear loads. IEEE Std C57.110™-2018 will be considered on the ground that it makes recommendations regarding losses in transformers when their load has harmonic content. In this research, the authors simulate the behavior of a real system using Power Factory software to know the values of current THDs that are compared to those obtained with a power quality device. For case studies, it is obtained that the K factor for TRAFO2 transformer is 1505 and the K factor for TRAFO1 transformer is 1283; although the load of TRAFO1 is higher than the load of TRAFO2.

34.1 Introduction

Currently, electrical power systems are mostly linear and nonlinear loads. Linear loads produce sine current waves at the network frequency, and nonlinear waves introduce other non-sinusoidal waves at different frequencies known as harmonic currents [1].

Harmonics cause many problems in electrical systems, including possible resonances between transformer windings and insulation problems due to temperature variations and small vibrations in the core. The temperature is one of the parameters that reduces or limits the load capacity due to the increase in losses in the transformer,

D. Contreras Ramírez (✉) · L.-G. Juan
Universidad Politécnica Salesiana, Chambers, 227, 090109 Guayaquil, Ecuador
e-mail: dcontreras@ups.edu.ec

Á. Rocha et al. (eds.), *Communication, Smart Technologies and Innovation for Society*, Smart Innovation, Systems and Technologies 252,
https://doi.org/10.1007/978-981-16-4126-8_34

so it is necessary that in a transformer the presence and effect of the harmonics must be considered so that the life of the transformer is not compromised [2, 3].

A variety of techniques has been presented for harmonic removal or attenuation. One of the most used techniques is active filters for their results; However, the drawback is higher cost and complexity, active filters create and control digitally to cancel harmonics in an installation. Calculating the K factor is another technique which is ideal for transformers where individual harmonic components are used, another effective method for removing or attenuating harmonics is through the use of phase change transformers [4].

As a result of the reduction in the lifespan of transformers operating at non-sinusoidal loads, the manufacturers of distribution equipment have presented K-factor transformers, this equipment has the possibility to operate safely when handling non-sine loads, the transformer is able to withstand harmonic load currents, the value of the K factor in an electrical power system is indicative value of the amount of harmonics in the system, the value of the K factor is directly proportional to the harmonic currents [5].

In studies [6], the researchers present K-factor calculations, without load and transformer losses with loads under the effects produced by harmonics. The calculations have been based on two strategies; the first is maximum non-sinusoidal load losses using losses by parasite currents (nominal with and without harmonic passive filter). The second strategy is maximum non-sine load losses using maximum losses by parasite currents with and without harmonic passive filter. The results show that additional losses are reduced in both strategies when using a harmonic passive filter F_{HL}.

In the case of transformers that are equipment involved in all functional stages of the PMI, the presence of harmonic currents increases losses in joule windings, parasitic losses and other additional losses. One of the most commonly used procedure for calculating harmonic distortion in a network using the K-factor index [7].

This article focuses on simulating a power system with actual data from individual harmonic currents provided by a local distribution company, measured in the low voltage die of single-phase transformers that feed residential electric loads, determining the K-factor and actual power of the transformer providing the service.

34.2 Background

Distribution lines connecting elements such as induction furnaces, discharge or fluorescent lamps, induction motors, computers and communications devices, among others, create problems for business owners or businesses and customers in general connected to such networks. Switched power supplies and nonlinearity models generate harmonic currents resulting in additional winding losses resulting in overheating, decreasing the actual capacity and lifespan of transformers.

34.2.1 Factor K

New designs in transformers have been developed over time, where loads containing a high harmonic content are fed, one of the solutions presented is to withstand the increase in temperature and increase in losses caused by $\boldsymbol{F}_{\mathbf{HL}}$ those at the rated frequency, these distribution transformers have a factor $K = 1$ [8, 9].

Using Eq. (34.1), you can calculate the K-factor:

$$K = \sum_{h=1}^{\infty} I_h^2(\text{pu})h^2 \tag{34.1}$$

In which we have $I_h(\text{pu})$, it represents the harmonic current for the order (h) in (value per unit of nominal load current RMS). It is important to know that the impact that harmonics can have on the transformer will depend on the individual current harmonics rather than the amount of load, it will depend on nature rather than the magnitude of the connected load [10].

34.2.2 Total Harmonic Distortion (THD)

Harmonic distortion is a technical parameter that is used to identify the signal leaving a system which is different from the signal that entered it, and this is due to how harmonics are multiples of the input signal as a result of equipment connected to the system that introduce some distortion or nonlinearity [11]. THD values can be current and voltage and tend to increase or decrease by nonlinear equipment or system reconfiguration, reconfiguration leads to an increase in the number of transformer energization resulting in the second harmonic distortion, transients are likely to occur due to transformer activation due to the change in load toward a lower resistance [12].

Several harmonics indices for measuring and characterizing harmonic distortions exist. Equation (34.2) presents the most common harmonic index which is total harmonic distortion [13].

$$\text{TDH} = \frac{\sqrt{\sum_{h=2}^{h\,\max} X_h^2}}{X_1} \tag{34.2}$$

where, X_1 is the RMS value of the fundamental wave, and X_h is the RMS value of the harmonic component h. Typical values for acceptable waveforms are 5% THD for current and 2% THD for voltage at the client connection point.

34.2.3 Software Power Factory

Power factory is a software used for engineering services in electrical power systems, comprising the phases of generation, transmission, distribution, industrial plants and the simulation and integration of renewable energies with connection to the grid [14].

By software can obtain harmonic load flow information, harmonic voltage and current indexes (IEC 61000-3-6, BDEW 2008), unbalanced harmonic sources, multiple harmonic injections: current and voltage sources, thyristor rectifiers, PWM converters, SVS, nonlinear loads, various harmonic distortion rates such as THD, HD, HF, THF, TAD, TIFmx, total RMS currents and voltages, loads and losses (defined according to IEEE and DIN/IEC standards), calculation of K factors and loss factors for 2 winding transformers, therefore adapted for the objectives set out in document [15, 16].

The power factory has a DPL programming language that provides an interface for machining tasks in the software. The DPL provides commands for decision and flow, allows the determination and use of programmer-defined data or variables, has a flexible interface for both input and output [17].

34.2.4 Transformer Losses Due to Harmonic Currents

IEEE Std C57.110™-2018 international standard sets recommendations for oil-submerged transformers undergoing nonlinear loads. Total losses are equal to vacuum losses plus load losses [18].

Vacuum losses can be considered unactioned by harmonics, and they remain virtually constant, which is why we focus on load losses [19]. Load losses (P_{LL}), in turn, correspond to losses in medium and low-voltage windings ($P = I^2R$), plus parasitic losses in windings are also known as Eddy (P_{EC}) plus other losses (P_{OSL}). In Eq. (34.3), the losses are expressed in Watt.

$$P_{\mathrm{LL}} = P + P_{\mathrm{EC}} + P_{\mathrm{OSL}} \tag{34.3}$$

whereas the currents measured in the wiring are taken at the rated current of the transformer, the load losses may be represented in per unit on the basis of power losses in the windings (P), as indicated in the following Eq. (34.4).

$$P_{\mathrm{LL\text{-}R}} = 1 + P_{\mathrm{EC\text{-}R}} + P_{\mathrm{OSL\text{-}R}} \tag{34.4}$$

where, $P_{\mathrm{IL\text{-}R}}$ are the load losses in per unit, they represent Eddy's losses on the windings in per unit under nominal conditions and means other additional losses in per unit under nominal conditions.

In practice it is rare for this to occur, therefore it is necessary to describe the parasite losses of the current measured at the frequency of the system. In addition to the basic premises in relation to transformer equivalent capacity, three assumptions are important for clarifying the use of this term [18].

- Parasitic losses are proportional to the square of the operating frequency.
- Parasitic losses depend on the current flowing in the conductors.
- Induced losses overlap, which in turn allows direct addition of losses induced by various harmonics.

For analysis, it is very beneficial to specify a unique value that is used to establish the capacity of the transformer that supplies power to the load. The factor used for harmonic losses in parasitic current windings (FHL) is the proportional ratio of total losses of parasitic currents in winding due to harmonics (PEC), divided for losses of parasitic currents from winding to nominal frequency (PEC-O), whose expression in terms of the fundamental load current (I_1) is:

$$\mathrm{FHL} = \frac{P_{\mathrm{EC}}}{P_{\mathrm{EC\text{-}O}}} = \frac{\sum_{h=1}^{h=h\max}\left(\frac{I_h}{I_1}\right)^2 h^2}{\sum_{h=1}^{h=h\max}\left(\frac{I_h}{I_1}\right)^2} \tag{34.5}$$

Whose terms represent, the loss factor in the windings by parasite currents, losses of parasite currents or Eddy in winding due to harmonics (watt), loss of parasite currents or Eddy in winding at nominal frequency (watt), harmonic order, is the highest harmonic number, is the rms current to harmonic h (amps) and is the fundamental load current (amps). Similarly, the standard recommends defining the factor [18].

$$\mathrm{FHL\text{-}STR} = \frac{\sum_{h=1}^{h=h\max}\left(\frac{I_h}{I_1}\right)^2 h^{0.8}}{\sum_{h=1}^{h=h\max}\left(\frac{I_h}{I_1}\right)^2} \tag{34.6}$$

where $F_{\mathrm{HL\text{-}STR}}$, it is known as harmonic loss factor due to additional losses, but with the exponent in h of 0.8.

The effect of other additional losses is considered for transformers whose dielectric is oil, since there is a ratio quite similar to the harmonic loss factor for parasitic winding losses as indicated in Eq. (34.5), except that the losses due to the collector bar connections, structural parts, tank, etc., are proportional to load I_2 and harmonic order at power 0.8, as seen in Eq. (34.6).

In the case of oil-submerged transformers, all losses should be considered, as indicated in IEEE StdC57.91 and load losses can be estimated considering harmonics based on FHL (Eq. 34.5) and FHL-STR factors (Eq. 34.6). Load losses considering the loss factors for the P_{LL} harmonics are expressed in Eq. (34.7).

$$P_{\mathrm{LL}} = P + F_{\mathrm{HL}} * P_{\mathrm{EC}} + F_{\mathrm{HL\text{-}STR}} * P_{\mathrm{OSL}} \tag{34.7}$$

Whose terms represent harmonic load losses in (watts), P load I^2R losses (watts), harmonic loss factor in winding, Eddy losses in windings (watts), loss factor from other additional losses and other additional losses (watts). P_{IL}, F_{HL}, P_{EC}, $F_{HL\text{-}STR}$, P_{OSL}.

It is important to be clear that of the total of other losses by parasite currents, these are divided and of which, as specified by IEEE Std C57.110-2018 for ONAN transformers, depending on the capacity ranges of oil-submerged transformers, for which they are less than 300 kVA (as are case studies), losses mean 60% (55% of low-voltage winding and 5% of high-voltage winding) and 40% correspond to [18]. P_{EC}, P_{OSL}, P_{EC}, P_{OSL} .

If the load losses considered in the expression (34.7) are represented in per unit, based on the power losses in the windings (P), we will get the expression (34.8), which allows us to calculate the Imax(pu).

$$I\max(\text{pu}) = \sqrt{\frac{P_{IL\text{-}R}(\text{pu})}{1 + F_{HL} * P_{EC}(\text{pu}) + F_{HL\text{-}STR} * P_{OSL}(\text{pu})}} \quad (34.8)$$

Equation (34.8) is already defined in previous lines and allows us to know the maximum current (pu) that the transformer could support with nonlinear loads versus operation with linear loads, i.e., the percentage of the actual capacity for which it was manufactured. The numerator represents load losses under nominal conditions without harmonics and the numerator represents load losses under harmonic conditions [19].

34.3 Case Study

As part of the methodological process of the project, the effect of harmonics on distribution transformers caused by residential and street lighting loads has been analyzed, for which n 2 single-phase 50 kVA transformers are considered. The information collected is from the history of energy quality measurements made in CNEL EP Business Unit Guayas—Los Ríos, updated to 2020. The selected transformers are presented in Table 34.1, which are supplied by Feeders departing from

Table 34.1 Characteristics of case studies

	Case 1	Case 2
Transformer	TF-002	TF-007
Geographic location	625282, 9774180	625292, 9774030
Feeder	Stick of Iguana	Milan
Transformer code	41817	41731
Users served	14	7
Monthly consumption	6740 kWh	2113 kWh

the S/E Catalonia that has a power transformer of 16/20 MVA with a voltage level of 69/13.8 kV [20].

The information of the harmonic currents is obtained from the electric distribution company in electronic sheets, which are imported by the power factory software, by creating a DPL with reading command "xls" or "xlsx" and declaring the variables type "set" and "object." Internally, read sub-permissions and scans of the created objects are performed. The Software provides us with K factor, FHL and FHL-STR harmonic loss factor calculations.

The one-line diagram of the system under study (see Fig. 34.1), which starts from the output bar of the S/E Catalonia that has an electrical power transformer of 16/20 MVA OA-FA, at a voltage level of 69/13.8 kV [20]. From the output bar of the S/E Catalonia, the Palo de Iguana and Milan Feeders are connected, whose characteristics are:

Feeder Palo de Iguana, mostly with conductor aluminum 4/0 ACSR, with a length of approximately 600 m to the site where the transformer of 50 kVA monophasic is supplied (TRAFO1).

Milan feeder, 4/0 ACSR aluminum conductor, with a length of 450 m to where the 50 kVA single-phase transformer is supplied (TRAFO2).

At the output of the transformers the bars in low voltage will appear, where harmonic current measurements have been made.

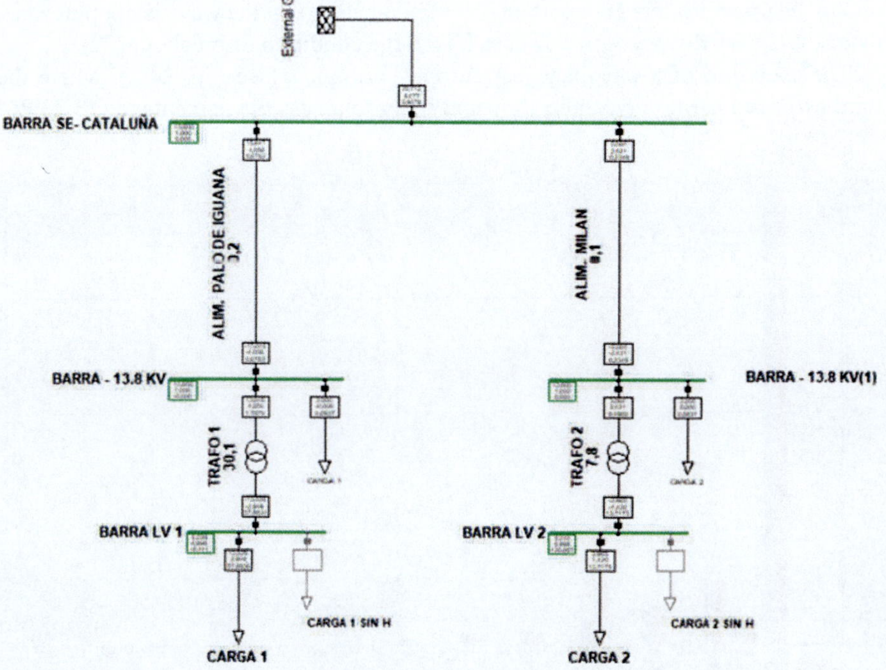

Fig. 34.1 One-line diagram in power factory with case studies

The data of the measurements used correspond to data on low voltage bars of the indicated processing centers, which were carried out in compliance with the requested in the Ecuadorian Energy Quality Regulation ARCONEL 005-18 [21]. Field measurements were performed with FLUKE 1744 power quality logger power quality analyzers, which comply with IEC 61000-4-30.

Current measurements made with their respective harmonic components are received via Excel files and uploaded to the power factory, by creating a code in DPL, which allows us to analyze measurements made in the distribution transformers that are the subject of the study and the analysis of the effect of the K-factor increase on transformers by harmonic distortion due to nonlinear residential loads and street lighting, by using the power factory and comparing them with the data obtained with the Fluke energy quality recorder.

34.3.1 Presence of THD in Low Voltage Bar

The transformer can be energized from the primary or secondary side, so the torque meters on both sides of the equipment must be considered for the calculation of %THD. It is considered a range between 5 and 30% for THE THD to discriminate normal operating conditions. The selection of 5% is due to some inherent harmonics are present in the supply by associated nonlinear loads, while if the harmonic distortion in the upper limit is greater than (30%), indicating that the wave is not pure sine, that is, it is very distorted, in this case it is in the condition of breakout [22].

The spectrum of low-voltage bar current harmonics (see Fig. 34.2), where the third extended harmonic can be shown in order to assess the percentage (13.339%)

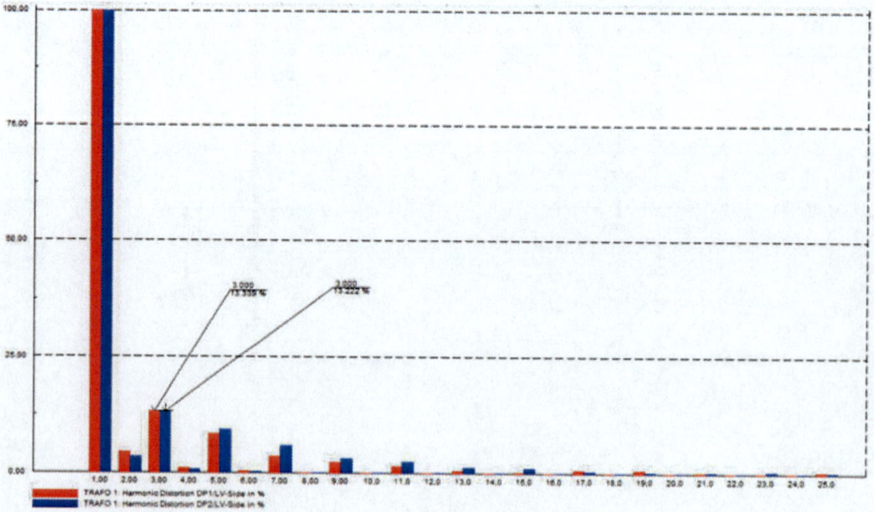

Fig. 34.2 Spectrum of low-voltage current harmonics of the transformer

phase 1 and (13.222%) phase 2. Several investigations select a range of 5–30% for the normal cycle, considering that if the upper limit is exceeded.

34.4 Case Simulation

In Fig. 34.3, there is a representation of the current signal on the high-voltage side with harmonics for TRAFO1. As can be seen, harmonics distort the signal, which is sinusoidal.

34.4.1 Simulation and Determination of Factor K

With the data obtained, K Factor is presented with the use of power factory. The location of the harmonic measurement is in the common coupling of individual loads and is located at the output of the transformers 50 kVA, i.e., on the low-voltage bars at 240 V between phases.

The results obtained with the simulation in the software, for the 2 case studies, as regards Factor K, FHL and FHL-STR, are shown below in Table 34.2.

The harmonic data for each or transformers have been considered from order 1 to 25, considering that they are the ones that influence the most.

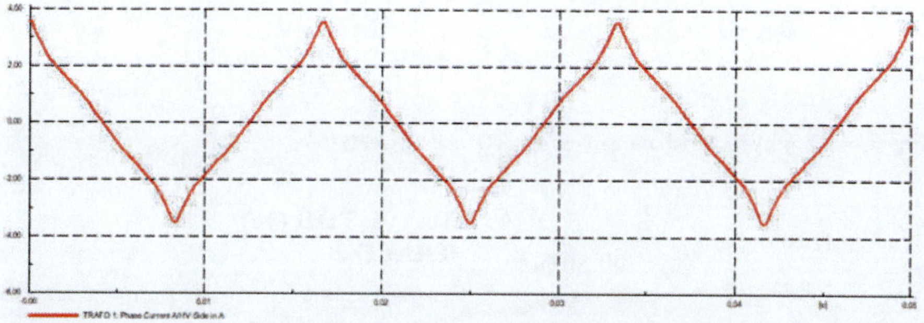

Fig. 34.3 Current signal on the medium voltage side of TRAFO1 with harmonics

Table 34.2 Power factory results for the transformers under study

Transformer	THD		K		F_{HL}		$F_{HL\text{-}STR}$	
	MT	BT	BT		BT		BT	
			L_1	L_2	L_1	L_2	L_1	L_2
1	16.67	17.18	1.28	1.21	1.28	1.21	1.03	1.02
2	10.09	11.14	1.50	1.72	1.50	1.72	1.06	1.07

34.4.2 *Loss Simulation and Analysis*

Losses in a transformer are increased with the presence of nonlinear loads. The standards recommend that they be analyzed considering the harmonic loss factors and, which were already defined. The parameters entered to simulate the single-phase transformers of 50 kVA 7967-120, 240 V, are set out in Ecuadorian Regulations NTE INEN 2112-2004 [23].

Being single-phase transformers with residential loads, where single-phase loads predominate, we have unbalanced systems that allow us to have different current values in each of the phases in low voltage and therefore have loss factors for each of the phases. The values obtained from the Power Factory when running the flow of harmonics, determine that the presence of harmonics represent an increase in the occupied capacity or the return of their actual capacity of the transformers. For TRAFO1, see Fig. 34.4. For TRAFO2, see Fig. 34.5.

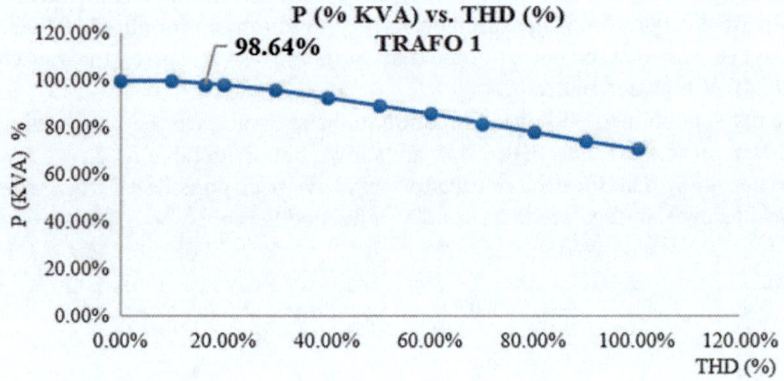

Fig. 34.4 P(% kVA) based on % of current THDs in TRAFO1

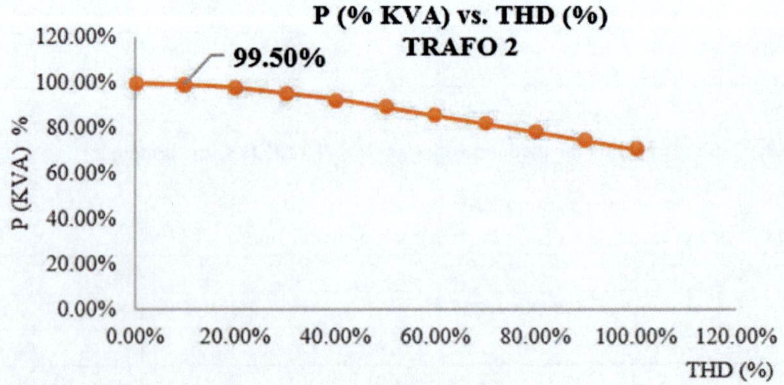

Fig. 34.5 P (% kVA) based on % of current THDs in TRAFO2

With the values obtained from THDI, if we make an estimate for cases where transformers operate to the maximum of their capacities, we would have to support only 95.31% on average per phase, while TRAFO2 would be 89.61%.

34.5 Case Discussion

Research on harmonics introduced by nonlinear loads in distribution transformers is largely part of the implementation of the recommendations given by IEEE [18], with most studies applied to three-phase transformers. It is very important to be able to obtain a reliable estimate of the real state of charge of the distribution transformers, which will allow knowing the opportune moment to increase the capacity or replace the transformer with one with an appropriate design considering the K factor [5]. Many researches are consistent in considering the importance of the transformer in power systems, so it transcends its studies and treatments of the consequences of harmonic currents on the capacity and life of these equipment and in some cases even if it has not reached its face value the transformer has manifest overheating [19].

If we compare the frequency spectrum of current readings with processed harmonics with the Fluke analyzer and compare them with those processed with power factory, it is obtained that in both the harmonics that have the most involvement are those of order 3, 5, 7, 9 and from there the percentages are less significant in both cases. The error between the two for harmonic 3 that has the most affected is of the order of 4.14%.

34.6 Conclusions

This article used power factory software to analyze the K factor in distribution transformers, starting with current THDs. The case studies are single-phase distribution transformers that are in operation in electric distribution company CNEL EP Business Unit Guayas-Los Ríos and that serve the Residential area. The information processed has been taken in the field in the period from September 2nd to September 10th 2020, according to the quality requirement of the Regulatory Agency.

Case studies conclude that factor K depends on the contribution percentages of each of the harmonic orders and not on the load ability of the transformer. TRAFO1 has a higher load chargeability than TRAFO2, however the K factor of TRAFO2 (1505) is greater than that of TRAFO1 (1283). The reduction in the actual capacity of the transformer is whether it will be proportional to the value of K. The higher the K value, the greater the capacity decrease. For case studies, the maximum currents at which they can work on average per phase for TRAFO1 must be at the top of their load capacity, while TRAFO2 will be 89.61%.

References

1. J.D. Arcila, http://www.ieb.com.co. Last accessed 2021/01/14
2. M.S. Taci, A. Domijan, The effects of linear and nonlinear operation modes in transformers, in *11th International Conference on Harmonics and Quality of Power* (2004), pp. 244–249
3. G. Ueta, T. Tsuboi, S. Okabe, Y. Shimizu, E. Hino, K-factor value and front time related characteristics of UHV-class air insulation for positive polarity lightning impulse test. IEEE Trans. Dielectr. Electr. Insul. **19**(3), 877–885 (2012)
4. S. Torabzad, M. Davudi, B. Ojaghi, Analysis of harmonics and harmonic mitigation methods in distribution systems. Aust. J. Basic Appl. Sci. **5**(11), 996–1005 (2011)
5. E.F. Fuchs, D. Yildirim, M. Grady, Corrections to measurement of eddy-current loss coefficient Pec-r-derating of single-phase transformers, and comparison with K-factor approach. IEEE Trans. Power Deliv. **15**(4), 1357 (2000)
6. M.A. Taher, S. Kamel, Z.M. Ali, K-factor and transformer losses calculations under harmonics, in *18th International Middle East Power Systems Conference* (2016), pp. 753–758
7. B. Ruggero Ríos, M.E. Sánchez Quintana, Revista Científica de la UCSA **1**(1), 33–51 (2014)
8. J.M. Frank, Harmonics, How they affect the development and design of ‘K’ factor transformers, in *IEEE Conference Record of Annual Pulp and Paper Industry Technical Conference* (1996), pp. 204–208
9. J.M. Frank, Origin, design and development of ‘K’ factor transformers (2002), pp. 2273–2274
10. P. Bagheri, W. Xu, K. Shaloudegi, New indices to evaluate the impact of harmonic currents on power transformers, in *Proceedings of International Conference on Harmonics and Quality of Power, ICHQP*, vol 2018 (2018), pp. 1–6
11. J. Rodriguez Maldonado, Total harmonic distortion estimation, minimization inter harmonic amplitude and expanding bands rejection in TKF filters. IEEE Lat. Am. Trans. **14**(2), 652–656 (2016)
12. A.R.A. Jerin, N. Prabaharan, N.M. Kumar, K. Palanisamy, S. Umashankar, P. Siano, Smart grid and power quality issues, in *Hybrid-Renewable Energy Systems in Microgrids, Integration, Developments and Control* (Elsevier, Amsterdam, 2018), pp. 195–202
13. R.R. Bhoyar, S.S. Bharatkar, S.A. Khadtare, Measurement of harmonics at HT end users in power distribution system, in *1st International Conference on Automation, Control, Energy and Systems* (2014)
14. DIgSILENT, https://www.digsilent.de/en/powerfactory.html. Last accessed 2020/12/21
15. Liuli, Zhaoyi, Yepeng, Wanggang, Luxiaoxu, Influence on power quality by wind turbine’s isolating operation, in *China International Conference on Electricity Distribution* (2014), pp. 1207–1210
16. P. Pachanapan, S. Premrudeepreechacharn, Coordinated voltage control between wind power plant and shunt capacitors in weak distribution networks, in *International Electrical Engineering Congress* (2014)
17. I. Netzberechnungs, Software power factory, vol 15 (2012), pp. 0–26
18. IEEE Power and Energy Society, *IEEE Recommended Practice for Establishing Liquid-Filled and Dry-Type Power and Distribution Transformer Capability When Supplying Nonsinusoidal Load Currents* (2008)
19. J. Gómez-Sarduy, E. Quispe, R. Reyes-Calvo, V. Sousa-Santos, Viego-Felipe, Influencia de los armónicos de corriente sobre las pérdidas en los transformadores de distribución monofásicos con derivación central. *El Hombre y la Máquina*, Cali, Colombia, N° 45, pp. 33–43, (2014)
20. Geoportal CNELEP, https://geoportal.cnelep.gob.ec/cnel/. Last accessed 2021/01/14
21. ARCONEL, https://www.regulacionelectrica.gob.ec/Regulacion-de-Calidad-Suscrita. Last accessed 2020/12/30

22. M. Raichura, N. Chothani, D. Patel, K. Mistry, Total harmonic distortion (THD) based discrimination of normal, inrush and fault conditions in power transformer. Renew. Energy Focus **36**(1), 43–55 (2020)
23. C. P. E. NEC, Ecuadorian Electrification Institute (2019), pp. 16–931

Chapter 35
Lead Contamination in the City of Cuenca

Lisenya Requene, Geovanna Saguay, and Fredi Portilla

Abstract Urban agriculture has emerged as an alternative to the lack of connection between cities and the environment; however, on certain occasions, the presence of heavy metals has been found in urban crops, which implies a risk for consumers. The objective of this study was to determine the level of lead contamination in the city of Cuenca-Ecuador, using vegetables such as cabbage (*Brassica viridis*), broccoli (*Brassica oleracea italica*), leaf lettuce (*Lactuca sativa var. Crispa*), and cabbage lettuce (*Lactuca sativa var. capitata*), produced in the context of urban agriculture in educational institutions in the north district, historic center, and south district of the city of Cuenca as indicators of contamination. The methodology used started from the deductive method with the information obtained, passing to the comparative method, analyzing the general data using descriptive statistics. The results showed a high concentration of lead in vegetables in the historic center with 115.98 mg/kg, followed by the south district where a concentration of 22.48 mg/kg was found and finally the north district with 6.98 mg/kg, giving a total lead concentration of 145.44 mg/kg in the city of Cuenca-Ecuador.

35.1 Introduction

Agricultural industrialization, demographic change, and urban expansion are part of a generational change in the new century. Traditional agricultural practices have been left behind by mechanization, the use of agrochemicals, and the excessive dependence on improved and transgenic varieties in agribusiness (agricultural model based on selective species of mass commercialization) [1].

In food resources, the challenge faced by humanity is indisputable, considering the loss of fertility of surfaces destined for cultivation and the continuous detriment of the environment that risk the food security of the population. Faced with this, urban agriculture has been taking place, which is the set of agricultural practices that

L. Requene · G. Saguay · F. Portilla (✉)
Environmental Engineering, Salesian Polytechnic University, Calle Vieja 12-30, Elia Liut Street, Cuenca, Ecuador
e-mail: fportilla@ups.edu.ec

Á. Rocha et al. (eds.), *Communication, Smart Technologies and Innovation for Society*, Smart Innovation, Systems and Technologies 252,
https://doi.org/10.1007/978-981-16-4126-8_35

are carried out within a city and is proposed as an alternative to the lack of connection between cities, the environment, and natural processes that facilitate food processing, in addition to providing spaces for education and environmental awareness [2].

There is a latent concern about the quality of crops in areas with high vehicular traffic, because the increase in the concentration of all heavy metals near the road indicates that vehicles are the source of emissions; thus, the overall traffic load increases the content of trace metals in biomass [3]. Heavy metal contamination in both water, soil, and air is one of the most severe problems affecting food safety and public health worldwide.

Several studies have confirmed the presence of heavy metals, including lead, in vegetables such as lettuce, broccoli, squash, cabbage, and potatoes. The main route of entry of heavy metals into plants is through air, water, and soil; thus, plants become the link for the transfer of heavy metals to consumers [4]. Because the toxicity of heavy metals is directly proportional to the ease of absorption by living beings.

Several studies have confirmed the presence of heavy metals, including lead, in vegetables such as lettuce, broccoli, squash, cabbage, and potatoes. The main route of entry of heavy metals into plants is through air, water, and soil; thus, plants become the link for the transfer of heavy metals to consumers [4]. Because the toxicity of heavy metals is directly proportional to the ease of being absorbed by living beings.

Most of the vegetable production areas are located in the Ecuadorian highlands [5]. In the city of Cuenca, the environment has been contaminated by soil and air quality derived from the needs of transportation, communication, and in general by the pace of life within the city [6]. Therefore, several investigations have been carried out on the concentration of lead in various sectors of the city, obtaining that in certain places, the concentration of lead is above the permissible levels established by the FAO Codex Alimentarius standard, which represents a risk to human health [7].

Currently, Ecuador does not have a specific standard for foods contaminated with heavy metals, so the studies used as a reference in this research were based on international standards such as the FAO Codex Alimentarius and the European Union. For the case of lead, the European Union and FAO have established permissible limits that vary according to the type of food, referring to leafy vegetables the maximum permissible of 0.3 mg/kg has been established.

The objective of this research is to determine the level of lead contamination affecting the city of Cuenca-Ecuador, by comparing data from studies conducted on vegetables produced in the context of urban agriculture in the northern district, southern district, and historic center of Cuenca. In this way, the present study becomes a tool that provides information to the population on the environmental quality in which they live and on whether or not the vegetables grown in these areas are suitable for human consumption.

35.2 Method

The following methodological components were considered and developed in this research. We started from the deductive method with the information obtained from historical data and bibliographic reviews of theses, scientific articles, and other sources, passing to the comparative method of the data of interest found, analyzing the general data through the use of descriptive statistics for which the Microsoft Excel tool was useful where the data were ordered and tabulated allowing their analysis and comparison through statistical graphs, until reaching the inductive method through which it is induced what has happened and conclusions are obtained. The methodology applied consists of the following stages: (1) search for information, (2) organization of the information, and (3) data analysis.

35.2.1 Search for Information

Several studies have quantified the levels of lead present in vegetables grown through urban agriculture, so different sources of information on the topic of interest were consulted, including scientific articles, scientific journals, and university theses obtained from the Institutional Repositories of Universities such as University of Cuenca, Universidad Politécnica Salesiana, and Catholic University of Cuenca, among other sources of information.

35.2.2 Organization of the Information

The most relevant documents for the comparison of lead levels present in vegetables in the north district, historic center, and south district of the city of Cuenca were six theses obtained from the Institutional Repository of the Universidad Politécnica Salesiana.

35.2.3 Data Analysis

The clusters from which the vegetable samples were collected are selected educational institutions located in the northern district, historic center, and southern district of the city of Cuenca. In the case of the northern district, five educational institutions studied by Quizphi and Sarango [8], and the Universidad Politécnica Salesiana analyzed by Fernández [9] and Galarza [10] were considered. In the historic center, five educational institutions studied by González and Portocarrero [11] and a house located in Av. 3 de Noviembre analyzed by Mora [6] were taken into account. Finally,

Table 35.1 Coordinates of the educational institutions under study

Zone	UTM X	UTM Y
Cristo Rey School of Basic Education	722,189.86	9,680,735.33
Federico Proaño School of Basic Education	721,519	9,680,498
Francisca Dávila Educational Unit	722,013.87	9,680,145.49
Julio María Matovelle Educational Unit	722,188.56	9,680,185.8
Tres de Noviembre Educational Unit	721,413.09	9,680,216.31
Corazón de María Educational Unit	722,264.73	9,678,999.24
María Auxiliadora Educational Unit	723,027.66	9,680,241.29
Luisa de Jesús Cordero Educational Unit	721,095.75	9,679,341.62
Santa Mariana de Jesús Educational Unit	721,978.06	9,680,363.77
Fiscomisional School of Basic Education Padre Carlos Crespi	721,601.88	9,680,300.17
Técnico Salesiano Educational Unit	720,564.64	9,677,484.24
Av. 3 de Noviembre	721,127.76	9,679,988.25
Nuestra Familia Educational Unit	720,890.51	9,678,054.62
Bilingüe Interamericano Educational Unit	721,147.77	9,677,771.88
Particular Borja Educational Unit	716,784.11	9,676,885.03
La Asunción Educational Unit	722,313.75	9,677,125.92
Universidad Politécnica Salesiana (UPS)	723,509.27	9,680,707.77

in the southern district, five schools studied by Calle and Zhindón [12] were taken as cluster, and this being the only study found on lead contamination of vegetables in the southern district of the city of Cuenca-Ecuador. Table 35.1 shows the coordinates of the educational institutions under study, and Fig. 35.1 shows their respective geographical location.

As indicators of lead contamination, vegetables such as cabbage (*Brassica viridis*), broccoli (*Brassica oleracea italica*), leaf lettuce (*Lactuca sativa var. Crispa*), and cabbage lettuce (*Lactuca sativa var. capitata*) were used, since data on these vegetables were found in the six theses mentioned above, being useful for the comparative analysis of lead levels between the north district, south district, and historic center of the city of Cuenca-Ecuador.

35.3 Results

The fulfillment of the objective of this research involves the comparison of lead concentrations present in the different districts of the city of Cuenca. The results are presented in order starting with the comparison of lead concentrations between the educational institutions corresponding to each district, and finally, the comparison is made in general between the north district, historic center, and south district.

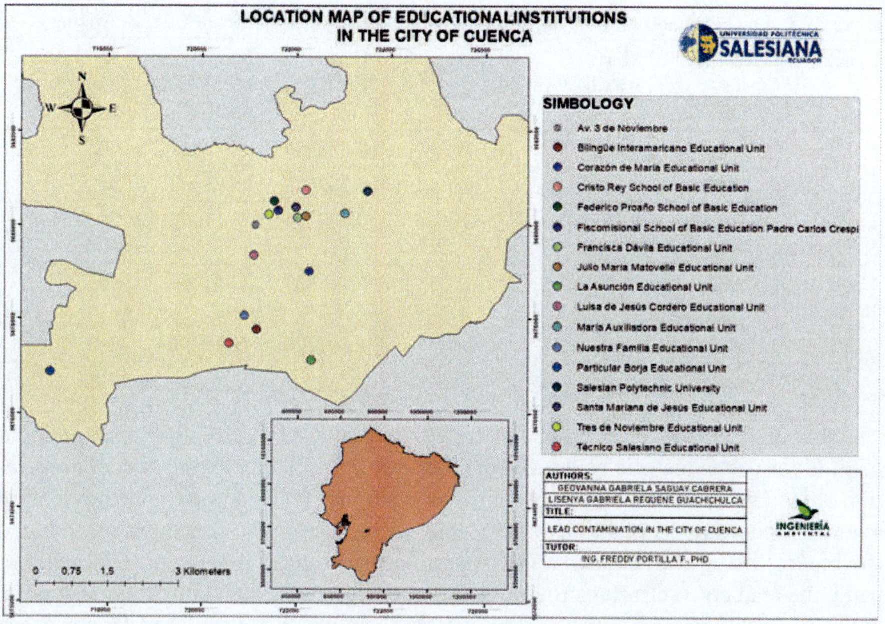

Fig. 35.1 Geographical location of the educational institutions under study

Table 35.2 shows that in the northern district, a high concentration of lead was found in the Julio María Matovelle Educational Unit, also standing out compared to the other institutions is UPS; in these two locations, the level of lead in all the vegetables grown exceeds the established permissible limit, highlighting cabbage lettuce as the vegetable with the highest concentration of lead 4.87 mg/kg in the Julio María Matovelle institution; while in the UPS, it was broccoli with a concentration of 3.27 mg/kg. In the rest of the educational institutions, lead concentrations in all vegetables were found to be below the established permissible limit.

Table 35.2 Lead concentration in vegetables grown in the northern district of Cuenca (mg/kg)

Species	Julio María Matovelle	Federico Proaño	Tres de Noviembre	Cristo Rey	Francisca Dávila	UPS
Cabbage	2.15086	0.07175	0.14502	0.14573	0.07223	2.34
Broccoli	2.14271	0.07256	0.07141	0.07207	0.0728	3.27
Leaf lettuce	4.29225	0.073	0.0729	0.07217	0.01434	2.02
Cabbage lettuce	4.87277	0.07173	0.0719	0.07205	0.07117	1.37
Total	13.45859	0.28904	0.36123	0.36202	0.23054	9

Table 35.3 Lead concentration in vegetables grown in the historic center of Cuenca (mg/kg)

Species	Luisa de Jesús Cordero	María Auxiliadora	Corazón de María	Carlos Crespi	Santa Mariana de Jesús	Av. 3 de Noviembre
Cabbage	0.324098	0.343419	0.017931	0.288659	0.338461	60.33
Broccoli	0.214307	0.146794	0.268960	0.387672	0.425693	67
Leaf lettuce	0.217900	0.352898	0.321820	0.240575	0.211048	56.78
Cabbage lettuce	0.293978	0.458292	0.315863	0.188833	0.344050	49
Total	1.0502833	1.301403	0.924574	1.105739	1.319252	233.11

Table 35.3 allows us to compare the lead concentration among the educational institutions of the historic center. As can be seen in this table, the highest concentration of lead was found in vegetables grown in a house located in Av. 3 de noviembre where broccoli stood out as the vegetable with the highest concentration of lead (67 mg/kg) being above the maximum permissible limit of 0.3 mg/kg. On the other hand, the lead concentrations found in the other five educational institutions studied in this district were negligible, indicating that there was no presence of lead in those vegetable crops.

Table 35.4 shows the lead concentration among the educational institutions of the southern district. There is possible to observe that all the vegetables grown in the five educational institutions studied in this district presented lead concentrations above the maximum permissible, highlighting the Bilingual Educational Unit followed by the Salesian Technical Educational Unit where a higher concentration of lead was found. However, there is a difference between the type of vegetables that obtained the highest concentration; in the case of the Bilingual Educational Unit, it was broccoli with 12.17 mg/kg; while in the Salesian Technical Educational Unit, it was the leaf lettuce with 12.24 mg/kg.

With respect to the concentration of lead in the different districts is possible to observe in Table 35.5, there is a high concentration of lead in the vegetables of the historic center, obtaining a total concentration of 115.98 mg/kg. This was followed by the south district where a concentration of 22.48 mg/kg was found. The lowest concentration of lead was found in the north district with 6.98 mg/kg, although the

Table 35.4 Lead concentration in vegetables grown in the southern district of Cuenca (mg/kg)

Species	Nuestra Familia	Técnico Salesiano	Bilingüe	Borja	La Asunción
Cabbage	5.9	8.09	8.48	3.53	4.33
Broccoli	6.79	3.89	12.17	1.92	4.03
Leaf lettuce	1.73	12.24	6.99	2.24	3.1
Cabbage lettuce	5.92	6.41	8.91	1.9	3.85
Total	20.34	30.63	36.55	9.59	15.31

Table 35.5 Lead concentration in vegetables in the northern, historic center, and southern districts

Species	North district (mg/kg)	Historical center (mg/kg)	Southern district (mg/kg)
Cabbage	1.7340	30.0337	6.066
Broccoli	2.3421	33.3557	5.76
Leaf lettuce	1.6450	28.2556	5.26
Cabbage lettuce	1.2590	24.3399	5.398
Total	6.9801	115.9849	22.484

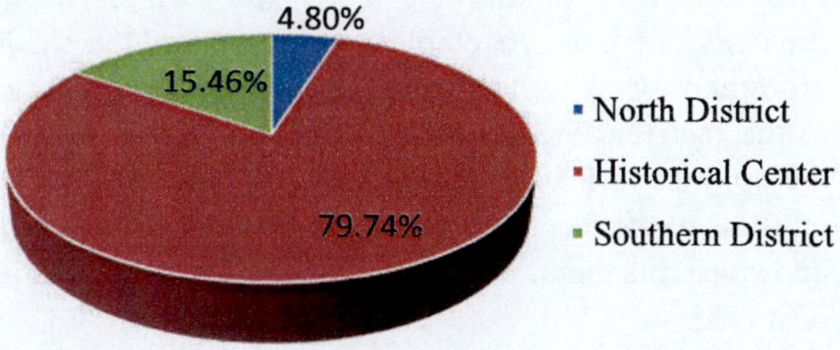

Fig. 35.2 Percentage of lead concentration by district in the city of Cuenca

vegetables grown in two educational institutions in this district were found to be unfit for human consumption.

To accentuate the implication of these results, we show in Fig. 6 that of the 145.44 mg/kg obtained as total lead concentration in the city of Cuenca 79.74% corresponds to the historic center, 15.46% to the south district and only 4.8% corresponds to the north district, being evident that the highest concentration of lead is in the historic center (Fig. 35.2).

35.4 Discussion

In this research, it can be observed in Table 35.3 that the concentration of lead is higher in the Julio María Matovelle Educational Unit, finding all the vegetables cultivated in this institution with Pb levels above the established permissible limit, showing a great difference with respect to the other four educational institutions studied by Quizphi and Sarango [8]; since in all the vegetables cultivated in the other four institutions, there were acceptable lead levels below the permissible limit. The difference is due to the type of substrate used for each crop; since in the case of the Julio María Matovelle school, the crop was grown in a field assigned by the school; while in the other schools, the vegetables were grown in boxes with black soil brought from the hill. The above is related to that mentioned by Hallat et al. [13] and Brewer et al. [14], who comment that the higher concentration of Pb at the plant level is usually related to contamination by that metal in the soil, and Pb contamination in the soil may possibly be produced by remains of paints in construction and demolition waste or by the proximity to factories. Also state that the presence of Pb in plants comes from soil absorption, irrigation water, and atmospheric pollutants.

In relation to the northern district, it is also observed that in UPS, a high concentration of Pb exceeding the maximum permissible limit in all vegetables is grown there, and it is worth mentioning that in the study conducted by Galarza [10], a higher concentration of Pb was found in broccoli, keeping relation with the study conducted by Fernández [9] where broccoli was also the vegetable with the highest concentration of Pb this is because the level of lead present in the different vegetables is related to the development of each species in both studies broccoli was the species that developed more, and this development is usually determined by several factors such as the amount of irrigation water, the frequency of irrigation, care of the land where the vegetables are grown and the shade cast on it. The authors of the aforementioned studies state that the high concentrations of Pb in the UPS are due to the fact that the vegetables were grown in the parking area where they receive the amount of Pb emanating from the exhaust of vehicles entering the campus, in addition to the fact that Pb has a permanence time in the environment of 15–20 years, and since this metal was used as an additive for gasoline, there must still be remnants of it [15].

In the comparison of lead concentration in the historic center, a great difference can be seen between two studies carried out in that place. The study conducted by Mora [6] in a house located on Av. 3 de Noviembre showed lead levels much higher than the study conducted by González and Portocarrero [11] where lead concentrations were almost zero. This notable difference in the studies may be due to the fact that the location of the crop on Av. 3 de Noviembre in the historic center is considered a high traffic area were located inside each institution in the back part, and these same authors mention that the barriers of buildings between the cultivation areas and the roads reduce significantly the contamination of crops in urban areas, corroborating what was published by Calle and Zhindón [12] who mention that the contamination of urban orchards depends on several factors such as type of crop, area, ventilation, humidity, wind direction, and the predisposition to contamination of the city. Codex Alimentarius [16] also suggests that "Those who cultivate private or small commercial gardens should take measures to reduce lead contamination. They should avoid planting near roads."

Regarding the southern district in all the institutions under study, the vegetables grown by Calle and Zhindón [12] exceeded the permissible limit of Pb concentration, being Bilingüe Interamericano, Técnico Salesiano, and Nuestra Familia the educational institutions where the highest concentration of Pb was found, this because the vegetables were grown in parking lots, or next to roads with high vehicular traffic, Reyes mention that atmospheric deposition is the main cause of Pb contamination. In the case of Técnico Salesiano, the crop was grown next to the mechanical workshop, so it received emissions from vehicles entering there. The educational institutions with lower Pb concentrations were Particular Borja and La Asunción, since the crops were grown at the back of these institutions; however, the Pb concentrations found also exceeded the established limit.

35.5 Conclusions

A total lead concentration of 145.44 mg/kg was found in the city of Cuenca-Ecuador, distributed in the three study zones as follows: north district with 6.98 mg/kg (4.8%), historic center with 115.98 mg/kg (79.74%), and south district with 22.48 mg/kg (15.46%). The highest concentration was found in the historic center with the species *Brassica oleracea itálica*, specifically on Av. 3 de Noviembre, due to the high vehicular traffic in the area.

In the northern district, the Julio María Matovelle Educational Unit followed by UPS was the institution in which a high concentration of lead was found, exceeding by far the maximum permissible limit established by the European Union and FAO of 0.3 mg/kg, so that all the vegetables grown in these two institutions were found to be unfit for human consumption as they can affect the health of consumers. However, even if vegetables are unfit for consumption, they serve as a defense mechanism against climate change by contributing to the extension of plant mass, absorption of lead, and other pollutants.

The vegetables with the highest lead uptake were cabbage lettuce (4.87 mg/kg), leaf lettuce (4.29 mg/kg) and broccoli (3.27 mg/kg) in the north district, broccoli with a content of 67 mg/kg in the historic center, and leaf lettuce (12.24 mg/kg) together with broccoli (12.17 mg/kg) in the south district, showing that these vegetables are good indicators of environmental contamination and are useful in addressing climate change.

The institutions that presented low levels of lead contamination in the north district were Federico Proaño, Av. 3 de Noviembre, Cristo Rey and Francisca Dávila, with an average of 0.07768 mg/kg, and in the historic center: Luisa de Jesús Cordero, María Auxiliadora, Corazón de María, Carlos Crespi, and Santa Mariana de Jesús, with an average of −0.28506 mg/kg, which is below the permissible limits established in the regulations. Although the study shows that lead levels are below the established limits, the importance of monitoring and regulatory studies to ensure food safety is highlighted.

References

1. D. Calderón, Rescate de los saberes tradicionales del adulto mayor a través de agricultura urbana, para determinar concentraciones de plomo y coliformes totales en hortalizas en el Vecino-Cuenca. Obtenido de: http://dspace.ups.edu.ec/handle/123456789/18171 (2019)
2. J.L. Toledo Q, Recuperación de márgenes de quebrada a través de Agricultura Urbana (2019). https://dspace.ucacue.edu.ec/handle/ucacue/1685
3. J. Vásquez, C. Sangurima, M. Alvarez-Vera, Concentraciones de plomo (Pb) en cultivos de lechuga (Lactuca sativa) en Azuay, Ecuador, Septiembre 2019. http://www.scielo.org.pe/scielo.php?pid=S2077-99172019000300013&script=sci_arttext
4. J.L. León, K.E. Ulloa, Implementación de educación ambiental por medio de "Agricultura Urbana" y valoración de plomo, carbono y coliformes totales en cinco centros educativos urbanos del cantón Paute, Agosto 2019. https://dspace.ups.edu.ec/handle/123456789/17834

5. S. Escobar, Determinación de la presencia de plomo y cadmio en frutilla (Fragaria ananassa) y tomate (Solanum lycopersicum) en El Quinche, Quito (2016)
6. C. Mora, Determinación de la concentración de elementos contaminantes y bacterias patógenas presentes en un sistema de agricultura urbana compuesto por hortalizas, implementado en el centro histórico de Cuenca, Octubre 2017. http://dspace.ups.edu.ec/handle/123456789/14858
7. D.S. Ochoa C, Transformación urbana. Gentrificación Rural, el caso de Challuabamba-Cuenca entre el 2000 y el 2017 (2018). https://dspace.ucacue.edu.ec/handle/ucacue/1754
8. M.J. Quizphi, T.A. Sarango, Determinación de plomo, bacterias patógenas (E. Coli y Coliformes) y captura de carbono en la agricultura urbana (hortalizas y ornamentales) emplazadas en cinco instituciones educativas ubicados en el Distrito Norte de la ciudad de Cuenca, Julio 2020. https://dspace.ups.edu.ec/handle/123456789/18812
9. J.V. Fernández, Agricultura urbana y su aporte contra el efecto invernadero en la Universidad Politécnica Salesiana sede Cuenca, Octubre 2016. https://dspace.ups.edu.ec/handle/123456789/12815
10. J.L. Galarza, Determinación de la concentración de elementos contaminantes y bacterias patógenas presentes en un sistema de agricultura urbana compuesto por hortalizas, implementado en la Universidad Politécnica Salesiana sede Cuenca, Octubre 2017. https://dspace.ups.edu.ec/handle/123456789/14777
11. D. González, L. Portocarrero, Educación ambiental bajo los principios IAR-FAO en cinco instituciones educativas del centro histórico de Cuenca y determinación de contaminantes (PB, E. Coli y CO2) en la agricultura urbana (hortalizas y ornamentales), Diciembre 2019. http://dspace.ups.edu.ec/handle/123456789/18365
12. D.C. Calle, J.A. Zhindón, Evaluación de la presencia de plomo a través de huertos hortícolas agricultura urbana en cinco colegios ubicados en el distrito Sur de Cuenca. Obtenido de: http://dspace.ups.edu.ec/handle/123456789/17317, Mayo 2019, https://dspace.ups.edu.ec/handle/123456789/17317
13. J. Hallat, A. Castro, P. Burgos, R. López, Contenido de metales pesados en legumbres y hortalizas procedentes de huertos urbanos (2016). https://digital.csic.es/handle/10261/152362
14. L.J. Brewer, S. Angima, D. Sullivan, P. Deol, Cómo evaluar y reducir el peligro del plomo en los huertos y jardines (2017). https://catalog.extension.oregonstate.edu/ec1616s
15. M.H. Noriega, D.J. Simbaña, D.E. Torres, Estudio tribológico en función de la temperatura utilizando aditivos no sólidos (2018). http://repositorio.uide.edu.ec/handle/37000/2508
16. CODEX, Código de prácticas para la prevención y reducción de la presencia de plomo en los alimentos (2004). http://www.fao.org/fao-who-codexalimentarius/codex-texts/all-standards/es/

Chapter 36
Impact of Urban Agriculture on the Environmental Education of Children and Young People of the City of Cuenca

Dayanna Vargas, María Fernanda Vázquez, and Fredi Portilla

Abstract The urban agriculture at educational institutions is a strategy to improve and promote education, cultural exchange, social relations and recreation by allowing students to gain experience with nature and become more sensitive to the future of the environment and the planet since the lack of environmental education of children and young people leads to a lack of knowledge and implementation of environmental issues. The objective of the research was to assess the impact of urban agriculture on the environmental education of children and young people in the educational institutions of the city of Cuenca, and in turn, learn the importance of applying experimental works in the institutions. The methodology of this research is: comparative, inductive and deductive which includes: information search, information organization and analysis of the impact matrix of the three theses, the same ones that applied urban agriculture in the environmental education of children and young people, obtaining as a result from the impact matrix a percentage of 80% representing a positive impact. The research achieved in the students an active participation and interest in environmental education through urban agriculture allowing these activities to be well received in the institutions, where the experiments were carried out.

36.1 Introduction

For some years, now and today the practice of urban agriculture (AU) has increased markedly in different countries of the world due to population growth affected by the development of cities [1] mainly due to the increase in the poor population, migration and human mobility, natural disasters, environmental degradation, dese mpleo and the lack of resources that as it extends the food needs of urban families [2].

With the majority of the population living in urban areas, agriculture is increasingly spreading to cities large and small as an increasing number of people are engaged in practicing it globally [3], this is because it provides healthy foods since

D. Vargas · M. F. Vázquez · F. Portilla (✉)
Universidad Politécnica Salesiana, Cuenca, Ecuador
e-mail: fportilla@ups.edu.ec

Á. Rocha et al. (eds.), *Communication, Smart Technologies and Innovation for Society*, Smart Innovation, Systems and Technologies 252,
https://doi.org/10.1007/978-981-16-4126-8_36

the quality of the products consumed daily is known without the danger that they contain agrochemicals or are curded with black water; it also provides nutrients because these crops are managed in an organic way [4] and above all generate sources of employment that form a space of recreation and work that can be shared in the family nucleus and those close; in addition, green belts are created that strengthen the resilience of cities to climate change [3].

Similarly, it mitigates negative impacts on circumstantial and distant biodiversity as it is more sustainable than rural biodiversity because it depends on fewer agro-chemicals and is less biologically harmful [5]. That is why the AU as a set of systems ranging from primary production to food processing at the family and artisanal level, to positioning itself in differentiated markets where the added value generated [2] is recognized seeking to be an instrument of environmental, economic and social sustainability.

On the other hand, promoting urban agriculture through environmental education is an important point as it promotes training actions where people discover new ways to meet and relate to the environment [6] by leveraging natural resources such as soil, water, air, flora and fauna without losing their natural properties and recovering not only the environment but also food security [7].

The objective of this research is to assess the impact of urban agriculture on the environmental education of children and young people in the city of Cuenca, by collecting information from the ins held in the city's educational institutions that have been raised in order to resume agriculture in the urban part as aid for environmental education programs for children and young people in schools that in turn serves as a contribution to them by instilling responsibilities about the care of the planet.

36.1.1 Environmental Education and Its Contribution

According to Labrador Herraiz and López [8] environmental education (EA) is a permanent process in which individuals and communities become aware of their environment learn the knowledge, values, destrezas, experience and determination that empowers them to act, individually and collectively, in solving present and future environmental problems. Therefore, while it is true, it is the new generations that bring important environmental attitudes since they bring as their main consequence a balanced environment between the human being and the ecosystem that surrounds them [9]. It is, therefore, worth mentioning that in several countries environmental education was used on educational basis since this terminology is not new because it had already been used in schools since the 1970s. Where the Ministry of the Environment of Chile [10] states that it was the teachers who carried out the first EA trials, in subjects such as natural sciences or as activities of knowledge, care and conservation of nature. So is Novo [11]. He points out that ecological education is the first method of uniting education with the environment as people seek to promote environmental education at different levels of education whether formal or informal.

According to Valdés et al. [12], it should be taken as a reference to countries such as France, England, Poland, Russia and Spain that have already implemented in their mesh of study subjects focused on environmental education, incorporating different approaches such as technology, culture, economy and even politics. It is therefore of great importance that all countries around the world are involved and committed to a model of environmental education in each school and institutions in general.

36.1.2 Impact of Urban Agriculture

Agriculture within cities is not recent, although in the late 1980s urban agriculture has already been heard about urban agriculture globally as this term has been developed mainly for people with greater food shortages and that generates a grievance instrument for healthier eating and a less polluted environment [13].

It is then that the term urban agriculture according to FAO [14] refers to "small areas such as plots, orchards, margins, terraces, containers located within a city and intended for the production of crops and the rearing of minor livestock or dairy cows for own consumption or for sale in neighborhood markets." In another context, AU's expression refers to small areas located on the urban perimeter for intensive cultivation and breeding of small domestic animals where this production is mainly carried out on empty plots, patios and terraces that are transformed into community and family huer coughs, made exclusively by people who live and work in cities [13].

In urban agriculture, several benefits and advantages have been recognized, such as that orchards become at best, fifteen times more productive than rural farms and in an area of one square meter provides 20 kg of food annually [15].

In addition, it provides benefits to the environmental, economic, social and cultural dimension by improving food security, modifying adequate diet and nutrition, saving at home and reducing poverty, minimizing waste generated in household cuisine, combining buildings and green spaces, contribution to improving quality of life and urban development [16]. It also offers great development possibilities and greater environmental benefits in aspects related to waste management and treatment, renewable energy production and efficient water management [17].

The following specific fundamental objectives were therefore raised in the study, such as knowing the impact urban agriculture had on the environmental education of children and young people in the three study areas of the city of Cuenca, as well as analyzing the impact of experimental work on the implementing of educational institutions and showing that urban agriculture can contribute significantly to the environmental education of children and young people.

36.2 Methods

This research is of an inductive and deductive type as the information was based on the consultation of research documents already carried out found on websites and databases that are related to the subject of this work. The variables that were analyzed to understand the impact of urban agriculture on the environmental education of children and young people in the city of Cuenca were also determined. The method used consists of the following three stages: (1) search for information, (2) information organization and (3) analysis of information.

36.2.1 Search for Information

The research was carried out based on the study of thesins prepared by students of the Salesian Polytechnic University of the career of Environmental engineering applied in educational institutions of the city of Cuenca, as well as in the revision of other bibliographic documents.

For the search for information of the location coordinates of educational institutions, the Google Earth Preprogram was used.

36.2.2 Information Organization

The most relevant documents in the review of the information were those made by the thesins with the following topics presented in Table 36.1. These being the most important as research focused mainly on comparing mental experimentation.

Table 36.1 Documents relevant to the review in the research topic

Theme	Authors
"Assessment of the presence of lead through horticultural orchards urban agriculture in five schools located in the southern district of Cuenca" [18]	Daisy Carolina Calle Loja Jorge Andrés Zhindón Rodríguez
"Environmental education under IAR-FAO principles in five educational institutions in the historic center of Cuenca and pollutant determination (PB, *E. COLI* Y) in urban **CO_2** agriculture (Vegetables and ornamentals)" [19]	Daniela Gonzáles Pesántez Laura Catalina Portocarrero Marquez
"Determination of lead, pathogenic bacteria (E. COLI and Coliforms) and carbon capture in urban agriculture(Vegetables and ornamentals) located in five educational institutions located in the northern district of the city of Cuenca promoting environmental education under IAR-FAO principles" [9]	Maria José Quizhpi Pintado Tania Alexandra Sarango Guaman

Subsequently, other bibliographic documents obtained from databases were searched and reviewed systematically for further ideas on the research topic that provided sufficient information to learn more about agriculture urban and environmental education.

36.2.3 Analysis of Information

Using software. In the analysis of the thesis information, only one of them had the location points of the educational institutions where the study was carried out, and for the following, the Google Earth Pro tool was used which were subsequently saved in a Microsoft Excel sheet.

ArcMap 10.3 was then used where Excel points were exported to locate educational institutions in the three areas of the city of Cuenca on the map.

Microsoft Excel software that enabled statistical analysis and histograms to know the results of the research were used for the impact matrix performed by its own authorship.

Study variables. It was taken as significant research variables to educational institutions in the southern, central and northern part of the city of Cuenca (Table 36.2) that were subsequently placed in the impact matrix.

Impact matrix. Through the impact matrix, it was possible to focus the points of greatest interest on the issue of urban agriculture in environmental education where they were evaluated on a scale of assessment in order to know the impact that young people and children of educational institutions had. The steps to perform the matrix were as follows:

Identification of the aspects to be evaluated

In the first step for the matrix, the aspects to be evaluated of the thesins related to the issues of urban agriculture and environmental education and educational institutions were selected.

Determination of the impact level

The second step was based on a three-level valuation scale where level 1 is low indicating a negative impact, level 2 is average indicating a moderately positive impact and level 3 high indicating a positive impact as you can see in the Table 36.3.

Impact calculation

In the third step calculated the aspects to be evaluated with the valuation scale, since knowing the information and results of the experimental work assessed the impact of the experimental work, then averaged the results obtained from educational institutions in the north, south and center that were subsequently released in percentage range (Table 36.4).

Table 36.2 Study variables

Areas	Educational Institutions	Courses
South of the city	UEP Borja	5 to 6 to 7 mo Basic
	EU La Asunción	9th Baccalaureate "C"
	UEP Nuestra Familia	1st Baccalaureate
	EU Técnico Salesiano	2nd Baccalaureate
	EU Bilingüe Interamericano	1st Baccalaureate
City center	EU Santa Mariana de Jesús	7th and 10th Basic
	EU Salesiana María Auxiliador	7th and 10th Basic
	EMU Corazón de María	7th and 10th Basic
	UEBF Padre Carlos Crespi	7th and 10th Basic
	EU Luisa de Jesús Cordero	7th and 10th Basic
North of the city	EU Julio Maria Matovelle	5th Basic
	BSE Federico Proaño	8th basic
	EU Tres de Noviembre	6th basic
	EEB Cristo Rey	6th basic
	EU Francisca Dávila	6th basic "A" and "B"

Table 36.3 Valuation scales

Valuation scale		description
Low	1	Negative impact
Middle	2	Medium-positive impact
High	3	Positive impact

Table 36.4 Percentage range

Percentage (%)	Description
1–34	Negative impact
35–70	Medium-positive impact
71–100	Positive impact

36.3 Results

36.3.1 *Location of Educational Institutions*

We are in the city of Cuenca in southern Ecuador in the province of Azuay, with geographical coordinates 2° 39 s at 3° 00° south latitude and 78° 54° to 79° 26° of west longitude, with an altitude of 2560 m above sea level (Appendix 1).

36.3.2 *Impact of Urban Agriculture on Environmental Education*

A summary and comparative table of the results obtained from the three study areas assessed in the impact matrix (Appendix 2) was presented. Also, histograms were presented for the analysis of the same which determined the impact of urban agriculture on the environmental education of children and young people (Table 36.5).

Table 36.5 Results of the evaluation of the impact matrix

Aspects to be evaluated	North	Center	South	Average
Theists applied the IAR-FAO principles	3	3	1	2.3
Opening institutions to experimental work	3	3	3	3
Previous knowledge on Environmental Education and urban agriculture	2.6	3	3	2.9
Schools have a "TINI" project	1.8	1	1.4	1.4
Assimilation of information on the topics covered in the talks	2.8	3	2.8	2.9
Delivery of support material to students	3	3	2	2.3
It has land intended for an orchard	1.4	1	2.4	1.6
The students were involved in planting the plants	3	1	3	2.3
The students were involved in the irrigation of the plants	3	1	3	2.3
Students were involved in comprehensive pest management	3	1	3	2.3
Students showed interest in replicating an orchard at home	2	2.4	2.6	2.3
Support of teachers in the activities carried out	3	3	3	3
Support of parents in the activities carried out	3	1	2,6	2,2
Overall average	2.7	2.0	2.5	2.4
Percentage (%)	90	66.67	83.33	80

Analysis of results. Below are the results obtained from the impact of educational institutions according to the study area (Fig. 36.1).

As can be seen in the histogram, the institutions of the northern area had a positive impact since 10 of the 13 aspects evaluated demonstrated high valuations, since the theists implemented the IAR-FAO principles by encouraging participants to take positive action on food security, nutrition and improving sustainable livelihoods through urban gardens where the participants involved demonstrated full performance in each of the activities carried out.

Likewise, two aspects were observed thoroughly positive since 3 of the 5 institutions do not have the project "TINI" promoted by the Ministry of Educación of Ecuador being this relevant since its objective is to promote, generate, compromise and strengthen the environmental awareness, in terms of replicating an orchard at home the students showed interest when the theists imparted the importance of it but did not apply it. And as a negative impact, it has to have only one of the 5 educational institutions had a land destined for the orchard while the others used drawers previously designed.

With all aspects evaluated, a total average of 2.7 dam was obtained, with 90% giving a positive impact on educational institutions in the north (Fig. 36.2).

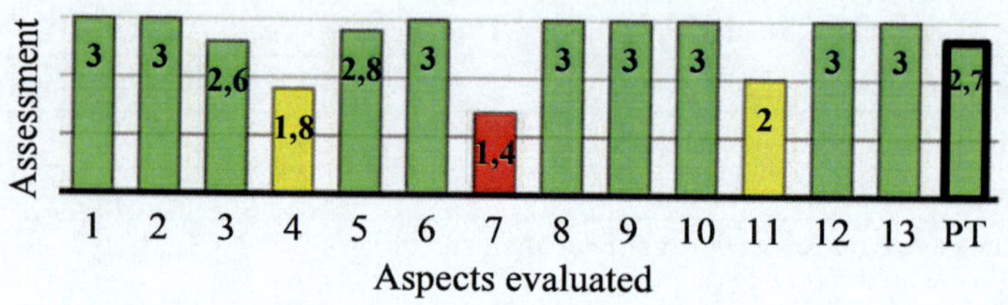

Fig. 36.1 Histogram of results obtained in the north zone impact matrix

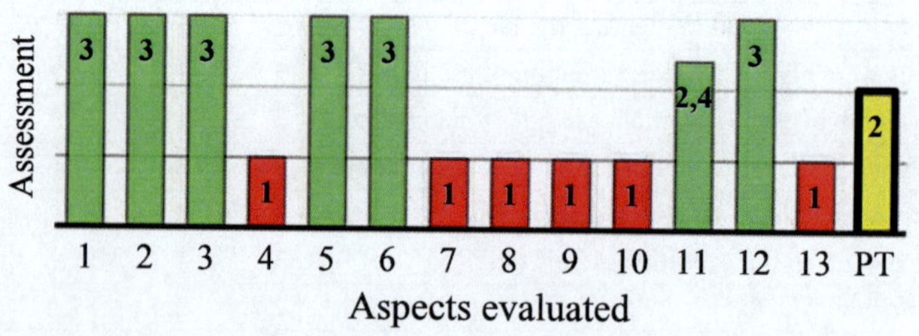

Fig. 36.2 Histogram of results obtained in the center zone impact matrix

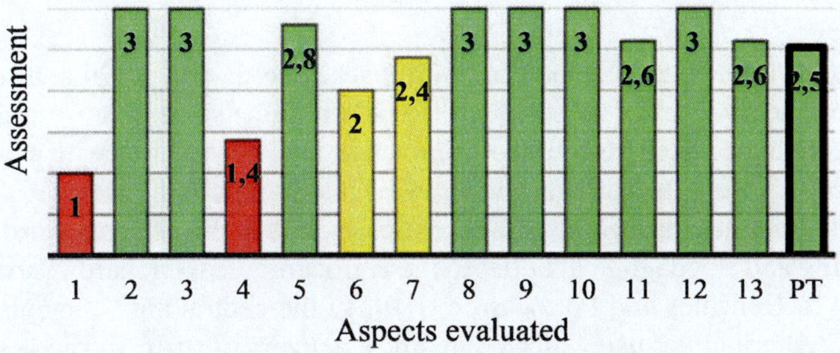

Fig. 36.3 Histogram of results obtained in the south zone impact matrix

The histogram shows that of the 13 aspects evaluated 7 they had positive impacts as the theists implemented the IAR-FAO principles through talks, dynamics, garden visits and questionnaires to evaluate participants' knowledge on the subject of environmental education and urban agriculture thus demonstrating interest and assimilation in the topics covered.

In the institutions of this area, the interest of replicating an orchard at home was moderately positive as few showed evidence of having applied in the meantime others were interested indicating that they would apply at some point. And as negative impacts 6 this was obtained because in aspect 4 none of the institutions had the project "TINI," and the other aspects that encompass the activities of crop management did not have land intended for an orchard, so also the students were not involved in the planting, irrigation and integrated management of pests, and in the end, those activities were carried out by theists.

The analysis determined a total average of 2 representing 66.67%, thus making a positive impact on the educational institutions of the central area (Fig. 36.3).

It is notorious of the 13 aspects evaluated 9 had positive impacts since all the participants involved demonstrated knowledge in urban agriculture and environmental education also assimilated the information provided by the instructors in the different activities they carried out, showing an active participation in the integral management of the crop. In aspect 6, videos on the topics covered were presented as support material, and in aspect 7, three of the institutions in the south had land intended for an orchard and the others used different spaces with a moderately positive impact. And as negative impacts, the theists did not apply the IAR-FAO principles and did not have the "TINI" project.

From the parameters studied, a total average of 2.5 representing 83.33% was determined, thus giving a positive impact on educational institutions in the southern area.

36.4 Discussion

According to the results obtained in the impact zone, it was possible to determine that among the three areas the results were not the same since the institutions of the north and south had a positive impact compared to the central area with a moderately positive impact, this is because the theists in the institutions of the north [9] applied the IAR-FAO principles through playful activities such as lucky roulette, word soup, EA card making and a pedagogical outlet to the Agroecological Orchard "San Joaquín"; compared to Gonzáles and Portocarrero [19]. In the central area, they also applied the IAR-FAO principles using talks, dynamics, delivery of a triptych and evaluations but unlike the previous one did not make exits to other orchards, and the institutions of the southern area the theists Calle and Zhindón [18]. They implemented activities such as talks, games, videos and evolutions but did not apply the IAR-FAO principles, and did not make a pedagogical outing thus showing that the students who performed it had better results in the au and EA topics. Since "Pedagogical outings turn out to be an enhancer for learning that places students at the center of the teaching process managing to be in direct contact with the environment by performing the learning process in an imminently practical way." All these activities are achieved thanks to the accompaniment of the professionals in charge.

The assimilation of the information of the talks given to the students had a high impact since according to Quizhpi and Sarango [9]. They showed an 80% understanding of the topics covered, while Gonzáles and Portocarrero [19] they say that the percentage of assimilation obtained was 91.49%, in terms of Calle, C. and Zhindón [18]. They do not have a percentage of assimilation but mention that to encourage student participation, talks and trainings were given provided by students in the sixth cycle of the UPS Environmental engineering career expressing a harmonious coexistence of the students of the school with the university students.

With regard to the TINI project, the participating institutions were the EU "Julio María Matovelle" and BSE "Federico Proaño" belonging to the northern area, as did the EU "La Asunción" in the southern area and in the s of the central area none had it; (c) to emphasize that this project aims to strengthen the knowledge, skills, values and self-esteem of students to live in armonía with the environment, that is why within the institutional curriculum hours has an established time to develop environmental issues since at the time of the activities, these hours were very useful unlike Gonzáles and Portocarrero [19] they had drawbacks due to the interruption of their class hours.

Likewise, in the processes of crop management, the institutions of the north and south [9, 18] were supported by students, teachers and parents in the activities carried out, compared to the institutions of the center [19] indicated that the students were not involucrate in the process because by their class hours they did not grant them the respective permits and also because of the number of holidays that were held in the school year that they performed the experimental work; it is then that theists performed the management of the crop being an aspect against it since the students did not implement what they theoretically learned. In this process, it was also known that few institutions had land destined for an orchard so that theists sought other alternatives for the application such as reusable drawers, pallets and planters.

It should be noted that the students showed interest in replicating an orchard at home but did not apply it, so the institutions that if they showed evidence of having applied in their home were the EU Salesiana María Auxiliadora, the EU Luisa de Jesús Cordero, UEP Nuestra Familia, EU Técnico Salesiano and EU Bilingüe Interamericano.

36.5 Conclusions

In relation to the research, it can be concluded that the impact of urban agriculture on the environmental education of children and young people in the city of Cuenca participating in the experimental work or had an overall average of 2.4 representing 80% indicating a positive impact from the evaluation of the impact matrix.

According to the results, 90% were achieved in the north area demonstrating a positive impact as the theists achieved a satisfactory methodology that resulted in a high level of learning and collaboration in the participants; followed by the southern area with 83.33% demonstrating a positive impact since in their methodology they involved sixth-cycle students of the Environmental engineering career providing greater knowledge resulting in the active participation of students. As for the downtown area with 66.67%, there is a moderately positive impact since the theists had some difficulties in executing their methodology in which it involved student participation in crop management because of the lack of availability of hours but it is worth mentioning that the students showed positive attitudes and actions in the topics of agriculture to urban and environmental education.

With regard to the implementation of experimental work in educational institutions linked to environmental education and urban agriculture issues, it was analyzed that these projects are of high impact, so it must continue to be implemented since they achieved in students the strengthening of their environmental values so that they can be applied with a vision in the future.

It was possible to show that urban agriculture can contribute significantly to the environmental education of children and young people since through learning they develop productive skills in the environmental, food and social field which allows them a dynamic between their environment that are involved in the processes of crop management and the maintenance of it generating awareness to the environment and a comprehensive and group participation among the participants and achieving in turn formative learning or with the projects carried out.

References

1. P. Velasquez Torres, D. Flores Murillo, *Social Project 'Huertos Urbanos* (Universidad Católica de Santiago de Guayaquil, 2012)
2. A. Rodríguez Dueñas, I. Proaño Rivera, *Quito sows: Urban Agriculture* (2016) [Online]. Available: http://www.conquito.org.ec/
3. FAO, *Urban Agriculture: Cultivating Soils in the City* (2015). http://www.fao.org/soils-2015/news/news-detail/es/c/332731/. Accessed 20 Jan 2021
4. A.V. de Guadalajara, *Urban Agriculture Manual* (2009)
5. RUAF, The urban agriculture magazine. *La Revista Agricultura Urbana* (2001)
6. J.A. Martínez Mejía, The school garden and urban agriculture as strategies to strengthen environmental culture in students of the 8th grade-2 , of the I.E. Santo Tomás de Aquino of the Municipality of Titiribí, Antioquia (Fundación Universitaria los Libertadores, 2020)
7. L.A. Portilla Insuasti, J.N. Sañudo Díaz, Urban agriculture and environmental awareness in the communities of: Corazón de Jesus, Emilio Botero, Belén, Miraflores y el Pilar, Municipality of Pasto-Nariño (University of Manizales, 2016)
8. C. Labrador Herraiz, D.V. López, Environmental education in international documents. Notes for a comparative study. Rev. Complut. Educ. **6**, 17 (1995) [Online]. Available: https://dialnet.unirioja.es/servlet/articulo?codigo=150176&orden=1&info=link
9. M.J. Quizhpi Pintado, T.A. Sarango Guamán, *Determination of Lead, Pathogenic Bacteria (E. coli and Coliforms) and Carbon Capture in Urban Agriculture (Vegetables and Ornamentals) Located in Five Educational Institutions Located in the Northern District of the City of Cuenca Promoting* (Salesian Polytechnic University, 2020)
10. M. del M. A. MMA, *Environmental Education A look from Chilean environmental institutionality.* Chile (2018)
11. M. Novo, Formal and non-formal environmental education: two complementary systems. Rev. Iberoam. Educ. (1996). Accessed: 11 Nov 2020. [Online]. Available: https://www.researchgate.net/publication/28075969_La_Educacion_Ambiental_formal_y_no_formal_dos_sistemas_complementarios
12. O. Valdés Valdés, A. Rodríguez Salvá, M. Llivina Lavigne, A. Betancourt Blanco, I. Santos Abreu, Environmental education and sustainable development: strategies of curricular and institutional interdisciplinary integration into programs, projects and good practices in universities, schools, families and communities in Cuba. *Educ. Cuba.* (2013) [Online]. Available: http://www.unesco.org/
13. M.-H. Zaar, Urban agriculture: some reflections on its current origin and importance. *Bibliographic Mag. Geogr. Social Sci.* (2011). http://www.ub.edu/geocrit/b3w-944.htm. Accessed 11 Nov 2020
14. FAO, *Urban and Peri-Urban Agriculture* (1999) http://www.fao.org/unfao/bodies/COAG/COAG15/X0076S.htm. Accessed 11 Nov 2020
15. FAO, "Creating Greener Cities," 2010, [Online]. Available: http://www.fao.org/ag/agp/greenercities/pdf/ggc-es.pdf.
16. Huertoencasa, *Manual of Urban Agriculture* (2012). https://www.ecoagricultor.com/wp-content/uploads/2013/04/Manual-Agricultura-Urbana.pdf. Accessed 26 Jan 2021
17. A. López, Environmental benefits of urban and peri-urban organic agriculture. *Inst. Recur. NAT. Agrobiol. Seville* (2014) [Online]. Available: https://digital.csic.es/bitstream/10261/116301/3/Beneficios_ambientales_agricultura_ecológica_urbana_periurbana_2014_Com_Congr.pdf
18. D.C. Calle Loja, J.A. Zhindón Rodríguez, *Evaluation of the Presence Of Lead Through Horticultural Orchards Urban Agriculture in Five Schools Located in the Southern District of Cuenca* (Universidad Politécnica Salesiana, 2018)

19. G. Pesántez, L.C. Portocarrero Márquez, *Environmental Education Under IAR-FAO Principles in Five Educational Institutions in the Historic Center of Cuenca and Determination of Pollutants (PB, E. coli and* CO_2*) in Urban Agriculture (Vegetables and Ornamentals)* (Universidad Politécnica Salesiana, 2019)

Chapter 37
Estimates of the Carbon Capture Potential in Urban Parks and Vehicle CO_2 Emissions in Cuenca, Ecuador

María Elisa Muñoz, Edison Vásquez, and Fredi Portilla

Abstract Nowadays, the adverse effects of global warming are evident due to the increase in greenhouse gas emissions, with urban areas being one of the areas that emits large amounts of CO_2; however, urban trees act as an important sink for carbon within the city; so in this work, the estimates of the potential of carbon capture in 28 urban parks and vehicular CO_2 emissions in Cuenca, in a study area comprising 64.51 ha. The non-destructive method is applied which, from the diameter to the chest height (DAP) and tree height, estimates carbon sequestration. The capture potential is 11,418.88 t and an average capture of 8.81 kg CO_2/m^2/year. On the other hand, the estimated CO_2 emissions from fuel consumption in Cuenca were a total of 828.81 Gg of CO_2. In reference to the economic valuation, in the compliance market, the SENDECO2 entity that pays $ 24.81 per ton of CO_2 would generate an income of $ 283,302.5 and in the voluntary market an income of $ 229,176.93 with the CDM Gold Standard entity that pay $ 20.07 per ton. It is concluded that the 28 sampled parks can mitigate 1.37% of the total CO_2 emissions emitted by the car park, demonstrating the importance of this environmental service and that the sale of carbon generates a relevant benefit, demonstrating the feasibility of carbon markets.

37.1 Introduction

Currently, it is necessary to maintain an environmental-economic balance, since the adverse effects of climate change are evident, with two main causes standing out: an increase in greenhouse gases (GHG), carbon dioxide being the predominant gas, and a significant loss of natural carbon sinks [1].

Ecuador has a national GHG catalog divided into five sectors, where the energy sector accounts for more than 40% of these emissions, when talking about this sector we refer to the transport sector generated by the burning of fuel [2].

M. E. Muñoz · E. Vásquez · F. Portilla (✉)
Environmental Engineering Researchers. Faculty of Life Sciences, Universidad Politécnica Salesiana, Calle Vieja 12-30 Elia Liut, Cuenca, Ecuador
e-mail: fportilla@ups.edu.ec

Á. Rocha et al. (eds.), *Communication, Smart Technologies and Innovation for Society*, Smart Innovation, Systems and Technologies 252,
https://doi.org/10.1007/978-981-16-4126-8_37

In response to this problem, alternatives for the preservation of natural resources have been proposed, such as carbon sequestration, which is one of the most efficient alternatives to mitigate the effect of global warming, since trees contribute to the reduction of greenhouse gas emissions [3]. The trees contribute to the reduction of carbon dioxide emissions, thus guaranteeing us a healthy environment with a healthy climate. The trees contribute to the reduction of greenhouse gas emissions, thus guaranteeing a healthy environment with good air quality [4, 5].

Therefore, the objective is to estimate the CO_2 capture potential of the parks that correspond to the urban area of Cuenca and compare it with the CO_2 emissions from the automobile fleet, highlighting the importance of protecting, restoring, financing and, if necessary, increasing green areas in the city.

37.2 Materials and Methods

37.2.1 Location of the Parks Under Study

The 28 parks are located in the 15 parishes that make up the urban area of the city of Cuenca, in the province of Azuay [6]. They occupy 64.51 ha of its surface area. The perimeter and area of each park were defined by georeferencing, using geographic information system tools and satellite images.

37.2.2 Park Selection

Two or more parks have been selected for those parishes that have a considerable surface area; while in parishes with small surface areas, one park was selected in addition to the most representative parks of the city and is located in areas considered to have high pollution indexes, so their capture potential could be high [7].

37.2.3 Forest Inventory and Estimation of Captured CO_2

Diameter at breast height (DBH) was measured at a height of 1.30 m from the base of the tree, while height was measured with the help of a hypsometer [8, 9]. The measurement was performed on all individuals in the selected parks.

For the estimation of captured CO_2, the indirect or non-destructive method was used, through which it was possible to estimate biomass, carbon content, and captured carbon dioxide. References [10, 11] were used to estimate biomass, carbon content, and carbon dioxide captured.

Biomass estimation. The calculation of the biomass of a green space or area is essential to determine its carbon sequestration potential (FAO 2000). For this estimation, it was necessary to estimate the aerial green biomass and the root biomass, and the former is estimated through a mathematical model [12].

$$\text{Bva} = a * D^b * H^c \tag{37.1}$$

where

- Bva: Aerial green biomass, kg.
- *D*: Diameter at breast height, cm.
- *H*: Total height, m.
- *a*: 0.026.
- *b*: 1.529.
- *c*: 1.747.

While the green root biomass was taken as 20% of the green biomass area [10].

The sums of these data make up the total biomass possessed by an individual [13].

Finally, the amount of dry biomass was estimated which can be obtained from the total green biomass in a natural way with a humidity lower than 60% [13].

$$\text{Bs} = \text{Bvt} - \frac{\text{Best} * 40}{100} \tag{37.2}$$

where

- Bs: Biomass in dry weight, kg.
- Bvt: Total green biomass, kg.

Estimation of carbon in parks. For the calculation of the aerial carbon, the fraction of carbon that exists in the biomass is used, whose factor is 0.5, and this factor belongs to the dry biomass that was obtained through an interspecific and specific weighting of carbon content factors of branches, leaves, and roots [14].

$$\text{CAT} = B_s * 0.5 \tag{37.3}$$

where

- CAT: Total airbone carbon, kg.
- Bs: Dry biomass, kg.

Root carbon is calculated through an indirect method where aerial carbon is related to the average ratio that exists between soil biomass which has a value of 0.24 [15, 16]. Using the following equation.

$$\text{CR} = \text{CAT} * 0.24 \tag{37.4}$$

where

- CR: Root carbon, kg.
- CAT: Root aerial carbon, kg.

To determine the amount of total carbon, total aerial carbon and root carbon are added together [15, 17]. The following equation was used for this purpose.

$$CT = CAT + CR \quad (37.5)$$

where

- CT: Total carbon, kg.
- CAT: Total airbone carbon, kg.
- CR: Root carbon, kg.

Estimated CO_2captured in parks. For the calculation of dioxide, an equation was used [18], from the product of total carbon and a carbon to carbon dioxide conversion factor.

$$CO_2 = \frac{CT * 3.6663}{1000} \quad (37.6)$$

where

- CO_2: Carbon captured in tons of carbon dioxide, t.
- CT: Total carbon stored, kg.
- 3.6663: Carbon dioxide conversion factor, which is the result of the quotient of the molecular weights of carbon and carbon dioxide.
- 1000: Conversion factor from kilograms to tons.

37.2.4 *Species Sampling and Identification*

The collection of in situ information (samples) was carried out through numerical coding, in order to determine the characteristics of the forest individual to be inventoried, using the Azuay Digital Herbarium and the book of Trees and shrubs of the rivers of Cuenca, both published by the University of Azuay, as a database.

37.2.5 *Importance Value Index*

The importance value index is calculated from the sum of the relative density, relative dominance, and relative frequency for each species, divided by three in order to obtain the proportion with respect to the percentage (100%) [19].

This value shows us the relative ecological importance of each species sampled in a plant group (Cueva 2015). That is, the higher a species has the IVI means that it is ecologically dominant [20].

$$\text{IVI} = \frac{\text{DR} + \text{DmR} + \text{fr}}{3} \quad (37.7)$$

where

- IVI: Importance value index [%].
- DR: Relative density [%].
- DmR: Relative dominance [%].
- fr: relative frequency [%].

37.2.6 Estimated Vehicle CO_2 Emissions

For this a methodology, was used that is a function of fuel consumed, which is equal to fuel sold [21–23] according to IPCC guidelines. For this it was necessary to obtain the volume of sales of super gasoline, extra gasoline and premium diesel, main fuels used in the city, then they were transformed from volume units to mass units to subsequently multiply it by the calorific value and obtain the fuel consumption in energy units which are Tera joules (TJ) [24]. Then the CO_2 emissions were obtained using the equation:

$$E = \sum_a [\text{Combustible}_a * \text{EF}_a] \quad (37.8)$$

where

- $E = CO_2$ emissions (kg)
- Combustible_a = fuel sold (TJ)
- EF_a = emission factor (kg/TJ).
- a = fuel type (gasoline, diesel, natural gas, etc.).

Finally, the emissions of each fuel are added together to obtain the total emissions of the vehicle fleet.

37.3 Results and Discussion

37.3.1 Tree Composition of the Parks

In the parks sampled, which occupy a total area of 64.51 ha, 3953 individuals were recorded, grouped into 48 species and 29 families (Table 37.1), including native and introduced species for the reforestation programs of green areas in the city.

Table 37.1 Species inventoried by family

Number	Common name	Scientific name	Family
1	Acacia	*Acacia dealbata Link*	Fabaceae
2	Acacia	*Acacia melanoxylon R. Br*	Fabaceae
3	Blue acacia	*Acacia Baileyana F. Muell*	Mimosaceae
4	Poplar	*Populus balsamifera* L	Salicaceae
5	White poplar	*Populus alba* L	Salicaceae
6	Alder	*Alnus acuminata Kunth*	Betulaceae
7	Arupo	*Chionanthus pubescens Kunth*	Oleaceae
8	Cañaro	*Erythina edulis Triana*	Fabaceae
9	Capulí	*Prunus serótina Ehrh*	Rosaceae
10	Casuarina	*Casuarina equisetifolia L*	Casuarinaceae
11	Cedar	*Cedrela montana Moritz ex Turcz*	Meliaceae
12	Cepillo	*Callistemon lanceolatus (Sm.)*	Myrtaceae
13	Cepillo blanco	*Callistemon salignus (Sm.) Sweet*	Myrtaceae
14	Magenta cherry	*Syzygium paniculatum (Lam.) DC*	Myrtaceae
15	Cypress	*Cupressus sempervirens*	Cupressaceae
16	Citrus	*Citrus aurantifolia*	Rutaceae
17	Coco	*Parajoubaea cocoides Burret*	Arecaceae
18	Eucalyptus	*Eucalyptus citriodora Hook*	Myrtaceae
19	Ficus	*Ficus spp.*	Moraceae
20	Fresno	*Tecoma stans* (L.)*Juss. ex Kunth*	Bignoniaceae
21	Guaba	*Inga insignis Kunth*	Fabaceae
22	Guabisay	*Podocarpus sprucei Parl*	Podocarpaceae
23	Guaylo	*Delostoma integrifolium D. Don*	Bignoniaceae
24	Guineo or Banana	*Musa paradisiaca*	Musaceae
25	Higuerón	*Ficus* spp.	Moraceae
26	Huahual o Arrayan	*Myrcianthes rhopaloides (Kunth) McVaugh*	Myrtaceae
27	Huesito	*Pittosporum undulatum Vent*	Pittosporaceae
28	Jacaranda	*Jacaranda mimosifolia D. Don*	Bignoniaceae
29	Jiguerón	*Aegiphila ferruginea Hayek & Spruce*	Lamiaceae
30	Magnolia	*Magnolia grandiflora* L	Magnoliaceae
31	Molle	*Schhus molle L*	Anacardiaceae
32	Mulberry	*Morus alba* L	Moraceae
33	Walnut	*Juglans neotropica Diels*	Juglandaceae
34	Palm	*Cycas* sp.	Cycadaceae

(continued)

Table 37.1 (continued)

Number	Common name	Scientific name	Family
35	Pink stick	*Tipuana tipu (Benth.) Kuntze*	Fabaceae
36	Pena pena	*Fuchsia boliviana Carrière*	Onagraceae
37	Peral	*Pyrus communis*	Rosaceae
38	Pine	*Pinus radiata D. Don*	Pinaceae
39	Pine	*Araucaria araucana*	Araucariaceae
40	Pintador	*Prunus opaca (Benth.) Walp*	Rosaceae
41	Banana	*Platanus occidentalis* L	Platanaceae
42	Quishuar	*Buddleja americana* L	Scrophulariaceae
43	Ramo de novia	*Yucca guatemalensis Baker*	Asparagaceae
44	Roble australiano	*Grevillea robusta A. Cunn. ex R. Br*	Proteaceae
45	Willow	*Salix humboldtiana Willd*	Salicaceae
46	Weeping Willow	*Salix babylonica* L	Salicaceae
47	Siete cuero	*Tibouchina lepidota*	Melastomataceae
48	Tilo	Sambucus mexicana C. Presl ex DC	Adoxaceae
49	Urapán	*Fraxinus excelsior* L	Oleaceae

There is a wide variety of individuals per species, such is the case of Eucalyptus citriodora Hook with 1554 individuals (39.2%) and Salix humboldtiana with 464 individuals (11.77%), being the most representative species of the total number of individuals, followed by Callistemon salignus (4.95%), Jacaranda mimosifolia (4.29%), Fraxinus excelsior (3.88%), Acacia dealbata (3.66%), Schhus molle and Populus alba (2.69%), and Tecoma stans (2.89%). On the other hand, 39 species (24.01%) had less than 100 individuals.

37.3.2 Importance Value Index by Species

With respect to the importance value index (IVI), the most dominant species is *Eucalyptus citriodora* Hook with an equivalent IVI of 35%, followed by *Salix humboldtiana* Willd with an IVI equal to 10.74%, *Jacaranda mimosifolia D. Don* with an IVI of 3.89% and *Acacia dealbata* Link with an IVI of 3.71, the rest of the species do not exceed in IVI equal to 3%. It should be noted that of these 4 species, 3 are considered introduced species, with only *Salix humbold* and *Acacia dealbata* Link having an IVI of 3.71%. [25]. The only dominant introduced native species is *Salix humboldtiana* Willd, which demonstrates the predominance of introduced species over native species (Table 37.2).

Da: Absolute density, Dr: Relative density, Fa: Absolute frequency, Fr: Relative frequency, DmR: Relative dominance.

Table 37.2 Dominant species in parks

Value of importance index (VII)							
Scientific Name	Total of individuals	Da (ind/m^2)	Dr (%)	Fa (%)	Fr (%)	DmR (%)	IVI (%)
Acacia dealbata Link	143	0,002,217	3,62	55,17	5,33	2,192	3,71
Jacaranda mimosifolia D. Don	169	0,000,262	4,28	68,97	6,67	0,725	3,89
Salix humboldtiana Willd	464	0,000,719	11,75	24,14	2,33	18,127	10,74
Eucalyptus citriodora Hook	1.544	0,002,393	39,1	27,59	2,67	63,343	35,04

37.3.3 Tree Clasification According to DBH

Considering all the individuals sampled, the average DBH is 38.63 cm with a standard deviation of 27.86 cm. The largest number of trees is grouped in a range of 5–55 cm, within which there are 2981 individuals. This means that most are young trees whose growth is developing, as well as their carbon sequestration capacity (Fig. 37.1).

37.3.4 CO_2 Captured by Sampled Species

In the parks sampled, the total CO_2 stored is 11,418.88 t, with an average capture rate of 8.81 kg/m^2. The species with the largest amount of CO_2 stored is *Eucalyptus citriodora* Hook with 9,928.70 t, representing 86.95% due to the large number of individuals in the parks, with the Tarqui-Guzho megapark having around 924 individuals of this species, followed by *Salix humboldtiana* Willd, which stores 607.53 t, representing 5.32%, while the other species do not capture more than 2% of the total (Fig. 37.2).

Referring to the average carbon dioxide capture and total capture at the urban level, Cuenca has an average of 88.1 t CO_2/ha and a value of 11,418.88 t of total carbon dioxide capture, being higher in comparison to other cities such as Tijuana, Mexico [24], Junín and Loreto, Perú [26, 27]. It is worth mentioning that the above-mentioned works may differ in the degree of accuracy in data collection, different methodologies applied, and the extent of the study area.

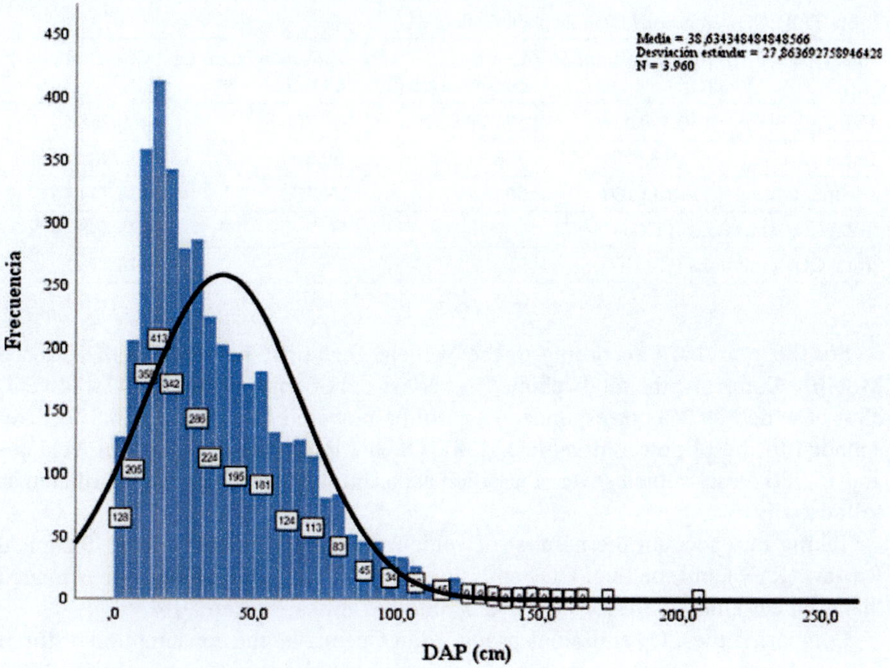

Fig. 37.1 DBH histogram of the species sampled

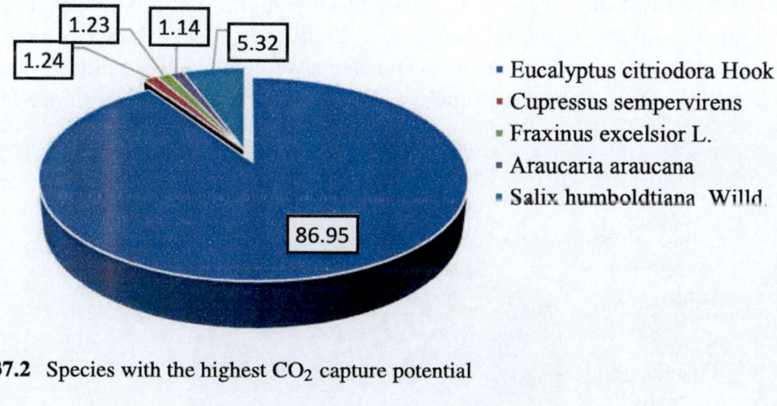

Fig. 37.2 Species with the highest CO_2 capture potential

37.3.5 CO_2 Estimates for Fuel Consumption

Estimated CO_2 emissions from mobile sources (vehicles) in Cuenca for 2014 were 828.81 Giga grams of CO_2, and this according to the methodology outlined above, with the consumption of extra gasoline and premium diesel being responsible for 93% of emissions (Table 37.3).

Table 37.3 CO_2 emissions from the vehicle fleet

Fuel type	Fuel consumption (t)	Energy consumption (TJ)	Emission factor kg CO_2/TJ	CO_2 emitted (t)
Extra gasoline	130,380.156	5775.84	69.300	400,265.71
Super gasoline	17,523.957	776.31	69.300	53,798.28
Premium diesel	117,611.091	5057.27	74.100	374,744.45
Total CO_2 emissions (t CO_2)				828,808.44
Total CO_2 emissions (Gg CO_2)				828.81

For the year 2014 according to the Vehicle Technical Review and the Cuenca Mobility Company, the total vehicle fleet consisted of approximately 147,484 vehicles, of which 89.2% corresponded to gasoline-powered vehicles and 10.8% corresponded to diesel-powered vehicles [28]. Of this total, 131,488 gasoline vehicles and 15,966 diesel vehicles were classified according to their class and distributed as follows (Fig. 37.3).

Taking into account the number of vehicles registered in the Vehicle Technical Review (RTV) and the total CO_2 emissions from fuel consumption, it is estimated that each car emitted 5629.76 kg of CO_2 into the atmosphere in 2014.

Comparing the CO_2 emissions produced in Cuenca by the consumption of fossil fuels in the transportation sector with other cities in the country, it is obtained that Cuenca produces 38.92% less emission than Guayaquil [22] and 23.24% less than Quito [29]. The highest emissions are produced by gasoline vehicles (super and extra), since most of the car parks are made up of vehicles that run on this fuel.

In contrast to other cities in America, it emits more CO_2 than Cochabamba, Bolivia, despite having fewer vehicles, but because Cuenca has a higher altitude above sea level, the combustion process is more incomplete and therefore higher

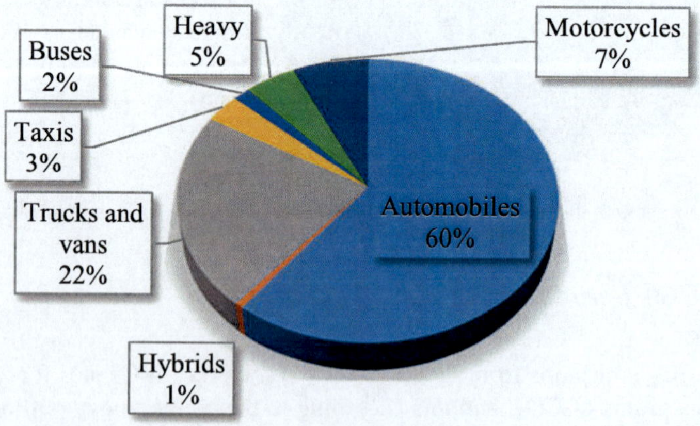

Fig. 37.3 Composition of Cuenca's vehicle fleet in 2014

Table 37.4 Economic valuation of all sampled species

Scenario	Entity	Total CO_2 captured (t)	Price (USD)	Total value (USD)
Market volunteer	CDM Gold Standard	11.418,88	20.07	229,176.93
Compliance market	SENDECO2		24.81	283,302.5

emissions [30]. However, because Cuenca has a higher altitude above sea level, the combustion process is more incomplete and therefore higher emissions. On the other hand, it emits 30.52% and 51.96 less CO_2 than Ibagué, Colombia and Tijuana, Mexico, respectively, which are cities with larger populations and larger numbers of vehicles.

37.3.6 Economic Valuation of Carbon Sequstration

The economic valuation for the sale of one metric ton of captured CO_2 was carried out for both the voluntary market and the regulated compliance market, with prices offered by the following entities, SENDECO2, and CDM Gold Standard. The revenues for the two markets do not have a significant difference, however, it is recommended to execute the second proposal because the monetary revenues are higher (Table 37.4).

According to Gallegos (2012), Ecuador is involved in the carbon market, in addition to carrying out emission reduction projects, about 92 projects have been presented, of which 43% correspond to hydroelectric plants, 12% waste treatment, 11% forestry, 10% energy efficiency, 10% gas recovery, 7% wind energy, 3% agriculture, 3% geothermal, and 1% transportation, demonstrating that it is viable to implement this type of carbon sale project in both voluntary and compliance markets; however, it is advisable to work with the second market because it develops CDM projects.

37.4 Conclusions

According to the forest inventory carried out in the 28 parks that correspond to the 15 urban parishes defined, 3953 trees that meet the selection criteria were recorded, with 20 native species and 29 introduced species belonging to 29 families. The most representative species were *Eucalyptus* sp., (39.2%) and *Salix humboldtiana* with (11.77%), being the most representative species of the total number of individuals.

The CO_2 capture of the sampled parks was 11,418.88 t during 2019, and an average capture of 8.81 kg CO_2/m^2/año considering the total area of the parks. Native species captured 669.08 t and introduced species captured 10,709.8 t, showing that introduced species capture 93.79% of the total. The parks that captured the greatest amount were Tarqui-Guzho, Parque Lineal Av. Cumanda, and El Paraíso, due to their larger surface area and greater number of individuals.

The species that captured the most CO_2 were *Eucalyptus* sp. (9.928.696 kg), *Salix humboldtiana* (607.305,42 kg) y *Cupressus sempervirens* (141.361,18 kg).

Estimated CO_2 emissions from fuel consumption in Cuenca during 2014 were 400,265.71 t by extra gasoline vehicles, 53,798.28 by super gasoline vehicles, and 374,744.45 t by premium diesel vehicles, totaling 828.81 Gg of CO_2.

In reference to the economic valuation, in the compliance market, the SENDECO2 entity paying $24.81 per ton of CO_2 would generate an income of $283,302.5 and in the voluntary market an income of $229,176.93 with the CDM Gold Standard entity paying $20.07 per ton.

The capture potential of the sampled parks can capture 1.37% of the total CO_2 emissions emitted by vehicles in the urban area of the city, demonstrating that it is a methodology that can be considered to combat climate change.

References

1. H. Benavides, G. León, Información técnica sobre gases de efecto invernadero y el cambio climático (Instituto de Hidrología, metereología y estudios ambientales, 2007)
2. Ministerio del Ambiente Ecuador, Inventario Nacional de Gases de Efecto Invernadero del Ecuador (Quito, 2012)
3. G. Avila, F. Jimenez, J. Beer, M. Gomez, M. Ibrahim, Almacenamiento, fijacion de carbono y valoracion de servicios ambientales en sistemas agroforestales en Costa Rica (Agroforesteria en las Americas, 2001), pp. 32–35
4. FAO, Los bosques y el cambio climático (2006)
5. P. Chambi, Valoracion economica de secuestro de carbono mediante simulcion aplicado a la zona boscosa del Rio Inambari y Madre de Dios (Tacna, 2001)
6. P.F. Vintimilla, *Análisis de Resultados de La Medición de Emisiones de Gases Contaminantes de Fuentes Móviles a Partir de La Implementación de La Revisión Técnica Vehicular En El Cantón Cuenca* (Universidad Politénica Salesiana, Cuenca, 2015)
7. M. Saldaña, K. Vera, *Diagnóstico de Plagas y Enfermedades Presentes En Las Plantas de La Zona Urbana de La Ciudad de Cuenca* (Univeridad Politécnica Salesiana, Cuenca, 2019)
8. J. Romo, *Evaluación Del Carbono En La Biomasa de 3 Especies Forestales Nativas (Shiripe-Myrsine Dependens, Rañas-Viburnum Triphyllum, Yugyug-Miconia Theaezans) En El Bosque Aguarongo* (Universidad Politécnica Salesiana, Cuenca, 2016)
9. M. Gutiérrez, *Evaluación Del Carbono En La Biomasa de 3 Especies Forestales Nativas (Miconia Aspergillaris, Vallea Stipularis, Oreocallis Grandiflora) En El Bosque Aguarongo* (Universidad Politécnica Salesiana, Cuenca, 2016)
10. M.P. Morales, M.P. Vasquez, *Valoracion Economica de La Captura de Carbono En Las Especies Podocarpus Spucei y Oreocallis Grandiflora En El Bosque Protector Aguarongo* (Universidad Politecnica Salesiana, Cuenca, 2019)
11. R. Russo, Guía práctica para la mediciónde la captura de carbono en la biomasa forestal (2009)
12. N. Higuchi, J. Carvalho, Fitomassa, e conteúdo de carbono de espécies arbóreas da Amazonia (Scielo, 1994), pp. 125–153
13. A. Barriga, L. Toscano, Análisis de los parámetros y selección de hornos para la combustión de biomasa (aplicación a biomasas locales típicas) (Espol, 2016)

14. L. Toscano, Análisis de los parámetros de hornos para la combustión de biomasa (Escuela Superior Politécnica del Litoral, Guayaquil, 2009)
15. N. Guerra, Valoración económica del secuestro de $C0_2$ y stock de carbono en plantaciones de Simarouba amara (UNAP, 2013)
16. D. Sánchez, Evaluación del carbono almacenado en la biomasa, necromasa y carbono orgánico del suelo de tres diferentes hábitats en la Península de Osa, Costa Rica (Universidad Nacional, 2016)
17. J. Torres, V. Mena, E. Álvarez, Carbono aéreo almacenado en tres bosques del Jardín Botánico del Pacífico, Chocó, Colombia. Entramado **13**(I), 200–209 (2017)
18. J. Sosa, Valoración económica del secuestro de CO_2 en tres tipos de bosque en el distrito del Alto Nanay, Loreto-Perú (UNAP, 2016)
19. Z. Aguirre, Guía de métodos para medir la biodiversidad (Universidad Nacional de Loja, 2013)
20. C.A. Jumbo, C.D. Arévalo, L. Ramirez Cando, Medición de carbono del estrato aebóreo del bosque natural Tinajillas-Limón Indanza, Ecuador (Revista de Ciencias de la Vida, La Granja, 2018)
21. IPCC, capitulo 3: Combustión móvil (2006)
22. J.A. Sagñay, Bases para inventario de emisiones del parque automotor en la ciudad de Guayaquil (Guayaquil, Guayas: Escuela Superior Politécnica del Litoral, 2016), pp. 15–23
23. S.N. Chiriboga, *Análisis de Emisiones de CO_2 Del Transporte En La Ciudad de Ibarra* (Universidad Técnica del Norte, Ibarra, 2018)
24. A.Y. Dominguez, Estimaciones de captura de los parques y emisiones de CO_2 vehicular en tijuana B.C. (El Colegio de la Frontera Norte, Tijuana, Baja California, 2016)
25. D. Minga, A. Verdugo, Árboles y arbustos de los ríos de Cuenca Azuay-Ecuador, Azuay: Serie Textos Apoyo a la Docencia Universidad del Azuay (Imprenta Don Bosco, 2016)
26. M.A. Chamorro, S. Falconi, Potencial de secuestro de carbono por los árboles en los parques urbanos de los Distritos de El Tambo, Huancayo y Chilca (Universidad Nacional del Centro de Perú, 2019)
27. K. Coquinche, Secuestro de CO_2 y producción de oxigeno en árboles urbanos de la Av. Abelardo Quiñones-Distrito San Juan Bautista (Universidad Nacional de la Amazonia Peruana, 2017)
28. R. Parra, Inventario de emisiones atmosfericas de la ciudad de cuenca año 2014 (Azuay, 2016)
29. D. Vega, L. Ocaña, R. Parra, Inventario de emisiones atmosféricas del tráfico vehicular en el Distrito Metropolitano de Quito. Año base 2012. Avances en Ciencias e Ingenierías **VII**(2), C86–C94 (2015)
30. A. Pareja, M. Hinojosa, M. Luján, Inventario de Emisiones Atmosféricas Contaminantes de la Ciudad de Cochabamba, Bolivia, año 2008. Acta Nova **V**(3), 344–373 (2012)
31. A. Dominguez, Estimaciones de captura de los parques y emisiones de CO_2 vehicular en Tijuana, B.C. (Tijuana, 2016)
32. GAD Municipal de Cuenca, Plan de desarrollo y ordenamiento territorial del cantón Cuenca-actualizacion 2015 (2015)

Chapter 38
Flow Maps in the Boiling Process in Domestic Refrigerators

Fernando Toapanta-Ramos, César Nieto-Londoño, Elizabeth Suquillo-Goméz, and William Quitiaquez

Abstract Forced flow convection boiling and flow patterns are analytically investigated in this study to determine the behaviour of R600a within domestic refrigeration systems when copper oxide nanoparticles are added or not to this refrigerant. The flow maps are analysed, for this, the mass flows and the heat flows are of great importance since according to the numerical value of each of these the region where the flow will be found during the boiling process is established. Flow maps are presented using established correlations for halogenated refrigerants or zeotropic mixtures in the forced convection boiling process. The results showed that the heat transfer coefficient increased in mass flow while it decreased with increasing quality. On the other hand, when metallic nanoparticles are used, the heat transfer increases while maintaining the mass flow. In the same way, R600a has greater heat transfer when adding nanoparticles and decreases when oil is added, this is noticeable in the interface between intermittent and annular flow. Therefore, adding nanoparticles makes nanorefrigerants more convective compared to base fluids. All this generates an improvement in domestic refrigeration, despite having a flammable refrigerant (R600a).

38.1 Introduction

There is an urgent need for green refrigerants in the refrigeration industry with increasing attention to environmental issues. As a natural refrigerant, R600a not only has zero ozone depletion potential and relatively low global warming potential, but it also has advantages over other refrigerants in terms of thermodynamic properties [1].

F. Toapanta-Ramos (✉) · E. Suquillo-Goméz · W. Quitiaquez
The Research Group on Renewable Energies and Mechanical Implementation of SMEs, Universidad Politécnica Salesiana, GIERIMP, Quito, Ecuador
e-mail: ltoapanta@ups.edu.ec

C. Nieto-Londoño
Universidad Pontificia Bolivariana, Medellín, Colombia

Á. Rocha et al. (eds.), *Communication, Smart Technologies and Innovation for Society*, Smart Innovation, Systems and Technologies 252,
https://doi.org/10.1007/978-981-16-4126-8_38

The heat transfer characteristics of R600a play an essential role in evaluating and optimizing the performance of refrigeration cycles [2]. Simultaneously, the flow pattern is expected to influence heat transfer and flow characteristics during the phase change the heat transfer process. Therefore, it is necessary to understand flow patterns and predicts the type of flow accurately. Therefore, research on two-phase flow patterns is essential. During the last decades, the study of the two-phase flow regime of different refrigerants was still attractive, and many flow pattern transitions have been developed to predict two-phase flow patterns in tubes [3].

The flow maps of a fluid circulating inside pipes and changes its phase from liquid to vapour have been studied for several decades. Throughout history, several horizontal pipe flow maps have been developed, among the most prominent are Taitel and Dukler [4], Kattan et al. [5, 6], Zürcher et al. [7], Wojtan et al. [8, 9], Moreno-Quibén and Thome [10, 11], Bandarra et al. [12].

Taitel and Dukler [4] in their investigation of flow patterns for gas and liquid, consider that, when performing the analysis of the transition of fluid in a liquid phase, there are five necessary flow regimes, smooth stratified (SS), wavy stratified (SW), intermittent (I) (slug and plug), annular with dispersed liquid (A-D) and dispersed bubbling (DB). Besides, they do not distinguish between other regimes such as slug, plug or elongated bubbles.

The study carried out by Hatamipour and Akhavan-Behabadi [13] to the flow patterns in forced boiling for smooth horizontal pipes found that there are three flow regimes: stratified-wavy, wavy annular and annular flow. Furthermore, they argue that the layered wavy flow pattern's characteristic is the complete gravitational separation of the liquid and vapour phases via a wavy interface. The liquid flows through the bottom and the vapour through the top throughout the entire tube. Due to the difference in speed between the two phases, the vapour stream disturbs the liquid–vapour interface and becomes wavy.

On the other hand, the flows with high vapour qualities, the shear at the interface of the high-speed vapour on the liquid film becomes dominant over gravity, so the liquid is expelled from the centre of the tube and flows as a thin layer in the wall of the tube. In contrast, the gas flows as a continuous phase to the centre of the tube; this flow is known as annular, and also, it is characterized by the complete separation of the two phases. The tube wall's entire circumference is continually moistened with a non-uniform liquid film; some small droplets may be trapped in the core of the vapour, and this flow regime is incredibly stable [14].

Wojtan et al. [9] interpret that after performing an iterative method to the geometric variables to the analysis tubing, and the transition curves for the flow maps and the regimes were established in the following: the transition between stratified and wavy stratified flow designate it as (S-SW), between stratified-wavy and intermittent/annular flow as (SW-I/A), between intermittent and annular flow as (I-A), between annular and dew flow (AM).

The research carried out by Yang et al. [1] presents various flow patterns for R600a with different heat fluxes at a pressure of 0.215 MPa in the format: mass flow vs steam quality. Besides, they note four main flow patterns: plug, slug, stratified-wavy and annular flow.

In the same way, they show that the beginning of the annular flow regime occurs early when the mass flow increases; with a high mass flow, the small bubbles quickly join in elongated bubbles that lead to annular flow. Likewise, the intermittent flow regime takes place in a narrow range of steam quality, and the annular regime occupies a wide range in steam quality [1].

This study aims to identify, establish and develop flow maps for the R600a refrigerant, when it is pure and mixed with oil or nanoparticles, with variable heat and mass fluxes. Also, the conditions that the intersection boundaries must meet for each regime are placed within the flow map.

38.2 Methods

The refrigeration system in domestic refrigerators consists mainly of long cylindrical pipes in what are called condensers and evaporators. This research work has a cylindrical copper tube with a 7 mm internal diameter and a length of 1.2 m, which is made to flow R600a refrigerant with nanoparticles and oil. The mass flows vary between 100 and 400 kg/m^2 s, while the heat flow changes from 10 to 20 kW/m^2.

The dimensionless geometric variables established by Taitel–Dukler are taken to elaborate the flow maps of any fluid. The same maps were modified in the work of Kattan et al. [5]. The latter includes a method to predict the onset of drying at the top of the tube for non-adiabatic annular flow.

On the other hand, the boiling transfer model of Kattan et al. [6] using Steiner's version [15] model is derived from Rouhani and Axelsson [16] for horizontal tubes in an empty fraction of cross section ε.

$$\epsilon = \frac{x}{\rho_V}\left[(1+0.12(1-x))\left(\frac{x}{\rho_V}+\frac{1-x}{\rho_L}\right)+\frac{1.18(1-x)[g\sigma(\rho_L-\rho_V)]^{0.25}}{G\rho_L^{0.5}}\right]^{-1} \tag{38.1}$$

The vacuum fraction model is easy to use and provides an explicit function of the total mass flow, and Taitel–Dukler does not consider this. Therefore, it is feasible to use the same void fraction model in the flow map. So, in the surface section of tube A, the values of A_{LD} and A_{VD} are directly determinable.

$$A_{LD} = \frac{A(1-\epsilon)}{D^2} \tag{38.2}$$

$$A_{VD} = \frac{A\epsilon}{D^2} \tag{38.3}$$

To avoid any iteration method, the geometric expression for the stratified angle θ_{strat} can be calculated from an approximate expression, evaluated in terms of empty fraction, by Biberg [17] as follows:

$$\theta_{\text{strat}} = 2\pi - 2\left\{(1-\epsilon) + \left(\frac{3\pi}{2}\right)^{1/3}\left[1 - 2(1-\epsilon) + (1-\epsilon)^{\frac{1}{3}} - \epsilon^{\frac{1}{3}}\right]\right.$$
$$\left. - \frac{1}{200}(1-\epsilon)\epsilon[1 - 2(1-\epsilon)]\left[1 + 4\left((1-\epsilon)^2 + \epsilon^2\right)\right]\right\} \quad (38.4)$$

The transition curves between dry annular flow (A-D) and dry mist flow (D-M) are displayed below:

$$G_{\text{dryout}} = \left[\frac{1}{0.235}\left(\ln\left(\frac{0.58}{x}\right) + 0.52\right)\left(\frac{D}{\rho_V \sigma}\right)^{-0.17}\right.$$
$$\left. \times \left(\frac{1}{g D \rho_V (\rho_L - \rho_V)}\right)^{-0.37}\left(\frac{\rho_V}{\rho_L}\right)^{-0.25} \cdot \left(\frac{q}{q_{\text{crit}}}\right)^{-0.70}\right]^{0.926} \quad (38.5)$$

$$G_{\text{mist}} = \left[\frac{1}{0.0058}\left(\ln\left(\frac{0.61}{x}\right) + 0.57\right)\left(\frac{D}{\rho_V \sigma}\right)^{-0.38}\right.$$
$$\left. \times \left(\frac{1}{g D \rho_V (\rho_L - \rho_V)}\right)^{-0.15}\left(\frac{\rho_V}{\rho_L}\right)^{0.09} \cdot \left(\frac{q}{q_{\text{crit}}}\right)^{-0.27}\right]^{0.943} \quad (38.6)$$

wherein:

$$q_{\text{crit}} = 0.131 \rho_V^{0.5} h_{LV} (g(\rho_L - \rho_V)\sigma)^{0.25} \quad (38.7)$$

The SW-I/A flows' transition curve is determined by the relationships established by Kattan et al.

$$G_{\text{wavy}} = \left\{\frac{16 A_{VD}^3 g D \rho_L \rho_V}{x^2 \pi^2 \left(1 - (2h_{LD} - 1)^2\right)^{0.5}} \cdot \left[\frac{\pi^2}{25 h_{LD}^2}\left(\frac{We}{Fr}\right)_L^{-1} + 1\right]\right\}^{0.5} + 50 \quad (38.8)$$

wherein:

$$Fr_L = \left[\frac{G^2}{\rho_L^2 g D}\right] \quad (38.9)$$

$$Fr_V = \left[\frac{G^2}{\rho_V^2 g D}\right] \quad (38.10)$$

$$We_L = \left[\frac{G^2 D}{\rho_L \sigma}\right] \quad (38.11)$$

$$We_L = \left[\frac{G^2 D}{\rho_L \sigma}\right] \tag{38.12}$$

The transition between flows (S-SW) and flows (I-A) in the same way as the previous one is calculated using the original work of Kattan et al.

$$G_{\text{strat}} = \left\{\frac{226.3^2 A_{LD} A_{VD}^2 \rho_V (\rho_L - \rho_V) \mu_L g}{x^2 (1-x) \pi^3}\right\}^{1/3} \tag{38.13}$$

$$x_{IA} = \left\{\left[0.34^{1/0.875}\left(\frac{\rho_V}{\rho_L}\right)^{-1/1.75}\left(\frac{\mu_L}{\mu_V}\right)^{-1/7}\right] + 1\right\}^{-1} \tag{38.14}$$

The limits for each of the flow regimes are established as follows:

- The limits in the transition of flows (A-D) are calculated using Eq. 38.5.
- The boundary for the transition flow (D-M) is determined from Eq. 38.6.
- The stratified-wavy region is subdivided into three zones:
  $G > G_{wavy}(x_{IA})$ gives a slug zone. $G_{strat} < G < G_{wavy}(x_{IA})$ and $x < x_{IA}$ gives a slug / stratified-wavy zone. $G > G_{strat}$ and $x \geq x_{IA}$ gives a stratified-wavy zone.
- The transition from annular to intermittent flow occurs when: $G > G_{wavy}$ and $x \geq x_{IA}$.
- Intermittent flow occurs when: $G > G_{wavy}$ and $x < x_{IA}$.

38.3 Results and Discussion

38.3.1 Flow Maps for R600a

In this section, reference is made to the flow maps for the R600a refrigerant, which are determined by the equations written in this study that are taken from the work done by Wojtan et al. To check the veracity of the flow maps, they are compared with the work of Hu et al. [18] studying the flow maps for the refrigerant R410A in its pure state and R410A with lubricating oil. Figure 1a shows the flow map of Hu et al. Besides, each zone of flow regime for the mentioned refrigerant is appreciated. Figure 1b indicates how the flow map varies when the refrigerant is mixed with 5% oil.

After confirming the established correlations, for flow maps in refrigerants and refrigerant-lubricating oil mixtures, Wojtan et al. and Hu et al., respectively. It has been noted that the vapour quality at the border between intermittent and annular flow for mixing with oil is higher by 2.78% concerning the refrigerant alone. In other words, adding oil slows down the phase change process in the refrigerants.

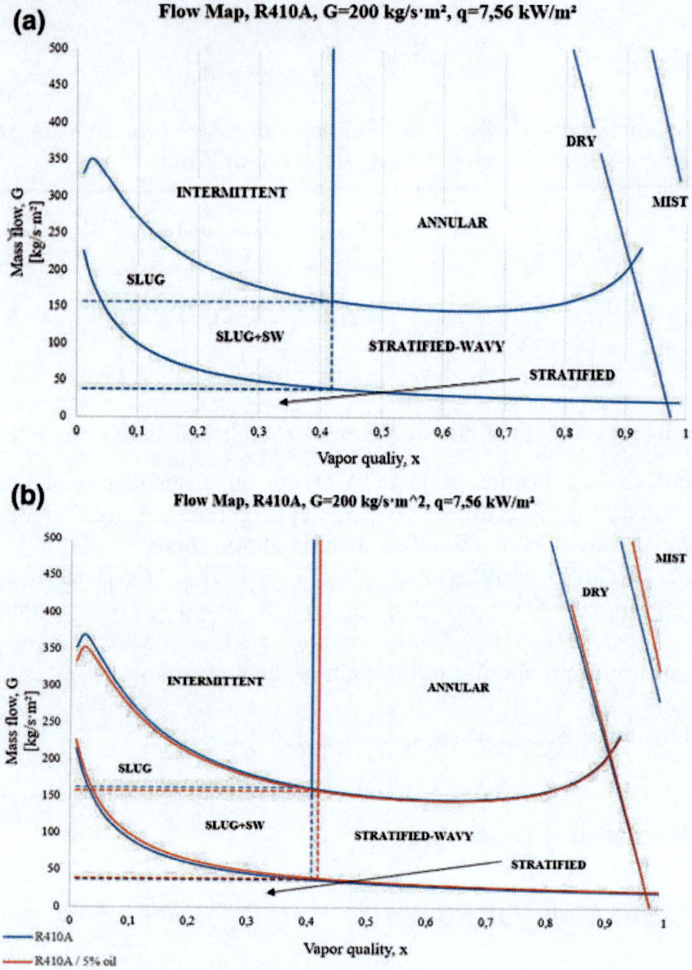

Fig. 38.1 Flow map for R410A, **a** refrigerant pure, **b** refrigerant with 5% oil [18]

Next, the flow maps for R600a, the base fluid of this investigation, are presented; likewise, flow maps are placed with modification of the mass flow or heat flow.

Figure 38.2 corresponds to the flow regime map for R600a, at a saturation temperature of 8 °C and an internal tube diameter of 7 mm, with heat and mass flows of 20 kW/m^2 and 100 kg/s·m^2, respectively. This map shows that steam quality at the border (I/A) is 0.2721, which varies drastically from the previous maps since the others are with R410A.

Furthermore, the stratified and intermittent flow regions are smaller than with a zeotropic mixture (R410A), the annular region is also shortened, and the drying and haze zones are wider for R600a.

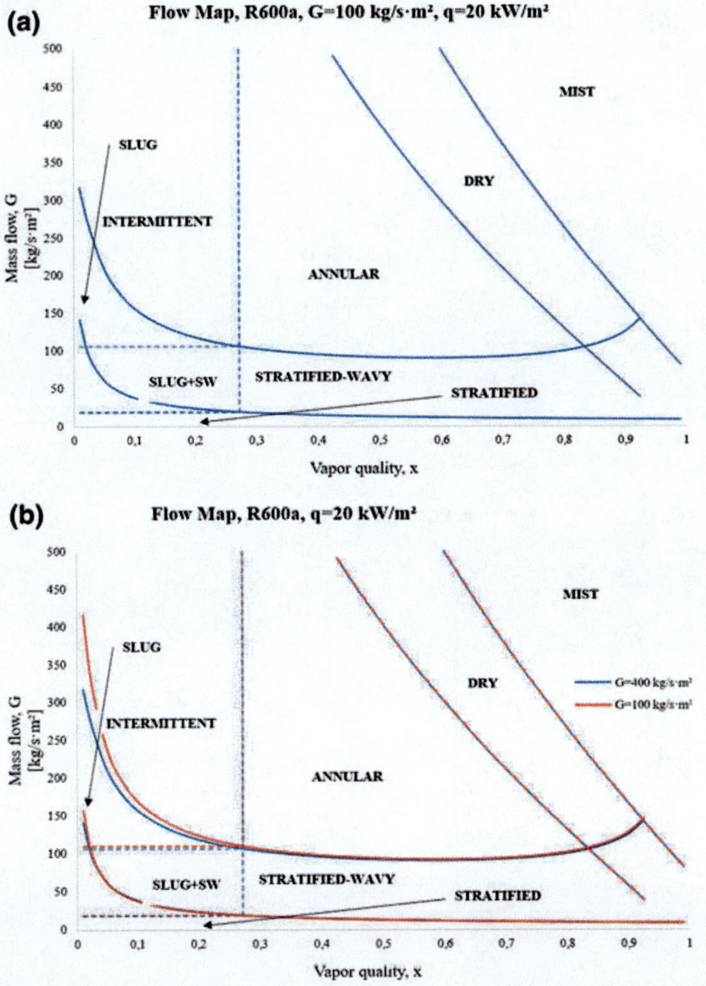

Fig. 38.2 Flow map for R600a refrigerant, $q = 20$ kW/m^2, **a** G $=$ 100 kg/s m^2, **b** $G = 100$–400 kg/s m^2

In Fig. 2b is observed that, as the mass flow rate increases, from 100 to 400 kg/s·m^2, the flow map changes in the intermittence, wavy stratified, slug and stratified sections. However, not variation in steam quality is reported for intermittent flow to the annular interface and the drying and mist zones.

Several modifications can be made to the process variables. Among them is the heat flow that, for the two previous figures, remained constant. In Fig. 3a, b, the flow map is displayed for a heat flow that is reduced to 10 and 6 kW/m^2, respectively, there the transitions for intermittent, annular, wavy stratified and stratified flow remain constant. However, there are appreciable changes in the dry and mist flow regimes, and these are reduced when the heat flow also decreases.

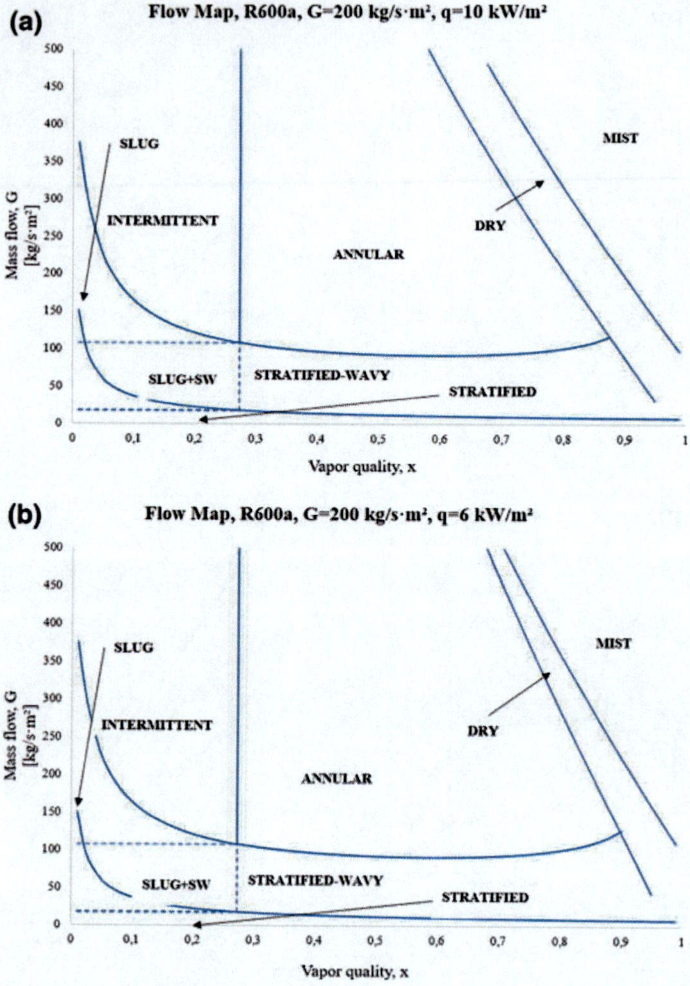

Fig. 38.3 Flow map for R600a refrigerant, $G = 200$ kg/s·m^2 **a** $q = 10$ kW/m^2, **b** $q = 6$ kW/m^2

38.3.2 Flow Maps for R600a with Oil

As shown in Fig. 1b, the flow maps can be made for a mixture between refrigerant and oil, in this case, miscible lubricating oil PAO; in the flow maps, it can see the flow regimes, and which are the variants that have relative to that of refrigerant pure.

The refrigerant R600a will be combined with the maximum amount of oil allowed to observe extensively, which is 5%. In Fig. 4a, we have R600a with a saturation temperature of 8° C flowing through a 7 mm internal diameter pipe. The mass flow is 200 kg/s m^2, for a heat flow of 20 kW/m^2 and the amount of oil added is 5%.

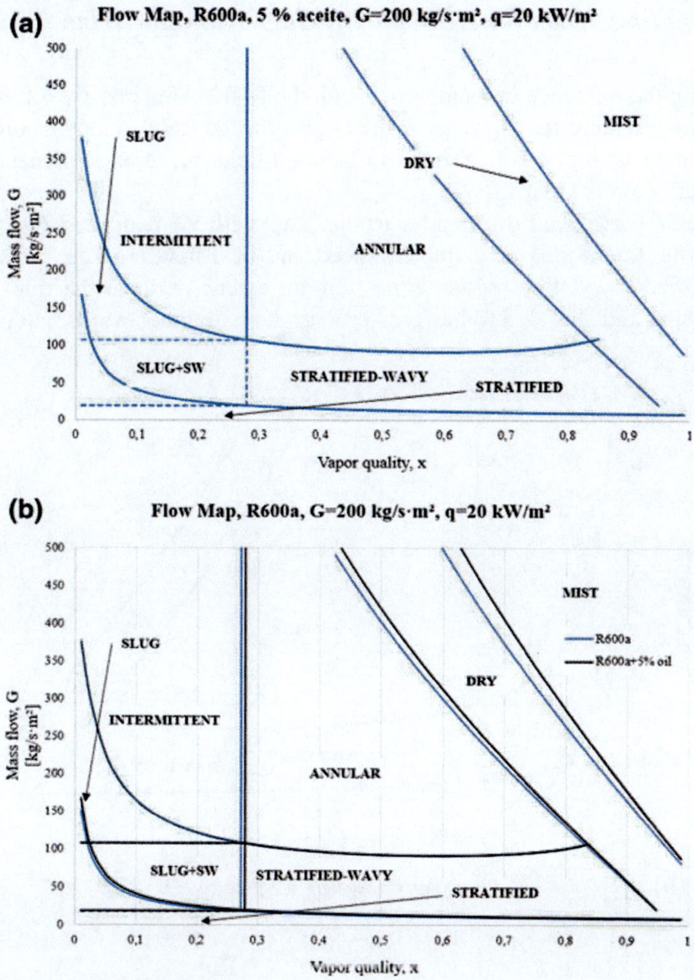

Fig. 38.4 Flow map for R600a refrigerant, $G = 200$ kg/s·m^2, $q = 20$ kW/m^2, **a** 5% oil, **b** R600a and R600a refrigerant with 5% oil

In Fig. 4b, the comparison of the flow maps for pure R600a and R600a + PAO is performed under the same applicability conditions. In this figure, the quality of steam is higher in the flow interface (I/A); reaching the drying regime is more time consuming than for the pure refrigerant, in the same way as for the zone of fog. Adding oil hinders the phase change in the fluid.

38.3.3 Flow Maps for R600a with CuO Nanoparticles

Identifying the influence of nanoparticles on the R600a flow maps is a fundamental part of this research; for this reason, flow maps for the R600a/CuO nanorefrigerant are presented below to observe the influence of the percentage of nanoparticles concentration on the maps.

Figure 5a shows the flow map for R600a/CuO with 5% nanoparticles in concentration. The dimensions of diameter, speed and heat flow are kept constant. The slug/stratified-wavy flow regime zone is more extensive since the transition line between slug and slug + SW has greater mass flow. In other words, adding a solid

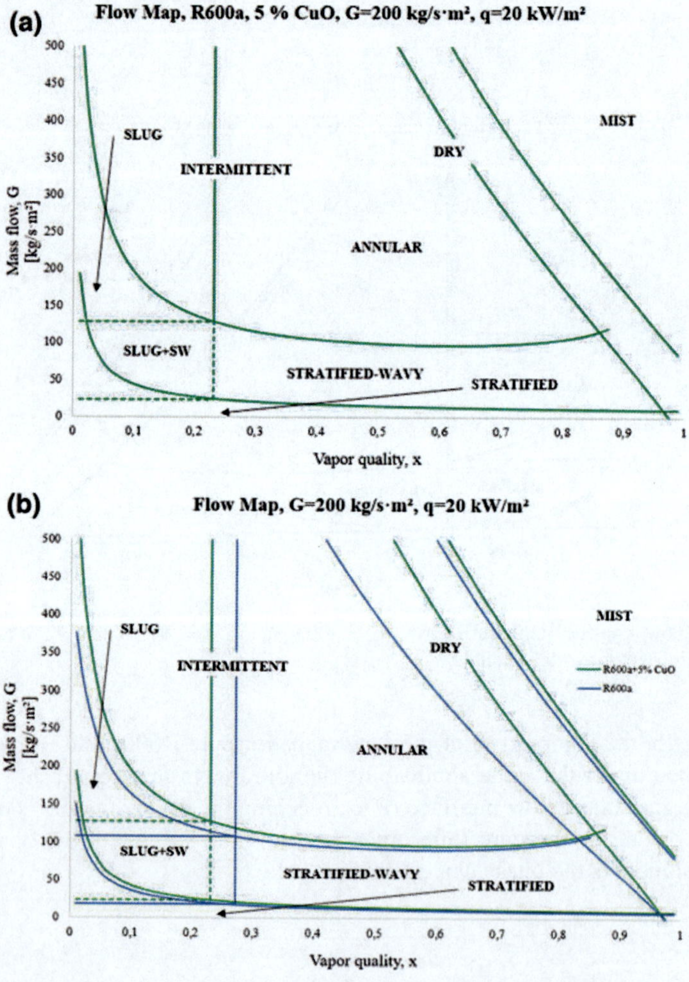

Fig. 38.5 Flow map for: **a** R600a/CuO nanorefrigerant, **b** R600a and R600a/5% CuO

substance to a moving fluid causes the flow to ripple for a broader range of mass flows.

On the other hand, the comparison between pure R600a and R600a/CuO at 5% is observed in Fig. 5b. The interface between the intermittent flow zones is annulled, for the nanorefrigerant is produced at a lower vapour quality (0.2318) compared to pure R600a (0.2721). This causes the intersection between these zones to be 14.81% by adding particles to the fluid, due to the heat transfer is more significant when you have a solid inside the flow. Likewise, the interface between annular and dry flow decreases as each flow regime's mass flow increases.

38.3.4 Flow Maps for R600a with CuO Nanoparticles and PAO Oil

Finally, the flow maps for the mixture of refrigerant, nanoparticles and PAO lubricating oil are prepared. The purpose of these is to know the zones and regimes that the R600a refrigerant has when it is mixed with oil and nanoparticles and determine how the borderlines move for each transition. Similarly, as the previous flow maps were constructed, the Wojtan et al. Correlations, the study variables and the properties established in the last chapter are taken.

Figure 6a is the flow map for the R600a/5% PAO/5% CuO mixture; this is like that of pure R600a for all its regimes and transitions.

Figure 6b represents the three case studies' flow maps, R600a in its pure state, R600a with copper oxide nanoparticles and R600a with lubricating oil. It can be noted that the boundary line for the intersection of the intermittent flow towards the annular has a horizontal transfer; when the fluid is mixed with nanoparticles, the transition occurs faster, if this is related to the quality of pure refrigerant vapour transition increases 14.79%. There is also a significant change in the transition from annular flow to dry flow. This is slower than the base refrigerant, producing an annular flow in a considerable section of the tube.

In the same way, the refrigerant mixed with 5% PAO oil is displayed, in this case, there is a movement of the transition line between intermittent and annular; however, it is directed to the right, delaying the process of change of phase, despite mixing a significant amount of oil, the quality only varies by 2.72%. Likewise, it is observed that the transitions of all the flows are like the base refrigerant.

38.4 Conclusions

At the end of this study, the following conclusions are obtained:

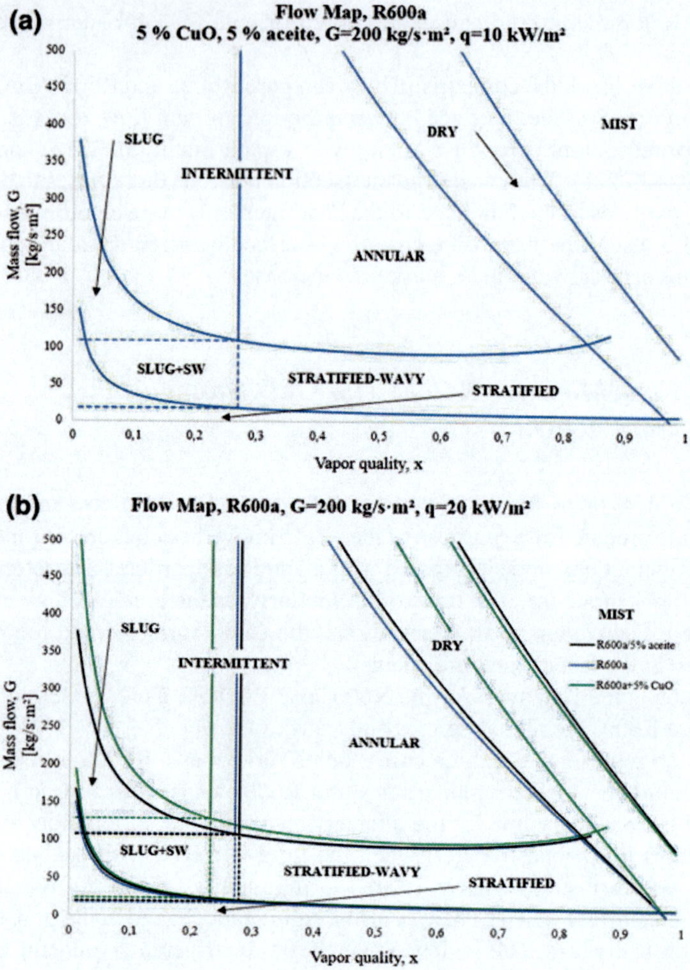

Fig. 38.6 Flow map for. **a** R600a/5% CuO/5% oil, **b** R600a, R600a/5% CuO and R600a/5% PAO

1 With the correlations cast in this study, the flow maps for the natural refrigerant R600a were elaborated. Before this, the results were verified with previous works presented in high-impact scientific articles.

2 When the refrigerant is mixed with nanoparticles and/or miscible oil, the thermal transport properties that were already set in the previous chapter were used to establish the flow maps.

3 It was determined that when adding oil to the refrigerant, the transition between the flow regimes and the quality of vapour is higher, while adding nanoparticles this quality is lower. Thus, it is inferred that its heat transfer increases and that oil decreases.

4 Finally, it is recommended to carry out an experimental study with the use of R600a/CuO nanorefrigerant since this research is analytically focussed on the boiling process; therefore, it is essential to compare these results with experimental data.

Acknowledgements The authors of this research thank the mechanical engineering career of the Salesian Polytechnic University, the Research Group on Renewable Energies and Mechanical Implementation of SMEs GIERIMP, the Branch ASHRAE UPS-QUITO and the Universidad Pontificia Bolivariana, Medellín-Colombia, for their generous contributions in this study.

References

1. Z. Yang, M. Gong, G. Chen, X. Zou, J. Shen, Two-phase flow patterns, heat transfer and pressure drop characteristics of R600a during flow boiling inside a horizontal tube. Appl. Therm. Eng. **120**, 654–671 (2017). https://doi.org/10.1016/j.applthermaleng.2017.03.124
2. X.R. Zhuang, M.Q. Gong, S. Wang, X. Zou, G.F. Chen, J.F. Wu, Two phase flow pattern for propane in a horizontal smooth tube. Phys. Procedia **67**, 661–666 (2015). https://doi.org/10.1016/j.phpro.2015.06.112
3. Z.Q. Yang, M. Gong, G. Chen, Z. Lin, H. Huang, H. Feng, A new diabatic two phase flow pattern transition model of R600a. Int. J. Refrig. **99**, 138–144 (2019). https://doi.org/10.1016/j.ijrefrig.2018.12.025
4. Y. Taitel, A.E. Dukler, A model for predicting flow regime transitions in horizontal and near horizontal gas-liquid flow. AIChE J. **22**(1), 47–55 (1976). https://doi.org/10.1002/aic.690220105
5. N. Kattan, J.R. Thome, D. Favrat, Flow boiling in horizontal tubes: Part 1-development of a diabatic two-phase flow pattern map. J. Heat Transfer **120**(1), 140–147 (1998). https://doi.org/10.1115/1.2830037
6. N. Kattan, J.R. Thome, D. Favrat, Flow boiling in horizontal tubes: part 3-development of a new heat transfer model based on flow pattern. J. Heat Transfer **120**(1), 156–165 (1998). https://doi.org/10.1115/1.2830039
7. O. Zürcher, D. Favrat, J.R. Thome, Development of a diabatic two-phase flow pattern map for horizontal flow boiling. Int. J. Heat Mass Transf. **45**(2), 291–301 (2001). https://doi.org/10.1016/S0017-9310(01)00146-6
8. L. Wojtan, T. Ursenbacher, J.R. Thome, Investigation of flow boiling in horizontal tubes: Part II - Development of a new heat transfer model for stratified-wavy, dryout and mist flow regimes. Int. J. Heat Mass Transf. **48**(14), 2970–2985 (2005). https://doi.org/10.1016/j.ijheatmasstransfer.2004.12.013
9. L. Wojtan, T. Ursenbacher, J.R. Thome, Investigation of flow boiling in horizontal tubes: part I—a new diabatic two-phase flow pattern map. Int. J. Heat Mass Transf. **48**(14), 2955–2969 (2005). https://doi.org/10.1016/j.ijheatmasstransfer.2004.12.012
10. J. Moreno Quibén, J.R. Thome, Flow pattern based two-phase frictional pressure drop model for horizontal tubes, Part II: New phenomenological model. Int. J. Heat Fluid Flow **28**(5), 1060–1072 (2007). https://doi.org/10.1016/j.ijheatfluidflow.2007.01.004
11. J. Moreno Quibén, J.R. Thome, Flow pattern based two-phase frictional pressure drop model for horizontal tubes. Part I: diabatic and adiabatic experimental study. Int. J. Heat Fluid Flow **28**(5), 1049–1059 (2007). https://doi.org/10.1016/j.ijheatfluidflow.2007.01.003
12. E.P. Bandarra Filho, L. Cheng, J.R. Thome, Flow boiling characteristics and flow pattern visualization of refrigerant/lubricant oil mixtures. Int. J. Refrig. **32**(2), 185–202 (2009). https://doi.org/10.1016/j.ijrefrig.2008.06.013

13. V.D. Hatamipour, M.A. Akhavan-Behabadi, Visual study on flow patterns and heat transfer during convective boiling inside horizontal smooth and microfin tubes. World Acad. Sci. Eng. Technol. **45**(9), 701–707 (2010)
14. J.R. Thome, *Engineering Data Book III* (Wolverine Tube, Inc., 2006), pp. 1–34
15. D. Steiner, *Heat Transfer to Boiling Saturated Liquids* (VDI-Warmeatlas, VDI Heat Atlas, 1993)
16. S.Z. Rouhani, E. Axelsson, Calculation of void volume fraction in the subcooled and quality boiling regions. Int. J. Heat Mass Transf. **13**(2), 383–393 (1970). https://doi.org/10.1016/0017-9310(70)90114-6
17. D. Biberg, An explicit approximation for the wetted angle in two-phase stratified pipe flow. Can. J. Chem. Eng. **77**(6), 1221–1224 (1999). https://doi.org/10.1002/cjce.5450770619
18. H. Hu, G. Ding, W. Wei, Z. Wang, K. Wang, Heat transfer characteristics of R410A-oil mixture flow boiling inside a 7 mm straight smooth tube. Exp. Therm. Fluid Sci. **32**(3), 857–869 (2008). https://doi.org/10.1016/j.expthermflusci.2007.10.003

Chapter 39
Design of a Photovoltaic System for Self-consumption in Buildings at High-Altitude Cities Located in the Equator

Marlon Quinde-Abril, John Calle-Siguencia, and Julio Amador Guerra

Abstract In recent years, many countries have opted for the use of non-conventional and environmentally friendly renewable energies as an alternative to gradually reduce the use of conventional energies, specifically those coming from fossil fuels. In Ecuador, a country located at the Equator, where the solar radiation is perpendicular throughout the year, an alternative to be considered is obtaining electrical energy using photovoltaic solar panels, especially at high-altitude cities such as Cuenca. This research presents the design of a photovoltaic installation in a building of the Universidad Politécnica Salesiana, at Cuenca, Ecuador, based on a consumption profile corresponding to a typical day, it is analyzed the architectural condition of building, identifying in the rooftop the elements that can produce shadows, to determine the minimum separation of the photovoltaic modules from these obstacles and define an effective surface for the installation, such that at midday the installation is free of shadows. The performance ratio (PR) obtained for the installation was 83.4%, which is considered satisfactory from the point of view of energy and based on the building demand conditions. The contribution of the facility was 36.35%, significantly reducing the dependency from the network.

39.1 Introduction

In Ecuador, in recent years policies have allowed the change of the energy matrix, with the main objective of gradually reducing the use of fossil fuels over time and promoting the energy development of non-conventional renewable resources and

M. Quinde-Abril (✉)
Ingeniería Mecánica, Universidad Politécnica Salesiana, Cuenca, Ecuador
e-mail: mquinde@ups.edu.ec

J. Calle-Siguencia
Ingeniería Mecatrónica, Universidad Politécnica Salesiana, Cuenca, Ecuador

J. Amador Guerra
Ingeniería Eléctrica, Universidad Politécnica de Madrid, Madrid, Spain

Á. Rocha et al. (eds.), *Communication, Smart Technologies and Innovation for Society*, Smart Innovation, Systems and Technologies 252,
https://doi.org/10.1007/978-981-16-4126-8_39

friendly with the environment [1]. However, contributing to reduce the environmental pollution is not achieved by only changing the energy sources, as indicated in [2], the energy consumption increases every day because there are parameters such as population growth that increase the demand, the development experienced by the electrical energy has enabled supplying approximately 65% of the worldwide energy production; the most commonly used renewable energies are: solar, wind, hydraulic, hydrogen, and fuel cells [2].

Solar energy, which is the object of study in this research work, corresponds to the largest energy source existing in nature and above all renewable; it comes from the reactions occurring in the hydrogen and helium nucleus within the sun [3]. Ecuador, since it is located at the Equator, receives throughout the year a solar incidence almost perpendicular to the surface, the angle of incidence does not vary as in other latitudes according to the season of the year, which favors that the reception of solar energy is larger and, above all, constant [4].

The solar energy used in photovoltaic installation has a high reliability, a low maintenance, and its relatively low operational cost compensates a high cost of implementation, and its use enables a sustainable economic growth, independent of the price of fossil fuels [2, 3].

39.2 Methods

This research is quantitative non-experimental since specific data will be used for its design and simulation, various methodical activities are integrated in its development to obtain required information about the topic and the problem to be solved.

39.2.1 Definition of the Location

The building taken as model for this study is the CIMA, Automotive Mechanical Engineering Degree located at the Universidad Politécnica Salesiana, at Cuenca, Ecuador, it is analyzed the orientation, the consumption profile for a typical day, the incident solar radiation and the solar path in the study area shown in Fig. 39.1. Table 39.1 shows the geographic location and the main meteorological data [5].

39.2.2 Data of the Building and of the Surroundings

The CIMA building has 4 floors, 21 classrooms each with a capacity of 35 students, 1 elevator, 1 projector per classroom, and 9 restrooms distributed in the first, second a third floor; each floor has 1 restroom for men, for ladies and for people with

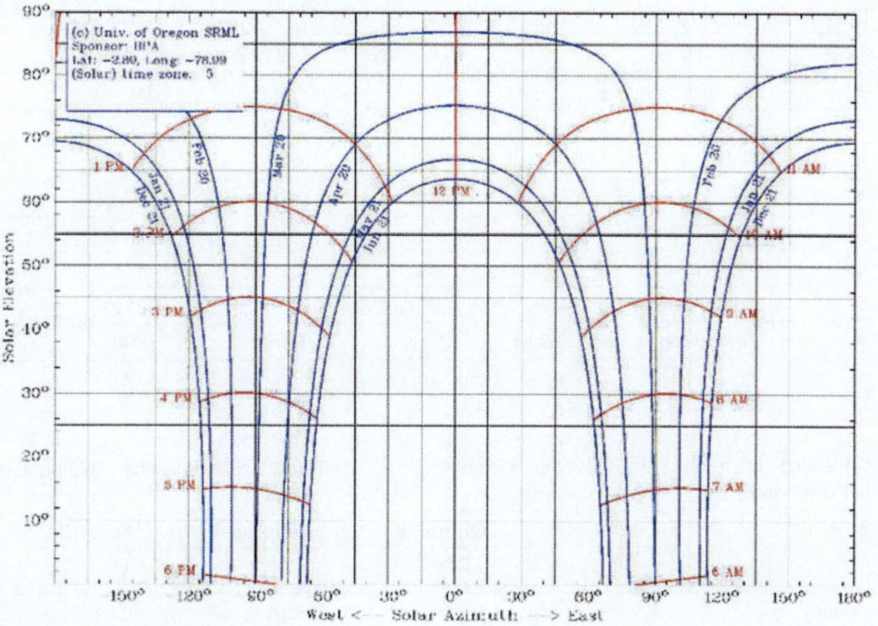

Fig. 39.1 Solar path in the study area

Table 39.1 Geographic location and main meteorological data of the CIMA building

Parameter	Description
City	Cuenca–Azuay–Ecuador
Latitude	2.886176 S
Longitude	78.988730 O
Altitude	2529 masl
Average sunshine hours	12 h
Global Radiation	4548 Wh m^{-2} day^{-1}
Average ambient temperature	16 °C
Maximum ambient temperature	30 °C
Minimum ambient temperature	3 °C
Orientation (Azimuth)	50°
The meteorological data were measured	

disabilities; Table 39.2 shows the load installed in the building, and Table 39.3 shows the daily average values measured.

Table 39.2 Load installed in the CIMA building

Item	Description	Quantity	Load (W)	Subtotal (W)
1	External lightning	6	750	4500
2	Elevator	1	7500	7500
3	Projector	21	250	5250
4	Recessed light 3 × 17	648	24	15,552
5	Hand dryer	9	1500	13,500
6	WIFI router	6	8	48
7	Laptop computers	50	65	3250
8	Other (mobile, tablets, etc.)	2	1000	2000
			Total W	51,600

Table 39.3 Daily average values of the variables: global and diffuse radiation, ambient temperature and wind speed

Month	Global_R	Diffuse_R	Ambient_T	Wind_S
	$\frac{\text{Wh}}{\text{m}^2\ \text{day}}$	$\frac{\text{Wh}}{\text{m}^2\ \text{day}}$	°C	$\frac{\text{m}}{\text{s}}$
January	4326	1082	15.61	2.03
February	5429	1303	16.29	2.22
March	4391	1142	15.78	1.76
April	4344	1303	15.63	1.92
May	4490	1115	15.70	1.95
June	3825	803	14.46	2.22
July	4131	943	14.67	2.07
August	3975	835	14.77	2.26
September	4980	1245	15.22	2.25
October	4290	1073	15.66	1.97
November	5206	1302	16.25	1.89
December	5190	1298	16.25	2.32
Yearly average	4548	1120	16	2

39.2.3 *Analysis of the Available Surface*

The analysis building shown in Fig. 39.2 and has a rooftop (1) in which there is a meteorological station (2), a structure that supports the elevator (3), a structure for accessing the rooftop (4), and a sloping roof (5). Table 39.4 shows the available surface, taking into account the flat roof and the sloping roof shown in Fig. 39.2.

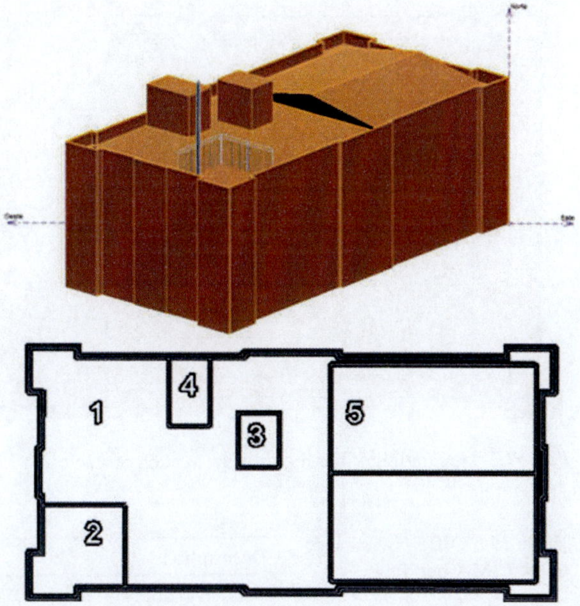

Fig. 39.2 3D representation of the CIMA building and plant view of the flat roof

Table 39.4 Surface available atop the terrace of the CIMA building

Item	Description	Surface m^2
1	Rooftop	632.22
2	Meteorological station	– 40.47
3	Structure that supports the elevator	– 12.09
4	Access structure	– 15.12
5	Sloping roof	– 272.54
	Top of the sloping roof	+ 243.6
	Total surface available	535.6

39.2.4 Analysis of Shadows

To determine the effective surface, an analysis of the shadows that the existing elements can cause is carried out, the spring equinox (September 22nd) and autumn (March 21st) and the summer solstice (December 21st) and winter (June 21st) are identified; due to the fact that the building is located toward the southern hemisphere, the analysis was carried out in the effective hours of sun from 10:00 to 14:00. Figure 39.3 shows the areas near each obstacle that would affect the photovoltaic installation. Table 39.5 shows the total effective area available for the photovoltaic installation, taking into account that it is guaranteed on the most unfavorable day of the year (June 21) at least 4 h of solar collection without shadows, while for the other

Fig. 39.3 Separation of the modules from each obstacle

Table 39.5 Effective surface of the CIMA building

Description	Surface m^2
Total available surface	535.6
Surface of separation of obstacles	216.43
Total effective surface	319.17

days that are indicated for example December 21, get 5 to 6 continuous hours of free solar collection without shadows.

39.2.5 Daily Consumption Profile

The daily consumption profile of the building is shown in Fig. 39.4, and this building works from September to February and from April to August, considering only the weekly working days from Monday to Friday [6].

39.3 Results

39.3.1 Dimensioning of the Photovoltaic Facility

Analysis of the optimal orientation and inclination. The dimensioning and simulation of the installation are carried out in the PVsyst software; since the building is located at a latitude very close to the Equator (2.89 S), the orientation will be to the

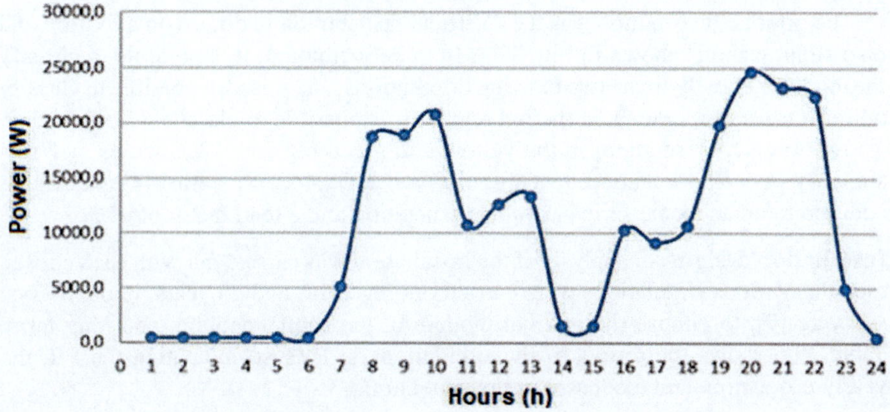

Fig. 39.4 Consumption profile of a typical day in the CIMA building

north while the inclination of the modules will be between 0° and 10°, considering the latter as the optimal inclination because for inclinations larger than 10° there are optical losses [7], and in addition, it will be avoided the accumulation of existing dust.

Since there is a flat roof and a sloping roof in the terrace of the CIMA building, it is considered an arrangement in branches and "dome" shaped taking as the symmetry plane one of the sides of the modules.

Distribution of the photovoltaic modules. With the definition of the effective surface, the best distribution of modules is analyzed and the best performance of the installation is obtained [8]; the polycrystalline modules of 60 cells of the brands Suntech (16.3%), Yingli Solar (18%), REC (16.1%) and Solar World (15.51%) of 260 Wp, with similar dimensions, were analyzed. For the series and parallel distribution, the number of modules per row (1) and the number of rows (2) that can be distributed on the effective surface. Considering length (l_m) and width (w_m) of the modules and length (L_r) and width (W_r) of the rooftop.

$$N_r = \frac{L_r}{w_m} \tag{39.1}$$

$$N_m = \frac{W_r}{l_m} \tag{39.2}$$

The analyzed distributions were:

- Vertical distribution in branches on rooftop and in dome on sloping roof.
- Horizontal distribution in branches on rooftop and in dome on sloping roof.
- Vertical distribution in domes on rooftop and on sloping roof.
- Horizontal distribution in domes on rooftop and on sloping roof.

The optimal distribution was the "Vertical distribution in domes on a rooftop and on a sloping roof," shown in Fig. 39.5. In this distribution, it is possible to classify the modules equally in the two existing orientations, and it is also possible to classify uniform rows of modules, in the red region, it is possible to classify 12 modules in 4 rows in each orientation, in the yellow and green regions, 12 modules in 1 row similarly in each orientation, reaching a power of 37,440 Wp with 144 modules, in addition to an adequate series–parallel association and a load factor of 75%.

Installation design. An analysis of the installation was carried out with each chosen module. Various simulations were carried out for 1, 10, and 25 years, in the PVsyst software [9], to choose the optimal module in the short, medium, and long term. Table 39.6 shows the results of the simulations in PVsyst, indicating the PR, the yearly production and the losses in the generator.

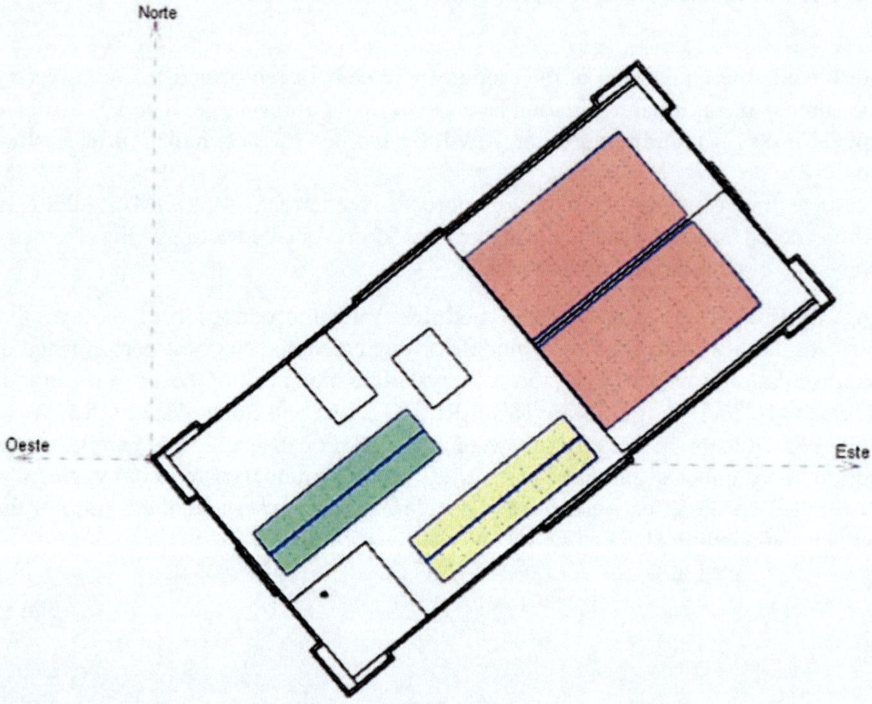

Fig. 39.5 Optimal dome shaped arrangement of modules

Table 39.6 Results obtained

Modulus	PR			Yearly production (kWh / year)			Generator losses (kWh/kWp/year)		
	1 year	10 years	25 years	1 year	10 years	25 years	1 year	10 years	25 years
Suntech	82.5	78.2	70.3	51,790	49,069	44,142	0.70	0.90	1.27
Yingli solar	83.4	79.2	70.9	52,353	49,699	44,517	0.66	0.86	1.24
REC	82.7	78.7	70.8	51,895	49,423	44,441	0.69	0.88	1.24
Solar world	82.3	78.3	70.5	51,657	49,144	44,232	0.71	78.3	1.26

39.4 Discussion

With the results obtained, the module that provides the best performance to the installation is the Yingli Solar module. Table 39.7 shows the summary of the results obtained with this module.

When analyzing the results, it is mainly observed a performance ratio (PR) of 83.4%, considering the performance of the installation as very good, since due to the losses and factors that affect the facility, the average PR is at 80% [10].

In addition, it may be seen that the energy generated in the facility during one year was 52,353 kWh, this energy corresponds to 36.34% of the building demand, 22,335 kWh (42.66%) of this supply of solar energy is consumed by the building, while 30,018 kWh (57.34%) is injected to the electric network; this is because there is a greater energy consumption in the building at evening hours.

Table 39.7 Summary of the optimal system

Summary of results	
Module	Yingli solar 260 Wp
Inverter	SMA Sunny Boy 9000 W
Arrangement of modules	Vertical dome shaped arrangement
Number of modules	144
Number of inverters	4
Nominal power	37,440 Wp
Power of inverters	36,000 W
Yearly production	52,353 kWh/year
PR (1 year)	83.4
Generator losses	0.66 kWh/kWp/year

39.5 Conclusions

This research has analyzed four brands of photovoltaic modules of similar features, both physical and technical, with the purpose of dimensioning the best facility according to the load of the building under analysis; the Yingli Solar module with an efficiency of 18% (with respect to modules Solar World (15.51%), Suntech (16.3%), Rec Solar (15.8%)) was the one that provided the best characteristics to the system, the simulations carried out in the PVsyst software, enabled validating the efficiencies showed by each module.

Considering the total effective surface of 317.19 m^2 in the CIMA building rooftop, and considering the separations for minimizing the shadows produced by every existing element, 144 solar panels were placed in the best dome shaped arrangement; thus, it has been possible to obtain a facility with a performance ratio (PR) of 83.4%, which is considered very satisfactory, and besides indicates an excellent performance and quality of the facility; from the point of view of energy and based on the demand conditions of the CIMA building, the contribution of the facility is 52,353 kWh that corresponds to 36.35%, of which the building consumes 42.66% (22,335 kWh) of solar energy, and the remaining 57.34% (30,018 kWh) is injected to the electrical network.

In addition, these results enable knowing with certainty the contribution that a photovoltaic system may give in a facility for self-consumption of a building located in high-altitude cities in the Equator, significantly reducing the dependency from the network, and even injecting to such network part of the energy generated, or else using it for other requirements.

References

1. Universidad Politécnica Salesiana, *Investigación, Desarrollo e Innovación en Ingeniería, Grupo de Investigación en Energía*, [En línea]. Available: https://bit.ly/3try76j
2. A. Beltrán-Telles, M. Morena-Hernández, F. López-Monteagudo, R. Villela-Varela, Prospective of wind and solar photovoltaic energy for electricity production. CienciaUAT **11**(2) (2017). Available: https://bit.ly/30HgUZS
3. R. Pereira Micena,O. Llerena, W. Queiroz Lamas, J. Silveira, A technical study of solar and biogas energy usage in electric vehicles in Ilhabela, Brazil. Ingenius **20**, 58–69 (2018). https://doi.org/10.17163/ings.n20.2018.06
4. J. Calle, O. Tinoco, Obtaining of SHW with solar energy in the canton Cuenca and analysis of environmental pollution. Ingenius **19**, 89–101 (2018). https://doi.org/10.17163/ings.n19.2018.09
5. G. Guarderas Burbano, *Revisión y mejora de procedimientos de dimensionado de sistemas fotovoltaicos conectados a la red para aplicación a localidades de la región intertropical* (Universidad Politécnica de Madrid, Instituto de Energía Solar, Madrid , 2013), pp. 42–58
6. O. Vanegas Guillén, «studio de la Viabilidad Técnico Económica de un Sistema Fotovoltaico para la autogeneración del Aeropuerto "Mariscal Lamar" de la ciudad de Cuenca Ecuador, Madrid (2014)
7. S. Báez, Análisis del Consumo Energético-Eléctrico de la Universidad San Francisco de Quito, Quito (2011)

8. J. Fernández Ferichola, Caracterización de módulos fotovoltaicos con dispositivo portátil, Madrid (2009)
9. D. Thevenard, A. Driesse, D. Turcotte y Y. Poissant, *Uncertainty in Long-term Photovoltaic Yield Predictions*, report # 2010–122 (RP-TEC) (Varennes Research Center, Natural Resources Canada: CanmetENERGY, 2010)
10. N. Martin, J.M. Ruiz, Calculation of the PV modules angular losses under field conditions by means of an analytical model. Solar Energy Mater. Solar Cells **70**, 25–38 (2001)

Chapter 40
Techno-economic Feasibility to Produce Electrical Energy Considering a Hybrid Photovoltaic—Wind System

Cabanilla José, Moisés Villacreses, Juan Lata-García, and Christopher Reyes-Lopez

Abstract The complexity in the geography for the distribution of energy in certain sectors and the misuse of renewable resources for the production of electrical energy leads to consider the implementation of systems capable of storing clean energy to ensure the constant supply of the required demands. This manuscript proposes a technical–economic analysis of a wind-photovoltaic system through the use of software for optimization of electrical models of renewable energies. As a result of the simulation, it allowed to obtain a detailed report of different variants of technological and economic criteria.

40.1 Introduction

Latin America possesses vast and underexploited renewable energy potential. In addition, several countries in the region have developed robust electricity markets, presenting an attractive destination for renewable energy project developers and investors seeking geographic diversification and clean investments [1].

The Ecuadorian energy growth rate was 3.60% in the period from 2000 to 2016, with an electricity fee higher than current values. The growth rate of Ecuadorian energy consumption was higher than the South American average with a value of 2.22%, while for the Andean community was 2.86% and the bordering countries, which have similar economies, reached 3.46 and 1.85%, for Peru and Colombia, respectively [2].

Ecuador has an estimated population of more than 17.5 million in 2020 [3]. It is a country full of natural resources, and for decades, it has been able to use them for the production of electricity [4]. There are twenty-four provinces, and the total electricity coverage for 2018 is 97.05% [5]. The country's government promotes "The National

C. José · M. Villacreses · J. Lata-García (✉) · C. Reyes-Lopez
Universidad Politécnica Salesiana, 090109 Guayaquil, Ecuador
e-mail: jlatag@ups.edu.ec

C. José
e-mail: jcabanillac@est.ups.edu.ec

Á. Rocha et al. (eds.), *Communication, Smart Technologies and Innovation for Society*, Smart Innovation, Systems and Technologies 252,
https://doi.org/10.1007/978-981-16-4126-8_40

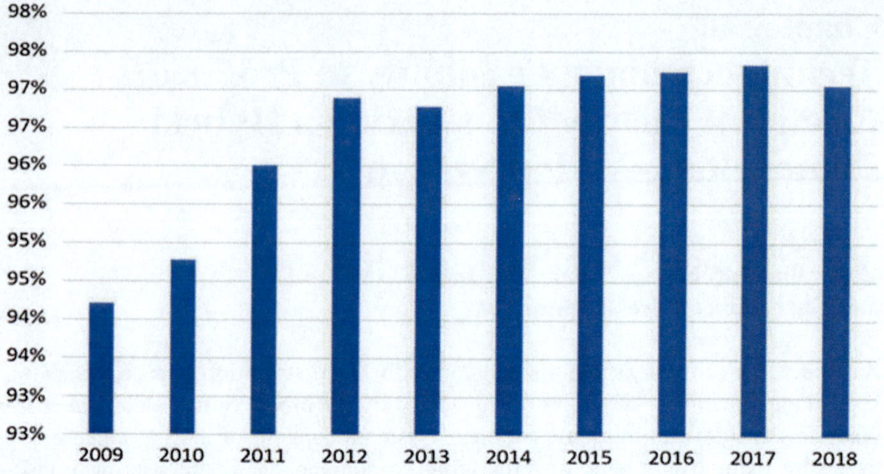

Fig. 40.1 Evolution of electrical demand coverage [3]

Plan of Good Living," which have as an objective to increase renewable energy sources from 68.8 to 90% by the end of 2021 [6].

Figure 40.1 shows the evolution of electricity service coverage in the country, in percentage, from 2009 to 2018. It can be noticed a mean growth until 2012 of 2.68%; for the following year, it decreased by 0.13%. In the following 4-year period (2013–2017), it increased by 0.56%, and finally, it decreased the following year by 0.28%.

In [7], the authors show a combination of both forms of electrical generation, photovoltaic generators, and hydrokinetic turbine to configure a hybrid system together with a group of batteries to guarantee the electrical supply of a set of isolated loads. The study includes the optimal location of the turbine and the technical economic analysis of all the components of the hybrid system for an efficient use of renewable energy resources. The results show that the hybrid system provides the power demanded by the loads, and the hydrokinetic and solar photovoltaic generation is optimized, and the state of charge of the batteries is maintain. The increase in the urban population over the next few years is well known to governments and scientists around the world, where its percentage changes from 55.3 to 68.4% [8].

In [9], the researchers present a simulation to find the feasibility of the implementation of an isolated system comprising photovoltaic panels, wind turbine, and a battery bank, and the technical economic parameters are analyzed to obtain the lowest cost of energy to be considered competitive, and the levelized cost of energy is directly related to the optimal cost of each of the generation sources; a considerable increase in the introduction of this type of hybrid systems produces the disenchantment of using fossil fuel-based generators.

This study seeks to obtain a complete report on the feasibility of the hybrid system in the community of Patul, with the use of HOMER PRO software, which is commercial software that is used to model, simulate and optimize any power microgrid system [10] which is a platform for optimizing the design of microgrids in all sectors. The economic models used to calculate the average grid value are the total cost price (NPC) and the levelized cost of energy (LCOE). The NPC is calculated as the total cost minus revenue over the lifetime of the project, while the LCOE is defined as the average cost per kWh of electrical energy in the system [11].

The study seeks to supply the electricity demand in isolated communities, where the conventional power grid does not meet the requirements for their development and progress. A study of the feasibility of a hybrid system is carried out, and the lowest possible COE is obtained through the use of electrical network optimization software, as well as the establishment of community parameters based on official government data, and projections of the system in techno-economic factors for power generation.

The paper is structured as follows. After the introduction, the methods are found in Sect. 40.2. Section 40.3 shows the simulation results and the implementation and investment costs. Finally, Sect. 40.4 shows the respective conclusions of the study performed.

40.2 Methods

40.2.1 Location

The current study takes the community of Patul belonging to the parish Molleturo in the province of Azuay is located at 3460 m above sea level. The climate, although there are mild days, is cold, something typical of the altitude, with an average temperature of 8–13 °C (degrees Celsius). Currently, 7 families live there; there is a mixed school with 10 students and a teacher; it has no road, and the village is accessed only by an old and busy bridle path, 3 h on foot or on horseback from the Illincocha sector, at the height of kilometer 36, thus being the study of the techno-economic feasibility of a hybrid photovoltaic wind system to supply the entire demand of the community taking into account the usual consumption of the inhabitants of that province by customs and similar activities [12].

The demographic profile in the study area is constructed based on official information from the National Institute of Statistics and Census. According to the 2010 Population and Housing Census, the population of Molleturo parish is 6,996, while the population of the census sector (010157999038), including the community of Patul with 103 inhabitants, with a slight increase in the male population (52.43%) [13].

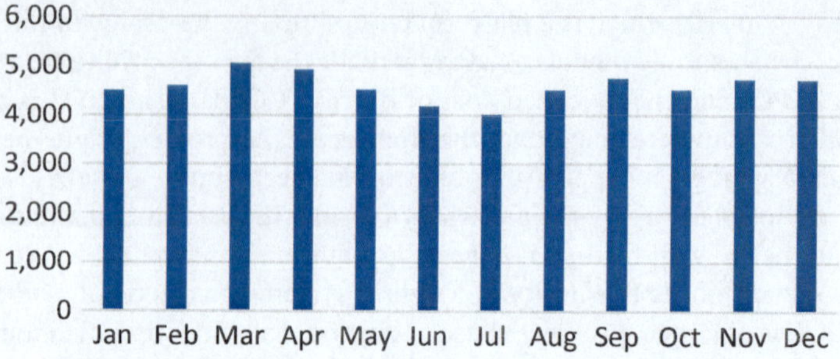

Fig. 40.2 Daily radiation graph

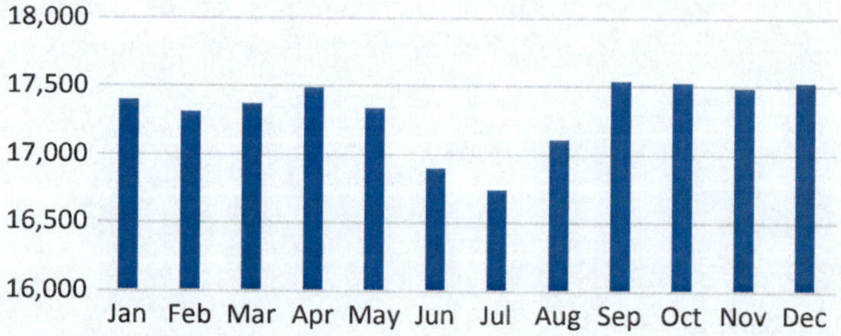

Fig. 40.3 Monthly temperature graph

40.2.2 Parameters and Variables

The solar resource and temperature were extracted from the NASA database, indicating daily radiation (kWh/m2/day). Figure 40.2 contains the solar radiation data, given in kWh/m^2/day, and the annual percentage is 4.58, but it can reach maximum values of 5 kWh/m^2/day.

Figure 40.3 contains monthly temperature data for the province of Azuay, showing the temperature in °C, which currently averages 17.31 °C, but can reach maximum values such as 17.54 °C.

Figure 40.4 contains data from the 2013 Wind Atlas of Ecuador showing the wind speed. Annually it is 7.63 m/s, with an average for the province of Azuay between 7 and 8.5 m/s.

40.2.3 Electrical Load Profile

For this scenario, the selected load profile presents the energy consumption of the province of Azuay in 2019, which is 1041.11 GWh for a population of 867,239 habitants. This corresponds to an annual per capita consumption of 1200.48 kWh/habitant.

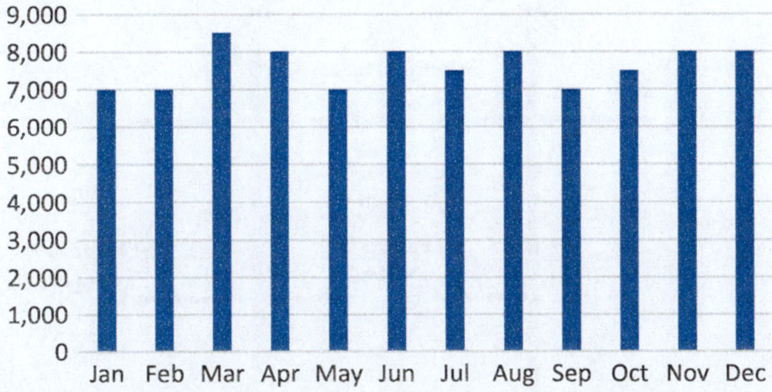

Fig. 40.4 Annual wind speed

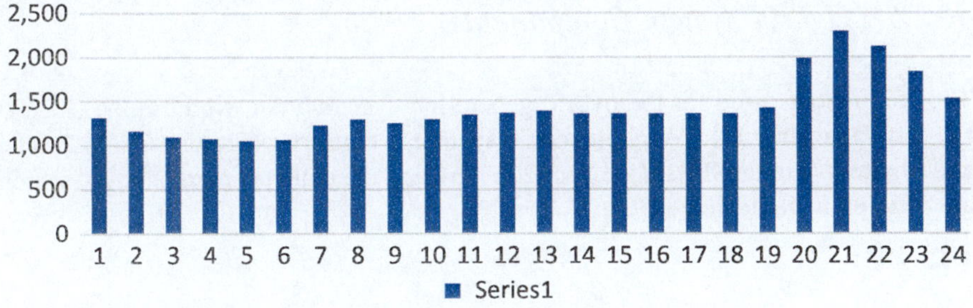

Fig. 40.5 Demand curve of the Patul Community

Therefore, it indicates that the per capita consumption in the province in one day is 3289 kWh/habitant.

According to the National Electricity Council (CONELEC), the characteristic profile corresponding to the residential sector under the Power Company "Centro Sur" for the Patul community [14] (Fig. 40.5).

40.2.4 Schematic of the Hybrid Generation System

The wind-photovoltaic hybrid generation system consists of a system (PV) connected to the DC bus, battery, converter, and a wind turbine connected to the AC bus. Figure 40.6 shows the scheme to be simulated to supply the load demand. The wind turbine and the PV panels supply the electrical load that is connected to the AC bus.

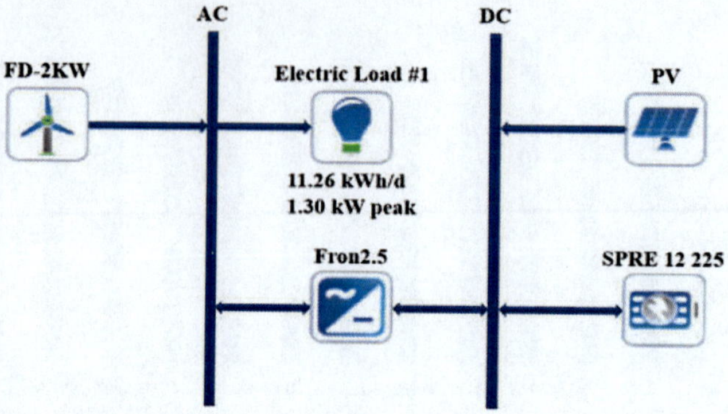

Fig. 40.6 Hybrid system scheme

40.2.5 Hybrid System Components

The components used for the hybrid generation system are a wind turbine (ZEYU FD-2 kW), a generic 1 kW photovoltaic system, a converter (Fronius Galvo 2.5-1), and a battery (Trojan SPRE 12 225 of 226 Ah), and the main characteristics of which are detailed in Table 40.1.

40.3 Results

Table 40.2 shows the results of the hybrid system developed at HOMER. The study was carried out by entering the values of the inhabitants' demand, the wind speed, and the economic and technical parameters (capacity, power, storage autonomy, annual yield, and expected lifetime) of the components included in the system. Among the results, it provides information on the base case with the optimization of the model based on the requirements of the community, as well as comparative data between your systems. The batteries have an expected life of approximately 9 years, and an autonomy of 27.8 h, for 6 units connected in parallel, as follows, the state of charge in Fig. 40.7.

The value of the initial investment in the implementation of the system under study, once the simulation has been carried out, provides us with the average cost per kWh of energy (COE) produced by the system, as well as the hours of autonomy due to possible maintenance or disconnection.

The solar radiation and wind speed data were beneficial to the study because the simulation showed a high amount of energy production in kWh/year, which would not be the same if the systems were simulated independently.

Table 40.1 Component characteristics

Photovoltaic PV	
Name:	Generic flat plate PV
Rated capacity (kW)	8
Capital ($)	$900
Replacement ($)	$200
O&M ($/year)	$30
Time (years)	25
Derating factor (%)	80
Wind turbine	
Name:	ZEYU FD-2KW
Rated capacity (kW)	1
Capital ($)	$1000
Replacement ($)	$500
O&M ($/year)	$50
Lifetime (years)	24
Hub height (m)	17
Storage	
Name:	Trojan SPRE 12 225
Maximum capacity (Ah)	226
Throughput (kWh)	2172
Nominal voltage (V)	12
Roundtrip efficiency (%)	80
Capital ($)	$399
Replacement ($)	$10
O&M ($/year)	$10
Converter	
Name:	Fronius Galvo 2,5–1
Capital ($)	$350
Replacement ($)	$50
O&M ($/year)	$5
Lifetime (years)	10
Efficiency (%)	95

40.4 Conclusions

This study allows the community of Patul to supply the electricity demand through the generation of renewable energy based on a hybrid wind-photovoltaic system with a low cost and high sustainability, supporting its development and progress.

Table 40.2 Simulation results

NPC ($)	COE ($)	Operating cost ($/year)	Production (kWh/year)	Autonomy (h)
$12.945	$0.244	$262.26	11,549	27.8
$14.389	$0.271	$183.04	6674	83.3
$15.515	$0.292	$243.50	9344	64.8
$15.210	$0.286	$302.03	11,549	27.8
$16.876	$0.318	$251.02	$7.545	69.5
$18.267	$0.344	$331.78	$15.088	60.2
$17.345	$0.327	$240.43	$8.879	88.0
$18.761	$0.353	$371.79	$12.884	27.8

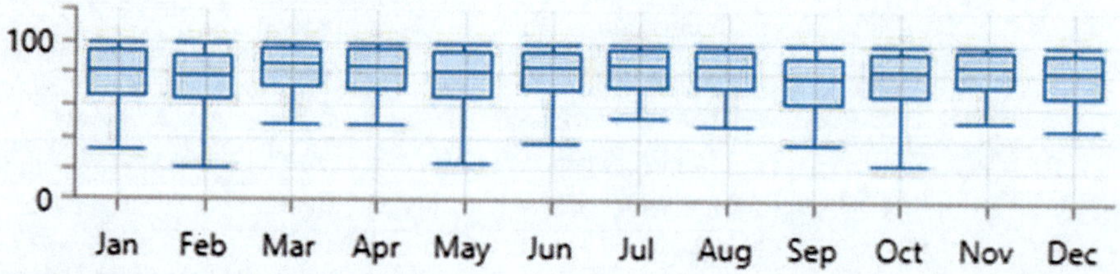

Fig. 40.7 State of charge

An optimal location of wind turbines and solar panels was determined to obtain maximum generation, supplying electricity to a set of isolated loads.

The study showed that the designed hybrid system is technically feasible; i.e., it is able to meet the demand by taking advantage of renewable sources; also, a correct management is needed in the storage devices to constantly supply the generated energy, the maximum wind speed was 8.5 m/s, and the maximum solar radiation is 17.31 °C. The optimization of the proposed hybrid system was carried out using HOMER, where the total cost price (NPC) is $12,945, the COE is 0.244 $/kWh.

The study of the feasibility of hybrid systems isolated from the conventional grid through the data obtained concludes that the values provided depend on the power generation capacity, the cost of the storage system, and the value of the maintenance required within a year.

References

1. OLADE e IRENA, http://www.olade.org/noticias/olade-e-irena-ponen-a-las-energias-renovables-en-el-corazon-de-la-recuperacion-economica-post-pandemia-de-america-latina-y-el-caribe/#:~:text=Los países de América Latina, inexplotado potencial de energ. Last accessed 02 Feb 2021
2. P.L. Castro Verdezoto, J.A., Vidoza, W.L.R. Gallo, Analysis and projection of energy consumption in Ecuador: energy efficiency policies in the transportation sector. Energy Policy **134**

(2019). https://doi.org/10.1016/j.enpol.2019.110948
3. Instituto Nacional de Estadisticas y Censos del Ecuador (INEC): ¿Cómo crecerá la población en el Ecuador? https://www.ecuadorencifras.gob.ec/documentos/web-inec/Poblacion_y_Demografia/Proyecciones_Poblacionales/presentacion.pdf (2011). Last accessed 03 April 2021
4. J. Cevallos-Sierra, J. Ramos-Martin, Spatial assessment of the potential of renewable energy: the case of Ecuador. Renew. Sustain. Energy Rev. **81**, 1154–1165 (2018). https://doi.org/10.1016/j.rser.2017.08.015
5. A.N. Constituyente, L.A. Paredes, M. Cáceres, A. Busso, L. Vera, A. Firman, G. Leiva, L. Luque, Nacional, T., ARCONEL, No, N.L., Oficial, R., Suplemento, N., Estadisticas Anuales Y Multianual Del Sector Eléctrico Ecuatoriano 2018. Rev. Bras. Energ. Sol. **16**, 47–54 (2019)
6. D. Icaza, D. Borge-Diez, S.P. Galindo, Proposal of 100% renewable energy production for the City of Cuenca—Ecuador by 2050. Renew. Energy. **170**, 1324–1341 (2021). https://doi.org/10.1016/j.renene.2021.02.067
7. J. Lata-García, F. Jurado, L.M. Fernández-Ramírez, H. Sánchez-Sainz, Optimal hydrokinetic turbine location and techno-economic analysis of a hybrid system based on photovoltaic/hydrokinetic/hydrogen/battery. Energy **159**, 611–620 (2018). https://doi.org/10.1016/j.energy.2018.06.183
8. J. Litardo, M. Palme, M. Borbor-Cordova, R. Caiza, J. Macias, R. Hidalgo-Leon, G., Soriano, in Urban heat island intensity and buildings' energy needs in Duran, Ecuador: simulation studies and proposal of mitigation strategies. Sustain. Cities Soc. **62**, 102387 (2020). https://doi.org/10.1016/j.scs.2020.102387
9. M J.P. Medina, , Lata-García, J.: Optimal model of a hybrid electrical system photovoltaic panel /wind turbine/battery bank, considering the feasibility of implementation in isolated areas. J. Energy Storage. **36**, 102368 (2021). https://doi.org/10.1016/j.est.2021.102368
10. A.K. Raji, D.N. Luta, Modeling and optimization of a community microgrid components. Energy Procedia **156**, 406–411 (2019). https://doi.org/10.1016/j.egypro.2018.11.103
11. L. Uwineza, H.G. Kim, C.K. Kim, Feasibilty study of integrating the renewable energy system in Popova Island using the Monte Carlo model and HOMER. Energy Strateg. Rev. **33**, 100607 (2021). https://doi.org/10.1016/j.esr.2020.100607
12. El Telégrafo: Patul: Un paraje de incomparable belleza aislado de El Cajas, https://www.eltelegrafo.com.ec/noticias/2015/1/patul-un-paraje-de-incomparable-belleza-aislado-de-el-cajas (2015)
13. Gobierno parroquial Molleturo: Diagnóstico ambiental línea base medio socio económico y cultural (2019)
14. CONELEC (Consejo Nacional de Eléctricidad): Estudio y Gestión de la Demanda Eléctrica. Plan Maest. Electrif. 2013–2022. 2. 206 (2013)

Chapter 41
Sustainable Management Model of Rural Electrification Project Isolated by Renewable Photovoltaic Energy

Edison Mora-Villagómez and Christopher Reyes-López

Abstract The electrification of remote communities based on renewable energy has achieved the development of the inhabitants, reducing poverty; however, the lack of tracing and management of renewable energy systems becomes a short-term problem. This work presents a proposal for an alternative sustainable management model for the electrification of isolated communities, considering the objectives for the sustainable development of the United Nations and the legislation of Ecuador. To complete the work, the economic technical study of a photovoltaic system for the electrification of the parish Pílalo is carried out, where the initial costs, operation, and maintenance are considered. The cost of energy (COE) is 0.216 USD/kWh, with which the value can be established by the energy supply with the aim of raising funds and sustaining the project during the operating time.

41.1 Introduction

Achieving access to electricity is considered one of the fundamental factors for the development of rural areas where it has energy, social, and economic constraints; in these localities, the rates of development are low. Even the International Energy Agency (IEA) estimates that around 980 million people do not have electricity service, sustainable development goals (SDGs) seek access to affordable, reliable, sustainable, and modern energy for all by 2030 removing barriers and driving sustainable development recognized by the global community [1].

Experiences of institutions and governments have brought to the fore relevant aspects of the problem of rural electrification; in Ecuador, several projects were developed including the Eurosolar program that has as its main objective; the development of the most disadvantaged populations of the Ecuadorian East through access

E. Mora-Villagómez · C. Reyes-López (✉)
Universidad Politécnica Salesiana, Chambers 227 y 5 de Junio, 090114 Guayaquil, Ecuador
e-mail: creyesl@ups.edu.ec

E. Mora-Villagómez
e-mail: emorav@est.ups.edu.ec

Á. Rocha et al. (eds.), *Communication, Smart Technologies and Innovation for Society*, Smart Innovation, Systems and Technologies 252,
https://doi.org/10.1007/978-981-16-4126-8_41

to basic services powered by photovoltaic solar energy, the project benefits 39,629 inhabitants from 7 provinces of the country, the project is integral complemented by computer and Internet training for the use of info centers, and technical training for the maintenance of photovoltaic systems [2].

Several rural renewable energy (ER) electrification projects have been implemented around the world; however, they have not been sustainable over time under different circumstances; for this reason in Ref. [3], researchers present a sustainable management model for the electrification of Brazilian communities in the Amazon region in accordance with the legislation of the country, and taking into account the isolation of populations, the proposal of the model is complemented by the sizing of a renewable energy system formed by a biomass plant considering the high costs per operation and maintenance due to being isolated, and these costs directly influence the price of kWh.

In developing countries, micro-resistance electrification is increasingly well received to electrify isolated and hard-to-reach populations, and the solution for supply is complex due to the intermittent primary energy sources; however, modeling, simulation, and sizing techniques achieve the solution for reliable supply, and these tools collaborate in decision-making to evaluate the design of the renewable energy system taking into account the cost of energy. However, the social or environmental part of the design of the system is not considered, and these considerations are key to achieving the long-term sustainability of the project.

In Ref. [4], they present a multi-criteria procedure where they compare electrification designs with different technologies considering different aspects, integrated into a methodology divided into two phases, the first phase allows to generate an optimization model for the economic technical design of open source, and the second phase is considered 12 criteria that meet technical, socio institutional, and environmental aspects. The proposed methodology is implemented and validated in 26 case studies in Plateau—Nigeria.

The document present aims to establish a management model to promote the sustainability of isolated electrification projects with isolated renewable energies; in the first part, we review the sustainability of projects implemented in Ecuador; in the second part, the parameters are defined to ensure long-term sustainability; in the third part, a management model is formulated to be applied in the rural area; in the last part, a system is sized according to the management model in the Pilalo village.

41.2 Background

The United Nations Education Organization for Science and Culture in 2017 presented the goals for sustainable development. Objective7 of the sustainable development goals (SDGs) seeks affordable and non-polluting energy in a safe and modern way for all; today the world is heading toward achieving the goal, and in recent years, it has been reduced from 1.1 trillion to 980 million people without access to electricity through electrification plans; in the poorest countries, the implementation of renewable energy systems is accelerated by improving energy efficiency and the use of renewable technology, however, there is a need to improve the sustainability of projects to achieve the 2030 SD targets [5].

The basic purpose of human development is to increase the opportunities of the population and transform them into their own benefit, rather than increase economic growth, thereby improving living conditions. Access to energy reliably and with sustainable projects has to be governed by five principles that are the managers of the management model grouping several factors such as the electricity company, users, and the state. Below are the pillars of the management model for sustainable renewable energy projects.

Access to energy. Ensuring access to feasible, reliable, and modern energy services in remote or isolated communities, without the search for economic benefits, the purpose of these projects is the well-being and social development of the inhabitants.

Equity and diversity. The inhabitants of a population must have equal opportunities for access to energy, training, and maintenance. Avoiding inequity among end users, breaking down barriers of high upfront costs, property right over the dwellings they live in, approval of credits, etc.

Sustainability. The term does not have to be understood only as the preservation of the project, instead sustainability must be understood as the conservation of the project with the same degree of satisfaction of energy delivery during the lifetime of the system. The term is quite broad where environmental, human, and financial aspects are involved among others.

Empowerment. It is a collective effort where users are not classified as objects but as people of development, who under a confessor have the opportunity to make free decisions without affecting living conditions, having full participation in the elaboration and implementation of improvements on the project.

Productivity. Economic growth is based on community development, and residents would harness electricity for productive tasks, conserve food, and expand public service hours; so people are not only considered the means but the purposes of development.

41.2.1 Management Model

In this work, the management model to be proposed is based on the purpose of human development fulfilling the objectives of financing efficiency with the lowest cost of energy and technical efficiency during the lifetime of the system [6], and the case study is in the Pilalo community belonging to the canton Pujili.

The management model consists of the following actors: owner of photovoltaic systems, association of inhabitants, and users. The owner of the photovoltaic systems may be a state entity or non-profit organizations, delivering the equipment well in comfort, i.e., delivering them free of charge so that the community can use them or in concession, ceding the exploitation to the association sharing responsibilities over the equipment and if necessary the replacement of the components.

The association consists of users who will choose policy, among its functions is the responsibility for good financial management, taking as revenue the collection of the fixed monthly rate of users, the difference against egress will be deposited in a bank account through a reserve fund that will be used to replace the system when the end of the working time arrives, and the state or non-profit entities are not interested in electrifying remote or isolated areas. The association's egress will be intended for payment of operation, predictive maintenance, and corrective maintenance of system equipment.

The association would convene its members to regular meetings for financial review and provide an annual balance sheet. Users are obliged to pay a monthly fee, belong to the association and be taken care of in case of equipment failures or maintenance, to guarantee the service the user can submit suggestions or complaints.

Figure 41.1 presents the management model to be implemented in the rural electrification project, which is subdivided into three phases. The first phase assesses the availability and potential of the community's renewable energy resources such as solar, wind, biomass, etc. With the assessment of the energy potential is presented the project of the ER system together with the technical sizing to know the number of components to use and meet the energy demand of the load, the lack of this data together with the analysis results in the power outages per discharge of the storage bank and the distrust of the user.

The economic evaluation of the system will indicate the cost of kWh, which will be agreed with the directory of the future association, and the cost of the fee in any case will be less than the expense incurred by families when buying candles, diesel, batteries, etc., ensure payment of energy consumption.

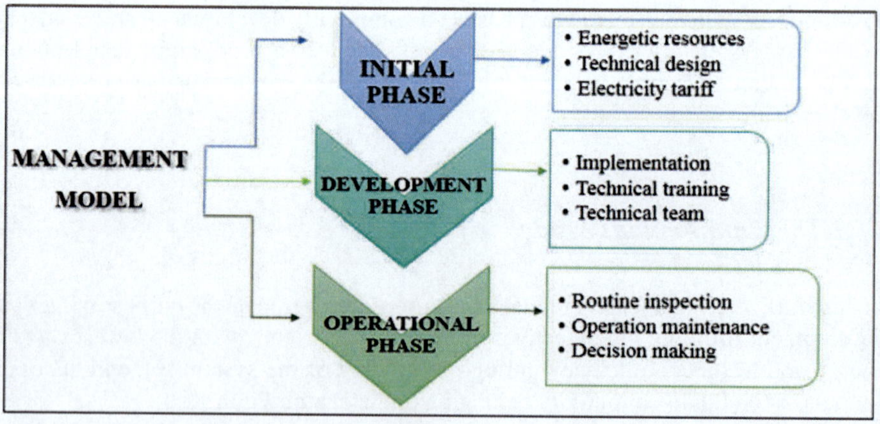

Fig. 41.1 Management model to be implemented

In the second phase or development, the model indicates the implementation of the system where according to the parameters indicated by the project design, system location, optimal angle of photovoltaic panels, wiring, etc.

Training to the association is essential for the efficient use of energy as well as the monitoring and proper functioning of the system, and additional with the inhabitants is selected on merit candidates to form a technical team that will have a comprehensive training to provide preventive, corrective maintenance, and operate the different renewable energy systems.

41.2.2 Ecuador's Legislation

The Agency for Regulation and Control of Electricity (ARCONEL) according to the organic law of electricity, and according to incentives for distributed generation, renewable energies, energy efficiency, and demand management, the board approved regulation 057/018 [7] micro photovoltaic generation for self-supply of end consumers of electricity.

The regulation was analyzed technically by the CNRSE (national coordination of regulation of the electricity sector) indicating the characteristics to be met by consumers to implement such systems and be enabled by the local electricity company.

However, the legislation only covers photovoltaic systems with network connection leaving isolated systems unmoderated, and the particularity of regulation 057/018 indicates that the owner of the property can install only one SF to reduce the power consumption of the conventional grid with a maximum capacity of 100 kW for residential consumers and 500 kW for industrial consumers with an operation of 20 years.

41.3 Case Study

41.3.1 Pilalo's Village

This town has documentation from 1852 and is located on the slopes of the Western mountain range; the community is located approximately at km 90 of the Latacunga—La Mana road, and its main activity is livestock. It has a population of approximately 2640 people but in recent years have emigrated in a significant number mainly to Spain. The climate is semi-arid with an average temperature of 12 C.

41.3.2 *Organization and Dynamics of the Population*

The community is organized through a decentralized autonomous government (GAD) parish, having a president elected for 4 years of management, has a budget to execute a series of works for the improvement of the quality of life of the inhabitants.

The chair of the GAD may be the coordinator of the association for the energy management of the system to be implemented, which must be socialized and approved by members to avoid inequity in energy use and decision-making. The priority of the systems is to feed the lighting and basic appliances.

Being a purely agricultural parish, the inhabitants spend most of their time in the countryside and in the grazing of cows. In recent years, community tourism and tourist attractions such as the waterfall la Chorrera are promoted.

41.3.3 *Data Collection*

The parish Pílalo is made up of 528 families; for the most part, they live in a dispersed way, and the per capita consumption of the inhabitants of the province of los Ríos is 738 kWh/year the inhabitants of a temperate climate with a monthly average of 14.2. The daily consumption of a family consisting of 5 people is 2.62 kWh/day, with a maximum load of 0.45 kW, and a minimum load of 0.04 kW.

Because the population is in a valley surrounded by mountains, the average wind speed is 3.6 m/s, so it is not possible to operate a wind turbine, which need a speed of about 12 m/s to reach the rated power, and the highest speed is reached in July with 4.79 m/s; in this month, the wind turbine produces less than 10% of the rated power.

Solar radiation varies throughout the year and is affected by mist and clouds, and the energy potential produced by the photovoltaic panel depends on the angle of inclination, temperature, and system efficiency, and the Meteonorm base is used for the study through the Pvsyst [8] (−0.94, −0.78) are located or through the interactive map where the data is located and imported, and the magnitude of the monthly average solar irradiation is 147 kWh /m^2/month.

Table 41.1 shows the monthly weather variables that include solar radiation, wind speed, temperature, and relative humidity.

41.4 Software and Economic Viability

41.4.1 *Pvsyst*

It is a software designed to be used by engineers and researchers; the procedure of use is quick and simple; among the advantages offered by this import weather data from

Table 41.1 Weather variables in the Village Pilalo take it

	Irradiation global (kWh/m²/month)	Diffuse irradiation (kWh/m²/month)	Temperature (°C)	Wind speed (m/s)	Relative humidity (%)
January	143.9	78.3	14.6	3.39	75.2
February	138.5	74.3	14.5	3.4	75.4
March	160.2	85.5	14.4	3	78.5
April	147.9	62.7	14.2	2.89	80.1
May	152.2	67	14.4	3.5	77.8
June	142.6	60.9	13.6	4.19	76.2
July	139.7	70.8	13.7	4.79	72.2
August	155.4	64.8	13.8	4.8	70.5
September	160.2	70.2	14	4.29	71.6
October	151.7	72.8	14.8	3.41	72.8
November	135.2	76.6	14.4	2.89	75
December	146.8	66.5	14.7	2.99	76
Year	1774.4	850.4	14.2	3.6	75.1

different sources, you can place the operating hours of the loads, and the components of the trading system are commercial, allows to perform a preliminary simulation, has warning messages for oversizing certain components, etc.

One of the strong engines that the program has against others is the economic evaluation where it allows to enter the values of all the materials used in the implementation including transport costs, taxes, etc., also sales policies, connection taxes, and distribution of investment funds.

The parameter that defines the economic viability of the system is the level cost of kWh energy (LCOE), and other parameters help decision-making such as net present value (VPN) and profitability analysis, recovery period, and return on investment rate.

41.4.2 Cost of Energy

LCOE is the cost or price of the kWh produced by the photovoltaic system. Pvsyst considers the present value of future cash flows using a shielding, and the program uses Eq. 1 for the calculation of LCOE.

$$\mathrm{LCOE} = \frac{\sum_{t=1}^{n} \frac{I_t + M_t}{(1+r)^t}}{\sum_{t=1}^{n} \frac{E_t}{(1+r)^t}} \quad (41.1)$$

From where, I_t it is the investment and expenses of the year (t), it is the cost of operation and maintenance, it is the energy production of the system and n is the lifetime of M_t the system E_t.

41.4.3 System Sizing

The photovoltaic system is proposed to install the closest to the load, analyzed studies implement centralized systems due to a series of improvements that they present, among which a lower cost of energy stands out. The selected equipment is chosen according to local availability, the solar panels, batteries, and charge controller are connected to a DC bus, while an inverter converts from DC to AC to connect consumer loads. The operation of the system is supported by the battery bank to satisfy the load at night or when there is low radiation.

Figure 41.2 shows the scheme used for the technical sizing of the photovoltaic system, and each of the components is shown in the graph.

According to the per capita consumption of the inhabitants of the town is proposed a series of loads that are used in a conventional house, the refrigerator to be used has a capacity of 300 L, with an energy efficiency A and an annual consumption of 310.3 kWh/year, and Table 41.2 details the appliances and the daily energy it consumes.

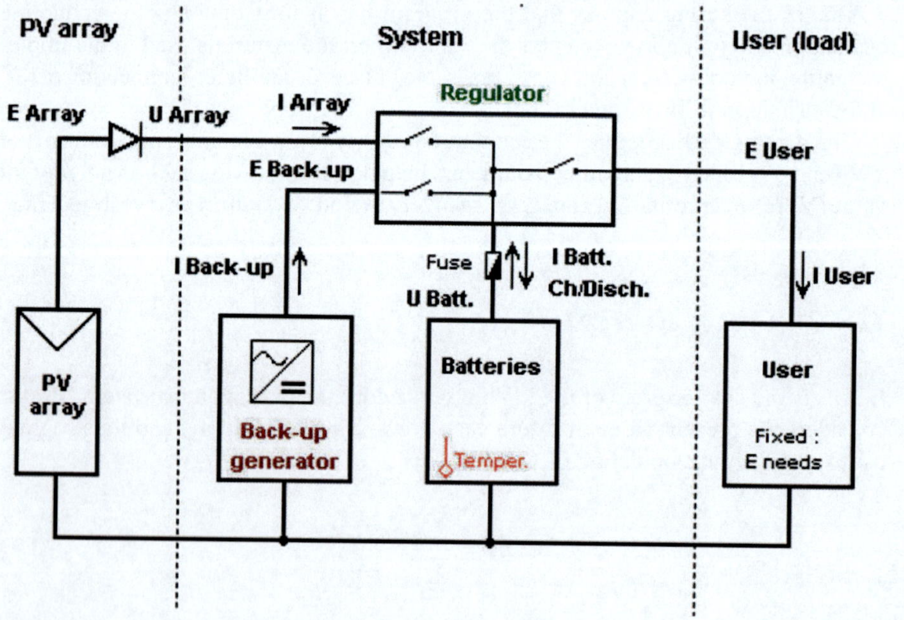

Fig. 41.2 Layout of a stand-alone system

Table 41.2 Domestic consumption

Number	Appliance	Power (W)	Daily use (h)	Daily energy (Wh)
6	Luminarias	9	6	324
1	TV	80	4	320
2	Appliances	100	2	800
1	Fridge	0.86 kWh/day	24	860
1	Several	100	3	300
Total daily energy				2628 Wh/day
Monthly energy				78.8 kWh/month

The photovoltaic panels used in the simulation of the system are from the manufacturer Yingli Solar with a power of 260 W polycrystalline type, the maximum voltage is 30.3 V, the open circuit voltage is 37.7 V, the maximum current is 8.59 A, and the efficiency is 15.9%. The load controller is selected from the manufacturer Victron Energy 30A/24 V type MPPT, the input voltage range is up to 100 V, and the self-consumption is less than 10 mA.

The batteries employed are bought from a local manufacturer: Dural Ltd. They have a voltage of12 volts, using a type of acid lead, and built to support a minimum discharge of 20% and an energy storage of 8.6 kWh. When the batteries are charged to 100%, they can reach a voltage of 28.8 V, while in discharge the voltage is 23.6 V.

The photovoltaic generator for each family has a power of 1kWp, consisting of 4 photovoltaic panels of 260 W polycrystalline type, covering an area of 6.5 m 2, in addition to 6 batteries of 12 V of acid lead and 150 Ah and 2 charge controllers Victron model 100/30-24 V. The operating voltage of the system is 24 V; so the batteries will have serial-parallel connection.

The system achieves a power production of 1501 kWh/year, of which 959 kWh/year is used, the unused power is 460 kWh/year and is due to the lack of connected charge certain generation hours, and the batteries are fully charged. The system performance percentage is 52.93%, and battery usage status cycles are 93.6%.

Normalized production and loss factors are shown in Fig. 41.3, the energy produced and used is 52.9%, and the loss of the system and by the photovoltaic assembly is 7.7 and 15%, respectively, when the storage bank is full, the unused energy is 24.3% of the total generated.

41.4.4 Economic Assessment

After technical sizing, the program allows to carry out the economic evaluation of the system; taking into account all the economic parameters involved in the implementation of the system, the tool allows to limit the initial cost of the project and the costs per operation and maintenance; with this data, the calculation engine presents

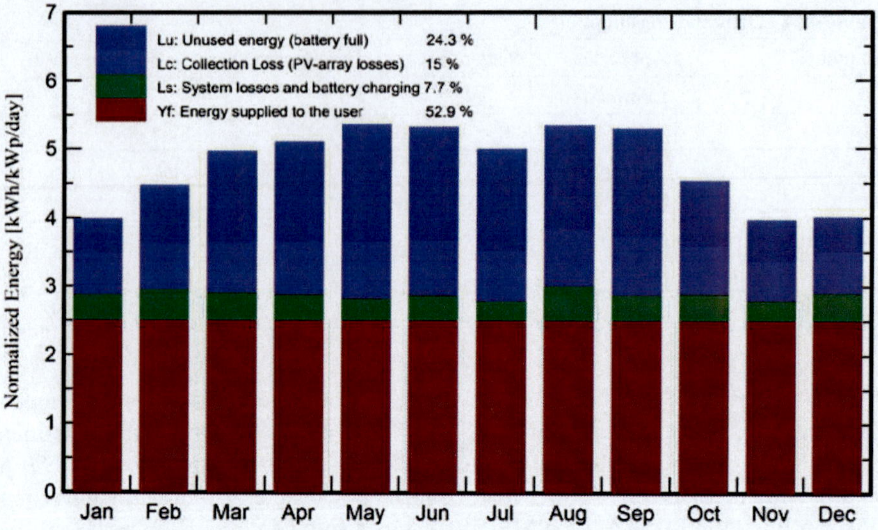

Fig. 41.3 Energy monthly supplied to the user and losses system

the cost of the energy produced LCOE, and the software estimates long-term productivity based on various financial parameters. The cost of energy produced by hybrid systems ranges from $0.10 to $0.35/kWh, and the factor involved in this variation is due to taxes on the components that each country has and the demand for components, low demand increases the cost [6, 9, 10]

The 4 photovoltaic panels cost $840, the module support costs $100, the 6 batteries cost $2634, being the most expensive component, the charging controller is priced at $92, and an estimated $20 cost for consumable materials such as the connection box, fasteners, etc. The engineering analysis is worth $50, the cost is representative as they are 528 similar systems, and the installation is included with the cost of the elements. For operating costs, annual maintenance is considered at a cost of $20, because it does not contain moving parts, the tasks to be performed are cleaning and tightening terminals. Cost per land rental, bank taxes, insurance, etc., are null. The total cost of the installation is 3736 USD, and the cost of energy is 0.216 USD/kWh, with a recovery period of 17 years.

41.5 Conclusions

Power supply to isolated communities or frequent power outage problems is a challenge for the state. The complexity of access means that the costs per operation and maintenance are significantly increased, in addition that Ecuadorian legislation does not contemplate or regulate the mechanisms of supply of photovoltaic systems without connection to the grid, therefore, having an internal management system where it conforms and uses equipment for maintenance, and operation will reduce the dependence and costs represented by the local power company's workforce. Through the organization of the community and the three phases of the model, it leads to the sustainability of the project and the possibility of replication in other communities reducing energy costs and quality of supply.

The cost of electric power service must be discussed among all actors; in the case of study applied for this document, each family must cancel 17.02 USD/month; at 0.021 kWh/USD, the energy generated by the photovoltaic system is 1501 kWh/year.

References

1. World Energy Outlook 2020, in *Analysis—IEA*. https://www.iea.org/reports/world-energy-outlook-2020. Accessed 19 Feb 2021
2. Electrificación rural con energías renovables – Ministerio de Electricidad y Energía Renovable. http://historico.energia.gob.ec/electrificacion-rural-con-energias-renovables/. Accessed 21 Feb 2021
3. G. Pinheiro, G. Rendeiro, J. Pinho, E. Macedo, Sustainable management model for rural electrification: Case study based on biomass solid waste considering the Brazilian regulation policy. Renew. Energy **37**, 379–386 (2012). https://doi.org/10.1016/j.renene.2011.07.004
4. M. Juanpera, P. Blechinger, L. Ferrer-Martí, et al., Multicriteria-based methodology for the design of rural electrification systems. A case study in Nigeria. Renew. Sustain. Energy Rev. **133**, 110243 (2020). https://doi.org/10.1016/j.rser.2020.110243
5. L. Ferrer-Martí, B. Domenech, A. García-Villoria, R. Pastor, A MILP model to design hybrid wind-photovoltaic isolated rural electrification projects in developing countries. Eur. J. Oper. Res. **226**, 293–300 (2013). https://doi.org/10.1016/j.ejor.2012.11.018
6. H.S. Das, C.W. Tan, A.H.M. Yatim, K.Y. Lau, Feasibility analysis of hybrid photovoltaic/battery/fuel cell energy system for an indigenous residence in East Malaysia. Renew. Sustain. Energy Rev. **76**, 1332–1347 (2017). https://doi.org/10.1016/J.RSER.2017.01.174
7. ARCONEL, Estadística anual y multianual sector eléctrico 2016 (Quito, 2017)
8. PVsyst. http://www.pvsyst.com/en/. Accessed 8 Nov 2017
9. T.T.D. Tran, A.D. Smith, Incorporating performance-based global sensitivity and uncertainty analysis into LCOE calculations for emerging renewable energy technologies. Appl. Energy **216**, 157–171 (2018). https://doi.org/10.1016/j.apenergy.2018.02.024
10. R. Syahputra, I. Soesanti, Renewable energy systems based on micro-hydro and solar photovoltaic for rural areas: a case study in Yogyakarta, Indonesia. Energy Rep. **7**, 472–490 (2021). https://doi.org/10.1016/j.egyr.2021.01.015

Chapter 42
Experimental and Analytical Studies of the Wear of CVD and PVD Cutting Tools in AA7075 Turning at Different Operating Conditions Through ANOVA

Leonidas Ramírez, Cristian Guilcaso, Willan Castillo, and Diego Molina

Abstract The cutting tool wear for the turning machining of aluminum alloy 7075 was analyzed by experimental and analytical studies considering two different cutting tool types at different operating conditions through analysis of variance (ANOVA). An orthogonal array was obtained by using a software, and for cutting speed, feed rate, depth of cut, and cutting tool type, with two levels, were generated 16 experiments for these combinations. Material specimens were prepared according ISO 2540:2016 standard, and machining was carried out by using a CNC turning machine; meanwhile, cutting tool wear at the flank was analyze with ISO 3685:1993 standard. A scanning electronic microscope was utilized to obtain wear measures on the cutting tool surface and flank. ANOVA determined the influence for each material on the required response, cutting tool wear, presenting that cutting tool type had an influence of 82.14% and the PVC cutting tool registered the highest wear value with 121.09 μm.

42.1 Introduction

Manufacturing processes provide raw material and final products carrying out metal forming and machining specific pieces. The productivity is focused to manufacture as many pieces as a machine/operator can make to obtain a high-quality product. As turning operation is one of the most common machining processes, it must be optimized to serial production and reducing costs. Rotating speed of the turning machine has the major influence on the machining, however, other operating parameters, such as depth of cut, also increases material removal rate [1]. Previous investigations have studied how to increase the material remove, by combining operating parameters

L. Ramírez (✉)
Universidad Politécnica Salesiana, Quito 170702, Ecuador
e-mail: lramirezg@ups.edu.ec

C. Guilcaso · W. Castillo · D. Molina
Universidad Técnica de Ambato, Ambato 180207, Ecuador

Á. Rocha et al. (eds.), *Communication, Smart Technologies and Innovation for Society*, Smart Innovation, Systems and Technologies 252,
https://doi.org/10.1007/978-981-16-4126-8_42

and different cutting tool geometry; nevertheless, cutting tool wear has an important influence on the process, and it has not the corresponding attention.

It is difficult to predict the influence of each parameter on the response exactly; hence, different methods, such as Taguchi and ANOVA are utilized. The tool wear was investigated by Kuntoglu and Saglam [2] to determine the optimized machining parameters in turning. The experiment used AISI 1050 steel and three different cutting tool types, all of them with the same geometry but with different coatings. The parameter with the major influence on the cutting tool flank wear was the cutting speed with 44% and the cutting tool coating was an influence of 34%.

Laghari et al. [3] studied the optimization of tool wear in turning of Al/SiCp using response surface methodology. The experiment was carried out by using tungsten carbide tools at cry cutting conditions and the combination of operating parameters with four different levels. The more influenced parameters were cutting speed and feed rate, obtaining and influence of 45% in the cutting tool wear. The authors recommend a cutting speed, feed rate, and depth of cutting of 6.28 mm·min^{-1}, 0.01 mm rev^{-1}, and 0.2 mm, respectively, to maximize the cutting tool life to machine the analyzed material.

A cutting tool can be affected by different types of wear, such as abrasive and adhesive wear, presented on the flank mainly. Suresh et al. [4] developed an experimental investigation on tool wear on turning AISI H13 steel by using response surface methodology and ANOVA. Three different cutting tools were utilized, tungsten carbide, and ceramic ones, with coating and without coating. The experiments consisted in four different operating parameters at five levels, where the cutting speed had the major influence on the response with 22.5, 14, and 12.6% for each cutting tool, respectively.

Dutta and Mukherjee [5] studied the machining parameter for tool wear of steel during turning by ANOVA to optimize them. The authors analyzed several cutting parameters, such as rotating speed, feed rate, and depth of cut, obtaining 27 different combinations of parameters. The parameter with major influence on the tool wear was rotating speed with 49.51%, followed by feed rate and finally the depth of cut with 25.79 and 23.87%, respectively. The author presented the following parameters to reduce cutting tool wear at wild steel turning, 250 rpm, 0.08 mm rev^{-1}, and 0.4 mm for rotating speed, feed rate, and depth of cut, respectively.

Aluminum alloys (AA) are used in different applications; hence, it is important to establish the optimum operating parameters for the machining processes. ANOVA and Taguchi methods were used by Niranjan et al. [6] to propose an optimization of cutting process parameters on AA6061. The required surface roughness had to be less than 1 μm; therefore, different combinations of cutting speed and depth of cut were experimented. For this alloy, authors recommend to configure operating parameters at 429 m min^{-1} and 2 mm for cutting speed and depth of cut, respectively.

Turning process parameters optimization on aluminum alloy AA6082 by using Taguchi and ANOVA was presented by Palaniappan et al. [7]. The orthogonal array was 127 combinations by using tungsten carbide (WC) cutting tool. The optimum parameters to machining this alloy was 0.25 mm rev^{-1} and 1.0 mm for feed and depth of cut, respectively, at 1600 rpm. The authors evaluated the process according to the surface roughness, establishing that rotating speed is the most significant parameter on it.

An optimization of turning parameters of AA6082 by using Taguchi method was proposed Aryan et al. [8]. Two different cutting tools were utilized for the machining process, HSS, and tungsten carbide tools. The aim of the investigation was to determine the optimum conditions to get the highest material remove; therefore, by using HSS tool, ANOVA presented that rotating speed had an influence of 43% on the response. AA6082 machining was done at 900 rpm with 1.1 mm and 8° for depth of cut and flank angle, respectively, for HSS tool. Meanwhile, the tungsten carbide cutting tool had minor flank wear, persevering the rotating speed with a 62.7% of influence, but depth of cut and flank angle had a negligible influence, with 9.8 and 7.4%, respectively.

An analysis for the optimization of operating parameter for flank wear and cutting forces in turning of AISI 5140 was presented by Aslan [9]. The author chooses an orthogonal array of four parameters and two levels, cutting tool angle, cutting speed, feed rate, and depth of the cut, needing 16 experiments. The speed increasing generated more friction, and it affected the cutting tool wear, where the highest influence on the wear was obtained by the feed rate, with 53.2%. The cutting tool angle had an influence of 22.4% and cutting speed 18.5%. The author recommends to machine this steel at 500 m mm^{-1}, 0.06 mm rev^{-1}, and 1.5 mm for cutting speed, feed rate, and depth of cut, respectively, by using a 60° cutting tool angle.

The aim of this study was to provide information about cutting parameters for AA7076 CNC turning machining by presenting the influence of them in this process. According to this information, it is going to be possible to select cutting parameters adequately; therefore, productivity shall increase by reducing cutting wool wear. This document id is distributed as follows. Materials and methods expose the analytical process to select parameters and levels, as well as the formulas and experimental development according to standards. Results section presents the analysis of photos taken with the scanning electronic microscope and the ANOVA responses obtained by using a software. Finally, conclusions discuss about the obtained information, and the combination of parameters and levels to obtain the minimal cutting tool wear is proposed.

42.2 Materials and Methods

42.2.1 Selection of Factors and Levels

Factors and levels have been selected according to the influence of each cutting parameter on the cutting tool wear. ISO 3685:1993 standard [10] indicates the process to evaluate a cutting tool wear; therefore, the operating conditions were chosen considering close values to the limits. This standard presents different conditions for the cutting process, with a tolerance of ± 5% according to the parameters and gears

Table 42.1 Experimental levels and factors

	Factors	Levels	
A	Cutting speed (S_c) [m·min^{-1}]	200	400
B	Feed rate (f) [mm·rev^{-1}]	0.1	0.3
C	Depth of cut (D_c) [mm]	1.5	3
D	Cutting tool type	CVD	PVD

of the turning machine. Two different types of cutting tools were analyzed, both of them with the same geometry DCGX 11T3, but with different coating, physic vapor deposition (PVD) and chemical vapor deposition (CVD), class N for aluminum and aluminum alloys. Table 42.1 presents the parameters and levels to carry out the experimental process.

42.2.2 *Taguchi Method*

An orthogonal array is required to organize the defined factors and levels, according to their degrees of freedom (*DF*). By using Taguchi method, it is possible to demonstrate that *DF* of the chosen orthogonal array must be greater than or equal to the total *DF* of the experimental process. Therefore, Dao and Huang [11] calculated *DF* of the process by:

$$DF = n_f \cdot (n_1 - 1) + n_i \cdot (n_1 - 1) \cdot (n_1 - 1), \tag{42.1}$$

where n_f, n_i, n_l are the number of iterations, factors, and levels of the experimental process. The experiment considered two levels and 4 factors, which generated an orthogonal array with 16 experimental tests by using a software. Table 42.2 contains the combination of parameters and levels for each experiment in such a way that factors happen the same times.

42.2.3 *Wear Measurement*

To support the experimental process and to complement the analytical study, cutting tools wear was seen through a Tescan Vega$_3$ scanning electron microscope (SEM). By this visual inspection, it was possible to determine the loss of material in the flank wear of the tool. ISO 3685:1993 standard was utilized to carry out the analysis of the cutting tool wear by studying the lack of the material in the cutting tool after machining.

Table 42.2 Relation between parameters and levels for number of experiments

No. of experiment	Cutting speed (S_c) [m min^{-1}]	Feed rate (f) [mm rev^{-1}]	Depth of cut (D_c) [mm]	Cutting tool type
1	200	0.1	1.5	CVD
2	200	0.1	1.5	PVD
3	200	0.1	3.0	CVD
4	200	0.1	3.0	PVD
5	200	0.3	1.5	CVD
6	200	0.3	1.5	PVD
7	200	0.3	3.0	CVD
8	200	0.3	3.0	PVD
9	400	0.1	1.5	CVD
10	400	0.1	1.5	PVD
11	400	0.1	3.0	CVD
12	400	0.1	3.0	PVD
13	400	0.3	1.5	CVD
14	400	0.3	1.5	PVD
15	400	0.3	3.0	CVD
16	400	0.3	3.0	PVD

42.2.4 *Turning Machining*

The analyzed material was aluminum alloy (AA) 7075, and every test required a cylindrical piece of 60 × 370 mm of diameter and length, respectively. According to ISO 2540:2016 [12] standard, these specimens must be prepared with face turning machining and a central drilling of 6.3 mm at 120°. It indicates that, machining length must be 300 mm; hence, 70 mm is the subjection length inside the chuck of the turning machine. ISO 3685:1993 has the purpose to provide guidelines to analyze the change of shape of the cutting tool. The cutting process was carried out by using a CNC turning machine, with maximum of 10 times through the length and a rotating speed around 1100 rpm for a diameter of 60 mm.

42.2.5 *Analysis of Variance*

Analysis of variance (ANOVA) is utilized to analyze the variability of data, getting a reliability percentage measured by the sample variance. The total summation for the obtained squares for each factor is calculated by [13]:

$$S_{\mathrm{T}} = \sum_{i=1}^{n} Y_i^2 - \left[\left(\sum_{i=1}^{n} Y_i \right) / n \right], \tag{42.2}$$

where Y_i is the response to ith iteration. Savkovic et al. [14] defined the variance (V) for each factor in function of S_T and DF by applying:

$$V = \frac{S_T}{DF} \tag{42.3}$$

Fisher's test (F) compares the relation between obtained values for variance and for the error (e), and it is obtained by:

$$F = \frac{V}{e} \tag{42.4}$$

This analysis lets to know the percentage influence (P) of each factor by using this equation:

$$P = \frac{V}{S_{\mathrm{T}}} \cdot 100\% \tag{42.5}$$

42.3 Results

A CNC cycle was programed to machine the AA7075 specimens until the estimated length was obtained. Figure 1a presents the material specimen in the supply conditions; meanwhile, Fig. 1b, c shows the material after turning machining with S_c of 400 mm·min^{-1}, f of 0.3 mm rev^{-1}, D_c of 3 mm, by using CVD, and PVD tool coating, respectively.

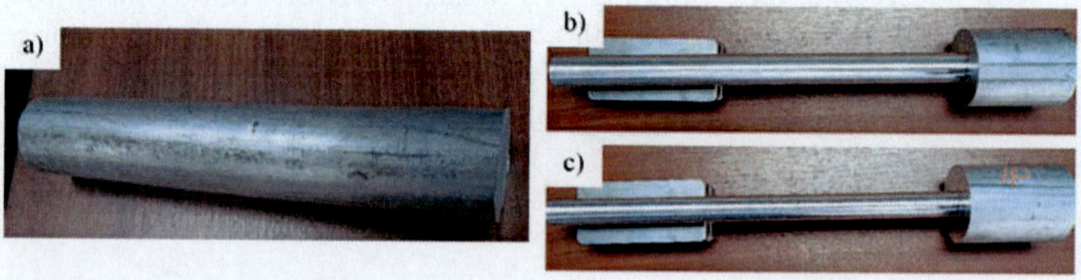

Fig. 42.1 AA7075 **a** supply conditions, **b** CVD machining, **c** PVD machining

42.3.1 Cutting Tools Wear Values

Figure 42.2 represents the measurements of cutting tools wear taken for each experiment by using SEM. There are several differences between measurements due to the influence of chosen parameters and levels. V_{B} indicates the central flank wear in the cutting tool; meanwhile, V_{M} and V_{S} are the shortest and the deepest flank wear, respectively.

Figure 3a shows that experiment 8 obtained the highest values for cutting tool wear, with values for VM, VB, and VS of 104.76, 122.05, and 136.45 μm at 200 m min^{-1}, 0.3 mm rev^{-1}, and 3 mm for S_{c}, f, and D_{c}, respectively. The minimum cutting tool wear was 0.64 μm in the experiment 13 by using a S_{c}, f, and D_{c} of 400 m mm^{-1}, 0.3 mm·rev^{-1}, and 1.5 mm, respectively. This information is presented in Fig. 3b.

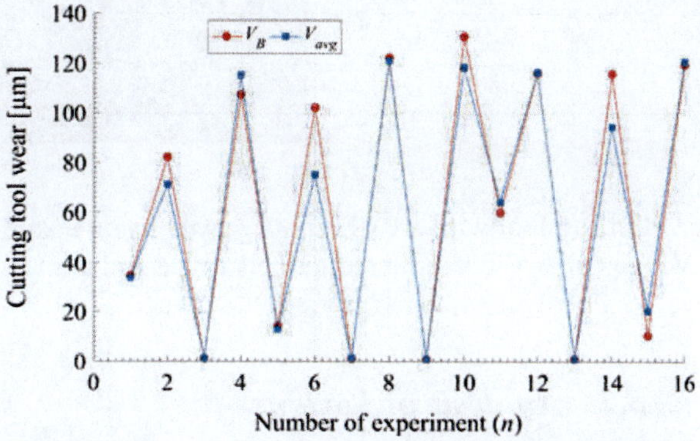

Fig. 42.2 Cutting tool wear for each experiment

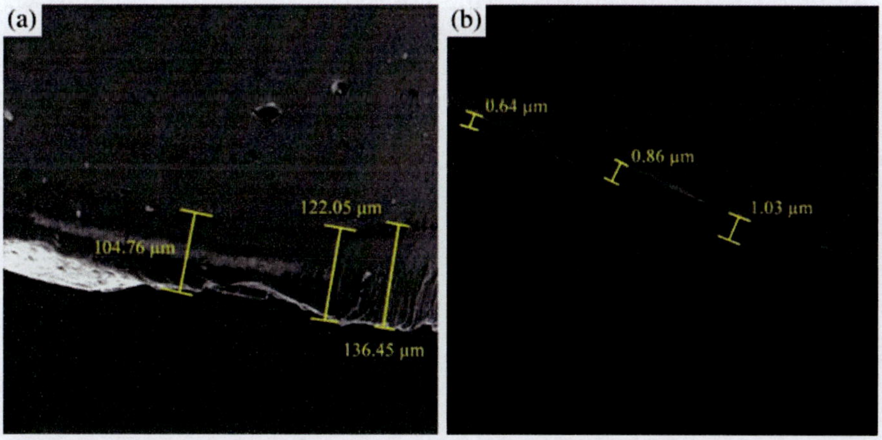

Fig. 42.3 Cutting tool wear looked through SEM at 50 μm **a** maximum, at experiment 8, **b** minimum, at experiment 13

Fig. 42.4 PVD cutting tool wear looked through SEM at 2 mm

Cutting tool with a PVD coating even had a naked eye wear as shown Fig. 42.4, where cut AA7075 adhered to the surface and a lot of notches are noticeable.

42.3.2 *Analysis of Variance*

Fisher's test (F) considered the geometric distribution of the number of experiments, and it determined the factor that had the most influence in the cutting tool wear after turning machining. F value was 81.25 for the cutting tool type, having an influence about 82%; meanwhile, S_c, f, and D_c had a negligible influence in the final response, 1.84, 0.95, and 3.95%, respectively. Table 42.3 contains the most important information that was calculated with ANOVA.

Table 42.3 Analysis of variance results

Source	DF	S	V	F	[%]
S_c	1	682.78	682.78	1.82	1.84
f	1	350.06	350.06	0.94	0.95
D_c	1	1464.34	1464.34	3.91	3.95
Cutting tool type	1	30,414.78	30,414.78	81.25	82.14
Error	11	4117.83	374.35		
Total ST	15				

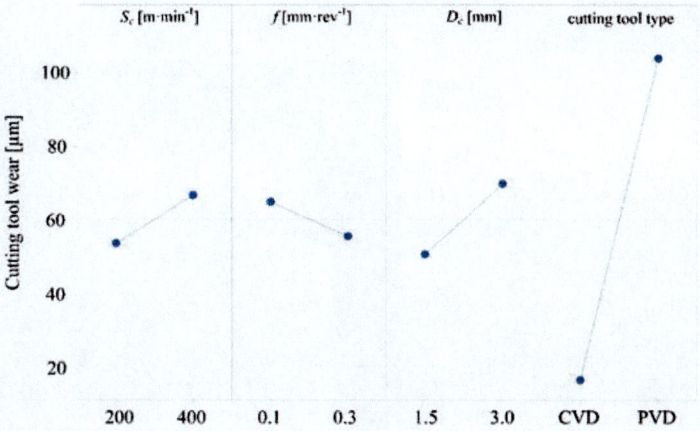

Fig. 42.5 Effect of the factors and levels in the cutting tool wear

Figure 42.5 presents the main effects for cutting tool wear according to the chosen factors and levels. The largest difference is observed in the cutting tool type; therefore, it is obvious that this parameters is going to be the most influential factor on the response. After analyzing the obtained results by the analytical process, the other factors, such as Sc, f, and Dc had a minimal influence.

Figure 42.6 contains the iterations for each cutting parameter in the turning machining process, S_c, f, and D_c. These factors, which were considered for the orthogonal array, did not have a significant difference between them. However, there

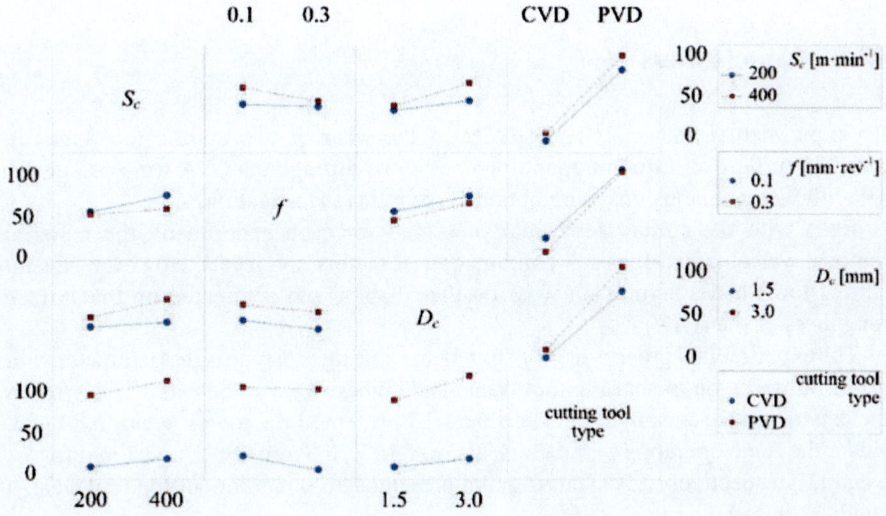

Fig. 42.6 Interaction between factors and levels for cutting tool wear

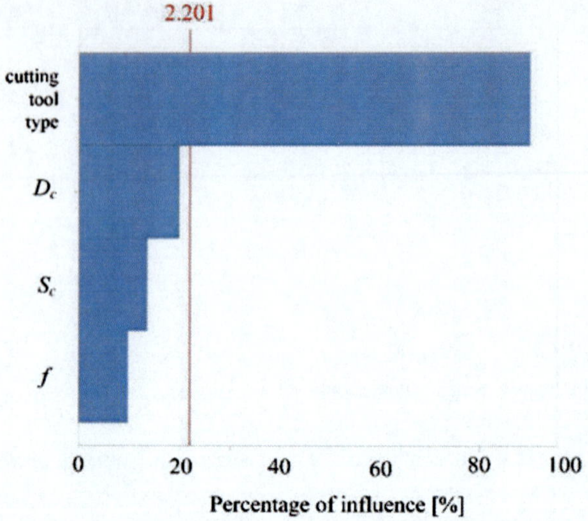

Fig. 42.7 Pareto chart for standardized effects

is a large range with the values for the different cutting tool types; hence, this factor is going to influence into the response.

Figure 42.7 analyzes the standardized effects according to the reference line, with a value of 2.2% provided by the software. The operational factors, such as S_c, f, and D_c are below the line; therefore, they do not have a significant influence on the response as long as the cutting tool type is clearly the most important factor, reaching an 82.14% of the influence during the machining process.

42.4 Conclusions

An experimental and analytical studies of the wear of two cutting tool types in AA7075 turning at different operating conditions through ANOVA were presented. The following conclusions were obtained from this investigation.

Even with the cutting tool wear presented for each experiment, the material removal was obtained for each combination of factors and levels. However, several cutting tools had a significant wear on their flanks, and it affected on the surface roughness of the AA7075.

The experimental process demonstrated that cutting tool type is the parameter with major influence on the cutting tool wear. The highest wear value was 121.09 μm in the experiment 8; meanwhile, experiment 13 presented the lowest wear, 0.84 μm. Under the same operating parameters, 200 m min^{-1}, 0.3 mm rev^{-1}, and 3 mm for S_c, f, and D_c, respectively, CVD cutting tool presented 1.20 μm for cutting tool wear, a hundred times lower than PVD.

The analysis of obtained Figures by using ANOVA and a software allowed to analyze the parameters and levels in an easy visual manner. The minimum value of

2.201 μm to consider a parameter as negligible determined that the only variable to consider was the cutting tool type, with an influence of 82.14% on the response.

Hence, the following values are proposed as the optimum parameters for turning machining of AA7075, CVD as cutting tool type, cutting speed of 400 m·min^{-1}, feed rate of 0.3 mm·rev^{-1}, and 1.5 mm for the depth of the cut.

References

1. M. Modi, G. Agarwal, V. Patil, A. Kare, S. Sukla, A. Sankhala, Modeling and analysis of turning process on lathe machine by Taguchi and Anova approach. Int. J. Sci. Technol. Res. **8**(10), 1466–1470 (2019)
2. M. Kuntoglu, H. Saglam, Investigation of progressive tool wear for determining of optimized machining parameters in turning. Measurement **140**, 427–436 (2019)
3. R. Laghari, J. Li, Z. Xie, S.-Q. Wang, Modeling and optimization of tool wear and surface roughness in turning of Al/SiCp using response surface methodology. 3D Res. **9**, 1–13 (2018)
4. R. Suresh, A. Joshi, M. Manjaiah, Experimental investigation on tool wear in AISI H13 die steel turning using RSM and ANN methods. Arab. J. Sci. Eng. **46**, 2311–2325 (2020)
5. S. Dutta, S. Mukherjee, Optimization of machining parameters for tool wear of mild steel during turning by S/N ratio and ANOVA. Int. J. Res. Anal. Rev. **46**(1), 662–666 (2019)
6. D. Niranjan, G. Shivashankar, R. Sreenivas, R. Praveen, Optimization of cutting process parameters on AL6061 using ANOVA and TAGUCHI method. Mater. Today: Proc. **4**, 10845–10849 (2017)
7. S. Palaniappan, K. Muthukumar, R. Sabariraj, S. Kumar, T. Satish, CNC turning process parameters optimization on aluminum alloy AA6082 by using Taguchi and ANOVA. Mater. Today Proc. **21**, 1013–1021 (2020)
8. R. Aryan, F. John, S. Kumar, A. Kumar, Optimization of Turning parameters of AL-alloy 6082 using Taguchi method. Int. J. Adv. Res. Innov. **5**(2), 268–275 (2017)
9. A. Aslan, Optimization and analysis of process parameters for flank wear, cutting forces and vibration in turning of AISI 5140: a comprehensive study. Measurement **163**, 107959 (2020)
10. ISO, IN *ISO 3685:1993 Tool-life testing with single-point turning tools* (1993). https://www.iso.org/standard/9151.html. Last accessed 2 Nov 2020
11. T. Dao, S. Huang, Optimization of a two degrees of freedom compliant mechanism using Taguchi method based grey relational analysis. Microsyst. Technol. **23**, 4815–4830 (2017)
12. ISO, *ISO 2540:2016 Centre drills for centre holes with protecting chamfer* (2016). https://www.iso.org/standard/64194.html. Last accessed 2 July 2020
13. T. Chen, M. Xu, J. Tu, H. Wang, X. Niu, Relationship between omnibus and post-hoc tests: an investigation of performance of the F test in ANOVA. Shanghai Arch. Psychiatry **30**(1), 60–64 (2018)
14. B. Savkovic, P. Kovac, A. Stoic, B. Dudic, Optimization of machining parameters using the Taguchi and ANOVA analysis in the face milling of aluminum alloys AL7075. Tech. Gazette **27**(4), 1221–1228 (2020)

Chapter 43
Secure Data Model for the Healthcare Industry in Ecuador Using Blockchain Technology

Guillermo Melendrez-Caicedo and Joe Llerena-Izquierdo

Abstract Patients' and doctors' mistrust of data security in third-party information storage is a latent and relevant problem in the healthcare sector in Ecuador. This paper presents a data model to improve information security in Ecuador's healthcare sector by using Blockchain technology. A descriptive research is conducted using an exploratory-deductive, qualitative methodology to analyze relevant studies referencing Blockchain and Smart Contract technologies in the healthcare industry. The results of the study present the definition of participants and roles for a Blockchain-based healthcare model in Ecuador, a conceptual architecture to improve information security in the healthcare sector and impact factors. It is concluded that Blockchain technology allows sharing information securely; the proposed model is developed in the Hyperledger platform with 56% of information analyzed remaining as a private network.

43.1 Introduction

In health care, professionals such as doctors and nurses work in physical infrastructures where clients come for medical services. These places provide the necessary spaces to offer specialized care that is funded both privately and publicly, including hospitals, clinics, primary care centers, medical care points, and doctors' offices. Patient data is stored by each provider according to its own internal processes. Public health data is stored in a database accessed only by public sector providers. In Ecuador, a segment of the healthcare industry has replaced handwritten records with computerized medical recording systems [1], improving security and increasing information efficiency while forming part of a strategic policy [2].

G. Melendrez-Caicedo · J. Llerena-Izquierdo (✉)
Universidad Politécnica Salesiana, Guayaquil, Ecuador
e-mail: jllerena@ups.edu.ec

G. Melendrez-Caicedo
e-mail: gmelendrezc@est.ups.edu.ec

Á. Rocha et al. (eds.), *Communication, Smart Technologies and Innovation for Society*,
Smart Innovation, Systems and Technologies 252,
https://doi.org/10.1007/978-981-16-4126-8_43

Blockchain technology is suggested as a solution in medical data management [3]; in health care, it is used to maintain patient identity, manage medical inventory, clinical reviews and investigations, medical adulterations, and healthcare systems, along with other tasks [4]. Blockchain is a distributed database resource (Ledger) that contains participants' records; it uses cryptography and distributed computing to reach consensus among participants; data is stored in blocks chained by hash values and is in chronological order; all data blocks are copied to participating nodes that are distributed in the network [1]; transactions are kept immutable among nodes. Each block contains data, timestamp, hash, and keys and is divided in three categories: public, private, and consortium [4]; an immutable data block is distributed across participating nodes, and there is no central authority [5].

Some of the data security and storage challenges analyzed in previous studies include: non-integration of computer applications, management challenges, demand for multiple accesses, interoperability drawbacks, low levels of security, and privacy of information [1]; possible attacks on medical information stored on providers' servers or in cloud-based services; lack of medical data privacy [5]; the use of centralized systems to store data and reports; poor filtering or exploitation of information; inconsistencies such as patients without control of their data; and slow systems [6]. IT applications have the following problems: interoperability, i.e., exchange of information between different applications; asymmetric access, i.e., healthcare facilities have access to data and not patients; data visibility violations [2]. The lack of reliable data storage solutions generates doubt in patients and physicians, and the problem of information security is highly relevant when the storage is in third-party sites.

Why is it necessary to secure health data in Ecuador using Blockchain technology?

To understand the importance of medical data, it is necessary to review the information security models in the medical area, understand this new technology applied in a healthcare system, examine the benefits of the technology, and consider the challenges.

This research aims to propose a data model to improve information security in the healthcare industry in Ecuador through Blockchain technology.

43.2 Materials and Methods

A descriptive research study is conducted using an exploratory-deductive, qualitative methodology to analyze studies related to Blockchain and Smart Contract technologies and their application in health care.

In Ethereum's Blockchain technology platform for healthcare providers [1], each node is a provider, and the provider is responsible for the Blockchain; data from each provider is brought to the Blockchain for interoperability between healthcare systems. Ethereum's system is based on three modules; users are nurses, doctors, patients, and administrators each with specific access levels. In the literature review, some authors make a qualitative comparison of Ethereum's architecture with other Blockchain proposals in access, storage, integrity, and scalability

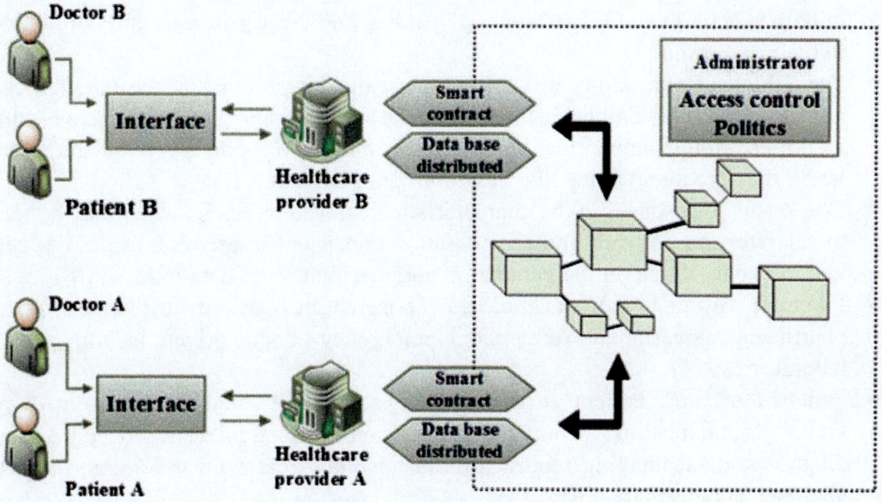

Fig. 43.1 Healthcare architecture in Blockchain

[2]. The Hyperledger-based architecture contains Smart Contract and access policies to maintain data traceability and logging; data capture is completed with medical sensors. The system is comprised of the following components: patients, healthcare providers, sensors, data, node, and Blockchain; the processing time of test runs of 5 to 320 patients is 40 ms [3]. A 4-layer architecture and tiered authentication is proposed [4], for access, storage, and sharing of data between healthcare providers (see Fig. 43.1); doctors and patients have access to their data, and data is protected against attacks through Blockchain's safeguards.

The system's architecture is composed of six major components and access policies that allow data to be generated from medical devices connected to an IoT network, while giving patients control over their data [5]. The proposal contains a contract structure and two participating nodes. The structure may contain patient data such as names, address, ID, date, healthcare provider, costs, and signatures; the components of the contract are patient registration and signature of the document [6]. In the data protection model with the aim of increasing levels of patient privacy, data source certification, and patients' control over their data, a similar model in Ethereum is used. Simulations are performed with more than 1.5 million records with maximum times reaching 122 s [7]. Attacks are classified into five types: network attacks, user wallet, Smart Contract attacks, transaction verification, and exploit groups [4].

The components of the Blockchain model are:

- Ledger: Ledger is the data structure distributed across the nodes where each network participant contains a copy of the transactions; moreover, these data blocks are kept consistent and immutable [8].
- Smart Contract: Smart Contract is a software code that is implemented in the Ledger, it automates the processes of updating nodes to the Blockchain, it is used

in both Ethereum and Hyperledger platforms, and these processes are immutable in their execution [1].

- Permissions: Permissions are the characteristics that define permissioned blocks and permissionless blocks. In the former, the administrator gives permissions with identities through authentication certificates like the Hyperledger platform; in the latter, they are anonymous like the Ethereum platform [9].
- Consensus: Consensus is the characteristic achieved by nodes when they adhere to rules for the network and thus obtain a consistent transaction in the Ledger, i.e., the nodes agree on the generation and distribution of data blocks [10].
- Potential participants or stakeholders: Those entities such as hospitals, doctors, pharmacies, government, researchers, emergency outlets, private insurance, and laboratories [11].
- Patient attributes: Patient attributes are characteristics that are determined by vital signs, medications, laboratory results, comments, fluid data, procedure data, diagnostic data, imaging report, time in hospital, treatment, vaccination times, allergies, and discharge data [11].

Among the advantages of this technology, lies that Blockchain is not centralized, nor managed by a single person/administrator; it guarantees high levels of data security; the security of block transactions is distributed throughout the network; it allows a very particular distributed data repository [12]; it creates identities for users; its data is immutable; it authorizes interested parties to save a transaction [7]; it does not use intermediaries in the transactions; it does not have layer dependency; it has non-updatable data; and it provides traceability of transactions [13].

As for the disadvantages, Blockchain has a slow digital signature verification; a company does not have control of the network; high costs in time and knowledge to adopt this technology; legacy networks compromise the project; excessive data growth; energy consumption [13].

The considerations for the model are: (i) to appoint public and private healthcare providers; (ii) to adopt Blockchain to generate a secure model for healthcare data; (iii) to propose a network in Hyperledger for the private part of the Blockchain network; (iv) to propose Smart Contract processes; (v) to propose that the model only include the following services for public users: appointments, diagnosis, and prescriptions; and (vi) to classify the references used between public and private platforms.

43.3 Results

43.3.1 Definition of Participants and Roles for the Blockchain-Based Healthcare Model in Ecuador

In Ecuador, according to the Organic Health Law [14], the Ministry of Public Health of Ecuador (MSP) is the governing authority in health. The proposed participants that manage data in the Blockchain network are the MSP, public hospitals, health centers, private insurance businesses, the Ecuadorian Institute of Social Security (IESS), and the Ecuadorian Red Cross.

The MSP is responsible for starting the network, configuring the network, creating the consortium, defining the network policies, providing access credentials, creating the nodes, and creating the structure for the transactions. The public hospital is part of the consortium and attends to patients for admissions or care that lasts for days or consultations with specialists. Health centers attend outpatient consultations. Private insurance companies attend to people who receive medical care through payment by an employer or through private payment. IESS for its affiliates who have general insurance through his or her job. Ecuadorian Red Cross for consultation of patient data in case of donation or extraction of blood or another vital organ.

43.3.2 Conceptual Architecture to Increase Data Security in the Healthcare Industry Through Blockchain Technology in Ecuador

The model proposes a conceptual architecture developed in Hyperledger which is a Blockchain platform with the following components:

The network administrator is from the first entity, i.e., the MSP, which defines the ordering services, network configuration, and authentication certificate; furthermore, the administrator adds the public hospital with its authentication certificate, and both entities are administrators. The MSP inputs the primary data of the citizens, using the civil registry database to immediately create a profile for each living person and newborn so that the other participants can only consult the primary data and can generate new data such as appointments and diagnoses. This information is stored in the Ledger.

The consortium contains the canal administrators and is comprised of the MSP and the public hospital. The Blockchain channel is created by the administrator through the consortium. This channel is managed with channel configuration (see Fig. 43.2).

All network participants are a node and have a specific authenticity certificate, software application, Smart Contract copy, and Ledger copy. They are also connected to the channel; each node depends on the Ledger status, and the Ledger is encrypted;

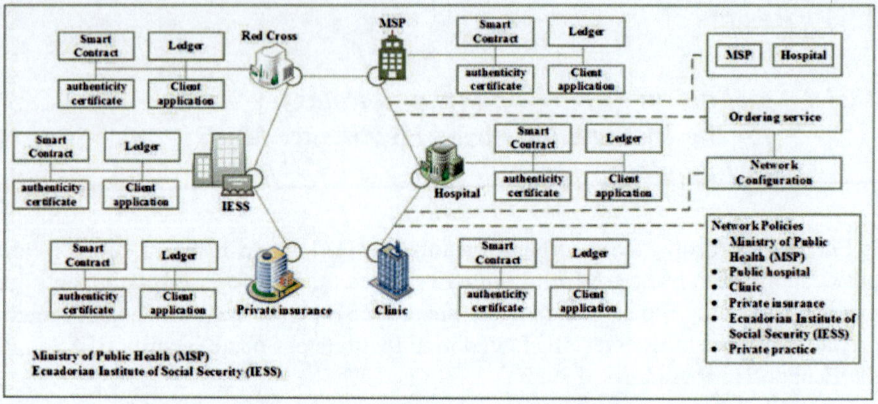

Fig. 43.2 Proposed conceptual healthcare architecture in Blockchain

access is granted according to an authentication certificate that controls and restricts access to the patients' data.

The membership provider specifies that each participant has a local membership for local collaboration rights, and a channel membership for channel collaboration rights; this provider issues and validates the certificates and applies cryptographic protocols; there can be several membership providers, facilitating scalability.

Ordering services allow nodes to connect to the channel and send node transactions, and these transactions are received by all nodes to ensure delivery and delivery reliability; each node can be created, restarted, stopped, excluded, or configured [15].

The Smart Contract must be considered within the architecture by including read and write processes in the Ledger. These processes are invoked by the participants or when performing an activity in the Ledger and contain the access policies. The processes are presented in Table 43.1.

In the architecture, it is also considered a data model; for a better understanding, the data model of the Ledger is represented in a similar way to a database; the Smart Contract procedures write and read the data blocks. The model and the relationships between the data are represented (see Fig. 43.3), the Users table contains the data of the authorized persons with their credentials; the table Pacient contains the data of the citizens with their details that can be obtained from the Civil Registry.

The Appointments table contains the medical calls that are delivered to the citizens, it is linked to the hospital table and the doctor table, in addition the calls have the medical details and the medicines recommended by the doctor; the Medicines table contains the details of the drugs that doctors can use; the Digital Certificate table contains the identification of the users; the Encrypted Key table contains the users' keys and the blockchain network applies the AES algorithm.

Table 43.1 Smart Contract procedures

Processes	Description	Responsible
Update participants	Create and update network participants	Ministry of Public Health
Update policies	Create and update network and access policies for the dataset	Ministry of Public Health
Create certificates	Create and update certificates of authenticity	Ministry of Public Health
Create patients	Create the data of the citizens registered in the civil register	Ministry of Public Health
Create appointments	Register a new appointment for a citizen	All except Red Cross
Create diagnostics	Register a new diagnosis for a citizen	All except Red Cross
Create recipes	Register a new prescription for a citizen	All except Red Cross
View patient data	Citizen data selection	All
Consult appointments	Selection of citizen appointments	All
Consult diagnostics	Selection of citizen diagnostics	All
Consult perscriptions	Selection of citizen perscriptions	All

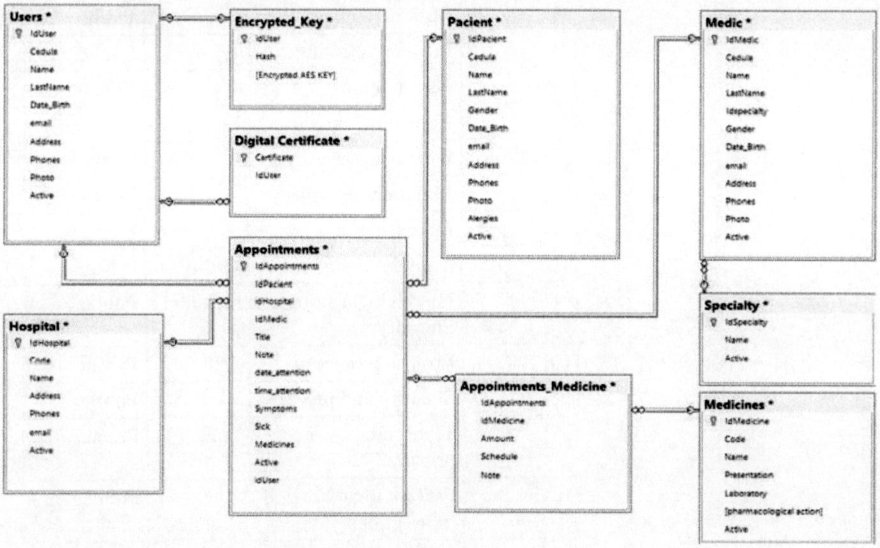

Fig. 43.3 Blockchain data model for architecture

43.3.3 Impact of Blockchain Technology in the Healthcare Sector in Ecuador

A seen in the studies included in the bibliography, Blockchain technology is relevant when applied to the healthcare industry. This small sample shows which platforms can benefit from this technology (Table 43.2).

When analyzing the objectives of the references included in this study, it is determined that three focused on security directly, six focused on security components (confidentiality, integrity, and availability), and nine focused on privacy.

Of the references in public platforms 7, corresponding to 39%, use Ethereum; of the references in private platforms 10, corresponding to 56%, use Hyperledger; 1 reference corresponding to 5% uses a hybrid platform employing both Ethereum and Hyperledger in its architecture. The general objectives of this group of references: 3 focused on security directly, 6 focused on security components (confidentiality, integrity, and availability), and 9 references focused on privacy.

Table 43.2 Utility of Blockchain in healthcare

References	General objective	Platform
[1]	Minimize manipulation	Public
[2]	Increasing confidentiality	Public
[3]	Data tracking	Private
[4]	Protection against attacks	Private
[5]	Increasing privacy	Private
[6]	Maintaining privacy	Private
[7]	Increasing privacy	Public
[8]	Confidentiality and scalability	Private
[9]	Hierarchical storage to maintain integrity	Public
[10]	Privacy protection	Private
[11]	Security and privacy	Public
[15]	Transparency of access and privacy	Private
[16]	Prevent unauthorized transactions	Public
[17]	Improve privacy and reliability	Public
[18]	Increasing protection	Private
[19]	Security and scalability	Private
[20]	Maintaining privacy and confidentiality	Private
[21]	Privacy and security	Public/private

43.4 Discussion

In the proposed conceptual model scalability is completely viable because a new participant obtains a copy of the Ledger and the Smart Contract. Transaction blocks are in the Ledger; access control is relevant in the management of health data because the data is extremely sensitive. Policies preside over access and authorizations contributing to the security and privacy of medical data; high levels of data confidentiality are guaranteed because Blockchain does not allow users outside the network to enter and extract data, only authorized users; regarding system integrity, the immutability of the data over time is guaranteed, and patient data is added to the transaction block; regarding availability, the data is copied or distributed in all nodes. This conceptual proposal is an alternative to increase data security focused on blocks of transactions generated by Blockchain technology; among its properties, the data remains immutable and guarantees the traceability of updates.

Only public or private entities are considered as participants in the private network based in Hyperledger; this model does not consider patients as participants because it would extend the Blockchain network to public networks and consider many elements for the Ethereum network.

This research can serve as a basis for other researchers to continue the design of data, such as the attributes or columns of entities such as patients, diagnoses, and prescriptions; it can also serve to continue the design of the public part made up of patients, or to store data from other services provided to patients. This research agrees that Blockchain architecture is relevant in the healthcare industry based on Refs. [1–4, 7, 10, 11, 15, 16] and [19]; knowledge and experiences are adopted to propose a conceptual architecture. References [5, 6, 8, 17, 18] present Blockchain models in health care; reference [9] is a proposal for Blockchain with IoT; reference [12] is a general model in health care; references [20] and [21] present logical designs in Blockchain.

Of the 18 references, 50% focused on security and 50% focused on privacy; 39% are public networks in Blockchain; 56% are private networks; and 6% are mixed networks.

43.5 Conclusions

It is concluded that Blockchain technology allows sharing information securely; the model proposed in this study uses the Hyperledger platform to remain as a private network; the review of rigorous analysis on previous works of relevance obtained 39% of entities use Ethereum, 56% use Hyperledger, and 5% are hybrids. This determines the prevalence of a type of technology that has had a high penetration rate in the environment compared to others. The security and efficiency of health records are especially important for interoperability, exposure time, sharing between entities, access by various levels of professionals, among others.

The functions of the Smart Contract allow access to participants through a Ledger. A distributed and consistent solution is suggested; moreover, Blockchain technology is useful in the healthcare industry because medical records are kept with confidentiality, integrity, and availability which are components of information security.

References

1. G. Yang, C. Li, A design of blockchain-based architecture for the security of electronic health record (EHR) systems, in *Proceedings of International Conference on Cloud Computing Technology and Science (CloudCom)* (2018), pp. 261–265. https://doi.org/10.1109/CloudCom2018.2018.00058
2. A. Shahnaz, U. Qamar, A. Khalid, Using Blockchain for electronic health records. IEEE Access. **7**, 147782–147795 (2019). https://doi.org/10.1109/ACCESS.2019.2946373
3. H. Guo, W. Li, M. Nejad, C.C. Shen, Access control for electronic health records with hybrid blockchain-edge architecture, in *Proceedings of 2019 2nd IEEE International Conference on Blockchain (Blockchain 2019)* (2019), pp. 44–51. https://doi.org/10.1109/Blockchain.2019.00015
4. B.L. Radhakrishnan, A. Sam Joseph, S. Sudhakar, Securing blockchain based electronic health record using multilevel authentication, in *2019 5th International Conference on Advanced Computing and Communication System (ICACCS 2019)* (2019), pp. 699–703. https://doi.org/10.1109/ICACCS.2019.8728483
5. K.M. Hossein, M.E. Esmaeili, T. Dargahi, A. Khonsari, Blockchain-based privacy-preserving healthcare architecture, in *2019 IEEE Canadian Conference on Electrical and Computer Engineering (CCECE 2019)* (2019). https://doi.org/10.1109/CCECE.2019.8861857
6. V.M. Harshini, S. Danai, H.R. Usha, M.R. Kounte, Health record management through blockchain technology, in Proceedings of International Conference on Trends in Electronics and Informatics (ICOEI 2019) (2019), pp. 1411–1415. https://doi.org/10.1109/icoei.2019.8862594
7. Y. Zhuang, L.R. Sheets, Y.W. Chen, Z.Y. Shae, J.J.P. Tsai, C.R. Shyu, A patient-centric health information exchange framework using blockchain technology. IEEE J. Biomed. Heal. Inform. **24**, 2169–2176 (2020). https://doi.org/10.1109/JBHI.2020.2993072
8. S.M.T. Toapanta, , F.G.M. Quimi, M.G.T. Espinoza, L.E.M. Gallegos, Proposal of a model to apply hyperledger in digital identity solutions in a public organization of Ecuador. In: 2019 Third World Conference on Smart Trends in System Security and Sustainable (2019), pp. 21–28. https://doi.org/10.1109/WorldS4.2019.8903981
9. G. Wang, Z.J. Shi, M. Nixon, S. Han, ChainSplitter : towards blockchain-based industrial IoT architecture for supporting hierarchical storage, in *2019 IEEE International Conference on Blockchain (Blockchain) ChainSplitter* (IEEE, 2019), pp. 166–175. https://doi.org/10.1109/Blockchain.2019.00030
10. L. Xu, A. Bagula, O. Isafiade, T. Chiwewe, Design of a credible blockchain-based e-health records (CB-EHRS). Presented at the (2019). https://doi.org/10.23919/ITUK48006.2019.8995905
11. N. Poonguzhali, A. Deebika, A framework for electronic health record using blockchain technology. In: *IEEE ICSCAN 2020* (2020). https://doi.org/10.1109/ICSCAN49426.2020.9262369
12. F.M. Enescu, A. Cirstea, Blockchain technology applied in health the study of blockchain application in the health system (I), in *ECAI 2018—International Conference*, 10th Edn. (IEEE, 2018), pp. 10–13

13. B.A. Kumar, T. Mohanraj, S. Shahulhammed, R. Santhosh, A study of blockchain technologies and health care systems, in *Proceedings 4th International Conference on IoT in Social, Mobile, Analytics and Cloud (ISMAC 2020)* (2020), pp. 265–267. https://doi.org/10.1109/I-SMAC49090.2020.9243529
14. Nacional, E.L.C., Salud, L.E.Y.O.D.E., Preliminar, T.: Ley organica de salud del Ecuador (2015)
15. S. Naveen Kumar, M. Dakshayini, Secure sharing of health data using hyperledger fabric based on blockchain technology, in *2020 International Conference on Mainstreaming Block Chain Implementation (ICOMBI 2020)* (2020). https://doi.org/10.23919/ICOMBI48604.2020.9203442
16. Ajayi, O., Abouali, M., Saadawi, T.: Secured Inter-Healthcare Patient Health Records Exchange Architecture. Proc. - 2020 IEEE Int. Conf. Blockchain, Blockchain 2020. 456–461 (2020). https://doi.org/10.1109/Blockchain50366.2020.00066.
17. M.T. Quasim, A.A.E. Radwan, G.M.M. Alshmrani, M Meraj, A blockchain framework for secure electronic health records in healthcare industry, in Proceedings of International Conference on Smart Technologies in Computing, Electrical and Electronics (ICSTCEE 2020) (2020), pp. 605–609. https://doi.org/10.1109/ICSTCEE49637.2020.9277193
18. Y. Liang, Identity verification and management of electronic health records with blockchain technology, in *2019 IEEE International Conference on Healthcare and Informatics* (ICHI 2019) (2019), pp. 2019–2021. https://doi.org/10.1109/ICHI.2019.8904712
19. S. Hasavari, Y.T. Song, A secure and scalable data source for emergency medical care using blockchain technology, in *Proceedings of 2019 IEEE/ACIS 17th International Conference on Software Engineering, and Research, Management and Application (SERA 2019)*, pp. 71–75 (2019). https://doi.org/10.1109/SERA.2019.8886792
20. L. Hirtan, P. Krawiec, C. Dobre, J.M. Batalla, Blockchain-based approach for e-health data access management with privacy protection, in IEEE International Workshop on Computer Aided Modeling and Design of Communication Links and Networks (CAMAD 2019) (2019). https://doi.org/10.1109/CAMAD.2019.8858469
21. U. Goel, R. Ruhl, P. Zavarsky, Using healthcare authority and patient blockchains to develop a tamper-proof record tracking system, in *Proceedings of 5th IEEE International Conference on Big Data Security and Cloud, BigDataSecurity 2019, HPSC 2019, 4th IEEE International Conference on Intelligent Data Security* (2019), pp. 25–30. https://doi.org/10.1109/BigDataSecurity-HPSC-IDS.2019.00016

Chapter 44
Prototype of an Integrated System with IoT Technology for Physiotherapeutic Rehabilitation in Older Adults Through Assisted Technology in a Gerontological Center

Steven Pogo-Ateaga, Cindy Ortiz-Araujo, Milton Escobar-Sánchez, and Carlos Andrade-Villacís

Abstract This research is focused on creating a physiotherapeutic prototype through four activities: (1) structural design; (2) embedded system; (3) wireless communication; and (4) web application. Taking the data from the field survey conducted by ESPE branch Latacunga in collaboration with MIES Cotopaxi in the gerontological center "Sagrada Familia" canton La Maná, it was determined that the residents suffer mostly from osteoarthritis, arthritis and tendinitis, in addition, it was determined that the gerontological center does not have the necessary equipment to treat these diseases. The use of this prototype will help to reduce muscle aches and pains caused by stiffness in the joints of the elderly and will elevate their mood through a multimedia system that encourages them to continue with the exercises, and it will also facilitate the physiotherapist staff in the monitoring of their patients through the use of the web application via the IoT. The prototype was validated through technical tests performed in each activity of the established methodology; it was found that the structural design is functional and stable to support the weight of the patient, the embedded system controls the multimedia system effectively, allowing an adequate wireless connection and the transmission of the information (physiotherapeutic exercise of the patient) to the web application, storing those records in a database.

44.1 Introduction

In the province of Cotopaxi, the population as of 2020 was estimated at 488,716 inhabitants, with 7.92% (38,714) of the population being elderly. Outstanding is the fact that 3% of the total population lives in poverty and 0.4% in extreme poverty [1]. The entity in charge of caring for the vulnerable elderly population is the Ministry

S. Pogo-Ateaga · C. Ortiz-Araujo · M. Escobar-Sánchez (✉) · C. Andrade-Villacís
Universidad de Las Fuerzas Armadas ESPE, Latacunga, Ecuador
e-mail: meesacobar1@espe.edu.ec

Á. Rocha et al. (eds.), *Communication, Smart Technologies and Innovation for Society*, Smart Innovation, Systems and Technologies 252,
https://doi.org/10.1007/978-981-16-4126-8_44

of Economic and Social Inclusion (MIES), which, through its gerontological centers and personnel specialized in the care of the elderly, provides medical care, accommodation and rehabilitation [2]. There are two modalities of care: (1) day care (354 people) and (2) residential care (139 people) with a total of 493 people in all its cantons [2].

It should be noted that there is an active agreement between the Ministry of Economic and Social Inclusion (MIES) Cotopaxi branch [2] and the University of the Armed Forces ESPE Latacunga branch, which consists of the following stages: (1) Research for data collection, (2) Methodological development of the project, (3) Design, implementation and execution of field projects, (4) Impact assessment in gerontological care centers in their different modalities, according to the methodology developed in the research phase and (5) Publication of results.

Currently, stages 1 and 2 have been completed. The following results were obtained for stage 2: (a) lack of specialized machines and equipment (95% average—Cotopaxi), (b) lack of interest in performing exercises (physiotherapy), (c) existence of joint, muscle and related diseases (38% shoulder tendinitis, 35% knee osteoarthritis, 18% back pain, 6% spastic hemiparesis, 3% neuropraxia of the sciatic nerve) [2].

In addition, it is important to note that the elderly are legally protected through a law approved and in force in Ecuador since July 19, 2018 [3]. Granting them as main rights the claim to food and a specialized and integral health. According to [4], technology has shown great progress over time, and it has been used in different areas, such as the solution of social problems and for the benefit of vulnerable communities.

In the same way [5] indicates the importance of developing special equipment around the rehabilitation of the elderly, such equipment must comply with a technical–economic planning, however, its prototype of rehabilitation machine does not have the appropriate technology that allows the physiotherapist to monitor and motivate the patient's performance. According to [6] determines that the Internet of Things (IoT) allows to improve the patient's quality of life by collecting and analyzing important data establishing a connection between the physical environment (sensors, actuators and indicators) and the virtual environment (applications, web pages, among others) allowing the creation of an integrated system between a rehabilitation machine, mobile networks and embedded systems. This is how [7] exposed the idea of creating a prototype of an ergonomic chair for patients using an Arduino Mega, biometric sensor and a database.

According to [8], gerontological centers are special institutions that guarantee physical and emotional stability and improve the lifestyle of the elderly, highlighting the importance of having specific areas and equipment that accompany the daily life and increase the self-esteem of the elderly. Furthermore, according to [9] a clear example of this proposal is an electronic prototype which is equipped with intelligent devices and technological equipment for the rehabilitation of patients with knee injuries. Controlled through mobile devices, allowing the physiotherapist to access data through a local database that has the patient's data with their respective rehabilitation routine.

Another important aspect to emphasize is the constant monitoring of the data obtained by the patients when performing physiotherapeutic exercises, helping to obtain an effective and controlled recovery of the treatment as mentioned by [10] who proposes a wireless modular structure for monitoring the functional status of patients with cardiovascular diseases during cardiac rehabilitation, highlighting that the physiotherapeutic procedures must be supervised by a doctor and must be properly organized.

Taking into consideration the preceding paragraphs, the activities of the methodology are developed, which are summarized in a functional and stable structural design that provides the patient a comfortable support to perform the exercises with his hands on the pedals, in addition to connecting all the elements of the embedded system, which integrates the hardware (sensors, multimedia components, etc.) and the wireless connection with the software (web application and database).

According to the analysis made in the field survey where the reality of older adults suffering from physical illnesses and not having the necessary equipment for their treatment is exposed, this research focuses on creating a physiotherapeutic prototype (integrated system—IoT) that facilitates the monitoring and analysis of data in the rehabilitation of older adults of the gerontological center "Sagrada Familia" canton La Maná province of Cotopaxi (diurnal mode) [2].

To validate this prototype, neither the design methodology nor the results obtained from the research proposed according to [9, 10]. However, the validation was based on the three stages proposed according to [9] involving the following areas: mechanical, electronic and software design. In the electromechanical field, a force analysis was performed using a free body diagram to visualize the points with the highest load, while in the electronic field, the degrees of freedom were calculated by testing the embedded system, as well as a test of the wireless communication. It should be noted that within the software area, black box tests were performed in a similar way according to [9]. That is, the sending of data from the prototype to the web application is evaluated.

44.2 Materials and Methods

This section provides a brief description of the general methodology used for the implementation of the prototype developed in four activities. The first activity is the structural design, which consists of a general scheme in 3D format made up of the following main parts: structure and general base of the bicycle, support for the sensors and screen support; then with the previous design made in the first activity, the second activity called embedded system is built and adapted, which integrates: hardware (sensors, components, etc.) and the connection with software (web applications, database), in addition to a multimedia system that fulfills the task of increasing the self-esteem and interest of the elderly in continuing with the physiotherapeutic rehabilitation, then we proceed to the third activity which is the connection of the prototype with the wireless module, thus allowing us to perform the fourth and

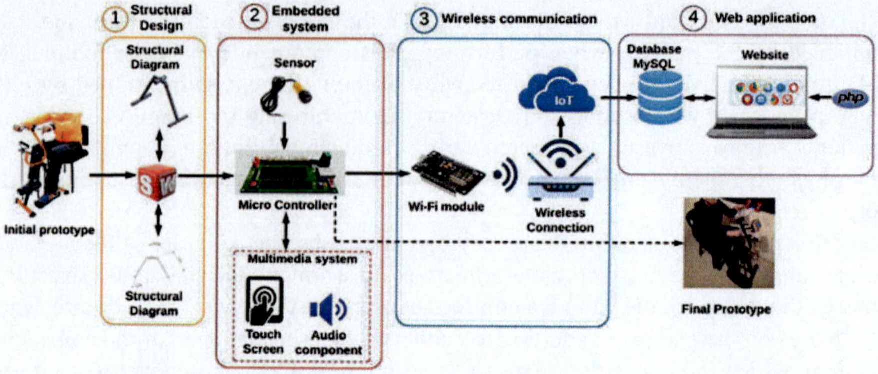

Fig. 44.1 General scheme of the development of the prototype

final activity which is the connection of the prototype with the web application (see Fig. 44.1).

Figure 44.1 explains a general approach of the functionalities and integration of the parts contained in the prototype proposed in this research (see Fig. 44.1). The structural design allows to schematize the functionality and connection of all components to the Arduino (microcontroller), serving as the basis for the embedded system which integrates the hardware with the software composed of sensors, multimedia system (touch screen and audio component) and web applications with database, respectively.

It should be noted that the wireless module facilitates the interaction of the prototype with the web application that is developed in PHP language through the use of MySQL database which stores and manages important information such as patient data and physiotherapeutic exercises, allowing the physiotherapist to monitor the activity performed by the elderly. The processes necessary for the development of the prototype according to the parts that make up the system and its multidisciplinary approach are divided into: (a) electromechanics (structural design, construction and integration), (b) electronics (electronic design, wireless communication) and (c) software (web application) (see Fig. 44.2).

The activities performed in each of the previously mentioned disciplines are detailed below (see Fig. 44.2).

1. Electromechanics—Discipline responsible for the structural manufacturing process for the prototype, taking into consideration a previous analysis allowing the combination of electrical and mechanical parts (corresponding mechanism). For which two activities are performed:

   (a) Structural design—Activity that allows the design of the prototype. This activity required the following tasks: 1) Dimensioning (modeling of parts); and 2) 3D modeling and assembly (preview of the prototype).

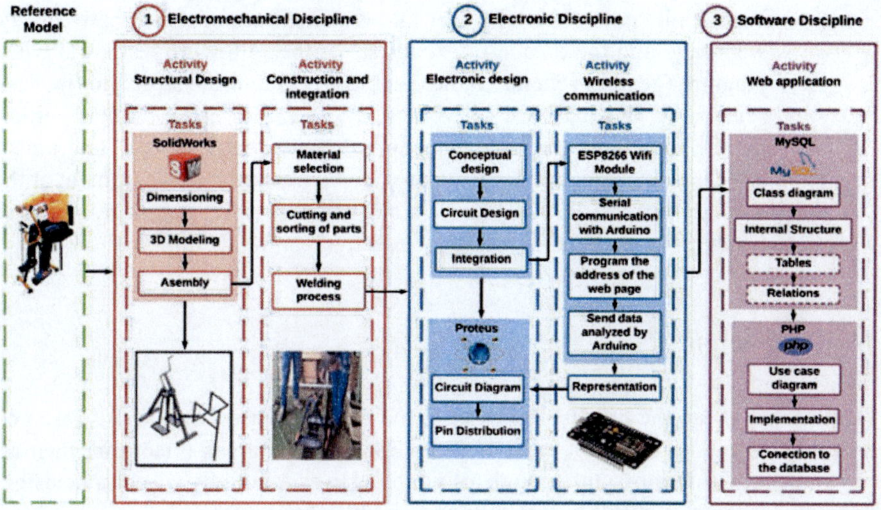

Fig. 44.2 Multidisciplinary scheme of the necessary processes of the prototype

(b) Construction and integration—Activity that allows the construction of the prototype requiring the following tasks: (1) Selection of the material (stainless steel); (2) Cutting and classification of parts (ranges and appropriate sizes); and (3) Welding process (welding components).

2. Electronics—Discipline that allows the coupling of multimedia technology and wireless communication to the prototype through two activities:

   (a) Electronic design—Activity that allows the creation and design of the electronic circuit for the development of the proposed prototype. Using the following tasks: (1) Conceptual design (analysis of user requirements); (2) Circuit design (exploration and selection of components); (3) Integration (individual modules to the microcontroller); and (4) Proteus (circuit diagram and pin layout).
   (b) Wireless communication—Activity that allows sending the data obtained from the prototype to the web application [11]. It consists of the following tasks: (1) ESP8266 Wi-Fi module (essential for internet connection); (2) Serial communication with Arduino (necessary step to receive the data processed by Arduino); (3) Programming (the connection to the Wi-Fi network and the address of the web page); and (4) Sending (the processed data to the address of the web page).

3. Software—Discipline that in conjunction with several logical components allows the development and implementation of the web application. It consists of the following activities:

(a) Web application—Activity that allows the communication between the web page and the database. It consists of the following tasks: (1) Class diagram (general schema of the database); (2) Internal structure (internal schema of the database in MySQL); (3) Tables (entities that participate); (4) Relationships (connections between tables); (5) Diagram of use cases (system requirements); (6) Implementation (development of the web application in the Atom framework with PHP language); and (7) Connection with the database (connection chain).

44.3 Results

The followings are the results that validate the prototype proposed in this research according to the approach of each of the proposed disciplines (electromechanics, electronics and software) and through tests applied in each activity (structural design, embedded system, wireless communication and web application).

44.3.1 Electromechanics

Validation of the structural design. This is done through the analysis of the components and a study of the evaluation of the forces with their corresponding reactions, which are used to check the structural design of the base and seat of the physiotherapeutic prototype together with its corresponding free body diagram (see Fig. 44.3) [12]. For the calculation of the forces, Newton's second law is used, which is the

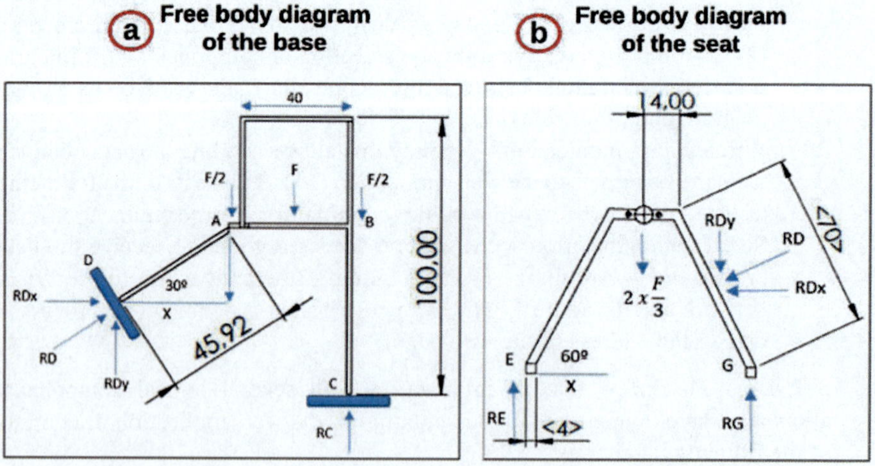

Fig. 44.3 Free body diagram of the base and seat of the physiotherapeutic prototype

proportionality between force and acceleration where the mass is constant [13]. This law is represented in the following equation:

$$\text{Force} = \text{mass} \times \text{gravity} \tag{44.1}$$

The following equation is obtained to check the structural design of the physiotherapeutic prototype base:

$$-\left(2 \times \frac{F}{3}\right)(35 + 14) - RD(35 + (14 \times 2) + \frac{35}{2}$$
$$+ RG\ ((2 \times 35) + (2 \times 14)) = 0 \tag{44.2}$$

where: F = Force applied; RD = Resultant at point D; RG = Resultant at point G.

By adding the sum and clearing RE (Resultant at point E) of the previous equation as shown in figure, we obtain: (see Fig. 44.3).

$$RE - \frac{2F}{3} - RD + RG = 0; \quad RE = 234.42\ \text{N}$$

For the calculation of the force analysis used to test the seat of the physiotherapeutic prototype, we start from equation number one and obtain the following equation:

$$-\frac{F}{2}(X) - \frac{F}{2}(X + 40) + RC(X + 40) = 0 \tag{44.3}$$

Adding the sum and clearing RC (Resultant at point C) from the previous equation we obtain:

$$RD - \frac{619.922}{2} - \frac{619.922}{2} + 464.55 = 0; \quad RD = 155.442\ \text{N}$$

Applying the free body diagram [14], the base and the seat of the prototype are analyzed, obtaining for the base that the greatest weight is at point G with 334.35 N (see Fig. 44.3a), while the support at point E of the structure is lower with a value of 234.42 N. That is to say that point G is the normal force exerted by the ground on the support of the structure which will support the average weight of the older adult which is equivalent to 619.992 N (63.2 kg) [15]. And for the seat, the greatest weight that will be exerted is at point C with a value of 464.55 N (see Fig. 44.3b), while at point D there will be very little support, since it will only support 155.442 N. In other words, point C is the normal force exerted by the floor on the leg of the seat, which will support the average weight of the older adult, equivalent to 619.992 N (63.2 kg).

44.3.2 Electronics

To determine the movements performed by the patient with the prototype, the degrees of freedom were calculated using the Kutzbach–Grübler criterion, separating the mobile mechanism into two parts, upper (arms) and lower (legs), the links N_i are marked in (see Fig. 4b) as well as the pairs of revolution (J_I).

Upper mechanism:

$$GDL = 3(N-1) - 2P_I - P_{II}$$
$$N = 4 \text{ Links; } P_I = 3 \text{ lower pairs; } P_{II} = 0 \text{ upper pairs}$$
$$GDL = 3(4-1) - 2(3) = 3$$

Lower mechanism:

$$N = 2 \text{ Links; } P_I = 1 \text{ lower pair; } P_{II} = 0 \text{ upper pairs}$$
$$GDL = 3(2-1) - 2(1) = 1$$

The upper mechanism has 3 GDL so that the patient can make a 360° turn with his hands, the photoelectric sensor (S) is placed to detect a complete turn and count how many revolutions each patient makes during the exercise, the lower mechanism has 1 GDL that facilitates the movement of the lower joints, however its movement is not measured.

Validation of the embedded system. Testing is performed on the sound system controlled by the touch screen, reception of the sensor signal mounted on the prototype and serial communication with the different modules, implementing the schematic diagram shown in (see Fig. 44.4a).

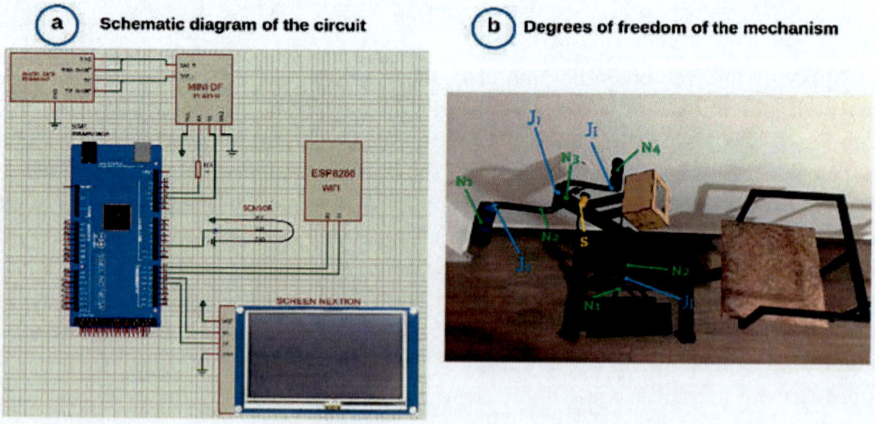

Fig. 44.4 Prototype testing according to the electronic approach (embedded system)

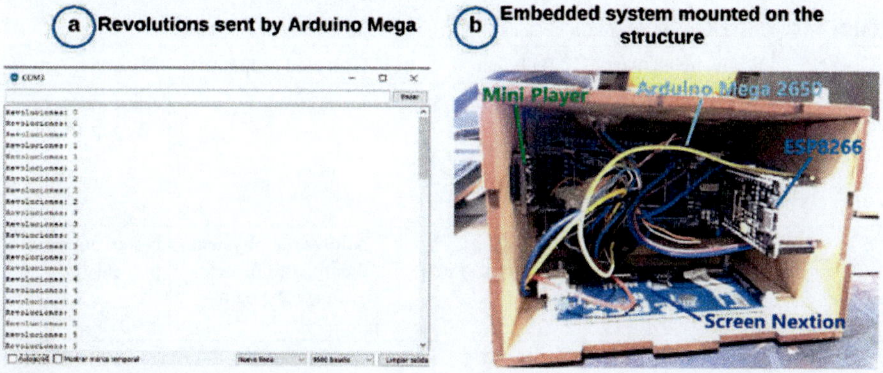

Fig. 44.5 Connection test

Table 44.1 Connection status between device and router

N°	Distance (m)	Connection
1	0.3	Si
2	0.9	Si
3	5	Si
4	8	Si

Validation of the wireless communication. A functional test of the data sent by the Arduino Mega to the ESP8266 Wi-Fi module (see Fig. 44.5a) that is mounted on the prototype (see Fig. 44.5b) is performed. This test obtains the data processed by the Arduino and sent through the serial port to the module, which sends it to the cloud. Distance tests were also performed with barriers taking into account that the module is inside the embedded system box as shown in Table 44.1.

44.3.3 *Software*

Validation of the Web application. The evaluation of the functionalities proposed in the requirements established for the development of the web application is performed [16]. For which the functional test model was taken into consideration, which consists of four phases: (1) specification, (2) planning, (3) execution and (4) evaluation [17]. The efficiency and quality of the developed web application were ratified as shown in Table 44.2.

Table 44.2 Test cases of the main classes

Case No.	Equivalence class	Test data	Expected output	Observed output
1	Invalid equivalence type: Data_bool = character string; Payload_blob = 70 bytes	pedro, 54 + 6sdf, dfgd6 × fgd-{ 80 bytes of data	Error	Alert message (Wrong format)
2	Valid equivalence type: Data_bool = 0 or 1; Payload_blob = 65,535 bytes	0, 1 40 bytes stored	Allows the physical therapist to further evaluate the data	No errors are present

44.4 Discussion

The following Table 44.3 shows the important characteristics, through which a comparison is made between the designed prototype and the comparison prototype [9]. The following relevant data were obtained for each of the disciplines:

1. Electromechanics—The validation of the physiotherapeutic prototype designed in this research (see Fig. 44.3) was performed through an analysis of forces and reactions of the base and seat of the prototype, using a free body diagram. The strengths and weaknesses of the structure were observed. Unlike the bending analysis in [9] which does not allow us to observe these points. It should be noted that the mass value used was 63.2 kg (average weight of an older adult) [13].
2. Electronics—The validation of the circuit (see Fig. 44.4a) integrates the components of the multimedia system and wireless transmission through IoT. This

Table 44.3 Comparative table of prototype characteristics and validations versus other research

Characteristics	Prototype designed	Comparison prototype
Author / Year	Pogo, Ortiz, Escobar, Andrade (2021)	Mercy Dalila Rea Yanez. (2019)
Communication Device	ESP8266 Wi-Fi module	Bluetooth HC-05
Degrees of freedom	Upper mechanism 3 GDL Lower mechanism 1GDL	Upper mechanism 3 GDL
Hardware	MEGA 2560 Arduino board DFplayer mini Tft nextion 2.8´´ Screen Photoelectric Sensor E 18-D80NK	MEGA 2560 Arduino board Incremental rotary encoder MCT38A
Software	Design: SolidWorks Development: Web application (Atom) with database (Mysql)	Design: SolidWorks Development: Mobile application (Android Studio) with database (SQLite)

improves the data transmission compared to [9] which uses a Bluetooth module limited to a maximum distance of 7.5 m. The prototype has an upper mechanism (3 GDL) and lower mechanism (1 GDL) for each leg for the movement of all extremities, while [9] is focused on the movement of one joint (1GDL).

3. Software—The validation is performed through tests of delivery and reception of data as mentioned in [9] demonstrating that the expected output satisfactorily meets the requirements set out in the equivalence classes, determining that the web application is functional and obtains an adequate communication with the prototype. In addition, a database hosted in the cloud is used which, unlike the local database proposed in [9] facilitates the storage of information and provides an improvement to the proposed prototype and by developing a web application demonstrates an improvement in usability with the mobile device proposed in [9].

44.5 Conclusions

The prototype of an integrated system and wireless networks for physiotherapeutic rehabilitation focused on reducing pain caused by stiffness in the joints of older adults through assisted technology has been developed at the "Sagrada Familia" gerontological center in the province of Cotopaxi.

The results of the structural analysis that were obtained through the free body diagram allowed to evaluate and identify the points with higher and lower resistance, focusing on determining the appropriate applied load of the physiotherapeutic prototype, taking into account the deflection of the material and its resistance. The embedded system allows the prototype to integrate IoT technology that assists the physiotherapist in monitoring and recording the patient's performance managed through a web application. In addition, it integrates a multimedia system which raises the self-esteem of the elderly not to abandon their rehabilitation.

Among the weaknesses of the prototype are the following observations: (1) semi-rigid structure, i.e., it maintains adequate ergonomics, but is not totally perfect, (2) connection stability depends on the type of router and quality of internet available and (3) exposure to risks that threaten the integrity, availability and confidentiality of patient information. The followings are proposed as future works: (1) To access and control the prototype remotely, (2) To include alarm controls to determine the start and end of the physiotherapeutic activity and (3) To implement a multiplatform service to the prototype.

References

1. INEC, Proyecciones Poblacionales, https://www.ecuadorencifras.gob.ec/proyecciones-poblacionales. Last accessed 09 May 2021

2. Proyectos de Vinculación 2019 archivos - Departamento Eléctrica y Electrónica, https://deee-el.espe.edu.ec/category/proyectos-de-vinculacion-2019/. Last accessed 02 Mar 2021
3. MIES: Servicios y Programas, https://www.inclusion.gob.ec/wp-content/uploads/downloads/2013/06/folleto-servicios-del-mies.pdf. Last accessed 02 Mar 2021
4. S. Clara, H. García Garcés, L. Navarro Aguirre, M. López Pérez, M. de Fátima Rodríguez Orizondo, The information and communication technology in health and medical education. EDUMECENTRO **6**, 253–265 (2014)
5. Repositorio Digital Universidad Técnica de Cotopaxi: Implementación de una máquina ejercitadora ergonómica para la rehabilitación física de personas con parapléjia y adultos mayores, http://repositorio.utc.edu.ec/handle/27000/4393. Last accessed 02 Mar 2021
6. C. Viloria Núñez, P. Sanmartín Mendoza, K. Avila Hernández, D. Jabba Molinares, Internet de las cosas y la Salud centrada en el Hogar. Rev. Científica Salud Uninorte **32** (2016)
7. Repositorio Institucional de la Universidad Politécnica Salesiana: Desarrollo de una silla estática para rehabilitación de las rodillas de personas parapléjicas de la Fundación "Ayudemos a Vivir," https://dspace.ups.edu.ec/handle/123456789/15290. Last accessed 05 Mar 2021
8. Repositorio PUCESA: Diseño de la zona fisioterapeutica para adultos mayores del servicio de gerontología del Hospital Regional Docente Ambato, https://repositorio.pucesa.edu.ec/handle/123456789/856. Last accessed 08 Mar 2021
9. DSpace ESPOCH.: Desarrollo de un prototipo de robot para rehabilitación de pacientes con lesiones en rodillas controlado mediante dispositivos móviles., http://dspace.espoch.edu.ec/handle/123456789/13679. Last accessed 08 Mar 2021
10. A.D. Sutyagina, A.A. Anisimov, E.K. Lyakhova, The device for monitoring of functional status of patients during cardiac rehabilitation, in *Proceedings of the 2019 IEEE Conference of Russian Young Researchers in Electrical and Electronic Engineering, ElConRus 2019* (Institute of Electrical and Electronics Engineers Inc., 2019), pp. 1331–1335. https://doi.org/10.1109/EIConRus.2019.8657152
11. S. Rajesh, A. Gehlot, L.R. Gupta, B. Singh, P. Tyagi, Introducción a Internet de las cosas con Launch Pad y ESP8266, Libros de River Publishers, IEEE Xplore, https://ieeexplore.ieee.org/document/9227080. Last accessed 10 Mar 2021
12. C. Mora, D. Herrera, Una revisión sobre ideas previas del concepto de fuerza. Latin-Am. J. Phys. Educ. **3**, 72–86 (2009)
13. P.M. Romero, *Mecánica: Estática y Cálculo Vectorial* (Universitat Politécnica de Valéncia, Valencia, 2016)
14. R. Norton, *Diseño de Máquinas*, 4th edn. (Pearson Educación, Naucalpan de Juárez, 2011)
15. R. Budynas, K. Nisbett, *Diseño de Ingeniería Mecánica*, 9th edn. (McGraw-Hill, México, 2011)
16. M. Torres, *Desarrollo de aplicaciones web con PHP y MySQL*, 1st edn (Marcombo, Lima, 2016)
17. M.E. Escobar-Sánchez, W.M. Fuertes-Díaz, Modelo formal de pruebas funcionales de software para alcanzar el Nivel de Madurez Integrado 2. Rev. Fac. Ing. **24**, 31 (2015). https://doi.org/10.19053/01211129.3549

Chapter 45

Time Series in Sensor Data Using State-of-the-Art Deep Learning Approaches: A Systematic Literature Review

Luis-Roberto Jácome-Galarza, Miguel-Andrés Realpe-Robalino, Jonathan Paillacho-Corredores, and José-Leonardo Benavides-Maldonado

Abstract Internet of things (IoT) and artificial intelligence (AI) are becoming support tools for several current technological solutions due to significant advancements of these areas. The development of the IoT in various technological fields has contributed to predicting the behavior of various systems such as mechanical, electronic, and control using sensor networks. On the other hand, deep learning architectures have achieved excellent results in complex tasks, where patterns have been extracted in time series. This study has reviewed the most efficient deep learning architectures for forecasting and obtaining trends over time, together with data produced by IoT sensors. In this way, it is proposed to contribute to applications in fields in which IoT is contributing a technological advance such as smart cities, industry 4.0, sustainable agriculture, or robotics. Among the architectures studied in this article related to the process of time-series data, we have: long short-term memory (LSTM) for its high precision in prediction and the ability to automatically process input sequences; convolutional neural networks (CNN) mainly in human activity recognition; hybrid architectures in which there is a convolutional layer for data pre-processing and recurrent neural networks (RNN) for data fusion from different sensors and their subsequent classification, and stacked LSTM autoencoders that extract the variables from time series in an unsupervised way without the need of manual data pre-processing. Finally, well-known technologies in natural language processing are also used in time-series data prediction, such as the attention mechanism and embeddings obtaining promising results.

L.-R. Jácome-Galarza (✉) · M.-A. Realpe-Robalino · J. Paillacho-Corredores
Facultad de Ingeniería en Electricidad y Computación, Escuela Superior Politécnica del Litoral, ESPOL, CIDIS. Gustavo Galindo Km, 30.5 Vía Perimetral, Guayaquil, Ecuador
e-mail: lrjacome@espol.edu.ec

J.-L. Benavides-Maldonado
Universidad Nacional de Loja, La Argelia, Loja, Ecuador

Á. Rocha et al. (eds.), *Communication, Smart Technologies and Innovation for Society*, Smart Innovation, Systems and Technologies 252,
https://doi.org/10.1007/978-981-16-4126-8_45

45.1 Introduction

The Internet of things (IoT) is formed by many connected devices and transmit vast amounts of data [1]. It is currently being applied in fields like agriculture [2, 3], smart cities [4], smart homes [5, 6], health care [7, 8], and human activity recognition [9, 10]. As time goes by, these IoT sensors are used, and more insights can be obtained from them that may help to predict the behavior of systems which is useful for systems maintenance [11, 12], yield performance [13, 14], resource allocation [15], or business planning [16]. Deep neural networks have achieved state-of-the-art results in complex tasks like image recognition [17] using CNN [18], human language understanding [19] using LSTM [20], or gaming [21] with reinforcement learning [22]. It is relevant to understand how well deep learning algorithms can extract patterns in time series and what trends those algorithms can find. When designing a deep learning architecture for time-series forecasting, RNN and LSTM networks are considered, but CNN networks are also used to report good results [23]. Accordingly, this paper aims to analyze what deep learning architectures are being utilized to forecast time series, their applications, efficiency, challenges, and trends in time.

The rest of this paper is structured as follows: Sect. 45.2 describes the research methodological framework; Sect. 45.3 shows the results obtained; and Sect. 45.4 concludes the article.

45.2 Materials and Methods

We used Barbara Kitchenham's methodology [24], which includes formulation of research questions, search process, inclusion and exclusion criteria, data extraction, and data analysis and classification. The following research questions were posed to attain this purpose:

- RQ1: What deep learning architectures are being used in projects involving time series in sensor data, and what are their application and efficiency?
- RQ2: What challenges are found in projects with time series in sensor data?
- RQ3: What trends are found in the applied methodologies in projects with sensor data and time series?

A manual search on the Science Direct, Google Scholar, and Springer databases was conducted. The search was performed in June 2020. The search string was the following: "deep learning" AND "sensor data" AND "time series." Additionally, only scientific papers written in English and later than 2016 were considered.

From the first search process, 8033 articles were extracted (532 Science Direct, 61 Springer, 7440 Google Scholar). After that, the extraction process resulted in 40 articles with the following year distribution: 2016 (3 articles), 2017 (11 articles), 2018 (10 articles), 2019 (10 articles), and 2020 (6 articles).

The data analysis and classification steps are explained below.

1. Reading the abstracts and conclusions.
2. Searching for each criterion within the complete content of the articles.
3. Reading the whole article when necessary.
4. Classifying articles by criteria.

45.3 Results

45.3.1 First Query Results

Table 45.1 is presented to answer the RQ1. This table includes the list of analyzed papers, their application, the deep learning architecture, and efficiency.

Deep learning architectures. It can be seen in Fig. 45.1 that LSTM is the most used architecture. LSTM networks can automatically learn high-level representative features containing the long-range temporal relationships. Stacked LSTM autoencoders extract the features from time series in an unsupervised manner avoiding the manual feature engineering. CNN are preferred for human activity recognition. Human activities are hierarchical because complex activities are composed of basic actions. Human activities are also translation-invariant because different people perform the same activity differently. A fragment of an activity can be performed at different points in time. CNN can address raw temporal signals in machine health monitoring (MHM) tasks without any time–frequency transformation. Convolutional kernels can be thought of as signal filters and remove the need for manual signal processing; in more complex architectures, CNN is used for feature extracting, while RNN is used for time-series classification.

Embeddings (numerical vectors that represent features of a system) can be used for prediction. The embeddings of two machines (P020) with similar operational behavior are close to each other. Heart rhythm embedding (P016) for both Atrial Fibrillation (AFib) and Normal Sinus Rhythm (NSR) vary between patients, so unsupervised clustering applied to heart rhythm embeddings can improve the model's accuracy. Furthermore, paper P015 explains that fully connected structures of DAE and DBN may lead to high computation costs and overfitting problems caused by huge model parameters. To avoid those disadvantages, they use a hybrid Conv LSTM architecture for machine health monitoring.

Multi-modal data fusion architectures. In paper P004, the authors present a multi-modal data fusion model for car tracking with motion sensors (accelerometer, gyroscope, magnetometer, and GPS for the ground truth), heterogeneous human activity recognition (accelerometer, gyroscope), and user identification with biometric motion analysis (accelerometer, gyroscope). The Fourier transform is used to obtain the frequency dimension.

Table 45.1 List of analyzed papers

Paper id	Application	Year	Deep learning architecture	Efficiency
P001	Water quality monitoring network	2019	LSTM	69.2%, 70.3%, 98.3%, and 76% improvement in RMSE, MAE, MAPE, SMAPE
P002	Generate sensory data for a deep learning-based discriminator model	2017	LSTM—MDN	Discriminator distinguishes 50%
P003	Indoor air quality analysis	2017	GRU LSTM	84.69% accuracy
P004	Heterogeneous human activity recognition, user identification with motion analysis, car tracking with motion sensors	2017	CNN GRU	99.7% accuracy
P005	Real-time activity classification	2016	CNN spectrogram	85% accuracy
P006	Human emotion classification	2019	CNN LSTM	95% accuracy
P007	Machinery fault diagnosis	2017	DNN	100% accuracy
P008	Predictive monitoring of production processes	2017	LSTM FFNN	87% accuracy
P009	Human activity recognition	2016	CNN	94.79% accuracy
P010	Human activity recognition based on mobile sensor data	2020	DRN CNN GAF	96% accuracy
P011	Driving behavior identification	2019	CNN, LSTM, GRU	97% accuracy
P012	Marine sensor data prediction	2020	DBEN	91% accuracy
P013	Predicts if participants would attend to a notification within 10 min	2017	RNN	40% improvement respect to random baseline
P014	Forecasting the future use of energy for the home appliances	2017	GMDH	63.23 RMSE
P015	Machine health monitoring (MHM)	2018	CONV LSTM	8.39 RMSE
P016	Atrial fibrillation detection using wearables	2018	CONV LSTM	False positive and false negative rates below 2 × 10 − 3
P017	Aircraft hard-landing prediction	2018	LSTM	1.1 × 10 − 3 MSE

(continued)

Table 45.1 (continued)

Paper id	Application	Year	Deep learning architecture	Efficiency
P018	Multi-modal sensor fusion for human activity recognition	2019	LSTM	94.46%, accuracy
P019	Anomaly detection	2017	LSTM	97%, 95%, 97% accuracy
P020	Predicting remaining useful life of machines	2017	GRU RNN	466 MSE
P021	Monitor the performance degradation of commercial aircraft air conditioning systems (ACS)	2020	LSTM	0.0466 error
P022	Construction equipment activity recognition	2019	LSTM	79.9%, 96.7% accuracy
P023	Aircraft landing speed prediction	2018	LSTM	3.5 RMSE
P024	Data augmentation using synthetic data for time-series classification	2018	CNN	97% accuracy
P025	Remaining useful life RUL prediction model	2019	CNN	2.5, 6.41, 3.25E-03, 10.72, 6.26, 3.12, 19.02 RMSE
P026	Time-series classification with multivariable data	2018	CNN	97.4%, 97.7% accuracy
P027	Human activity recognition with inertial sensors	2016	CNN	97.01% accuracy
P028	Human activity recognition from inertial sensor time series	2018	LSTM	92% accuracy
P029	Multi-time series anomaly detection	2019	CNN RNN	Anomalies reduced by 3%
P030	Landslide hazard prediction	2018	LSTM, BPNN	81.2%, 62% accuracy
P031	Data-driven anomaly detection for UAV sensors	2019	LSTM	99% accuracy
P032	Recovering missing air quality data	2018	LSTM	0.63, 2.90, 1.87 MAE
P033	Missing sensor data prediction in IoT	2020	LSTM	0.07, 0.08, 0.17 RMSE
P034	Time-series data for equipment reliability analysis	2020	FFNN	75%, 82%, 85%, 79%, 80% accuracy

(continued)

Table 45.1 (continued)

Paper id	Application	Year	Deep learning architecture	Efficiency
P035	Recognizing and forecasting the underlying high-level states from raw sensory data for activity recognition	2017	CNN LSTM	87.47%, 80.21% accuracy
P036	Sleep apnea severity detection	2017	LSTM-RNN	100%, 99.9% accuracy
P037	Activity recognition via multi-channel sensor data	2020	CNN LSTM	0.755, 0.933, 0.918 F1 score
P038	Mobile application usage prediction	2019	LSTM	89.55%, 95.53% accuracy
P039	Wearable-based Parkinson's disease severity monitoring	2019	FCN	84.3% accuracy
P040	Classify the dangerous levels of methane concentration	2019	RBM DBN	97.6%, 99.2%, 96.3% accuracy

Bi-LSTM Bidirectional Long Short-Term Memory; *GRU* Gated Recurrent Units; *FCNN* Fully Connected Neural Network; *GAN* Generative Adversarial Networks; *DFNN* Deep Feedforward Neural Networks; *LSTM* Long Short-Term Memory; *DBEN* Deep Belief Echo State Network; *RNN* Recurrent Neural Networks; *BPNN* Back Propagation Neural Networks; *DRN* Deep Residual Networks; *CNN* Convolutional Neural Networks; *DAE* Deep Autoencoder; *FCN* Fully Convolutional Net; *DBN* Deep Belief Networks

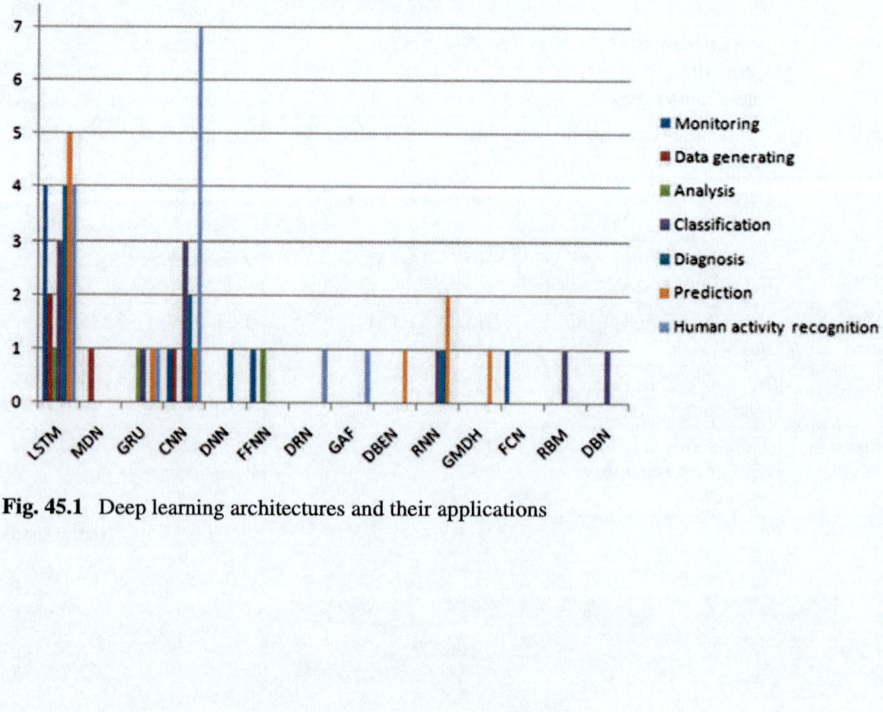

Fig. 45.1 Deep learning architectures and their applications

A CNN extracts local features within each sensor modality and merges the local features into global features hierarchically; a stacked GRU extracts temporal dependency. In paper P006, they achieve motion detection fusing various sensor modalities: on-body, environment, and location data. Two convolutional layers analyze the concatenated data. After that, a fully convolutional layer is used, then a LSTM layer processes the temporal information, and a softmax layer is used for classification. The model gets better performance than single CNN and LSTM. In paper P010, they use multi-modal data fusion for human activity recognition. The model encodes sensor data as images using Gramian Angular Fields (GAF), representing time-series data in polar coordinates instead of the Cartesian coordinates.

A CNN analyzes the image representation of each sensor's data. The model performs better than random forests, SVM, restricted Boltzmann machines, FCNN, and LSTM. In paper P018, the authors use multi-modal data fusion for human activity prediction. The data from on-body IMU devices (tri-axial accelerometer, gyroscope, and magnetometer) are analyzed individually by an LSTM classifier; the results from each sensor modality are used as features for a model that integrates the results, applying simple soft-voting based on the weighted average of class probabilities (soft-voting gets higher overall performance than hard-voting because it can give more weight to highly confident votes). This model uses random forests, kNN, and SVM for the final classification. In P029, they present an industrial case where three types of anomalies detection are considered: point, context-specific, and collective.

CNN blocks individually process sensors; then, the resulting feature maps are fed to an RNN. A multi-channel CNN architecture is used to process multiple time series. A single feature map containing the main features of all the time series is obtained. The number of extracted features per window is much higher for the multi-head convolution, which performs better than the multi-channel CNN architecture. In paper P030, they use multi-modal data fusion for landslide hazard prediction. They combine digital elevation model (DEM), high-resolution remote sensing (HR-RS) images, 1:50,000 GM (geologic maps), and meteorological data. A CNN processes the images, the build-up index, and the stream power index. The DEM, HR-RS, GM, and meteorological data are fed into a LSTM for classification. The model performs better than SVM, decision trees, and backpropagation neural networks.

45.3.2 Second Query Results

In order to answer the RQ2, Fig. 45.2 shows that complexity is the main challenge reported in manufacturing projects since the data has high-dimensional characteristics. Missing or noisy data is also found as a challenge in all kinds of projects. In projects with wireless sensor data, low bandwidth can be a problem. The lack of computation power and energy consumption is common problems in projects that use sensors in mobile devices. Imbalanced classes can be an issue since more errors are found on samples from minority classes. Data privacy is a limitation in some driving datasets. The lack of performance metrics is a limitation in unsupervised learning

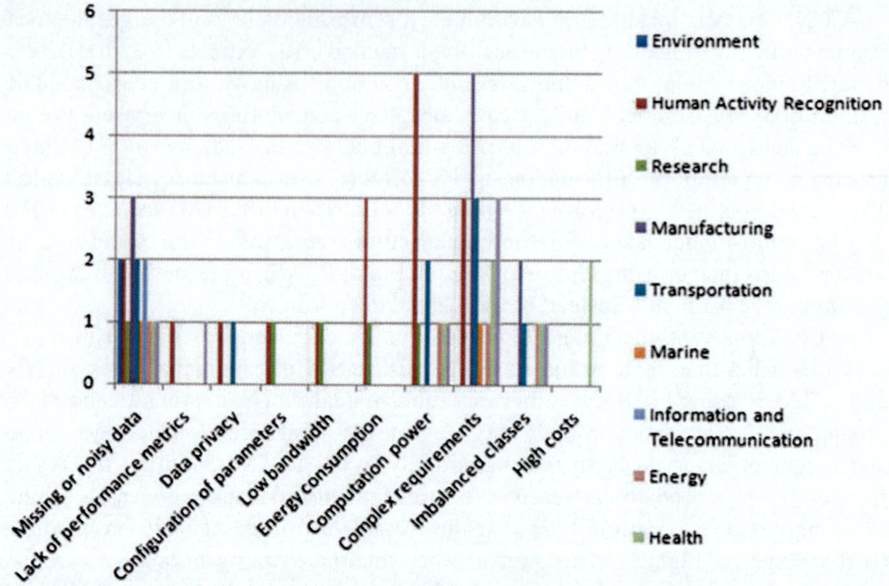

Fig. 45.2 Challenges detected in the projects of time-series and sensor data

and generative adversarial networks. Marine data (P012) has fluctuations, outliers, noise, and complex characteristics. The intermittent nature of heart conditions such as AFib (P016) is a challenge for data collection. Human movements are encoded in a sequence of time, and the activities are not defined by one small window of data alone, and multiple sensor devices are required (P028). There is much more data related to normal behavior in anomaly detection than anomalous data (P029). Sensor-generated data is expected to be noisy, uncertain, erroneous, and missing due to the lack of battery power, communication errors, and malfunctioning devices (P033). Healthcare projects report high costs for acquiring medical datasets.

Regarding the mitigation of the impacts of the detected challenges, the following measures are explained:

Sampling frequency. If all sensors have the same sampling rate, 2D spatio-temporal sequence data is formed. Down-sampling reduces the number of dimensions in the original time series and makes it easier to learn patterns; in contrast, up-sampling can be done by the Lagrange interpolation method.

Time-series data augmentation. Some techniques are slicing window, jittering, scaling, time warping, and permutation (in paper P024, it is reported that data augmentation improved the performance of the CNN model considerably). Some algorithms incur a heavier loss for errors on samples from minority classes; putting more focus on the minority classes during training is needed. Over-sampling and under-sampling techniques are usually used to balance the number of instances of both classes.

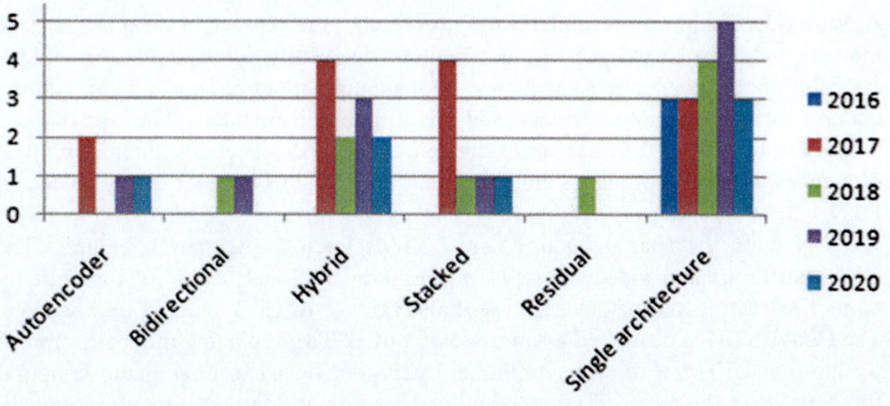

Fig. 45.3 Deep learning architectures and their applications

Sliding window. Time-series data is processed in a window-based manner. The sliding window length affects the number of parameters of each layer. Allowing overlap between segments avoids creating gaps in the output signal.

Other tools and techniques. Other techniques such as batch normalization are commonly used, and transfer learning in time-series data is proposed, but it needs further research. Likewise, InfluxDB, an open-source distributed database of time-series events and metrics, is popular for storing real-time IoT data; the Ray v0.7.6 distributed application framework is used for hyperparameter search. In paper P037, they evaluate the training and architectural parameters on the mentioned framework.

45.3.3 Third Query Results

In order to answer RQ3, Fig. 45.3 shows that the use of a single architecture is still the most utilized approach for time-series forecasting and LSTM networks as the most popular choice. LSTM is proven to be the most stable and robust RNN model to learn long-range temporal patterns in practical applications.

Bi-LSTMs were used mainly for recovering missing sequence data. Papers P018, P019, and P028 used stacked LSTM, which is built instead of using a higher number of memory blocks in a standard recurrent hidden layer. One more recurrent hidden layer is stacked, providing better generalization and robust temporal pattern learning. In P019, the model gets 97, 95, and 97% accuracy in anomaly detection performing better than traditional methods like cumulative sum and exponentially weighted moving average.

In P028, the model gets 92% accuracy in human activity recognition having better results than SVM, deep convolutional LSTM, FCNN, and single LSTM. They state that batch normalization can obtain similar accuracy with fewer training epochs.

Autoencoders explore the underlying structure in the dataset in an unsupervised manner; they can be categorized as simple or deep fully connected autoencoders, convolutional autoencoders, sequence-to-sequence autoencoders (LSTM autoencoders), contractive autoencoders, and variational autoencoders. The encoder can be used to read from the time-series data, and a decoder generates the missing data sequences. Hybrid approaches are also found in P006, P011, P015, P016, P035, and P037.

CNN learn the spatial features and LSTMs learn the temporal features. The CNN can be treated independently, forming convolutional heads for individually feature extracting sensor data, having the advantage of easily adding new sensors. The ConvLSTM is described as an extension of LSTM, replacing the matrix multiplication in LSTM with a convolutional operation for performing spatio-temporal forecasting. In the paper, P035, the authors propose an long-term recurrent convolutional LSTM network (LR-ConvLSTM) model for time-series classification and high-level state prediction. LR-ConvLSTM is a hybrid deep learning representation that has a stacked convolutional LSTM component to learn complex temporal features and an LSTM component that learns external temporal features.

Most of the projects use the "many-to-one" approach (the input is divided into fixed-length overlapping windows, then a model processes each window individually, generating a class prediction for each one, and then the predictions are concatenated). Researchers use the "many-to-many" approach in the paper P037. The entire output time series is generated with a single model evaluation; they propose a GPU-optimized (cuDNN) LSTM implementation for improving the model. In P029, scholars generate an already-trained model using transfer learning (Transfer Knowledge-Based—TKB) instead of training a model from scratch. Authors use transfer learning to improve their model for monitoring Parkinson's disease with wearable devices in paper P039. An optimal mechanism for time step searching is needed to improve learning models was found in P003.

45.4 Conclusions

Deep learning has achieved impressive improvement in many research fields. This paper presents a systematic literature review about the use of DL in time-series data produced by IoT sensors. Its insights may be beneficial for optimization in many industrial processes. The fusion of multi-modal sensor data is a challenging task suitable for deep learning networks since they can find patterns from raw and massive data. Findings revealed that multi-modal approaches increase the accuracy of the predicting models.

Well-known technologies in natural language processing are also being used in time-series prediction. Thus, the attention mechanism and embeddings disclosed promising results. The combination of CNN and LSTM networks is a good approach for improving time-series data prediction. The CNNs are used for the pre-processing of the sensor data, and the RNNs are used for processing time patterns.

CNN networks can also obtain spatial patterns. Moreover, stacked LSTM autoencoders are used for extracting variables from time series in an unsupervised way without performing manual data pre-processing. The challenges found in time-series prediction are missing and noisy data, privacy restrictions of datasets, high costs of datasets, especially in the health area, high computational demand, and high wireless speed requirements for mobile devices. Applying transfer learning to time-series prediction still needs research. Other relevant topics for time-series prediction studies are temporal convolutional networks and sequence-to-sequence learning.

The current paper was limited by its focus on deep learning techniques and restraint of classical approaches. As future work, we intend to extend this analysis to the projects that utilize the ConvLSTM and temporal convolutional network architectures toward verifying their advantages. Additionally, tools for automatic hyperparameter search need further research.

Appendix

The list of the analyzed papers is available in the Appendix available in: https://www.dropbox.com/s/1hlpy6i1n0vb7rk/appendix.pdf?dl=0

References

1. M. Mahdavinejad, M. Rezvan, M. Barekatain, P. Adibi, P. Barnaghi, A. Sheth, Machine learning for Internet of things data analysis: a survey. Digital Commun. Networks **4**(3), 161–175 (2018)
2. N. Gondchawar, R. Kawitkar, IoT based smart agriculture. Int. J. Adv. Res. Computer Commun. Eng. **5**(6), 838–842 (2016)
3. J. Muangprathub, N. Boonnam, S. Kajornkasirat, N. Lekbangpong, A. Wanichsombat, P. Nillaor, IoT and agriculture data analysis for smart farm. Comput. Electron. Agric. **156**, 467–474 (2019)
4. T. Kim, C. Ramos, S. Mohammed, *Smart city and IoT* (2017)
5. S. Samuel, A review of connectivity challenges in IoT-smart home, in *2016 3rd MEC International Conference on Big Data and Smart City (ICBDSC)* (IEEE, 2016), pp. 1–4
6. Y. Kim, Y. Park, J. Choi, A study on the adoption of IoT smart home service: using value-based adoption model. Total Qual. Manag. Bus. Excell. **28**(9–10), 1149–1165 (2017)
7. A. Ukil, S. Bandyoapdhyay, C. Puri, A. Pal, IoT healthcare analytics: the importance of anomaly detection, in *2016 IEEE 30th International Conference on Advanced Information Networking and Applications (AINA)* (IEEE, 2016), pp. 994–997
8. S. Tyagi, A. Agarwal, P. Maheshwari, A conceptual framework for IoT-based healthcare system using cloud computing, in *2016 6th International Conference-Cloud System and Big Data Engineering (Confluence)* (IEEE, 2016), pp. 503–507
9. A. Rghioui, S. Sendra, J. Lloret, A. Oumnad, Internet of things for measuring human activities in ambient assisted living and e-health. Network Protocols Algorithms **8**(3), 15–28 (2016)
10. C. Shi, J. Liu, H. Liu, Y. Chen, Smart user authentication through actuation of daily activities leveraging WiFi-enabled IoT, in *Proceedings of the 18th ACM International Symposium on Mobile Ad Hoc Networking and Computing* (2017), pp. 1–10

11. Y.K. Al-Douri, H. Hamodi, J. Lundberg, Time series forecasting using a two-level multi-objective genetic algorithm: a case study of maintenance cost data for tunnel fans. Algorithms **11**(8), 123 (2018)
12. M. Baptista, S. Sankararaman, I. de Medeiros, C. Nascimento Jr., H. Prendinger, E. Henriques, Forecasting fault events for predictive maintenance using data-driven techniques and ARMA modeling. Comput. Ind. Eng. **115**, 41–53 (2018)
13. E. Kamir, F. Waldner, Z. Hochman, Estimating wheat yields in Australia using climate records, satellite image time series and machine learning methods. ISPRS J. Photogramm. Remote. Sens. **160**, 124–135 (2020)
14. O.D. Adeniyi, A. Szabo, J. Tamás, A. Nagy, Wheat yield forecasting based on Landsat NDVI and SAVI Time Series (2020)
15. F. Kadri, F. Harrou, S. Chaabane, C. Tahon, Time series modelling and forecasting of emergency department overcrowding. J. Med. Syst. **38**(9), 1–20 (2014)
16. E. Demir, S. Dincer, Place and solution proposals of data mining in production planning and control processes: a business application. PressAcademia Procedia **11**(1), 189–193 (2020)
17. M. Wu, L. Chen, Image recognition based on deep learning, in *2015 Chinese Automation Congress (CAC)* (IEEE, 2015), pp. 542–546
18. H. Qassim, A. Verma, D. Feinzimer, Compressed residual-VGG16 CNN model for big data places image recognition, in *2018 IEEE 8th Annual Computing and Communication Workshop and Conference (CCWC)* (IEEE, 2018), pp. 169–175
19. G. Tur, A. Celikyilmaz, X. He, D. Hakkani-Tür, L. Deng, Deep learning in conversational language understanding, in *Deep Learning in Natural Language Processing* (Springer, Singapore, 2018), pp. 23–48
20. T. Wen, M. Gasic, N. Mrksic, P. Su, D. Vandyke, S. Young, Semantically conditioned lstm-based natural language generation for spoken dialogue systems (2015). arXiv preprint arXiv:1508.01745
21. D. Silver, J. Schrittwieser, K. Simonyan, I. Antonoglou, A. Huang, A. Guez, D. Hassabis, Mastering the game of go without human knowledge. Nature **550**(7676), 354–359 (2017)
22. O. Vinyals, T. Ewalds, S. Bartunov, P. Georgiev, A. Vezhnevets, M. Yeo, A New Challenge for Reinforcement Learning (2017). arXiv preprint ArXiv:1708.04782
23. R. Liu, G. Meng, B. Yang, C. Sun, X. Chen, Dislocated time series convolutional neural architecture: an intelligent fault diagnosis approach for electric machine. IEEE Trans. Industr. Inf. **13**(3), 1310–1320 (2016)
24. B. Kitchenham, O. Brereton, D. Budgen, M. Turner, J. Bailey, S. Linkman, Systematic literature reviews in software engineering—a systematic literature review. Inf. Softw. Technol. **51**(1), 7–15 (2009)

Chapter 46
LoRaWAN: Application of Nonlinear Optimization to Base Stations Location

Erwin J. Sacoto-Cabrera, Gabriel León-Paredes, and Walter Verdugo-Romero

Abstract This paper analyzes the application of nonlinear optimization in the localization of LoRaWAN base stations at a point equidistant from the sensors located within the base station coverage. Specifically, in the proposed model, we obtain the optimal location of the base station based on the geographical location of the sensors, the coverage parameters specified for urban and rural areas established in the LoRaWAN technical specifications, and a coverage error factor set out in previous LoRaWAN studies. We conclude that such a model is feasible and allows optimizing the location of a LoRaWAN base station.

46.1 Introduction

Nowadays, low power area networks (LPWAN) is the most popular low-cost, long-battery lifetime, and long-range communication technology for developing the wireless sensor network (WSN). LPWAN uses protocols to facilitate communication as LoRaWAN [1, 2], NB-IoT [3, 4], and Sigfox [5, 6], which have reduced the complexity of developing new IoT devices.

About the above, several technologies, platforms, and protocols have been developed for WSN, as described in [7–10]. The authors in [11] compare many characteristics of technologies as a balance between power efficiency, latency performance, quality of service (QoS), cybersecurity, coverage, and conclude that LoRaWAN is a technology designed for the Internet of Things (IoT) [11–13]. Specifically, LoRaWAN wireless radio frequency technology permits the use of minimum sensor power, i.e., battery-powered sensor nodes can reliably send data over long distances with minimal power draw by adopting LoRaWAN technology. With LoRaWAN, these devices can stay active for many years due to a low data bit rate and low power draw during device sleep states, as described in [14–16].

The authors in [17, 18] described that the LPWAN technologies ensure long transmission range, low energy consumption, and low-cost deployment solution. It

E. J. Sacoto-Cabrera (✉) · G. León-Paredes · W. Verdugo-Romero
GIHP4C, Universidad Politécnica Salesiana, Cuenca, Ecuador
e-mail: esacoto@ups.edu.ec

Á. Rocha et al. (eds.), *Communication, Smart Technologies and Innovation for Society*, Smart Innovation, Systems and Technologies 252,
https://doi.org/10.1007/978-981-16-4126-8_46

allows up to 40 km as a communication range in rural zones and 10 km in urban zones.

Several studies carried out on the coverage of base stations described below. The authors in [19–24] described the path-loss (PL) models applicable to outdoor environments for the 800–1800 MHz and 2.5–5 GHz, e.g., Okumura–Hata, Cost 231-Hata, Bertoni-Walfisch, International Telecommunications Union (ITU) Advanced, WINNER+, and 3GPP spatial channel model. Although many of these PL models are widely used today, they are not intended for long-range LoRaWAN network. For this reason, the authors in [25] presents an in-depth study of the radio propagation characteristics and proposed PL models fit measurements with more accuracy and simplicity compared to other models. The mentioned article concludes that, in a dense urban area, a coverage range up to 9 km was attained, whereas in the rural case a coverage range up to 47 km was reached using a single deployed gateway (GW). The quality of transmission in high shadowing and blocking environments (buildings and elevation profile) can be improved by the installation of additional GWs and the optimization of GW locations.

In the literature, some researches are focused on the study of the local positioning systems, such [26, 27], as well as propagation and coverage measurements [28–30]. For instance, in the context of local positioning systems, in [31–33], the algorithms to improve the energy efficiency of the global positioning system (GPS) are analyzed. In the same way, there are some studies that analyze the propagation and coverage measurements, based on simulation and experimental evaluation of the capabilities of the LoRaWAN networks, as described in [34]. However, the mentioned studies do not analyze the optimization using nonlinear disciplines for the LoRaWAN base stations (BS) location from a mathematical viewpoint.

In this paper, we analyze two models to optimization LoRaWAN station's base location: urban and rural zones. This analysis is conducted by means of nonlinear optimizations. Our main contribution is to prove that models analyzed are technical viable to optimize the points for stations base location.

The rest of this paper is organized as follows. In Sect. 46.2, we describe in detail the model. In Sect. 46.3, we analyze and solve the model. In Sect. 46.4, we present the results. Finally, in Sect. 46.5, we present the conclusions.

46.2 Model

We consider the IoT scenario that is depicted in Fig. 46.1 with a LoRaWAN BS to provide connectivity to sensors. The sensor nodes are grouped into clusters. Each cluster has a large number of sensing nodes connected through a LoRaWAN BS [35].

This section presents the mathematical model to optimize the location of LoRaWAN BSs. The model allows us to find the center of the circle which is the smallest that contains all the points representing the sensors, where set $S = \{s_1, s_2, s_3, \ldots, s_n\}$ represent the location of the sensors.

To pose the problem in mathematical terms, we make the following considerations:

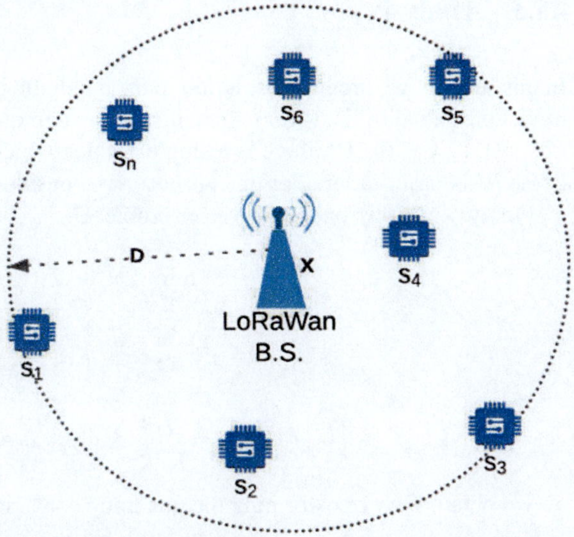

Fig. 46.1 LoRaWAN base station

Signal quality decreases with distance as described in [19–24]. We assume circular areas of 9 km of LoRaWAN base stations coverage for urban areas, as described by authors in [28–30]. We assume circular areas of 40 km of LoRaWAN base stations coverage for rural areas, as described by authors in [28–30].

The mathematical model is represented by:

$$\min_{(x,\delta)\in R} \delta + D_z - \epsilon_z$$

$$\text{s.t} \quad |x - s^i| \leq \delta, \quad i = 1, 2, \ldots, n. \tag{46.1}$$

The notations in Eq. (46.1) are specified as follows:

- δ: coverage area radius, $\delta > 0$
- s: geographic position of sensors
- x: optimal location LoRaWAN base station
- i: total sensors in the cell
- D: LoRaWAN cell maximum coverage radius
- ϵ: coverage correction factor, is a small positive value, as described in [36]
- z : rural zone (r) or urban zone (u).

The same problem for three points in the plane and solution used in this work is proposed in [37–39].

46.3 Analysis

In this section, we propose a solution to the quadratic programming problem, for the model presented in Eq. (46.1). Specifically, we demonstrate that, if any fixed values of s_i, D, ϵ, $x \in (0, D]$, there is a single x value which optimize the location of the LoRaWAN base station. For this purpose, we consider the following steps:

Firstly, if we rely on the Sylvester problem.

$$\min_{(x,\delta)\in R} \; |\omega|^2 + 2$$

$$\text{s.t.} \quad |x - s^i|^2 \leq \delta^2, \quad i = 1, 2, \ldots, n \tag{46.2}$$

$$T = \left\{(x, \delta)\, \mathbb{R} | \left\|x - s^i\right\|^2 \; \leq \delta^2 \,, \quad i = 1, 2, \ldots, n\right\}. \tag{46.3}$$

We obtain T by transforming the quadratic constraints into linear as follows:

$$m = x \tag{46.4}$$

$$\alpha = \frac{1}{2}(\delta^2 - |x|^2) \tag{46.5}$$

$$H = \left\{(m, \alpha) \in R \mid \alpha + m' \, s^i \; \frac{1}{2} \left|s^{\,i}\right|^2, \quad i = 1, 2, \ldots, n\right\} \tag{46.6}$$

Secondly, if we consider $T \neq 0$, $H \neq 0$, then $T \subset H$ as a direct consequence of Eq. (46.5) and $\left(0, \max\left\|s^i\right\|, i = 1, \ldots, k\right)$ belongs to T since the finite set of points is bounded. $\Gamma(T)$ is an injective function, as follows: given (x, δ) and $\left(\hat{x}, \hat{\delta}\right)$ of T, where $x = \hat{x}$ and $\frac{1}{2}\left(\delta^2 - |x|^2\right) = \frac{1}{2}\left(\delta^2 - |\hat{x}|^2\right)$, then $\delta^2 = \widehat{\delta^2}$ and $\delta = \hat{\delta}$ therefore (x = y) and are non-negative.

Finally, to show that problem raised has a solution, we check that Eq. (46.2) is a bounded function. The function described in Eq. (46.2) is lower bounded at 0 as follows: if $|\omega|^2 + 2\alpha \leq \frac{1}{2}\left|s^i\right|^2$, then $|\omega|^2 + 2\alpha \leq \frac{1}{2}\left|s^i\right|^2$ then $|\omega|^2 + 2\alpha \leq \left|m^i\right|^2 - 2m's^i + \left|s^i\right| = \left|m - s^i\right| \leq 0$.

The problem proposed in Eq. (46.2) is a convex quadratic problem, where the Karush–Kuhn–Tucker (KKT) conditions are necessary and sufficient to determine a global optimum. We conclude that the central point of the circular area is contained in set $S = \{s_1, s_2, s_3, \ldots, s_n\}$.

We assume that $D = 9$ with $\epsilon = 273.03$ m and $D = 49$ with $\epsilon = 273.03$ m (where ϵ is as coverage correction factor, is a small positive value, as described in [36]) for urban and rural areas, respectively, and we have proceed numerically in the analysis to solve Eq. (46.2).

46.4 Results

In this section, we will present some numerical results obtained with the above model. First, we apply the model in the city of Cuenca, Ecuador. We obtain the geographic information of the 10 points where the LoRaWAN sensors will locate in an urban zone, as shown in Table 46.1, and 10 points where the LoRaWAN sensors will locate in rural zone Table 46.2. Second, we will calculate numerically the location of the LoRaWAN base station, to provide coverage to the points indicated above. Finally, we implement the numerical solution in MATLAB. The input points introduced in the function are the columns in the matrix, and the outputs are x and δ.

Table 46.1 LoraWAN sensors location—urban zone

Latitude	Longitude
– 2.8963533907518335	– 79.00300052318362
– 2.8975596186788906	– 79.00433153655682
– 2.8850470585514105	– 79.00639343919428
– 2.9078662328614935	– 79.03862815724942
– 2.9228987044519146	– 79.03902046053604
– 2.920588174149547	– 79.00221753007577
– 2.9094584198910955	– 78.98181270428492
– 2.898328555753347	– 78.9962530425369
– 2.898328555753334	– 79.0158730673358
– 2.8991123525901834	– 78.9963315226361

Table 46.2 LoraWAN sensors location—rural zone

Latitude	Longitude
– 2.7966587092209174	– 79.2780783944704
– 2.8755262168401092	– 79.35317579914539
– 2.904030174408659	– 79.29325956911605
– 2.930533178769365	– 79.23576187699562
– 2.8812324250285246	– 79.25010989930803
– 2.9189353765234483	– 79.20703588579389
– 2.890819083808092	– 79.17133031622328
– 2.847614644397509	– 79.14592443018267
– 2.7948070201477884	– 79.16927037951738
– 2.7701169234102814	-79.22832189842252

46.4.1 Base Station—Urban Zone

The results of this section were obtained based on the parameter values shown in Table 46.1, with $D = 9$ km and $\epsilon = 273.03$ m.

Table 46.3 shows the x , δ and with $D \leq 9$ is less, the coverage of the LoRaWAN base station is assured.

Figure 46.2 shows the location of the base station, which is equidistant from the farthest points (3, 4, 7); i.e., the model optimizes the location of the base station LoRaWAN for urban zones.

Table 46.3 LoRaWAN base station location—urban zone

Latitude	Longitude	δ
– 2.9142	79.0109	2.95

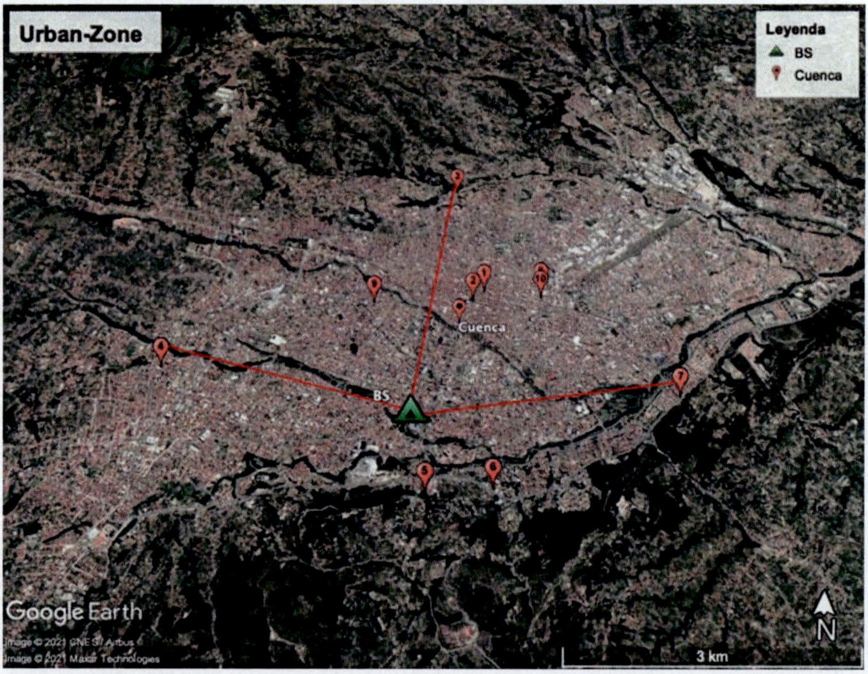

Fig. 46.2 Urban zone

Table 46.4 LoRaWAN base station location—rural zone

Latitude	Longitude	δ
– 2.8615	– 79.2496	1.046

46.4.2 *Base Station—Rural Zone*

The results of this section were obtained based on the parameter values shown in Table 46.1, with $D = 40$ km and $\epsilon = 273.03$ m.

Table 46.4 shows the x, δ and with $D \leq 40$ is less, the coverage of the LoRaWAN base station is assured.

Figure 46.3 shows the location of the base station, which is equidistant from the farthest points (2, 4, 9), i.e., the model optimizes the location of the base station LoRaWAN for rural zones.

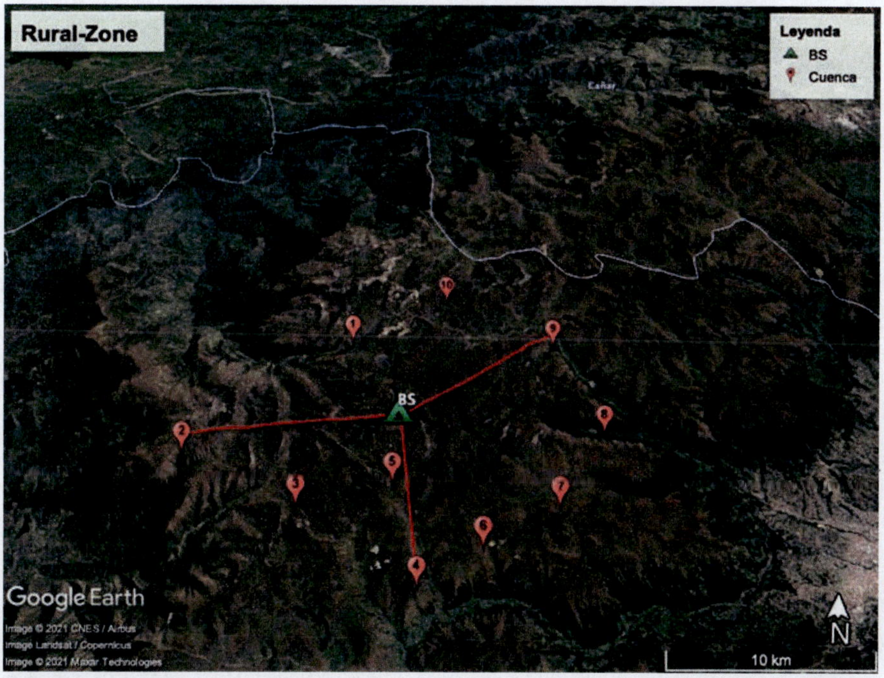

Fig. 46.3 Rural zone

46.5 Conclusions

In this paper, we have studied the application of nonlinear optimization to base stations LoRaWAN location from a mathematical point of view. Our main results suggest that the nonlinear optimization problem proposed in Sect. 46.2, and the analysis developed in Sect. 46.3 allows us to calculate the optimal point to locate the LoRaWAN base station within the coverage area, i.e., at a point equidistant between the location of the sensors and the base station, as shown in Figs. 46.2 and 46.3. Also, the proposed model optimizes the use of sensors batteries by minimizing their distance from the base radio, which is an important pre-requisite in LPWAN. Additionally, LoRaWAN coverage conditions for urban and rural areas are considered in this study. We conclude that such a model is feasible and allows optimizing the location of a LoRaWAN base station.

Future research could focus on the implementation of a LoRaWAN network in other contexts and improve the model with field tests.

References

1. P. Basford, F.M. Bulot, M. Apetroaie-Cristea, S.J. Cox, S.J. Ossont, LoRaWAN for smart city IoT deployments: a long term evaluation. Sensors 648 (2020)
2. N. Sornin, M. Luis, T. Eirich, T. Kramp, O. Hersent, *Lorawan Specification*. LoRa Alliance (2015)
3. R.S. Sinha, Y. Wei, S.H. Hwang, A survey on LPWA technology: LoRa and NB-IoT, in *Ict Express* (2017)
4. R. Ratasuk, B. Vejlgaard, N. Mangalvedhe, A. Ghosh, NB-IoT system for M2M communication, in *IEEE Wireless Communications and Networking Conference* (2016)
5. J.C. Zuniga, B. Ponsard, Sigfox system description. LPWAN, in *IETF97* (2016)
6. C. Gomez, J.C. Veras, R. Vidal, L. Casals, J. Paradells, A sigfox energy consumption mode. Sensors **19**(3), 681 (2019)
7. E. Sacoto-Cabrera, J. Rodriguez-Bustamante, P. Gallegos-Segovia, G. Arevalo-Quishpi, G. León-Paredes, Internet of things: informatic system for metering with communications MQTT over GPRS for smart meters, in *CHILEAN Conference on Electrical, Electronics Engineering, Information and Communication Technologies (CHILECON)*, Pucón (2017)
8. V. Vimos, E. Sacoto-Cabrera, D. X. Morales, Conceptual architecture definition: implementation of a network sensor using Arduino devices and multiplatform applications through OPC UA, in *IEEE International Conference on Automatica (ICA-ACCA)* (2016)
9. V. Vimos, E. Sacoto-Cabrera, Results of the implementation of a sensor network based on Arduino devices and multiplatform applications using the standard OPC UA. IEEE Lat. Am. Trans. **16**(9), 2496–2502 (2018)
10. E.J.S. Cabrera, S. Palaguachi, G.A. León-Paredes, P.L. Gallegos-Segovia, O.G. Bravo-Quezada, Industrial communication based on MQTT and Modbus communication applied in a meteorological network, in *International Conference on Advances in Emerging Trends and Technologies* (2020), pp. 29–41
11. K. Mekki, E. Bajic, F. Chaxel, F. Meyer, Overview of cellular LPWAN technologies for IoT deployment: Sigfox, LoRaWAN, and NB-IoT, in *International Conference on Pervasive Computing and Communications Workshops* (2018)
12. P.J. Basford, F.M. Bulot, M. Apetroaie-Cristea, S.J. Cox, S.J. Ossont, LoRaWAN for smart city IoT deployments: a long term evaluation. Sensors **20**(3), 648 (2020)

13. A. Sanchis-Cano, J. Romero, E.J. Sacoto-Cabrera, L. Guijarro, Economic feasibility of wireless sensor network-based service provision in a duopoly setting with a monopolist operator. Sensors **17**(12), 2727 (2017)
14. O. Khutsoane, B. Song, A.M. Abu-Mahfouz, IoT devices and applications based on LoRa/LoRaWAN, in *IECON 2017–43rd Annual Conference of the IEEE Industrial Electronics* (2017)
15. A.J. Wixted, P. Kinnaird, H. Larijani, A. Ahmadinia, N. Strachan, Evaluation of LoRa and LoRaWAN for wireless sensor networks, in *IEEE Sensors* (2016)
16. J. de Carvalho Silva, J.J. Rodrigues, A.M. Alberti, P. Solic, A.L. Aquino, LoRaWAN—a low power WAN protocol for Internet of Things: a review and opportunities, in *2nd International Multidisciplinary Conference on Computer and Energy Science (SpliTech)* (2017)
17. M. Centenaro, L. Vangelista, A. Zanella, M. Zorzi, Long-range communications in unlicensed bands: the rising stars in the IoT and smart city scenarios. IEEE Wirel. Commun. **23**(5), 60–67 (2016)
18. D. Patel, M. Won, Experimental study on low power wide area networks (LPWAN) for mobile Internet of Things, in *Vehicular Technology Conference (VTC Spring)* (2017)
19. Y. Okumura, Field strength and its variability in VHF and UHF land-mobile radio service. Rev. Electr. Commun. Lab. **16**, 825–873 (1968)
20. P.E. Mogensen, J. Wigard, COST Action 231: digital mobile radio towards future generation system, final report, in*Section 5.2: On Antenna and Frequency Diversity in GSM. Section 5.3: Capacity Study of Frequency Hopping GSM Network* (1999)
21. Series, M. ITU, *Guidelines for Evaluation of Radio Interface Technologies for IMT-Advanced*, Report ITU, 72
22. Y.D.J. Bultitude, T. Rautiainen, IST-4–027756 WINNER II D1. 1.2 V1. 2 WINNER II Channel Models. EBITG, TUI, UOULU, CU/CRC, NOKIA, Tech. Rep (2007)
23. M. Juha, K. Pekka, H. Lassi, WINNER+ final channel models, deliverable D5. 3 V1. 0. 2007111130) [2014104116]. http://wenku.baidu.com/link
24. ACCESS, E. U. T. R., Further advancements for E-UTRA physical layer aspects, Further advancements for E-UTRA physical layer aspects (2019)
25. R. El Chall, S. Lahoud, M. El Helou, LoRaWAN network: radio propagation models and performance evaluation in various environments in Lebanon. IEEE Internet Things J. **6**(2), 2366–2378 (2019)
26. A. Vazquez-Rodas, F. Astudillo-Salinas, C. Sanchez, B. Arpi, L.I. Minchala, Experimental evaluation of RSSI-based positioning system with low-cost LoRa devices. Ad Hoc Netw. **105**, 102168 (2020)
27. R. Zekavat, R.M. Buehrer, *Handbook of Position Location: Theory, Practice and Advances* (Wiley, 2011)
28. H.T. Nastiti, I.E. Praditya, I.W. Mustika, Evaluation of XBee-Pro transmission range for wireless sensor network's node under forested environments based on received signal strength indicator (RSSI), in *2nd International Conference on Science and Technology-Computer (ICST)* (2016)
29. T. Addabbo, A. Fort, M. Mugnaini, E. Panzardi, A. Pozzebon, V. Vignoli, A city-scale IoT architecture for monumental structures monitoring. Measurement **131**, 349–357 (2019)
30. J. Petajajarvi, K. Mikhaylov, A. Roivainen, T. Hanninen, M. Pettissalo, On the coverage of LPWANs: range evaluation and channel attenuation model for LoRa technology, in *4th International Conference on ITS Telecommunications (ITST)* (2015)
31. T. Hadwen, V. Smallbon, Q. Zhang, M. D'Souza, Energy efficient LoRa GPS tracker for dementia patients, in *39th Annual International Conference of the IEEE Engineering in Medicine and Biology Society (EMBC)* (2017)
32. B.C. Fargas, M.N. Petersen, GPS-free geolocation using LoRa in low-power WANs, in *In 2017 Global Internet of Things Summit (Giots)* (2017)
33. P. Kietzmann, D. Kutscher, T.C. Schmidt, M. Wählisch, Long-range IoT: Is LoRaWAN an option for ICN? in *In Proceedings of the 7th ACM Conference on Information-Centric Networking* (2020)

34. R.S. Alonso, I. Sittón-Candanedo, O. García, J. Prieto, S. Rodríguez-González, An intelligent edge-IoT platform for monitoring livestock and crops in a dairy farming scenario. Ad Hoc Netw. **98**, 102047 (2020)
35. P.V. Mekikis, E. Kartsakli, A. Lalos, A. Antonopoulos, L. Alonso, C. Verikoukis, Connectivity of large-scale WSNs in fading environments under different routing mechanisms, in *In 2015 IEEE International Conference on Communications (ICC)* (2015)
36. M. Aernouts, R. Berkvens, K. Van Vlaenderen, M. Weyn, Sigfox and LoRaWAN datasets for fingerprint localization in large urban and rural areas. Data **3**(2), 13 (2018)
37. S. Garrod, G. Ferrier, S. Campbell, In and on: investigating the functional geometry of spatial prepositions. Cognition **72**(2), 167–189 (1999)
38. M.J.C. Cánovas, M.S. Orts, V.H. Navarro, Optimización matemática aplicada. Enunciados, ejercicios y aplicaciones del mundo real con MATLAB., Editorial Club Universitario (2012)
39. W. Verdugo-Romero, L. González-Delgado, E. Sacoto-Cabrera, A. Verdugo-Cabrera, O. Verdugo-Cabrera, Mathematical and statistical analysis for the simulation of the vehicular flow in a specific sector of the city of Cuenca-Ecuador, in *In 2018 IEEE Third Ecuador Technical Chapters Meeting (ETCM)*, pp. 1–4

Chapter 47
Wireless Sensor Network Applied to Precision Agriculture: A Technical Case Study at the Technical University of Manabí

Marely Rosario Cruz del Felipe, Maikel Leyva Vázquez, and Jorge Luis Piguave Bermello

Abstract Today, the use of technologies in agriculture, which is called precision agriculture, is increasingly necessary to improve crop yields. One of the most widely used technologies is wireless sensor networks (WSN). This article deals with the development, implementation, and evaluation of a network of wireless sensors applied to precision agriculture at the Faculty of Agriculture of the Technical University of Manabí to study environmental phenomena in crops. The sensor network collects data on environmental variables such as soil moisture, temperature, and ambient humidity. This data is sent wirelessly to a head node responsible for uploading the information to an Internet of Things (IoT) platform. The tools used were Arduino Mega Rev3 as main coordination node board, Arduino Pro mini, XBee S2c devices for wireless communication, DHT-22 and YL-69 sensors to obtain environmental information and rechargeable batteries with solar panels. Custom Arduino libraries were used to enable Arduino sleep mode and achieve power savings.

47.1 Introduction

Since ancient times, man has carried out the agricultural activity, obtaining results based on the physical effort invested, the methods applied, the environment in which are located and the climate. Currently, some technologies facilitate and improve most of the processes carried out manually in any field. An improvement in the level of crop production can be achieved if the technological tools corresponding to the needs of the crop are correctly applied.

M. R. C. del Felipe · J. L. P. Bermello
Universidad Técnica de Manabí, Portoviejo 130105, Ecuador
e-mail: marely.cruz@utm.edu.ec

M. L. Vázquez (✉)
Computer Science Department, Universidad Politécnica Salesiana Guayaquil, Guayaquil, Ecuador
e-mail: mleyva@ups.edu.ec

Á. Rocha et al. (eds.), *Communication, Smart Technologies and Innovation for Society*, Smart Innovation, Systems and Technologies 252,
https://doi.org/10.1007/978-981-16-4126-8_47

Agriculture in Ecuador is one of the primary sources of employment and is considered a strategic activity to ensure food security and sovereignty; the agricultural sector is one of the most relevant tasks in Ecuador because this country has great biodiversity and, due to its geographical location, has advantages to grow several products for export [1].

The Faculty of Agricultural Engineering is located at the Technical University of Manabi, Ecuador; there are several land areas for different crops or nurseries within it. This faculty's objective is to increase or improve their efficiency levels in crop production, which would require more time and expense if done through conventional procedures. With the help of technology, different methods can be used to make this more economically effective; farming and technology are called precision farming [2].

The use of technology is essential today in all fields of life and industry. Particularly in the agricultural case, wireless sensor networks or WSNs are a good example; nodes in these networks can provide information of several climatic variables to measure through the software on desktop, web, or mobile platforms, facilitating data collection and monitoring of environmental variables [3].

In Ref. [4], an exhaustive analysis of the main techniques, standards, and tools used in WSN is made, where there are works related to the use of motherboards that are variants of Arduino, Waspmote or Raspberry Pi, as well as for communication protocols or standards such as IEEE 802.15.4, ZigBee or LoRa. It also reviews the selection of sensor models that depend on the type of variables to be measured, such as temperature and humidity, soil moisture, and others which was exposed in [5].

The parts of a sensor network are (1) Sensors: of different types and technologies that take information from the medium and convert it into electrical signals. (2) Sensor nodes take the sensor data through their data interfaces and send the information to the base station. (3) Gateways: Elements for interconnection between the sensor network and a TCP/IP network. (4) Base station: data collector. (5) Wireless network: usually based on the ZigBee 802.15.4 standard [6].

Sensor networks generate data that can be treated and processed in remote servers; currently, IoT platforms are of great use for this processing. The use of sensor networks with IoT platforms is discussed in [7], where they are widely used in various applications. Among the IoT platforms used are Cayenne, Thingboard, and ThingSpeak [8, 9].

Given that WSN plays an essential role in agriculture and given the need to increase the crop efficiency levels of farmers through the use of technology without the need for excessive economic investment, the need to improve resource management and also the need for environmental sustainability; this paper presents a WSN design that takes into account the following aspects: First, the design of an environmental sensor network and second, the use of IoT platform for data monitoring.

The remainder of the paper is organized as follows: Sect. 47.2 shows the design requirements for data acquisition, the design of the sensor network, and the energy savings achieved by the network. Section 47.3 discusses the research results and their interpretation, showing how the data is brought to an IoT platform. Section 47.4 presents conclusions and future work.

47.2 Materials and Methods

In this research, different sensor network designs were carried out, and the technologies studied were tested to select the best design.

Technologies such as Arduino Pro mini, Arduino Micro, Arduino Uno, Arduino Mega Rev3, XBee S2c, DHT-11, DHT-22, and YL-69 sensors were analyzed.

Tests were carried out to determine the battery consumption with the different technologies and the performance of the system, for which the consumption was compared at no load or when the node is "asleep" term used to define when the sensor node is not transmitting or receiving, and algorithms are used to ensure that the sensor node remains most of the time in a state of "sleeping" and thus achieve a lower battery consumption.

47.3 Results

This section continues as follow (3.1) system requirements, (3.2) electrical design of the sensor network, (3.3) sensor network operation with IoT platforms, (3.4) energy-saving tests on nodes.

47.3.1 System Requirements

The functional requirements of the system are the services provided by the wireless sensor network:

- The information of different environmental variables will be collected through the sensors connected to the Arduino board.
- The network will have a wireless connection using XBee devices.
- The Zigbee communication standard will be used.
- The data collected can be consulted on a web platform and from a mobile application.
- The network nodes will be powered by an external rechargeable battery with solar panels.
- The coordination node will connect to the Internet to upload data to the IoT platform.

The non-functional system requirements are the statement of the general characteristics and constraints of the wireless sensor network, which are detailed below:

- The end-user will have access to add alarms according to the data reported by the sensor network.

- Network nodes should be no more than 40 m apart in indoor environments and no more than 200 m apart in a line-of-sight configuration.
- It is not possible to change the pins on the Arduino board to which the sensors are connected.
- A mid-range cell phone device is recommended for using the mobile version of the IoT platform where the information will be displayed.

47.3.2 Electrical Design of the Sensor Network

The diagram in Fig. 47.1 shows the functionality of the sensor network, where an external agent called "User" interacts with the IoT platform and, through widgets, can visualize the information collected by the sensor network nodes. He also has access to other functions of the IoT platform, such as the creation of alarms. Figure 47.1 shows the functions that the system must fulfill to meet the user's requirements.

Figure 47.2 shows the final device node showing the Arduino Pro mini's connection with the XBee module mounted on the XBee Explorer, the DHT-22 sensor, the YL-69 sensor, and the rechargeable batteries with solar panels.

The coordination node's general schematic can be seen in Fig. 47.3. The Arduino Mega Rev3 is connected to an Ethernet Shield W5100 module, an XBee mounted on the XBee Explorer, and a 110–5 V adapter.

Figure 47.4 shows the location of all the nodes that are part of the wireless sensor network, and nodes 1, 2, 3, and 4 are the nodes that have the sensors to measure the environmental variables that are located within a hectare of crop, node 6 is the coordinating node located in the Faculty of Agricultural Engineering because node 6 has no line-of-sight with the other nodes, node 5 helps to expand and improve wireless communication between the other nodes.

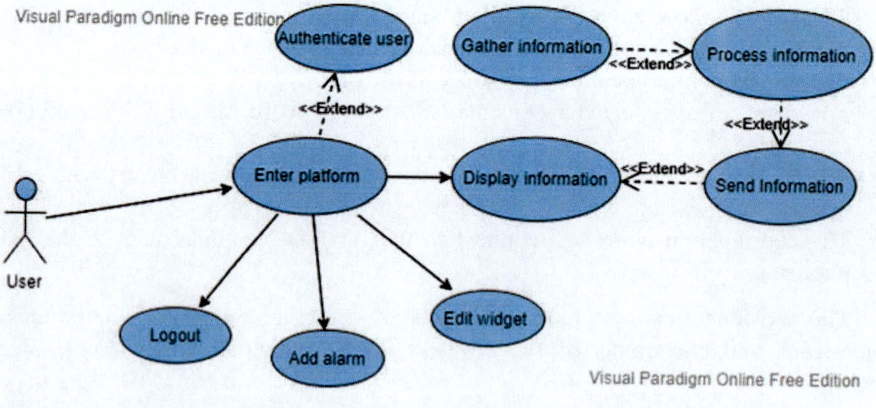

Fig. 47.1 User use case diagram

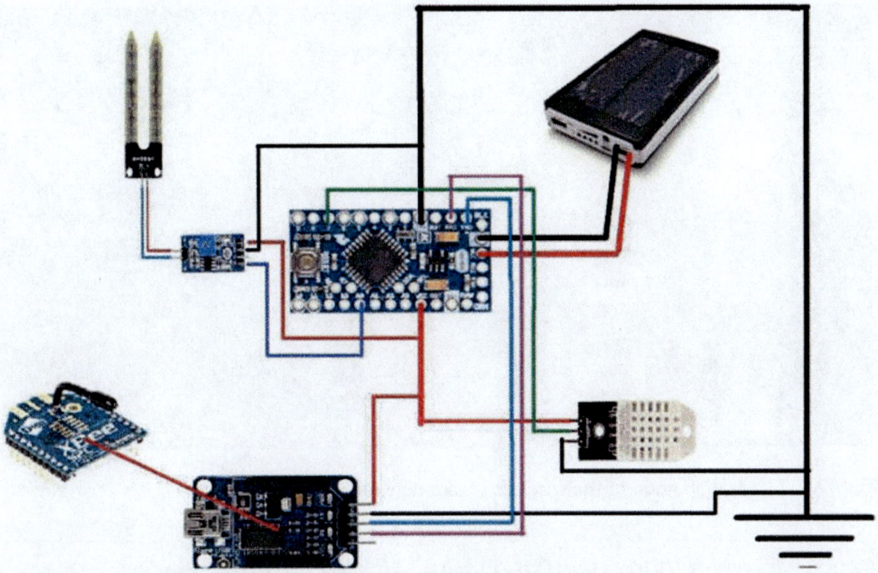

Fig. 47.2 General diagram of the final device nodes

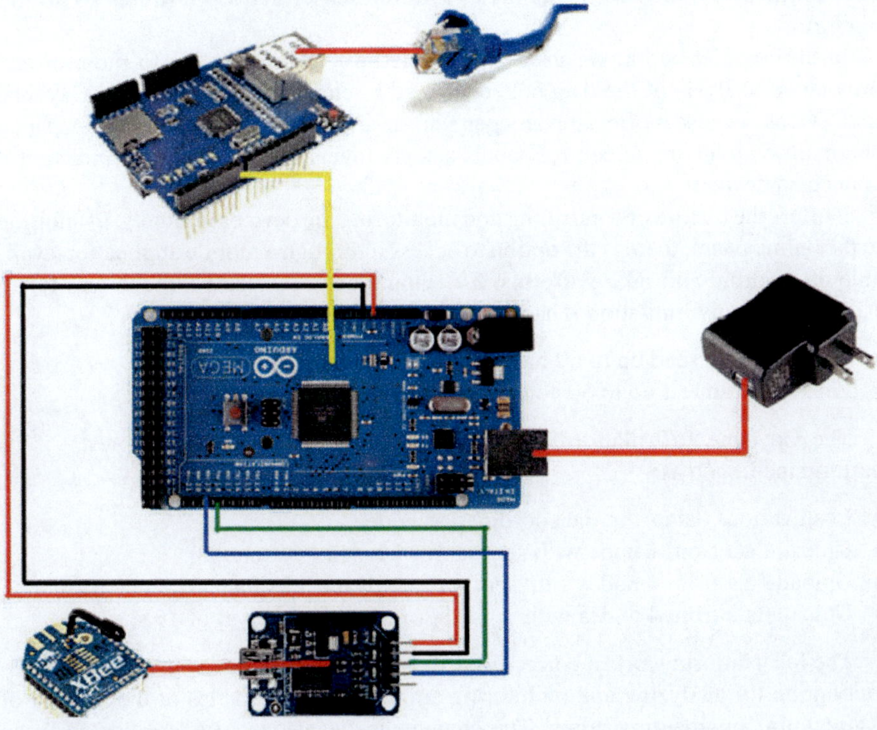

Fig. 47.3 General scheme of the coordinating node

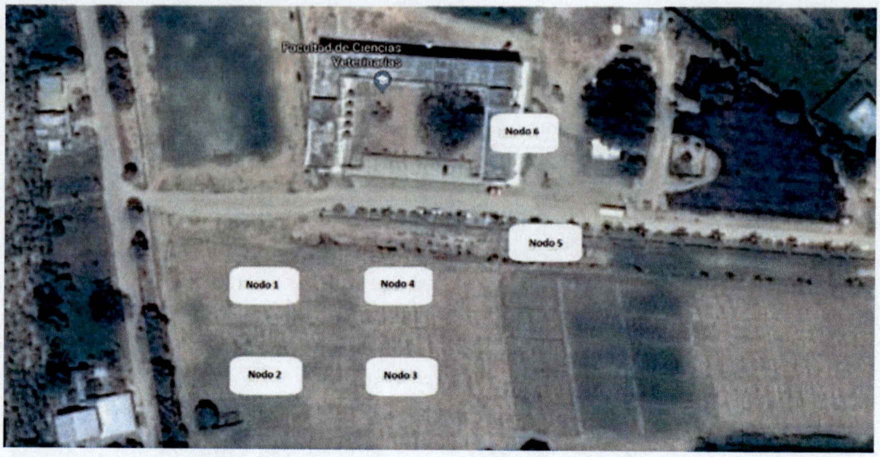

Fig. 47.4 General diagram of the wireless sensor network

47.3.3 Sensor Network Operation with IoT Platforms

The results of the measurements made with the sensor network are sent to an IoT platform.

In the research work, we used an IoT platform that allows us to monitor and provide an analysis of the data collected by the sensors; we chose to use Cayenne myDevices because it offers advantages such as an easy-to-use administration panel, and it allows to create of control panels using widgets to view, manage, and control connected devices.

It offers the feature of controlling and monitoring the devices remotely. In addition to the online panel, there is the option to access it through a mobile application available on Android and IOS platforms. Everything communicates through an MQTT API, and the only limitation it has are the API calls:

- Each user can send up to 60 messages per minute per IP.
- Each user can test up to 50 connections every 10 min per IP.

The Cayenne IoT platform was evaluated by performing several tests of its performance, such as:

- Load various data from the coordination node.
- Upload data from 1 node with sensors for 1 h at a 15 s interval.
- Upload data from 4 nodes with sensors for 1 h at a 15 s interval.
- Upload data from 4 nodes with sensors for 6 h at an interval of 15 s.

The IoT platform worked correctly in 50 tests performed, proving to be an excellent option for analyzing and monitoring environmental variables of the Faculty of Agricultural Engineering crops. The changes in the widgets were reflected in an average of 10 s in the web platform, considering that its mobile version sometimes

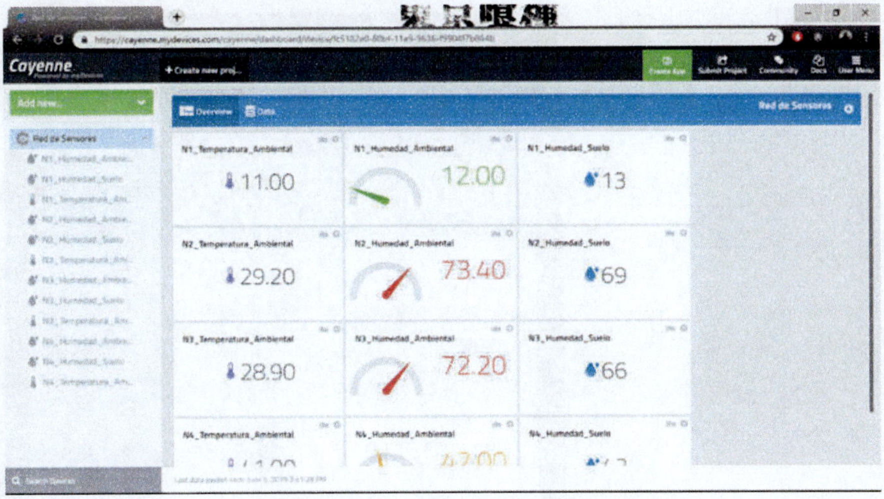

Fig. 47.5 Main panel of the sensor network in Cayenne web version

did not reflect the changes, so it had to abandon and re-enter the project; on all occasions, there were no problems with the IoT platform server. Figure 47.5 shows the Cayenne main panel in its web version.

Figure 47.6 shows the data analysis of a widget in Cayenne in its mobile version.

For the use of alerts in the IoT platform, an event with specific information must first occur, e.g., if the ambient temperature exceeds 27 °C, an alert is sent by text message or email. Figure 47.7 shows the panel for managing triggers or alerts in Cayenne in its web version.

47.3.4 Energy-Saving Tests

Tests were performed to put the Arduino to sleep with the "LowPower," "JeeLib," and "SnoozeLib" libraries, opting for the "LowPower" library because it presented no problems when it came to putting the Arduino Pro micro-board to sleep. With this library using the Watch Dog timer, user can put the Arduino to sleep in a maximum time of 8 s without the need for an external interrupt. However, by adding a loop, uses can extend the time as needed.

Also, in order to save the energy consumption of the final device nodes, several Arduino models were tested.

The Arduino Nano Rev3 resulted in a 48 mAh consumption at idle. The Arduino Pro micro resulted in 15 mAh. However, it went into a complete hibernation mode when programmed in sleep mode, not recognizing the COM port. So it is discarded the use of this model of Arduino.

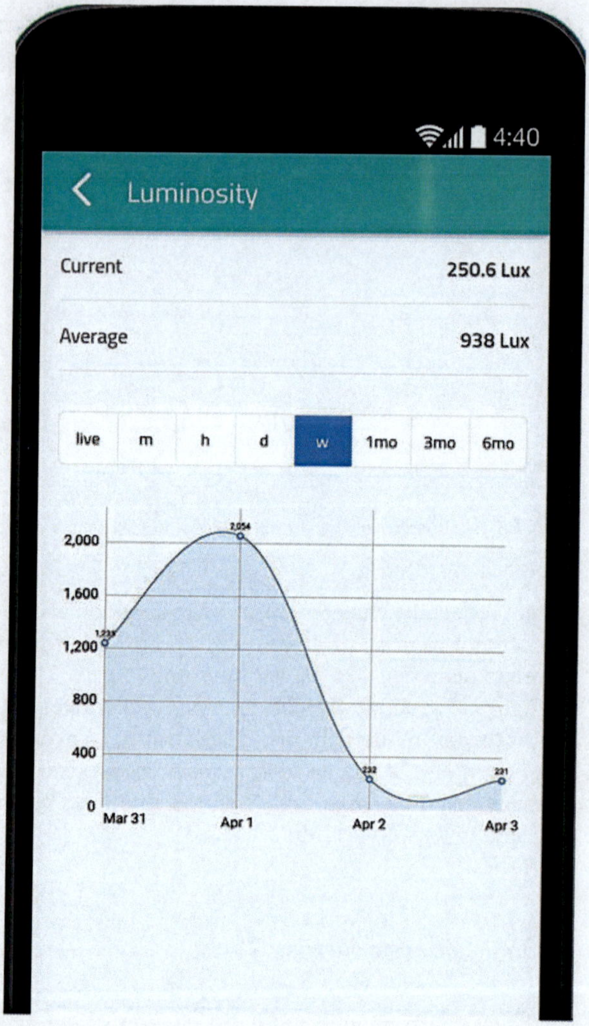

Fig. 47.6 Data analysis of a widget in the mobile version of Cayenne

The research was done on the use of different libraries such as "LowPower," "JeeLib," and "SnoozeLib" to sleep the Arduino Pro micro-board. However, there was still the drawback that the COM port does not work correctly in all tests performed.

The Arduino Pro mini showed excellent results of consumption at idle with 11 mAh and in sleep mode 5 mAh, also presents no problems with serial communication or sleep mode, choosing this model of Arduino as the main board of all nodes end devices along with the library "LowPower" to enter sleep mode to the Arduino.

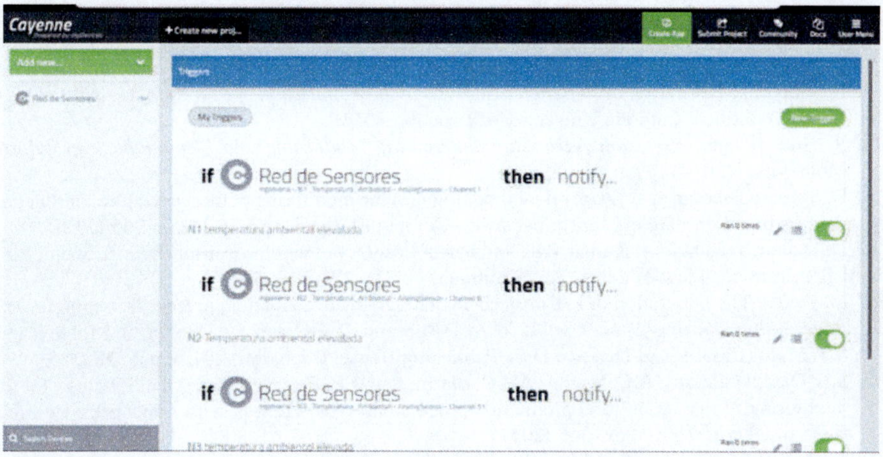

Fig. 47.7 Chayenne trigger panel web version

47.4 Conclusions

By investigating different technologies that are applied to implement sensor networks in precision agriculture, it was determined that the optimal design of the wireless sensor network for the proposed project should employ technologies such as: Arduino Pro mini, Arduino Mega Rev3, XBee S2c, DHT-22, and YL-69.

A wireless sensor network was implemented in the Faculty of Agronomic Engineering of the Technical University of Manabi, in a hectare of crops, achieving efficient wireless communication between the nodes to collect data from the relative humidity, ambient temperature, and soil moisture sensors.

The IoT platform "Cayenne" was used, which is optimal for this project, allowing functions such as adding widgets to visualize the information in a user-friendly way, the option of providing node statistics and adding alerts when a specific event occurs.

Several tests were performed to improve the wireless sensor network's performance, such as the power supply of the nodes opting to build a rechargeable battery system with solar panels, and the power consumption of various Arduino boards giving the Arduino better results. The Arduino Pro mini showed excellent results with different libraries at the sleep mode, being the "LowPower" the more efficient, for sending and receiving data in wireless communication via XBee, using the configuration in transparent mode on the Arduino hardware serial port.

It is recommended in future work to perform a correlation analysis of the variables used in the sensor network to optimize an irrigation system.

References

1. R. Villamil, M. Ricardo, *Cost-Benefits Analysis of the Application of Precision Agriculture Case Cocoa* (Pontifical Catholic University of Ecuador, 2012)
2. R. Bongiovanni, *Precision Agriculture Integrating Moderning and Sustainable Agriculture*. Montevideo (2014)
3. D. Soledad Escolar, et al., A novel methodology for the monitoring of the agricultural production process based on wireless sensor networks. Comput Electron Agri **76.2**, 252–265 (2011)
4. D. Thakur, Y. Kumar, A. Kumar, P.K. Singh, Applicatibily of wireless sensor network of precision agriculture: a review. Wireless Pers Commun. **107**(1), 471–512 (2019)
5. M. Oprea, The integration of IoT projects in undergraduate education, in *New Technologies and Redesigning Learning Spaces*, vol I, ed. by I. Roceanu, D. Belgian, I.A. Stefan, A. Moldoveanu, S.T. Matu (Carol I Natl Defence Univ Publishing House, Bucharest, 2019), pp. 318–325
6. S.E. Diaz, J.C. Perez, A.C. Mateos, M.-C. Marinescu, B.B. Guerra, A novel methodology for the monitoring of the agricultural production process based on wireless sensor networks. Comput. Electron. Agric. **76**(2), 252–265 (2011)
7. I. Bhakta, S. Phadikar, K. Majumder, State-of-the-art technologies in precision agriculture: a systematic review. J. Sci. Food Agric. **99**(11), 4878–4888 (2019)
8. K.E. Lakshmiprabha, C. Govindaraju, Hydroponic-based smart irrigation system using Internet of Things. Int. J. Commun. Syst. e4071 (2019)
9. B. Jiang, M.T. Iqbal, Open source data logging and data visualization for an isolated PV system. Electronics **8**(4), 424 (2019)

Chapter 48
Benchmarking of Supervised Machine Learning Algorithms in the Early Failure Prediction of a Water Pumping System

Gerardo Herrera and Paulina Morillo

Abstract Preventing failures in water supply systems is of vital importance for the development of a population, especially when its economic engine is the agricultural sector. Therefore, it is important to apply new control techniques, which incorporate machine learning and allow prediction failures effectively. This paper performs a comparative analysis of three classification algorithms, random forest, support vector machines, and artificial neural networks, to predict failures in a water pumping system. The methodology employed considers the selection of a training dataset, data preprocessing, training, and evaluation of each model, and its subsequent performance comparison. According to the results, the lowest average accuracy was obtained by the SVM algorithm (83.24%), while RF obtained the highest accuracy (99.98%), closely followed by ANN (86.94%). According to the hypothesis tests, there are significant differences between the SVM, RF, and ANN algorithms, showing that the latter two achieve better performances than SVM, but without significant differences between them, so that to select one of them, it is necessary to consider other aspects such as training time and interpretability. The results show that supervised learning algorithms can reach values higher than 80% of accuracy in the detection of system failures, which evidences their usefulness in control systems.

48.1 Introduction

According to the United Nations (UN), water supply is a service and a fundamental right [1]. A continuous and reliable service serves the development and economy of a population, especially when its most important economic activities are agricultural and industrial [2]. Pumps, tanks, and pipelines are part of the water supply systems of large and small cities. Their correct operation guarantees the availability of this vital element, for many activities of daily life. Therefore, failures in water systems can cause serious consequences on health, agriculture, and in general on the development of the affected populations [3].

G. Herrera · P. Morillo (✉)
Universidad Politécnica Salesiana, Quito, Ecuador
e-mail: pmorillo@ups.edu.ec

Á. Rocha et al. (eds.), *Communication, Smart Technologies and Innovation for Society*, Smart Innovation, Systems and Technologies 252,
https://doi.org/10.1007/978-981-16-4126-8_48

In rural areas, agriculture is one of the most important economic activities, and its development requires large quantities of water, especially for plantation irrigation [4]. In the case of Chaiyaphum, province of Thailand, a region whose economic engine is agriculture, the water supply service is essential to sustain the sector. However, this service has been interrupted seven times in one year due to system failures that could not be detected by the water company, even though they have a classic control in the tanks [5]. This may be because only 10–20% of the pumps reach their useful life without preventive maintenance, as indicated by [6], and premature damage to the equipment is prevented if there is proper maintenance. In addition, when a pump failure occurs, it can take several days for the water service to be restored, since the replacement of a component can take a long time. This causes the service to be suspended, causing losses to the company and the citizens. With an adequate failure prediction system, the service suspension time could be reduced. According to [7], it will be possible to reduce different parameters that affect the company's assets and productivity, such as the number of stops in the work field, breakdowns in primary components, maintenance costs, and contingency costs.

Sometimes, the traditional control does not obtain the desired results, because it does not consider wear and tear and failures in the elements of the system. For such reasons, the new paradigms of control include the use of machine learning to predict critical failures [8]. In [9], for example, the prediction of failures in a water system using the Kaplan–Meier method and regression models to estimate the service life of water pipes, although other factors such as flow rate and temperature are not considered. In [10], it addresses the detection of aqueduct pipeline failures, considering static and dynamic factors, using a clustering algorithm implemented in MATLAB. This model considers only the temperature and the pipe manufacturing static data.

Another work [11] analyzes a neuro-fuzzy model to forecast flow demand and detect possible failures in water supply systems. On the other hand, the work of [12] has had successful results in monitoring hydraulic systems using multivariate statistical analysis and using linear discriminant analysis (LDA), managing to classify the failure condition satisfactorily, but without considering the tank level as a study variable. In [13] random forest (RF) and artificial neural networks (ANN) algorithms are used to predict the hourly throughput of a ground source heat pump (GSHP) system and use it to schedule its maintenance.

As evidenced in the literature, the use of supervised machine learning algorithms is frequently used in the solution of fault detection problems, the models that stand out for their precision and accuracy are support vector machine (SVM) [14, 15], RF [16, 17] and ANN [18, 19]. This paper performs a comparative analysis of three machine learning (ML) techniques: SVM, RF, and ANN applied in the early prediction of faults in a water pumping system. Therefore, this article is organized as follows. Section 48.2 presents the methodology used to implement and compare the ML techniques. Section 48.3 presents the experiments and results obtained. Finally, Sect. 48.4 presents the conclusions and future works.

48.2 Methodology

This work is based on the classical machine learning methodology [20], from data processing to configuration, training, evaluation, and performance comparison of the models (Fig. 48.1). Since the problem to be solved is the prediction of the state, failure/non-failure, of a water pump in a supply system, the paradigm addressed was supervised machine learning, specifically a binary classification task. For this purpose, a dataset was used [21], which was the result of monitoring the water supply system of the population of Chaiyaphu/Thailand. This dataset originally had 52 sensors (attributes) and 220, 320 instances taken with a time interval of one minute. The target variable called machine status had three classes referring to the status of the pumps, *normal*, *recovery*, and *broken*, and the variables measured were pressure in psi, tank level in m, and flow rate in l/min. On the other hand, the models selected to compare their performance were random forest, support vector machine, and artificial neural networks, since they are the most widely used algorithms for predictive maintenance applications, according to [22].

The first phase of the methodology was the data preprocessing [23]. The following procedures were performed data rescaling, missing values analysis, inconsistent values analysis, class balancing, dimensionality reduction, and label coding (Fig. 48.2).

After data cleaning, we moved on to the training and evaluation phase of the models, for which we used the cross-validation technique with a number equal to ten [23, 24]. This technique guarantees that the results obtained by the model are independent of the participation of training and test data. The training partition was set up with 80% of the data, while the remaining 20% was used for the test set. Three different configurations (C_1, C_2, C_3) were used for each algorithm to evaluate parameters such as training time (*T_train*) and performance. The details of each configuration are shown in Table 48.1.

The metrics used to analyze model performance were accuracy (Acc), precision (Pr), sensitivity or recall (Re), specificity (Sp), and Fisher coefficient ($F1$). The calculation of the evaluation metrics was performed considering that the positive class corresponds to the *recovery* label and the negative class to the *normal* label. After training and validating the three algorithms with their respective configurations and to

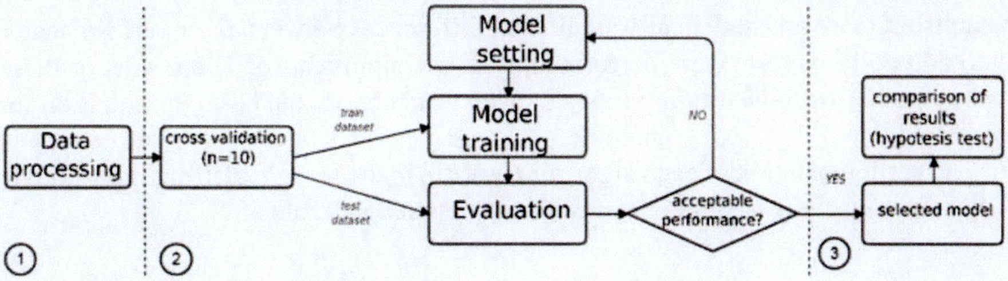

Fig. 48.1 Methodology diagram

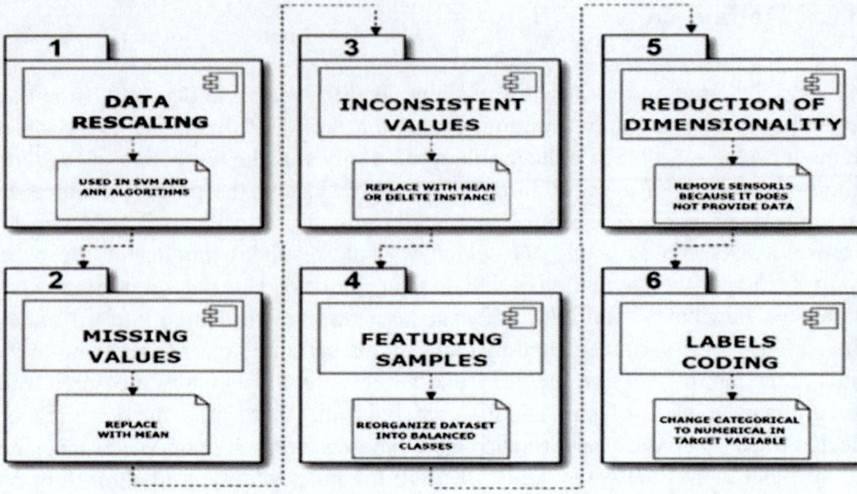

Fig. 48.2 Data preprocessing diagram

Table 48.1 Algorithm's parameters and settings

Types of models	Parameter	Value
RF_{C1}	Trees (n)	25
RF_{C2}	Trees (n)	100
RF_{C3}	Trees (n)	500
SVM_{C1}	Kernel (k)	Linear
SVM_{C2}	Kernel (k)	Radial, gamma = 0.1, $C = 1$
SVM_{C2}	Kernel (k)	Polynomic (poly)
ANN_{C1}	Hidden layers (L) y # hidden neurons(h)	$L = 1, h = 15$, epochs = 10, bs = 10, nn = 52
ANN_{C2}	Hidden layers (L) y # hidden neurons(h)	$L = 1, h = 30$, epochs = 10, bs = 10, nn = 79
ANN_{C3}	Hidden layers (L) y # hidden neurons(h)	$L = 3, h = 45$, epochs = 300, bs = 20,nn = 94

establish if there are statistically significant differences between their performances, paired hypothesis tests were performed, with 5% significance. The results of these tests, together with the training time, allowed us to choose the best configuration for each model.

The implementation of each algorithm was run in the Kaggle platform and Google Colab, in a machine with the characteristics detailed in Table 48.2.

Table 48.2 Hardware and software features of the machine used in the training process

	Parameter	Value		Parameter	Value
Hardware	Memory	12.71 Gb	Software	SO	linux
	Disk	107.77 GB		Kernel	4.19.112+
	CPU model	Xeon(R)		Python	3.6.9
	Velocity	2.30 GHz		Pandas	1.1.2
	# cores	2		Sklearn	0.22.2
	Marca	Intel		Joblib	0.16.0
				Seaborn	0.11.0
				Scipy	1.4.1

48.3 Experiments and Results

48.3.1 Data Processing

In this phase, we follow the steps shown in Fig. 48.2. Firstly, data rescaling was performed, which consisted of transforming the values of each attribute by values between zero and one, using the MixMaxScaler function (Eq. 48.1), which facilitates the learning and fast convergence of the models [25], the rescaled attributes were pressure (S01, S02, S03, S11), water level (S00, S18), and flow (S04, S14, S16, S17, S19–S23, S25–S28, S30, S31, S44, S50, and S51).

$$\text{MinMaxScaler} = (\text{Xstd})(\text{Max} - \text{Min}) + \text{Min} \tag{48.1}$$

$$\text{Xstd} = (X - X \cdot \text{min})/(X \cdot \text{max} - X \cdot \text{min}) \tag{48.2}$$

After the scaling, the missing data by each attribute were replaced by the corresponding average of the remaining values. In total, 17,033 missing values, were replaced. The attributes with the highest number of missing data were s01, s44, s50, and s51. Inconsistent values were also analyzed, of which a total of 476 were found in the *recovery* class. Inconsistent data refers to values outside the normal range (well above or below the mean), with non-valid formats (strings, characters), etc. Those instances with more than four inconsistent fields were removed. Since the number of *broken* instances was very low concerning the other classes, they were discarded, remaining only the instances with *normal* and *recovery* labels. The machine learning task thus became a binary classification, in which 14,001 instances of the *recovery* class and 14,002 instances of the *normal* class were randomly selected to achieve the balance between both classes. Then the attributes of the dataset were analyzed. This process was carried out manually and considering the meaning of each variable, its units, and its usefulness in the control process. In this way, 24 attributes were selected

Table 48.3 Original and cleaned datasets description

Dataset	Features	Normal	Recovery	Broken	Total
Original	55	205,836	14,477	7	220,300
Cleaned	24	14,001	14,002	0	28,003

from the original 55 attributes (S00–S04, S11, S14, S16–S23, S25–S28, S30, S31, S44, S50, and S51).

Finally, label coding was performed, which consisted of replacing the labels of the classes by zero for *normal* (N) and one for *recovery* (R) to facilitate their training. The final dataset used for model training consisted of 28,003 instances, 24 attributes, and two classes, showing Table 48.3.

The final version of the debugged dataset and the Python code used for its debugging is hosted at https://github.com/gherrera2k1/tesis_maestria_2020/blob/master/hardware%20_collab.ipynb.

48.3.2 *Training and Evaluation of Models*

For each model and configuration, we measured the average training time and performance on the classification task. For this purpose, cross-validation was performed, where the training partition had 22,402 instances and the test partition had 5601 instances. The results, in percentage, are summarized Table 48.4, in which the highest values by the model have been highlighted.

Likewise, the first two rows of Table 48.4 show the confusion matrices for each model and configuration. As can be seen in the three configurations of the random forest algorithm, the instances were almost entirely correctly classified, obtaining accuracies of 100%. While for the other models, the number of misclassified instances was higher. Thus, for the SVM model, the highest accuracy was 96.73% in C_2 and for ANN it was 95.69% in C_3. The configuration with the lowest accuracy was C_3 of the support vector machine model. On average, the three models reached values higher than 80% of accuracy. On the other hand, when observing the sensitivity, the RF algorithm stands out again, with values higher than 99%, which shows its high capacity to detect instances of possible failure. SVM also obtained high sensitivity values, especially in C_3, in contrast to ANN, whose sensitivity remains 79.79% for all their configurations. Similarly, when analyzing the average $F1$-score values, the high performance and robustness of the RF model are evident. The SVM and ANN models obtain similar average $F1$-Score values, higher than 80%.

Figure 48.3 shows the accuracy values achieved in the different cross-validation partitions. The dashed line shows the average accuracy of each setting. As can be seen, the performance of the RF algorithm is higher than that of the other two algorithms, reaching an average accuracy of 99.98%. Contrary, in the support vector machine model, there are partitions that considerably decreased their performance. In the case

Table 48.4 Performance comparison of the different algorithms

		RF						SVM						ANN					
Predicted classes		C1		C2		C3		C1		C2		C3		C1		C2		C3	
		N	*R*	*N*	*R*	*N*	*R*	*N*	*R*	*N*	*R*	*N*	*R*	*N*	*R*	*N*	*R*	*N*	*R*
Actual classes	*N*	2789	0	2828	0	2792	0	2561	235	2755	78	1526	1266	2557	233	2638	152	2689	101
	R	1	2811	1	2772	0	2809	579	2226	463	2305	195	2614	569	2242	569	2242	568	2243
Pr (%)		100		100		100		90.45		96.73		67.37		90.59		93.65		95.69	
Re (%)		99.96		99.96		100		79.36		83.27		93.06		79.76		79.76		79.79	
F1 (%)		99.98		99.98		100		84.54		89.50		78.16		84.83		86.15		87.02	
Acc (%)		99.98		99.98		100		85.46		90.34		73.91		85.68		87.12		88.05	
T_train (s)		16.54		65.66		295.09		1086.46		198.5		586.28		222.54		688.39		2181.65	
Pr Avg (%)		100						84.85						93.31					
Re Avg (%)		99.97						85.23						79.81					
Sp Avg (%)		100						81.17						94.19					
*F*1 Avg (%)		99.96						84.07						86.00					
Acc Avg (%)		99.98						83.24						86.95					
T_train Avg (s)		125.76						623.74						1030.86					

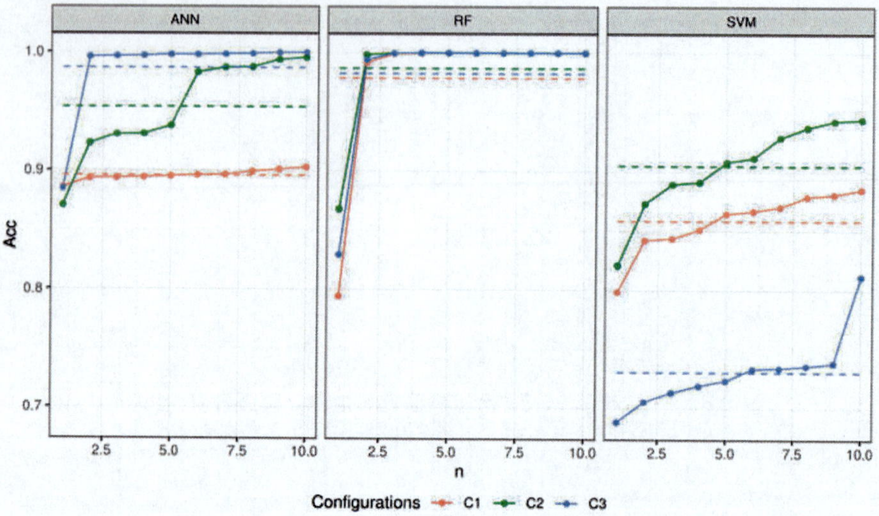

Fig. 48.3 Accuracy of each model and settings in the different cross-validation partitions

of artificial neural network models, the third configuration reached, for most of the partitions, 100% accuracy, achieving an average of 95.69%.

In general, in terms of training time, the models with the highest values were neural networks and support vector machines. Thus, the third ANN configuration obtained a longer training time, with 2181.65 s. It was mainly due to the increase in the number of neurons and hidden layers compared to the others. Compared to the SVM model, configuration one took approximately five times longer to train than configuration two and almost twice as long as configuration three. In this case, kernel selection (linear, radial, and polynomial, respectively) plays an important role. On the contrary, the random forest models obtained much shorter times. Despite the increase in parameters in their configuration, the training time was proportional to the number of trees. Therefore, the shortest time was achieved by RF_{C1} (16.54 s), with 25 trees.

48.3.3 Comparison of Model Performances

Once the evaluation phase of the algorithms was completed, a comparison of the results was carried out between the configurations of each model and then between the different models (using the best performing configuration). To this end, paired hypothesis tests of statistical significance [26] were performed with an $\alpha = 5\%$, comparing the accuracy obtained from the cross-validation of each configuration. The hypotheses proposed were the null hypothesis H_0: $\mu C_1 = \mu C_2$, and the alternative hypothesis H_A: $\mu C_1 \neq \mu C_2$ and the same was done between configuration two and

Table 48.5 Performance comparison of the different algorithms

	RF	SVM	ANN	
C_1 versus C_2 (%)	29.58	0.062	0.010	
C_1 versus C_3 (%)	29.58	5.26e−5	0.40	
Significant	None (C_2)	C_2	C_3	
RF versus (%)	N.A	0.08	44.03	
SVM versus (%)	0.08	N.A	3.68e−3	
ANN versus (%)	44.03	3.68e-3		N.A

three. These results are shown in the first two rows of Table 48.5. The configurations one and two of the ANN model have significant differences; therefore, the null hypothesis was rejected (*p*-value < *0.05)*. In this case, the second configuration had a significant improvement in their performance respect configuration one. In contrast to the RF and the SVM models, whose p-values (greater than 0.05) allowed the null hypothesis to be accepted. This indicated that there is no significant improvement between C_2 and C_1. Following this same analysis, the results of the tests between C_2 and C_3 show that for the SVM model. There are significant differences in their performance, while for the RF and the ANN model, there is no significant difference in their accuracy. Each model, according to the highest accuracy value were RF_{C3}, SVM_{C2}, and ANN_{C3}.

In the case of the random forest algorithm configurations, the number of principal components selected for each configuration was also analyzed. As shown in Fig. 48.4, the components with the highest weight and variability in all configurations were s50, s28, 25, s26, and s44. These sensors do not change position, regardless of the number of trees used in the model configuration. Comparing these results with the description of the author of the dataset [27], it was identified that most of the sensors (s25, s26, s28, s31, s44, s50, and s51) belong to the main pipe of the supply system (s25, s26, s28, s31, s44, s50, and s51).

Likewise, the performances of the different models were compared, using the configurations with the highest Acc. The results are shown in the last three rows of Table 48.5, and there are no significant improvements between the performances of the three models, except between the SVM model and ANN. Therefore, it can be considered that the difference between the performances of ANN with respect to RF can be considered statistically insignificant. In the case of the random forest algorithm configurations, we also analyzed the number of principal components selected by each configuration of principal components selected by each configuration was also analyzed. As shown in Fig. 48.4, the components with the highest weight and variability in all the configurations were s50, s28, 25, s26, and s44. These sensors do not change position, regardless of the number of trees used in the model configuration. Comparing these results with the author's description of the dataset [27], it was identified that most of the sensors (s25, s26, s28, s31, s44, s50, and s51) belong to the main pipeline of the supply system.

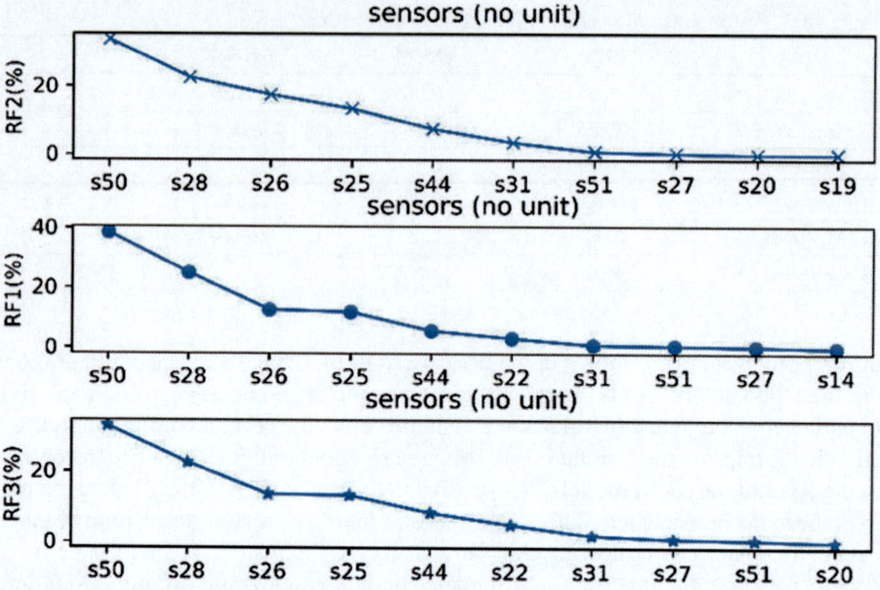

Fig. 48.4 Comparison of the main characteristics in RF

48.4 Conclusions and Future Works

In this paper, we compared three machine learning models in a binary classification task used in the failure control of a water pump. The results show that the use of machine learning models is effective in fault prediction when a robust training dataset is provided, and adequate data preprocessing is performed. The model with higher average accuracy and shorter training time was the random forest model, in all its configurations, which shows that it is possible to resort to alternative models for certain problems. The SVM model proved to be less efficient to address this task.

As future work, it is recommended to incorporate the trained ANN and RF algorithm in the control system and verify the results in a real environment. In addition, the failure control system could be evaluated with other algorithms not considered in this research.

References

1. N. Jakab, G. Melypataki, The right to water as a social fundamental right. J. Agric. Environ. Law **14**(26), 7–63 (2019)
2. R. Chuchird, N. Sasaki, I. Abe, Influencing factors of the adoption of agricultural irrigation technologies and the economic returns: a case study in Chaiyaphum province, Thailand. Sustainability **9**(9), 1524 (2017)

3. A.W. Bivins, T. Sumner, E. Kumpel, G. Howard, O. Cumming, I. Ross, K. Nelson, J. Brown, Estimating infection risks and the global burden of diarrheal disease attributable to intermittent water supply using QMRA. Environ. Sci. Technol. **51**(13), 7542–7551 (2017)
4. T.P.L. Nguyen, N. Inkong, N. Faysse, Role of local institutions in the transition towards sustainable agriculture: the case study of Thailand. WIT Trans. Ecol. Environ. **245**, 135–143 (2020)
5. R. Dorf, *Modern Control Systems* (13th edn.). Pearson. ISBN 1292152974 (2017)
6. J. Fausing Olesen, H.R. Shaker, Predictive maintenance for pump systems and thermal power plants: state-of-the-art review, trends and challenges. *Sensors* **20**(8), 2425 (2020)
7. A. Rastegari, M. Mobin, Maintenance decision making, supported by computerized maintenance management system, in *2016 Annual Reliability and Maintainability Symposium (RAMS)* (IEEE, 2016), pp. 1–8
8. M.M. Giraldo-Gonzalez, J.P. Rodriguez, Comparison of statistical and machine learning models for pipe failure modeling in water distribution networks. Water **12**(4), 1153 (2020)
9. K. Pietrucha-Urbanik, A.˙Zelazko, Approaches to assess water distribution failure. Period. Polytech. Civ. Eng. **61**(3), 632–639 (2017)
10. R. Farmani, K. Kakoudakis, K. Behzadian, D. Butler, Pipe failure prediction in water distribution systems considering static and dynamic factors. Procedia Eng. **186**, 117–126 (2017). https://doi.org/10.1016/j.proeng.2017.03.217
11. D. Vries, B. Van Den Akker, E. Vonk, W. De Jong, J. Van Summeren, Application of machine learning techniques to predict anomalies in water supply networks. Water Sci. Technol. Water Supply **16**(6), 1528–1535 (2016)
12. A. Robles-Durazno, N. Moradpoor, J. McWhinnie, G. Russell, A supervised energy monitoring-based machine learning approach for anomaly detection in a clean water supply system, in *2018 International Conference on Cyber Security and Protection of Digital Services (Cyber Security)* (IEEE, 2018), pp. 1–8
13. S. Lu, Q. Li, L. Bai, R. Wang, Performance predictions of ground source heat pump system based on random forest and back propagation neural network models. Energy Conv. Manag. **197**, 111864 (2019)
14. S. Porwal, S. Akbar, S. Jain, Leakage detection and prediction of location in a smart water grid using SVM classification, in *2017 International Conference on Energy, Communication, Data Analytics and Soft Computing (ICECDS)* (IEEE, 2017), pp. 3288–3292
15. A.K. Panda, J.S. Rapur, R. Tiwari, Prediction of flow blockages and impending cavitation in centrifugal pumps using support vector machine (SVM) algorithms based on vibration measurements. J. Int. Measure. Confederat. **130**, 44–56 (2018). ISSN 02632241. https://doi.org/10.1016/j.measurement.2018.07.092
16. A. Shirzad, M.J.S. Safari, Pipe failure rate prediction in water distribution networks using multivariate adaptive regression splines and random forest techniques. Urban Water J. **16**(9), 653–661 (2019)
17. J. Myrans, Z. Kapelan, R. Everson, Automated detection of faults in wastewater pipes from cctv footage by using random forests. Procedia Eng. **154**, 36–41 (2016)
18. A.M. Sattar,¨O.F. Ertu˘grul, B. Gharabaghi, E.A. McBean, J. Cao, Extreme learning machine model for water network management. Neural Comput. Appl. **31**(1), 157–169 (2019). ISSN 09410643. https://doi.org/10.1007/s00521-017-2987-7
19. M. Kuty Lowska, Comparison of two types of artificial neural networks for predicting failure frequency of water conduits. Period. Polytech. Civ. Eng. **61**(1), 1–6 (2017)
20. J.E.T. Akinsola, Supervised machine learning algorithms: classification and comparison. Int. J. Comput. Trends Technol. (IJCTT) **48**, 128–138 (2017). https://doi.org/10.14445/22312803/IJCTT-V48P126
21. N.K. Phantawee, *Pump Sensor Data—Kaggle* (2018). https://www.kaggle.com/nphantawee/pump-sensor-data
22. T.P. Carvalho, F.A. Soares, R. Vita, R.D.P. Francisco, J.P. Basto, S.G. Alcala, A systematic literature review of machine learning methods applied to predictive maintenance. Comput. Industr. Eng. **137**, 106024 (2019). ISSN 03608352. https://doi.org/10.1016/j.cie.2019.106024

23. W. McKinney, *Python for Data Analysis. Data Wrangling with Pandas, NumPy, and IPython.* O'Reilly 2nd edn. (2017). ISBN 978-1-491-95766-0. http://gen.lib.rus.ec/book/index.php?md5=6558C6F3EFA0F99E1A4B40356EF854C6
24. O. Kramer. Scikit-learn. in *Machine Learning for Evolution Strategies* (Springer, 2016), pp. 45–53
25. E. Bisong, Introduction to scikit-learn, in *Building Machine Learning and Deep Learning Models on Google Cloud Platform* (Springer, 2019), pp. 215–229
26. D. Consonni, P.A. Bertazzi, Health significance and statistical uncertainty. The value of p-value. MedLav **108**(5), 327–331 (2017)
27. N. Phantawee, kagglepumpschallenge/description.txt at master·n-phantawee/kagglepumpschallenge·GitHub (2020). https://github.com/nphantawee/kaggle_pumps_challenge/blob/master/description.txt

Chapter 49
Introductory Analysis of LoRa/LoRaWAN Technology in Ecuador

Alejandra M. Pinto-Erazo, Luis E. Suárez-Zambrano, Mario M. Mediavilla-Valverde, and Ronni E. Andrade-Guevara

Abstract This research paper presents an introductory analysis or state of the art of LoRa/LoRaWAN technology, and its development approach and application in Ecuador and worldwide, which are growing exponentially due to its focus on the Internet of Things (IoT). LoRa/LoRaWAN is a technology that allows all objects to connect to the Internet in a reliable and secure way. This study focuses on the technology within the current context that we live in Ecuador and the world, its advantages, application scenarios, and how this technology behaves and has its contribution in simulation environments for future implementations in both indoor and outdoor scenarios.

49.1 Introduction

The exponential increase of information and communication technologies in recent years has made the Internet of Things (IoT) a promising area for future service implementations. IoT brings the concept of a world where most objects can be connected to the Internet with the ability to share information through the interoperation of various communication systems, platforms, and protocols [1–3].

IoT technologies are emerging that operate in free spectrum, and that does not require a license from the government in which the technology is used. In general, the current growth rate is in short- and long-range technologies at low power since unlicensed operation must normally be tolerant to interference from other devices. LoRa technology uses the ISM (Industrial, Scientific, and Medical) bands below 1 GHz (433, 868, and 915 MHz depending on the version of the microchip used).

Sigfox and LoRa/LoRaWAN are license-free technologies that only require their own infrastructure, being a solution for networks with low-data throughput and low-latency requirements [4]. The Sigfox and LoRa/LoRaWAN technologies are also available for networks with low data rates and low-latency requirements.

A. M. Pinto-Erazo (✉) · L. E. Suárez-Zambrano · M. M. Mediavilla-Valverde · R. E. Andrade-Guevara
Universidad Técnica del Norte, Ibarra, Ecuador
e-mail: ampinto@utn.edu.ec

Á. Rocha et al. (eds.), *Communication, Smart Technologies and Innovation for Society*, Smart Innovation, Systems and Technologies 252,
https://doi.org/10.1007/978-981-16-4126-8_49

LoRa and LoRaWAN are not the same. LoRa uses spread spectrum modulation, being a physical implementation layer, and is independent of the upper layer implementations. Its main advantage is its long-range capability. LoRa modulation can be used as a physical layer by many different protocol architectures [1, 5].

The LoRaWAN protocol is created with low power consumption in mind and is designed to scale from a single gateway to large global networks with millions of devices. Innovative features of the LoRaWAN specification include support for redundant operation, low cost, and geolocation [6].

49.2 Methods

49.2.1 Technical Component

This section contains the important parameters of the technology, such as: architecture, security and activation modes, features, protocol.

LoRa. LoRa started in 2009, when the three founders (Nicolas Sornin, Olivier Seller, and François Sforza) were targeting the metering industry and aimed to add wireless communication capabilities for gas, water, and electricity meters. This led them to create Cycleo which was later acquired by Semtech in May 2012. In February 2015, LoRa Alliance was founded, and the network protocol was renamed "LoRaWAN." The objectives of the LoRa Alliance were, and remain, "to support and promote the global adoption of the LoRa-WAN standard by ensuring the interoperability of all LoRa-WAN products and technologies" [7].

LoRa, from the acronym Long Range, is the physical layer or wireless modulation used to create long-range links. LoRa modulation, patented by Semtech, is based on Frequency Shifting Keying (FSK) and a variation of chirp spread spectrum (chirp spread spectrum-CSS), in which the low power characteristics of FSK modulation are maintained, but achieving a more significant communication range; it also employs Forward Error Correction (FEC) [2, 5, 8].

The signal can make frequency hops in short periods of time, allowing simultaneous transmissions, also by using a wide spectrum, the power handled for transmission is small as shown in Fig. 49.1, which translates into minimal power consumption (see Fig. 49.1).

LoRa employs a variable data rate, driving orthogonal spreading factors (spreading factor SF), allowing the system designer to trade off data rate for range or power to optimize network performance over a constant bandwidth; it operates in the ISM (Industrial, Scientific, and Medical) bandwidths (USA: 915 MHz, EU: 433 MHz and 868 MHz), Fig. 49.1. The broadband nature allows compensating for a better signal-to-noise ratio (SNR) which enables LoRa to demodulate signals even when it is 19.5 dB below the noise floor [2, 5, 8].

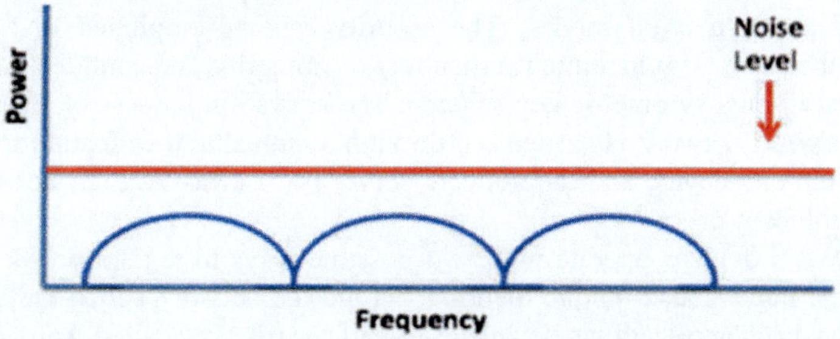

Fig. 49.1 Relationship between power and frequency as design metrics for IoT applications

A single gateway or base station can cover hundreds of square kilometers or even entire cities. The range is also highly dependent on the environment and the degree of obstructions present at a given location [2].

LoRa architecture. The basic architecture of a solution using a LoRa network includes three types of equipment: nodes, gateways, and application server. The nodes communicate with the gateway using LoRa. The gateway communicates with the application server via IP. The Infrastructure is presented in Fig. 49.2.

It is feasible to implement mesh type networks, where nodes relay messages from other nodes; but in this case, the nodes should be persistently receiving and transmitting messages, which generates high energy consumption. The alternative for transmission and reception is to consider a lower performance gateway.

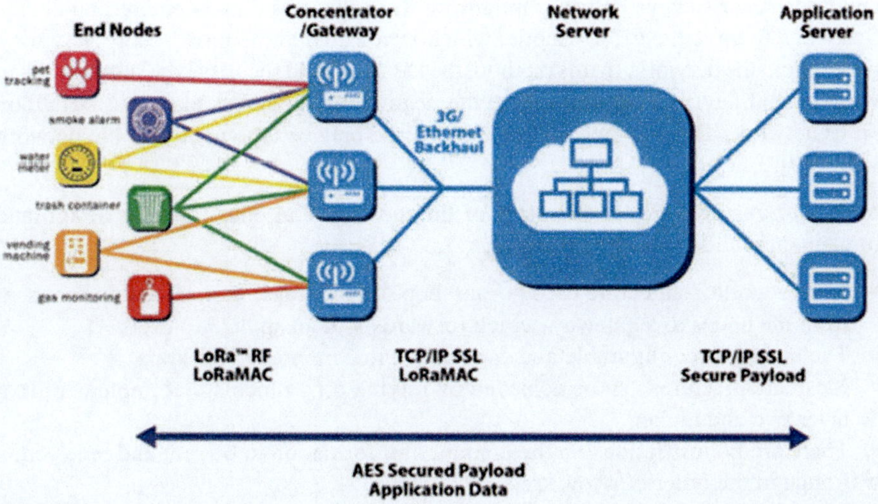

Fig. 49.2 LoRa architecture, with emphasis on the nodes, gateway, servers, and security context

Safety and activation modes. The security scheme employed in LoRaWAN provides the network with mutual authentication, integrity, and confidentiality mechanisms based on a symmetric key scheme. The activation process of an end device in a LoRaWAN network is carried out through a mutual authentication mechanism between the end device and the network server [9]. LoRaWAN devices operate in two possible ways.

LoRaWAN devices operate with two possible ways to register in the network; each of the nodes has a unique identifier within the network called DevAddr, and in addition, each application gets an end-to-end security key called AppSkey, which allows each node part of the network to send data to a certain system. Additionally, to make the network even more secure, there is also a second key at the network level that allows device access to the network named as NwkSkey [9].

- AES-128 network session key or NwkSKey: allows the message integrity code (MIC) to be calculated and verified. This process checks that the integrity of the packet has not been compromised. Both the end-node and the network server share the same NwKSKey [9].
- AES-128 application session key or so-called AppSKey: allows the payload of each message to be encrypted and decrypted, ensuring that neither the gateway nor the network server reads the payload of the transmitted packet. Both the end-node and the application server share the same AppSKey [9].

For a module to be part of the LoRaWAN network and be able to send messages up to the server, there are two modes that standardize access: Activation by Personalization (ABP) and Over-the-Air Activation (OTAA).

First, there is the ABP mode, which allows manual configuration of the DevAddr and node security keys. When configuring the node, this data is entered manually. Secondly, there is the OTAA mode, which is considered the most secure way for a node to join the network. In this method, the DevAddr and security keys are negotiated with the end-device in a manner like the negotiation that takes place via DHCP on an IP network. This key negotiation takes place between the node and the network server [9].

Main features of LoRa. A summary of the most general ones for their design and implementation is presented, such as:

- The network architecture used is "one-hop star topology," i.e., information is sent from the nodes to a gateway, which forwards it to an application server.
- The bit rate is configurable and can reach a maximum of 37.5 kbps.
- Mesh architectures can be achieved by relying on protocols that implement it at layer two and higher.
- There are no restrictions on the amount of information to be sent and received.
- Range in the order of a few km.
- It is possible to manipulate modulation characteristics, which vary the bitrate, power consumption, and the range obtained, among other parameters.
- Freedom to design customized wireless networks.
- Low cost per module ($30–$33 depending on quantity).

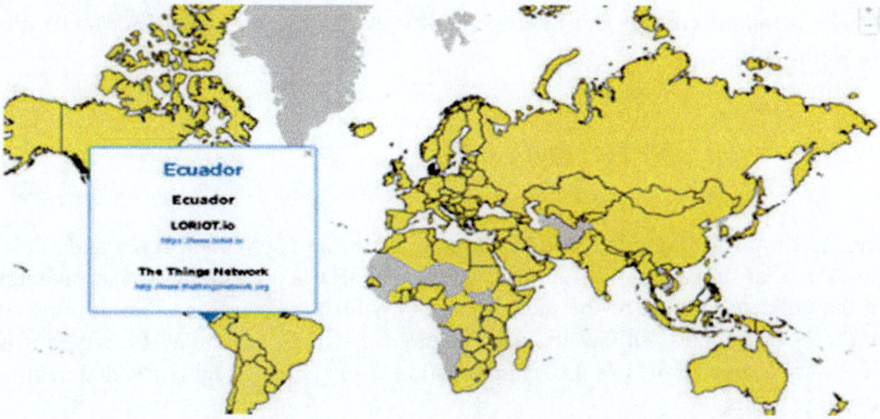

Fig. 49.3 Global coverage

- Integrated circuits manufactured only by Semtech.

LoRaWAN. The alliance is a non-profit association that has become one of the largest alliances in the technology sector, committed to the large-scale deployment of low-power wide area networks (LPWANs), through the development and advancement of the LoRaWAN open standard. Through standardization and the LoRa Alliance's accredited certification scheme, the LoRa Alliance brings interoperability to LPWAN networks at scale, making LoRaWAN a solution for global LPWAN deployments.

LoRa Alliance has standardized the MAC protocol called LoRaWAN. The physical layer, LoRa, in long-range communication links enables it, while LoRaWAN defines the communication protocol and the network system architecture, being fundamental aspects in the lifetime of a node (LoRa-Thing) with respect to the battery, including network capacity, security, quality of service (QoS), and the diversity of applications that can be served by the network [8, 10] (See Fig. 49.3).

LoRaWAN guarantees data rates from 0.3 to 50 kbps, which are considered tolerable for sending real-time sensor data, machine-to-machine (M2M), smart cities, and industrial applications. However, data transfer, real-time imaging, or any data that demands high bandwidth may not be suitable for LoRa networks.

This low data rate ensures low power consumption by end-devices, accessing the use of batteries with a lifetime of a few years, as LoRa/LoRaWAN is intended for sensors and applications that need to send low amounts of data over long distances, several periods per hour and from different environments [1, 2, 8].

LoRaWAN protocol features. The following parameters are considered, based on the standards of the same technology, provided by LoRa Alliance:

- It is a fully bi-directional protocol that enables reliable message delivery.
- It includes end-to-end encryption for security and data privacy.
- It allows over-the-air registration of end nodes and has multicast capability.

- The standard ensures the interoperability of the various LoRaWAN networks worldwide.

49.2.2 Applicable Regulations

International Standards. LoRaWAN belongs to the application layer and to the protocol that runs on LoRa radio modulation. LoRa is a technology that operates in the unlicensed band of the radio spectrum, which makes it possible to transmit in these frequencies without the need to pay for its use, unlike Wi-Fi, where it is necessary to pay for its use. LoRa modulation is subject to a regulation that defines its usage limits.

The generic regulatory standard at European level is defined by the European Telecommunications Standards Institute (ETSI). At present, the current regulation is set by the ETSI EN 300 220-1. ETSI EN 300 220-1 establishes how radio waves in the 25–1000 MHz range are to be used, i.e., much of this standard does not apply to LoRa, since its majority use in the European area is reduced to the free 868 MHz band [11, 12]. LoRa Alliance has created a document on the use of LoRa in Europe.

LoRa Alliance has created a document called LoRaWAN Regional Parameters which explains the specific standards for LoRa and LoRaWAN. This document explains the specific regulations that apply to LoRa modulation and LoRaWAN (modulation and application layer together) [13]. LoRaWAN is the only LoRaWAN modulation and application layer.

National Standards. The LoRa Alliance announced on November 10, 2020 the release of Regional RP2-1.0.23 Parameters, which include support for all versions of the LoRaWAN Layer 2 Specification. Support for LoRaWAN operation in Senegal, Montserrat, Mali, Guinea, Syria, and Vanuatu has been identified, and the RP now supports more than 200 countries and regions worldwide [6].

A significant change in regional parameters is the addition of support for new long-range frequency hopping spread spectrum (LR-FHSS4) data rates for the EU868585, US915, and AU915 regions. The use of LR-FHSS data rates is optional and, therefore, preserves compatibility with all previous revisions of the Regional Parameter Specification [6]. The LoRa Alliance with the LoRa Alliance summary of the new LR-FHSS4 data rates is available for the EU8685, US915, and AU915 regions.

The LoRa Alliance with the summary of the worldwide regulations place Ecuador in the Australian regulation and, therefore, look at those items titled AU915-928. While it is true that device manufacturers must be very attentive to this regulation, at the user level of a LoRaWAN network, what we must be aware of are the following limitations:

- Frequency bands: 915–928 MHz
- 3–4 channels, 8 with data rates between 250 and 5.5 kbps, 1 of 11 kbps, and 1 FSK of 50 kbps.
- Duty Cycle between 0.1 and 1% depending on the channel.

Table 49.1 Regional parameters provided by LoRa Alliance

ISO 3166–1 Country name (Code alpha-2)	Band/channels (MHz)	Channel plan	LoRaWAN certified devices
Ecuador (EC)	902–928	AU915-92834	
Egypt (EG)	433.05–434.79	EU433	
	865–868	IN865-867	
	863–870	EU863-870	
El Salvador (SV)	915–928	AU915-9283	
Finland (FI)	433.05–434.79	EU433	
	863–873	EU863-870	
France (FR)	433.05–434.79	EU433	X
	863–870	EU863-870	

- In private networks, there is no channel dwell time limitation.

Table 49.1 shows how the regional parameters are set up, considering the different LoRaWAN channel plans for a given country. The table also has an indication of the existence of known end devices that have LoRaWAN certification with regulatory approval in the country in question [6]. The table also shows the number of LoRaWAN channel plans for the country in question.

In Ecuador, there is no entity that regulates this wireless technology LoRa (not licensed), so the analysis of research in the country is ongoing; however, there are few companies that adopt LoRa to generate applications through low-cost transmissions.

Ministry of Telecommunications (MINTEL) has promoted the digital transformation of conventional cities toward smart cities as a style adopted in many cities in the world, thus, promoting a paradigm shift in business activities, health, education, industry, and transportation among others.

49.3 Results and Discussion

49.3.1 Presence of LoRaWAN

One of the major impacts generated by LoRaWAN is the introduction of technology into the IoT market, such as the ability to develop any type of business model in the implementation of privately owned networks to subscribe to LoRaWAN connectivity. In addition, the collaborative nature of LoRaWAN has been fostering the emergence of global open developer communities [12]. The coverage of the global network is very high.

The global network coverage is the aggregation of any type of LoRaWAN network. Figure 49.3 shows the countries that have global coverage, where Ecuador is present

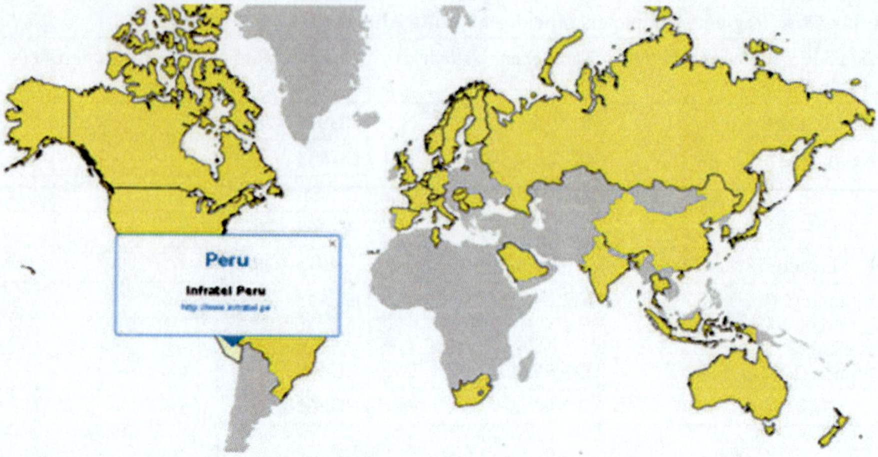

Fig. 49.4 Authorized supplier in Peru

under a commercial implementation model that is oriented to the open community, private, or generated by a network operator.

Currently, no mobile operator in Ecuador provides this service, resulting in openly monetizing connectivity or end-to-end services to third parties. The official network operators or providers, which are members of the LoRa Alliance, are considered LoRaWAN public network operators.

An analysis of how LoRaWAN is being developed at the operator level in South America, Peru, and Brazil, where this type of network is already being deployed. In the case of Peru (see Fig. 49.4), there is Infratel Peru that promotes the development of IoT-based solutions on its national LoRaWAN network for various industries [14], among them are:

- Cargo monitoring, truck tracking, and package location.
- Remote reading of water and gas meters, remote control of valves, etc.
- Soil quality monitoring such as humidity and chemical elements.
- Panic buttons and people tracking.
- Open different sensors such as: door, window, smoke, fire, flood, other.
- Intelligent parking, noise sensors, and traffic light management.

For the case of Brazil (see Fig. 49.5), the company American Tower which was named one of the most admired companies in the world in 2020 by FORTUNE magazine for its great technological development and approach to LoRaWAN networks at present [15].

LoRaWAN has open developer communities including two Ecuadorian companies such as LORIOT.io and The Things Network. LORIOT.io promotes this type of technology. LORIOT.io is a global IoT company that provides a growing distributed LoRaWAN infrastructure and an easy-to-use software platform that allows users to build, operate and scale a secure IoT network. LORIOT.io's goal is to ensure that

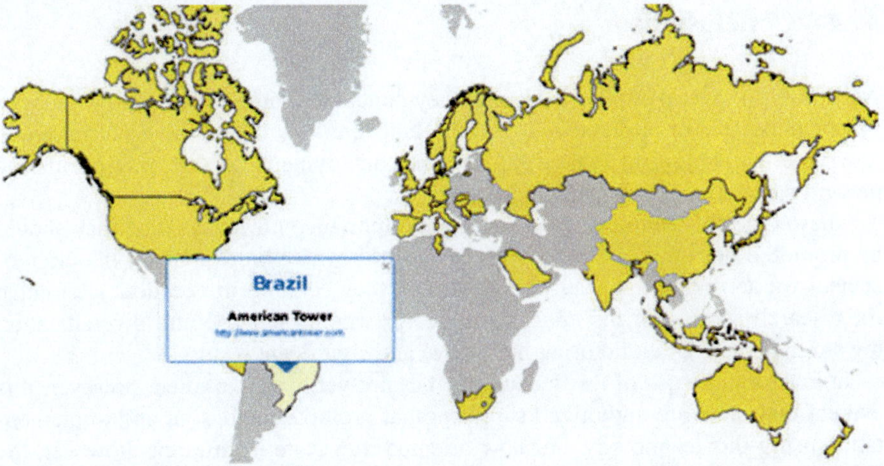

Fig. 49.5 Authorized supplier in Brazil

customers such as telecoms companies, utilities, system integrators, local/municipal governments, Bluechip companies, and multinational organizations have the highest standards for their enterprise-level implementations [15].

Table 49.2 shows the deployment of LoRaWAN public network operators in the Americas, compared to Europe, where more countries have this long-range technology.

LORIOT promotes solutions in the following cases: smart street lighting control, window and door management, cold chain monitoring, low power trailer tracking, flow monitoring, smart agriculture, and smart building.

Table 49.2 Authorized supplier in South America

Country	Authorized supplier	Web site
Canada	Eleven-x	http://www.eleven-x.com
United States of America	Machine, a Comcast Company	https://iot.machineq.com/lora-alliance
	Senet, Inc	http://www.senetco.com
	Everynet	http://everynet.com
Perú	Infratel Perú	http://ilwww.infratel.pe
Brazil	American Tower	http://www.americantower.com

49.4 Conclusions

According to several studies analyzed, they conclude that the metrics to evaluate a system simulated or implemented with LoRa technology fall on the performance in low power and long-range that would enhance deployments of wireless systems and networks based on these technologies.

These types of technologies are of great importance nowadays, and they should be promoted not only in Ecuador but also in the whole world. Given the circumstances we are living in because of Covid-19, they become an excellent alternative for research starting in the educational sector where their use and diversification toward the company could bring innovative and significant results.

In Ecuador, the use of this technology has not yet been exploited; however, it is evident that there are authorized suppliers that promote the design and implementation using this technology. For now oriented to private institutions, however, the field is open for research and future implementations not only at the private level but also at the public level.

This introductory analysis of LoRa/LoRaWAN technology suggests generating many possibilities for the design of wireless systems with LoRa network deployment, initially in indoor environments without major interference and later focused on outdoor environments with a variety of interferences.

In South America, there are few countries that have given rise to studies and implementations with these technologies, which means an open field for research and design of systems based on this type of deployments.

References

1. J.M. Marais, R. Malekian, LoRa and LoRaWAN testbeds: a review, in *2017 IEEE Africon* (2017), pp. 1544–1549
2. M. Saari, A.M. Baharudin, P. Sillberg, S. Hyrynsalmi, W. Yan, LoRa—a survey of recent research trends, in *2018 41st International Conference on Information Communication and Technology Electronics Microelectronic* (2018), pp. 872–877
3. J. Kölsch, C. Heinz, S. Schumb, C. Grimm, *Hardware-in-the-Loop Simulation for Internet of Things Scenarios* (2018), pp. 1–6
4. T. Opferman, A.K. Emarievbe, J. Koepp, *Emerging Communication Technologies Enabling the Internet of Things White Paper*, vol. 1, no. 2 (Rohde and Schwarz, 2016), p. 17
5. Semtech Corporation, *LoRa Modulation Basics* (2015), pp. 1–26, 2015.
6. LoRa AllianceTM, *LoRa Alliance* (2018). [Online]. Available: https://www.lora-alliance.org
7. SEMTECH, *Three Inventors Share Their Story* (2020). https://blog.semtech.com/a-brief-history-of-lora-three-inventors-share-their-personal-story-at-the-things-conference. Accessed 19 March 2021
8. LoRa-Alliance, *A Technical Overview of LoRa and LoRaWAN* (2015), pp. 1–20
9. I.O. Monfort, *Estudio de la arquitectura y el nivel de desarrollo de la red LoRaWAN y de los dispositivos LoRa* (2017). Available: http://openaccess.uoc.edu
10. R. Pérez García, *Avaluació de LoRa/LoRaWAN per an escenaris de Smart City* (2017), pp. 1–89
11. Erm, *EN 300 220-1—V3.1.1—Short Range Devices (SRD) Operating in the Frequency Range 25 MHz to 1 000 MHz; Part 1: Technical Characteristics and Methods of Measurement* (2017).

Accessed: 24 Feb 2021. [Online]. Available: https://portal.etsi.org/TB/ETSIDeliverableStatus.aspx
12. L. Alliance, *Otice of* (2020)
13. Logitek, *Parámetros regionales de LoRaWAN* (2020). https://www.m2mlogitek.com/parametros-regionales-de-lorawan/. Accessed 23 Feb 2021
14. INFRATEL, *Infratel, IoT Solutions* (2020). http://iot.infratel.pe/servicios
15. American Tower, americantower.com (Global Wireless Solutions, 2020). https://www.americantower.com/index.html

Chapter 50
Optimal Control Algorithms for WNCS Implemented with IoT Tools

Carlos Pillajo, Patricio Sulca, and Juan Escobar

Abstract In this work, the implementation, simulation and analysis of LQG and PID algorithms have been carried out; applied in a DC motor with brushes by means of a control system in a WNCS wireless network, where the network is part of the closed control loop, using the protocol MQTT for communication between the broker and the plant. The performance of each of the control algorithms was evaluated through simulations in the network, which allowed determining the dynamic response parameters of the system. With the data obtained, the algorithms were implemented using a development tool based on the flow for visual programming, with the purpose of controlling and monitoring wirelessly or even through a mobile device remotely in real time, thus becoming an IoT application. Optimum performance is corroborated through communication and control tests when the plant is subjected to disturbances.

50.1 Introduction

Industry 4.0 contains numerous technologies and associated paradigms, such as Internet of things (IoT) control through the cloud, artificial intelligence as described in [1, 2]. The objective of Industry 4.0 is to achieve a higher level of operational efficiency and productivity, a higher level of automation, for which, optimal control algorithms are needed, which allow greater adaptation and human–machine interaction (HMI) and automatic data communication. These characteristics are not only highly correlated with Internet of Things technologies and advanced algorithms, in this work, it is proposed to implement PID and LQG controllers over a network, with the help of MATLAB software and the True Time library, which allows simulation of the behavior of the dynamic system in the network.

For the implementation of WNCS, the MQTT protocol was used, which is defined as a lightweight messaging protocol ideal for M2M scenarios [3], as well as being an alternative for the output signal to continue without delay despite variations in the signal input, such as: motor speed variation and data exchange in real time, since it

C. Pillajo (✉) · P. Sulca · J. Escobar
Carrera de Ingeniería Electrónica, Universidad Politécnica Salesiana, Quito, Ecuador
e-mail: cpillajo@ups.edu.ec

Á. Rocha et al. (eds.), *Communication, Smart Technologies and Innovation for Society*, Smart Innovation, Systems and Technologies 252,
https://doi.org/10.1007/978-981-16-4126-8_50

allows the implementation of different qualities of service that identify the loss of messages or duplication in them. In Sect. 50.4, the results of the design of the system and the application of the algorithms are presented, and finally, in Sect. 50.5, the results are analyzed to determine the conclusions.

50.2 Background

In this work, the MQTT protocol was used, which uses an editor-subscriber model, which has a low overhead in the network, which allows its implementation in low-power devices, such as microcontrollers that could be used in remote IoT sensors. In addition to using the Mosquitto broker that is designed to be used in all situations, where there is a need for lightweight messaging, particularly on devices with limited resources [4]. It is complemented together with Node-RED, which is a flow-based open-source development tool for the combination of IoT hardware devices and online services developed by IBM Emerging Technology.

In addition to being free and JavaScript-based, it is built on the Node.js platform, which provides a browser-based visual flow editor [5]. For the design of the system, the broker that will be hosted on the Raspberry Pi is used. It is a small computer, the size of a credit card at a very low cost for any type of project, and it can be used practically as a PC. The software used in this card is open source, and it occupies an operating system adapted from Debian, called Raspbian, although the card can also work in other operating systems, including version 10 of Windows [6]. According to the aforementioned, it is summarized in wireless network control systems, based on the study of control systems through wireless networks, dynamic control systems require communication between sensors and actuators; therefore, establish a closed loop control system that uses a wireless network for the transmission of information [7]. It should be taken into account that due to the different failures that wireless networks present, the application of feedback controllers is discussed, that is, WNCS in the industrial field [8].

For PID control, there are several types of considerations. Due to its simplicity, it is one of the most used options at the moment in control applications, and due to its three-parameter structure, it is considered to use the PIDup or PID in parallel in this research, since each action of the components that make up the controller are separated in the equation to later be added [9]. On the other hand, the LQG control is based on the analysis of the intrinsic behavior of the systems, that is, on the variables that constitute the state spaces. Understanding the alteration of each of the variables that make up the interior of the control process in a system, reinforces control and solves the difficulties of classical control theory [10], where its main foundation is the analysis of input and output for dynamic systems with constant parameters.

The state space facilitates the understanding of the behavior of one or more variables with a minimum amount of information in a given time, which allows forecasting future instants in the behavior of the variables [11]. The mathematical model of the LQG control consists of an LQR control and a Kalman filter, being LQR, the one that observes the regulation problem, this solves it from the optimization of a quadratic index with a linear solution as shown in Eq. 50.1.

$$J = \int_{t0}^{tf} \left(x^T Qx + u^T Ru\right)\mathrm{d}t \tag{50.1}$$

where x represents the states of the system and the weight matrices Q, R are positive symmetric matrices. The Kalman filter can behave as an estimator of states that can be a filter or predictor, so it is understood as an optimal estimator due to the approximation obtained that seeks to have the lowest possible error. Assuming a dynamic plant, Eqs. 50.2 and 50.3 are presented [12].

$$x_{k+1} = Ax_k + Bu_k + W_k \tag{50.2}$$

$$y_k = Cx_k + Bu_k + V_k \tag{50.3}$$

where the state $x_k \in R^n$, control input $u_k \in R^m$, for the exit y_k and matrices A, B, C are known constant matrices of appropriate dimensions.

50.3 Simulation and Implementation Methodology

In this identification and experimental modeling section, you focus on the prototype shown in Fig. 50.1. It is built and instrumented with: encoder sensor for speed measurement, the actuator, which is a brushed motor and the control unit. Regarding

Fig. 50.1 Speed control system implemented

the location of the sensor, it must be in a place where the luminosity has no influence so that the data acquisition is obtained with the least amount of oscillations possible and the control does not become complex. Methodologically, assuming the constant variation of PWM to reach the maximum speed of the motor; the transfer function of the plant is obtained; simulated using MATLAB software to later discuss the performance of the PID and LQG control in the DC motor through the implemented IoT tools. The case of this system is to automatically control and compare the performance of each applied algorithms.

50.3.1 Operation

To manage the process in the experimental prototype, the ESP-8266 is used, which must be able to read the message from the MQTT broker that is hosted on the Raspberry Pi and then apply the PID and LQG controllers that must stabilize the plant speed on a local server. The response time to any disturbance in the closed-circuit must be minimal to stabilize the DC motor, since, according to the result, the MQTT protocol used in this research uses the Mosquitto corridor and the QoS level 0, understood as the way of managing robustness to send messages to the client in case they are not published and subscribed as a messaging service mechanism [13]. It is an event-based architecture, so each message is sent to the recipients who have subscribed to a specific publication, the broker is in charge of distributing the messages to the recipient, in this case to the engine. To install MQTT on a Raspberry Pi, you must use a server widely known as Mosquitto [14] (Fig. 50.2).

Let's see a linear dynamic system with n states, m inputs and l outputs, the reason is to determine that the control signal u is minimal in Eq. 50.2, where the separation theorem establishes that the global optimum is obtained by dividing the problem into sub-problems [15]. The feedback regulation of the state variables is obtained as presented in Eq. 50.3.

$$u = -K_c \hat{x} \tag{50.4}$$

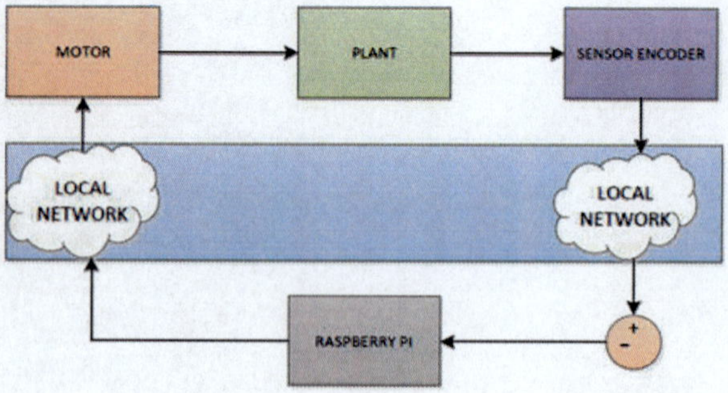

Fig. 50.2 WNCS block diagram

Being

$$K_c = R_c^{-1} B^T P_c \tag{50.5}$$

P_c is obtained from the Riccati equation:

$$A^T P_c + P_c A - P_c B R^{-1} B^T P_c + Q_c = 0 \tag{50.6}$$

Through the Kalman filter:

$$\frac{\mathrm{d}\hat{x}}{\mathrm{d}t} = A\hat{x} + B\hat{u} + K_o(y - C\hat{x}) \tag{50.7}$$

where

$$K_o = P_o C^T R_o^{-1} \tag{50.8}$$

And P_o is established of:

$$A P_o + P_o A^T + Q_o - P_o C^T R_o^{-1} C P_o = 0 \tag{50.9}$$

50.4 Implementation

With the help of the Arduino software, a sample of PWM versus speed data is obtained, which will later be entered in the Ident tool of MATLAB to find the transfer function of the DC motor, as can be seen in the Eq. 50.10.

$$\mathrm{FT} = \frac{0.4745}{s^2 + 1.071s + 0.2813} \tag{50.10}$$

Once the transfer function has been obtained with the assistance of the Node-RED tool, the PID control algorithm is carried out on our local server. For the method of tuning the controller parameters k_P, k_I and k_D, the PID Tuner tool was used of the MATLAB software that automatically calculates the controller gains in order to offer a fast and stable response, obtaining values as shown in the Table 50.1.

Table 50.1 PID controller gains

Gains	Values
k_P	0.24259
k_I	0.093188
k_D	0.15788

For the communication of the plant, the ESP8266 is programmed in Arduino, which is responsible for sending the data obtained from the encoder sensor and receiving the data to stabilize the speed at the set point established in the PID controller. For verification purposes, the simulation of the PID control loop was carried out with the Simulink tool as shown in Fig. 50.3.

On the other hand, for the simulation, the MATLAB software is used to obtain the state matrices of the transfer function; the system is discretized by means of the programming established for the LQG control. As presented in the following matrices obtained and, in the diagram, Simulink blocks in Figs. 50.4 and 50.5.

$$A = \begin{bmatrix} 0.9475 & -0.0137 \\ 0.0487 & 0.9997 \end{bmatrix} B = \begin{bmatrix} 0.0487 \\ 0.0012 \end{bmatrix} C = \begin{bmatrix} 0 & 0.4745 \end{bmatrix}$$

Once the physical connections of the plant have been made, the local server is created. For this, the Mosquitto Broker must be installed on the Raspberry Pi single board computer, where the PID and LQG control algorithm will be applied.

As can be seen in Fig. 50.6, using the Node-RED software, we proceed to carry out the programming of the PID algorithm that will be implemented in the broker hosted on the raspberry Pi.

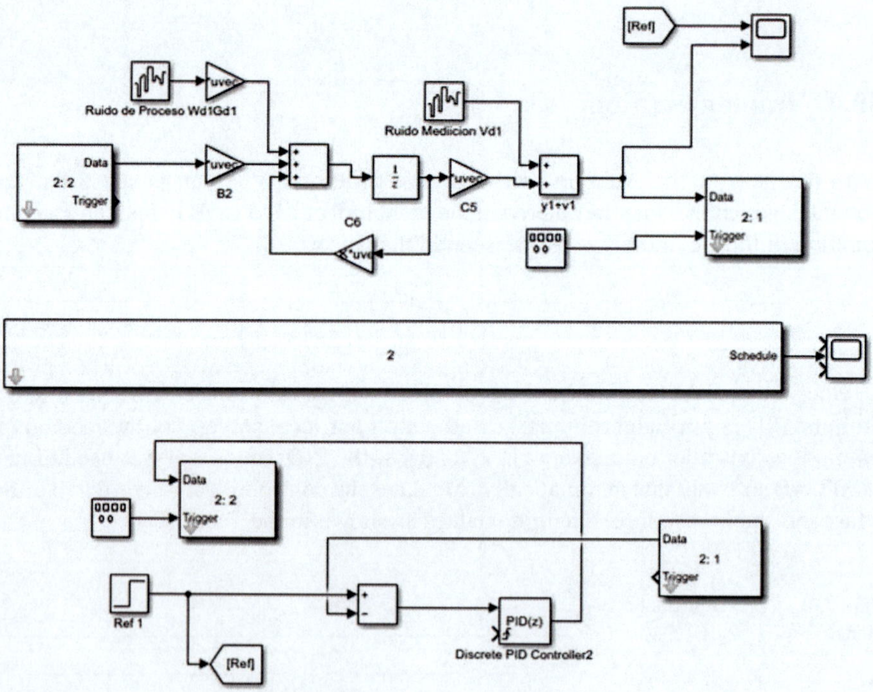

Fig. 50.3 PID control simulation in WNCS

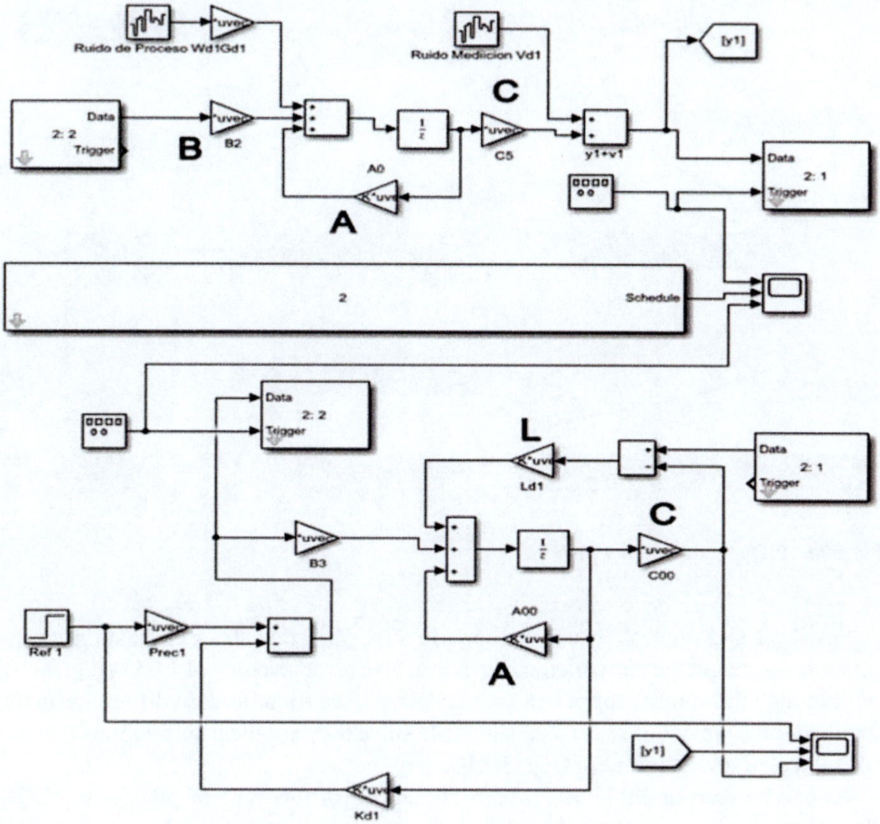

Fig. 50.4 LQG control simulation in WNCS

In the same way, we proceed to implement the LQG algorithm as presented in Fig. 50.7, which will be implemented in the broker hosted on the raspberry Pi.

50.5 Results and Discussion

Once the controls have been simulated and implemented in the local network, the following results are presented: Fig. 50.6 shows the stabilization of the simulated PID control signal in a local wireless network, and Fig. 50.4 corrects the error in a time of 2.55 s.

In Fig. 50.7, the performance of the PID control implemented in a local wireless network is shown, where by subjecting the system to a disturbance, it tends to correct the error in a time of 4 s. The delay that occurs in the PID control implemented in a practical way is considerable compared to the one carried out in the MATLAB Simulink simulation tool.

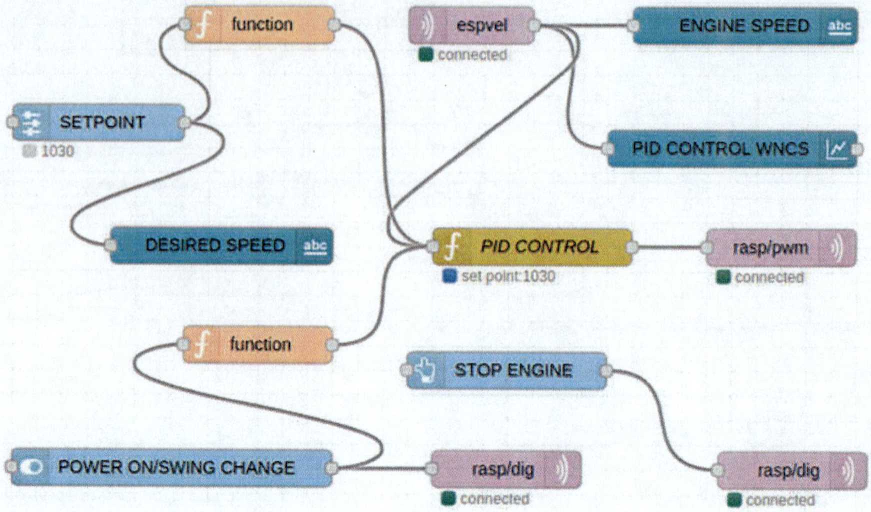

Fig. 50.5 PID algorithm block code

For the LQG control, it can be seen in Fig. 50.8 that the stabilization of the simulated signal in the network tends to correct the error in a time of 1.05 s (Fig. 50.9).

Although the control applied in the network presents a minimum difference in the stabilization time of 0.02 s, this is negligible since they are ideal simulations that are not subjected to disturbances (Fig. 50.10).

As can be seen in Table 50.2, the best control for this type of plant is the LQG algorithm, since it reduces the response time by more than half compared to the PID algorithm. In turn, the network control may have a delay due to the occupation of the channel of the devices connected to it.

50.6 Conclusion

Through the design methodology used, it can be seen throughout this work that, thanks to the analysis of the technical aspects of the control systems, the effect of delay, channel occupancy and the programmed communication protocol were evidenced. Controllers such as PID and LQG were applied for the manipulation of a DC motor, observing the LQG control a fast response to disturbances or to the change of the set point is obtained by 58.85%, minimizing the overshoot with respect to the PID control. In addition, an error of 35% was evidenced in the implemented PID control with respect to the simulated one, since it does not work with the disturbances that exist in the real control. Finally, a new work environment is proposed for the simulation and implementation of distributed systems control algorithms.

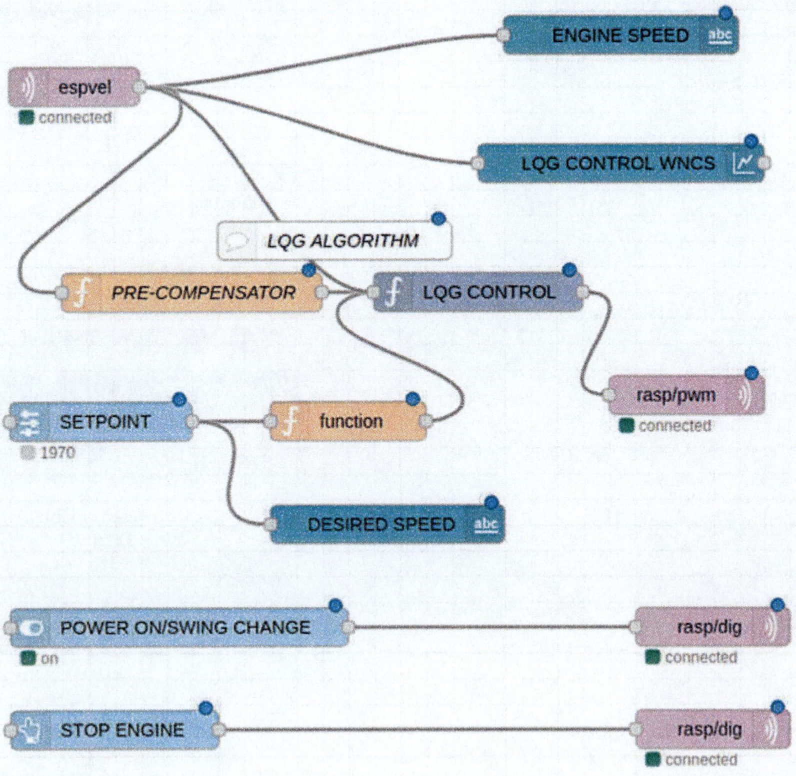

Fig. 50.6 LQG algorithm block code

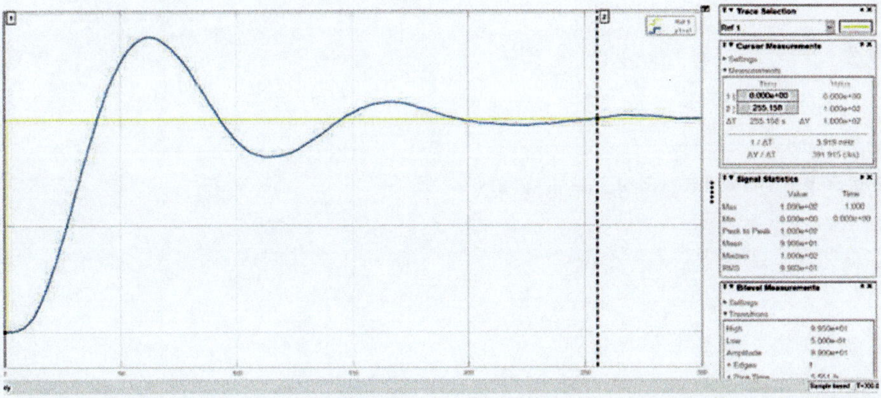

Fig. 50.7 Simulated PID control in WNCS

Fig. 50.8 Real PID control implemented in WNCS

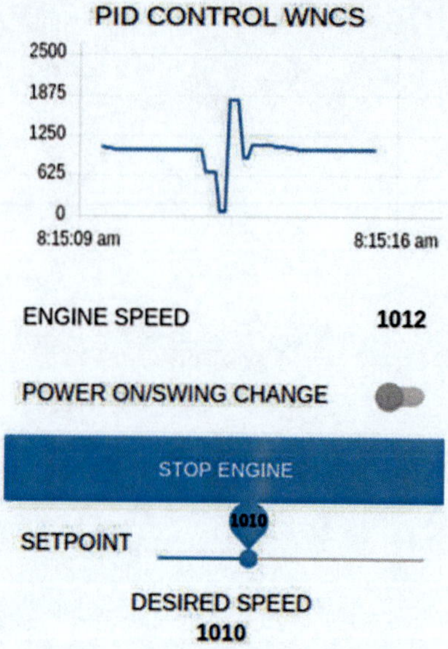

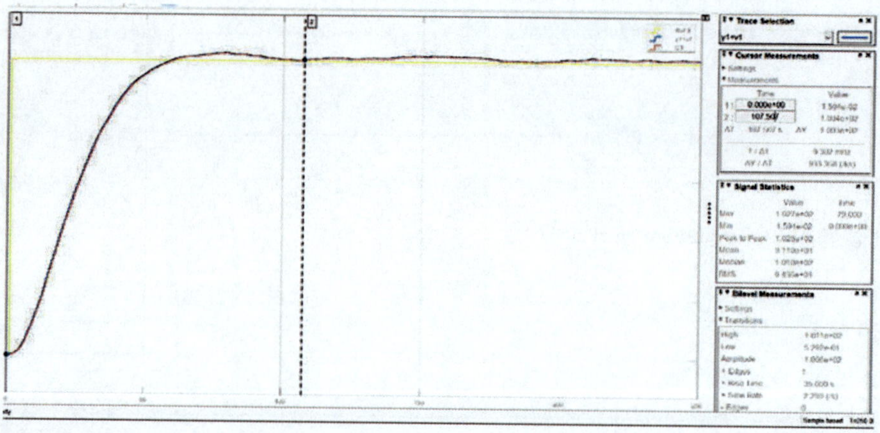

Fig. 50.9 LQG control simulation in WNCS

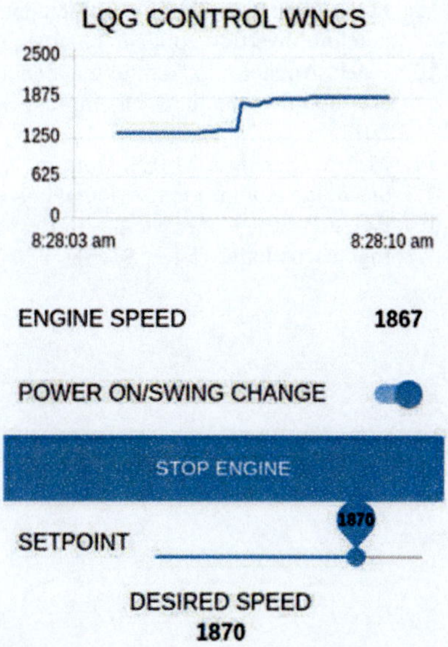

Fig. 50.10 LQG control in WNCS

Table 50.2 Comparison of response times

Control	Implemented with network (s)	Simulated with network (s)
PID	4	2.60
LQG	1.96	1.07

References

1. Y. Lu, *Industry 4.0: A Survey on Technologies, Applications and Open Research Issues*, vol. 6 (Springer, 2017), pp. 1–10
2. L.S. Dalenogarea, G.B. Beniteza, N.F. Ayalab, A.G. Franka, The expected contribution of Industry 4.0 technologies for industrial performance. Sci. Direct **204**, 383–394 (2018)
3. M. Kashyap, V. Sharma, N. Gupta, Taking MQTT and NodeMcu to Iot: Communication in Internet of Things. Sci. Direct 1611–1618 (2018)
4. R.A. Light, Mosquitto: server and client implementation of the MQTT protocol. JOSS 265 (2017)
5. M. Lekić, G. Gardašević, *IoT Sensor Integration to Node-RED Platform* (IEEE, 2018)
6. A.C.T. Martínez, J.F.A. Benavides, Diseño de un sistema de control para ascensores con análisis de sus variables energéticas en la nube utilizando machine learning, Bogotá
7. C. Pillajo, R. Hincapié, *Wireless Network Control Systems*. Abya-Yala, Quito (2018)
8. B.D.S. Pozo, A.E.T. Bastidas, Sistema de control en red inalámbrico (wncs) mediante visión artificial y herramienta de mensajería Telegram, para sistemas de nivel. UPS, Quito (2020)
9. R. Vilanova, V.M. Alfaro, Control PID robusto: Una vision panorámica. Sci. Direct **8**, 141–158 (2011)
10. J.R.C. Solís, Control de voltaje óptimo para sistemas de almacenamiento de energía de batería (BESS) acoplados a microrredes de AC utilizando algoritmos LQG (Quito, 2020)
11. C. Pillajo, R. Hincapié, *Stochastic Control for a Wireless Network Control System (WNCS)* (IEEE, 2020)

12. D.K.S. Peñafiel, diseño e implementación de un controlador óptimo LQG para un sistema de péndulo invertido aplicado en un equipo Lego Mindstorms (Quito, 2018)
13. P.A.F. Agudelo, IoT como tendencia, retos y beneficios generados por la implementación de este tipo de soluciones en empresas colombianas del sector de industria y comercio (Bogotá, 2019)
14. M.A.M. Sánchez, M.R.S. Herrera, A.M. Borrero, J.M.A. Márquez, Las ventajas del IoT en el diseño de laboratorios remotos: myGATEWAY (Universidad de Huelva, Huelva, 2019)
15. F.G. Álvarez, Contribuciones al problema del control óptimo (Escuela Técnica Superior de Ingenieros Industriales, Sevilla, 2004)

Chapter 51
Design and Implementation of a Prototype for Acquisition of Fluorescence Images for the Diagnosis of Non-melanoma Skin Cancer

Eduardo Ávalos, Silvia González, and Cristian Ponce

Abstract The fluorescence diagnosis is a non-invasive technique that allows obtaining information about the state of biological tissues and based on this determines the suffering of some type of pathology. The objective of this work was to design and implement a portable prototype that allows exciting biological tissues and acquiring the fluorescence image for the demarcation of different types of cancer of non-melanoma type. The prototype consists of four systems: an optical system that allows to direct the excitation light based on a matrix of LEDs toward the sample and direct the emission of fluorescence to a commercial camera. An electronic system that consists of a system of autonomy that uses a battery of Lithium ions as a power source and a control system for the intensity of the light source. An image acquisition system that uses a modified GoPro camera to acquire and transmit real-time fluorescence images to a PC, where open-source software is used to manage, process, and store these images. Finally, a mechanical system supports the optical and electronic elements that make up the prototype.

51.1 Introduction

Non-melanoma type is around 90% of skin cancer. In Quito, from 2006 to 2010, an average of 499 cases per year is reported. From them, at least 92 deaths have been registered on the National Registry of Tumors [1], conducted by "Sociedad de Lucha Contra el Cáncer" (SOLCA) as the national organization on the fighting against cancer is named.

Early diagnosis enables better conditions for the treatment of patients with this illness and prevents future complications. In Ecuador, a biopsy is the diagnosis technique applied. It consists of making an incision on damaged tissue to perform a study of histopathology and determine if it corresponds to cancer [2]. The disadvantages of this technique are related to his invasiveness and the lack of capacity of determining

E. Ávalos (✉) · S. González · C. Ponce
Escuela Politécnica Nacional, Quito, Ecuador
e-mail: eduardo.avalos@epn.edu.ec

Á. Rocha et al. (eds.), *Communication, Smart Technologies and Innovation for Society*, Smart Innovation, Systems and Technologies 252,
https://doi.org/10.1007/978-981-16-4126-8_51

the borders of the lesion. The lack of information about the borders of the lesion complicated treatment due to the odds of not taking off the complete cancer tissue or taking healthy tissue also, during surgery [3].

To overcome the problems derived from a biopsy, treatment based on fluorescence is an alternative and non-invasive technique. Its purpose is to mark cancer cells from a variety of skin cancer types, different from melanoma. This is possible thanks to the accumulation of protoporphyrin PpIX in cancer cells after the application of a photosensitivity element into the suspicious lesion.

Since protoporphyrin and some biomolecules of skin such as flavins and light pigments emit fluorescence, it is possible to distinguish between healthy and cancerous tissue. In that, biomolecules of skin when exposed to light on a spectrum of 400 ± 20 nm emit a characteristic fluorescence. Thus, protoporphyrin emissions correspond to a spectrum of 632 nm, associated with red color, and light pigments and flavins emit a spectrum around 500 nm corresponding to green color [4, 5].

This technique requires two important elements: a photosensitivity agent and an instrument to excite the biomolecules and gather information about fluorescence. Then, this research focuses on the development of the instrument to be used for the application of this technique. For that purpose, the gathered information about fluorescence will be shown through images. It will enable a qualitative analysis of the lesion and the use of the instrument for a clinical diagnosis of this disease.

The system created is of the kind of a wide optical field. Then, using a camera, it is possible to observe an image of the fluorescence of skin tissue. The prototype consists of an electronic optical system for directing both exciting and emission light, a system for image obtainment, and a mechanical system for the physical support of the elements of the system.

51.2 Methodology

The prototype is focused on the fluorescence generated by the biomolecules of skin tissue and the images obtained from it. It enables the identification of the state of the biological tissues. In that, there are four essential systems: an optical system, an electronic system, an image acquisition system, and a mechanical system; shown in Fig. 51.1.

Description of prototype design and implementation is presented as follows.

51.2.1 Optical System

The optical system of the prototype is similar to the one of a microscope of fluorescence. Thus, it takes advantage of the epifluorescent optical model that guarantees the efficiency of light exciting the sample and the fluorescence of light directed toward

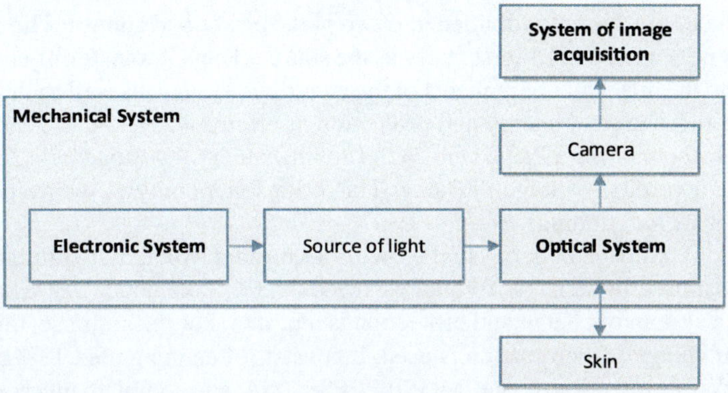

Fig. 51.1 Blocks of the general system of the prototype

Fig. 51.2 Trajectory and direction of light

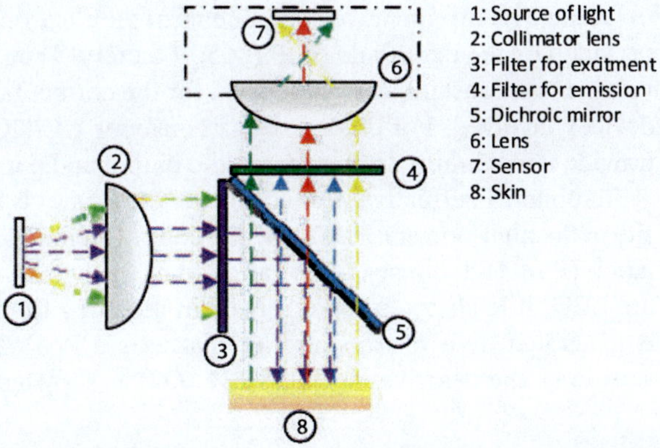

the detector [6]. This optical model consists of an optical filter for excitation, an optical filter for emission, and a dichroic mirror.

The source of light is a high-power LED matrix of 10 W that emits a spectrum around 400 nm. Its luminescence is collimated by a plano-convex lens of borosilicate toward the optical epifluorescent model, which directs light toward the camera as shown in Fig. 51.2.

51.2.2 *Electronic System*

The electronic system controls the intensity of the source of light and the energy supply system of the prototype. Furthermore, it allows the user to easily handle the prototype. It has an OLED screen for visualizing certain parameters of the prototype and push buttons for controlling the irradiance of the source of light.

The electronic system is divided into two parts: power and control. The electronic system of power uses an Li-ion battery as the source. Then, it transforms energy to be supplied to the different components of the electronic system, as well as the charging system of the source of energy and protection of electronic system in general.

This prototype uses a Panasonic NCR18650B battery. Its capacity is 3350 mAh, and it has a nominal voltage of 3.6 V. This brings autonomy to the system, being enough for its functioning.

Since this kind of battery charges with a constant voltage at different stages, they are exposed to damage if voltage exceeds 4.3 V, falls under 2.3 V, or exists a shortcut; a system of charge and protection is required. For this purpose, the TP4056 module of charge and protection is used. It consists of an integrated TP4056 charge device, DW01 protection device, MOSFET FS8205A, and complementary electronic components. Altogether, these devices enable a constant charge current of 1 A, a constant 4.2 V voltage, and bring protection to the battery. This guarantees the good performance of the instrument.

Voltage requirements of components of the electronic system are superior to those provided by the Panasonic NCR18650B battery. Then, it is necessary to increase the voltage to a constant and stable level for the correct functioning of all the electronic devices involved. For this reason, a converter DC/DC elevator MT3608 is used to provide a maximum 2 A current at the output and a maximum 28 V voltage.

To obtain an effective irradiance over the skin, it is necessary to control the intensity of the high-power LED. Then, the control technique applied is pulse-width modulation (PWM). It consists of a variation of the work cycle of the voltage supplied to the LED. It leads to a proportional variation of LED irradiance. Signal-type PWM is generated by a microcontroller that uses a PNP (Q1) transistor as a driver for activating and deactivating MOSFET (Q2) as shown on the circuit of Fig. 51.3.

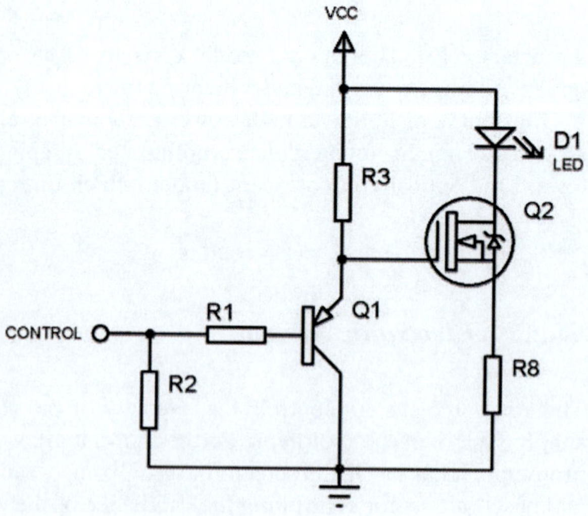

Fig. 51.3 Driver for controlling LED intensity

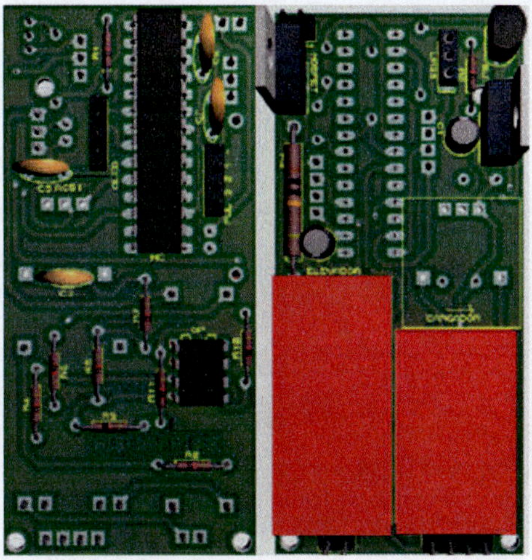

Fig. 51.4 3D diagram of electronic board

The electrical system of control uses a microcontroller as the central element of control. It receives signals from sensors to visualize battery status and signals from push buttons to handle the source of light and LED intensity. Also, it must be able to operate the communication protocol of the OLED screen which controls and receive signals from different sensors within the electronic system.

To fulfill these requirements, a low-consumption Atmega328p controller (AVR, 8 bits, RISC architecture) is used.

Furthermore, the implementation of the electronic system is made using Proteus Design Suite software, and ARES PBC layout is used for designing electronic boards. Figure 51.4 shows a 3D diagram of the board.

51.2.3 System of Image Acquisition

It brings the user a computational tool to visualize and capture in real time an image of the fluorescence. Also, it allows images to vary in characteristics such as brightness, contrast, and color. Then, saving on a database the diagnosis information of every patient and the image.

Figure 51.5 represents the general system of image acquisition which uses a small digital camera that captures images and videos in high resolution (GoPro-Hero4 Silver). The lens of the camera was replaced by one of a lower visual field since samples to be analyzed are small. A no-distortion-fisheye lens of 7.2 mm offers an angle of vertical vision of 37° and the possibility of macro or close caption photography. Then, this modified camera captures images and sends them to a computer through wireless communication. The software captures images of fluorescence and transmits them in real-time. Also, it provides editing tools, databases, and a graphic interface for the user to manage it correctly.

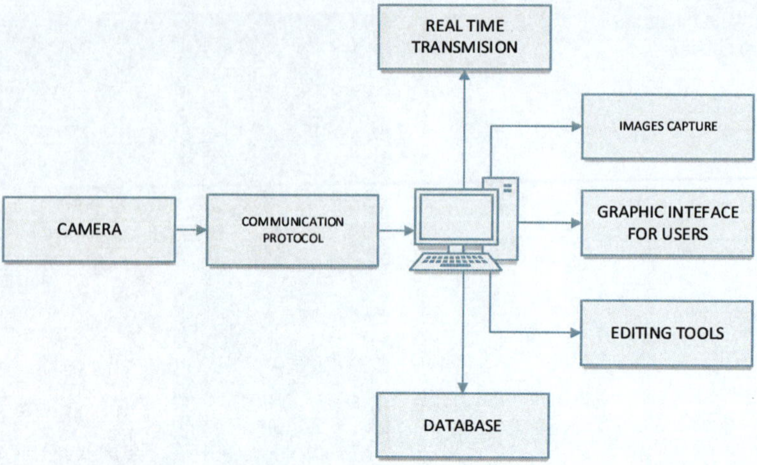

Fig. 51.5 General system of image acquisition

The software of the prototype was developed using Python to be compatible with operating systems such as Windows and Ubuntu. Additionally, GTK + 3 was used to develop the graphic interface, FFmpeg for reproducing and transmitting images in real-time, Goprocam for the capture of images and download, PIL for tools of image processing, and SQLite as the database engine. Then, Fig. 51.6 shows an example of the graphic interface for users.

Fig. 51.6 Graphic interface for users

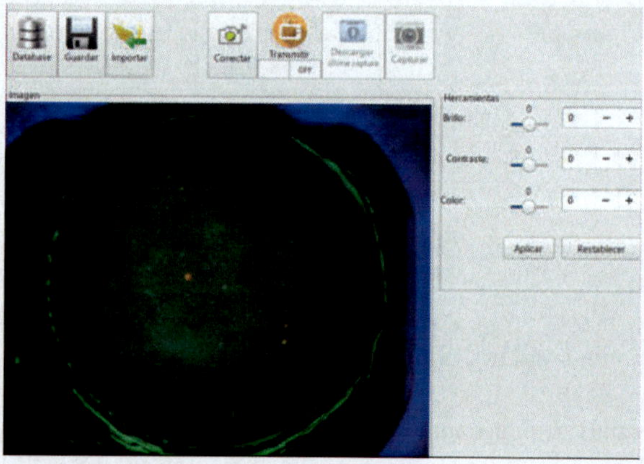

51.2.4 Mechanical System

It supports optical and electronic components of the prototype. Some of the most important mechanical parts of the prototype are the camera shooting, camera focus, cube for fluorescence filters and principal support. Also, in order to preserve time life of the equipment, it has an integrated heat sink to keep the heat of LED under 60°.

Principal mechanical support brings a robust coupling to optical, electronic, and mechanical elements of the device. This support is divided into an upper and a lower part. Figure 51.7 shows assembled parts of the prototype.

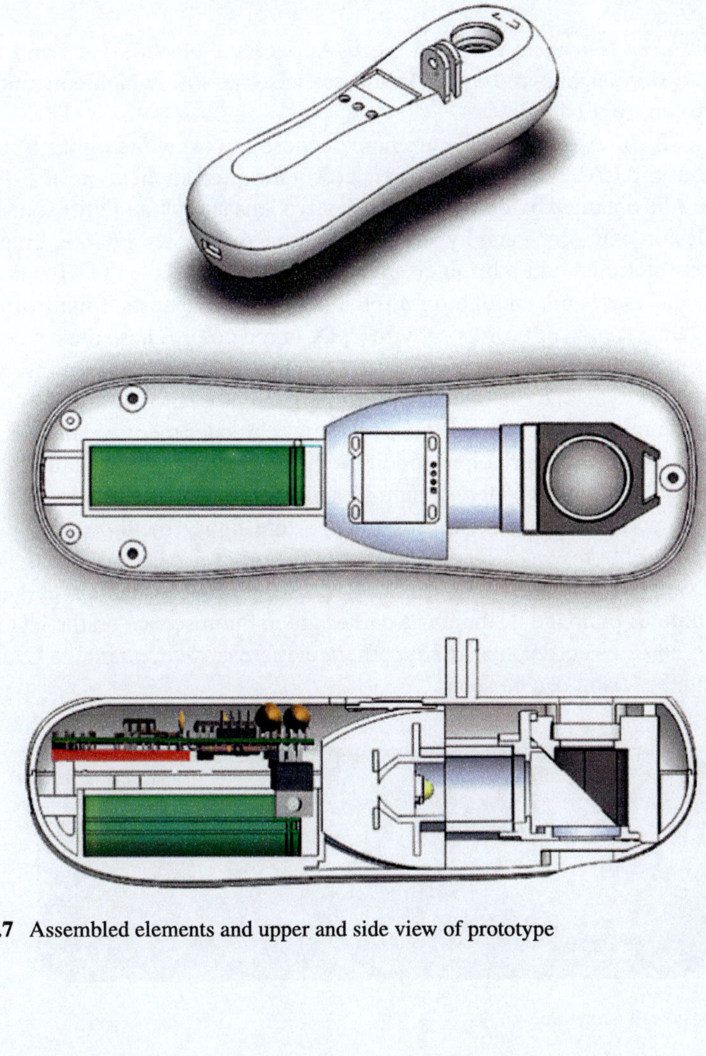

Fig. 51.7 Assembled elements and upper and side view of prototype

51.3 Results and Discussion

51.3.1 *Validation of Results*

The system so-called Fluor-Vision is validated by observation of fluorescence of biomolecules on skin and identification of the presence of protoporphyrin PpIX in cells of certain types of cancer when a photosensitive element is applied.

In total, 42 skin lesions were analyzed. Among them, cases of basal cell carcinoma and actinic keratosis were identified. These types of skin cancer are different from melanoma type as stated by Marks and Motley [7]. In order to discriminate among them, other no-cancerous skin lesions such as verruca vulgaris and nevus were analyzed.

The photosensitive agent used is 5-ALA (Aminolevulinic Acid) at 5 and 10%. It is applied to skin lesions as a topic. Irradiance is set at 40% which corresponds to 20 mW/cm^2 approximately.

Figure 51.8 shows a basal cell carcinoma exposed to (a) white light, (b) fluorescence before 5-ALA; and, (c) fluorescence 60 min after application of 5-ALA at 5%. Image A is obtained by dermatoscopy (macro lens and white light) and shows a pink skin lesion with scabs and hyperkeratosis. Image B shows a darker green fluorescence on the lesion and a brighter green on hyperkeratosis areas. However, this qualitative analysis is not enough for a reliable clinical diagnosis. Finally, image C shows the fluorescence of protoporphyrin PpIX in red color. These are cancer cells. Notice that there are red areas unidentified on images A and B. In consequence, without the prototype and the diagnosis technique applied, the identified borders of the skin lesion might be neglected. Those require a broader medical analysis.

Actinic keratosis is the principal condition that leads to squamous cell carcinoma on the skin [7], and it is considered a non-melanoma skin cancer. Figure 51.9 shows a lesion with actinic keratosis exposed to (a) white light, (b) fluorescence before 5-ALA; and, (c) fluorescence 60 min after application of 5-ALA at 5%.

When exposed to white light (A), a skin lesion in pink, scabs, and hyperkeratosis can be identified. In image B, there is a darker green fluorescence on the lesion, and a brighter green area corresponds to hyperkeratosis areas. And, on image C, a minor red color on the lesion can be seen.

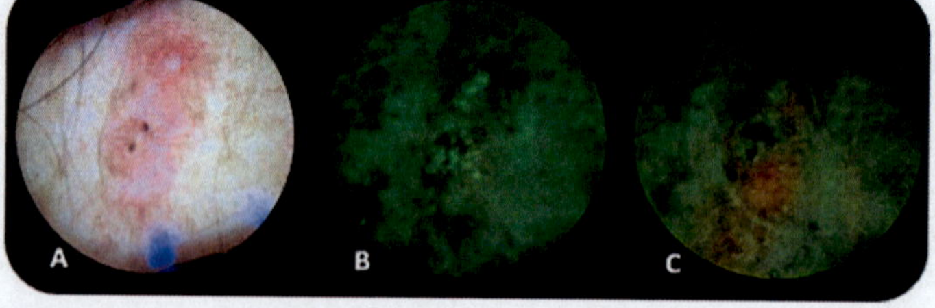

Fig. 51.8 Basal cell carcinoma

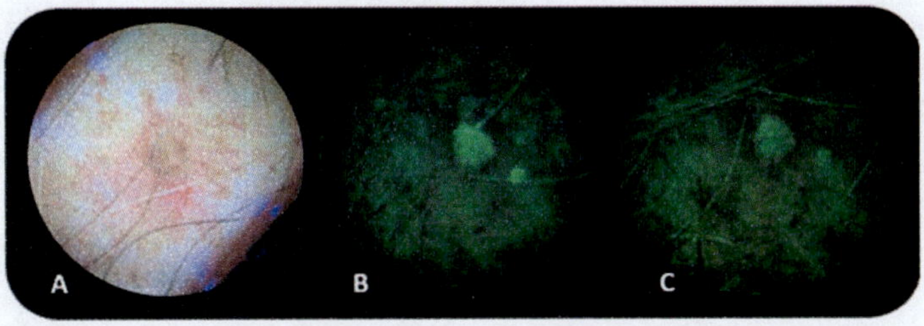

Fig. 51.9 Actinic keratosis

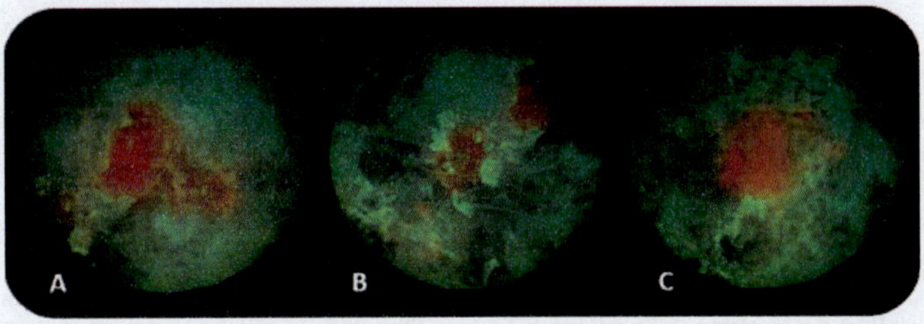

Fig. 51.10 Different cases of actinic keratosis after 90 min of 5-ALA at 10% application

As the accumulation of protoporphyrin PpIX varies according to time and concentration, the fluorescence of three different cases of actinic keratosis is analyzed to discard the lack of irradiance. Then, as shown in Fig. 51.10, after 90 min of application of 5-ALA at 10% and 40% of irradiance, lesions characterized by an intense red color are identified. This means that irradiance at 40% is enough to generate fluorescence of protoporphyrin PpIX, and 5-ALA is difficult to be absorbed by areas with hyperkeratosis. This might require a test by curettage or scraping.

51.4 Conclusions

Prototype for acquisition of fluorescence images constitutes an effective tool for visualization of fluorescence of different groups of biomolecules from biological tissue. It enables discrimination between healthy and cancerous cells due to the presence of protoporphyrin PpIX on skin cancer different from melanoma type when applying a photosensitive agent.

Irradiance around 20 mW/cm^2 is enough to generate fluorescence of protoporphyrin PpIX. It brings to the prototype a longer time of autonomy and a reduced temperature when operating. This contributes to its requirements of energy and time of life.

An optical model of epifluorescence is optimal for visualizing images of fluorescence from sample tissues. Also, the prototype is portable due to the Li-ion battery and wireless communication. Then, it becomes easier to use and more comfortable to visualize and acquire images of fluorescence.

Furthermore, the system of acquisition of fluorescence images has potential application on acne treatment and dental diagnosis. Then, this innovative prototype has the potential to be used on clinical diagnosis within a variety of medical fields. For instance, besides oncology, it can be used for dermatologic and dental diagnosis of health conditions and pathologies.

References

1. P. Cuevas, J. Yépez, Epidemiología del cáncer en Quito 2006–2010, in *Registro Nacional de Tumores*, SOLCA (Quito, Ecuador, 2014). Obtenido de: http://www.estadisticas.med.ec/Publicaciones/PUBLICACION-QU-2006-2010.pdf
2. A. Barros, Estudio Descriptivo: Diagnóstico Clínico e Histopatológico de Cáncer De Piel no Melanoma de Pacientes que Acudieron al Servicio de Dermatología del Hospital "Carlos Andrade Marín", Quito". Rev Med HJCA 2015 **7**(2), 123–127 (2014)
3. C. Andrade, Identificação de lesões de pele por detecção fotodinâmica mediada por ácido aminolevulínico. (Biblioteca Digitais de Teses e Dissertações da USP, São Carlos, 2012)
4. C. Costa, Identificação de lesões de pele por detecção fotodinâmica mediada por ácido aminolevulínico. (Biblioteca Digitais de Teses e Dissertações da USP, São Carlos, 2012)
5. M. Costa, Estudo de viabilidade de um protótipo integrado para o diagnóstico e tratamento da acne por técnicas ópticas. Tesis Doctoral en Processamento de Sinais de Instrumentação (Universidad de São Paulo, Escola de Engenharia de São Carlos, São Carlos, 2015)
6. M. Costa, *Desenvolvimento de Um Sistema Por Imagem de Fluorescência Óptica Para Uso Médico-Odontológico [Disertatión]* (Universidad de São Paulo, Escola de Engenharia de São Carlos, São Carlos, 2010)
7. R. Marks, R. Motley, *Dermatología* (El manual modern, México, 2012)

Chapter 52
Performance Analysis of Protected Next-Generation Optical Access Networks

Estefany Alpusig-Endara and Germán Arévalo

Abstract This manuscript analyzes the penalty of using different survivable topologies for next-generation PON, taking into account three scenarios: a non-protected topology, a topology with redundancy in the CO hardware and a topology with redundancy in the CO hardware and the feeder optical fiber. For comparing the performance of the aforementioned topologies we obtained, for each scenario, the bit error rate (BER) versus received optical power, using Optsim® optical systems' simulation software. To achieve accurate results, we employed in our simulation parameters' values retrieved from real optical components and devices. It were considered in this study two different lengths of ODN: 20 and 40 km. And, to compare the performance of current generation and next-generation PON, the simulations were made for two different bit rates: 10 and 40 Gb/s. Results show that with respect to a non-protected topology, there is almost no penalty in a topology with redundancy for a bit rate of 10 Gb/s. Nonetheless, there is a considerable penalty of using redundancy for a bit rate of 40 Gb/s.

52.1 Introduction

The use of protected topologies is a research topic that has been increasingly covered in recent years [1, 2]. In fact, the improvement of the reliability of the high-capacity next-generation (NG) optical access networks is mandatory to avoid the loss of very high amounts of data and the disconnection of a large number of users, especially for NG high and ultra-high bit rate passive optical networks (PON) [3, 4]. In this context, the fault-recovery times are crucial [1] as well as the deployment costs

E. Alpusig-Endara
Maestría en Tecnologías de La Información, Pontifica Universidad Católica del Ecuador, Av. 12 de Octubre 1076, Quito, Ecuador
e-mail: gealpusig@puce.edu.ec

G. Arévalo (✉)
Carrera de Ingeniería en Telecomunicaciones, Universidad Politécnica Salesiana, Rumichaca Sn Y Morán Valverde, Quito, Ecuador
e-mail: garevalo@ups.edu.ec

Á. Rocha et al. (eds.), *Communication, Smart Technologies and Innovation for Society*, Smart Innovation, Systems and Technologies 252,
https://doi.org/10.1007/978-981-16-4126-8_52

[2]. Most optical networks' protection techniques are based on the diagnostics and redundancy of key devices of the central office (CO) hardware and in the redundancy of key components of the optical distribution network (ODN). In the ODN of a PON the most fault-sensitive component is the feeder optical fiber (OF), due to it is shared by all the PON users and, in case of damage, all the users end up disconnected from the network.

A popular proposal for NG-PON is the wavelength division multiplexed PON (WDM-PON), due to the possibility of assigning dedicated high bandwidth to many users with no need of using media access control techniques. Nonetheless, there are not any current NG-PON standard which includes redundancy in its specifications [5]. Thus, it can be found a series of proposals for protected PON, like the reported in [6], where authors evaluate three different protected topologies utilizing redundancy in the feeder OF and fast-switching between the faulty and the recovery fiber. In that work, it is considered also the redundancy in the OLT employing optical couplers instead of switches.

In [7], it is covered the redundancy of hybrid TDM/WDM PON through the use of a ring-topology and optical coders for reducing the fault detection times, increasing the network reliability, by means of segmenting the topology in smaller pieces. Research works [8, 9] employ feeder OF and distribution OF protection by duplicating the amount of required OF. Such solutions have the drawback of using a big amount of idle fibers with high-deployment costs. The research work reported in [10] proposes a 1:N schema based on the cyclic property of arrayed waveguide gratings (AWGs) and on the use of a specific connection pattern among the optical network units (ONU) at the end user's side. This solution requires the use of a complex routing control unit.

In the design proposed in [11] it is employed a series of switches and optical filters in the optical line terminal (OLT) at the CO side, as well as the redundancy of AWG in the first remote node (RN). This topology provides protection to the feeder OF and the AWG at the same time. In [12], it is detailed a couple of WDM-PON using multiple sub-rings in order to achieve a protected topology. However, extensive use of monitoring circuits and switches is required, in the CO and the user premises, for the network operation which increases the costs of deployment and operation.

The research work in [13] presents a survivable WDM-PON which employs an optical circuit suppression (OCS) technique which includes the use of centralized protection of colorless ONU. It requires the use of many electro-optical devices, which increases the complexity and the cost of the solution. As well, in [14] it is proposed a network topology design that uses remote optical nodes and ONU with auto-restoration capabilities. Such a solution employs a 2 × 1 coupler and a 2 × 4 optical multiplexer in the RN, and in each ONU it uses an optical coupler besides an optical filter and an optical switch.

In this research work, we compare the performance of different topologies for next-generation PON, taking into account three possible scenarios: a non-protected topology, a topology with redundancy in the CO hardware and a topology with redundancy in the CO hardware and the feeder optical fiber. To the best of our knowledge, no research work compares, in the same scenario of hardware and users' demands, the performance of the proposed protected topologies in comparison with the non-protected solution. The remainder of this manuscript is organized as follows. Section 52.2 describes the network scenarios covered in the simulations. In Sect. 52.3, it is detailed the simulation setups and in Sect. 52.4 the main results of the simulations are presented. Finally, Sect. 52.5 concludes the paper.

52.2 Network Scenarios

For the network analysis, we used a traditional splitter-based PON topology with two levels of protection, as shown in Fig. 52.1: (a) no protection, (b) protection only on CO hardware (i.e., protection of the OLT), and (c) protection of the CO hardware and the feeder OF.

Notice that the Type A network corresponds to a traditional PON with ODN based on a tree-topology. In such case any failure of the OLT or in the feeder OF will result in a disconnection of all the users that were serviced by the PON (typically, 32 or 64 users), who will have to wait until a qualified technician repair the damage in the network, which may take several hours, to become connected again.

In the Type B network, the topology is almost the same as a non-protected PON, except for the use of a redundant OLT and an optical switch in the CO for each PON.

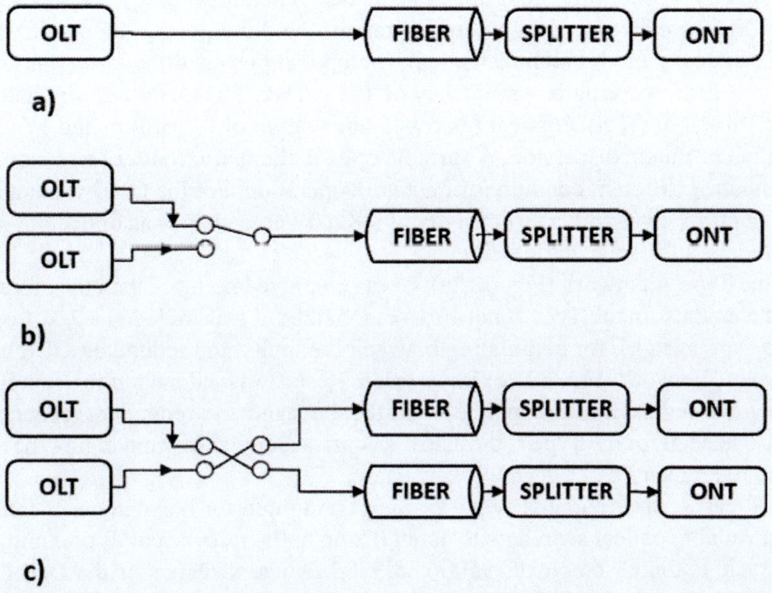

Fig. 52.1 Diagram of the three different topologies considered in the study: **a** Type A: non-protected PON, **b** Type B: protection of the CO hardware (redundancy of the OLT), **c** Type C: protection of the CO hardware and feeder OF

Alternatively, a redundant OLT could be shared by many active OLT, but from the performance point of view, it practically makes no difference (of course from the deployment cost perspective it does makes a great difference [2]).

Type C is a more complex and much more expensive solution given that it includes, besides the OLT redundancy, a redundancy in the feeder OF. This is due to the cost of installation of the optical fiber is the most expensive component of a PON deployment [2]. Of course, we considered a scenario where the redundant feeder OF is routed in a different path than the active feeder OF, and that is the reason for the big addition in the cost of deployment. This solution uses not only one but also two optical switches, which has an almost negligible impact on the cost of deployment, but it does have a non-negligible impact on the network performance.

52.3 Simulation Setup

Figure 52.2 presents the general simulation setups employed for this research, using the optical transmission systems simulation software Optsim®. We chose this software, besides to the licensing availability reasons, because of its versatility and reliability. The OLT was simulated by: A Lorentzian continuous-wave laser (laslor—CW) and an intensity optical modulator (Ioamod), which received electrical NRZ bits at a rate of 10 and 40 Gb/s from a modulator driver (recnrz). The driver received the data generated by a pseudo-random binary sequence generator (pnseq). The OLT transmitted NRZ optical bits, at the same bit rate of the driver, using the on–off keying (OOK) intensity modulation technique. The optical power of the laser was configured in 3 dBm, working at a frequency of 193.2 THz. The ODN was simulated for 20 and 40 km of G652D optical fiber with attenuation of 0.2 dB/km and 17 ps/(nm * km) of chromatic dispersion. A variable optical attenuator (oatten) was employed for emulating different conditions of network operation. For the ONU we employed a PIN photoreceiver with a responsivity of 0.8 A/W preceded by an optical filter and, followed by a Bessel electrical filter.

In the Type A network (Fig. 52.2a), it was employed no more than the aforementioned elements. In the Type B network (Fig. 52.2b), it was included a 2×1 optical switch (opt_switch), for commuting between the faulty and redundant OLT, and in the Type C network (Fig. 52.2c) it was used a 2×2 optical switch for commuting between the redundant OLT and between the damaged and redundant feeder OF. It was also needed for the Type C the use of an optical coupler for converging the active and the redundant OF.

In all cases, the simulated switches were configured for behaving as a commercially available optical switch, with about 0.2 up to 0.4 microseconds of commuting time, such as the Cisco Nexus 9500 and 9300 optical switches or the Dell Power Connect 6000 optical switch [15].

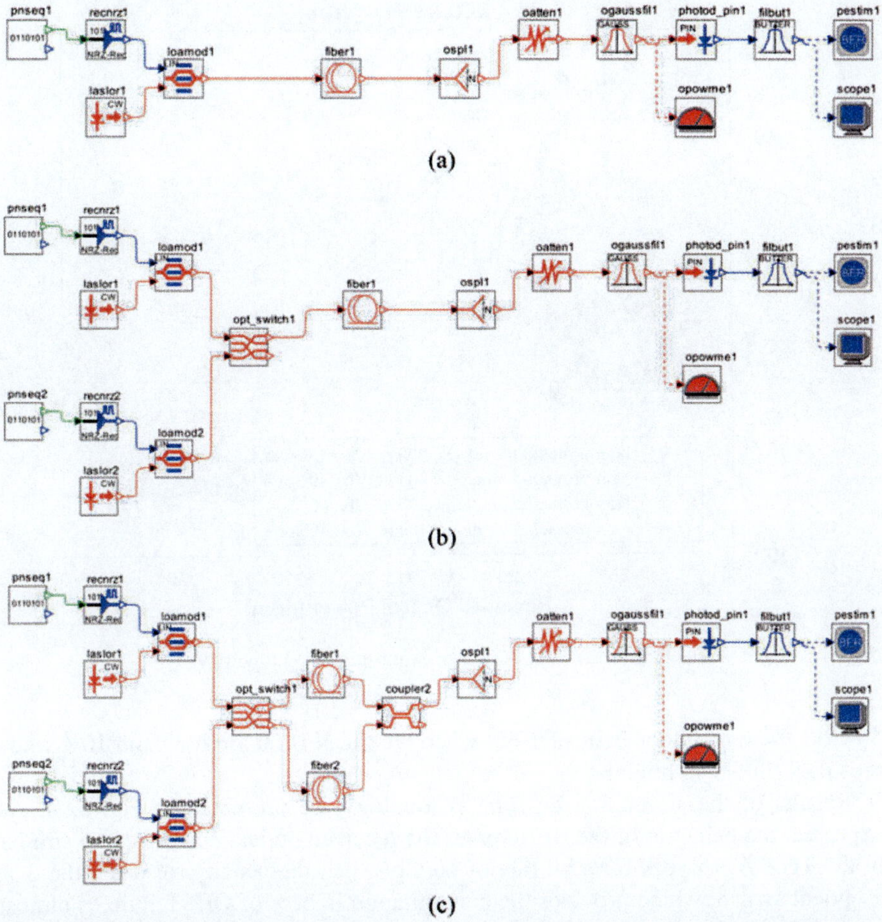

Fig. 52.2 Simulation setups for the three network topologies: **a** Type A, **b** Type B, **c** Type C

52.4 Results and Analysis

For each topology, it was evaluated the bit error rate (BER) for different values of received optical power (Prx). In Fig. 52.3, it can be seen the BER versus Prx curve for the Type A network. This is the reference performance to be employed in comparison with the results of the other two topologies. An important result that can be appreciated is the fact that when the bit rate is 10 Gb/s the Prx (i.e., the sensitivity required for the photodiode) for achieving a BER of 10^{-12} is close to –20 dBm for the network with a 20 km link, while the Prx for the network with 40 km of optical fiber link, for achieving the same BER, is − 18.5 dBm. Thus, there is a penalty of about 1.5 dB for increasing the link from 20 to 40 km. In the case of a bit rate of

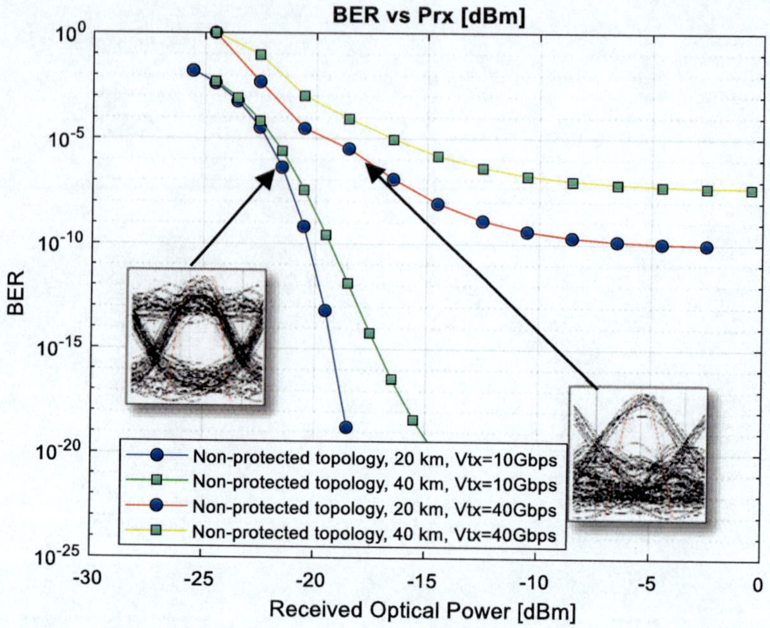

Fig. 52.3 BER versus Prx for the Type A network (non-protected topology)

40 Gb/s there is a floor BER of 10^{-10} when the ODN is 20 km long and 10^{-7} when the ODN link is 40 km long.

Results of the simulation of Type B topology are presented in Fig. 52.4. As expected, the behavior of the BER versus the received optical power is very similar to the Type A network for a bit rate of 10 Gb/s, i.e., the penalty for using the 2 × 1 optical switch, which provides the commutation in case of OLT failure, is almost negligible in this case. But, for a bit rate of 40 Gb/s there is a penalty of about 5 dB for achieving a BER of 10^{-9} when the ODN has a length of 20 km. Moreover, for a ODN of 40 km, the floor of the BER decreases to a value in the order of 10^{-3}.

Figure 52.5 shows the BER versus received optical power for the Type C network. It is evident that the use of a 2 × 2 switch and an additional optical coupler do make an impact on the network performance. In fact, for a bit rate of 10 Gb/s, using as reference a BER of 10^{-12}, it can be seen that the penalty for achieving that BER is close to 0.5 dB for the network with 20 km of optical link as well as for the network of 40 km of optical link. In the case of having a bit rate of 40 Gb/s, the floor BER for the network of 20 km significantly decreases to 10^{-5} and for 40 km, the BER stays degraded to 10^{-3}.

Table 52.1 presents the OSNR values of a received optical signal, with BER = 10^{-12}, for an ODN of 20 km. All values are very similar, which demonstrates that the optical devices employed for providing protection in Type B and C networks make a very low contribution in the degradation of the transmitted signal.

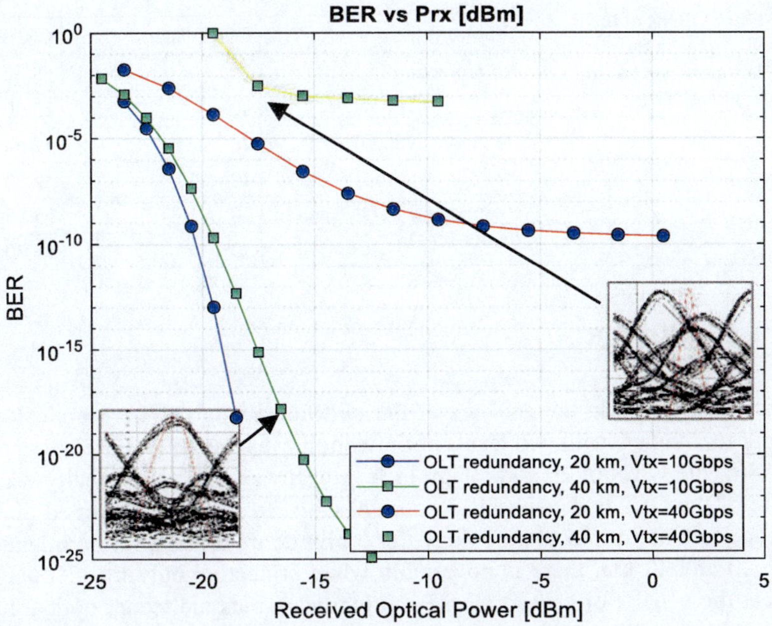

Fig. 52.4 BER versus Prx for the Type B network (OLT redundancy)

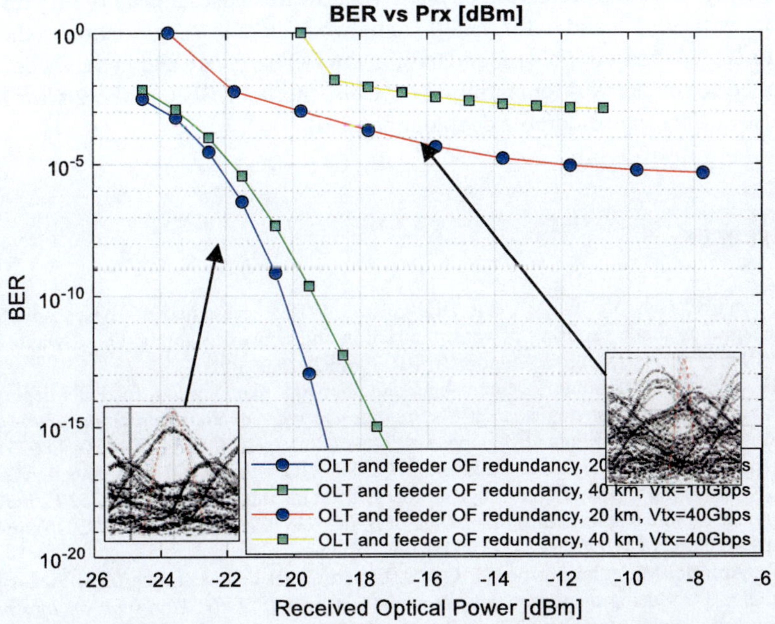

Fig. 52.5 BER versus Prx for the Type C network (OLT + feeder OF redundancy)

Table 52.1 OSNR of the received signals in the different topologies for an ODN length of 20 km

Topology	ODN distance (km)	OSNR [dB]
No protection	20	55.33
	40	55.30
Redundancy in OLT	20	52.34
	40	52.30
Redundancy in OLT and in the feeder OF	20	55.75
	40	55.70

52.5 Conclusion

In this research work, the analysis of the performance of different optical access network topologies, with two levels of protection, was presented for passive optical networks with ODN of 20 and 40 km of length. It was tested the penalty when the bit rate of the OLT transceiver is 10 and 40 Gb/s. Results demonstrated that with reference to a non-protected topology, for a bit rate of 10 Gb/s for both lengths of ODN, 20 and 40 km, there is no penalty when protecting only the CO hardware and that the penalty of protecting the CO hardware plus the feeder optical fiber is only close to 0.5 dB. In the case of a bit rate of 40 Gb/s, the non-protected topology presents a floor BER of 10^{-10} and 10^{-7} for 20 and 40 km of optical fiber length, respectively. And, the protection of the CO hardware causes a penalty of 5 dB for a network with 20 km, and in for the 40 km network there is a degradation in the floor BER to 10^{-3}. When a double redundancy, in the CO hardware and in the feeder fiber, is applied to the network operating at 40 Gb/s, the floor BER of the 20 km length link also suffers a considerable degradation to 10^{-5}.

References

1. C. Feng, C. Gan, Y. Gong, C. Wu, High-reliable WDM access network based on alternate transmission mode featuring power optimization and network stability. Opt. Switch. Netw. **18**(P1), 11–19 (2015). https://doi.org/10.1016/j.osn.2015.02.001
2. G.V. Arévalo, R. Gaudino, A techno-economic network planning tool for PON deployment including protection strategies, in *2017 19th International Conference on Transparent Optical Networks (ICTON)*, Girona (2017), pp. 1–4. https://doi.org/10.1109/ICTON.2017.8025122
3. P. Torres-Ferrera, M. Tipán, R. Gaudino, D. Cardenas, G.V. Arévalo, Direct-detection 25 Gb/s PON: PROs and CONs of digital signal processing at the transmitter side, in *2020 22nd International Conference on Transparent Optical Networks (ICTON)*, Bari, Italy (2020), pp. 1–4. https://doi.org/10.1109/ICTON51198.2020.9203535
4. G.V. Arévalo, M. Villavicencio, G. Flores, Experimental demonstration of a PAM-4 based 20 Gb/s PON using dispersion pre-compensation, in*2019 XVIII Workshop on Information Processing and Control (RPIC)*, Salvador, Brazil (2019), pp. 283–285. https://doi.org/10.1109/RPIC.2019.8882165

5. A.O. María, M. Yacelga-Pinto, G.V. Arévalo,Experimental evaluation of a NG-PON2 network, in *2018 IEEE Third Ecuador Technical Chapters Meeting (ETCM)*, Cuenca, Ecuador (2018), pp. 1–4. https://doi.org/10.1109/ETCM.2018.8580339
6. E. Wong, Survivable architectures for time and wavelength division multiplexed passive optical networks. Opt. Commun. **325**, 152–159 (2014). https://doi.org/10.1016/j.optcom.2014.03.084
7. M.A. Esmail, H. Fathallah, Fiber fault management and protection solution for ring-and-spur WDM/TDM long-reach PON, in *GLOBECOM—IEEE Global Telecommunication Conference* (2011), pp. 1–5. https://doi.org/10.1109/GLOCOM.2011.6133689
8. A. Chowdhury, H.C. Chien, S.H. Fan, C. Liu, C. Su, G.K. Chang, A survivable protection and restoration scheme using wavelength switching of integrated tunable optical transmitter for high throughput WDM-PON system. Opt. InfoBase Conf. Pap. **1**, 30–32 (2011). https://doi.org/10.1364/ofc.2011.othk
9. Y. Qiu, Z. Liu, C.K. Chan, A centrally controlled survivable WDM-PON based on optical carrier suppression technique. IEEE Photonics Technol. Lett. **23**(6), 386–388 (2011). https://doi.org/10.1109/LPT.2011.2107317
10. S. Chuan Ching, 1:N Protection scheme for AWG-based WDM PONs, in *GLOBECOM—IEEE Global Telecommunication Conference* (2006), pp. 2–6. https://doi.org/10.1109/GLOCOM.2006.358
11. X. Sun, C.K. Chan, L.K. Chen, Erratum: a survivable WDM-PON architecture with centralized alternate-path protection switching for traffic restoration. IEEE Photo. Technol. Lett. **18**(4), 631–633; **18**(6), 794–795 (2006). https://doi.org/10.1109/LPT.2006.872195
12. X. Cheng, Y.J. Wen, Z. Xu, Y. Wang, Y.K. Yeo, Survivable WDM-PON with self-protection and in-service fault localization capabilities. Opt. Commun. **281**(18), 4606–4611 (2008). https://doi.org/10.1016/j.optcom.2008.06.018
13. Z. Zhou, S. Xiao, M. Bi, T. Qi, P. Li, W. Hu, Survivable wavelength-division multiplexing passive optical network system with centralized protection routing scheme and efficient wavelength utilization. Opt. Eng. **52**(9), 096109 (2013). https://doi.org/10.1117/1.oe.52.9.096109
14. C.H. Yeh, C.M. Luo, Y.R. Xie, C.W. Chow, Y.W. Chen, T.A. Hsu, Survivable and reliable WDM-PON system with self-protected mechanism against fiber fault. IEEE Access **7**, 165088–165092 (2019). https://doi.org/10.1109/ACCESS.2019.2952944
15. Cisco, "Switches Cisco Nexus 9300," no. Figura 2, pp. 1–16, 2013, [Online]. Available: https://www.cisco.com/c/dam/global/es_mx/assets/C78-729405-00_Cisco_9300_DS_v4c.pdf

Chapter 53
Vermicomposting: Production of Humus and Biol

Alisson Patricia Bolaños Villarreal, Diana Elizabeth García Tumipamba, and Freddy Vicente Cuarán Sarzosa

Abstract This research was carried out in the city of Quito during the health emergency due to the COVID-19 pandemic, with the aim of exploring the vermicompost process that involves different stages, such as the construction of a vermicompost station, using a compost worm (*Eisenia foetida*); and the production of the final product (humus and biol), using dry leaves, dry material (cardboard) and organic waste. Californian red worms (*E. foetida*) can be used to treat solid organic waste. Many studies have reported the effect of temperature, humidity, and pH value on the growth, reproduction and survival of worms. This paper analyzes the influence of these parameters on the growth and development of earthworms in household organic residues for 6 weeks, monitoring pH values, moisture percentage, and $C:N$ ratio. Each value was recorded one day a week, except for the $C:N$ ratio, which is related to worm feeding. Worms were fed every 15 days. The control of these parameters had allowed the reproduction of worms, initially with a weight of 1 kg of worm, and after 6 weeks, 1.24 kg of Californian red worms and 400 mL of biol were recorded. For that purpose, 9.98 kg of organic waste were processed.

53.1 Introduction

In Ecuador, about 11,200 tons of solid waste are collected daily, of which only 10% are collected from the source in a differentiated way between organic and inorganic waste. Of this percentage, more than 50% correspond to organic waste [1]. Many of these organic wastes are thrown away, ending up in landfills and occupying a large amount of space after having emitted greenhouse gases, which contribute to climate

A. P. B. Villarreal
Environmental Engineering, Universidad Politécnica Salesiana, Av. Rumichaca y Morán Valverde s/n, Quito, Ecuador

D. E. G. Tumipamba (✉) · F. V. C. Sarzosa
Research Group in Ecology and Management of Natural Resources, Universidad Politécnica Salesiana, Av. Rumichaca y Morán Valverde s/n, Quito, Ecuador
e-mail: dgarcia@ups.edu.ec

Á. Rocha et al. (eds.), *Communication, Smart Technologies and Innovation for Society*, Smart Innovation, Systems and Technologies 252,
https://doi.org/10.1007/978-981-16-4126-8_53

change. When these materials are composted, they are abundant mixture of valuable nutrients that can benefit soil and plants.

Currently, the declaration of international health emergency due to coronavirus has been announced, so a large amount of waste has been generated around the world. There has been an increase in 600 ton/day of solid waste in Quito, i.e., the figure has increased in recent days to approximately 2800 tons/day [2].

In accordance with the United States Environmental Protection Agency (USEPA), there are five types of composting: on-site composting, vermicomposting, aerated (turned) windrow composting, aerated static pile composting and in-vessel composting. Vermicomposting or composting with worms is a low-cost biotechnological process that allows biodegrading and stabilizing organic waste under aerobic and mesophilic conditions by the action of certain species of earthworms, capable of feeding on the waste, while accelerating its microbial degradation. Therefore, in this process, the detritus capacity of worms is exploited, which ingest, crush and digest the organic residues, breaking them down by the action of their digestive enzymes and the aerobic and anaerobic microflora present in the interior of their intestine [3].

The vermicomposting is the process in which worms are used with the objective of converting organic residues in a final product that is stabilized, fine and homogenous, denominated earthworm humus or vermicompost, an organic product characterized by its agricultural value and/or as a soil floor covering agent. The worms that are responsible for accelerating the decomposition and humidification of organic material are known as Californian red worm or *Eisenia foetida*. Earthworm humus is an organic fertilizer obtained by the digestion process that worms make to organic material, resulting in humic and fulvic acids, which can be handled in solid form such as humus or in liquid form such as humus leachate or worm humus tea. Both earthworm humus and liquid worm humus are one of the best organic fertilizers due to their high content of Nitrogen, Phosphorus, Potassium, Calcium and Magnesium; these elements are essential for the development of plants.

53.2 Methodology

53.2.1 Site Description

The research was carried out in the city of Quito, Ecuador. The vermicomposting process lasted 6 weeks (42 days), taking place during June–August 2020. Vermicompost process started from the implementation and processing of vermicompost to the harvesting of worms. The organic solid waste was obtained from house-holds.

53.2.2 Elaboration of the Vermicompost

For the production of vermicompost, the boxes were defined as: box 1 (base box), which stores the leachate, and box 2, which stores the worms and the food supply.

As a first step, marks were done in box 2 to pierce the base of the box with a screwdriver, with the aim of allowing the leachate to pass and similarly, perforations were made both on the top, the lid, and on the sides of the container for an adequate distribution of ventilation. In box 1, holes were made with a screwdriver on the top, i.e., on the lid of the container to allow the passage of leachate coming out of box 2. It should be taken into consideration that a cut of 5 cm was made on each side of the lid of the container, to avoid unnecessary leakage of percolated liquid, this cut was made using a knife that has been subjected to heat to facilitate cutting. Box 1 was the recipient of the percolated liquids.

Subsequently, the valve was pressed into the bottom of box 1 to prevent leachate leaks.

53.2.3 Feeding of the Worms

The first layer was prepared, which corresponded to the bed for the worms. Initially, 0.10 kg dry leaves were placed in a container with water, and once the excess of water has been removed, it was placed in box 2. It should be taken into consideration that the implementations of a wet bed makes the worm habitat more comfortable and it keeps the container properly ventilated. Subsequently, 400 kg of soil, and 0.5 kg of organic waste were placed on the bed of worms, the amount is little at the beginning of the process because it is important to "acclimatize" worms in their new habitat, and finally worms were added and were covered with 0.01 kg dry sheets and 0.01 kg cardboard to keep the flies isolated. The order of layers was: dried leaves (worm bed), soil, worm, organic waste, dried leaves and cardboard.

After a week, worms continued being fed, maintaining the ideal *C*:*N* ratio which is ± 30 of the entire vermicompost process. In this way, its formation process is controlled [4].

In each food step, a pre-compostation of the organic residues was carried out for 15 days; but these wastes were previously washed very well and cut into very small pieces to facilitate the digestion of worms [5]; in addition, the cardboard and dried leaves were cut and crushed into small pieces with the same purpose.

The food was weighed on a scale to later perform the respective calculations, and in that way maintain the *C*:*N* ratio (Table 53.1). The following table shows the calculation used to determine the *C*:*N* ratio. The feeding process was repeated every 15 days during 6 weeks.

The percentage (%) by weight of the product is calculated as follows:

Table 53.1 Calculation to determine the *C*:*N* ratio of materials

Layers	*C*:*N* ratio	Weight		% × (*C*:*N* ratio)
		kg	%	
Product	*C*:*N* ratio of product	Product weight		% product × *C*:*N* ratio of product
Total (Summation)		Total weight	1.000	*C*:*N* = ±30

$$\frac{\text{Product } 1 \times 100}{\text{Total product}} \Big/ 100 \qquad (53.1)$$

The calculation, to determine the *C*:*N* ratio of the materials, was the same for the whole worm feeding process, the only variant was the amount of food added (Product).

During each feeding, an extra material was added to the worm diet. Initially, worms were fed with 1 kg of banana and 4 kg of potato peeling. During the next feeding, 1.20 kg of fruit remains, 0.01 kg of cardboard and 2.50 kg of potato peeling were added, and finally 0.50 kg of strawberry, 0.15 kg of avocado, 0.63 kg of banana and 0.01 kg of cardboard were added. It should be mentioned that the only constant products during the whole feeding process were: 0.01–0.15 kg of soil and 0.01–0.05 kg of dry leaves. However, quantities were varying to maintain the ideal *C*:*N* ratio which is ± 30.

53.2.4 Humidity and pH Measurement

Worms show interest if they live at a neutral pH. *E. foetida* can live perfectly in soils where the pH is between 7 and 8 [6].

As a first step to measure pH, water was boiled to remove the chlorine and then it was set still for 3 days, to ensure that the chlorine evaporates quickly and there is no alteration during the pH measurement. It was verified that water had a neutral pH. Then, a small sample was taken from the substrate, verifying that worms are not present in all their stages. The sample is added to the glass container with a few drops of water, because it must not be muddy. It was mixed for 1 min, and after that time, the glass vessel was tilted slightly in a 45° orientation so that pH strip is in contact with water (Fig. 53.1). It was expected that water rises through capillary action. The pH indicator strip was removed from water, and the new color was compared to the graph provided by the manufacturer to check the value. The pH measurement was taken to each day of the week during the 6 weeks of the study.

To maintain an adequate humidity of 80%, 250 mL of dechlorinate water was added twice a week with the help of an atomizer. The control of humidity percentage was done with the technique known as "clenched fist technique." [4] A quantity of the vermicompost was collected with the fist of the hand, and force was applied until

Fig. 53.1 Humidity and pH measurement in vermicompost

it was observed that 0.4–0.5 mL of water [7] were falling. As a photographic record, wooden sticks were inserted into the vermicompost for 1 min to verify moisture.

53.2.5 Worm Harvesting and Humus Production

In the last feeding of worms, the last layer of dry leaves was removed and a 5-mm plastic mesh of 50 cm × 40 cm was placed on the whole substrate; on this mesh, the food for worms was placed, maintaining the C:N ratio, so that worms ascend to food or, in turn, they remain in the first 10 cm of the substrate; in that way, the worm can be harvested and the weight of worms can be compared at the beginning and at the end. Worm population was observed and classified into: cocoons, clitelate worms and non-clitelate worms (Fig. 53.2).

Worms are photophobic, therefore, to separate worms from the humus, they were exposed to sunlight, so that they are induced to move toward areas with less light or that they are submerged to the depths. Subsequently, a plastic bag was placed on the floor and the vermicompost found in box 2 was sieved to obtain humus from the worms. Materials that were not able to decompose during 6 weeks were separated from the final product. In the same way, the biol that was in box 1 was collected in a glass container.

Fig. 53.2 Red Californian worms life cycle and stages

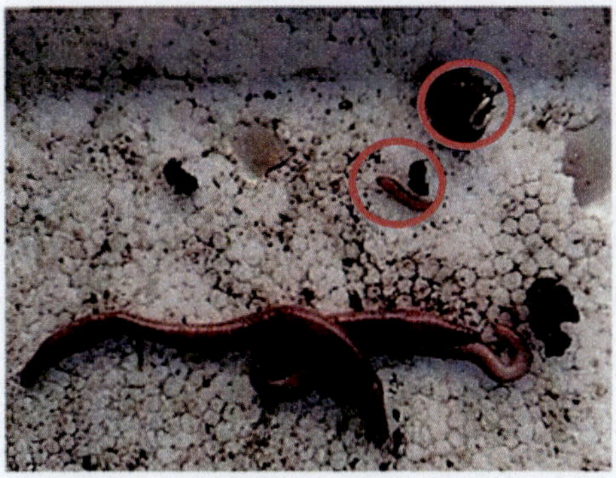

53.3 Results and Discussion

Table 53.2 summarizes the humidity and pH records throughout the process.

Week 3 and 6 recorded an average neutral pH, i.e., during those weeks, worms were in optimal conditions for their development. Citrus foods were avoided, for that reason, the average value of optimal pH was in the range of 7–8. A lower value was recorded during the first week, for this reason, more wet material was added to stabilize vermicompost. Therefore, a progressive increase in humidity is evident during the course of the weeks. In addition, the highest percentages of humidity recorded were during week 5 and 6 with a humidity value of 80%. It is relevant to notice that an average value of 71.7% was recorded. Both pH and humidity were related due to the increase in the number of worms. Initially, a value of 1 kg of worms was reported, after 6 weeks a total value of 1.24 kg of worms was recorded. Therefore, the humidity and pH conditions were optimal because they allowed the development and reproduction of Californian red worms. In addition, the average total optimal pH and humidity in this study were 7.7 and 71.7%, respectively.

Another priority parameter for obtaining compost is the *C*:*N* ratio, which was monitored with a frequency of fifteen days throughout the process. The process should maintain a *C*:*N* ratio at ± 30. According to authors [8], an ideal *C*:*N* ratio for composting is 30:1; therefore, it is observed that the development and quality of this vermicompost is ideal, and it is in the range of 30.024–30.477 (Table 53.3). The *C*:*N* ratio of the materials used were values obtained through tables that are established in research carried out [4, 9–12]. Each of the materials were weighed individually, and then, the total weight was obtained by taking the weight percentage of each material. The *C*:*N* ratio was calculated by multiplying each individual percentage by the *C*:*N*

Table 53.2 pH and humidity variables during the vermicomposting process

Week	pH	Humidity (%)
1	8.0	55
2	8.0	65
3	7.0	75
4	8.0	75
5	8.0	80
6	7.0	80
$\bar{x}$	7.7	71.7

Table 53.3 *C*:*N* ratio during the vermicomposting process

Week	% × (*C*:*N* ratio) total
1	30.083
3	30.477
5	30.024
$\bar{x}$	30.195

ratio consulted in tables. Subsequently, all these values were added and the value of the *C:N* ratio of the mixture used was obtained.

The harvest of worm humus and the collection of biol were carried out in week 6. It was observed as the worm humus change color; initially, it presented dark brown color and at the end a black color, with a pleasant smell similar to a forest. The collected biol had a dark appearance. pH as measured and its value was 9. In addition, 400 mL of leachate were collected.

The pH is one of the important factors in the vermicompost process. A pH value of 8 was recorded during the first and second week. At the third week, a stabilization of the pH was observed to neutral, however, during the following two weeks, this value was 8 again, due to the action of fungi and microorganisms involved [13]. In addition, the pH is an indicator of the humification degree of the organic matter and therefore of its maturity [14].

It is also noted that worms created aerobic conditions in material waste by their excavation actions, inhibiting the action of anaerobic microorganisms that release hydrogen sulfide and mercaptans [15].

The results in this study indicated that the average humidity percentage of vermicompost was 71.7%; however, in the research [16], it is mentioned that the range obtained during the elaboration of its vermicompost was 80–90%, ensuring quick growth of their worms.

The average pH of this study was slightly neutral in the vermicompost. Several researchers have reported that most species of earthworms prefer a pH of approximately 7 [17, 18]. The optimal pH ensures a growth and stability of the worms throughout the vermicompost process.

Edwards et al. [19], described that 0.5 kg of earthworms can process about 0.5 kg of organic material (at humidity 75–85%) and produce about 0.25 kg of vermicompost per day (40–50% conversion rate), which is a satisfactory population of earthworms, as defined by authors [20]. A good indicator of quality and stability in substrates is reflected in the survival of worms [21]. In this context, approximately twenty times more organic waste was processed compared to the above-mentioned study, i.e., 9.98 kg total with twice the worms, obtaining humus with neutral pH and presenting a pleasant smell similar to wet soil.

According to author [22], given the optimal temperature and humidity conditions, approximately 0.5 kg of adult worms (approx. 1000) or 5 kg of worms (approx. 10,000) can produce vermicompost from 10 kg of waste or 1 ton of waste, respectively, in just 30 days.

The pH fluctuated from 7 to 8, until it remained neutral in week 3 and 6, just when the compost was ready for harvest, which is similar to findings [16, 23], who recorded a pH range of 5–9 during the process. The values came almost to neutrality, when vermicompost was ready for harvest; this can occur due to the production of CO_2 and organic acids produced during microbial metabolism.

The combination of different plant materials used as substrate, mainly combinations of herbaceous and woody, generate a substrate of high biological value stabilized by the action of microorganisms and oligochaete annelids [24].

Romero et al. [14] state that the *C:N* ratio during the vermicompost process influences the rate of degradation and mineralization of organic material.

According to [25], the production of earthworms with horse substrate presented the best *C:N* ratio, pH and in count of worms. However, the cow substrate does not have favorable characteristics, so it is the least suitable as a soil improver.

Comparing the pH value of the study conducted by Morales et al. [25], with the present paper, it is concluded that with the substrates of animal origin of the horse, pig and hens, a similar value is obtained. However, when comparing *C:N* ratio in the different substrates, it is evident that the values do not exceed $C{:}N = 18$; for this reason, a deficient *C:N* ratio could present problems with pH, humidity, temperature, harming worms and the quality of vermicompost. During the conduction of this research, the *C:N* ratio was controlled and an optimal value was maintained in order not to prejudice the process.

Worms have the real potential to accelerate and improve the natural biodegradation and decomposition of organic materials from 60 to 80% promoting the growth of "beneficial aerobic decomposing bacteria" [26].

53.4 Conclusions

Earthworms are omnivorous, therefore, the food supplied must include, for the most part, organic residues. In order to make better use of the waste, a "pre-composting" process must be carried out before worms are fed. The vermicomposting with dry leaves, organic waste and dry materials was successful. The vermicompost product had a dark color, it smelt pleasantly as earth and it was homogeneous. *E. foetida* adapted to the different diets implemented during the 6 weeks. In implementing the last feed, it was observed that the detritus capacity of worms accelerated the decomposition process of the organic residues. The last diet implemented was made up of avocado and banana, both have a high-nutritional content and a high percentage of water, which facilitated that the worms transform into fertilizer for shorter periods. The food provided during the last week managed to stabilize the compost, at the same time, the ideal humidity parameter was obtained.

The parameters as pH, humidity and the *C:N* ratio were essential for the reproduction and development of *E. foetida*, with average values: pH 7.7, $C{:}N$ ratio $=$ 30.195 and humidity 71.7%. These values are within the recommended ranges so that the growth and reproduction of worms are not affected. The pH values recorded during the vermicompost remained stable. In addition, during process, the number of worms increased by 0.24 kg and 400 mL of biol was generated with a basic pH ($\text{pH} = 9$).

References

1. INEC, Estadística de Información Ambiental Económica en Gobiernos Autónomos Descentralizados Municipales, (2014). [Online]. Available: https://www.ecuadorencifras.gob.ec/documentos/web-inec/Encuestas_Ambientales/Municipios_ConsProvinciales_2014/Municipios-2014/201412_GADSMunicipalesDocumentoTecnicoDeResultados.pdf
2. A. Martínez, Quito: recomiendan colocar desechos en doble funda y rociar con agua y jabón (Metro Ecuador, Quito, 2020)
3. V.M. Cornelio Villegas, J.R. Laines Canepa, Vermicompostaje: I avances y estrategias en el tratamiento de residuos sólidos orgánicos. Rev. Mex. Ciencias agrícolas **8**(2) (2017). https://doi.org/10.29312/remexca.v8i2.59
4. P. Román, M. Martínez, A. Pantoja, Manual de Compostaje del Agricultor. Organ. las Nac. Unidas para la Aliment. y la Agric., p. 56 (2013) [Online]. Available: http://www.fao.org/3/a-i3388s.pdf
5. C.M. Acosta Durán, O. Solís Pérez, L. Villegas Torres, O.G.C. Vigueros, Precomposteo de Residuos Orgánicos y su Efecto en la Dinámica Poblacional de *Eisenia foetida*. Agron. Costarric **37**(1), 127–139 (2012) [Online]. Available: http://www.mag.go.cr/rev_agr/v37n01_127.pdf
6. R. Rostami, Vermicomposting. Integr. Waste Manag. **2**, 133 (2011). https://doi.org/10.5772/16449
7. Universidad de Magallanes, Manual de compostaje para zonas frías (Punta Arenas, 2018) [Online]. Available: https://educacion.mma.gob.cl/wp-content/uploads/2020/03/Manual-de-Compostaje.pdf
8. J.O. Ryan, M. Riffo, El compostaje y su utilización en agricultura. Fund. para la Innovación Agrar (2007) [Online]. Available: http://bibliotecadigital.fia.cl/bitstream/handle/20.500.11944/1907/1El_compostaje_y_su_utilizacion_en_agricultura.pdf?sequence=1&isAllowed=y
9. J.E. Simo, L.A. Ruiz, R. Rivera, O.M. Morales, D. Carvajal, T. Ramírez, Estudios integrales para el manejo y producción in situ de alternativas de fertilización en el cultivo del plátano (Instituto de Investigaciones de Viandas Tropicales, 2004)
10. C. Gamarra, M.I. Díaz Lezcano, M. Vera de Ortíz, M.D.P. Galeano, A.J.N. Cabrera Cardús, Relación carbono-nitrógeno en suelos de sistemas silvopastoriles del Chaco paraguayo. Revista Mexicana de Ciencias Forestales **9**(46). https://doi.org/10.29298/rmcf.v9i46.134
11. J.J.González-Fernández, J.M. Alvarez, Z. Galea, E. Guirado, J.M. Hermoso, I. Hormaza, R. López, Reciclaje de residuos de la producción de guacamole mediante compostaje, in *VII Congreso Iberico de Agroingenieria y Ciencias Horticolas, Ref. N^o C0463* (2013)
12. L.A. Sepúlveda, J.A. Alvarado, Manual de compostaje doméstico. Manual de aprovechamiento de residuos orgánicos a través de sistemas de compostaje y lombricultura en el Valle de Aburrá. in *Área Metropolitana del Valle de Aburrá*, 1st edn. (2013)
13. A. Camacho, L. Martínez, H. Ramírez, R. Valenzuela, M. Valdés, Potencial de algunos microorganismos en el compostaje de residuos sólidos. Terra Latinoam. **32**(4), 291–300 (2014) [Online]. Available: http://www.scielo.org.mx/pdf/tl/v32n4/2395-8030-tl-32-04-00291.pdf
14. J. Romero, J. Sánchez, M. Rodríguez, M. Gutiérrez, Producción de vermicompost a base de rastrojo de maíz (*Zea mays* L.) y estiércol de bovino lechero. AgroProductividad **8**(3), 52–59 (2015) [Online]. Available: https://biblat.unam.mx/hevila/Agroproductividad/2015/vol8/no3/9.pdf
15. R.K. Sinha, S. Agarwal, K. Chauhan, V. Chandran, B.K. Soni, Vermiculture technology: reviving the dreams of Sir Charles Darwin for scientific use of earthworms in sustainable development programs. Technol. Invest. **01**(03), 155–172 (2010). https://doi.org/10.4236/ti.2010.13019
16. Y. Indrani Ramnarain, A. Adil Ansari, L. Ori, Vermicomposting of different organic materials using the epigeic earthworm *Eisenia foetida*. Int. J. Recycl. Org. Waste Agric. **8**, 23–36 (2018). https://doi.org/10.1007/s40093-018-0225-7

17. S. Suthar, Development of a novel epigeic-anecic-based polyculturevermireactor for efficient treatment of municipal sewage. Environ. Waste Manag. **2**, 84–101 (2008). https://doi.org/10.1504/IJEWM.2008.016994
18. S. Panday, N. Karmegam, Effect of vermicompost amended alluvial soil on growth and metabolic responses of rice (*Oryza sativa* L.). Eco-friendly Agric. **4**(1), 35–47 (2009)
19. C.A. Edwards, S. Subler, N. Aracon, *Quality Criteria for Vermicompost* (CRC Press, 2011), pp. 287–301. https://doi.org/10.1201/b10453-19
20. C. Edwards, N. Aracon, *Interactions Among Organic Matter, Earthworms and Microorganisms in Promoting Plant Growth* (CRC Press, 2004), pp. 327–376. https://doi.org/10.1201/9780203496374.ch11
21. J.J.E.D. Santoya, R. Gómez Álvares, A. Jarquín Sánchez, G. Villanueva López, Caracterización de vermicompostas y su efecto en la germinación y crecimiento de Capsicum chinense Jacquin. Ecosistemas y Recur. Agropecu. **5**(14), 181 (2018). https://doi.org/10.19136/era.a5n14.1465
22. C. Visvanathan, J. Trankler, K. Jospeh, R. Nagendran, Vermicomposting as an Eco-Tool in Sustainable Solid Waste Management. Asian Inst. Technol. (2005)
23. J. Domínguez, C. Edwards, Biology and Ecology of Earthworms Species used for Vermicomposting (CRC Press, 2011), pp. 27–40 [Online]. Available: https://www.taylorfrancis.com
24. F. Salinas-Vásquez, L. Sepúlveda-Morales, G. Sepúlveda-Chavera, Evaluación de la calidad química del humus de lombriz roja californiana (*Eisenia foetida*) elaborado a partir de cuatro sustratos orgánicos en Arica. Idesia (Arica) **32**(2), 95–99 (2014). https://doi.org/10.4067/S0718-34292014000200013
25. J. Morales Munguia, M. Fernández, A. Montiel, B. Peralta, Evaluación de sustratos orgánicos en la producción de lombricomposta y el desarrollo de lombriz (*Eisenia foetida*). Biotecnia **11**(1), 19–26 (2009). https://doi.org/10.18633/bt.v11i1.49
26. R. Kuma, S. Herat, S. Agarwal, Vermiculture and waste management: study of action of earthworms *Eisenia foetida*, *Eudrilus euginae* and *Perionyx excavatus* on biodegradation of some community wastes in India and Australia. Environmentalist **22**(3), 1–10 (2002). https://doi.org/10.1023/A:1016583929723

Chapter 54
Evaluation of the Microbiological Quality of Iceberg Lettuce Commercialized in the Markets of the City of Cuenca

Angélica G. Zea, Marcelo E. Flores, Tony Viloria, and Ernesto Delgado

Abstract The consumption of vegetables free of pathogenic bacteria is of utmost importance since the absence or presence of these has a direct impact on the health of the population. Among the most frequently consumed vegetables is lettuce, due to the number of nutrients it has. We must consider that this vegetable is consumed raw, and therefore, in the environmental factors to which this type of crop is exposed, implying an undeniable risk of transmitting intestinal-type diseases if the proper measures are not taken when preparing them for human consumption. The present work focused on conducting a microbiological study to determine the presence of *Escherichia coli*, total Coliforms, Salmonella, and Shigella by the pour plate method in the culture medium Chromocult Agar and SS Agar as indicator bacteria of fecal contamination to determine the food safety of these lettuces at the different points of commercialization. As a result, it was found that the degree of contamination was high, independent of the sampling point.

54.1 Introduction

Foodborne diseases (FBD) constitute important mortality worldwide since, until 2010, it was estimated that 600 million cases caused by FBD caused 420,000 deaths [1]. Outbreaks of these diseases are attributed to the consumption of fresh vegetables, mainly leafy vegetables such as spinach, alfalfa sprouts, and lettuce [2], and since they are consumed without any type of cooking, they are potentially dangerous in case of contamination [3, 4].

According to INEC statistics, in Ecuador gastrointestinal diseases represent the second cause of morbidity in the total population and are one of the main causes of infant mortality [5].

The presence of *Escherichia coli* indicates fecal contamination in water and/or food since this microorganism is a normal inhabitant of the digestive tract of warm-blooded animals and is rarely found in water or soil that has not suffered some type of

A. G. Zea (✉) · M. E. Flores · T. Viloria · E. Delgado
Universidad Politécnica Salesiana, Cuenca, Ecuador
e-mail: azea@ups.edu.ec

Á. Rocha et al. (eds.), *Communication, Smart Technologies and Innovation for Society*, Smart Innovation, Systems and Technologies 252,
https://doi.org/10.1007/978-981-16-4126-8_54

fecal contamination; therefore, is considered a universal indicator. The identification of total coliforms is more difficult as these can come from soil, and surface freshwater; they are not always intestinal [6].

The control of the sanitary quality of food can be carried out by identifying bacteria that fecal contamination. These bacteria can be used to assess the quality of food, sediments, and water intended for human consumption, agriculture, industry, and recreation. There is no universal indicator, so the most appropriate one must be selected for the specific situation under study [7, 8].

The most widely used indicators of fecal contamination are total and thermotolerant coliforms, *E. coli* and enterococci [9]. According to González et al. and Méndez et al.

Although Salmonella and Shigella are not indicator bacteria for fecal contamination, contamination with Shigella and Salmonella is very common from contaminated water and food. In the USA, in the period 1996–2006, Salmonella and Shigella were pathogens frequently associated with foodborne illness (FBD), with outbreaks of shigellosis associated with consumption of lettuce, parsley, and onions and salmonellosis with tomato, Brussels sprouts, watermelon, and cantaloupe [10].

Salmonella was responsible for 83–87% of the recorded ATE outbreaks. Its presence was detected in fresh export vegetables (lettuce, mustard, basil, tubers, etc.,) with a frequency of 1.9–8.0% [13]. From the 1990s to the present, Salmonella is considered to be one of the most persistent pathogens causing TADs in the world [12]. Raw foods, such as lettuce, present a higher risk of transmission of these and other enteropathogenesis since there is not a post-processing step that removes the initial microbial loads [12].

Given the lack of studies on bacterial contamination by consumption of lettuce in our locality, it was decided to conduct a microbiological study in order to detect the presence of pathogenic bacteria in lettuce commercialized in different markets and supermarkets in the city of Cuenca with the aim of providing data to contribute to the prevention of public health.

54.2 Methods

54.2.1 Selection and Sampling

Lettuce was selected among other vegetables, due to the great demand for its consumption in raw form, and for the ease of production and marketing in our country; selecting iceberg lettuce because it is the most common.

The sampling points were established as four markets and three supermarkets most frequented by the community of the city of Cuenca, where most of the vegetables are marketed for raw consumption by the population.

The lettuces were collected from the four markets and three supermarkets in Cuenca city according to INEN 1975:2003 standards for sampling [13]. Two units were taken randomly per supermarket market. They were then taken to the Life Sciences Laboratory of the Salesian Polytechnic University for microbiological analysis.

54.2.2 Sample Preparation

To perform the microbiological analysis and calculate the Colony Forming Units (CFU), the pour plate technique is used [13], where for the microbiological analysis, it was necessary to perform decimal dilutions in test tubes contained in peptone water; 20 g of each sample (previously chopped) was weighed, then liquefied with 100 mL of distilled water, and the homogenized samples were placed in sterile flasks to take 1 mL of each sample and proceed to perform the respective dilutions according to the technique.

With a sterilized pipette, 1 mL of each of the preprepared dilutions was transferred to empty and properly labeled sterilized Petri dishes, and then 10–15 mL of the melted and moderately cooled culture medium was added to each Petri dish previously inoculated with its respective sample. Once the culture medium was solidified, the petri dishes were inverted and incubated at 37 °C for 24 h.

54.2.3 Selection of the Culture Medium

For the present work, it was required to isolate the following bacteria from lettuce samples: total coliforms, *E. coli* as fecal indicator bacteria in lettuce, and Salmonella and Shigella as pathogenic bacteria present in food.

Chromocult culture medium was selected as a selective agar for the growth of total coliforms and *E. coli* in water and food samples [11]. Chromocult agar is a selective agar for the growth of total coliforms and *E. coli* in food and water samples.

Salmonella Shigella Agar (SS Agar) is a selective and differential medium widely used in sanitary bacteriology to isolate Salmonella and Shigella. [12]. The Chromocult agar is a selective agar for the growth of Salmonella and Shigella in food and water samples (Table 54.1).

Table 54.1 Microbiological test (Incubation conditions: 35 ± 2 °C/18–24 h)

Microorganisms	Agar	specification	characteristic reaction
Coliformes Totales	Chromocult	Good growth	Red to pink colognes
Escherichia coli	Chromocult	Good growth	Dark blue to violet colognes
Salmonella	Agar SS	Good growth	Colorless cologne
Shigella	Agar SS	Good growth	Cologne colorless with black center

After the proposed time has elapsed, we proceed to observe the growth of the colonies and count them in the colony counter.

54.3 Results

A sampling of 48 lettuces taken during four consecutive months from different markets and supermarkets in the city of Cuenca was carried out as shown in Table 54.2. The samples were analyzed in the Microbiology Laboratory of the Salesian Polytechnic University, applying the same conditions that were explained in the methodology section, obtaining the results of the percentage of positive samples for the different bacteria as shown in Tables 54.3, 54.4, 54.5, and 54.6.

Table 54.2 Number of samples taken from the different markets and supermarkets in the City of Cuenca

Markets	Sampling 1	Sampling 2	Sampling 3	Total
Popular markets				
Feria Libre	2	2	2	6
3 de Noviembre	2	2	2	6
10 de Agosto	2	2	2	6
12 de Abril	2	2	2	6
Supermarkets				
Supermaxi	2	2	2	6
Coral Hipermercados	2	2	2	6
Mega Tienda del Sur	2	2	2	6
Gran Akí	2	2	2	6
Total samples				48

Table 54.3 Presence of total coliforms in the lettuce samples

Markets	Analyzed	Positive	Percentage
Popular markets			
Feria Libre	6	6	100
3 de Noviembre	6	3	50
10 de Agosto	6	4	67
12 de Abril	6	3	50
Supermarkets			
Supermaxi	6	2	33
Coral Hipermercados	6	4	67
Mega Tienda del Sur	6	4	67
Gran Akí	6	2	33

Table 54.4 Presence of *Escherichia coli* in the lettuce samples

Markets	Analyzed	Positive	Percentage
Popular markets			
Feria Libre	6	5	83
3 de Noviembre	6	3	50
10 de Agosto	6	3	50
12 de Abril	6	4	67
Supermarkets			
Supermaxi	6	2	33
Coral Hipermercados	6	4	67
Mega Tienda del Sur	6	3	50
Gran Akí	6	2	33

Table 54.5 Presence of Salmonella in the lettuce samples

Markets	Analyzed	Positive	Percentage
Popular markets			
Feria Libre	6	4	67
3 de Noviembre	6	3	50
10 de Agosto	6	2	33
12 de Abril	6	3	50
Supermarkets			
Supermaxi	6	1	17
Coral Hipermercados	6	2	33
Mega Tienda del Sur	6	3	50
Gran Akí	6	0	0

Table 54.6 Presence of Shigella in the lettuce samples

Markets	Analyzed	Positive	Percentage
Popular markets			
Feria Libre	6	4	67
3 de Noviembre	6	2	33
10 de Agosto	6	2	33
12 de Abril	6	2	33
Supermarkets			
Supermaxi	6	0	0
Coral Hipermercados	6	1	17
Mega Tienda del Sur	6	2	33
Gran Akí	6	1	17

54.4 Discussion

Formerly in the city of Cuenca, the food chain was very short since it was customary to consume what the community produced from their farms, home gardens, farm animals, and nearby markets, without worrying about the contamination by microorganisms. However, nowadays, the protection of public health plays a very important role in our society, directly involved with the prevention of environmental contamination. Foodborne diseases one of the biggest problems of public health, changing the focus of the concept of health that is currently directed to prevention and not to treat the diseases.

Since lettuce is one of the vegetables most consumed because of its quick and easy preparation, it was easy to have the necessary samples for the execution of this research and to collect sufficient data for future work. When analyzing the results globally, a percentage of each of the bacteria analyzed was found in all samples. Although in the markets, the percentage is higher compared to supermarkets, it is evident that there is a risk to people's health when consuming this type of vegetables regardless of their origin, if hygienic sanitary measures are not applied in their homes before their consumption.

When analyzing the incidence percentage of total coliforms, clearly the highest percentage of the presence of this fecal indicator bacteria is in the samples taken from the popular markets; however, we cannot rule out that although the percentage is lower in the supermarkets than in the popular markets, in both cases, there is no food safety regarding the lettuce sampled. Therefore, since there is a deficient hygienic and sanitary condition of the vegetables due to their fecal coliform counts, they become a source of propagation of bacterial diarrhea, according to the literature mentioned [13, 14].

Regarding the analysis of *E. coli*, it was found in all sampling points. Its presence, as well as the total coliform count, clearly indicates an unavoidable food risk, and as a result of the presence of this bacterium in lettuce, if people do not take the proper disinfection measures at the time of consumption, they would be exposed to the risk of diarrhea, as indicated above.

The presence of Salmonella and Shigella clearly shows that the risk exists. It is not possible to tell with naked eyes if a lettuce is infected with some type of bacteria but only by performing a microbiological analysis, but what we can do is to apply mandatory disinfection measures of food for consumption. In this case of any type of vegetable that is consumed raw, and thus ensure food safety at the household level, applying home disinfection techniques and that one of them is the use of water suitable for human consumption since the type of water used at the household level plays a key role in preventing foodborne diseases. Food safety is aimed precisely at that to ensure maximum possible food safety.

54.5 Conclusions

In conclusion, there are high counts of fecal coliforms and *E. coli*, and less presence of Salmonella and Shigella in the analyzed lettuces commercialized the points indicated above. Since they are above the maximum values acceptable by the standard, and these are identified as positive in the table of results, it is a risk to public health. It is important to provide environmental training on hygienic and sanitary measures for the consumers of these products, as well as the need for a strict and constant sanitary control of the vegetables consumed by our population.

Although the study covered four popular markets and four supermarkets in the city of Cuenca, since there are many more, the results obtained give us a general hygienic approach to vegetables, specifically lettuce, which was our object of study, and although in the popular markets the percentage was higher compared to the supermarkets, in no case and at no sampling point we can talk about was food safety guaranteed. It should be noted that the causes of contamination were not part of the objectives of this work, however some causes can be mentioned such as: quality of water used in vegetable crops, type of fertilizer, handling by people during harvest, transport, marketing, etc.

The information obtained in this work will allow us to raise awareness in the community in general about the importance of proper disinfection of vegetables and vegetables that are consumed at home, especially those that are consumed raw.

References

1. World Health Organization (ed.), *WHO Estimates of the Global Burden of Foodborne Diseases* (World Health Organization, Geneva, Switzerland, 2015)
2. Z. Soto Varela, L. Pérez Lavalle, D. Estrada Alvarado, Bacterias causantes de enfermedades transmitidas por alimentos: una mirada en Colombia. Rev. Salud Uninorte **32**(1), 105–122 (2016). https://doi.org/10.14482/sun.32.1.8598
3. A.F. López Camelo, *Manual for the Preparation and Sale of Fruits and Vegetables: From Field to Market* (Food and Agriculture Organization of the United Nations, Rome, 2004)
4. K.R. Matthews, Microorganisms associated with fruits and vegetables, in *Microbiology of Fresh Produce* (Wiley Ltd, 2005), pp. 1–19
5. A.P. Vélez Bravo, J.E. Ortega González, Determinación de coliformes totales y *E. coli* en muestras de lechuga expendidas en cuatro mercados de la ciudad de Cuenca (2013). Accedido: 01 May 2021. [En línea]. Disponible en: http://dspace.ucuenca.edu.ec/handle/123456789/4301
6. C.M. Arboleda-Baena, L.M. Arismendi-González, M. Sepúlveda-Sánchez, M. Rodríguez-López, J. Betancur-Urán, Verificación de la metodología colilert para la determinación y cuatificación de coliformes totales y *Escherichia coli* en una matriz de agua natural. Rev. Politécnica **12**(22) (Art. no 22) (2016)
7. D.S. Bachoon, S. Markand, E. Otero, G. Perry, A. Ramsubaugh, Assessment of non-point sources of fecal pollution in coastal waters of Puerto Rico and Trinidad. Mar. Pollut. Bull. **60**(7), 1117–1121 (2010). https://doi.org/10.1016/j.marpolbul.2010.04.020
8. S.P. Luby, S.K. Gupta, M.A. Sheikh, R.B. Johnston, P.K. Ram, M.S. Islam, Tubewell water quality and predictors of contamination in three flood-prone areas in Bangladesh. J. Appl. Microbiol. **105**(4), 1002–1008 (2008). https://doi.org/10.1111/j.1365-2672.2008.03826.x

9. J.A. Larrea-Murrell, M.M. Rojas-Badía, B. Romeu-Álvarez, Bacterias indicadoras de contaminación fecal en la evaluación de la calidad de las aguas: revisión de la literature **44**(3), 12
10. C. Monge, C. Chaves, M.L. Arias, Comparación de la calidad bacteriológica de la lechuga (*Lactuca sativa*) producida en Costa Rica mediante cultivo tradicional, orgánico o hidropónico, p. 5
11. E. Merck, Manual de medios de cultivo Merck. Merck, E., (1985)
12. S-S-AGAR-WITH-SODIUM-DEOXYCHOLATE.pdf. Accedido: 01 May 2021. [En línea]. Disponible en: https://www.tmmedia.in/sites/default/files/S-S-AGAR-WITH-SODIUM-DEOXYCHOLATE.pdf
13. L. López, J. Romero, F. Duarte, Microbiological quality and effect of washing and disinfection of pre-cut Chilean vegetables. Arch. Latinoam. Nutr. **53**(4), 383–388 (2003)
14. A.M. Gonzalez, R. Paranhos, M.S. Lutterbach, Relationships between fecal indicators and pathogenic microorganisms in a tropical lagoon in Rio de Janeiro, Brazil. Environ. Monit. Assess. **164**(1–4), 207–219 (2010). https://doi.org/10.1007/s10661-009-0886-9

Chapter 55
Clustering: A Fundamental Axis in the Competitiveness of the Textile Industry in the Post-pandemic Era

Franyelit Suárez, Luis Rosales, Omar Flor, and Christian Chimbo

Abstract This article presents a set of strategies to study the levels of technological innovation of the Ecuadorian textile companies in the post-pandemic period. As a secondary information source, data will be taken from national and international institutions and organizations which have current information about economic indices for Ecuador, in particular from the textile sector. Other interesting information sources are surveys and interviews applied to owners and managers for many companies randomly selected as primary sources. This methodology is based on surveys and the TASI2 method for measuring the levels of technological innovation in the Ecuadorian textile companies and establishing the influence of the textile clustering after the pandemic. Grouping several textile companies in clusters allows establishing common goals and projects for getting finance easily, setting strategies of technological innovation to step up competitiveness and productivity to develop levels of internationalization to compete at the post-pandemic period. A preliminary result suggests that the Ecuadorian textile industry should be agglomerated in clusters of focused subsectors, and that technological innovation together with the injection of capital are part of its future challenges, to be more competitive. This requires the training of the human resources necessary for its implementation. This work is fundamental for the Ecuadorian government in terms of the public policies to be implemented to have a more competitive textile company. Finally, clustering is directly related to competitiveness and innovation, hypotheses are raised about the challenges that the Ecuadorian textile companies to survive in a highly affected post-pandemic economy.

F. Suárez (✉) · O. Flor
Facultad de Ingeniería y Ciencias Aplicadas, Carrera de Ingeniería Industrial, Universidad de Las Américas, Quito, Ecuador
e-mail: franyelit.suarez@udla.edu.ec

L. Rosales · C. Chimbo
Polytechnic University of Venezuela, UNEXPO, Puerto Ordaz, Venezuela

Á. Rocha et al. (eds.), *Communication, Smart Technologies and Innovation for Society*, Smart Innovation, Systems and Technologies 252,
https://doi.org/10.1007/978-981-16-4126-8_55

55.1 Introduction

The textile industry is one of the fundamental pillars upon which modern economies rest. In this work, technological challenges and clustering as an organizational model for a stage are raised. This document presents technological challenges and clustering as an organizational model for a post-pandemic stage for Ecuador, a country with great socioeconomic potential. Technological innovation, location, competitiveness, and their current interrelationships together with their future challenges for the post-pandemic stage are the main aspects discussed in this work.

The health crisis that the country is going through due to the COVID-19 pandemic impacted the textile sector. Since the health emergency, the sector has recorded USD 150 million in losses and several Colombian textile companies have closed in Ecuador. According to the Association of Textile Industries of Ecuador (AITE), activities have been reduced both in production and sales. In this context, companies have reduced or suspended working hours. In other cases, the option has been to lay off workers to maintain sustainability. The textile sector continues to operate at 15 or 20% of its capacity. It is forecasted that manufacturing and commercial activities are resumed in a period of between three and six months after the quarantine is lifted [1]. As in other countries such as Italy and Spain, in Ecuador the textile industry could agglomerate geographically, forming clusters and industrial districts agreement on this activity [2].

It should be noted the business maturity of the textile sector and improve them to take responsibility for current production challenges, to generate new business opportunities, produce with quality, diversify their products and contribute to the expansion of markets [3]. Hence, the proposal to use clustering and technological innovation in the Ecuadorian textile industry in times of pandemic and post-pandemic.

Several authors [4] investigated the relationship between clustering and competitiveness in emerging economies and they found that clustering and innovation in its hotels. Therefore, the link between clustering and competitiveness is mediated through innovation.

This paper aims to study how the cluster organization influences the trends of technological innovation in the Ecuadorian textile companies in the post-pandemic period, in such a way that it is possible to propose industrial clusters as an agglomeration model. The objective of this work is to study the technological challenges that the Ecuadorian textile industry must assume to be more competitive and how cluster organization could influence technological innovation trends in the post-pandemic era. This will make it possible to prepare to compete with imported products, mostly from China.

Research questions:

What are the technological challenges that the Ecuadorian textile industry must assume to be competitive in a post-pandemic stage?

What are the indicators to be improved in the Ecuadorian textile industry that will allow them to be competitive in the post-pandemic stage?

What are the advantages offered by the clustering of the Ecuadorian textile industry in the post-pandemic stage?

In Latin America, the textile industry is a means of production mainly for the local and marginal market for export. The development of the industry has been favored by technology and innovation from foreign investment and by the interrelationships of companies through industrial clusters. However, this has not been enough to counter the Chinese boom. Chile, Brazil, Ecuador, Panama, Mexico, and Peru are the countries most affected in their trade with China in the clothing sector [5].

Some Italian researchers identify the determining factors of the economic dynamics of small companies, finding in external economies a good point of support and theoretical basis for the analysis Bellandi [6]. Sforzi [7] suggests that different aspects have contributed to the massification of the concepts of industrial district and cluster: (a) the resurgence of new alternative industrial models; (b) the renewed interest of researchers in the use of business systems as analysis unit, (c) the economic development that these areas have been showing, and (d) as in Italy, its recognition by the Public Administration for Industrial Policy purposes. In short, the territorial dimension and location are potential sources of competitive advantages.

A preliminary analysis suggests few policies that drive the productive sector across clusters, fostering the exchange of knowledge and innovation. However, most Central American countries have great strength in the clothing sector. The existing industrial clusters have been maintained through associations of industrialists. Many companies do not apply strategies to increase productivity and seek to benefit from tax incentives and other privileges that the government grants them to stay in the market through export [5]. Leading countries in this sector located in Asia, Europe, and Latin America work through clusters and global value chains through which they are grouped and work together, in this way, they acquire more strength and positioning in the markets.

The term "cluster" refers to a group of companies and institutions associated with a particular and geographically close field. These companies use common and complementary practices. Clusters take variable forms, depending on their depth and sophistication. Most include companies that make end products or services. Likewise, clustering implies another form of agglomeration and has generated excellent results in the textile industry, which uses synergy and improves competitive capacity [5].

Porter [8] considers clusters as geographic concentrations of interconnected companies, specialized providers, service providers, companies in related industries, partner institutions (e.g., universities, trade associations) that compete but also cooperate [8, 9]. Four main elements of the grouping have emerged from theory and evidence. It is about geographic agglomeration, ties, integration, and competition with cooperation. A fifth element refers to a shared goal or vision perceived by group members.

In the past two decades, the literature on industrial clusters has grown rapidly. The diagnosis and improvement of these are some of the most favored forms of competitiveness policies [10–12]. However, the conceptual foundations of clusters have remained weak. This is partly because their theory has been based on their absolute advantages and not on a comparative framework, which frames them as an alternative form of economic organization, such as markets and other modes that imply cooperation. Taiwan is recognized by the great number of clusters on its territory. It ranked in the first position in the world in 2007, since it had more than seventy industrial clusters, according to the 2008–2009 report on global competitiveness, issued by the World Economic Forum between 2015 and 2016, occupies the fifth position. Furthermore, Taiwan has the highest density of industrial clusters. Became the global supply base for the latest technology products [13].

There are a few industrial clusters in Latin America. However, they have little relevance in terms of structure and development. According to the 2015–2016 annual report of the World Economic Forum, Colombia is ranked number 66 internationally; Brazil at 36; Guatemala at 58; Ecuador in 98; Honduras in 48; El Salvador in 71; Nicaragua in 113; Costa Rica in 42 and Peru in 104. Some examples of the clusters in Latin America are the Medellín textile-clothing, design and fashion cluster, and the Bogotá fashion cluster. Guatemala has a textile cluster, considered the most integrated into the region, coordinated by the Association of the Clothing and Textile Industry (Vestex). The Choloma processing industrial zone is located in Honduras, directed by the Honduran Maquila Association; in Ecuador, the Atuntaqui cluster; and in El Salvador, the clothing and textile cluster. The latter is the only country that has developed a complete synthetic cluster [14].

There was an analysis of the textile sector to obtain conclusions about clustering, technological innovation, and its influence on competitiveness to face technological challenges in a post-pandemic era.

55.2 Methodology

The objective of this work is to study the effect that technological innovation generates on competitiveness through clustering in the post-pandemic period, and the methodology proposed for this study comprises a mixed targeting. The method, the collection, and the analysis of both qualitative and quantitative data, determined from the data to be analyzed, will be presented. The former objective is a suitable proposal for setting contributory strategies for the development of the competitive textile industry in another country so that to enhance the macroeconomic indicators of Ecuador through clusters groups. The research was non-experimental, exploratory type because information collection tools were used from secondary sources such as databases and national and international entities. It is also evaluative since as primary sources of information, surveys and selective interviews were applied to owners and/or managers of companies in the textile sector.

The research has also taken into consideration the primary sources such as technical journals, articles, and textile companies published, together with publications whose content has opinions and analysis on the subject. Also, the investigation is documentary since data from existing sources was investigated, interpreted, and presented, and a field investigation was carried out that had direct contact with the facts or phenomena that affect the investigation.

To determine the study sample, the directory of companies belonging to the Association of Textile Industries of Ecuador in the Sierra Region was used. Through a simple probabilistic sample and under the criteria that the companies are active and have their main headquarters within the limits of the metropolitan area of the Sierra, a total of 16 textile companies in the Sierra region of Ecuador were evaluated.

The descriptive component for the sample was:

- Active industries that interact with others belonging to the same branch, the relationship between region of location and productivity, legal status and subscribed capital, size of the industry, relationship between capital and age of the company, relationship between competitiveness and internationalization. These theories will be based on and evaluated with a survey applied to business leaders, shareholders, directors and managers of textile companies, taking into account the information in the published databases.

To carry out the data collection, surveys were conducted to obtain, in a systematic and orderly manner, information on the variables involved in this study. The survey was organized into categories: general company data, management and administration, economic and commercial data, human resources, innovation, design and manufacturing, and quality and maintenance system. In addition, the following aspects were determined: (a) Technological innovation and current status of the company (b). Machinery—number of employees—productivity (c). Personnel training, product quality, internationalization, and competitiveness.

About the information from the companies, the following was followed:

Part 1: Direct communication with human talent, knowledge of the production process (control techniques for the elements involved during production: raw material, labor), research, development and innovation activities, management and leadership capacity, Technological level, Personalized customer service, Industrial safety, Interaction with industries in the same branch.
Part 2: Market segments, new markets, credit facilities, exploitation of new technologies, fashion, and new trends.
Part 3: Inadequate distribution of machinery and equipment, excessive waste in the manufacture of products, training of operators in machinery and equipment, inadequate time, method of work, and sales pricing.
Part 4: Sales price, competition, taxes, fixed suppliers, customer liquidity, environmental policies.

Once the diagnosis of the Ecuadorian textile company has been made, its technological challenges are proposed in the post-COVID-19 stage, for example, cluster agglomeration models for the Ecuadorian textile company.

55.2.1 TASI Method

One way to evaluate the technological level of a company is the Technological Situation and Adequacy Index, TASI, developed by Torres [15], based on the International System of Universal Categories, SICU. It is a methodology for analyzing complex systems based on systems theory that explains the structure and operation of systems through the materials flow.

The evaluation of the technological level of the Ecuadorian textile sector was developed through the application of the Company Technological Situation and Adequacy Index (TASI2) model, proposed by Arzola [16], which quantitatively defines each of the variables considered in the partial indicators which weighting of 1–100% is applied, depending on the number of variables, degree of application, and importance.

TASI2 method arising from the lessons learned of its application in the academic and industrial field of researchers from the University of Zaragoza, as an adaptation to the characteristics of the technological environment, modifying the content of some indicators and adding others that include fundamental aspects in the management of the process of technological innovation in companies, given the current competitive dynamics, such as comprehensive information systems, the quality system, innovative product, organizational processes and actions, management of technical knowledge, strategic management of research and development activities, and innovation, among others.

55.2.2 Textile Companies

The basis for this study corresponds to a representative sample of textile companies in Ecuador. Through online surveys, the current status of the Ecuadorian textile industry was diagnosed and strategies on types of agglomeration and competitiveness were proposed [4, 17] for the post-pandemic stage. An updated diagnosis of the capabilities and levels of technological innovation of the Ecuadorian textile industry was obtained. The technological level was evaluated using the TASI2 method [16, 18]. The survey was organized into categories: general company data, management, and administration, economic and commercial data, human resources, innovation, and size (large, medium, and small). The objective of the surveys is to have a primary source of information, which allows for the validation of the information.

55.2.3 Limitations

To have a primary source, surveys were sent to 16 Ecuadorian textile companies in the highlands. The delay in responses was one of the limitations. Due to the type of variables involved in the study, part of the information was obtained from secondary sources: public, private, international organizations, and scientific journals. This work suggests additional lines of research such as technological challenges of the Ecuadorian textile industry, technological innovation through business clusters, productivity, and competitiveness of the Ecuadorian textile industry in the post-COVID-19 era.

55.3 Results

The textile sector from the year 2000 presented a significant development in the country's economy due to the adaptation of the sector to the new dollarized monetary scheme, allowing it to contribute with employment, capital, and innovation [19]. That year, there was an increase in exports of 8.14% increase over the year before, which marks a trend that began to be normal during this new millennium. Only in 2002, there was a decrease in textile exports, quickly recovered in the following two years, reaching exporting close to 90 million dollars in 2004, surpassing the highest peak in the last 10 years (1997–82 million dollars exported).

For the period 2006–2015, the textile sector contributed an average of 0.41% to the national PIB, indicating the importance it has in the evolution of the Ecuadorian economy since the value-added of this sector has the same behavior as the national PIB. However, in 2011, the PIB was affected due to the contraction of production and the economic situation of the country, the world recession, and the generalized growth of unemployment reducing national consumption [20].

Textile PIB only exceeded the percentage of growth of manufacturing PIB in 2007 with a positive variation rate of 4.91%, which indicates that even though the sector contributes significantly to the country's manufacturing, it presents a potential gap in your demand. One way to generate sustainable growth in the sector is to group in clusters and thus boost exports and reduce dependence on imports of yarns, clothing, and raw materials, among others.

The value-added of the national production of the textile industry represents between 0.85 and 1.04% of the total value of PIB as illustrated in Table 55.1. In 2014, about 607 thousand dollars or 0.86% out of 70 million, the dollar of total value-added of the National industry, represent the textile industry.

For the year 2012, the production of the textile sector decreased by 25.6% compared to last year. This is attributed to the increase in the prices of imported raw materials and the scarce-trained labor force. In addition, the entry of foreign companies into the local market, bringing with them a portfolio of products imported from China, the United States of America or Spain, at a lower cost than those produced locally, affected Ecuadorian production, which was put at a competitive disadvantage [21].

On a global level, during the period 2006–2016, the Ecuadorian trade balance for textile and clothing products showed a deficit trend, since imports exceeded exports. Despite the measures desired by the government in defense of national production,

Table 55.1 Textile exports in millions of dollars

Year	Tons	Total Millions of dollars
2008	18.708,95	150.70
2009	17.310,10	169.88
2010	21.127,37	220.59
2011	24.881,68	191.29
2012	26.478,90	143.21
2013	27.683,63	146.22
2014	28.229,06	145.25
2015	28.774,49	142,85
Total	193.194,18	1.309,99

Source National Information System and AITE [23]

the textile sector is still facing competition from imported products and internationally has the challenge of entering new markets, which requires more competitive products [22]. Clustering contributes to textile companies to march together toward a competitive economy after COVID 19.

During 2016, textile imports were $608,249, which corresponds to 82.41% more than exports, which were $131,931, which registered a trade deficit of $486,318. Similarly, 2015 recorded a steadily rising negative trade balance of $753,754. However, in 2014, he has presented a greater deficit within the study period, and these being 835,176 dollars since imports were 1,016,580 dollars, 83.52% more than exports [23].

Society is constantly evolving and strongly demand-driven by innovations in technology. The textile industry behaves similarly, adapting to changes in the market and consumers. It was considered appropriate to analyze the trends that will affect the Ecuadorian textile industry in the coming years and how technological advances will affect the traditional patterns of the industry to be more competitive in the post-pandemic period.

Ecuadorian textile companies sell their products mainly in the local market. This occurs mainly due to factors such as obsolete technology, low level of sales due to high competition with low-priced imported products, low competitiveness, low levels of innovation, not opening to markets international, among others. However, it is important that given the nature of the sector, its export potential is feasible, which would be reflected in an increase in sales to international markets, and therefore, higher income for the textile sector and the country.

The bibliographic review on technological innovation, studied with different approaches, among others by [4, 16, 17], that promotes the success of business clusters, through geographical location, among others, in competitiveness.

Some researchers consider that it is not correct that each company within a cluster benefits equally [12], although others have argued that the differences are a reason for the study. Martin and Sunley [11] showed that the impact can be positive as groupings, while in each company it causes a heterogeneous impact.

Regarding the role of territory and proximity in improving the competitiveness of companies [24, 25] to study their advantages and disadvantages, it should be taken into account that in most of the In EU countries, the manufacturing industry tends to be geographically concentrated in industrial conglomerates [26].

Cluster-based textile companies can capture changes in customer needs faster than non-clustered companies. The groups could also accelerate the uptake and implementation of new technologies, production systems, and marketing methods.

After the pandemic caused by COVID-19, the clustering of the textile company is expected to lead to the sharing of resources for innovation and related experiences, achieving a positive impact on the competitiveness of the textile company. In other words, in emerging economies, clustering will have a direct effect on competitiveness and an indirect effect on competitiveness through innovation. Consequently, the actions of the Ecuadorian textile company for the post-pandemic stage should be focused based on the following hypotheses.

Hypothesis 1: Textile companies that are organized in clusters will be more competitive than those that are not.

Hypothesis 2: The textile companies that are organized in clusters will be more technologically innovative than those that are not.

Clustering has a positive direct impact on the competitiveness and innovation of textile companies, and there is a positive effect of innovation on competitiveness. The textile company will actively participate in agglomerations, and the implementation of the innovations will lead to higher levels of competitiveness. Public policies should promote the creation of strategic collaboration networks in textile companies from emerging economies such as Ecuador.

Ecuadorian textile companies contribute enormously to the growth of the manufacturing sector. Exports of articles related to this industry have shown significant growth in recent years; however, they face the challenge of competing inside and outside the country with articles of foreign origin, particularly those of Chinese origin. Improving competitiveness through the location is the main challenge for the textile company after the pandemic, doing so will allow entry and positioning in foreign markets, incentivize production, and generate a greater number of jobs.

The cluster in Atuntaqui–Ecuador arose to seek an improvement in the competitiveness of companies through the promotion of grouping and productive integration [27]. The project includes the development of clusters and production chains aimed at micro, small, and medium-sized enterprises with the potential to contribute to the economic development of the regions where they are located.

Atuntaqui was promoted by investors, national and international entities, and public and private companies since the dollarization in Ecuador in 2000 when Ecuadorian companies faced great challenges to compete locally and globally. This cluster was selected for the "Development of Competitiveness through cluster formation" for the following reasons:

(1) Entrepreneurs were aware of the need to innovate and improve their performance to confront the international competition, mainly from China, in addition to the loss of markets due to non-competitive prices as the Ecuadorian economy dollarized.
(2) Atuntaqui is a part of sectors that can supply a large part of the local market and generate great advantages for national companies.
(3) The textile apparel industry represents practically 70% of the canton's employment, and therefore, an improvement in the industry leads to improvements for the population.

Based on the experience of the year 2000, it is expected that after the pandemic, other clusters will be established that will contribute to the strengthening and competitiveness of the Ecuadorian textile company.

55.4 Discussion of Results

Results were obtained on the levels of technological innovation and its relationship with clustering and competitiveness, which is intended to be useful to overcome the challenges of the Ecuadorian textile industry in the post-pandemic period.

Aware that the development of the sector is directly related to exports, textile industrialists must be agglomerated in a cluster in such a way that allows them to be more competitive in the face of a post-pandemic economy. Logically, this effort to be competitive must be shared.

It was determined that textile companies should work from an organizational perspective of strategic planning that considers the aspects of location, competitiveness, and efficiency, based on a model of organization in clusters, which allows them to join efforts and achieve in the medium term joint processes of innovation, quality, and competitiveness that leverage their permanence in the market.

There is no doubt that, in the future, medium and large companies will have to face technological challenges. Also, some small and few companies dedicated to low-quality products may join, however, only those that can get ahead with their means will coexist (not only financial but also technical and human resources) to compete and face the world market. In the survival of the Ecuadorian textile industry, it is not the price that will determine the competitiveness, but it will be the cluster organization model, in which large companies can subsidize small ones with a well-planned state policy.

Based on this study, it is determined that the textile industry should focus on the following aspects:

(1) Generation of greater public and private investment, internal and external investment, the flexibility of financial conditions in banking to allow development, research, and innovation of companies. (2) Adopt as an organization model the clusters with Ecuadorian textile industries to attract foreign markets as has been implemented in other countries with excellent results. This type of agglomeration will

allow the development of common projects with the support of public and private banks for research, development, and innovation activities. (3) Ecuadorian government support for the Ecuadorian textile industry as a state policy. (4) Coordination between the small, medium, and large textile industries, the state, and the private banks for access to sources of financing.

An open debate remains in the authorized literature whether or not the benefits of the cluster are contextually independent. McDonald et al. [27] argued that the effects of forming groups can be detected by comparing companies from different territorial areas within the same sector. However, the effects are neither always positive nor have the same intensity for all companies that share the same location within the company studied. In the case of the Ecuadorian textile company, everything leads to the fact that certainly not all companies benefit in the same way, as factors such as age, size, export, and import levels, external financing, and GDP influence factors. However, it is necessarily the model to implement for the post-pandemic stage.

55.5 Conclusions

In this work, the technological challenges of the Ecuadorian textile company for the post-pandemic stage were studied. This development was based on theoretical assumptions of technological innovation and its effect on competitiveness. Despite the extensive literature that links technological innovation with competitiveness in different contexts, this research focused on the role of clusters.

A preliminary result suggests that the Ecuadorian textile industry should be agglomerated in a cluster of targeted subsectors. The capital destined for the purchase of machinery, expansion of plants and innovation, will be generators of direct employment and catalysts of indirect employment. Technological innovation together with the injection of capital influences the competitiveness of the textile product. This will impact the level of sales, entry into new markets, and the exchange in textile trade with other countries.

It requires the training of human resources necessary to implement technological innovation for the addition of value in the field of design, machine programming, preparation, and spinning of textile fibers, weaving, finishing of textile and clothing products, as well as manufacturing of articles made of textile materials, finishing of garments, among others. In this sense, the academic offer should be increased in the coming years to strengthen the aforementioned areas. Until now, analyzes have converged on the convenience of creating specific production clusters or poles not to reduce common costs and share technology, but to improve quality and efficiency in the post-pandemic period.

It is not simply a matter of companies acquiring and coupling new technologies to their production. Once grouped in clusters, research and development areas must be created that allows them to advance in the matter of strategies for the improvement of productive processes and innovations that allow progress in business development, productivity, and competitiveness issues [28].

This work is essential for the Ecuadorian State based on the public policies to be implemented to have a more competitive textile company; for the Association of Textile Industries of Ecuador, for all its goals set for the next few years, including the clustering of the textile sector and for the academy. Since from this work, it could be possible to obtain conclusions about how the location influences the levels of innovation and resilience of the textile industry.

Currently, companies operate in highly dynamic environments, causing them to be exposed to several adverse situations, such as the COVID 19 pandemic, that alter their level of operation. In many cases, companies cannot guarantee a quick, effective, and efficient reaction to unexpected situations, which is why they need tools that allow them to prepare for the post-pandemic period.

References

1. M. Ortega-Vivanco, Efectos del COVID-19 en el comportamiento del consumidor: Caso Ecuador. Rev. Retos. **10**(20) (2020)
2. C. Molina, Análisis de los determinantes de la competitividad de la industria manufacturera en la Frontera Norte: 1994, 1999 y 2004. 2010, 104 (2010)
3. O. Rodríguez, F. Puig, M. Gonzalez, Clustering, innovation, and hotel competitiveness: evidence from the Colombia destination. J. Int. J. Contemp. Hosp. Manag. (2017)
4. J. Batanero Rodríguez, Concentración y Productividad en la Industria Textil: Comparativa España vs Italia. [Trabajo Fin de Grado] (2019)
5. P. López, P. Rodríguez, Tla-Melaua, El liderazgo de los países asiáticos en el sector del vestido: repercusiones para América Latina año 10, No 40, pp. 152–175 (2016)
6. M. Bellandi, Italian industrial district: an industrial economics interpretation. Eur. Plan. Stud. **10**(4), 425–437 (2002)
7. F. Sforzi, The industrial district and the 'New' Italian economic geography. Eur. Plan. Stud. **10**(4), 439–447 (2002)
8. M. Porter, *On Competition* (Harvard Business School Press, Boston, 1998)
9. M. Porter, Location, competition, and economic development: local clusters in a global economy . Econ. Dev. Q. **14**(1), 15–34 (2000)
10. A. Malmberg, P. Maskell, The elusive concept of localization economies: towards a knowledge based theory of special clustering'. Environ. Plan A **34**, 429–449 (2002)
11. R. Martin, P. Sunley, Deconstructing clusters: chaotic concept or policy panacea? J. Econ. Geogr. **3**(1), 5–35 (2003)
12. A. Potter, H.D. Watts, Evolutionary agglomeration theory: increasing returns, diminishing returns, and the industry life cycle. J. Econ. Geogr.. **11**(3), 417–455 (2011)
13. K. Schwab, The Global Competitiveness Report 2015–2016, Foro Económico Mundial. Disponible en (2016).http://www3.weforum.org/docs/gcr/20152016/Global_Competitiveness_Report_2015-2016.pdf
14. J. Cabrera, M. López, Análisis de la Competitividad de las Exportaciones de Honduras utilizando la metodología can. Economía y Administración (E&A) **5**(2), 105–125 (2015). Disponible en: http://iies-unah.org/Revista/index.php/EyA/article/view/281
15. F. Torres, Análisis y diseño funcional de sistemas de fabricación. Automatización, pp. 65–72 (1992)
16. M. Arzola, Métodos para evaluar el proceso de innovacion tecnológica, vol. 5, nro 17, pp. 15 (2001)
17. F. Puig, H. Marques, Effect in the textile industry. Eur. Plan. Stud. **19**(8), 1424–1437 (2011)

18. M. Arzola, M. De Armas, Evaluación del nivel tecnológico del sector metalmecánico del estado Bolivar, **6**(23), 135 (2002)
19. ProEcuador, Análisis Sectorial de textiles y confecciones. Recuperado de (2012). http://www.proecuador.gob.ec/wpcontent/uploads/2013/11/PROEC_AS2012_TEXTILES.pdf
20. AITE Asociación de Industriales Estadísticas. recuperado de: http://www.ahite.com.ec/descargas/category/52-estadisticasdiciembre-2014.html recuperado de: https://www.bce.fin.ec/index.php/component/k2/item/756 (2014)
21. C. Espinoza, R. Sorhegui, Rafael Antonio Sorhegui Ortega. Revista Científica ECOCIENCIA 1–28 (2016)
22. H. Torre, Análisis de Comportamiento y Oportunidades del sector Sistema Moda (SURA, Bogotá, 2014)
23. AITE Asociación Latinoaméricana de Integación. ALADI. Recuperado el 2017, de Sistema de información de Comercio Exterior: http://consultawebv2.aladi.org/sicoexV2/jsf/comercio_exterior_Entrada.seam?cid=24876 (2017)
24. G. Becattini, G. Dei Ottati, The performance of Italian industrial districts and large enterprise areas in the 1990s. Eur. Plan. Stud. **14**(8), 1139–1162 (2006)
25. R. Demartino, D. Mchardy Reid, S.C. Zygliodopoulos, Balancing localization and globalization: exploring the impact of firm internationalization on a regional cluster. Entrepreneurship Reg. Dev. **18**(1), 1–24 (2006)
26. S. Goglio, Introduction: the industrial district as a proving ground. Eur. Plan. Stud. **10**(4), 421–424 (2002)
27. F. McDonald, Q. Huang, D. Tsagdis, H.J. Tüselmann, Is there evidence to support portertype cluster policies? Reg. Stud. **41**(1), 39–49 (2007)
28. C. Arias, Estrategias empresariales para afrontar la competencia internacional: Caso del clúster de confecciones textiles de Atuntaqui (Univ. Andina Simón Bolivar, Ecuador, 2015)

Chapter 56
Optimization of the Economic Dispatch of a Hybrid Renewable Energy System Using CPLEX

Ronald Vallejo Duarte and Juan Lata-García

Abstract Currently the uncertainty of renewable energy and demand response has become one of the main problems in the shipments of hybrid energy systems. To optimize dispatch, different probability distribution functions and the load model with which it is associated are usually established. The different hypotheses on how to obtain the optimal solution for the energy system. For this problem, this article proposes to optimize the economic cost of a hybrid energy system using the GAMS computer tool. As a first approach, we have a power plant (thermal/photovoltaic) which has a total cost that represents a high economic impact. In second place, we have the application of the CPLEX solver whose main objective is to carry out an improvement to the plant, considering the capacity of generators and the respective restrictions within the objective function. Third, we emphasize the importance of improving an energy system, given the relevance that these have over the years. We emphasize the importance of improving an energy system, given the relevance that these have over the years. The applied algorithm analysis and resolution methodology allowed to improve economically through simulations, concluding with a result of Total Cost (TC) = 5.4037×10^4, which verifies the validity of the decision making used for this problem.

56.1 Introduction

Currently, energy systems have a high impact on industries, cities, countries, which has led to the generation of various energy systems to improve and increase power production. Renewable energy is the most optimal option in several of these cases, because it has several positive implications on the impact on the environment,

R. V. Duarte · J. Lata-García (✉)
Universidad Politécnica Salesiana, Robles 107 Chambers, Guayaquil, Ecuador
e-mail: jlatag@ups.edu.ec

R. V. Duarte
e-mail: rvallejod@est.ups.edu.ec

Á. Rocha et al. (eds.), *Communication, Smart Technologies and Innovation for Society*, Smart Innovation, Systems and Technologies 252,
https://doi.org/10.1007/978-981-16-4126-8_56

reducing CO_2 production, several of these scenarios are possible thanks to the help of alternative energy sources [1, 2].

Power systems always try to find a suitable process that reduces the economic impact. Economic planning is complex, affecting financial, tariff, social, business, and environmental aspects. The electrical system process is a task that requires correct decision making that begins in the long term and takes place in the short term [3]. These decisions are based on highly complex resolution calculations, in general terms the economic output of loads from an energy system whose primary job is to reduce the cost of energy production [4].

56.2 State of the Art

Abdelkader [5], I develop a methodology to optimize the size of a hybrid photovoltaic/wind system with an energy reserve system (battery—supercapacitor) developed based on a discrete Fourier transformation algorithm (DFT). The results obtained demonstrate that the PV/Wind turbine design selected by the proposed algorithm is better than any other case, with a total cost of electricity (TCE) of $ 23.856 compared to $ 30,414.

Acuña and others [6] also presented a novel methodology to optimize the size of an autonomous photovoltaic/wind hybrid system using battery storage using a multi-objective function, considering this study and a reliability indicator called maximum energy expected not supplied (MENS). The non-mastered algorithm II (NSGA-II) has been used to find the best system solution economically and reliably; they have evaluated the performance of a hybrid photovoltaic/wind system with a battery bank in the Dhahran KSA region. They have taken into consideration economic aspects with the cost of electricity (EC), the percentage of excess energy (EEP), and the operating life cycle (OLC).

Mandal and others [7], have proposed a study of renewable energy available in non-electrified areas, to meet the demand through a photovoltaic/wind system with a diesel generator and battery bank. For this evaluation, six different studies of hybrid systems are considered, and they have selected the most optimal configuration based on emissions of (TCE) and pollutants. Based on the methodology they used, they concluded that the photovoltaic generator/batteries/diesel is the optimal configuration for the studied site.

Robert and others [8] have proposed a multi-objective optimization method, and the method raises two main objectives: the (NPC) and the (LPSP). Furthermore, this method has been used to design an EPS for a community located in the Amazon region of Brazil. The results show a confirmation with a high level of reliability that implied a higher economic cost. However, the proposed configurations gave as feasible robust solutions and guarantee a reliable generation source [9].

Marco Ledesma and Santiago Espinosa [10], developed an optimization plan using GAMS, which allows minimizing investment, operation, administration, and maintenance costs associated with the generation plants of different technologies.

Why is it important to optimize an economical energy dispatch? To reduce system operating costs, minimize resources, and establish greater profitability in your operation, the main reason for developing this article is the optimization of the economic dispatch of hybrid energy (thermal/photovoltaic), as well as their respective configurations [11]. The criterion taken into consideration is the techno-economic cost of operating the system. Previously, there was an energy system with a high operating cost, this excess is given from a bad dimensioning, and to minimize the total cost (TC), an algorithm inspired by the nature of the problem has been used. This concludes in the use of GAMS, which is widely used to solve various mathematical models efficiently and profitably. The methodology carried out allows to determine the optimal configuration of the system, minimizing the objective function for the proposed problem. Said optimization makes it possible to obtain the most suitable solution to meet the restrictions given by the hybrid system. The simulation carried out results in a reduction in the operating cost, including the values that must be modified to arrive at the most appropriate operation.

56.3 Problem Optimization Formulation

This study presents a hybrid power generation system based on techno-economic analysis; therefore, the configuration of the system comprises four thermal generators and a photovoltaic panel which stores the energy in a battery, these together feed the main network, the optimization approach proposed for this system can be used for any other approach of the same type, the mathematical resolution was proposed in GAMS.

This has different optimizers that can be chosen at the time of resolution, and for this study, we will take the CPLEX resolution that is the most suitable for the analysis of economic dispatches using hybrid energy systems [12, 13], the configuration scheme of the plants are outlined in Fig. 56.1. For a greater understanding of the approach, it has been proposed.

56.3.1 Table of Variables of the Thermal Power Plant

The hybrid system has four thermal generators, which represent costs in twenty-four hours, thus the generation values were expressed in different units as expressed as follows.

Table 56.1 represents the values taken for the simulation in the GAMS software, taking them as base values to perform the optimization in the objective function 1 [14]. The decrease in total cost (TC) is the objective function because it is attached to the respective restrictions that must be met in the resolution process. With the present data, the system has a high operating cost, which represents a high economic impact;

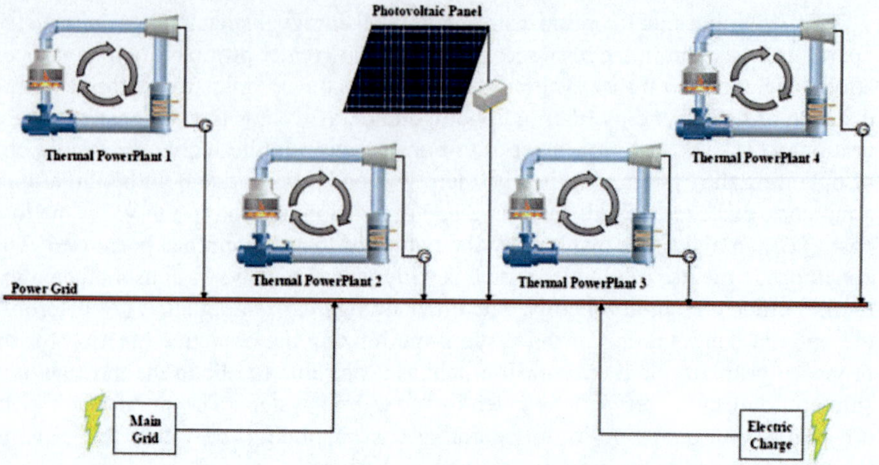

Fig. 56.1 Hybrid system configuration diagram (thermal/photovoltaic)

Table 56.1 Thermal power plant cost/power ratio

Unit	a_g (\$/MW2)	b_g (\$/MW)	c_g (\$)
Gen1	0.12	14.80	89
Gen2	0.17	16.57	83
Gen3	0.15	15.55	100
Gen4	0.19	16.21	70

the modification and adaptation of the proposed system is necessary to adjust the operating price, resulting in a timely use of resources.

56.3.2 *Analysis of the Objective Equation*

This section presents the different equations that will be the basis for solving the optimization, since they are the parameters and limits that the solution must have for it to be optimal.

Equation (56.1) is defined as being the multi-objective function to minimize, in this each of the restrictions, thermal units, and photovoltaic panel that are presented as problematic are represented. This comprises four thermal generators, which contain different emission factors, these factors are given by the execution time of the power plant.

Before this expression is defined in GAMS, there is a set of data, costs, and variables that must be exposed before proceeding to the resolution of Eq. (56.1). Because it presents different restrictions that must be expressed beforehand for their respective simulation.

$$\min_{P_{g,t},\mathrm{SOC}_t,P_t^d,P_t^c} \quad \mathrm{TC} = \sum_{g,t} a_g P_{g,t}^2 + b_g P_{g,t} + c_g \tag{56.1}$$

where $P_{g,t}$ is the power generated by the thermal unit g at time t (MW); SOC_t is the state of charge at time t (MW); P_t^d represents the emission generated by the thermal unit d at time t (MW); P_t^c stands for the emission generated by thermal unit c at time t (MW); a_g, b_g, c_g are the cost coefficients of thermal units ($); and TC represents the total cost of operations ($).

56.3.3 Restrictions

The simulation process also considers the parameters with their respective equivalences (Table 56.2).

The restrictions are vital to lead to an adequate optimization of any hybrid energy system, for which we have to apply them to the objective function, enforcing the operating limits of the system, taking into account these considerations, the equations to be fulfilled are presented as part of the solution [15].

$$P_g^{\min} \le P_{g,t} \le P_g^{\max} \tag{56.2}$$

$$P_{\min}^c \le P_t^c \le P_{\max}^c \tag{56.3}$$

$$P_{\min}^d \le P_t^d \le P_{\max}^d \tag{56.4}$$

$$\mathrm{SOC}_{\min} \le \mathrm{SOC}_t \le \mathrm{SOC}_{\max} \tag{56.5}$$

The values of the times taken into consideration for the resolution have been projected in Fig. 56.2 showing the load values for the twenty-four hours.

In Fig. 56.2, the relationship between power (kW) and time (hours) has been graphically represented, these data are obtained through the autochthonous repository that will help us as the residential load that will be used for the simulation, and these

Table 56.2 Simulation data for the hybrid power system

Parameter	Value
SOC_0	100 MW
$\mathrm{SOC}_{\max}$	300 MW
P^d max	0.2 $\mathrm{SOC}_{\max}$
P^d min	0
P^c max	0.2 $\mathrm{SOC}_{\max}$
P^c min	0

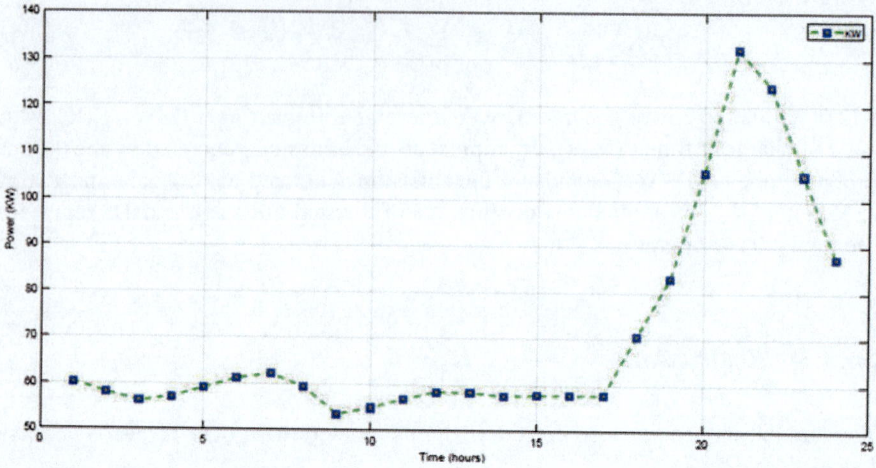

Fig. 56.2 System electrical load

are expressed in Eq. (56.1) of the document, which is the objective function to minimize.

The photovoltaic plant displays these data during a full day, so we can determine a limitation of its operation [16], which would be that its minimum power per day is not less than 50, and in hours its maximum peak is no greater than 140. With this, we can determine that its highest demand time is at 20 h a day, having a load of 132.30 kW.

Before proceeding with the simulations, the restrictions mentioned in Table 56.1 must be considered, since they are the limits established to execute the total cost minimization (TC) and not prejudice the solution of the objective-proposed function.

56.3.4 Algorithm Pseudocode

Next, the proposed solution scheme is described in a general way to address the problem by means of simulation and application of algorithms [17, 18]. To proceed to the resolution of the economic dispatch of the hybrid energy system, the development process in GAMS is presented through a flow diagram in Fig. 56.3.

A previous analysis of the system is carried out, using a database that contains the system, in which the high cost of production is confirmed. Requiring an improvement to reduce the economic impact, and significantly improve the profitability of its operation.

Once the problem is confirmed, the data is sorted through an indigenous repository (Excel), which through commands will be called into the program and the respective values entered automatically, emphasizing that the values can be modified at any time.

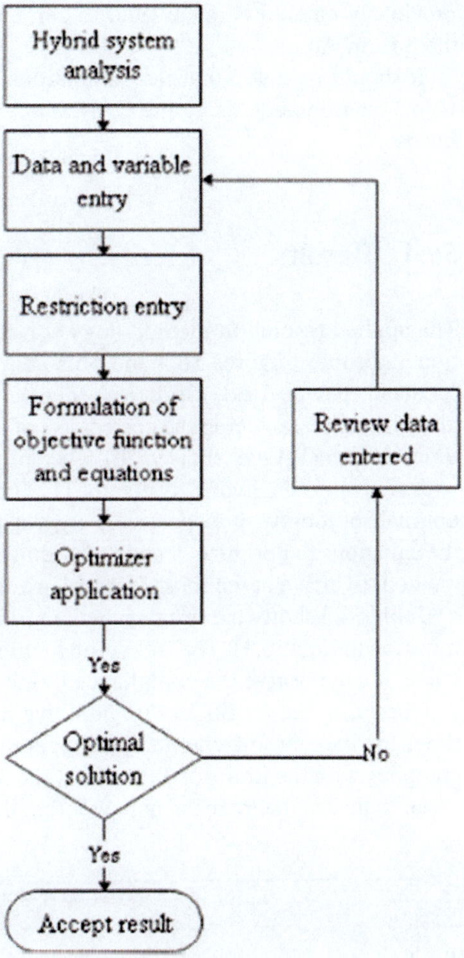

Fig. 56.3 Resolution methodology diagram

Before proceeding to the formulation of the objective function and equations, the operating limits of the system are considered, as well as the parameters these can be examined in Table 56.1, these data were also provided from the database.

The objective function is expressed from the formulated conditions, since it is the main basis for optimization as well as the equations, this can be denoted in the equations from (56.1) to (56.5). That must be expressed in an appropriate way in the software to avoid possible errors at the time of execution.

Once all the previous steps have been coordinated, the CPLEX optimizer is used through the solver command that allows to use the most efficient algorithm to carry out the different simulations, the main objective of which will be to find the most suitable option for the hybrid system.

Finally, we conclude with the final simulation that leads to the choice of the optimal result that must adhere to the restrictions set out above, otherwise the data previously entered is revalidated to correct possible errors at the time of inputting the information.

It should be noted that the simulations were carried out under the Windows 10 Home environment, using the GAMS software, which does not have a professional license.

56.4 Results

The applied resolution methodology solved efficiently through the CPLEX resolution algorithm. Figures 56.4 and 56.5 show the improvements it had in generation (demand, power, load, discharge), not only a cost reduction is guaranteed, but it also allows guaranteeing the correct use of resources, and their dynamic growth over time of the energy system [19, 20]. The optimization result presented by the software with respect to the hybrid system is $TC = 5.4037 \times 10^4$, which represents the most optimal solution with respect to energy dispatch, the results were ordered based on the equations to optimize described within the program using the CPLEX resolution method, to arrive at the most favorable result in the hybrid energy system.

Table 56.3 shows the results which must be applied to be able to perform the minimization in Eq. (56.1). The following figures graphically represent the data obtained, where it can observe the compliance of the data with the proposed constraints.

Observing the results in Fig. 56.4, we have the relationship between power and time, for the state of charge (SOCt) represented by the red color; we see that it complies with the limits of Table 56.1, the demand that is represented with the color green indicates the increase in power that the system has with the optimization with

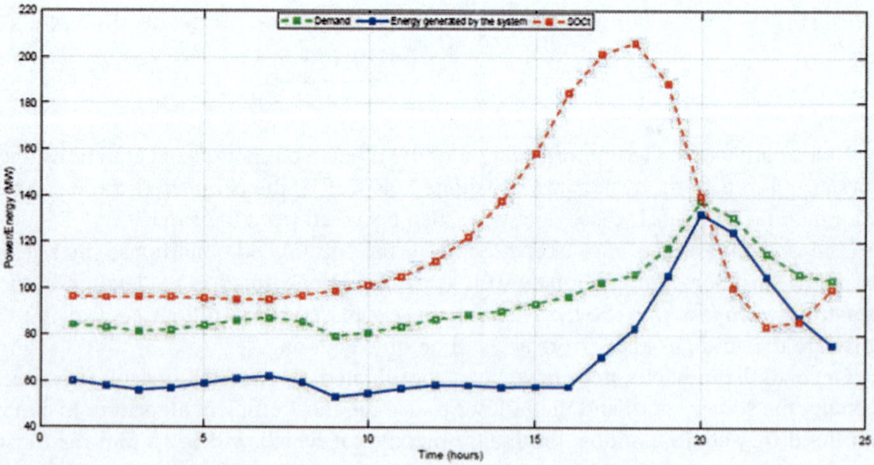

Fig. 56.4 Generation ratio over time

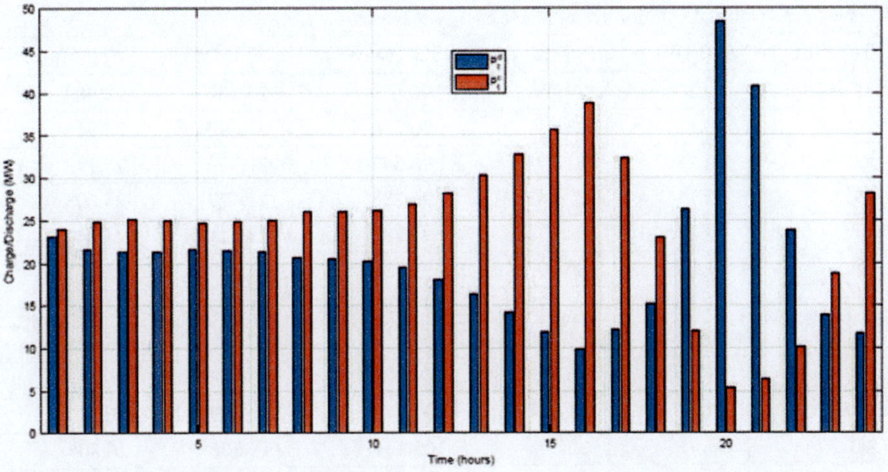

Fig. 56.5 Charge and discharge ratio in time

respect to the previous one, and finally, we have the total energy of the system that is represented by the blue color concluding with a notable improvement of the system.

In Fig. 56.5, it is observed that it is coupled to the restrictions named in Table 56.1, which indicate that the minimum emission status is 0 and the maximum is 60, complying with the operating conditions of the plant. It is denoted that the emissions were also reduced with the optimization, with respect to the previous one, which generates a minimization in the TC studied.

56.5 Conclusions

It was shown that the CPLEX solver identifies the problem through commands, which implies a simulation process prior to resolution. Unlike other algorithms, this one presents not several results, but the result in which the system must be adjusted to improve its total cost of operation.

It was concluded in a notable improvement of the economic dispatch studied considering all the restrictions for the thermal/photovoltaic system, and previously, there was a high economic impact, after having entered the various equations and variables, the program determined an improvement of $TC = 5.4037 \times 10^4$, highlighting the optimal values for the operation of the power plant.

Finally, CPLEX demonstrates superiority over several algorithms such as JAYA, ALO, GWO, and KH which provide several results in the resolution, but the algorithm used for the economic dispatch problem shows us a single result that indicates to be the most optimal solution.

Table 56.3 Optimal solution for efficient energy dispatch

Time (h)	Lt (MW)	$P_{g,t}$ (MW)	SOC_t(MWh)	P_t^d(MW)	P_t^c(MW)
$t1$	60.3	84.289	96.659	23.157	23.989
$t2$	58.12	83.037	96.230	21.690	24.917
$t3$	56.25	81.373	96.343	21.379	25.123
$t4$	57.14	82.143	96.288	21.427	25.003
$t5$	59.13	83.874	95.781	21.621	24.744
$t6$	61.23	86.003	95.336	21.582	24.773
$t7$	62.24	87.272	95.350	21.389	25.032
$t8$	59.5	85.548	97.204	20.602	26.048
$t9$	53.5	79.523	99.102	20.542	26.023
$t10$	54.9	81.027	101.542	20.219	26.127
$t11$	56.9	83.705	105.334	19.506	26.805
$t12$	58.45	86.614	111.928	18.145	28.164
$t13$	58.35	88.613	122.598	16.271	30.263
$t14$	57.65	90.375	137.892	14.215	32.725
$t15$	57.85	93.552	158.620	11.870	35.702
$t16$	57.85	96.685	184.546	9.871	38.835
$t17$	70.65	102.915	201.643	12.199	32.265
$t18$	83.15	106.176	206.656	15.176	23.026
$t19$	105.8	117.78	188.818	26.298	11.980
$t20$	132.306	137.773	140.141	48.483	5.467
$t21$	124.3	130.72	100.794	40.901	6.420
$t22$	105.2	115.362	83.989	23.812	10.162
$t23$	87.5	106.236	86.289	13.950	18.736
$t24$	75.6	103.706	100.00	11.691	28.106

References

1. M. Ikhsan, A. Purwadi, N. Hariyanto et al., Study of renewable energy sources capacity and loading using data logger for sizing of solar-wind hybrid power system. Proc. Technol. **11**, 1048–1053 (2013). https://doi.org/10.1016/j.protcy.2013.12.293
2. T.E. Boukelia, M.S. Mecibah, B.N. Kumar, K.S. Reddy, Investigation of solar parabolic trough power plants with and without integrated TES (thermal energy storage) and FBS (fuel backup system) using thermic oil and solar salt. Energy **88**, 292–303 (2015). https://doi.org/10.1016/j.energy.2015.05.038
3. S. Ferrari, F. Zagarella, P. Caputo, M. Bonomolo, Assessment of tools for urban energy planning. Energy **176**, 544–551 (2019). https://doi.org/10.1016/j.energy.2019.04.054
4. A. Kaabeche, Y. Bakelli, Renewable hybrid system size optimization considering various electrochemical energy storage technologies. Energy Convers. Manag. **193**, 162–175 (2019). https://doi.org/10.1016/j.enconman.2019.04.064

5. L.G. Acuña, M. Lake, R.V. Padilla et al., Modelling autonomous hybrid photovoltaic-wind energy systems under a new reliability approach. Energy Convers. Manag. **172**, 357–369 (2018). https://doi.org/10.1016/j.enconman.2018.07.025
6. S. Mandal, B.K. Das, N. Hoque, Optimum sizing of a stand-alone hybrid energy system for rural electrification in Bangladesh. J. Clean. Prod. **200**, 12–27 (2018). https://doi.org/10.1016/j.jclepro.2018.07.257
7. G. Prinsloo, A. Mammoli, R. Dobson, Discrete cogeneration optimization with storage capacity decision support for dynamic hybrid solar combined heat and power systems in isolated rural villages. Energy **116**, 1051–1064 (2016). https://doi.org/10.1016/j.energy.2016.10.026
8. L.K. Tartibu, B. Sun, M.A.E. Kaunda, Multi-objective optimization of the stack of a thermoacoustic engine using GAMS. Appl. Soft. Comput. **28**, 30–43 (2015). https://doi.org/10.1016/j.asoc.2014.11.055
9. P. Miao, Z. Yue, T. Niu et al., Optimal emission management of photovoltaic and wind generation based energy hub system using compromise programming. J. Clean. Prod., 124333 (2020). https://doi.org/10.1016/j.jclepro.2020.124333
10. P. Skworcow, D. Paluszczyszyn, B. Ulanicki et al., Optimisation of pump and valve schedules in complex large-scale water distribution systems using GAMS modelling language. Proc. Eng. **70**, 1566–1574 (2014). https://doi.org/10.1016/j.proeng.2014.02.173
11. D. Álvaro, R. Arranz, J.A. Aguado, Sizing and operation of hybrid energy storage systems to perform ramp-rate control in PV power plants. Int. J. Electr. Power Energy Syst. **107**, 589–596 (2019). https://doi.org/10.1016/j.ijepes.2018.12.009
12. M. Petkov, D. Markova, S. Platikanov, Modeling of electrical characteristics of photovoltaic power supply sources. IFAC Proc. **39**, 183–188 (2006). https://doi.org/10.3182/20061002-4-BG-4905.00031
13. M. Vidal, M. Martín, Optimal coupling of a biomass based polygeneration system with a concentrated solar power facility for the constant production of electricity over a year. Comput. Chem. Eng. **72**, 273–283 (2015). https://doi.org/10.1016/j.compchemeng.2013.11.006
14. A. Al-Sharafi, B.S. Yilbas, A.Z. Sahin, T. Ayar, Performance assessment of hybrid power generation systems: Economic and environmental impacts. Energy Convers. Manag. **132**, 418–431 (2017). https://doi.org/10.1016/j.enconman.2016.11.047
15. A. Soroudi, *Alireza Soroudi Power System Optimization Modeling in GAMS* (2019)
16. A. Naghiloo, M. Abbaspour, B. Mohammadi-Ivatloo, K. Bakhtari, GAMS based approach for optimal design and sizing of a pressure retarded osmosis power plant in Bahmanshir river of Iran. Renew. Sustain. Energy Rev. **52**, 1559–1565 (2015). https://doi.org/10.1016/j.rser.2015.08.018
17. M.G. Prina, M. Cozzini, G. Garegnani et al., Multi-objective optimization algorithm coupled to EnergyPLAN software: the EPLANopt model. Energy **149**, 213–221 (2018). https://doi.org/10.1016/j.energy.2018.02.050
18. S. Mehta, P. Agarwal, P. Shrivastava, J. Barlawala, Differential bond energy algorithm for optimal vertical fragmentation of distributed databases. J. King Saud Univ. Comput. Inf. Sci. (2018). https://doi.org/10.1016/j.jksuci.2018.09.020
19. M. Ćalasan, K. Kecojević, O. Lukačević, Z.M. Ali (2021) Testing of influence of SVC and energy storage device's location on power system using GAMS, in *Uncertainties in Modern Power Systems*, ed. by A.F. Zobaa, S.H.E. Abdel Aleem (Academic Press, 2021), pp. 297–342 (Chapter 10)
20. J. Lata-García, F. Jurado, L.M. Fernández-Ramírez, H. Sánchez-Sainz, Optimal hydrokinetic turbine location and techno-economic analysis of a hybrid system based on photovoltaic/hydrokinetic/hydrogen/battery. Energy **159**, 611–620 (2018). https://doi.org/10.1016/j.energy.2018.06.183

Chapter 57
Does Cooperation Boost a Transition to a Higher Innovation Intensity in Firms?

Grace Carolina Guevara-Rosero, Byron Acosta, Jaime Flores, and Jhonatan Velasteguí

Abstract The objective of this study is to determine the effect of R&D cooperation and non-R&D cooperation in the introduction of incremental and radical innovations. R&D cooperation tackles engineering, design, intellectual property, software development and database, and acquisition of tangible assets. Non-R&D cooperation is related to financing, training, and information. Using the National Survey of Innovation Activities of Ecuador 2009–2014, Tobit models are estimated to determine the effects of R&D and non-R&D cooperation on sales of non-innovative products, incremental innovations, and radical innovations are estimated. The control variables used in the model are firm's size, economic sector, R&D expenditure, age, corporate group, and staff qualifications. The results show that R&D cooperation boosts a transition from selling non-innovative products to selling incremental and radical innovations. This happens by increasing the percentage of new products, for both firm and market, and decreasing the percentage of sales of non-innovative products. Interestingly, some variables such as labor qualification, age, and R&D expenditure behave differently depending on the novelty of products.

57.1 Introduction

As the United Nations determines, an important channel to achieve sustainable development is investing in research, development, and innovation. However, in developing countries, efforts in research and development are scarce. In comparison with developed countries, developing ones invest much less in research and development. In Latin America, R&D expenditure reaches 0.54% of its GDP; meanwhile in Europe and the United States, this indicator reaches up to 1.84% and 2.6%, respectively [1]. It is then clear that firms in less developed countries need to widen their capabilities. Besides actions that can be taken internally as investment in human capital and R&D, innovation efforts can be capitalized through cooperation and knowledge exchange between firms and Science and Technology institutions.

G. C. Guevara-Rosero (✉) · B. Acosta · J. Flores · J. Velasteguí
Department of Quantitative Economics, Escuela Politécnica Nacional, Quito, Ecuador
e-mail: carolina.guevara@epn.edu.ec

Á. Rocha et al. (eds.), *Communication, Smart Technologies and Innovation for Society*, Smart Innovation, Systems and Technologies 252,
https://doi.org/10.1007/978-981-16-4126-8_57

By developing and managing a collaborative network based on strategic alliances and formal agreements with other organizations, firms can improve their innovative capacity [2]. It is a necessity for organizations to interact with each other to obtain, exchange, and develop new knowledge and useful information, as well as for obtaining funding and a wider variety of inputs. Companies that collaborate with others commonly manage to overcome the obstacles that arise when working independently [3]. Then, to make progress in innovation, companies need to engage in cooperation with other actors and not only rely on their own investments [4]. There are several reasons that support the interorganizational cooperation as an effective and efficient strategy for innovation. First, a collaborative network can increase accessibility to necessary resources to ensure the development of an innovative project, which could be more limited for an individual company [5]. Second, partnering with other organizations facilitates new knowledge exchange on the targeted market [6, 7]. Third, interorganizational cooperation also reduces the production costs of involved firms, or distributes them, optimally [8].

To succeed in terms of innovation, firms need to develop core competencies (management, market, and technical knowledge) and complement these with cooperation [9]. Firms use new sources to gain competitive improvement-knowledge, creativity, innovativeness, and cooperation [10]. Companies cooperate due to lack of financial resources, knowledge base, and technology limitations [10]. Cooperation is a critical complement to the internal generation of knowledge to innovate [9]. That is, efficient cooperation between firms will depend on the level of competence that the innovative firm has. Strategic cooperation agreements play an essential role in the firms' innovation management, as they contribute to innovation efforts [9, 11].

The lack of some capabilities generates the need to cooperate to fill this deficiency and execute successful innovation processes. Thus, cooperation is a strategy to acquire external knowledge to broaden a firm's internal knowledge base and, in turn, to generate innovation [9]. In this sense, the generation of innovations will depend on the level of internal knowledge base that the company has (*competence*) to absorb this new external knowledge (*cooperation*), and that these practices and processes are developed repetitively (*continuity*) [9].

Although cooperation can increase the level of innovation, in developing countries, it can lay the foundations for innovation; developing countries focus more on the absorption of knowledge, than on the generation of it [12]. This countries invest and cooperate mostly in non-R&D activities [13]. In fact, in Latin America, the percentage of cooperation in R&D is quite low. Based on this context, the following research question is posed: How R&D cooperation and non-R&D cooperation in Ecuadorian companies influence the introduction of incremental and radical innovations?

Firms find that cooperation partners gain external knowledge, share costs, technologies, and build up innovation collaboration [10]. For this study, *cooperation* is defined as agreements with partners (e.g., customers, suppliers, competitors, consultancies, government, and universities) [9, 10] for R&D, training and technical assistance, information and financing activities [11], and therefore create innovative outputs [10]. Two types of cooperation are analyzed in this study, R&D cooperation, and non-R&D cooperation. The first tackles cooperation objectives such as

R&D, engineering and design, and product testing [14]. The second addresses cooperation objectives such as training and technical assistance, information (marketing and brand equity activities), and financing (loans, seed capital, non-refundable funds) [11]. Thereby, the following research hypotheses are proposed:

H1: Cooperation improves the level of sales of incremental and radical innovations.
H2: Cooperation in innovation improves the level of sales of incremental and radical innovations.
H3: Cooperation, other than innovation, improves the level of sales of incremental and radical innovations.

To answer these hypotheses, Tobit models are estimated using data from the National Survey of Science, Technology and Innovation of Ecuador during 2009–2011 and 2012–2014. The results corroborate the hypotheses 1 and 2 because cooperation, specifically in R&D, is a main driver of sales in incremental and radical innovations. However, hypothesis 3 is not confirmed by the results because non-R&D cooperation increases only sales of incremental innovations, not those of radical innovations. The main contribution of this paper is to clarify the role of R&D and non-R&D cooperation in sales of different types of products: radical innovations, incremental innovations, and non-improved products. The main finding is that cooperation practices boost transition from selling non-improved products to selling radical innovations.

The remainder of this paper is organized as follows: Sect. 57.2 presents the data and methodology. Sections 57.3 presents the results. Section 57.4 discusses the results, and Sect. 57.5 presents the conclusions.

57.2 Methodology

Data come from the National Survey of Science, Technology and Innovation Activities Survey (ACTI, acronym in Spanish), for 2009–2011 and 2012–2014. Since dependent variable's measuring is the intensity of innovation [15] in terms of sales, only companies that carried out innovation activities are considered, resulting a pool data of 4254 observations.

Three econometric models with three different dependent variables are estimated. The first model considers the sales percentage of new products in the market (radical innovations). The dependent variable of the second model is the sales percentage of the firm's new products (incremental innovations). And finally, the dependent variable of the third model captures the sales percentage of non-improved products.

The Tobit methodology is used for estimations. There are two reasons for this choice: first, since dependent variables capture the firms' sales percentage, the predicted values require to be positive or equal to zero; second, dependent variables are expected to have a significant accumulation of zero values, because many firms in the sample do not sell innovative products.

With a simple linear regression, it is likely to obtain negative and not accurate predictions. Furthermore, since the distribution of the dependent variable can accumulate zeroes, it is clear that it will not have a conditional normal distribution. Fortunately, Tobit model corrects both problems using a latent variable y^* that depends linearly on x^*, through a parameter β that determines the relationship between the independent variable x^* and the latent variable y^* [16].

The specification of the econometric model is:

$$\begin{aligned} Y_i^* = \beta_0 &+ \beta_1 \,\text{R\&D cooperation}_i + \beta_2 \,\text{non R\&D cooperation}_i \\ &+ \beta_3 \,\text{corporative group}_i + \beta_4 \,\text{manufacture}_i \\ &+ \beta_5 \,\text{R\&D expenditure}_i + \beta_6 \,\text{graduated workers}_i \\ &+ \beta_7 \,\text{undergraduated workers}_i \\ &+ \beta_8 \,\text{workers with secondary education}_i + \beta_9 \,\text{age}_i \\ &+ \beta_{10} \,\text{age squared}_i + \beta_{11} \,\text{micro firm}_i \\ &+ \beta_{12} \,\text{small firm } f_i + \beta_{13} \,\text{medium-sized firm}_i \\ &+ \beta_{14} \,\text{year}_i + \mu_i \end{aligned}$$

where y_i* is a latent variable that represents the percentage of sales in products by type. Accordingly, three dependent variables are considered: (i) Percentage of sales in new products in the market (radical innovations), (ii) Percentage of sales of new products for the firm (incremental innovations), and (iii) Percentage of sales of non-improved products. The variables of interest are R&D cooperation and non-R&D cooperation of firm j that identify if a firm collaborates with other actors with the goal of innovating and with a different objective, respectively. The rest of control variables are specific characteristics of the firm such as size, economic sector, employees' qualification, market knowledge, R&D expenditure, and if the firm is part of a corporative group or not. The expected results of these control variables are presented in Table 57.1. B_0 is the constant coefficient. β_i is the coefficient of the independent variables. ui is the error term. The latent variable y^* satisfies the assumptions of the classical linear model, such as a normal distribution, homoscedastic with linear conditional mean.

Table 57.2 presents statistics of sales of new products in the market, of new products in the firm and in non-improved products of firms that cooperate in R&D aspects, in non-R&D aspects and firms that do not cooperate. The results show that the percentage of sales in radical innovations of firms that cooperate with R&D objectives is higher than that of firms that cooperate in non-R&D objectives. However, this difference is not significant. By contrast, the percentage of sales in radical and incremental innovations is higher for firms that cooperate, either in R&D or not, than firms that do not cooperate at all. These differences are statically significant. Likewise, the average R&D expenditure is much higher in firms that cooperate with R&D objectives and with non-R&D objectives than firms that do not cooperate at all.

Table 57.1 Expected results of independent variables

Variable	Description	Expected Sign	Author
Cooperation with an innovation objective	Dummy variable. 1 if the company cooperates with innovation objective; 0 if not	+	[14, 17]
Cooperation with another objective	Dummy variable. 1 if the company cooperates with an objective other than innovation; 0 if not	±	[14]
Experience	Quantitative variable. Years of experience of the company	+	[18]
Squared experience	Quantitative variable. Years of company's experience, squared	–	
R&D expenditure	Quantitative variable. Logarithm of the company's total R&D expenditure	+	[15, 19]
Corporate group	Dummy variable. 1 if the company belongs to a corporate group; 0 if not	+	[4]
Education level of workers	Quantitative variables. Logarithm of the number of workers in the company with postgraduate, undergraduate, or secondary academic training	+	[20–22]
Economic sector	Categorical variable. The category used is manufacturing sector	Ambiguous	[15, 22, 23]
Size of the company	Categorical variable. The categories used are microenterprise, small company, medium-sized company, and large company. The base category is large company	+	[4, 15, 19, 24–26]

57.3 Results

Table 57.3 summarizes the results on the effect of cooperation on the percentage of sales of new products in the market [shown in column (1)], the percentage of sales of new products in the firm [shown in column (2)] and the percentage of sales of non-improved products [shown in column (3)]. In columns (1.1), (2.1), and (3.1), the marginal effects of the respective models are presented.

Table 57.2 Differences in percentage of sales

	% of sales of new products to the market	% of sales of new products to the firm	% of non-improved products	R&D expenditure
R&D coop	9.27%	15.98%	74.74%	272,277.40
Non-R&D coop	8.44%	15.94%	75.62%	181,261.50
No cooperation	5.23%	12.22%	82.55%	30,144.84
Means difference R&D coop versus non-R&D coop	*t* stat = 1.484 *P*-value = 0.14	*t* stat = 0.066 *P*-value = 0.95	*t* stat = -1.129 *P*-value = 0.26	*t* stat = 1.457 *P*-value = 0.15
Means difference R&D coop versus no cooperation	*t* stat = 4.29 *P*-value = 0.00	*t* stat = 3.132 *P*-value = 0.00	t stat = -5.844 *P*-value = 0.00	*t* stat = 2.17 *P*-value = 0.03
Means difference non-R&D coop versus no cooperation	*t* stat = 3.558 *P*-value = 0.00	*t* stat = 3.232 *P*-value = 0.00	*t* stat = -5.388 *P*-value = 0.00	*t* stat = 1.73 *P*-value = 0.08

57.4 Discussion

The variable of interest, cooperation for innovation is highly significant to increase the sales of radical and incremental innovations and to decrease sales of non-innovative products. This indicates that cooperation invested in research and development, engineering, and design and product testing is well/addressed because it allows a transition from the production of non-innovative products to the production of innovative products. The cooperation in other aspects besides innovation does not increase sales in radical innovations; meanwhile, it increases the sales of incremental innovations. These results might suggest that cooperation with science, technology, and innovation (STI) actors conducts to radical innovations, whereas cooperation with partners that help doing, using, and interacting (DUI) conducts to incremental innovations [27, 28]. This study focuses on the objective of cooperation since firms can cooperate with DUI partners to innovate, and then, it is more likely to generate radical innovations than incremental ones, as suggested by the existing literature. Results also show that expenditure in R&D is the main driver of radical innovation [19, 24]. A 1% increase in R&D expenditures conducts to 0.03 percentage points in radical innovations. Investing in R&D allows firms to acquire new fundamental knowledge for innovation [24]. By contrast, R&D expenditure is not significant for incremental innovations, and it reduces the sales of non-innovative products.

When a firm is part of a corporative group, its sales in radical innovations are higher in 3.3% points than firms that are not part of a corporative group. Companies that belong to a business group have more probabilities to carry out innovation activities, because they have the support of other institutions [4].

More interestingly, the composition of labor in a firm is a key aspect for radical, incremental [20, 21], and non-innovative products. Our results show that graduated

Table 57.3 Tobit estimations of the cooperation on the percentage of sales products by novelty

	(1) Sales in new products to the market	(1.1) Marginal effects	(2) Sales in new products to the firm	(2.1) Marginal effects	(3) Sales in non-improved products	(3.1) Marginal effects
R&D coop	0.0521*	0.0114*	0.0462**	0.0205**	−0.0270**	−0.0269**
	(1.96)	(1.96)	(2.76)	(2.76)	(−2.90)	(−2.90)
Non-R&D coop	0.0290	0.00636	0.0841***	0.0373***	−0.0471***	−0.0469***
	(0.84)	(0.84)	(3.68)	(3.68)	(−4.16)	(−4.16)
Corp. group	0.154***	0.0338***	−0.0935***	−0.0415***	−0.00682	−0.00680
	(5.11)	(5.09)	(−4.13)	(−4.14)	(−0.57)	(−0.57)
Manufacture	0.134***	0.0294***	0.0117	0.00520	−0.0258**	−0.0257**
	(4.95)	(4.93)	(0.67)	(0.67)	(−2.67)	(−2.67)
R&D exp	0.0170***	0.00374***	−0.00169	−0.000748	−0.00205*	−0.00204*
	(7.07)	(7.03)	(−1.04)	(−1.04)	(−2.31)	(−2.31)
Graduated labor	0.0548***	0.0120***	0.00276	0.00123	−0.0141*	−0.0141*
	(3.54)	(3.54)	(0.23)	(0.23)	(−2.27)	(−2.27)
Undergrad labor	0.0399**	0.00877**	−0.0166	−0.00739	0.00532	0.00530
	(2.76)	(2.75)	(−1.71)	(−1.71)	(0.99)	(0.99)
Workers with secondary educ	−0.0373***	−0.00818***	−0.00802	−0.00356	0.0121**	0.0121**
	(−3.33)	(−3.32)	(−1.01)	(−1.01)	(2.81)	(2.81)
Age	0.000580	0.000127	−0.00219*	−0.000971*	0.00179***	0.00179***
	(0.39)	(0.39)	(−2.50)	(−2.50)	(3.31)	(3.31)
Age squared	0.0000003	5.39e−08	0.0000192**	0.000009**	−0.000015**	−0.00002**
	(0.02)	(0.02)	(2.72)	(2.72)	(−2.58)	(−2.58)
Micro firm	−0.0197	−0.00432	−0.0330	−0.0146	0.00680	0.00678
	(−0.20)	(−0.20)	(−0.47)	(−0.47)	(0.18)	(0.18)
Small firm	−0.0966	−0.0212	0.0200	0.00889	0.0200	0.0199
	(−1.64)	(−1.64)	(0.47)	(0.47)	(0.89)	(0.89)
Medium firm	−0.0391	−0.00858	0.000608	0.000270	0.0184	0.0183
	(−0.88)	(−0.88)	(0.02)	(0.02)	(1.07)	(1.07)
Year	−0.157***	−0.0344***	−0.0602***	−0.0267***	0.0469***	0.0467***
	(−6.25)	(−6.16)	(−3.70)	(−3.71)	(5.22)	(5.21)
Constant	−0.540***		−0.0199		0.727***	
	(−5.73)		(−0.30)		(20.36)	
Sigma	0.582***		0.456***		0.283***	
	(42.52)		(58.01)		(73.79)	

(continued)

Table 57.3 (continued)

	(1)	(1.1)	(2)	(2.1)	(3)	(3.1)
	Sales in new products to the market	Marginal effects	Sales in new products to the firm	Marginal effects	Sales in non-improved products	Marginal effects
No. observations	4254	4254	4254	4254	4254	4254

The standard errors are in parentheses, and the significance is represented by: ***$p < 0.01$, **$p < 0.05$, *$p < 0.1$

workers boost radical innovations, more than undergraduate workers. An increase of 1% of hired graduated workers increases sales in radical innovation in 1.2 percentage points, and an increase of undergraduate workers conducts to an increase in sales of radical innovations of 0.08 percentage points. For new knowledge to be correctly absorbed and used, hiring highly qualified employees are required [21, 22]. Workers with a high academic formation have a comparative advantage in adapting to new and changing technologies easier [20]. In contrast, an increase of graduated workers reduces the sales of non-improved products, constituting an expenditure, rather than an investment for firms that keep selling non-innovative products.

Contrary to what is expected [4, 25], the size of firms is not significant for sales in any type of products. However, firms' experience is important for sales of incremental and non-innovative products. On the one hand, in the case of sales in incremental innovations, the experience variable has a U-shape, meaning that young firms have decreasing sales of incremental innovations until a given maturity where their incremental innovation sales increase. On the other hand, the effect of experience in non-innovative sales has an inverted-U-shaped relationship, meaning that sales in non-innovative products increase until a given moment, in which sales increments start to decrease.

The different patterns show that a certain level of experience is required to increase innovative sales. The appropriate use of new information is conditioned by the firm's expertise. The more experience in the market a company has, the higher is its ability to use new knowledge [18]. By contrast, the increment of sales of non-innovative products is obtained by leveraging experience at first, but then, more experience is not enough to further increase sales. Non-innovative products cannot stay for long time in the market due to creative destruction competition [29]. Finally, the manufacturing sector proves to register higher sales in radical innovations than other economic sectors. The need and capacity to generate innovations depend on the economic sector, because not all industries need the same innovation intensity to be as competitive [22, 23].

57.5 Conclusions

This investigation contributes to the existing literature by establishing the different effects of R&D cooperation and non-R&D cooperation on sales of products by novelty, namely radical innovative products, incremental innovative products, and non-innovative products. The main finding of this study is that cooperation with innovation goals allows a transition from selling non-innovative products to selling innovative products, either incremental or radical innovations. In addition, cooperation in other aspects other than R&D increases incremental innovations, rather than radical ones. For radical innovations to increase, cooperation intended to innovate is required.

Moreover, the characteristics of firms influence differently on their sales depending on the type of products. For instance, the contribution of more qualified workers is effective to increase radical and incremental innovations but is not effective to increase non-improved products. In the same line, non-qualified workers increase sales of non-improved workers but lower the sales of incremental and radical innovations. In addition, R&D expenditure boosts incremental and radical innovations but reduces sales in non-improved products.

Given these results, some policy recommendations can be drawn. One of them is that firms need to increase R&D investment by partnering with universities and research institutions. The government can be a central actor that boosts these cooperative actions. Another governmental intervention is the attraction of technological intensive multinational enterprises. Another recommendation to firms is to hire more qualified workers to their production processes. In this line, governments and the private sector could increase scholarships' offer for masters and doctorates. In this way, there would be more availability of qualified workers in the labor market so firms can take advantage of it.

References

1. G. Anlló, D. Suárez, *Innovación: algo más que I + D*. Evidencias Iberoamericanas a partir de las Encuestas de Innovación: Construyendo las Estrategias Empresarias Competitivas
2. J. Rowley, Should your library have an innovation strategy? Libr. Manag. **32**(4/5), 251–265 (2011)
3. R. Rothwell, External networking and innovation in small and medium-sized manufacturing firms in Europe. Technovation **11**(2), 93–112 (1991)
4. B.S. Tether, Who co-operates for innovation, and why: an empirical analysis. Res. Policy **31**(6), 947–967 (2002)
5. T.K. Das, B.-S. Teng, A resource-based theory of strategic alliances. J. Manage. **26**(1), 31–61 (2000)
6. H.W. Chesbrough, *Open Innovation: The New Imperative for Creating and Profiting from Technology* (Harvard Business Press, 2003)
7. G. Ahuja, Collaboration networks, structural holes, and innovation: a longitudinal study. Adm. Sci. Q. **45**(3), 425–455 (2000)

8. J. Hagedoorn, Understanding the rationale of strategic technology partnering: interorganizational modes of cooperation and sectoral differences. Strateg. Manag. J. **14**(5), 371–385 (1993)
9. M. Kötting, A. Kuckertz, Short-, mid-, and long-term effects of innovation activities: a configurational analysis on continuity, competence, and cooperation. Int. J. Innov. Manag. **2150053**, 1–36 (2020)
10. V. Prokop, J. Stejskal, Different influence of cooperation and public funding on innovation activities within German industries. J. Bus. Econ. Manag. **20**(2), 384–397 (2019)
11. OECD and Eurostat, *Oslo Manual 2018: Guidelines for Collecting, Reporting and Using Data on Innovation*, 4th edn (Oslo, 2018)
12. C. Chaminade, C. Esquist, *Rationales for Public Policy Intervention in the Innovation Process: Systems of Innovation Approach* (Edward Elgar Publishing, 2010)
13. J. Fernández Sastre, C.E. Vaca Vera, Cooperation for innovation in developing countries and its effects: evidence from Ecuador. J. Technol. Manag. Innov. **12**(3), 48–57 (2017)
14. H.C. Kim, W. Yoon, Study on types of technology cooperation partner and innovation performance: focusing on incremental and radical innovation. Int. J. Innov. Manag. **23**(1), 1950005 (2019)
15. D. Faems, B. Van Looy, K. Debackere, The role of inter-organizational collaboration within innovation strategies: towards a portfolio approach. DTEW Res. Rep. **0354**, 1–33 (2003)
16. J. Wooldridge, *Introduction to Econometrics: A Modern Approach*, in *Ciudad de México, México, Edit. Cengage Learning*, pp. 1–865, 4th Spanish edn (2010)
17. C.A. Un, A. Cuervo-Cazurra, K. Asakawa, R&D collaborations and product innovation. J. Prod. Innov. Manag. **27**(5), 673–689 (2010)
18. N. López-Mielgo, J. Montes-Peón, C. Vázquez-Ordás, What does a company need to innovate? Research, experience and persistence. Eur. J. Bus. Manag. Econ. **21**(3), 266–281 (2012)
19. B. Ortiz, J. Bach, M. Donate, F. Guadamillas, Cooperation for external knowledge acquisition from interorganizational relationships as antecedent of product innovation: the mediating role of absorptive capacity. SSDC **180**, 43–57 (2019)
20. A. Bartel, F. Lichtenberg, The comparative advantage of educated workers in implementing new technologies. Rev. Econ. Stat. **69**(1), 1–11 (1987)
21. M. Buesa, T. Baumert, J. Heijs, M. Martínez, Determinants of innovation: an econometric analysis. Ind. Econ. **347**, 67–84 (2002)
22. X. González, D. Miles-Touya, C. Pazó, R&D, worker training and innovation: firm-level evidence. Ind. Innov. **23**(8), 694–712 (2016)
23. D. Nicolau-Juliá, M. Expósito-Langa, J. Tomás-Miquel, Exploration and exploitation of knowledge in the business environment. Scale validation in an industrial sector with a low technological profile. Eur. Res. Bus. Manag. Econ. **21**(3), 139–147 (2015)
24. M. Fritsch, R. Lukas, Who cooperates on R&D? Res. Policy **30**, 297–312 (2001)
25. A. Leiponen, Skills and Innovation. Int. J. Ind. Organ. **23**(5), 303–323 (2005)
26. S. Benito, M. Platero, A. Rodríguez, Determinants of innovation in Spanish microenterprises: the importance of internal factors. Univ. Bus. Rev. **33**, 104–121 (2012)
27. M. Jensen, B. Johnson, E. Lorenz, B.-A. Lundvall, Forms of knowledge and modes of innovation. Res. Policy **36**, 680–693 (2007)
28. D. Fitjar, A. Rodríguez-Pose, Firm collaboration and modes of innovation in Norway. Res. Policy **42**, 128–138 (2013)
29. J. Schumpeter, Creative destruction. Capital. Social. Democrat. **825**, 82–85 (1942)

Chapter 58
Intelligent White Cane for Visually Impaired People Based on an Ultrasonic Sensor

Nicolás Pozo, Fernando Yupa, and Edy Ayala

Abstract This article describes the development of a device adaptable to a white cane for guidance and mobility of visually impaired people. The proposal consists of an economical, portable, easy-to-maintain, and non-invasive device designed with a microcontroller which receives and processes signals obtained from an ultrasonic sensor. Similarly, it alerts obstacles' distances to the user based on ultrasonic measurements. The device is a compact prototype which is adapted to the handle of the white cane and can be removed for a comfortable transportation. The principle of operation is based on the detection of aerial obstacles that represent a threat for the user mobility. The ultrasonic sensor sends the signal to the microcontroller which generates output alerts and transmitted to the user by stimuli which are auditory or haptic. These signals respond to an algorithm implemented in the controller. The device has been designed not to interfere with the user's senses and has been programmed including different alert modes allowing user adaptation. The algorithm has been developed in a C-based programming language, and for the validation of the development, tests have been performed with visually impaired persons to verify the fulfillment of the proposed objectives in this research.

58.1 Introduction

Considering the high number of people with visual disabilities, the development of assistive devices has gained interest in this research field, with the main emphasis on aid proposals for the mobilization of blind people [1]. Among the most prominent devices, for instance, it can be mentioned: the smart cane presented in [2], eye replacement, and Global Position System (GPS), Fusion of Artificial Vision (FAV), Tongue Electro-tactile Device (TED) [3, 4]. The main purpose of assistive technology

N. Pozo · F. Yupa · E. Ayala (✉)
Universidad Politécnica Salesiana, Calle Vieja 12-30, Cuenca, Ecuador
e-mail: eayala@ups.edu.ec

N. Pozo
e-mail: apozoa@est.ups.edu.ec

Á. Rocha et al. (eds.), *Communication, Smart Technologies and Innovation for Society*, Smart Innovation, Systems and Technologies 252,
https://doi.org/10.1007/978-981-16-4126-8_58

Table 58.1 Some typical obstacles for the mobility of blind people

Obstacles	Anticipation (m)	Description
Aluminum post	2.0	Diameter: 13 cm
Dark shiny post	0.8	Diameter: 8 cm
Wooden post	2.5	Diameter: 20 cm
Big dustbin	3.2	0.45 m × 1 m × 1.5 m
Trash grill	3.0	0.60 m × 0.45 m
Big tree	2	Diameter: 35 cm

is to improve people's quality of life. One of the ways in which this technology assist people is through electronic devices easily accessible to the population. The design characteristics depend on their application, ergonomic considerations, and user requirements. In addition to vision, haptic sense also plays an important role for objects detection [5]. For instance, exploration such as "touching" allows a visually impaired person to comprehend the environment by understanding pressure, lateral movement, material properties, texture, hardness, temperature, and surface friction. In certain scenarios, vision has a secondary role for object recognition. Assistive technologies development for mobility involves multidisciplinary efforts and represent a challenge because the complex requirements [6]. In this way, active touch has become one of the techniques most used by blind people to obtain information through other senses [7], and therefore, the white cane cannot be easily replaced or avoided since it is an extension of the hand [8]. This sensory perception improves over time, and consequently, the user can generate a mental map where they can identify different objects. Table 58.1 presents some of the typical obstacles for the mobility of blind people [9].

It is very important to mention that these devices have a social impact when the blind person tries to interact and adapt, that is, they accept this technology; in this way, the person is coupled to the device, managing to mobilize with greater confidence. Table 58.2 presents some differences between a traditional white cane and a smart white cane [9].

The principle of the device operation for distance object measurement is echolocation, this technique was analyzed in the early 90s as mentioned in [10]. In this case, a similar recognition is proposed in [11] which states that for different intensities and differences of ultrasonic time waves reflected and transmitted by the sensors, it is possible to locate direction and size of an obstacle. This method is based on the Doppler effect which implies to process the time of emission and reception of an ultrasonic signal [12]. This same principle has been applied, so that the user is able to identify obstacles by means of an alert generated by the actuators distributed in the device. A representation of the operation of the sensor can be seen in Fig. 58.1.

The development of the proposal is based on the detection of obstacles with an intelligent white cane which provides support for the mobility. As the user approaches the obstacle, the system provides progressive vibrations to alert the presence and depth of an object. In the development of the device, the utilized ultrasonic sensor is the MB1040 model [12]. The implemented strategy can be seen in Fig. 58.2.

Table 58.2 Main differences between a traditional white cane and the smart cane

Characteristics	Traditional white cane	Smart cane
Weight	Light (0.45 kg)	Heavy (0.62 kg)
Length	120–155 cm	120–155 cm
Cost	Cheap (≈2200 USD)	Expensive (≈ 50,000 USD)
High obstacles detection	No	Yes
Wearing comfort	Comfortable	Slightly uncomfortable
Handle volume	Ergonomic	Bulky
Maintenance	No	Yes
Mobility	Limited	Higher reliability
Orientation	Limited	Higher reliability

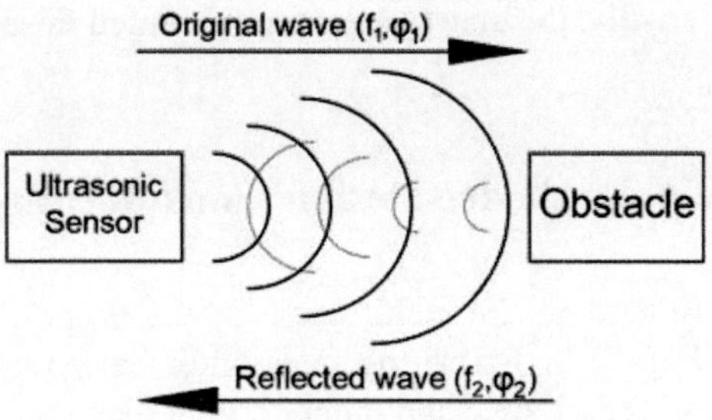

Fig. 58.1 Ultrasonic method for distance measurement

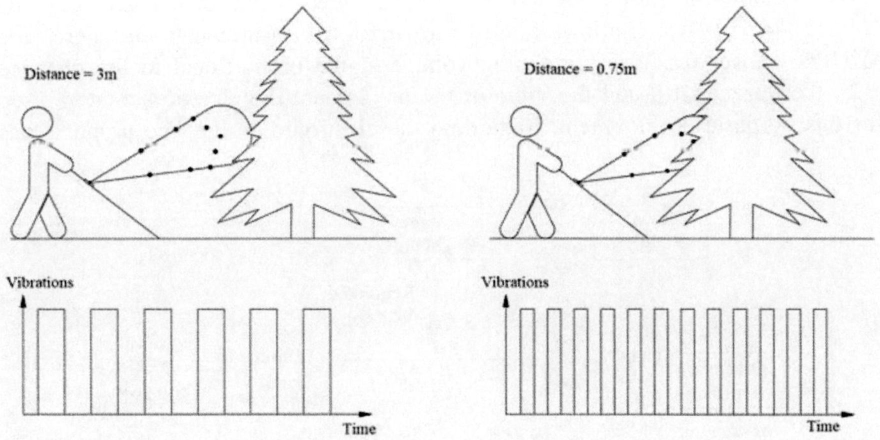

Fig. 58.2 Device operating conditions

The design has been developed in accordance with the recommendations for the implementation of a smart white cane shown in [8]. This proposal reduces the range of the scope to determine if the path is clear. For the proposed case, a maximum detection range has been configured: 0.75, 1.5, 2, and 3 m.

According to the characteristics of an ultrasonic sensor device detailed in [13], it is essential to understand the strategy for locating obstacles. Thanks to this development, various prototypes have been proposed, among them it can be highlighted: Brain Port [14], Ultra Cane [15], and Ray Device [16]. Similarly, there are devices developed in the academic as well as in [17] where basic mobility assistance equipment is detailed, in [18] an intelligent white cane is presented using ultrasonic sensors in an open-source platform, and finally in [1], the development of an intelligent white cane is shown infrared-based technology. Therefore, to other proposed devices that are able to detect aerial obstacles, this proposal uses basic materials and ultrasonic sensor. In section two, the design of the mechanical, electronic, and control components is detailed. Moreover, the prototype results analysis can be seen in section three. In section four, the discussion of results and subsequent analysis are described. Finally, the conclusions are presented in section five.

58.2 Device Design Considerations

Regarding the architecture of the proposed device, a basic structure is shown in Fig. 58.3, where the interaction between the microcontroller, sensors, actuators, power circuits, and input / output elements of the device can be observed.

The detailed electronic components implemented in the device are: the PIC 16LF15344 [19] which allows the user to perform measurements and alerts; the MB1040 sensor provides the analog voltage signal proportional to the distance [12], the buzzer [20], and the vibrator motor [21] are the electric actuators used for this proposal development. Regarding the electronic design, the proposal has

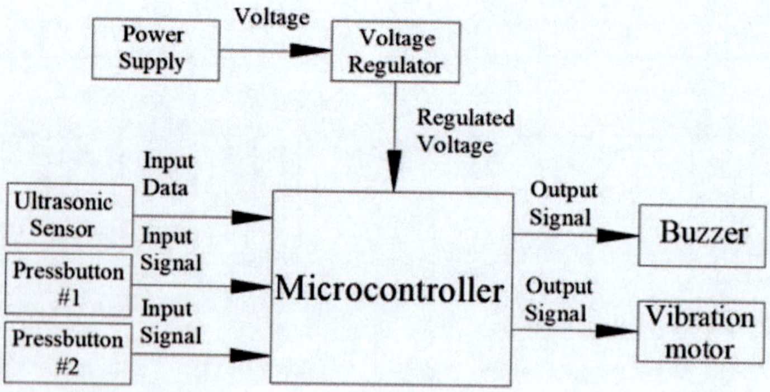

Fig. 58.3 Device architecture

been designed in Autodesk-Eagle and considers protection elements for the actuators, specifically transistors configured as common collector have been dimensioned to activate the warning signals elements.

Regarding to the voltage source, a plug located on the left side of the electronic board is presented in Fig. 58.4a. It has been incorporated to connect a 3.7 V lithium battery for powering the device. The microcontroller is in the middle part of the electronic board, and with respect to the other elements, they are distributed around it. Additionally, the device has multiple pins distributed, so that they allow the programming of the microcontroller for later versions improvements. Referring to the electronic board dimensions, it has been designed into a 60 × 35 mm total area using two layers printed circuit board (PCB). Finally, Fig. 58.4b presents a referential model of the electronic board implemented in the proposed device.

The mechanical design has been modeled on Autodesk-Inventor and developed following the design conditions outlined in previous sections. It has a distribution in such a way that the components take the least possible space including three buttons corresponding to selection of user mode and distance calibration. The assembly of the elements present in the device can be seen in Fig. 58.5a, while the referential assembly

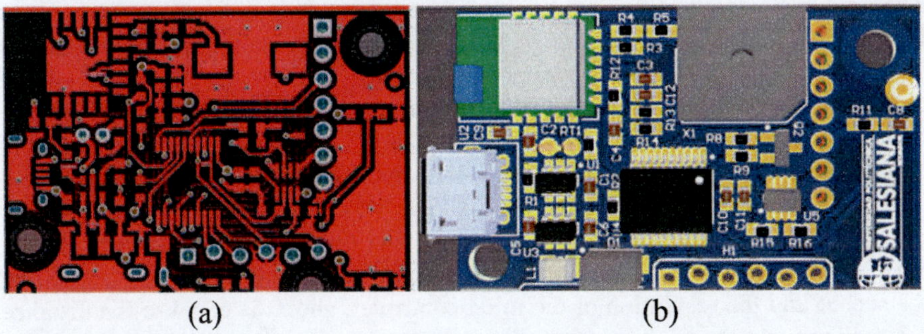

Fig. 58.4 Electronic board schematic of the device (**a**), and 3D view of the electronic board (**b**)

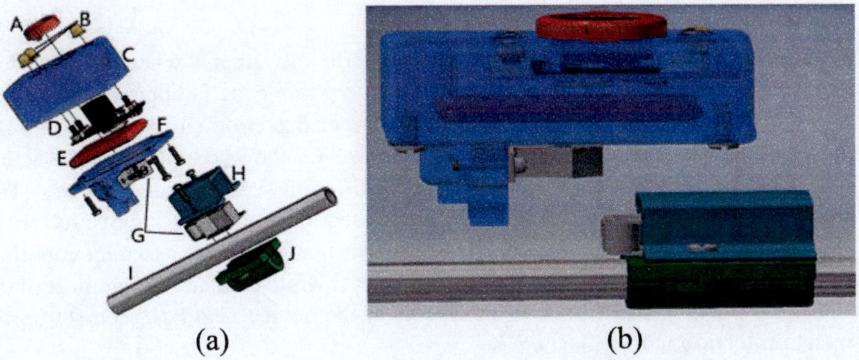

Fig. 58.5 Exploded diagram (**a**), and assembly mechanism for the white cane (**b**)

Table 58.3 Description of device elements

Parts	Description
A	Sensor cover
B	Calibration buttons
C	External case (cover)
D	Electronic board
E	Electronic board (subjection)
F	External case (top)
G	Grip
H	Grip
I	White cane

model can be seen in Fig. 58.5b. It is important to mention that the design of the device includes commercial elements such as which provides favorable conditions when it comes to replicating the proposal. Table 58.3 presents the details of the elements present in the device from Fig. 58.5a.

The proposed control algorithm consists of different subroutines: main program, user alert customization, and calibration of maximum detection ranges as shown in Fig. 58.6. It is important to mention that user mode and calibration functions are independent from the main program; however, they are necessary for the correct operation of the algorithm implemented in the microcontroller since they are the functions for providing the necessary information. The mathematical relationship to determine the distance between the sensor and the obstacle has been stabilized in Eq. (58.1) where V_{EZ4} is the analog voltage generated by the ultrasonic sensor proportional to the distance (v); V_{cc} is equal to the sensor voltage supply (3.7 v), $Kp = 1300.48$ is equal to a proportional constant based on the voltage provided by the sensor and the resolution of the microcontroller, and D is equal to the distance between the sensor and the obstacle in (cm) [9].

$$D = \frac{V_{EZ4}}{V_{cc}} Kp \tag{58.1}$$

On the other hand, blind people perform specific movements when using a white cane. Considering that the mobility range of the white cane is approximately 90°, the proposal has been considered to cover a larger detection surface thanks to the detection range of the sensors (Fig. 58.7). Figure 58.8 shows the proposed detection zone based on the movements of the user in different scenarios and devices [16]. Finally, the developed device integrates all the sections mentioned above to form a functional, compact, replicable, and non-invasive element which assist the mobility. This device can be incorporated into a commercial white cane including most of the available models. As can be seen in Fig. 58.8, the device also has been set up for impact protection with a plastic case.

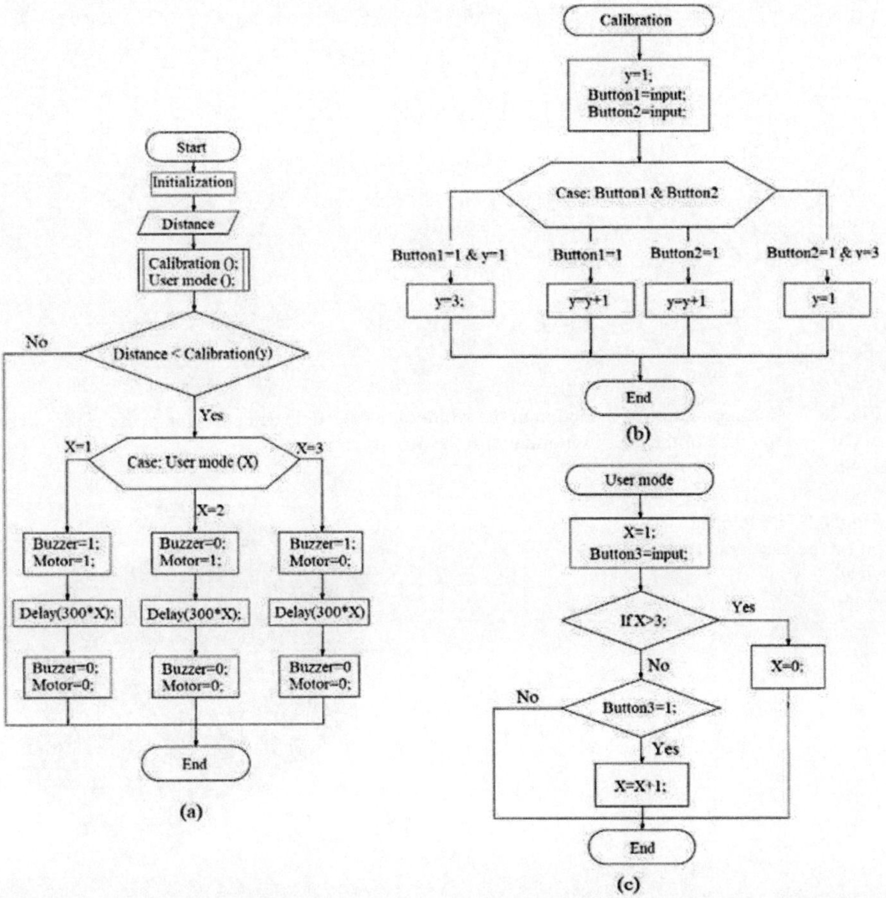

Fig. 58.6 Microcontroller algorithm (**a**), device alert calibration algorithm (**b**), and device detection range customization algorithm (**c**)

58.3 Prototype Results Analysis

For the validation of the proposal, two evaluation mechanisms have been used; the first consists in a study of distance measurement error by the device at different angles and heights, and the second is the validation by using TRIZ matrix.

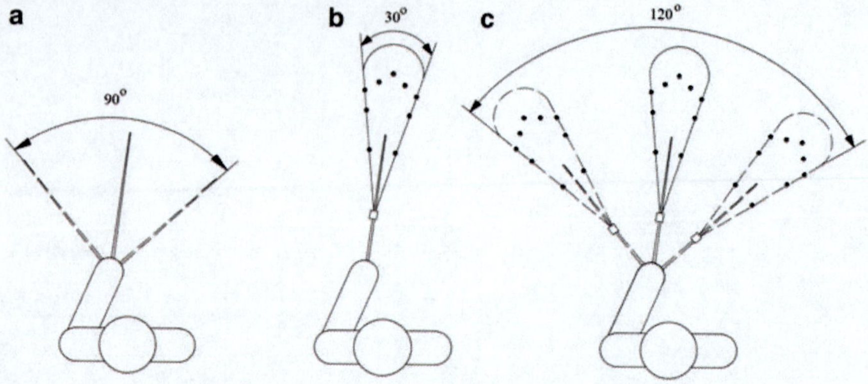

Fig. 58.7 Common range of detection of the white cane (**a**), static detection range of the prototype in (**b**), sweep range of the user in motion with the prototype in (**c**)

Fig. 58.8 Developed prototype case (external view)

58.3.1 Distance Error

The speed of sound varies depending on changes of media temperature which modifies the frequency at which the interactions between the particles that carry the vibration and change of propagation speed takes place. Consequently, the speed of sound is directly proportional to the temperature. The basic function at atmospheric pressure validates with enough precision for a range of temperature T (0–100 °C); the speed of the ultrasonic signal v is presented in Eq. (58.2):

$$v = 331\frac{\text{m}}{\text{s}} + \left(0.6\frac{\text{m}}{\text{s}}\frac{1}{^{\circ}\text{C}}\right)T \tag{58.2}$$

where the velocity increases by 0.6 m/s per each 1 °C increase of temperature. The results are approximated, because in the speed of the sound they are influenced by more aspects apart from the temperature, i.e., the pressure that has not been

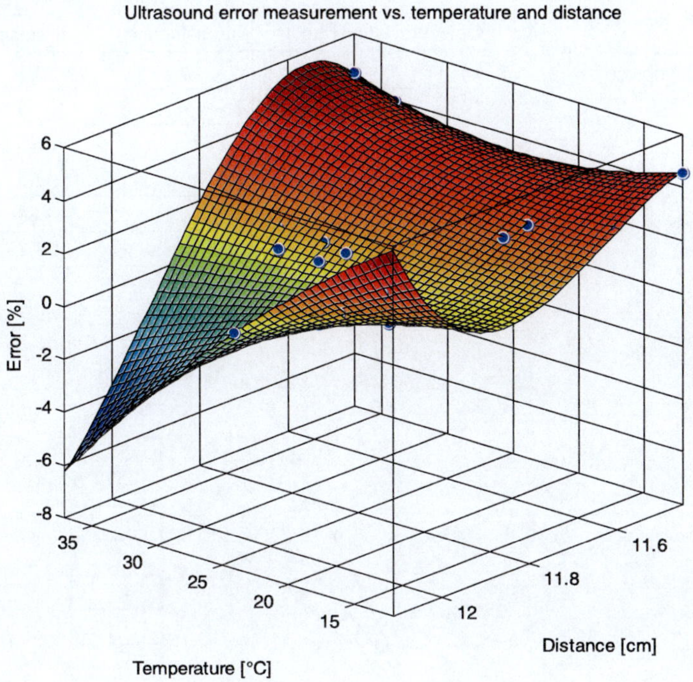

Fig. 58.9 Error measurement at different temperatures and distances

considered [22]. In conclusion, velocity and temperature are directly proportional because when one increases the other also does. At a constant speed of 340 m/s the measured distance with ultrasound varies between 3 mm of the original measurement, this with the sensor as the temperature increases there is an increase in frequency which causes the measured distance to decrease. Experiments using the EZ4 sensor have displayed an equation shown in Fig. 58.9 obtained also using Eq. (58.3) by taking multiple measurements:

$$e = -0.0017x^2 + 0.072x + 11 \tag{58.3}$$

where e is the obstacle distance error, and x is an independent variable representing the distance.

The ultrasonic waves rebound on a surface produces reflection in the same direction of the angle of the object´s surface. In Table 58.4, it is analyzed the effects of the measured distance depending on the pitch angle. For instance, when the surface´s angle increases the error also rises exponentially. The distance is affected considerably when the reflection surface angle of the object is higher than 30°. Figure 58.10 shows a representation of how the distance has been measured.

Table 58.4 Results of measurement at different angles of the object

Object distance (cm)	Pitch angle (°)	Distance (cm)
12	0	12,01
12	15	13,65
12	20	14,25
12	25	14.68
12	30	14.88
12	35	153.79

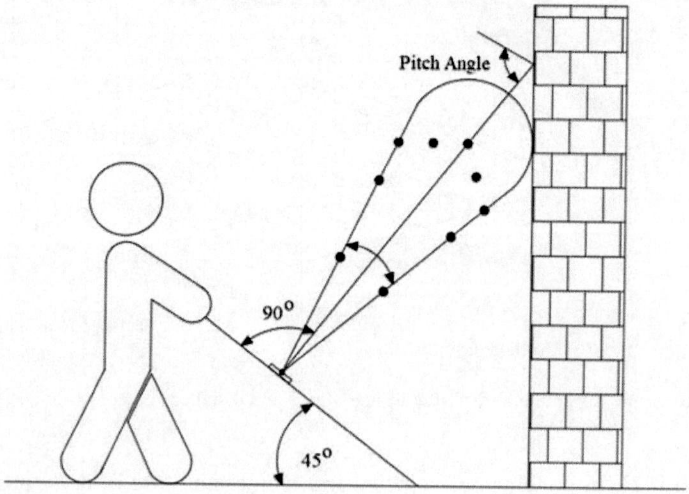

Fig. 58.10 Representation of data collection

58.3.2 TRIZ Matrix

The second mechanism for validation used during for this proposal consists of the characteristics analysis of the device considered for the final design resolved using the theory of the resolution of Invention-related Tasks (TRIZ) matrix methodology [23]. The result obtained once analyzed the TRIZ matrix has indicated that it is recommend maintaining the size and weight of the device or reduce as much as possible, since a larger device can hinder the mobility of people visually impaired. Moreover, depending on the socioeconomical characteristics of the population, but in general, it is important to reduce its cost and maintenance necessities. The resulting matrix can be observed in Fig. 58.11.

Fig. 58.11 TRIZ simplified resulting matrix for the device design

Features to improve / Features to worsen	27 Reliability	28 Measurement accuracy	32 Ease of manufacture	34 Ease of repair	35 Adaptability
1 Moving object (Weight)	1,3,11,27	28,27,35,26	27,28,1,36	2,27,28,11	29,5,15,8
7 Moving object (Volume)	14,1,40,11	25,26,28	29,1,40	19,35	10
12 Shape	10,40,16	28,32,1	1,32,17,28	2,13,1	1,15,29
19 Energy consumption	19,21,11,27	3,1,32	28,26,30	1,15,17,28	15,17,13,16
36 Device complexity	13,35,1	2,26,10,34	27,26,1,13	1,13	29,15,28,37

58.4 Discussion

The performance of this proposed intelligent white cane has been evaluated according to the documentation presented in the sensor manufacturer's data sheet. Different conditions such as: pitch angle and different distances, have been considered during the tests. As a result of this performance evaluation, it has been determined that the most appropriate angle for data collection is at 60º; however; a response analysis of the device in refractive elements has not been considered in the results validation stage.

Regarding the design in terms of comfort, the device has been proposed following the TRIZ matrix, specifically the users expressed disagreement in the fastening mechanism, and the implementation currently contemplates an easy removal mechanism; however, users state that the fastening system can generate limitations when assembling and disassembling the device, and this can cause possible loss or accidents with the device. On the other hand, the people surveyed expressed agreement with the autonomy, since being an adaptable device it can be used without a white cane; The surveyed users indicate that the prototype is easy to assemble; however, they mention that it is necessary a third person assistance for maintenance or repairs.

Regarding a commercial implementation, it is recommended for few devices production rapid prototyping using 3D printers. Moreover, the holding to the cane mechanism must be robust but also must not increase the cost of the device as well as its weight. It is important to note that this device can be personalized depending on the user necessities. Finally, regarding to the device's cost, the list of materials can be observed in Table 58.3 which includes basic electronic and assembly components for ease of construction. The device has been designed to be accessible for most countries where its manufacturing can be performed manually using conventional tools and materials. Moreover, similar proposals [14, 15] present some advantages because they allow the user to mobilize by detecting obstacles using also echolocation; however, the purpose of this research is to present a functional and easy to build device using low-cost and commercial elements. Although these proposals detect different obstacles, the need to incorporate this technology in a compact and portable device is highly recommended. The test results indicate that the device allows the user to detect aerial obstacles more rapidly compared with a sensor less white cane improving movement.

58.5 Conclusions

The proposal for the development of a smart white cane for blind people is based on an ultrasonic sensor. The compliance analysis and results have been presented in the present work providing support in mobility and allowing the user to identify obstacles rapidly. The customization established in the microcontroller allows the user to interact more friendly with the device, providing ease of use. Moreover, a lightweight, adaptable, easy-to-maintain, and customizable device becomes an attractive option for a developing assistive technology. Prototypes with similar characteristics are commonly limited by resources availability. Additionally, some devices restrict the user operation which may compromise the functionality, and the high-cost elements reduce its chance for commercialization in developing countries.

The development of this device allows a fast training for the user. The vibrations and sound alerts permit the user to add information to the already well-known white cane. This is also important since the white cane cannot be replaced or avoided for an independent mobilization of blind people. Consequently, adaptability is highly recommended as proposed in this work in order to provide an optional additional tool for mobility. This easy to use and cheap device allows to add information about the environment to the blind person to assist and improve his/her quality of life.

Regarding the design, thanks to the TRIZ methodology, this proposal has solved the inventive conflicts improving the concept from the first stages of development. The results also indicate an error rate about 6%. Nevertheless, the pitch angle produces a high error from 30° which must be considered.

For future works, it is recommended to consider performance evaluations of the assistive devices including specific characteristics for early age users providing them additional assistance during training. The capability to produce the device is related to the availability of resources such as electronic and mechanical components.

References

1. A. Nada, M. Fakhr, A. Seddik, Assistive infrared sensor based smart stick for blind people, in *Science and Information Conference* (2015), pp. 1149–1154
2. A. Sen, K. Sen, J. Das, Ultrasonic blind stick for completely blind people to avoid any kind of obstacles, in *2018 IEEE Sensors* (2018), pp. 1–4
3. W. Elmannai, K. Elleithy, Sensor-based assistive devices for visually-impaired people: current status, challenges, and future directions. Sensors (Basel) **17**(3), 565 (2017)
4. S. Ponnada, S. Yarramalle, M. Rao, A hybrid approach for identification of manhole and staircase to assist visually challenged, in *2016 IEEE International Conference on Robotics and Biomimetics* (*ROBIO*) (2016)
5. R. Klatzky, N. Giudice, C. Bennett, J. Loomis, Touch-screen technology for the dynamic display of 2D spatial information without vision: Promise and progress. Multisens. Res. **27**, 359–378 (2014)
6. J. Liu, J. Liu., L. Xu, W. Jin, Electronic travel aids for the blind based on sensory substitution, in *5th International Conference on Computer Science and Education* (2010), pp. 1328–1331
7. S. Lederman, R. Klatzky, Haptic perception: a tutorial. Atten. Percept. Psychophys. **71**, 1439–1459 (2009)
8. E. Ayala, Prototipo de un sistema de ultrasonido aplicado a un baston blanco, in *ABYA YALA* (Cuenca, 2013)
9. J. Villanueva, R. Farcy, Optical device indicating a safe free path to blind people. Instrum. Measur. **61**, 170–177 (2012)
10. S. Kim, G. Hwang, H. Bae, Bat-G2 net: bat-inspired graphical visualization network guided by radiated ultrasonic call. IEEE Access **8**, 189673–189683 (2020)
11. D. Dakopoulos, N.G. Bourbakis, Wearable obstacle avoidance electronic travel aids for blind: a survey. Syst. Man Cybern. Part C Appl. Rev. **40**(1), 25–35 (2010)
12. MaxBotix Inc., *LV-MaxSonar EZ4 High Performance Sonar Range Finder Datasheet* (MaxBotix, 2005)
13. D. Tudor, L. Dobrescu, D. Dobrescu, Ultrasonic electronic system for blind people navigation, in *5th IEEE International Conference on E-Health and Bioengineering* (2015), pp. 19–21
14. BrainPort Technologies, *BrainPort V100* (2015)
15. Ultracane, *Ultracane: Smart White Cane*. https://www.ultracane.com/.
16. J. Sanchez, C. Oyarzun, Mobile assistance based on audio for blind people using bus services, in *New Ideas in Computer Science Education* (2007), pp. 377–396 (in Spanish)
17. M. Mahmud, R. Saha, S. Islam, Smart walking stick—an electronic approach to assist visually disabled persons. Sci. Eng. Res. **4**(10), 111–114 (2013)
18. D. Gbenga, A. Shani, A. Adekunle, Smart walking stick for visually impaired people using ultrasonic sensors and Arduino, in *Engineering and Technology* (*IJET*) (2017), pp. 3435–3447
19. Microchip, *PIC 16(L)F15324/44 Datasheet and Instructions*. http://ww1.microchip.com/downloads/en/devicedoc/40001889a.pdf
20. TDK Corporation, *PS1240P02BT Datasheet and Instructions*. https://product.tdk.com/en/search/sw_piezo/sw_piezo/piezo-buzzer/info/print_pdf
21. Sparkfun, *310-101 Shaftless Vibration Motor Datasheet*. https://www.sparkfun.com/datasheets/Robotics/310-101_datasheet.pdf
22. P. Tipler, Movimiento ondulatorio, in *Fisica Para la Ciencia y Tecnologia*, vol. 1 (2006), p. 460
23. S. Savransky, *Engineering of Creativity: Introduction to TRIZ Methodology of Inventive Problem Solving* (CRC Press, Florida, 2000)

Chapter 59
Smart White Cane for Unevenness Detection Using an Optical Device for Visually Impaired People Assistance

Fernando Yupa, Nicolas Pozo, and Edy Ayala

Abstract This article presents the development of a device adaptable to a white cane to assist the mobility of blind people. The objective of the proposal is to detect unevenness that represents a risk. The device consists of an optical sensor that sends signals to a microcontroller which generates alerts to the user according to the depth. It includes a distance sensor based on laser imaging detection and ranging (Lidar) technology adapted by a portable and easy-to-maintain mechanism. The proposal has been designed to be non-invasive with the user improving hand posture and avoiding common diseases caused by inadequate postures. In addition, the device includes a wireless charging technology to reduce manipulation by the user. The device incorporates a microcontroller which is responsible for obtaining signals which are processed to generate alerts to the user. The controller algorithm can be customized by users of different builds to allow operation in different conditions. The algorithm controller can be adapted to different users' physical complexions allowing convenient and extended use. This proposal seeks for a comfortable design allowing the user to maintain the natural hand position, reducing risks of body posture-related disease. The Lidar sensor and the implemented algorithm permit to detect unevenness which is a major threat for mobility. The device has been subjected to various tests using different materials and slopes to verify the correct operation obtaining satisfactory results over refracting surfaces.

59.1 Introduction

One of the most important needs for people in the independent mobility. In most of the cases, blind people have to encounter multiple obstacles. Some of them are more dangerous than others, besides, common places for transit do not offer an inclusive

F. Yupa · N. Pozo · E. Ayala (✉)
Universidad Politécnica Salesiana, Calle Vieja 12-30, Cuenca, Ecuador
e-mail: eayala@ups.edu.ec

F. Yupa
e-mail: lyupar@est.ups.edu.ec

Á. Rocha et al. (eds.), *Communication, Smart Technologies and Innovation for Society*, Smart Innovation, Systems and Technologies 252,
https://doi.org/10.1007/978-981-16-4126-8_59

architecture. This it is very important to locate and find practical solutions when the environment cannot be improved. However, usually, it is very difficult to offer free-obstacle paths for blind people. In this context, it is important to find alternative solutions that allow users to detect at least the most significant threats, such as big holes or any unevenness where a blind person could experience an accident.

According to the World Health Organization (WHO), at least 2.2 billion people worldwide suffer partial or complete loss of sight which is a significant number considering the global population [1]. Considering the high rates of people with visual disabilities, the development of devices that contribute to the inclusion of people with severe loss of sight in society must be granted. There are support mechanisms for mobility and recognition of the environment using different technologies such as the traditional white cane. Nowadays, novel solutions have been proposed including sensors such as ultrasonic [2], radio frequency [3, 4], and equipment added to different parts of the body supported on belts [5], helmets [6, 7], or the white cane [8]. These devices present similar characteristics for usability; however, the difference is based on how the signals are acquired and processed varying also the device cost, weight, size, maintenance, and ergonomics [9, 10].

The basis of various mobility assistance devices is the difficulty of blind people to identify obstacles. In this work, it is considered an essential aspect for the development of the proposal in order to propose a convenient device. Moreover, in this work, it is considered a non-invasive for the haptic and auditory signals which may have an impact in the person's proprioception. The device seeks to improve the ability to recognize obstacles by means of alerts such as sounds or vibrations [9]. Figure 59.1

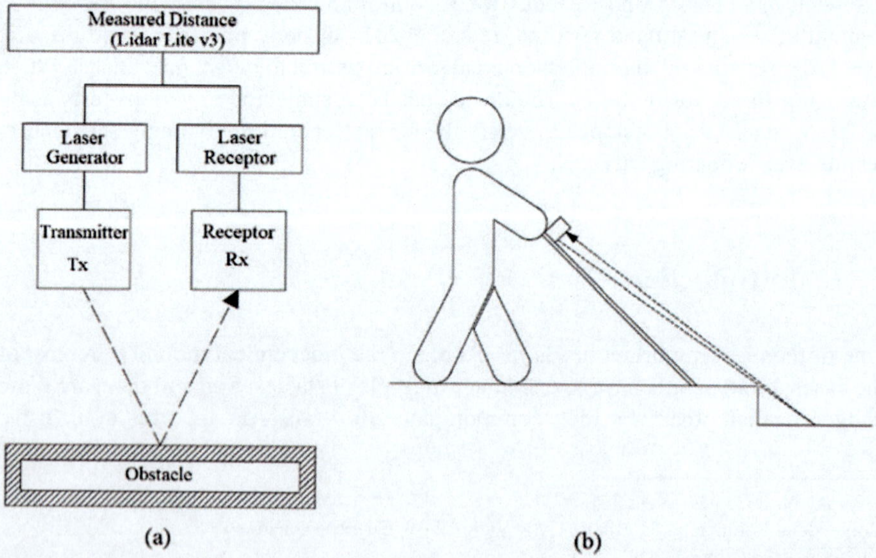

(a) (b)

Fig. 59.1 Lidar Lite v3 operation diagram (**a**) and obstacle detection with the sensor adapted to the smart white cane (**b**) shows the basics of the prototype operation principle. Traditionally, users utilize different assistive techniques such as white canes, trained dogs, among others [10].

The main objective in this work is to elaborate a device which improves people independence. In this research, different types of unevenness are considered to test representing real risk scenarios. Additionally, other factors are included for the design and test including different deeps and shapes for holes and stairs [11]. The operation of the equipment is based on an optical sensor that calculates the distance to an object thanks to the rebound time of the light signal by the means of the well-known Doppler effect. The Lidar sensor is capable to set a zero reference with respect to the surface in order to measure the distance to the obstacle. In Fig. 1a, it can be seen the sensor operation diagram [12, 13]. Consequently, the characteristics described require a microcontroller algorithm capable to generate precise obstacle alerts to the user.

The proposed device in this research is based on the obstacle detection with a device adapted to a white cane. Its main purpose is to assist the user's mobilization finding and alerting the presence of unevenness as shown in Fig. 59.1b. The system generates sound and vibration alerts when detecting objects or holes at a distance of 150 cm. The specific sensor is the Lidar Lite v3 operating range, and the processor is an Arduino [12], taking into account the characteristics and design conditions of a smart white cane as can be seen in Table 59.1 [14].

For the correct development of the presented device mechanism, several human factors have been considered that are presented in Table 59.2. During the development of this research, it has been included: anatomic-physiological factors based on biological requirements [15], anthropometric characteristics to observe general measurements of the body [16], psychical and mental limitations, and sociocultural factors [17].

In Sect. 59.2, the proposal of the device is developed, moreover, in Sect. 59.3, it shows the validation of the results of the tests carried out with the smart white cane. In Sect. 59.4, the results are discussed and finally, the conclusions are presented in Sect. 59.5.

Table 59.1 Characteristics of a smart white cane

Characteristics	Smart white cane
Weight	Heavy (0.62 kg)
Length	Customizable (120–155 cm)
Cost	Expensive (500$)
Comfort	Slightly uncomfortable
Mobility	Greater conformity
Orientation	Greater conformity
Maintenance	Yes

Table 59.2 Human factors for the proposed design of the smart white cane

Human factors	Environmental factors	Objective factors
Anatomic physiological	Temperature	Form
	Humidity	Volume
	Ventilation	Weight
Anthropometric	Illumination	Dimensions
	Color	Material
	Noise and sound	Finishes
Psychological	Vibration	Color
	Contamination	Technological texture
Sociocultural		Symbols and signs

59.2 Smart White Cane Design Considerations

The development of the proposal is based on the block diagram presented in Fig. 59.2, where the description of the smart white cane operation can be observed. For this system, the distance measured by the sensor provides a feedback signal which allows the correct operation of the algorithm. Moreover, a special consideration has been considered by including a wireless charging which has been proposed for better manipulation of the device and improves the user comfort.

Furthermore, the device uses the Atmel ATmega328 microcontroller detailed in [18], which allows to perform the tasks of measurement and processing of the signals obtained at the time of both measurement and self-calibration of the system. The main purpose is to provide appropriate alerts using signals generated by actuators similar to the ones used in mobile phones nowadays [19].

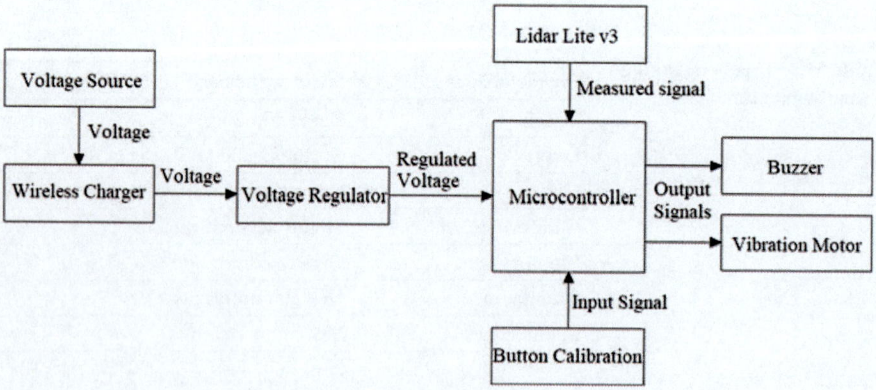

Fig. 59.2 Smart white cane architecture block diagram

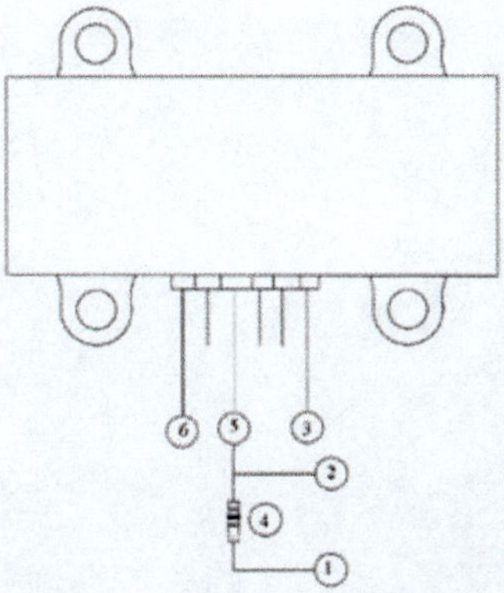

Item	Description	Notes
1	Trigger pin on microcontroller	Connect the other side of the resistor to the trigger pin on your microcontroller
2	Monitor pin on microcontroller	Connect one side of the resistor to the mode-control connection on the device, and to a monitoring pin on your microcontroller
3	Power ground (-) connection	Black Wire
4	1kΩ resistor	
5	Mode-control connection	Yellow wire
6	5 Vdc power connection	Red wire The sensor operates at 4.75 through 5.5 Vdc, with a max. of 6 Vdc

Fig. 59.3 Lidar wiring for PWM mode

For the development of the proposal, the electronic design of the device is proposed, where three development criteria are considered: power supply and load, data acquisition, and self-calibration. On the power supply, different elements are considered such as a 2000 mA lithium battery, responsible for supplying the operating voltage for the equipment, a wireless charging receiver which allows better handling by people with visual disabilities, also the voltage necessary for the operation of all the elements is obtained by a voltage regulator. For data acquisition, a pulse-width modulation (PWM) connection system is used with which the sensor works as detailed in [12], the printed circuit board (PCB) for the communication of the sensor with the microcontroller can be seen in Fig. 59.3.

On the other hand, in the electronic stage, the activation of the actuator and self-calibration is considered. The push-button in pull down programmatically sends the reference to the sensor which will allow a better detection of obstacles. For the protection of the actuator, a transistor configured as a common collector has been used. Once the design criteria fulfill the requirements listed above, an electronic board (11 × 4 cm) has been generated as can be seen in Fig. 59.4, respecting the criteria of easy maintenance, minimization of space, and easy access, the electronic board has been modeled in Ultiboard software.

Regarding the mechanical design, the proposal has a uniform distribution of elements, with protection from splashes of water and dust, as shown in Fig. 59.5. The device includes within the design a proposal for a handle adaptable to a white cane. This is an advantage for the user because it reduces the muscular effort generated in the wrist due to the use of a white cane daily. Due to the geometric distribution of a traditional white cane, the user is forced to generate pressure over the hand, wrist, and arm producing an unnecessary posture causing problems with long periods of use. These problems include inflammation, damage of joints, or ligaments, which causes diseases such as ulnar deviation [20]. The proposed design seeks to maintain the natural posture of the user's hand while manipulating the device. Thanks to these considerations from the initial phase of this research, the user will avoid possible

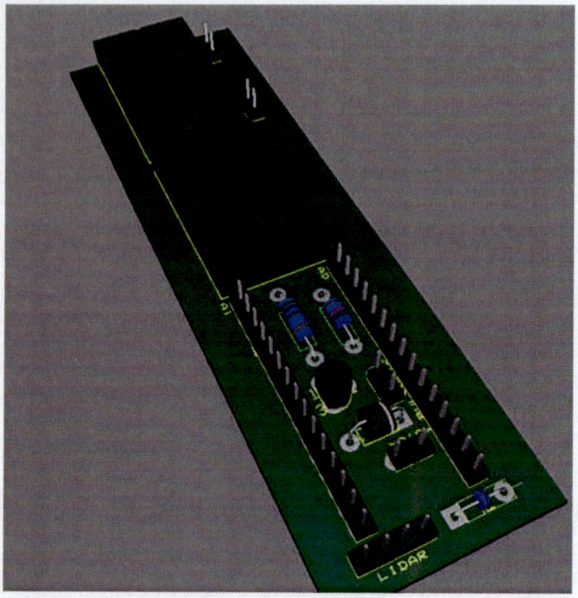

Fig. 59.4 Electronic board 3D view

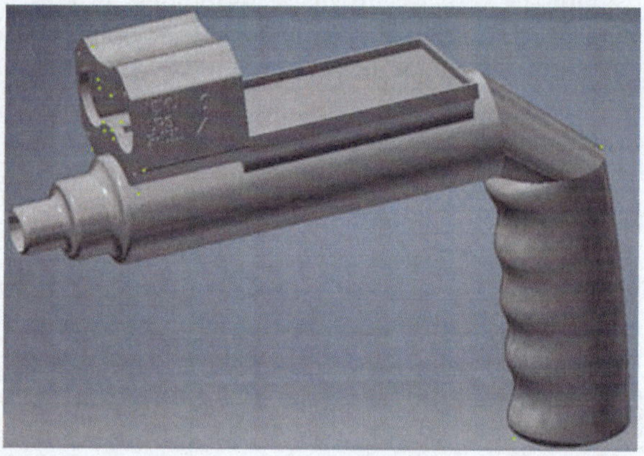

Fig. 59.5 Mechanical model design

risks and diseases related to the manipulation of the device with incorrect postures. A referential scheme displaying the incorrect and correct hand posture can be seen in Fig. 59.6. The mechanical model has been modeled in Autodesk Inventor software.

For the proposed control algorithm, the following flow chart has been proposed which defines the entire operating routine of the device and its self-calibration as shown in Fig. 59.7, the program logic is based on activating an actuator (Buzzer/Motor) to issue an alert signal to the user. The main program consists of an algorithm that compares a reference signal generated by the user against the continuous measurement of the device, the moment the distance is greater than the

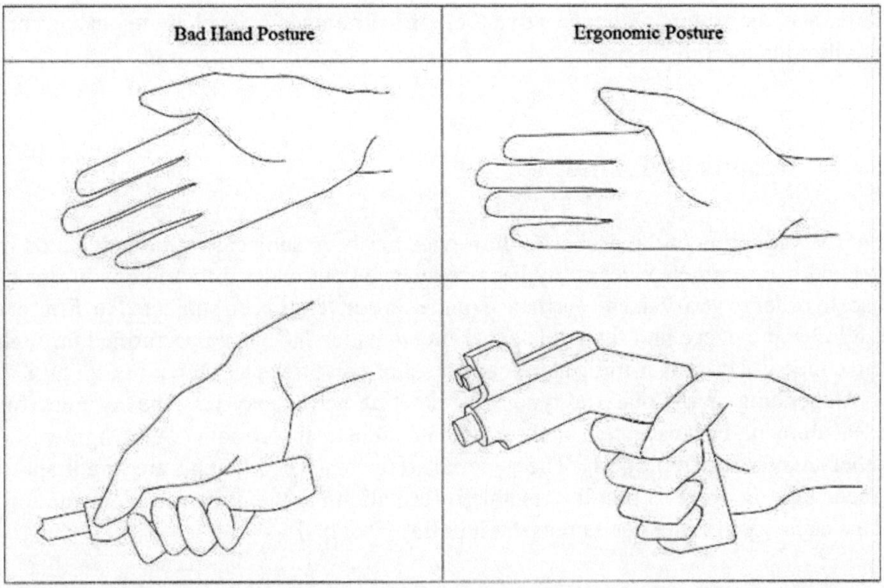

Fig. 59.6 Traditional withe cane bad hand posture vs smart white cane hand ergonomic posture

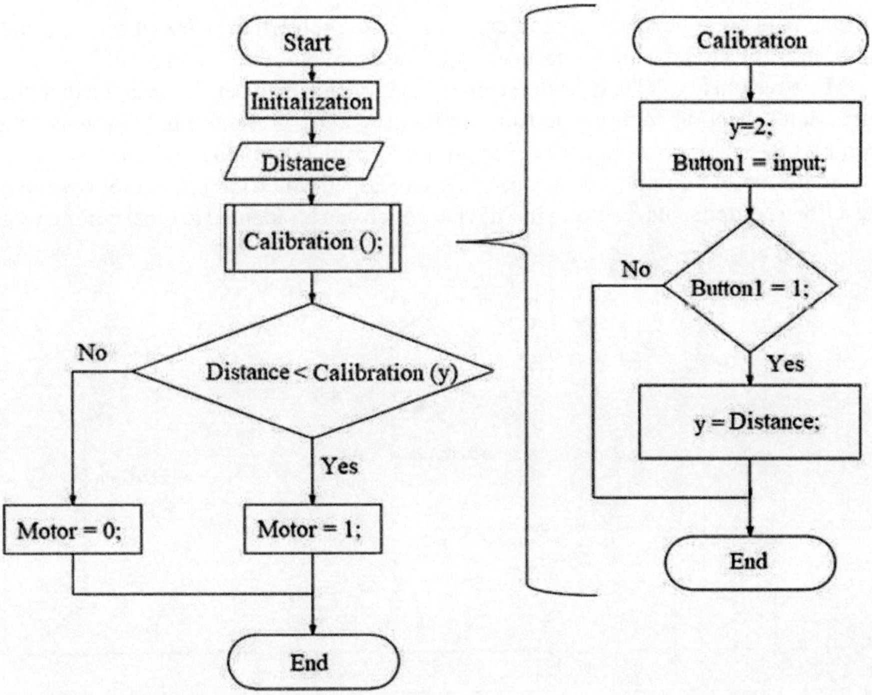

Fig. 59.7 Algorithm for the device operation diagram

reference, the system recognizes that there is a difference in level giving an auditory or vibration alert to the user.

59.3 Results and Analysis

For the validation of the proposal, the device has been subjected to tests as stated in [9] and measurement error study has been carried out under different conditions of use. In order to verify the performance of the sensor, it has been subjected to different angles of incidence and materials experiments which have been performed through the collection of sensor information and a set of conditions as shown in Fig. 59.8.

Depending on the material type, the reflecting waves may vary the accuracy for measurement. For instance, some surfaces are able to absorb some of the light while others may reflect it properly. The materials reflectance coefficients are important to understand in order to provide reliable information for the device implementation. This capacity of reflection is presented in Eq. (59.1):

$$R = \frac{(1-n)^2}{(1+n)^2} \tag{59.1}$$

where: R is the reflectance coefficient, and n is the refraction index of the material. The data collected during the tests are presented in Tables 59.3 and 59.4.

Moreover, in Fig. 59.9, it can be seen the relationship between pitch angle, distance error and reflection coefficients from a previous study presented in [13] where the complex behavior of the sensor could produce errors in the range of $\pm 10\%$.

Additionally, a compact device has been developed which integrates the previously described components resulting in a device which can be adapted to most commercial

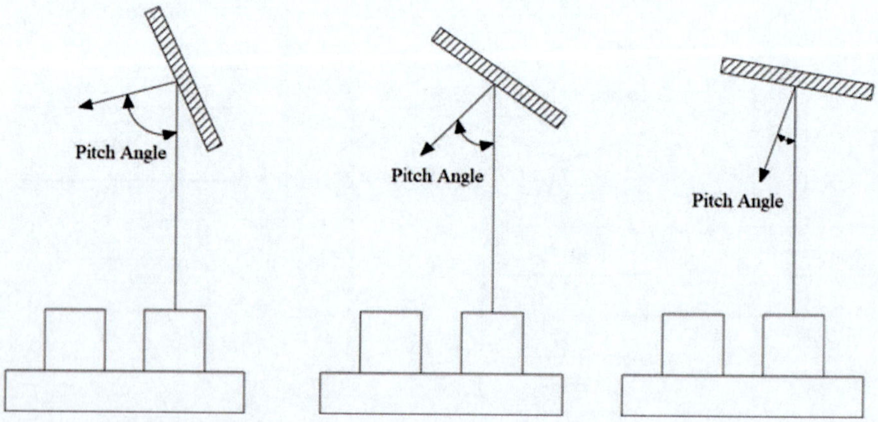

Fig. 59.8 Pitch angle test for the Lidar

Table 59.3 Measurement error with different angles

Angles (°)	Average measured distance (cm)	Real measurement (cm)	Percentage error (%)
20	93.74	93	0.789
25	98.76	98	0.769
30	99.58	100	0.422
35	110.66	112	1.211
40	127.54	128	0.361
45	133.09	132	0.826
50	155.34	156	0.425
55	156.95	158	0.665
60	250.38	250	0.151
65	261.18	262	0.314
70	265.68	268	0.873

Table 59.4 Measurement error with different types of materials with 45°

Material	Average measured distance (cm)	Real measurement (cm)	Percentage error (%)
Ceramics	95.71	91	4.921
Tile	109.62	102	6.951
Wood	110.42	115	4.147
Concrete	92.28	92	0.303
Porcelain	104.92	106	1.029
Asphalt	97.09	98	0.937
Paving stone	102.51	102	0.497
Grass	105.25	110	4.513
Cement	108.9	110	1.01
Sand	97.76	101	3.314

white canes easy to replicate. The manufacture of this prototype has been made by using a conventional 3D printer. This proposal has been elaborated using Polylactic Acid (PLA) which is characterized by its material strength as presented in Table 59.5 [21]. The final prototype built in PLA is shown in Fig. 59.10.

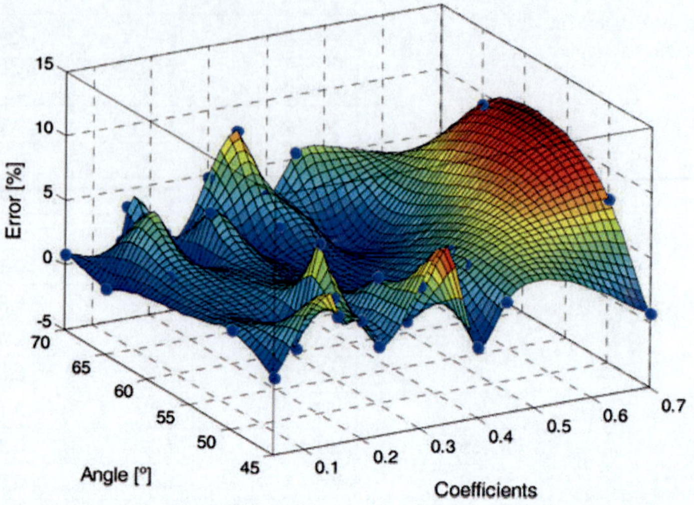

Fig. 59.9 Distance error depending on measurement at different angles and reflection coefficients presented in [13]

Table 59.5 PLA mechanical properties

Characteristic	Description
Density	1.4 g/cm^3
Modulus of elasticity (Young)	3.5 GPa
Elongation at break	6%
Flexural modulus	4 GPa
Flexural	80 MPa
Cutting module	2.4 GPa
Strength to weight ratio	40 kN-m/kg
Tensile strength (UTS)	50 MPa
Thermal conductivity	0.13 W/mK

59.4 Discussion

With the documentation raised and the tests carried out with the sensor, the device works adequately on non-refractory surfaces and for a certain range of angles. Moreover, the construction of the system has been possible in order to verify the adaptability to the white cane. However, it is recommended to carry out an exhaustive evaluation for more surfaces for complete validation.

On the other hand, 3D printing technology is used to generate the prototype and verify basic functionalities of the device, this being a long-lasting and expensive process, so it is recommended to use alternative manufacturing methods such as injection molding in order to reduce unitary cost by producing these devices in a

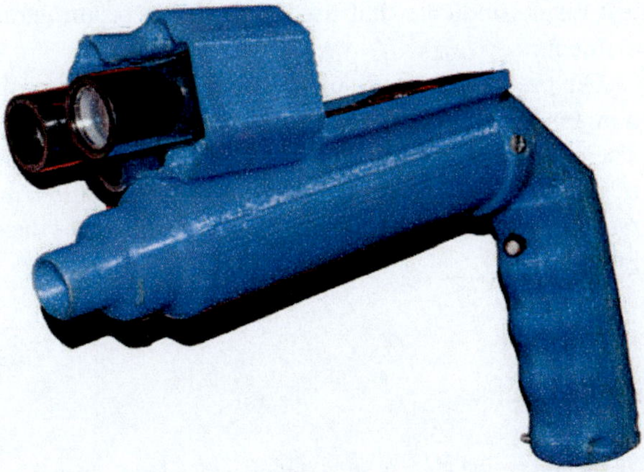

Fig. 59.10 Implementation of the proposed device

large scale. A factor that is not considered in the mechanical design is the calculation of normal, shear, and crush stress for the screws present in the device, which can generate long-term problems.

Regarding the qualitative part, the design of the equipment has been subjected to verification by a user who satisfactorily evaluated the autonomy, comfort, design, and adaptability criteria. In addition, commercial elements that are within the environment have been considered, this proposal is an alternative for other more expensive devices that do not have technical support or assistance, it should be noted that the proposal considers external help to maintain device.

Moreover, similar proposals [11] consider a low-cost device for the mobility of blind people, however, the design conditions are the differentiating factor of the device, which have been proposed methodologies [20] for easy-to-use and portable equipment.

59.5 Conclusion

The development of the device is to assist in the mobility of blind people generated adequate results. This proposal presented an easy adaptation to a commercial white cane in order to assist for unevenness location. This device has been designed for easy usage and maintenance. Moreover, it has fulfilled a set of tested conditions in order to provide reliable results of the system implementation. The proposal presents additional assistance through alerts in real time through the auditive and haptic signals produced by the device. It can be used by a greater number of users with different physical complexions improving ergonomics. In addition, using the Lidar sensor, it is possible to obtain a better response with respect to other measurement principles such as ultrasonic which provides greater security to the user. In this work, the presented test results indicate that this method can be implemented with relatively low-cost elements.

On the other hand, the mechanism has been designed considering multiple humans, environmental, and maintenance factors, to generate the greatest reliability. The proposed design improves the anatomical positions of the arms preventing risks of joints' diseases, ligaments, and muscles while using the device. Another important characteristic for this easy-to-handle device is the incorporated wireless charging module.

References

1. World Health Oragnization, *Informe Mundial Sobre la Visión* (*World Report on Vision*) (World Health Organization, Ginebra, 2020)
2. A. Sen, K. Sen, J. Das, Ultrasonic blind stick for completely blind people to avoid any kind of obstacles, in *2018 IEEE Sensor* (2018), pp. 1–4
3. L. Faggion, A. Graziano, Low-frequency RFID based mobility network for blind people, in *2011 IEEE International Conference on RFID-Technologies and Applications* (2011), pp. 364–369
4. M.E. Saaid, I. Ismail, M.Z.H. Noor, Radio frequency identification walking stick (RFIWS): a device for the blind, in *2009 5th International Colloquium on Signal Processing and Its Applications* (2009), pp. 250–253
5. L. Everding, L. Walger, V.S. Ghaderi, J. Conradt, A mobility device for the blind with improved vertical, in *2016 IEEE 18th International Conference on e-Health Networking, Applications and Services (Healthcom)* (2016), pp. 1–5
6. A. Brilhaut, S. Kammoun, O. Gutierrez, P. Truillet, C. Jouffarais, Fusion of artificial vision and GPS to improve blind pedestrian positioning, in *4th IFIP International Conference on New Technologies, Mobility and Security* (2011), pp. 1–5
7. J. Liu, J. Liu, L. Xu, W. Jin, Electronic travel aids for the blind based on sensory substitution, in: *2010 5th International Conference on Computer Science and Education* (2010), pp. 1328–1331
8. M. Hazzaz, R. Saha, S. Islam, Smart walking stick—an electronic approach to assist visually disabled persons. Int. J. Sci. Eng. Res. **4**(10) (2013)
9. W. Elmannai, K. Elleithy, Sensor-based assistive devices for visually-impaired people: Current status, challenges, and future directions. Sensors 17 (2017)
10. R. Gulati, GPS based voice alert system for the blind. Int. J. Sci. Eng. Res. **2**, 1–5 (2011)
11. S. Ponnada, S. Yarramalle, M. Rao, A hybrid approach for identification of manhole and staircase to assist visually challenged. IEEE Access **6**, 41013–41022 (2018)
12. S. Electronics, *LIDAR-Lite v3 hookup guide*. https://learn.sparkfun.com/tutorials/lidar-lite-v3-hookup-guide/all. Last accessed 17 Mar 2021
13. E. Ayala, J. Sotamba, B. Carpio, O. Escandon, Lidar Lite v3 module performance evaluation, in: *2018 IEEE Third Ecuador Technical Chapters Meeting (ETCM)*, pp. 1–5 (2018).
14. E. Ayala, *Prototype of an Ultrasound System Applied to a White Cane*, 1st edn (ABYA YALA, Cuenca, 2013) (in Spanish)
15. H. Rouviere, A. Delmas, *Descriptive, Topographic and Functional Human Anatomy. Volume 3: Members*, 11th edn (Masson, Barcelona, 2005), pp. 294–312 (in Spanish)
16. M. Vergara, M. Agost, *Anthropometry Applied to Product Design*, 1st edn (Universitat Jaume I, Spain, 2015) (in Spanish)
17. C. Flores, *Ergonomics for Design*, 1st edn (D. R. Libraría, Mexico, 2001) (in Spanish)
18. Atmel, *ATmega328P Datasheet*, https://ww1.microchip.com/downloads/en/DeviceDoc/Atmel-7810-Automotive-Microcontrollers-ATmega328P_Datasheet.pdf. Last accessed 12 Mar 2021

19. TDK Corporation, *PS1240P02BT Datasheet*, https://product.tdk.com/en/search/sw_piezo/sw_piezo/piezo-buzzer/info/print_pdf. Last accessed 12 Mar 2021
20. Daniels and Worthingham, *Muscle Balance Techniques*, 9th edn (Elsevier, Spain, 2018), pp. 156–170 (in Spanish)
21. PrintaLot Inc, *PLA Material Specifications*, https://www.printalot.com.ar/TDS/TDS-PLA.pdf. Last accessed 15 Apr 2021

Chapter 60
Functional Requirements for Management and Control of Public Transportation Vehicles, Applied to Sustainable Mobility in Medium-Sized Cities

Tomas Ramirez-Guerrero, Mauricio Toro, Gustavo Villegas López, and Leonel Castañeda

Abstract Public transportation systems stimulate the economy of cities, while affecting their infrastructure and influence the lives of citizens. For this reason, sustainable mobility is one of the most challenging areas for medium-sized cities. Factors such as delay in the technologies integration and increased motorization favor the increase of emissions, the rise of informal transportation, and the detriment of public transportation in these cities. To improve the operation and management of public transportation with the help of technology, it is essential to identify the technological needs to be covered in the public transportation systems of medium-sized cities. In this paper, an exploratory analysis was conducted to identify the functional requirements of the fleet management and control system for public transportation in the medium-sized cities of Colombia. Using qualitative analysis tools, the information gathered through interviews with public transportation administrators in different cities of the country was analyzed. Results show the functionalities required for the control of public transportation vehicles, the service delivery preferences in fleet management and control systems, and the challenges faced by public transportation systems of the medium-sized cities to be sustainable in time.

T. Ramirez-Guerrero (✉) · M. Toro · G. Villegas López · L. Castañeda
Grupo de Estudios en Mantenimiento (GEMI), Universidad EAFIT, Medellín, Colombia
e-mail: teramirezg@eafit.edu.co

T. Ramirez-Guerrero · M. Toro · L. Castañeda
Semillero de Investigación en Sistemas Inteligentes de Transporte (SiSIT), Universidad EAFIT, Medellín, Colombia

M. Toro
Grupo I+D+I en Tecnologías de la Información y las Comunicaciones (GIDITIC), Universidad EAFIT, Medellín, Colombia

Á. Rocha et al. (eds.), *Communication, Smart Technologies and Innovation for Society*, Smart Innovation, Systems and Technologies 252,
https://doi.org/10.1007/978-981-16-4126-8_60

60.1 Introduction

Public passenger transportation systems affect the infrastructure of cities, directly stimulate their economy, and influence the lives of their citizens. Given the densification of cities and the growing mobility needs of citizens, smart urban public transportation systems help to improve the quality of life and the environment in cities [1].

In recent years, motorization rates have increased steadily in Latin America, prioritizing the private vehicle to the detriment of public transportation. In the Colombian context, the annual increase of motorcycles and light vehicles is 14.7% and 6.6%, respectively [2]. An example in medium-sized cities is the case of Montería, Colombia, where the high number of individual trips reflects the high levels of use and possession of motorcycles. This has led to increased emissions, increased informal transport, and negative impacts on public transportation and road safety in the city [3].

To improve passenger transportation in the Latin American region, various reforms have been applied to transform the massive and regional public transportation systems [4, 5]. In Colombia, the Strategic Public Transportation Systems (SETP, in Spanish) are organizational structures designed to improve the provision of public transportation service in medium-sized cities. The objectives of SETPs are aimed at achieving safe, equitable, integrated, efficient, accessible, and environmentally sustainable mobility in the cities where they are implemented [6]. To achieve their objectives, SETPs must implement solutions based on information and communication technologies (ICT) to optimize the management and control of fleets, the electronic collection of rates, and the information delivery to users [7], while must integrate services of the Intelligent Transportation Systems (ITS) for vehicle fee collection, traffic lights, among others [8].

However, most SETPs have not started their operation due to delays in the adaptation of the road infrastructure and acquisition of the information systems that must be integrated for the management and control of fleets [7]. Added to these delays are the impacts of the COVID-19 pandemic on public transportation systems due to population confinement, social distancing, and fear of contagion. As part of the measures to contain the spread of the virus, the occupancy capacity in public transportation systems was limited to 35%, which generated a significant deficit in these systems [9].

As in other economic sectors, the transportation sector has benefited from advances in mobile communications, computer systems, and sensor technologies. Since the 1970s, technological advances in transportation have been the basis of the research fields addressed by ITS [10]. For this reason, it is necessary to work on the identification of IT services and technologies to be incorporated in the SETPs, to design cost-efficient solutions based on ICT and ITS. To achieve this, it is necessary to identify the functional requirements associated with fleet management and control in these transportation systems.

In this paper, an exploratory study was conducted to characterize the functional requirements of the fleet management and control systems for medium-sized cities in Colombia. In the development of this research, data was collected through interviews and observations made to the SETPs, focused on the existing technological needs for the management and control of public transportation vehicles. Results obtained show the prioritizations and functionalities for the fleet management and control systems in the SETPs, while reflecting the challenges currently facing public transportation systems in medium-sized cities.

60.2 Methods

In this work, the methodology design science research (DSR) in information systems [11] was used to conduct the exploratory study about the features and functionalities that the fleet management and control systems must have in medium-sized cities of Colombia. For the first phase of this research, the first two stages of the methodology were adopted: awareness of problem and solution suggestion.

60.2.1 Awareness of Problem

In this stage, a series of semi-structured interviews were conducted with the operation managers and technology directors of the public transportation systems in the medium-sized cities of Colombia, to know about the status of these systems.

The sample for this study corresponds to the medium-sized Colombian cities that have a SETP. A total of nine interviews were conducted: eight interviews were conducted in the medium-sized cities where the SETPs are currently implemented, and one additional interview was conducted in the city of Medellin, a Latin American reference for innovation and good practices in public passenger transportation. The questions and the steps followed in the interviews were elaborated following the interview design process of Portigal [12]. Due to the health emergency generated by the COVID-19 pandemic, the interviews were conducted through one-hour virtual sessions. Interviews included 12 basic questions and 6 supplementary questions aimed at identifying the status and technological needs of the public transportation systems in the cities studied. During the nine interviews conducted, 16 officers (operations managers and technology directors) were consulted. On average, each of the officers interviewed has been working with the entity for more than five years.

Figure 60.1 shows the geographic location of the medium-sized cities in Colombia where the interviews were conducted; the bubbles size is related to the number of inhabitants of each city interviewed, Santa Marta being the city with greater number of inhabitants, and Popayán the city with the least number of inhabitants.

Along with the literature review, with the information obtained from the interviews met the current state of implementation of the SETPs, the delays and limitations in

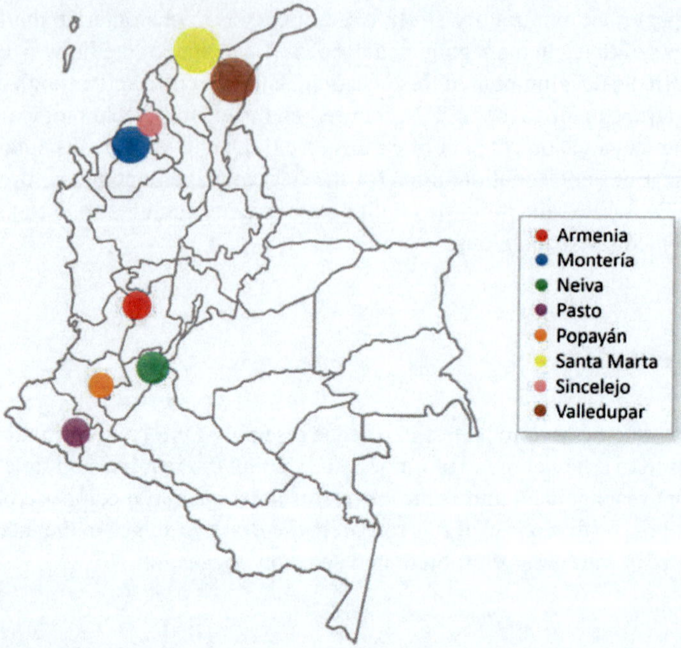

Fig. 60.1 Geographical location of the medium-sized cities of Colombia that were interviewed

the fully operational implementation of these systems, and the decrease in public transportation users caused by informal and illegal transportation and the COVID-19 pandemic [7, 13, 14].

60.2.2 Suggestion

The information collected in the interviews was analyzed using natural language processing techniques (NLP) and qualitative analysis tools, to identify the challenges, prioritizations, and expectations of each city for the management and control of fleets in its public transportation systems. The analysis included the categorization and interpretation of the information looking for similar patterns between the answers given by the operations managers and technology directors interviewed [15].

Based on the relations found, the functional requirements for IT services to be incorporated into public transportation in the cities which have a SETP are presented. From this, the requirements of interest for the design of a platform to support the operation and decision making based on data, for the studied public transportation systems, were selected.

60.3 Results

From the analysis and interpretation of the information obtained during the interviews, the challenges, the prioritizations, and the functional requirements associated with the public transportation systems of the medium-sized cities interviewed were identified.

60.3.1 *Current Challenges for Strategic Public Transportation Systems*

Strategic Public Transportation Systems must face several challenges to ensure that their operation is sustainable in time. These challenges are oriented toward the increase of passengers transported in the system and the technological adoption that must be integrated for the fleet management and control.

Decrease in users of the public transportation system in the medium-sized cities has been produced by factors such as the informal and illegal transport, the increase in the motorization rates, the COVID 19 pandemic, among others. This one of the operation managers interviewed reflects the consequences of the decrease of passengers:

> We design a system so that the number of people entering the system is sufficient to satisfy the financial model of the system... But what is happening now? The problems of motorcycle-taxis, lack of control of paratransit, motorization, and new elements such as the electric mobility, all of this does not allow us to reach that level of sustainability that we seek... More passengers are needed to close our financial model.

Table 60.1 shows the number of passengers transported by the SETP of each city, as well as the percentage of decrease in each system. It was observed that, even before the pandemic, in most medium-sized cities, the number of users of the

Table 60.1 Number of users in the public transportation systems of the cities interviewed

City	Passengers required by transport system	Passengers before COVID-19	Passengers after COVID-19	Percentage decrease (%)
Armenia	126,000	75,000	25,000	66.6
Neiva	98,000	74,000	12,000	83.4
Pasto	125,000	100,000	35,000	65.0
Popayán	127,000	109,000	10,000	90.8
Santa Marta	170,000	200,000	35,000	82.5
Sincelejo	24,000	6,500	5,400	16.9
Valledupar	58,000	25,000	5,800	76.8

Figures are for November 2020

public transportation system is less than the number required by the system. To compensate the economic limitations due to lack of passengers, the municipalities of each city contribute resources to the local public transportation system through the rate stabilization funds.

Otherwise, the technological adoption required for the public transportation systems of the cities interviewed must incorporate systems and services of the Intelligent Transport Systems. To that end, large investments must be made both in the acquisition and in the operation of the technology. Faced with this scenario, the comments made by several of the technology directors interviewed express that the technologies to be acquired must "*work with open communication protocols, be adjustable to each city, be interoperable, and be low-cost.*"

60.3.2 Prioritization of IT Services for Fleet Management and Control Systems

In the interviews, questions were asked about the functionalities that must be included in the fleet management and control systems of public transportation in the medium-sized cities of Colombia. Questions done to operations managers and technology directors interviewed included approaches such as the continuous monitoring and control of bus speed, route compliance, among others. Figure 60.2 shows the relevance of the functions associated with IT services in the public transportation system

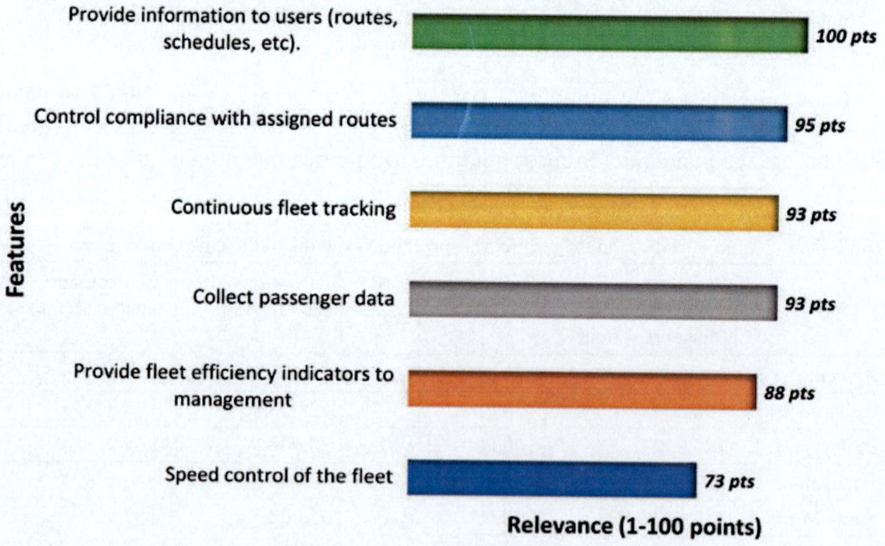

Fig. 60.2 Relevance of the functions associated with IT services in the SETPs

of the cities interviewed, on a scale of 1–100 points where 1 is nothing important and 100 is very important.

It was found that the IT services intended to provide information about the system to users are the most relevant, followed by the services intended to control the compliance with the routes assigned to public transportation drivers. One of the technological directors made the following comment about providing information to users:

> Part of all of this technological theme... is also to implement the user information system, this really makes it possible for us to have a greater demand capture, that the user feels happier in terms of service reliability... then this is one of the most important components be implemented in the SETP.

60.3.3 Functional Requirements for Fleet Management and Control System

To make a good description of the functional requirements for the public transportation system, the information provided by the stakeholders must be involved. Questions were asked about the requirements that the fleet management and control system must have in each medium-sized city, and diverse answers were obtained with various points of relation between them. The following are the most outstanding requirements for these systems, according to the SETP operation managers.

- Fleet scheduling
- Real-time bus location and traceability
- Cloud-based data storage
- Visualization and control of data in the operations center
- Passenger count
- Panic button and incident management
- APIs for connection with user information systems
- Interoperability with other information systems.

With these functionalities, the managers of the public transportation systems of the cities interviewed will be able to collect the necessary information to calculate the reporting indicators in the urban transportation information, monitoring, and evaluation system (SISETU, in Spanish) of the Ministry of Transport of Colombia [16]. Indicators related to the fleet management and control systems are number of kilometers executed, number of passengers, speed on routes during business day, total fuel consumption, among others.

60.3.4 Proposals for the Implementation of the Requirements of the Fleet Management and Control System

Current technological trends such as the Internet of Things, Cloud Computing and Fog Computing, Open Hardware and Open Software, Machine Learning, among others, allow us to integrate IT solutions that lead to the implementation of the requirements requested for fleet management and control systems in public transportation. Next, proposals for collection, processing, and visualization of data in public transportation are presented. Figure 60.3 represents the general flowchart for data handling in an Intelligent Transportation System, where the three stages mentioned above can be observed.

Data collection. Using mobile devices or onboard devices in the vehicle on board the vehicle, it is possible to capture public transportation data without resorting to specialized and high-cost technology. Proposals for capturing data through mobile devices identify the advantages and efficiency levels of using smartphones to collect user and bus data, their low cost compared to specialized devices, their facility to connect to mobile telecommunications networks, and the use of mobile apps created to provide information and evaluate aspects of public transportation [17–19]. On the other side, proposals focused on the use of onboard equipment present designs of automatic vehicle location (AVL) devices using LoRaWAN telecommunication networks, low-cost embedded systems, and geolocation through open standards global navigation satellite systems (GNSS) [20–22].

Data processing. Large volumes of data produced by public transportation systems can be processed using advanced trends in data analytics. Proposals aimed at the use

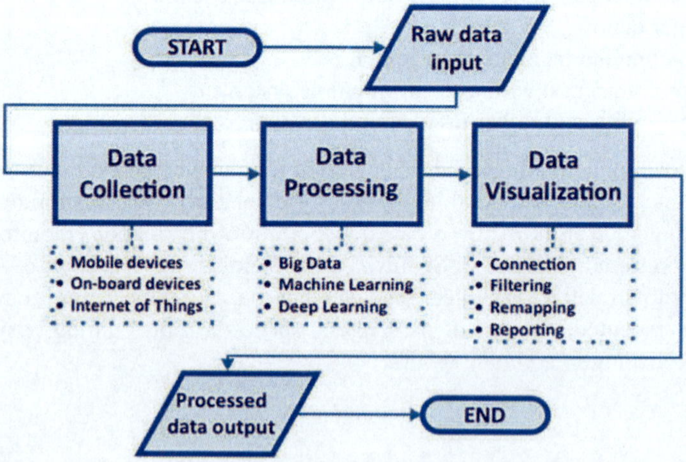

Fig. 60.3 Generic flowchart for data handling in an Intelligent Transportation System

Fig. 60.4 Mockups designed for the platform to support the operation of SETPs

of Big Data in public transportation indicate techniques for the integration, aggregation, fusion, and management of data from different sources; prioritization of data processing uses the support vector machines (SVM) algorithm; and the use of ontology-based methods to describe semantics and infer implicit relationships in data [23, 24].

Data visualization. New data visualization techniques in public transportation strategic plans support transportation system operators and managers in decision making. In the design of data visualization systems, strategies and techniques of data selection, connection, filtering, repairing, and remapping must be considered [25].

60.3.5 Work in Progress

In the next phase of this research, the development of a web platform that manages the processing and visualization of public transportation indicators is contemplated, aimed at assisting in making strategic decisions to management bodies of SETPs in medium-sized cities of Colombia. Figure 60.4 shows mockups designed for the platform based on the needs and functional requirements identified in this work.

60.4 Conclusions

In the medium-sized cities of Colombia, public transportation systems face different challenges related to the decrease of users, economic limitations, and the integration of technological solutions required by national legislation. This work identified the functional requirements for the fleet management and control systems requested by the managers of the public transportation in these cities. The interviews revealed a high level of interest from the operation managers in providing information to users based on data from the fleet management and control system, with the purpose of encouraging the use of public transportation and achieving the sustainability goals related to the number of transportation system passengers.

Based on the identified needs, it is possible to propose vanguard IT technologies and services to achieve the complete implementation of the fleet management and control systems in the eight medium-sized cities in Colombia that have a SETP. Interesting proposals in low-cost sensor technologies, cloud-based storage and processing data, predictive algorithms to support decision making, among others, can optimize the management of public transportation vehicles. Likewise, the recommendations presented serve as the basis for designing and integrating information systems in the four cities of Colombia that are structuring their SETPs.

The next phase of this research focuses on the design of a platform for fleet management and control in SETPs, adopting the functional requirements described in this work. Therefore, future research is proposed in public transportation systems aimed at the development, deployment, and evaluation of data-based decision support systems, transformational transportation strategies for the adoption of international standards, and development of low-cost IT solutions for public transportation.

References

1. R.M. Martinod, N. Rezg, L.F. Castaneda, O. Bistorin, Combined optimization methodology for operational planning and service policies: a case study of urban ropeway transport. Int. J. Transp. Econ. **42**(3), 337–363 (2020)
2. A. Jehanno, H. Niang, J. Ortíz, P. Laborde, P. López Camacho, *Desafíos Para la Integración de Sistemas de Transporte Masivo: Manual de Buenas Prácticas* (Banco de Desarrollo de América Latina, Caracas, 2019)
3. L.M. Quiñones, T. Van Laake, El estado del transporte urbano en América Latina, in *Transporte Urbano Sostenible en América Latina: Evaluaciones y Recomendaciones Para Políticas de Movilidad*, ed. by M. Moscoso, T. Van Laake, L.M. Quiñones, C.F. Pardo, D. Hidalgo (Despacio, Bogotá, 2019), pp. 11–18
4. A. Gómez-Lobo, Transit reforms in intermediate cities of Colombia: an ex-post evaluation. Transp. Res. Part A Policy Pract. **132**, 349–364 (2020)
5. GSD+, *Esquemas de Implantación de Tecnologías Inteligentes de Transporte en América Latina: Estudios de Casos y Recomendaciones* (Banco de Desarrollo de América Latina, Caracas, 2019).
6. Ministerio de Transporte, *Decreto 3422 de 2009 Reglamentación de los Sistemas Estratégicos de Transporte Público* (Ministerio de Transporte, Bogotá, 2009)
7. M. Rodríguez Porcel, J. Bernal González, Ó.M. Beltrán Real, *Programa de Sistemas Estratégicos de Transporte Público (SETP). Caso de estudio Colombia* (Banco Interamericano de Desarrollo, Washington, 2018)
8. Ministerio de Transporte, *Decreto 2060 de 2015* (Ministerio de Transporte, Bogotá, 2015)
9. *Sistemas de transporte público, entre crisis y cambios*, https://ieu.unal.edu.co/en/medios/noticias-del-ieu/item/sistemas-de-transporte-publico-entre-crisis-y-cambios. Last accessed 17 Mar 2021
10. C.R. García, A. Quesada-Arencibia, T. Cristóbal, G. Padrón, F. Alayón, Systematic Development of Intelligent Systems for Public Road Transport. Sensors, **16**(7), 1104 (2016)
11. V. Vaishnavi, B. Kuechler, S. Petter, *Design Science Research in Information Systems* (2004)
12. S. Portigal, *Interviewing Users: How to Uncover Compelling Insights* (Rosenfeld Media, Brooklyn, 2013)
13. A.M. Sánchez-Jabba, *La economía del Mototaxismo: El Caso de Sincelejo* (Banco de la República, Bogotá, 2011)

14. J. Valencia, T. Ramirez-Guerrero, L. Castañeda, M. Toro, Detección de infracciones y matrículas en motocicletas, mediante visión artificial, aplicado a Sistemas Inteligentes de Transporte. RISTI Rev. Iber. Sis. Tecnol. Inf. 37, 1–15 (2020)
15. A. Strauss, J. Corbin, *Bases de la Investigación Cualitativa. Técnicas y Procedimientos Para Desarrollar la Teoría Fundamentada* (Universidad de Antioquia, 2002)
16. Ministerio de Transporte, *Resolución 6609 de 2019 Reglamentación del Sistema de Información, Seguimiento y Evaluación al Transporte Urbano* (Ministerio de Transporte, Bogotá, 2019)
17. T. Ramirez-Guerrero, M. Toro, G.A. Villegas López, L.F. Castañeda, Low-cost computational systems applied to physical architectures in public transportation systems of intermediate cities. J. Phys. Conf. Ser. **1702**, 012018 (2020)
18. J. Buele, L.F. Salazar, S. Altamirano, R.A. Aldás, P. Urrutia-Urrutia, Plataforma y aplicación móvil para proporcionar información del transporte público utilizando un dispositivo embebido de bajo costo. RISTI Rev. Iberi. Sis. Tecnol. Inf. E17, 476–490 (2019)
19. E. Simonyi, Z. Fazekas, P. Gáspár, Smartphone application for assessing various aspects of urban public transport. Transp. Res. Proc. **3**, 185–194 (2014)
20. S. Morales, C. Pedraza, F. Restrepo-Calle, F. Vega, V. Bastidas, Analysis of requirements for secure automated vehicle location devices for public transportation systems in Colombia. Ing. Desarr. **36**(2), 298–320 (2018)
21. F. Jurado Murillo, J.S. Quintero Yoshioka, A.D. Varela López, R. Salazar-Cabrera, Á. Pachón de la Cruz, J.M. Madrid Molina, Experimental Evaluation of LoRa in Transit Vehicle Tracking Service Based on Intelligent Transportation Systems and IoT. Electronics, **9**(11), 1950 (2020)
22. H. Zhou, K.M. Hou, D. Zuo, J. Li, Intelligent urban public transportation for accessibility dedicated to people with disabilities. Sensors **12**(8), 10678–10692 (2012)
23. L. Zhu, F.R. Yu, Y. Wang, B. Ning, T. Tang, Big data analytics in intelligent transportation systems: a survey. IEEE Trans. Intell. Transp. Syst. **20**(1), 383–398 (2019)
24. S. Ftaimi, T. Mazri, Handling priority data in smart transportation system by using support vector machine algorithm. Adv. Sci. Technol. Eng. Syst. J. **5**(6), 1422–1427 (2020)
25. C.A. Steed, Interactive data visualization, in *Data Analytics for Intelligent Transportation Systems*. ed. by M. Chowdhury, A. Apon, K. Dey (Elsevier, Amsterdam, 2017), pp. 165–190

various branches of science and engineering; some examples include electromagnetic waves, sound waves, vibrations of strings, or membranes, seismic waves, traffic density waves, waves on the surface of a liquid, [1–3] among others. In this paper, a numerical solution of the Heaviside equation is presented. Mathematical and computational tools are used to obtain this solution, which allows using this example for didactic and research purposes.

The study and understanding of electromagnetism topics require high analysis and abstract reasoning processes due to its visual and mathematical complexity. Therefore, computational tools are used to simulate, analyze, and visualize the nature of electromagnetic phenomena in support of the theoretical framework used, which this research is the Heaviside equation.

The Heaviside equation is used to study the propagation of waves associated with the phenomenon of atmospheric discharges [1, 4, 5]; it is also known as the telegrapher's equation. The complexity of this natural phenomenon makes it difficult to find analytical solutions to the equations that describe it, so the need arises to use numerical models supported by increasingly advanced software and hardware. As an extension of the work [1, 6–8], the computational aspects of major relevance in a simulation are considered.

Oliver Heaviside developed the so-called telegraph equations, which form the theory for communication transmission lines but are fully applicable to the design of power transmission lines. The equations consist of two linear differential equations as a function of distance and time: one for $V(x, t)$ and one for $I(x, t)$. The model demonstrates that electrical energy can be reflected from the line and that known wave patterns can be formed [9–11].

The telegraph equations are simplifications of Maxwell's equations. For practical purposes, the conductor is assumed to be composed of bipoles (quadrupoles) [7, 12–15], each of which represents an infinitesimal segment of the transmission line. An infinitesimal segment of a transmission line has four distributed parameters, called primary parameters, which are mentioned below:

The distributed inductance [expressed in henries per unit length (PUL)] due to the magnetic field around the conductor is represented as a single coil in series L. This parameter models the process of energy storage in the form of a magnetic field occurring in the line [1, 16–18].

The distributed capacitive behavior (expressed in farads PUL) due to the electric field existing in the dielectric between the line conductors is represented by a single parallel capacitor C, located between "the forward conductor" and "the return conductor." The parameter C models the process of energy storage in the form of an electric field that occurs in the line.

The distributed resistance in the conductor (expressed in ohms PUL) is represented by a single series resistance R. This parameter models the energy dissipation due to the non-ideality of the conductors (ohmic losses).

The distributed conductance (expressed in mhos PUL or siemens PUL) is represented by a parallel conductance G, placed between "the forward conductor" and "the return conductor." The parameter G models the power dissipation that occurs due to the non-ideality of the dielectric medium (dielectric losses). This segment is

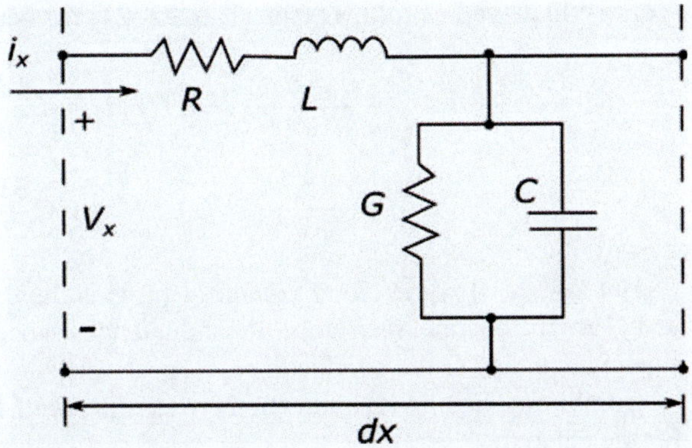

Fig. 61.1 Primary parameters of the transmission line

characterized by four distributed parameters (see Fig. 61.1), also commonly known as the primary parameters of the transmission line.

Where *ix* is the wave function of the transient current, and *Vx* is the wave function of the transient voltage. The inductance "*L*" models the energy storage process in the form of a magnetic field that occurs in the line, and it is expressed in henries per unit length (*h*/length). The capacitance "*C*" models the energy storage process in the form of an electric field that occurs in the line and is expressed in Farads per unit length (*f*/length). The resistance "*R*" models the power dissipation due to the non-ideality of the conductors (ohmic losses), and the conductance "*G*" models the power dissipation due to the non-ideality of the dielectric medium (dielectric losses).

It is in the transient regime where it will be observed how the electromagnetic wave evolves when it impacts with the transmission line defined as a function of time, position and initial conditions and the behaviour of the current (I) will be visualized. While for the steady state analysis, there is no time dependence, achieving a simpler and easier to understand visualization, however, its study leads to a set of mathematical models whose solutions are cumbersome and uninteresting. [11, 19–21]. The structure of the paper is as follows: Sect. 61.2 contains the description of the physical and mathematical models. The numerical model together with its implementation, convergence, and stability tests, and results are shown in Sect. 61.3. The most important conclusions are presented in Sect. 61.4.

61.2 The Physical Model and the Mathematical Model

The physical description of the system under study is as follows: The cloud-ground system forms a cylindrical capacitor whose separation is the discharge channel, which is assumed to be completely straight (no tortuosity). The cylindrical capacitor is 1 km [1, 6] apart, and the discharge channel is about 20 cm in diameter. In the simulation of the problem, a typical height of 1000 m is taken which is divided into 250 elements with a time step, Δt, of 0.5 μs. The system of equations that allows the study in this work is displayed as follows (see [1, 6, 22–24] for details).

$$-\frac{\partial V(x,t)}{\partial x} = RI(x,t) + L\frac{\partial I(x,t)}{\partial t} \tag{61.1}$$

$$-\frac{\partial I(x,t)}{\partial x} = GV(x,t) + C\frac{\partial V(x,t)}{\partial t} \tag{61.2}$$

Here, $V(x, t)$ is the cloud voltage, $I(x, t)$ is the discharge current, and R, L, C, and G are the channel resistance, the inductance associated with the discharge, C the capacitance, and G the conductance, respectively.

Combining Eqs. (61.1) and (61.2) yields the well-known telegrapher's equation for the current:

$$\frac{\partial^2 I(x,t)}{\partial x^2} - (RC+GL)\frac{\partial I(x,t)}{\partial t} - LC\frac{\partial^2 I(x,t)}{\partial t^2} - RGI(x,t) = 0 \tag{61.3}$$

It is easy to show that Eq. (61.3) has the same form for the lightning voltage. In this case, it is solved numerically to determine the return discharge current with appropriate initial and boundary conditions. Also, real parameters allow describing the atmospheric discharge accurately. It should be noted that this equation has been solved on other occasions, and the authors make assumptions that greatly simplify it [1, 10, 25–27].

61.3 Numerical Development and Implementation

For the discretization of Eq. (61.3), the basic idea of finite differences is to replace continuous space–time with a discrete set of points. The time step between two consecutive levels is denoted Δt and the distance between two adjacent points in space Δx [7]. The differential equations are then replaced by a system of discrete equations. This is achieved by approximating the differential operators by their corresponding finite differences taking into account the function values at adjacent points in the mesh. In this way, an algebraic equation is obtained at each point of the grid for each term of the differential equation [6, 28, 29]. These equations involve the values of the functions at the considered point and its nearest neighbors. Using the approximations obtained by Taylor's series development, each differential operator in Eq. (61.3) is replaced by its corresponding discretization, and Eq. (61.4) is obtained.

$$\begin{aligned} &\frac{I_{i+1,j} - 2I_{i,j} + I_{i-1,j}}{\Delta x^2} - (RC+GL)\frac{I_{i,j+1} - I_{i,j-1}}{2\Delta t} \\ &- LC\frac{I_{i,j+1} - 2I_{i,j} + I_{i,j+1}}{\Delta t^2} - RGI_{i,j} = 0 \end{aligned} \tag{61.4}$$

This equation has the property that it involves only two values of the wave function at the last time level, the value $I_{i,j+1}$, and then this value can be cleared in terms of values at previous times to obtain the discretized telegrapher equation, which will later be implemented in MATLAB. To implement this equation numerically, the initial and boundary condition must be specified.

The discretized equation describes the behavior of the current present in the transmission line, taking into account the different boundary conditions that may arise, such as a load that can be netted. To observe the time evolution of the current in the line using this method, it is necessary to create a structure or mesh as shown in Fig. 61.2 [7, 16, 30, 31], which intuitively points out the result of the discretization process mentioned above, observing a spatial and a temporal discretization for I. The current in the cloud is assumed to be a constant value that is taken as zero. Therefore, the boundary condition is:

$$I(h, t) = 0 \tag{61.5}$$

where h is the height of the cloud where the discharge starts.

At $t = 0$, the current distribution is assumed to be constant; therefore, the initial condition is as follows:

$$I(x, 0) = I_0 \tag{61.6}$$

An algorithm for solving the Heaviside equation is as follows: Clear the current from the discretized equation, define the computational mesh or grid, define the time step size between consecutive levels and the spatial length element, generate a plot zone showing the behavior of the output parameters, identify the equations to be used to obtain the output current, identify the input parameters, identify the output parameters, define the appropriate initial and boundary conditions, study the convergence of the code, guarantee the stability of the system with the CFL condition: $\Delta t \leq \Delta t \leq \Delta x\sqrt{LC}$ perform code runs for different months, vary initial and boundary conditions, study convergence and stability for different discretization schemes; centered

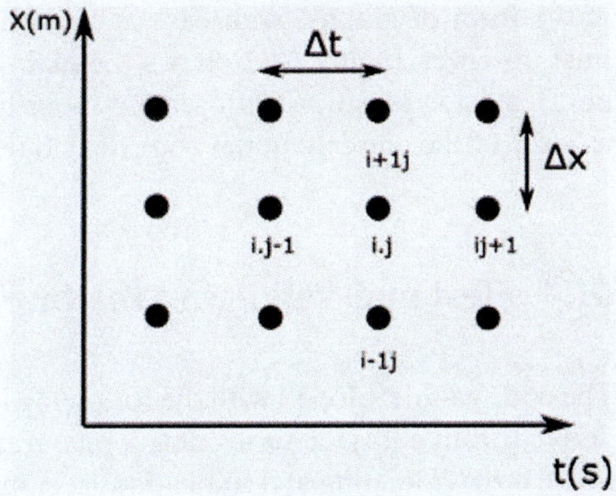

Fig. 61.2 Points of the discretized grid. The differential operators are replaced by their corresponding finite difference expression taking into account the function values at adjacent points on the mesh, where x is the spatial coordinate, and t is the time coordinate

Table 61.1 Electrical parameters of the discharge channel of an atmospheric electrical discharge (ATED)

Height (x)	1000 m
Inductance (L)	2.18 Mh
Capacitance (C)	6.95 Pf
Conductance (G)	5 Us
Resistance (R)	1000–5000 Ω

differences, forward and backward differences. Knowing the use of MATLAB and its different commands and subroutines in addition to the finite difference method, it is coded to find solutions to electromagnetic problems in the time domain, through the process of discretization of the partial derivatives present in the equations that describe the phenomenon [8, 32, 33]. Then, we proceed to study and synthesize the finite difference method for a single dimension since this procedure is used to represent the transients in the transmission line. As the resulting equations contain partial derivatives, by developing Yee's theory, they can be discretized, leaving mathematical expressions that a computational tool (or a proprietary code) can simulate [7, 8], which is shown in this work. Equation (61.4) is implemented numerically in MATLAB. Only the input parameters R, L, C, and G, which are assumed to be known in the discharge channel, are needed. The values used in this case are shown in Table 61.1.

The process to know the evolution of the current is obtained by solving the telegrapher's equation in all spatial samples for each time instant using the stored current values of the previous time. By recursion in solution I, the time evolution of the current in a transmission line is presented (see Fig. 61.5); the process culminates when the time spacing is terminated. MATLAB software offers powerful tools for numerical computation applied to science; it also has clear documentation to observe what this commercial software contains and has a large community that provides constant help for the subject [32, 33]. In MATLAB, the use of graphics makes it possible to show and analyze to the user the behavior of a phenomenon utilizing a curve with its coordinate systems and details that identify the origin of what is shown, and likewise that the generated figures can be stored in different formats for later study [8, 10, 14]. This offers the user alternatives when he needs to represent mathematical contents in the form of graphs, with ease and detail so that they can be understood by the final observer. Figure 61.3 shows the main computational code developed to solve the Heaviside equation with second-order finite differences. Figure 61.4 shows the module of the computational code developed to perform the convergence tests.

61.4 Test and Validity of the Implemented Code

The code was developed with the following in mind: finite difference expressions are derived from a Taylor series development up to second order; therefore, computational error (rounding and truncation) accumulates and must be controlled. There are

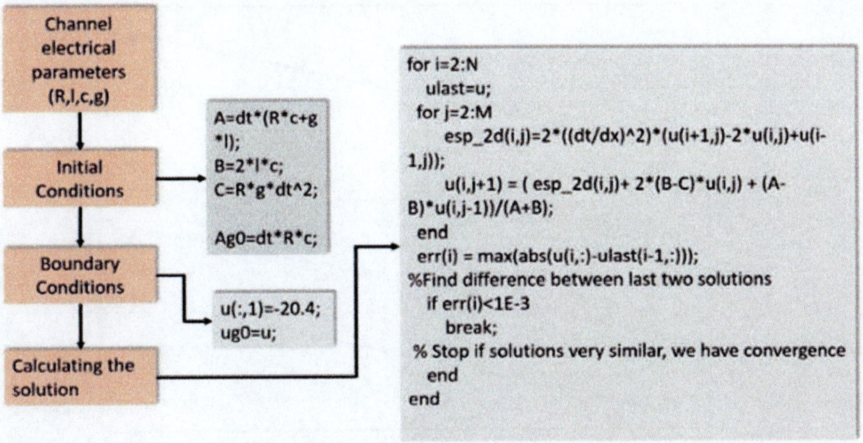

Fig. 61.3 Shows the computational code developed to solve the Heaviside equation with second-order finite differences

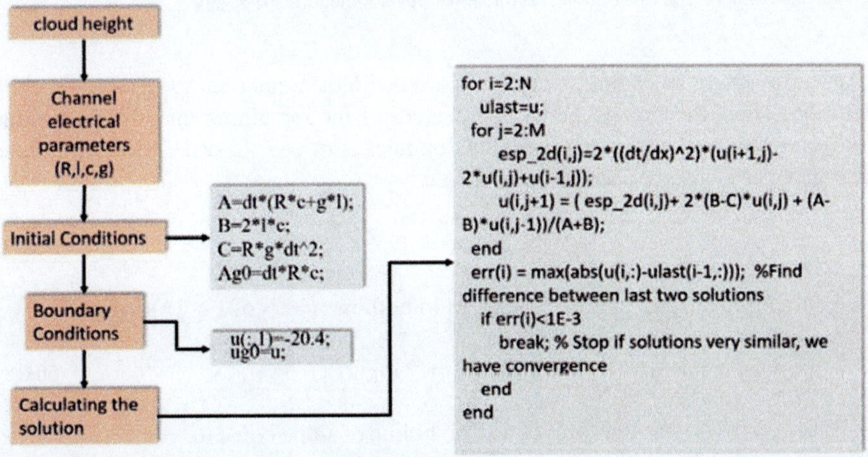

Fig. 61.4 Shows the module of the computer code developed to perform the convergence tests

errors inherent in the limitations of fitting the physical model to accurately describe the real physical system and the limitations of the boundary conditions. Convergence and stability tests ensure that the finite difference approximations used in the discretization are correct [8, 10, 14]. In this work, the electrical parameters are taken into account to have a sufficiently fine mesh to obtain a convergent code in as few iterations as possible. For this purpose, the CFL condition is used so that the meshing allows a solution closer to reality. To perform the convergence when the analytical solutions are known, the numerical value of the first iteration of the program is

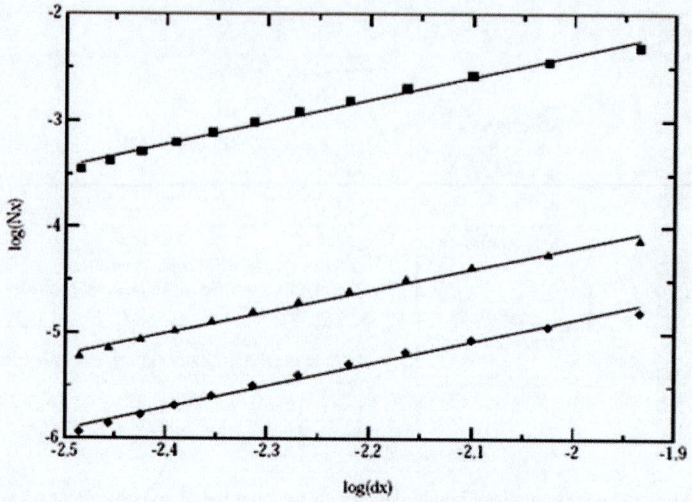

Fig. 61.5 Convergence of the MATLAB code for the three standards from top to bottom is shown L_1, L_2, L_∞, with slopes 1.9, 1.7, and 2, which is in agreement with the expected error for using second-order approximations in the Taylor series development, respectively

taken as the analytical solution and compared with the numerical value of the second iteration. Then, the same scheme is repeated for the remaining ones until the code converges. Since the Taylor series development is of second order, it is postulated that the error is also of second-order at Δx:

$$\text{Error} = \Delta x^2 \tag{61.7}$$

Applying logarithm and its properties to both members of Eq. (61.7):

$$\log(\text{Error}) = 2\log(\Delta x) \tag{61.8}$$

Equation (61.8) has the form of a straight line of slope equal to two. If *log (Error)* versus *log (Δx)* is plotted, the error between a known analytical solution and the numerical solution found for Eq. (61.4) is obtained. A slope approximately equal to two implies that the numerical solution converges to the analytical solution. In this case, a general expression for the convergence calculation was used, given by the following rules L_1, L_2, L_∞ [7, 8, 12].

The waveform generated with the Heidler function [2, 9, 14, 17, 18] was selected as the analytical solution and compared with the numerical solution of the telegrapher's equation to estimate the convergence error. Figure 61.5 shows a line of average slope $m = 1.9$ using least squares for linear regression, with differential steps of 3 microseconds. The numerical solution converges to the analytical solution to second-order approximation. When the parameters R and G are very small, their effects can be ignored, so that the transmission line can be considered an ideal, lossless structure.

In this case, the model depends only on the parameters L and C, from which we obtain a pair of partial differential equations, one for voltage and one for current.

These equations can be combined to form any of these exact wave equations for $G = 0$ (non-dissipative transmission line). The case of a lossless line ($R = G = 0$) leads to a differential equation identical to the string equation and whose algorithm is very easy to implement numerically:

$$\frac{\partial^2 I(x,t)}{\partial x^2} = LC\frac{\partial^2 I(x,t)}{\partial t^2} \tag{61.9}$$

$$\frac{\partial^2 V(x,t)}{\partial x^2} = LC\frac{\partial^2 V(x,t)}{\partial t^2} \tag{61.10}$$

It should be noted that the solution of Eqs. (61.9) and (61.10) can be used to verify the convergence of the code in the limits of $R = G = 0$. To verify the validity of the solution of the telegrapher's Eq. (61.3), in the codes where the equation is solved with $R = G = 0$ and the numerical solution of the current coincide with those of Eq. (61.11). The software developed for this particular case is the same for solving the wave equations for electric and magnetic fields. Besides, it is the same for calculating small displacements in materials mechanics. It should be noted that in each case what changes is the physical system and the propagation velocity.

Figure 61.6 shows the simulation of the discharge current. Mesh refinement was performed for $N = 1000$, 10,000, 200,000, 1,000,000, and the code remained stable and convergent for all meshes, only with the centered finite difference scheme.

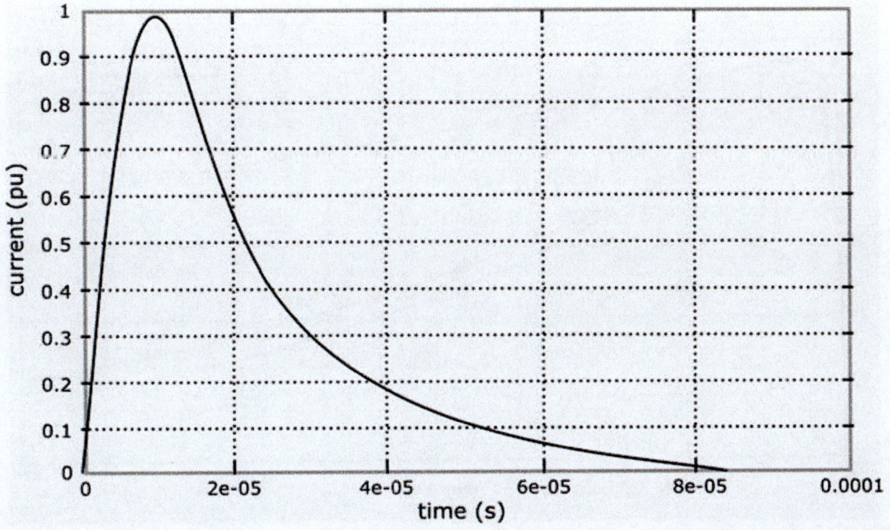

Fig. 61.6 Discharge current versus time for a discharge channel (a) Channel resistance of 20,000 Ω and cloud height of 3000 m

Another test and validation of the code developed in MATLAB was a series of runs of the code. A series of plots are shown for $G = 0$ and $G \neq 0$ for different numbers of points, that is, to vary the size of the spatial differentials and test the stability criteria previously stated. The fact that the conductance parameter is zero implies non-real solutions since in reality if there are real line losses. Meanwhile, when $G \neq 0$, the solutions are closer to reality.

Figure 61.7 shows a comparison between the waves obtained numerically with the code in MATLAB (both for $G = 0$ and $G \neq 0$) and the analytical solution. There it is visible the difference with $G = 0$, that is, because the dissipative characteristics of the discharge channel are not considered. The solutions closest to reality are obtained when $G \neq 0$.

Figure 61.8 shows a series of plots for different values of the discharge channel resistance, *R*, both for $G = 0$, to simulate the different resistivities present in the atmosphere by having areas with higher humidity and pollution than others.

Results to highlight are that the code was shown to be stable and convergent for different mesh types, with the best results being obtained with a count of nodes between 300 and 400. The magnitude of the return current of an EATD is affected inversely proportional to the value of the resistance of the conduction channel. The tail times are affected by the conductivity (*G*) present in the discharge channel. The code is oscillating and unstable for *G* values above 0.001 mhos. With wavefront derivatives (ramps) above 71kA/μs and attenuation time less than 100 μs, these values are presented in the double Exponential and Heidler Eq. (61.11), which were designed for non-tropical study areas, are markedly different from the data yielded by the numerical solution proposed in this work. Traditional commercial software packages, e.g., T-flash, do not present an exact solution of the phenomena associated

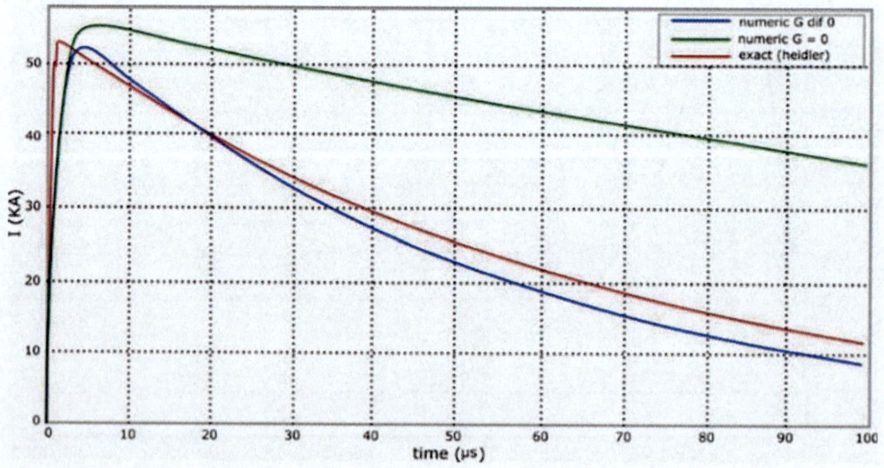

Fig. 61.7 Comparison between the numerical solutions for the conductance parameter $G = 0$ and $G \neq 0$, and the analytical solution (Heidler). This plot was obtained with a run of the program in MATLAB

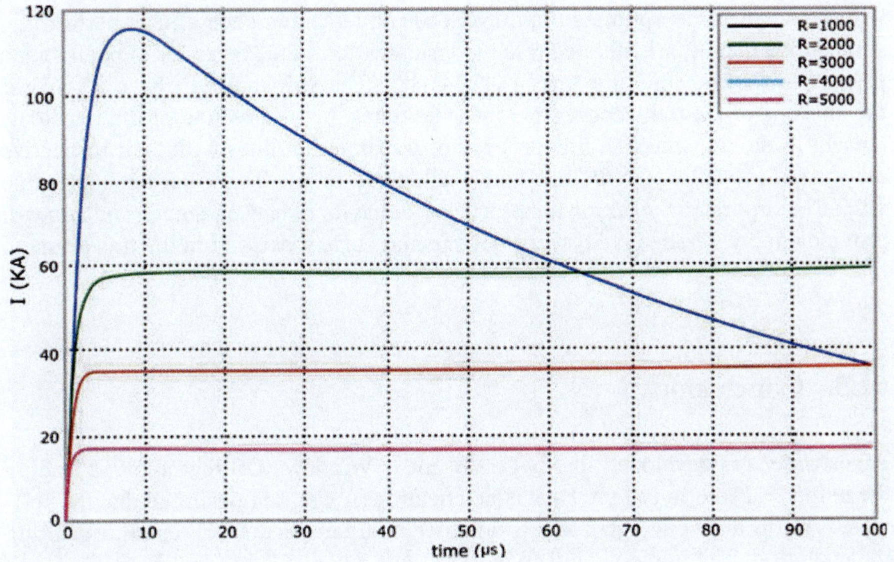

Fig. 61.8 Numerical solution with different values of channel resistance R and conductance parameter $G = 0$

with atmospheric discharges, but rather are based on probabilistic solutions.

$$h(t) = \frac{\text{Amp}}{\eta} \frac{\left(\frac{t}{T_f}\right)^n}{1 + \left(\frac{t}{T_f}\right)^n} e^{-t/\tau} \quad (61.11)$$

$$\eta = e^{\left(-\frac{-T_f}{\tau}\right)\left(n\frac{\tau}{T_f}\right)^{\frac{1}{n+1}}} \quad (61.12)$$

The parameters of this impulse type function are as follows: Amp = Current [Amp], T_f = Front duration in [seg], τ = Impact duration in [seg]. Interval between $t = 0$ and the point on the tail, n = Factor influencing the rate of increase of the function.

The implemented software offers support in the calculation of transient regime, stationary regime, and in the handling of equations [1, 6 34, 38]. However, with the technological advances, new advances and inventions are frequent that point to the development of science in multiple fields; and universities, being the cradle of knowledge and progress, do not escape from this reality, and should not be left behind. From this fact arises the need to maintain new computational tools and technological resources to expand the capabilities of understanding and analysis of the concepts of numerical methods and electromagnetism. In search of meeting the new challenges of technology, which impose the fact of having a software that allows solving different physical systems, and also be able to offer greater freedom and versatility to the user, with new and some optimized options. The implemented code allows performing simulations that offer better and reliable visualization of the behavior of phenomena present in the transmission lines [14, 34, 38]. The software will have a tool for the analysis of the transient regime that facilitates the observation of the transients present in the transmission line or array of transmission lines with their respective diagram of reflections, in the interactive calculation tools for the permanent regime will be incorporated the option to observe the behavior of new parameters, which will provide a more detailed analysis of voltage and current existing in the transmission line.

61.5 Conclusions

A software was developed in MATLAB under Windows OS that allowed finding the numerical solution of the Heaviside equation. It was demonstrated that the software is stable and convergent to second order, maintaining the CFL condition during the evolution. The implemented code represents a contribution for simulations and numerical solutions of partial differential equations, because, although it is implemented for the Heaviside equation, it can be applied to different problems, such as electromagnetic fields, material science problems, sound waves, vibrations of strings, or membranes, seismic waves, traffic density waves, waves on the surface of a liquid, among others. The code will allow computational applications to determine the effect on the control and protection systems of an electrical substation after being directly or indirectly impacted by atmospheric discharges. Development and improvement in the field of applied numerical methods, as well as diversifying the teaching of languages such as MATLAB. The implemented software can be extended in some cases for educational purposes, taking into account details such as more updated visual design, easy to learn, able to generate animations simply and visualize them in a graphical user interface. The code implemented in MATLAB, making use of the finite difference method, allowed to simulate the transients present in an array of transmission lines. The program offers the user complete content for the study and analysis related to transmission lines.

References

1. R. Álvarez, L. Rosales, Simulación de descargas atmosféricas y su efecto en líneas eléctricas de potencia. Rev. UCT **16**(64) (2011)
2. D. Agoris, V. Charalambakos, E. Pyrgloti, S. Grzybowski, A computacional approach on study of Franklin rod height impact on striking distance using a stochastic model. J. Electrostatics. **60**, 175–181 (2004)
3. L.C. Evans, *Partial Differential Equations, Graduate Studies in Mathematics*. American Mathematical Society (1998)

4. V. Rakov, M. Uman, *Lightning: physics and effects*. Cambridge University Press, Cambridge, UK (2003)
5. H. Torres El rayo en el trópico. Colección apuntes maestros, Ed. UN, Bogotá, Colombia. (2015)
6. P. Rodríguez, J. Toledo, E. Contreras, L. Rosales, Simulación de descargas atmosféricas mediante la ecuación de onda viajera. Rev. UCT **13**(53) (2009)
7. L. Rosales, W. Barreto, C. Peralta, B. Rodríguez, Nonadiabatic charged evolution in the postquasistatic approximation. Phys. Rev. D **40**, 24 (2010)
8. J. Thomas, *Numerical Partial Differential Equations-Finite Difference Methods* (Springer, New York Inc., USA, 1995), pp. 205–258
9. C. Hwang, Numerical modeling of lightning based on the traveling wave equations. IEEE Trans. Indus. Appl. **33**(2) (1997)
10. G. McCann et al., Lightning phenomena, in Chap. 12, *Electric Transmission and Distribution Reference Book*. Westinghouse Electric & Manufacturing Co., Pittsburg (1950)
11. M. Pinsky, *Introduction to Partial Differential Equations with Applications*. McGraw Hill (1984)
12. P. Cole, T.J. Krehbiel, G.M. Henebry, Web-enabled Landsat data time series for monitoring urban heat island impacts on land surface phenology. IEEE J. Sel. Topics Appl. Earth Obs. Remote Sens. **9**, 2043–2050 (2016)
13. J.D. Logan, *Applied Partial Differential Equations*. Springer, Undergradute Texts in Mathematics (1998)
14. J. Cooper, *Introduction to Partial Differential Equations with MATLAB*. Birkhauser (1998)
15. S. Farlow, *Partial Differential Equations for Scientists and Engineers*. Dover (1993)
16. H. Torres, El Rayo, Mitos, Leyendas, Ciencia y Tecnología, 2da Edición, Bogotá, UNIBIBLOS (2002)
17. F. Heider, Calculation of lightning current parameters. IEEE Trans. Power Deliv. **14**(2), pp. 309–404 (1999)
18. J. Herrera, *Nuevas Aproximaciones En El Cálculo de Tensiones Inducidas Por Descargas Eléctricas Atmosféricas* (Universidad Nacional de Colombia, Colombia, Tesis de Doctorado, 2006)
19. P. Duchateau, D. Zachmann, Applied partial differential equations. Dover (2002)
20. G. Ortigoza, Animaciones en MATLAB y Maple de Ecuaciones Diferenciales Parciales de la Física–Matemática. Rev. Mexicana de Física (2006)
21. E. Zauderer, *Partial Differential Equations of Applied Mathematics*. Wiley–Interscience Series in Pure and Applied Mathematics (1998)
22. L. Konstadia, T. Hallam, Traveling wave solutions of a nonlinear reaction–advection equation. J. Math. Biol. **38**, 346
23. C. Mejías, O. Hernández, Desarrollo de un Software Educativo Didáctico Interactivo para Sistemas de Ondas Guiadas Versión 2. Tesis de grado. Universidad de Carabobo. (2008)
24. A. Bianchi, Sistemas de ondas guiadas. Marcombo, D.L España (1980)
25. M. Lutz, *Programming python*, vol. 8. O'Reilly (1996)
26. G. Rossum, *History and License* (2001)
27. R. González, *Python Para Todos* (2006)
28. Matplotlib. *The Matplotlib Development Team* (2016)
29. G. McCann, Lightning phenomena, in Chap. 12, *Electric Transmission and Distribution Reference Book* (Westinghouse Electric & Manufacturing Co., Pittsburg 1950).
30. M. Uman, Natural lightning. IEEE Trans. Ind. Appl. **30**(3), 785–790 (1994)

31. V. Rakov, M. Uman, Review and evaluation of lightning return stroke models including some aspects of their application. IEEE Trans. Electromagn. Compat. **40**(4), 403–426 (1998)
32. C. Hwang, Numerical modeling of lightning based on the traveling wave equations. IEEE Trans. Ind. Appl. **33**(2), 1520–1523 (1997)
33. F. Heidler, Calculation of lightning current parameters. IEEE Trans. Power Deliv. **14**(2), 309–404 (1999)
34. P. Hevia, Fuentes de Impulso en el ATP, Argentina, Santa Fe, Universidad Tecnológica Nacional (2000), p. 10
35. PyQt. *Riverbank Computing Limited* (2016)

Chapter 62
Smart Contracts for User Registration on Ethereum Technology: Systematic Literature Review

Ramiro-Cristian Narváez, Jhonny Carrión, Leonardo Benavides Maldonado, and María del Cisne Ruilova

Abstract The security and application of new technologies will not always be easy to integrate them, due to different characteristics, especially the benefits they will offer and their cost, among others; due to this, the search for security will always be a difficult point to achieve. In the present systematic literature review (SLR), the way to develop smart contracts within the user registration module of a system is sought, in addition to detecting which are the tools that will provide a better application of the contracts. Barbara Kitchenham methodology was applied with all its phases and stages, which defined Remix IDE as one of the most used tools for the development, testing and debugging of smart contracts. This review also provided us with guidelines for the development of contracts in different environments such as cloud, fog and storage systems.

62.1 Introduction

Nowadays, the security of computer systems is of great importance, and the implementation of new technologies makes it more efficient every day. Smart contracts1 in conjunction with blockchain technology provide a new security proposal within these systems. By managing secure and decentralized systems using the emerging blockchain technology, the owner of an account can be offered security by registering and storing their private data of unique knowledge, preventing the theft of information, use of services and illegal access to the existing resources within an account by third parties [1].

This RSL is based on the methodology of Barbara Kitchenhamm, whose results were used to choose a correct model in the development of smart contracts for the Ethereum platform based on decentralized and independent data management, integration of registration and user control for information validation using Remix, Ethereum Geth client, truffle, solc-js and Solidity language tools.

R.-C. Narváez (✉) · J. Carrión · L. B. Maldonado · M. del Cisne Ruilova
Facultad de Ingeniería Electromecánica, Universidad Nacional de Loja, UNL, Av. Pío Alvarado y Reinaldo Espinoza, La Argelia, Loja, Ecuador
e-mail: cristian.narvaez@unl.edu.ec

Á. Rocha et al. (eds.), *Communication, Smart Technologies and Innovation for Society*, Smart Innovation, Systems and Technologies 252,
https://doi.org/10.1007/978-981-16-4126-8_62

Table 62.1 Systematic literature review process

Phases	Progress stages
Review planning	Identification of the need for a review Formulation of research questions Review protocol development
Conducting the review	Identification of the research Selection of primary studies Quality assessment of the study Data extraction and monitoring Data synthesis
Submitting a report	Discussion and analysis

The following sections of the study are as follows: The Scope, which determines the reason for conducting the SLR, the methodology where the process established by Barbara Kitchenham was defined within the revisions, the results where the SLR is applied based on the phases of the methodology, and finally, the conclusions based on the research carried out.

62.1.1 *Scope*

The purpose of this systematic literature review is to investigate and analyze the studies related to the development and implementation of smart contracts within the user registration module, in order to identify the necessary tools and guidelines to integrate these contracts to the user registration in control systems.

62.2 Methodology

Barbara Kitchenham's methodology [2] describes its process in three phases, which are as follows: planning the review, conducting the review and reporting as presented in Table 62.1.

62.3 Results

62.3.1 *Review Plannig*

Identification of the need for a review. A systematic literature review implementation allows to analyze knowledge that covers a certain research topic; in our case study, it allowed to choose all the necessary requirements to develop smart contracts, the tools to implement, the programming language and libraries to interact with the smart contracts within the user registration management of a system.

Formulation of research questions. In order to implement the above proposal and based on the subject matter, Table 62.2 presents two research questions that limit the course of the review.

Review Protocol Development. The protocol to be applied is the one proposed by Mark Petticrew and Helen Roberts [3], PICOC (Population, Intervention, Comparison, Outcome and Context), which helps to structure the search strings, and with the help of the Parsifal tool, the review can be easily performed.

In order to search and select the articles, the following scientific databases were used:

- IEEE digital library (https://ieeexplore.ieee.org/)
- Science@Direct (https://www.sciencedirect.com/)
- Scopus (http://www.scopus.com).

Keyword Selection. The selection of key words is based on some scientific articles detailed in Table 62.3.

Research Chains. For each virtual library, specific search strings formed are presented in the Table 62.4.

Eligibility Criteria. The following criteria will be taken into account in order for the studies to be included:

Table 62.2 Research questions

Research questions	
RQ1	Which studies discuss the components for the development of smart contracts in EVM user registration and control?
RQ2	Which studies present tools for the development of Ethereum-based smart contracts for user registration and control?

Table 62.3 Articles and keywords

Articles	Keywords
RBAC-SC: Role-based access control using smart contract [4]	Blockchain technology, role-based access control, smart contracts
A secure cloud storage framework with access control based on blockchain [5]	Cloud storage, access control, Ethereum, blockchain, smart contract
EIDM: An ethereum-based cloud user identity management protocol [6]	Cloud computing, identity management, blockchain, reputation, smart contract
Smart contract-Based access control for the Internet of Things [1]	Internet of Things, access control, blockchain, smart contract

Table 62.4 Research chain

Libraries	Chains
Scopus	TITLE-ABS-KEY (("Smart Contract" OR "Pactos Inteligentes" OR "SmartContract") AND ("ethereum" OR "EVM") AND (register% OR user%) AND NOT (bitcoin)) AND (LIMIT-TO (DOCTYPE, "ar")) AND (LIMIT-TO (PUBYEAR, 2020) OR LIMIT-TO (PUBYEAR, 2019) OR LIMIT-TO (PUBYEAR, 2018) OR LIMIT-TO (PUBYEAR, 2017) OR LIMIT-TO (PUBYEAR, 2016))
IEEE digital library	((((((("All Metadata":smart contract) OR "All Metadata":contratos inteligentes) OR "All Metadata":SmartContract) OR "All Metadata":smart contracts) AND "All Metadata":login) OR "All Metadata":register users) OR "All Metadata":ethereum) NOT "All Metadata":bitcoin)
Science@Direct	(("Smart Contract" OR "Pactos Inteligentes" OR "SmartContract") AND ("ethereum" OR "EVM") AND (register% OR user%) NOT(BITCOIN))inteligentes basados Ethereum para el registro y control de usuarios?

- Papers published between 2016 and 2020.
- The English language will be the most relevant in the documents and Spanish will also be accepted.
- The abstract should contain information on the application of smart contracts and the use of tools inside it.
- Articles that are classified as journals.

62.3.2 Conducting the Review

Identification of the survey. The purpose of the systematic literature review is to answer the questions raised by analyzing studies based on the implementation of blockchain technology and smart contracts according to [35] states that an organization must provide security when storing, retrieving and accessing services or data using current technologies to be at the forefront.

Managing secure and decentralized systems by applying the use of emerging blockchain technology can provide an account owner with security when recording and storing their private data of unique knowledge, preventing theft of information, use of services and illegal access to existing resources within an account by third parties [36].

Primary studies selection. Table 62.5 is the result of the primary studies obtained after applying the search strings and the inclusion and exclusion criteria in their respective virtual libraries.

Quality assessment. To define the quality of each article, the following questions were defined:

Table 62.5 Primary studies

Libraries	Found	Selected
IEEE digital library	87	52
Science@Direct	51	16
Scopus	89	58

- QA1: Article focuses on the development of smart contracts in the user registration module?
- QA2: This article describes the components of the smart contract?
- QA3: Article presents some tools for the development of smart contracts?

Data extraction and monitoring. Following the entire process, 18 studies containing related information were accepted and are detailed in Table 62.6.

Data Synthesis. From the accepted studies, each article was analyzed in which the diversity of tools for the different stages in the construction of a smart contract is evidenced: development, deployment, compilation, and testing with Ethereum. It is important to highlight the Solidity programming language for the development of smart contracts, these tools allow to observe simulation, in different test networks and thus validate the behavior in the blockchain network as well as its costs (gas consumption per transaction).

(a) *Which studies discuss the components for the development of smart contracts in EVM user registration and control?*

The implementation of smart contracts has a great breadth, given this in Table 62.7, studies are presented based on cloud services or fog (fog) and through Ethereum with the blockchain network try to make these systems decentralized and independent in the management of data, which integrate the registration and control of users to validate their information, also describe how the smart contract will act between the intervening parties and the blockchain network based on Ethereum, while in based on the presentation of information implement smart contracts and decentralize the registration system and user access control which significantly increases security and efficiency by creating users finally control access to their devices through smart contracts with a blockchain network (p2p) which gives to understand that these technologies have a great immersion in the different types of systems to decentralize them.

(b) *Which studies present tools for the development of Ethereum-based smart contracts for user registration and control?*

Different tools exist to link smart contracts to current systems with user registration modules, which have different functionalities for both security and quality, as shown in Table 62.8, detailed by study.

Table 62.6 Accepted studies

No.	Title	Year
D01	Combating Deepfake Videos Using Blockchain and Smart Contracts [7]	2019
D02	Blockchain-Based Proof of Delivery of Physical Assets with Single and Multiple Transporters [8]	2018
D03	Proof of Delivery of Digital Assets using Blockchain and Smart Contracts [9]	2018
D04	Smart Contract-Based Data Trading Mode Using Blockchain and Machine Learning [10]	2019
D05	A Secure Cloud Storage Framework With Access Control Based on Blockchain [11]	2019
D06	Integrated Application of Blockchain in the Electric Information Management System [12]	2019
D07	Monetization of Services Provided by Public Fog Nodes Using Blockchain and Smart Contracts [13]	2020
D08	EIDM: A Ethereum-Based Cloud User Identity Management Protocol [6]	2019
D09	On the Design of a Flexible Delegation Model for the Internet of Things Using Blockchain [14]	2020
D10	Smart contract-based access control for the internet of things [1]	2019
D11	Using Ethereum blockchain to store and query pharmacogenomics data via smart contracts [15]	2020
D12	IoT Public Fog Nodes Reputation System: A Decentralized Solution Using Ethereum Blockchain [16]	2019
D13	A blockchain-based framework for data sharing with fine-grained access control in decentralized storage systems [17]	2018
D14	Smart Contract-Based Review System for an IoT Data Marketplace [18]	2018
D15	Caterpillar: A business process execution engine on the Ethereum blockchain [19]	2019
D16	RBAC-SC: Role-based access control using smart contract [4]	2018
D17	Blockchain for Secure EHRs Sharing of Mobile Cloud-Based E-Health Systems [20]	2019
D18	Blockchain-based decentralized reverse bidding in fog computing [21]	2020

Table 62.7 Smart contracts developed in different systems

System	Studies
Based on cloud and fog	D05, D12, D17
Based on presentation of information	D06, D10, D16
Storage based	D13

62.3.3 *Presentation of the Report*

Discussion and analysis. The different articles analyzed during the review stage show that there is a diversity of tools for the different stages such as development, deployment, compilation and testing of smart contracts within the Ethereum platform

Table 62.8 Tools for the development of smart contracts on ethereum

Funcionality	Studies	Dev. tools
Development, testing and debugging	D01, D02, D03, D04, D05, D07, D08, D10, D12, D14, D18	Remix IDE
Deployment and interactions	D05, D09, D10, D17	Ethereum Geth Client
Testnet access and testing	D11, D12	Truffle
Compilation	D09, D15	Solc-js standard compiler Solidity

which through a programming language which allows you to create software to manage transactions and automate highlighted, and these tools allow to observe the simulation and behavior in the blockchain network, as well as the costs to be incurred due to gas consumption by transactions, and in addition, it can be highlighted that one of the most used tools that works based on a browser is Remix IDE which provides an interface and implements programming with solidity which is the base language and recognized by Ethereum for the development of smart contracts, and in contrast, you can use libraries within IDEs and other languages to perform some of the functions or connections to the blockchain network.

In studies D01, DO2, DO3, D04, D05, D07, DO8, D10, D12, D14, D18, it presents a powerful tool widely used Remix IDE currently used for the development, debugging, testing and implementation of smart contracts based on the solidity language, having all these capabilities makes it ideal for implementation in systems working with the Ethereum blockchain network. In addition, studies D05, D07, D08, D12, D18 are oriented to systems implemented in the cloud and fog that use this tool for the development of smart contracts, these intervene mainly to store and retrieve data, monitor and measure their services, in addition to the interaction between nodes of the blockchain form distributed access control schemes. Within the studies D04, D14 talk about data trading where there must be security a very important part, that is, where smart contracts intervene to verify the owner of this data and its quality avoiding the intervention of third parties; based on what was observed, the use of remix within the Ethereum platform is essential despite the fact that very little specification of the tool was found in the documents, the versions with which it was worked and the specifications in which it intervened are unknown, even so the capacity of Remix is robust to implement in all types of system and important for the development of this work.

The studies D11 and D12 use the well-known and implemented Truffle tool that serves as Frond-End, in addition to covering the testing framework and asset pipeline of the blockchain, in the document D12 I make the combination of the operation of Remix and Truffle to cover the development and testing within IOT within an architecture of the fog (Fog), with this is intended to see the breadth of tools that work in conjunction with Ethereum and define them to be applied in the work, despite the fact that Truffle has many capabilities Remix proves to be more robust and cover many phases that smart contracts entail and that is demonstrated in the number of studies in which it intervenes.

The methodology for implementing smart contracts within the Ethereum platform and its blockchain network is defined by its documentation, and it is a design by contract, which takes the design elements as participants in a relationship similar to a business contract. Although the studios couple their systems to their own methodologies for architecture, the writing and design of smart contracts are driven by this design-by-contract methodology, which is the one we will proceed to apply to program.

62.4 Conclusions

By analyzing the SLR, it can be concluded that:

1. The methodology for implementing smart contracts is not explained in any of the works, but based on the Ethereum platform and its documentation, it is known as design by contract, which takes the design elements as participants in a relationship similar to a business contract. Although in the studies they couple their systems to their own methodologies for architecture, the writing and design of smart contracts is driven by this design-by-contract methodology which is the one we will proceed to apply to program it.
2. The most widely used tools for programming smart contracts are Remix IDE and Truffle, whose functionalities cover the needs to meet the objectives of implementing smart contracts in current systems. It has been concluded that among the most used tools for the programming of smart contracts are Remix IDE and Truffle where the functionalities that allow us to cover the needs to meet the objectives of implementing smart contracts in current systems.
3. For the interaction and testing of smart contracts more information is needed, since to do so we need the intervention of another tool called MetaMask, which is not present in the studies carried out.
4. The authors in many of the related works talk about the application of smart contracts to parts of the system, which there is no reference of the implication to a complete system that works with them.

References

1. Y. Zhang, S. Kasahara, Y. Shen, X. Jiang, J. Wan, Smart contract-based access control for the internet of things. IEEE Internet Things J. **6**(2), 1594–1605 (2019)
2. B. Kitchenham, Procedures for performing systematic literature reviews, Jt. Tech. Report, Keele Univ. TR/SE-0401 NICTA TR-0400011T.1, vol. 33 (2004), p. 33
3. Systematic Reviews in the Social Sciences (2006)

4. J.P. Cruz, Y. Kaji, N. Yanai, *RBAC-SC: role-based access control using smart contract*, vol. 3536, no. c (2018)
5. S. Wang, X. Wang, Y. Zhang, A secure cloud storage framework with access control based on blockchain. IEEE Access **7**, 112713–112725 (2019)
6. S. Wang, R. Pei, Y. Zhang, EIDM: an Ethereum-based cloud user identity management protocol. IEEE Access **7**, 115281–115291 (2019)
7. H.R. Hasan, K. Salah, *Combating deepfake videos using blockchain and smart contracts*. IEEE Access, vol. 7, no. c (2019), pp. 41596–41606
8. H.R. Hasan, K. Salah, *Blockchain-based proof of delivery of physical assets with single and multiple transporters*. IEEE Access, vol. PP, no. 8 (2018), p. 1
9. H.R. Hasan, K. Salah, Proof of delivery of digital assets using blockchain and smart contracts. IEEE Access **6**(8), 65439–65448 (2018)
10. W.E.I. Xiong, L.I. Xiong, *Smart contract based data trading mode using blockchain and machine learning*. IEEE Access, vol. PP (2019), p. 1
11. S. Wang, X.U. Wang, Y. Zhang, A secure cloud storage framework with access control based on blockchain. IEEE Access, vol. PP (2019), p. 1
12. C. Xu, Y. Fang, Y. Ma, ScienceDirect Integrated Application of Blockchain in the Electric Information Management System. Procedia Comput. Sci., vol. 162, no. Itqm 2019 (2020), pp. 88–93
13. M. Debe, K. Salah, M. Habib, U.R. Rehman, *Monetization of services provided by public fog nodes using blockchain and smart contracts*, vol. 8 (2020)
14. S. Pal, T. Rabehaja, M. Hitchens, V. Varadharajan, A. Hill, On the design of a flexible delegation model for the internet of things using blockchain. IEEE Trans. Ind. Inform. **16**(5), 3521–3530 (2020)
15. G. Gürsoy, C.M. Brannon, M. Gerstein, Using Ethereum blockchain to store and query pharmacogenomics data via smart contracts. BMC Med. Genomics **13**(1), 1–11 (2020)
16. M. Debe, K. Salah, M. Habib, U.R. Rehman, IoT public fog nodes reputation system: a decentralized solution using ethereum blockchain. IEEE Access **7**, 178,082–178,093 (2019)
17. S. Wang, Y. Zhang, Y. Zhang, A blockchain-based framework for data sharing with fine-grained access control in decentralized storage systems. IEEE Access **6**, 38,437–38,450 (2018)
18. J.S. Park, T.Y. Youn, H. Bin Kim, K.H. Rhee, S.U. Shin, Smart contract-based review system for an IoT data marketplace. Sensors (Switzerland) **18**(10), 1–16 (2018)
19. O. López-pintado, L. García-bañuelos, I. Weber, A. Ponomarev, Caterpillar: a business process execution engine on the Ethereum blockchain. April, 1162–1193 (2019)
20. D.C. Nguyen, P.N. Pathirana, S. Member, Blockchain for secure EHRs sharing of mobile cloud based E-health systems. IEEE Access **7**, 66792–66806 (2019)
21. M. Debe, K. Salah, M. Habib, U.R. Rehman, *Blockchain-Based Decentralized Reverse Bidding in Fog Computing*, vol. 4 (2020)
22. A. Rosic, ¿Qué son los contratos inteligentes? [Guía definitiva para principiantes sobre contratos inteligentes] (2017) [Online]. Available: https://blockgeeks.com/guides/smart-contracts/. [Accessed: 23 Feb 2020]

Chapter 63
Encapsulation Component and Its Incidence into Scientific Software Performance

G. Omar Pizarro-Vasquez, Felix Barahona, and Miguel Botto-Tobar

Abstract Performance is considered an important feature rather than the application of programming techniques for better software design in most scientific software developers. Therefore, the problem arises if the software is developed without considering a specific paradigm or some programming technique when performing maintenance; tasks related to this activity are complicated, since almost no one would understand the source code. The goal of this research is to verify the performance of the software with or without an encapsulation component. An ex-post-facto experimental methodology has been implemented, carrying out a descriptive analysis of the data and then concluding by verifying the hypothesis by means of a robust test. This work was carried out by running algorithms written in the programming language Java by using three data groups in different conditions to analyze their behavior. Results show is that the application of the encapsulation component of the object-oriented paradigm does affect the execution of the scientific software performance.

G. O. Pizarro-Vasquez (✉) · F. Barahona
Universidad Politécnica Salesiana, Guayaquil, Ecuador
e-mail: gpizarro@ups.edu.ec

G. O. Pizarro-Vasquez
Research Group in Software Engineering and Knowledge Engineering (GIISIC), Guayaquil, Ecuador

M. Botto-Tobar
Eindhoven University of Technology, Eindhoven, The Netherlands

Research Group on Artificial Intelligence and Information Technologies (IATI), Universidad de Guayaquil, Guayaquil, Ecuador

Á. Rocha et al. (eds.), *Communication, Smart Technologies and Innovation for Society*, Smart Innovation, Systems and Technologies 252,
https://doi.org/10.1007/978-981-16-4126-8_63

63.1 Introduction

63.1.1 Object-Oriented Paradigm

Object-oriented paradigm (OOP) has been the design principle of many programming languages [1]. The idea behind OOP was derived mainly from the representation of knowledge in the human brain according to the real world. According to this paradigm, everything can be modeled as an object, which is composed of: identity, state and behavior. This allows us to make software design more accessible by information systems developers and architects.

Unhelkar [2] presents six fundamentals of Software Engineering, which are also those of OOP, such as: classification, polymorphism, abstraction, inheritance, association and encapsulation. *Classification* refers to the grouping of identified entities or potential objects; *polymorphism*, as the runtime feature with respect to an instantiated object when understanding a message sent by another object; *abstraction*, understood as the classification of objects that are identified as classes; *inheritance*, which results from classes that have been generalized; the *association*, as the characteristic that allows to relate classes; and ***encapsulation***, a feature that is taken into account in this research work, such as the one that locates data and prevents it from being directly exposed to the rest of the system, improving quality and reuse because the data is accessed through calls to operations (methods or functions) of a class and shows the set of "data and code" depending on its visibility (*public*, *private* or *protected*).

63.1.2 Software Scientific Development

Scientific software development refers to the analysis, design, implementation, testing and deployment of software applications for scientific research purposes, for example in the field of physics, biology, medical analysis, data science, among others. The need for continuous experimentation and validation of techniques (e.g., simulations) prior to the publication of scientific results has led to the emergence of the field of scientific software development as an important method for researchers to be successful in multiple fields [3].

According to Arvanitou [4], most of the code implemented for scientific software does not follow a guideline with respect to some paradigm that allows considering some non-functional requirement, such as, coupling, scaling, modularity, among others, since the efficiency of the execution of an algorithm prevails over design techniques.

Hypothesis: Encapsulation component within the object-oriented paradigm (OOP) impacts on the performance of a scientific software.

The paper is organized as follows: Sect. 63.2 presents the materials and methods: data source and runtime environment are mentioned. Section 63.4 describes the results obtained, and finally, Sect. 63.5 presents our conclusions and recommendations.

63.2 Materials and Methods

63.2.1 Experimental Design

This research is based on code of Pizarro's Master's Thesis [5], not yet published; in which, an ex-post facto experimental study was carried out, with simulated data for the generation of batches with orders (instances from [6] which contains index of place where are items) and their respective collection in a rectangular warehouse with one cross aisles (see Fig. 63.1).

The order grouping algorithms were: *Random* (batches are formed randomly), *First Come First Served (FCFS)* batches are formed according to the orders come up to the capacity of the cart, *Strict Order Picking (SOP)* a batch is formed with a single order, *Greedy 1 (G01)* are ordered from the highest to the lowest number of items of each order and batches are created, *Greedy 2* (G02) are ordered from the lowest

Fig. 63.1 Rectangular warehouse with one cross aisles

to the highest and batches are created and *Greedy 3* (G03) are grouped according to the closest orders and batches are created.

There is a heuristic that was applied to two groups of experimental data after having a set of solutions obtained with the algorithms explained in the previous paragraph, this heuristic is called **Local Search (LS)** with four variants: 1×0 two batches are taken randomly and a single-random order is taken out from each batch then exchanged if is into cart capacity; 1×1 two random batches are taken and a random order is taken from each batch then exchanged; 2×1 two random batches are taken and two random orders are taken from one batch and one random order from another batch is then exchanged; and, 2×2 two random batches are taken and two random orders are taken from one batch and another two random orders are taken from another batch, then exchanged.

It should be noted that exchanges are made if the verification of the cart's capacity is fulfilled; otherwise, the exchange is not made, another batch is sought until the exchange can be made.

Routing algorithms were: ***S-Shape*** (the route through the warehouse is like a letter S) and ***Largest Gap*** (the products are collected first at the top and then the products at the bottom, in general).

63.3 Simulations

Three simulations were run running the order grouping algorithms together with the routing algorithms in three groups:

- Group 1. Constructive Algorithms (Random, SOP, FCFS, Greedy 01—G01, Greedy 02—G02 and Greedy 03—G03) with S-Shape and Largest Gap.
- Group 2. Random constructive algorithm (Random) and heuristic Local Search (1×0, 1×1, 2×1, 2×2) with S-Shape and Largest Gap.
- Group 3. Greedy constructive algorithms (G01, G02 and G03) and heuristic Local Search (1×0, 1×1, 2×1, 2×2) with S-Shape and Largest Gap.

The specification of the detailed experimentation in the previous paragraphs can be seen in Fig. 63.2.

The source code of the warehouse configuration with several cross aisles for this research is implemented in Java [7] based on a Perl code from the research works of [8–10].

For data analysis, the statistical programming language named R was used, and RStudio [11] was used as IDE.

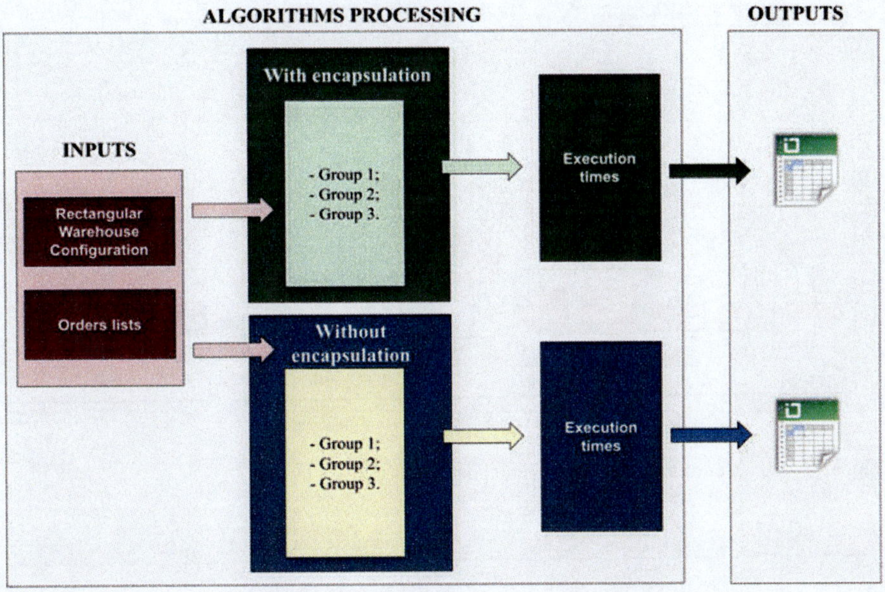

Fig. 63.2 Scientific research design

63.4 Results

63.4.1 Data Analysis Exploration

Prior to apply a hypothesis testing to the experimental data, it is necessary to verify them in a descriptive way and thereby check the statistical results of the hypothesis test.

In all boxplot plots (Figs. 63.3, 63.4 and 63.5), there have been applied the logarithm of base 10 with respect to the axis of the execution times (in nanoseconds), so that they can be displayed in an adequate way, as shown in graphs, since previously they could not be appreciated in a better way due to the amount of aberrant data obtained in experiments. In these three groups, it can be observed data and its execution times with encapsulation is slightly higher than execution times without encapsulation.

The following tables show the values of means, trimmed means to 10% and variances of execution times (in nanoseconds) of each group that were previously visualized. In Tables 63.1, 63.2 and 63.3, it can be seen how the experimental data where encapsulation was applied in the source code is greater than the experimental data in which encapsulation was not applied.

In figures of the density of execution times (Figs. 63.6, 63.7 and 63.8), both with encapsulation and without encapsulation, it is evident that they don't have a normal

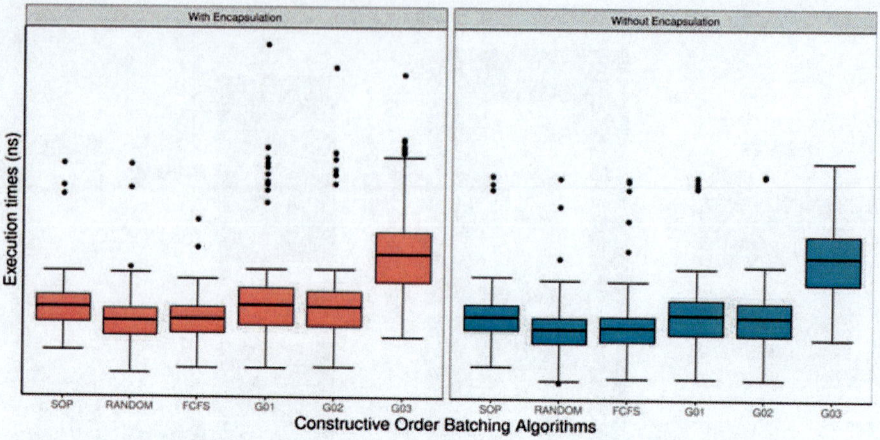

Fig. 63.3 Box plot of group 1 experimental data

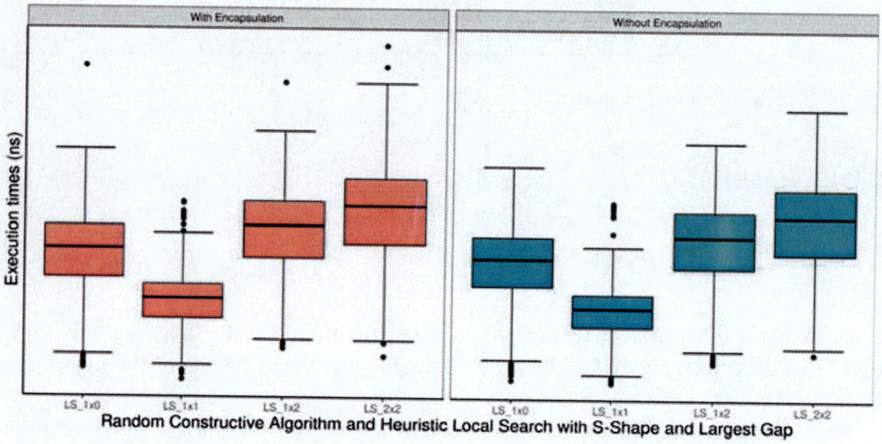

Fig. 63.4 Box plot of group 2 experimental data

distribution, which allows us to deduce that an alternative other than a parametric hypothesis testing.

63.4.2 Hypothesis Testing

Graphically, these three groups of experimental data don't have a normal distribution; which was verified in the three groups of experimental data, using Lilliefors normality test (Kolmogorov–Smirnov) [12], the null hypothesis being that "*the experimental data have a normal distribution*":

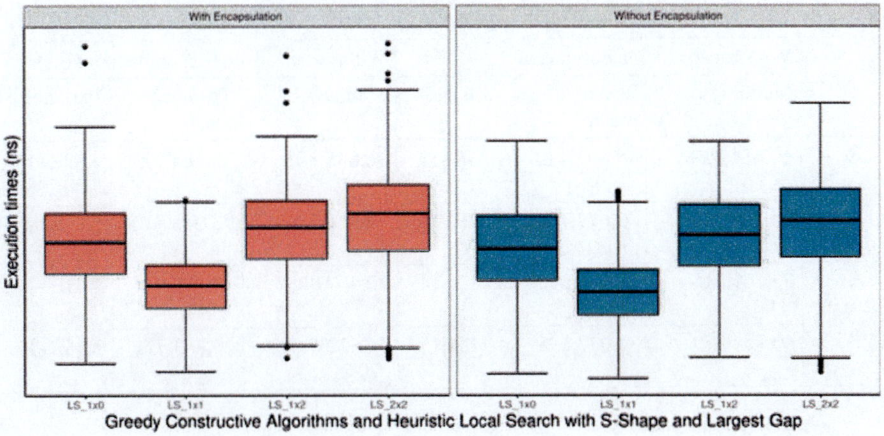

Fig. 63.5 Box plot of group 3 experimental data

Table 63.1 Descriptive experimental data from group 1

	With encapsulation component (ns)			Without encapsulation component (ns)		
	Media	Trimmed mean	Variance	Media	Trimmed mean	Variance
SOP	1,024,637.6	984,356.9	1.07E+12	862,458.2	804,322.5	1.52E+12
Random	718,507.1	686,736.8	8.47E+11	599,995.9	576,146.4	4.37E+11
FCFS	738,722.7	722,968.7	1.28E+11	647,044.9	605,959.2	8.28E+11
G01	1,555,259.4	1,087,353.7	3.56E+14	952,754.1	862,217.4	2.40E+12
G02	1,320,236.1	1,017,183.3	1.09E+14	848,957.1	814,291.2	8.64E+11
G03	5,504,069.7	4,530,229.5	1.20E+14	4,935,173.2	4,415,207.6	2.52E+13

Table 63.2 Descriptive experimental data from group 2

	With encapsulation component (ns)			Without encapsulation component (ns)		
	Media	Trimmed mean	Variance	Media	Trimmed mean	Variance
LS 1×0	9,557,378	8,472,821	2.69E+14	7,483,487	6,780,018	4.16E+13
LS 1×1	2,154,638	2,064,382	1.98E+12	1,801,292	1,707,401	2.00E+12
LS 2×1	18,266,779	15,728,281	4.20E+14	14,671,111	12,794,689	2.18E+14
LS 2×2	38,733,059	29,950,634	3.52E+15	30,546,922	24,013,919	1.71E+15

Table 63.3 Descriptive experimental data from group 3

	With encapsulation component (ns)			Without encapsulation component (ns)		
	Media	Trimmed mean	Variance	Media	Trimmed mean	Variance
LS 1 × 0	19,684,438	15,891,130	1.25E+15	16,845,381	13,864,973	4.58E+14
LS 1 × 1	4,103,574	3,677,788	1.32E+13	3,971,539	3,346,910	1.89E+13
LS 2 × 1	23,209,315	20,085,973	8.91E+14	20,067,883	17,582,944	4.41E+14
LS 2 × 2	40,754,879	31,027,295	4.21E+15	34,820,637	26,843,721	2.46E+15

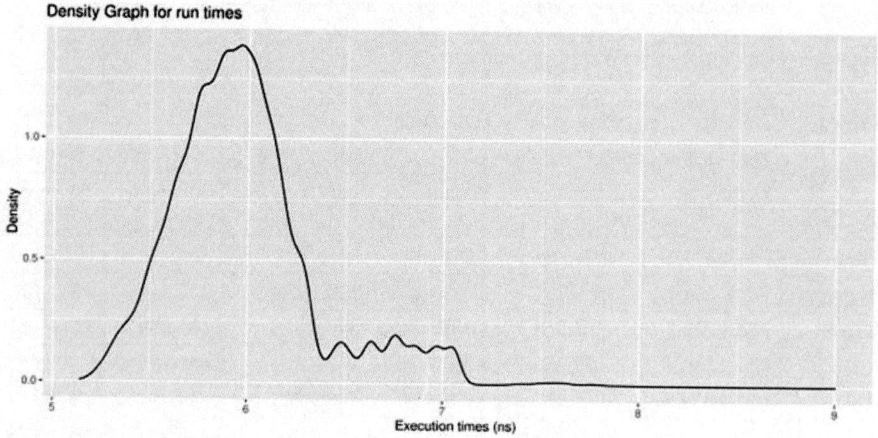

Fig. 63.6 Density diagram of the experimental data from group 1

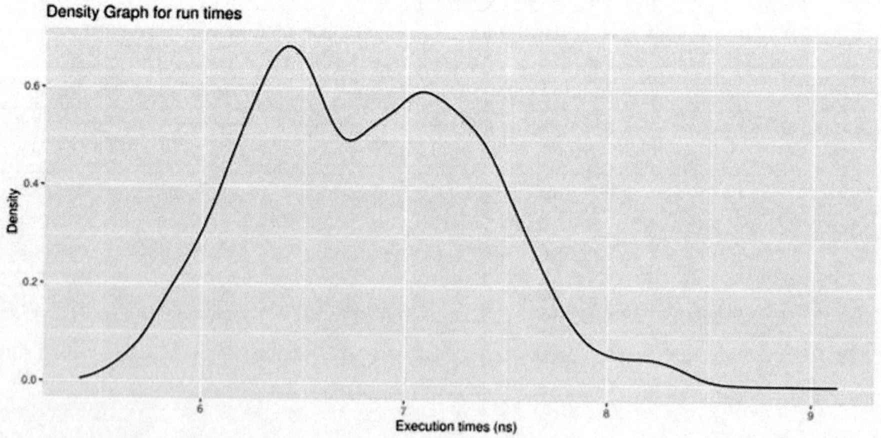

Fig. 63.7 Density diagram of the experimental data from group 2

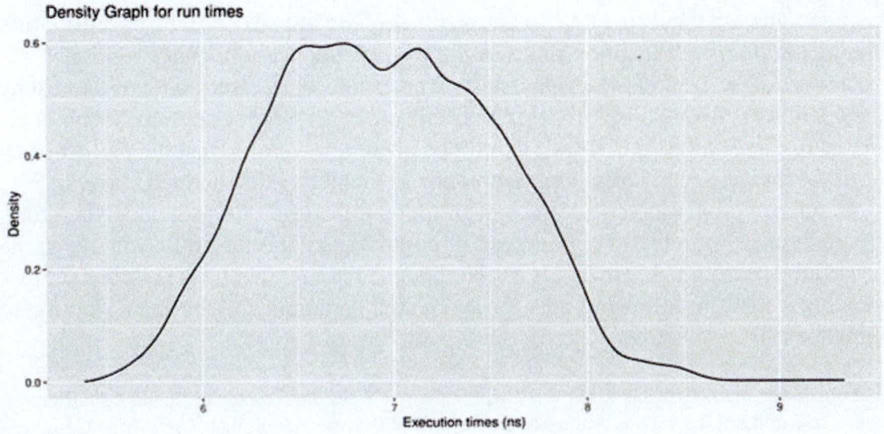

Fig. 63.8 Density diagram of the experimental data from group 3

In Groups 1, 2 and 3, the following conclusion is reached: "*with a value of significance close to zero (<2.2e−16), it can be concluded that the null hypothesis is rejected; therefore, the data do not have a normal distribution*".

Since these three groups of data do not follow a normal distribution; the requirement to apply the non-parametric hypothesis test must be proved, verifying the null hypothesis: "*homogeneity of the variance in the data*", using the Bartlett homoscedasticity test recommended in [13, 14].

In Groups 1, 2 and 3, the following conclusion is reached: "*with a value of significance close to zero (<2.2e−16), it can be concluded that the null hypothesis is rejected; therefore, the variance is not homogeneous*".

Since it does not meet the two previous requirements: neither the assumption of normality nor homogeneity in the variance, now it will be statistically verified if the mean of execution times with the code implemented with encapsulation is greater than the mean of the execution times with the source code without encapsulation, using function proposed by Yuend [15] robust hypothesis test of two dependent groups, with null hypothesis: "*the mean of the execution times of each group is equal*":

In Groups 1, 2 and 3, the following conclusion is reached: "with a significance value close to zero and a difference of means bounded with a positive value, it can be concluded that the null hypothesis is rejected; therefore, the bounded mean of each group is different and that the mean of the execution times with encapsulation is greater than the mean of the execution times without encapsulation".

63.5 Conclusions

The impact on how the source code is implemented, considering encapsulation or not, on the execution times of scientific software *(hypothesis about this research)*; It has been shown statistically that the execution times of the source code with encapsulation are greater than the execution times of the source code without encapsulation.

According to a study [4] where it has not been considered as a matter of interest by the scientific software development community to consider programming techniques; In this research work, it is shown that if considered, it would affect the performance of the software, specifically if encapsulation is included in all defined classes.

In scientific software one of the most important feature is performance; therefore, a paradigm more in line with this type of computer solution must be sought; as the procedural paradigm. From this, it is recommended to carry out future research implementing a solution following the object-oriented paradigm and the same solution to implement it considering the procedural paradigm.

Acknowledgements Thanks to Universidad Politécnica Salesiana, due to the financial contribution to this research work, which is within the framework of the research project: "ACISoft—Computational Analysis in Software Engineering" carried out in the Research Group of Software Engineering and Knowledge Engineering (GIISIC).

References

1. T. Rentsch, Object oriented programming. ACM SIGPLAN Notices **17**(9), 51–57 (1982). https://doi.org/10.1145/947955.947961
2. B. Unhelkar, *Software Engineering with UML*. CRC Press (2017)
3. H. Hourani, H. Wasmi, T. Alrawashdeh, A code complexity model of object oriented programming (OOP), in *2019 IEEE Jordan International Joint Conference on Electrical Engineering and Information Technology (JEEIT)* (IEEE, 2019), pp. 560–564. https://doi.org/10.1109/JEEIT.2019.8717448
4. E.M. Arvanitou, A. Ampatzoglou, A. Chatzigeorgiou, J.C. Carver, Software engineering practices for scientific software development: a systematic mapping study. J. Syst. Softw. **172**, 110848 (2021). https://doi.org/10.1016/j.jss.2020.110848
5. G.O. Pizarro-Vasquez, E.G. Pardo, *Resolution of the problem of optimization of the batching and routing of orders in warehouses with multiple transversal aisles* (unpublished)
6. S. Henn, S. Koch, K.F. Doerner, C. Strauss, G. Wäscher, Metaheuristics for the order batching problem in manual order picking systems. Bus. Res. **3**(1), 82–105 (2010). https://doi.org/10.1007/BF03342717 LNCS Homepage, http://www.springer.com/lncs. Last accessed 21 Nov 2016
7. G.O. Pizarro-Vasquez, Repository of source code of the TFM research project in Java. https://github.com/omarjcm/warehouse, [Online; accedido 25-Febrero-2021]
8. C.A. Valle, J.E. Beasley, Order batching for picker routing using a distance approximation. arXiv preprint arXiv:1808.00499 (2018)
9. C.A. Valle, J.E. Beasley, A.S. da Cunha, *Modelling and solving the joint order batching and picker routing problem in inventories*. Springer International Publishing, Cham (2016), pp. 81–97. https://doi.org/10.1007/978-3-319-45587-7
10. C.A. Valle, J.E. Beasley, A.S. da Cunha, Optimally solving the joint order batching and picker routing problem. Eur. J. Oper. Res. **262**(3), 817–834 (2017). https://doi.org/10.1016/j.ejor.2017.03.069
11. G.O. Pizarro-Vasquez, *Statistical analysis source code repository in R*. https://github.com/omarjcm/poo-research, [Online; accedido 25-Febrero-2021]

12. G.E. Dallal, L. Wilkinson, An analytic approximation to the distribution of Lilliefors's test statistic for normality. Am. Stat. **40**(4), 294–296 (1986). http://www.jstor.org/stable/2684607
13. M.S. Bartlett, R.H. Fowler, Properties of sufficiency and statistical tests, in *Proceedings of the Royal Society of London. Series A—Mathematical and Physical Sciences*, vol. 160, no. 901 (1937), pp. 268–282. https://doi.org/10.1098/rspa.1937.0109, https://royalsocietypublishing.org/doi/abs/10.1098/rspa.1937.0109
14. O. Gonzales, Parametric and non-parametric mathematical modelling techniques: a practical approach of an electrical machine identification. Ecuadorian Sci. J. **5**(1), 30–36 (2021). https://doi.org/10.46480/esj.5.1.86
15. P. Mair, R. Wilcox, Robust statistical methods in R using the WRS2 package. Behav. Res. Methods **52**(2), 464–488 (2019). https://doi.org/10.3758/s13428-019-01246-w

Chapter 64
Applying Maturity Models in Organizations for Digital Transformation: A Comparative Study

Juan Francisco Reyes, Villie Morocho, and Priscila Cedillo

Abstract Digital transformation and maturity models are being used by organizations, consulting providers, and academia. The objectives that maturity models pursue are optimizing and using technological functionalities and measuring the degree of maturity in defining an implementation path. These models allow the identification of dimensions that an organization needs to follow toward greater digital maturity. This paper presents the most relevant features that consider several maturity models developed from different sources and have been applied in the last ten years. Also, it is focused on reporting studies indexed in Scopus that consider case studies related to research performed around the world. Therefore, this study's primary goal is to provide a tool to guide information technology personnel during digital transformation processes. Moreover, surveys have been applied in several local companies to evaluate their involvement and assess the models' dimensions during a transformation process. Finally, the interpretation of a real implementation process has been carried out where several dimensions of the model are evaluated to verify its validity and subsequent analysis.

64.1 Introduction

Main Chief Executive Officers (CEOs) of the World who lead organizations use digital advances such as analytics, mobility, social networks, integrated intelligent devices, and other improvements like optimizing ERP [1]. Digital transformation is considered an essential strategy for organizations to renew businesses and stay competitive [1]. However, the establishment and implementation of digital plans could be challenging for managers because they sometimes do not know several, specific topics, and configurations that provide the best results [2, 3]. Investment in technology is not the same as doing it in digital transformation [4]. Most companies spend a lot of money purchasing new tools, services, or platforms. However, these

J. F. Reyes (✉) · V. Morocho · P. Cedillo
Department of Engineering, University of Cuenca, Ave. 12 de Abril, Cuenca, Ecuador
e-mail: juanf.reyes@ucuenca.edu.ec

Á. Rocha et al. (eds.), *Communication, Smart Technologies and Innovation for Society*, Smart Innovation, Systems and Technologies 252,
https://doi.org/10.1007/978-981-16-4126-8_64

investments do not mean that entities will change how they compete in a digital economy through transformation [3].

For this reason, there are several attempts to provide a global vision. Then, this paper reviews outstanding digital maturity models and frameworks, which identify different dimensions, by performing a comparative study among maturity models. Also, it depicts digital transformation strategies that enable the possible transition of companies from a "Business as usual" state to another where the innovation is linked [4, 5].

Consequently, this study identifies variables such as content, scope, and the organizations where models were implemented if they are open. Thus, it is possible to identify, as a result, the dimensions that are proposed as a reference for organizations [6]. Besides, Ecuador has been selected as a specific place to show the feasibility of applying the theories and guidelines presented in this contribution.

64.2 Literature Review Outcomes

64.2.1 Digital Transformation

Digital transformation has been a trendy subject during the last years, and it has attracted the attention of several researchers and professionals [5, 7]. Here, the need to implement and deploy new technologies in organizations, plus the interest of executive leaders in updating their companies, occur as an evolution toward changing the way of doing business [8]. The reasons take advantage of technological tools to become agile, optimize processes, and make effective use of business resources [9]. Emerging companies with a digital mindset attract customers through their platforms [7]. These organizations offer a variety and convenience in the use of products/services [3]. Finally, new working conditions and bio-security due to health issues, such as the current COVID-19 pandemics, establish urgent needs for technological implementations that contribute to the "new normality" industry.

In recent decades, the world has witnessed the collapse of large companies interrupted by insufficient innovative business models and the fault of new digital technologies, supported by infrastructure and platforms [8, 10]. Clear examples are Kodak, the first, giant in analog photography, and neglected the digital transition. Blockbuster LLC, which Netflix overtook, shifted from physical DVDs and video clubs to online streaming of digital content [6]. The digital transformation process is a challenging, long, and complex experience. Therefore, it is necessary to consider that it is not only oriented toward the technological field; the solutions implemented may not bring true value to the potential of a total transformation of the business model [7, 8].

The term transformation, in this context, refers to a fundamental change in business, which has an impact on organizational strategy and internal structures [11]. Therefore, even modifications related to the distribution of power require companies to align themselves and initiate a complete change process [11, 12]. This is concerning structures as well as business models, which are certainly challenging and require constant learning [8]. Transformation is a general concept and incorporates competitive strategies, which should be adopted in search of significant improvements in entrepreneurship performance [1, 7].

64.2.2 Main Models for Digital Maturity

Digital Culture Model. Being part of the digital culture is key to settling the different organizational transformations and hence its importance [13]. In this sense, it is defined as a set of behaviors and habits that are developed and applied by both managers and employees of a company to take advantage of the functionalities provided by the implemented technologies [9, 14]. Likewise, as part of this digital culture, there is the correct socialization in the organization, because transmitting the changes must establish a positive impact reflected by the end customer [15].

MIT Maturity Map. This model presents how organizations react and proceed to digital opportunities. It was born from a study in four hundred companies that reviewed the initiatives used by these organizations [10, 16]. The model is given under the combination of two dimensions. The first, **digital intensity**, which is defined as the level of investment in technological initiatives that adjust the way of operating [11, 12]. The second, **intensity in the management of transformation**, is the investment related to the leadership capabilities that facilitate transformation in the organization.

Capability Maturity Model (CMM). This model was created at Carnegie-Mellon University—CMM is the product of a study that examined organizations that intended to build software, with the US Department of Defense [11, 13]. Then, based on this model, some companies took it as a basis for creating other patterns aimed at different software branches, such as the guide for the acquisition of software, for the development of integrated products, among others [10].

McKinsey Digital Quotient. About five years ago, McKinsey conducted a study in several companies to understand the challenges organizations face in digitization. In this process, eighteen practices related to digital strategy, capabilities, and culture were taken into account, which shows a metric that seeks to measure companies' digital maturity. This resulted in the so-called McKinsey Digital Quotient [11].

Industry 4.0. It was born from the application of the IoT to the industry, which triggered the concept of Industry 4.0 [14, 17]. There is a high degree of connectivity and data generation, which by themselves do not present any value [18]. With due process and association, they become a source of analysis, which creates a differentiating factor. Moreover, issues, such as changes in the organization and control over the product cycle chain, are considered [13]. The Model of Degree of Preparation for Industry 4.0. was carried out in Germany [18]. It is mainly oriented to organizations in the engineering and manufacturing sectors, presenting four large areas where there is potential for its application [14].

64.3 Methodology

Organizations need capabilities to modify their processes; delivering solutions require effective methods and solid skills. These are the bases of a successful transformation process [1, 19].

An exploratory study was developed with a mixed approach, that is, qualitative and quantitative. This allows analyzing the models and maturity frameworks useful for the digital transformation (DT) of organizations in strategic management. The former approach allows the study to be substantiated using the systemic review that supports the exposed data [8]. As an initial approach, the qualitative character seeks the description of the characteristic features of the information collected without numerical measurement, (i.e., bibliographic foundation). Then, five fields were defined from the bibliographic analysis: (i) organization, (ii) culture, (iii) strategy, (iv) technology, and (v) internal personnel of the entities [7, 13]. Additionally, through the quantitative approach, designing and applying a survey, numerical data was collected, measurable instruments to identify the position of the companies' respect to five fields defined before.

In this study, the bibliographic-documentary technique was used, using a systemic review and later the application of a questionnaire directed at organization's strategic management. Thus, the TD models applied in organizations around the world were obtained [7, 8]. Exploratory research consists of the synthesis of data and information related to the subject of study in books, academic articles, scientists, and indexed journals. Data were extracted from Scopus and Science Direct.

During these activities, the study topic was identified and discussed. This action is related to aspects of the subject based on the reflections obtained in relevant documents of related research. Secondly, were the state of the art and theoretical framework on TD and the adoption of the digital maturity model, by national companies. This material allowed to identify better the research questions that allowed obtaining the key platforms and the dimensions to study. With this, the documentary information addressed in investigations of the last ten years was identified and selected. Subsequently, the information extraction was carried out, based on five dimensions that comprise a digital framework: organization, culture, strategy, technology, and personnel.

During the survey design, four questions were considered for each of the five dimensions. For the answers, a Likert scale was used. In addition, the design of the survey took into account essential suggestions described in the literature. To avoid bias or a central tendency to arrive at precise results, an even number of options were used. A trend of answers that do not clearly verify the respondent inclination toward the approach decreased. Likewise, the induction that the approach itself may have was reduced, to avoid directing the respondent toward one of the alternatives.

With these antecedents, data classified by type of study, author, date, and maturity frame were collected, which provided the research's main results and conclusions. The findings were critically discussed and served to explain, support, or expand previous theories on the subject. The following keywords were used as descriptors, both in Spanish and English: digital maturity model (DMM), maturity capability model, and TD maturity models. With these guidelines, 200 documents were selected in the Scopus indexation site and Science Direct library. Of which 69 were related to the theme, and the other results were ignored (i.e., results repeated, jobs already covered). That point was reviewed in detail 31 articles, the criteria, and context of this study. The selected papers are classified as scientific articles, conference papers, research reviews, consultancy publications, and report publications. The additional chosen variables were type of publication, country of origin, year of publication, model, area of application, and the fields of action proposed by the model.

The survey design was based on the study objectives. In addition, the defined dimensions have been added of the Digital Maturity Models: organization, culture, strategy, technology, and personnel [7, 13]. To measure the perception of certain questions and answers, it was validated through analysis performed with domain experts. The instrument's consistency is held that the content validity by expert judgment is a valuable method to identify the reliability of an investigation; it is understood as the opinion provided by people, whose trajectory is extensive in the subject, in addition to being recognized by other qualified professionals. On the other hand, scale validity is good (Cronbach $\alpha = 94$). The application of the survey was oriented to the management personnel of recognized companies in the city of Cuenca, Ecuador.

The questions were written to be oriented to measure the perceptions about this approach. It refers to the scale of the survey, which goes from a lower level that corresponds to "completely disagree", to the high level that indicates "totally agree". The survey was carried out using technological tools, using electronic mail, and telephone calls. The response time shows this topic's importance, motivation to perform related studies at national and regional levels.

64.4 Results

Among the findings of this study, it was identified that most studies were carried out at an international level, mainly in Europe. Therefore, it shows a weakness in digital transformation (DT) research in the Latin American region.

According to the literature reviewed, digital maturity models are used by various industries and economic sectors to measure TD progress [20]. The findings show the existence of various models with an emphasis on improving organizational capabilities. Among which 12 stand out: DMM Industry 4.0, DMM Peyman, DMM Berghaus, Digital Readiness Assessment, TD Structure, DMM Gill, DMM Manufacturing, Organizational DMM, DMM Newman, DMM Deloitte, and Road Map for TD [5, 21–23]. The surveys' results were tabulated to determine the value of

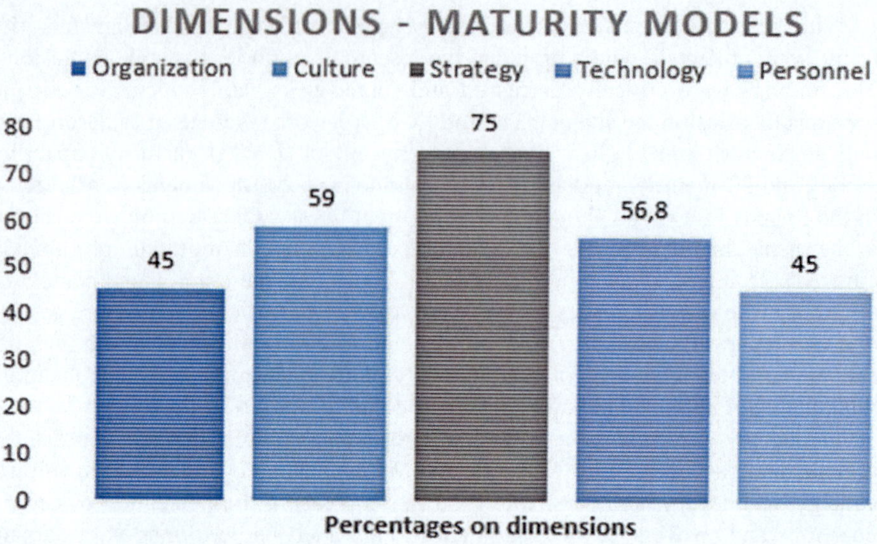

Fig. 64.1 Results on dimensions of maturity models

the quantitative variable that is more common among each group of answers. For this, the mode (statistics) was determined, but in the visualization, the average over groups of four questions, which correspond to one dimension, was used. In total, the survey has five groups of questions that correspond to five dimensions raised.

The percentages of the proposed dimensions were obtained for the analysis, as shown in Fig. 64.1. In specific sectors, the level of coincidence is high around the levels. Therefore, state a good strategy prior to the start of a DT is essential. It is visible in the results that the strategy stands out over the other dimensions, according to results obtained in the bibliographic review. Therefore, this dimension is taken as the—"central axis"—[7]. The reason is associated with the alignment that must exist between the organization strategy around the business line and the technology to be adopted [4]. Thus, leading organizations, with a high level of maturity, present a clear digital strategy that pursues long-term objectives. Meanwhile, companies with a relatively low maturity level of maturity continue to investigate the impact of digital disruption without a clear plan [8, 23].

Up to this point, general results have been presented that cover the sectors subject to analysis. However, each industry has peculiarities, and it is essential to identify them. Next, sectors such as banking, multinational organizations, and production organizations are specifically presented (see Fig. 64.1).

There are leading sectors in digital transformation; banking is one of them. In this sector, budgets are prioritized in technological projects providing reliability to those businesses (i.e., security, confidence); any investment is highly profitable [6, 15]. Therefore, the results obtained on this type of organization stand out, where

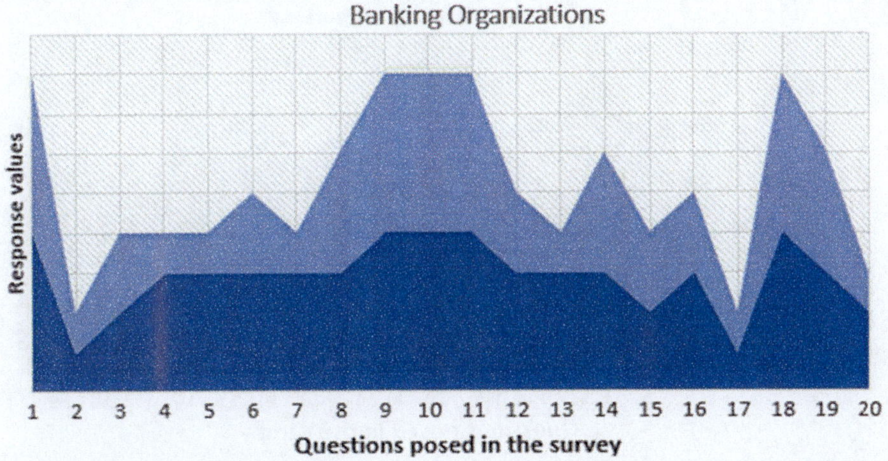

Fig. 64.2 Results on banking organizations

there is a high level of coincidence and shows why the sector has evolved successfully. Regarding the technology's implementations, it aims to achieve an objective marked by competitiveness and attract a population that is becoming more and more accustomed to technology [1]. Proof of this, a high number of services are provided through digital channels. In this way, it is possible to improve the customer experience, promoting multiple initiatives that leverage technological functionalities. In this sector, digital committees are preceded by the general management, who develop strategies to effectively introduce changes in customer services (see Fig. 64.2) [12, 16].

Finally, in the banking sector, the culture created by the staff and in the client stands out to be aware of new functionalities that improve interaction with the bank. Regarding multinational organizations with significant presence in the country, the guidelines are given from the headquarters in a clear way [1, 8]. They execute strategies that allow them to follow global standards [1]. This type of result is not a coincidence, and instead, they show a marked trend toward innovation and the incorporation of new technologies. Their implementation allows reaching highly competitive in an environment where there is great volatility due to the continuous appearance of changes [16]. The observations have clear slants. Figure 64.3 shows that in the central data, which obey the *strategy* dimension, there is a very high level of coincidence. Similarly, this trend is repeated in the group that belongs to the *culture*.

Not all of the results obtained are coincidence; the differences obtained in terms of the production organizations are an example of this. One of the studied companies is going through a process of DT in its entire back office, which denotes a beginning on the path of transformation [6, 24]. The objective is to have a better quality of information, allowing them to project based on the correct results. Another organization in the same sector had a similar process in previous years, so it is currently stable. This process seeks to consolidate the data collected with business intelligence

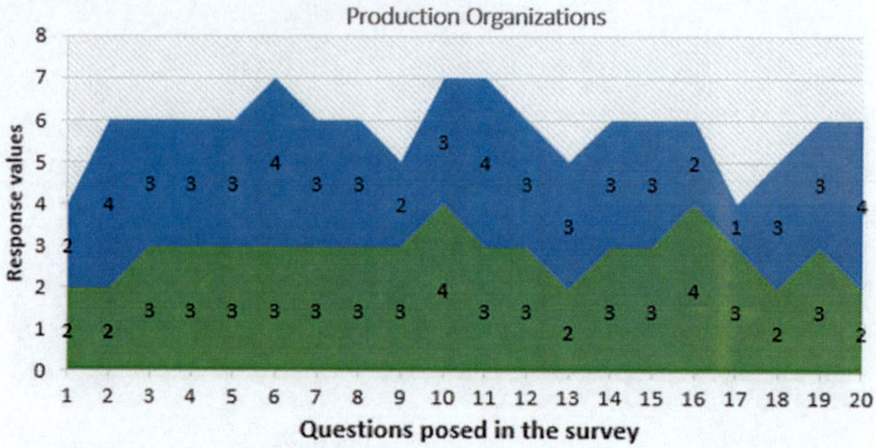

Fig. 64.3 Results production organizations

modules and other verticals [15]. The objective is to have a clearer picture, which will help them to be competitive in several of their lines [1].

This type of organization has a high level of agreement, especially in the dimensions of organization and culture. It differs in this case in the dimension of strategy, due to the realities as mentioned above. However, the technology dimension variation is noticeable, both trying to mature but in different timelines. All these results are shown in Fig. 64.3.

64.5 Discussion

Every digital transformation process has an identified beginning due to carrying out a technological implementation [4, 6]. That is why it is essential to present it in a way that the need raised is understood. The ultimate goal is to count on the collaboration and commitment of the organization managers, until they are aligned [7, 25].

According to the results obtained, in the application environment of the model of this study, clear guidelines must be prioritized between the technology and business areas. In addition, there must be a qualified technology provider in each process, and the organization must have the ability to follow an appropriate process to choose it [26]. It is recommended to consider, within the organization, variables such as the budget for the project and the profile required for the involved personnel [27]. Regarding the supplier, it is necessary to take into account the positioning in the technological environment. The possibility of support services at the national or local level, references from nearby organizations, where a case of success is evident, and above all who knows the market in which implementation is proposed [1, 2].

Several points are not part of this study but are undoubtedly relevant in the process of digital transformation. They may take a back seat, but they are a fundamental part, such as the case of change management [4, 12]. If adoption is desired an optimally, it must be planned, allowing the person to consider future changes. This point deserves to have a clear strategic plan where personnel profiles, monitoring methods, and capacity building to support operations are considered [6]. Among the lessons learned, the fact of making consensual decisions with managers, who are clear participants, stands out. Plan the load related to the tasks that are key in the project to avoid deviating from the initial planning [7].

Finally, it was possible to verify that the results when applying surveys are consistent with the plans proposed in the case study to achieve digital transformation. It can be mentioned that the model where these five dimensions prevail is feasible to carry out in the country. It should be taken into account that the proposed dimensions obey the specific needs of the organization, but it is these dimensions that encompass the best practices in the world.

64.6 Conclusions

The proposed model, as a result of the bibliographic analysis, served for the design and application of the survey, yielded the main axis, the strategy, which highlights the alignment with the business to achieve digital maturity.

As part of the findings in the case study, it was found that the digital transformation process requires having the skills and influence that leaders possess, creating a clear and convincing transformative vision to drive the organization. Then, from the bibliographic review of the models applied in recent years, it has been possible to determine that the maturity model has specific dimensions, depending on the sector where it is to be used.

The maturity models can be considered generic; their application depends solely on the current situation of the organization, and the scope that the transformation process is intended to give. Based on the bibliographic review and the application of the survey, it is demonstrated that the determining dimensions are used as the main axes in the implementation projects.

Pressure points for technological change are increasing from various angles. Globalization makes efficient integration mandatory that organizations can only achieve through digital processes and where digital collaborative tools would greatly support. As a result of this document, it can be indicated that digital transformation is an integration of technologies, which impact the main processes of an organization. This makes it possible to obtain a differentiation factor on your final products for the customer.

References

1. G. Westerman, M. Tannou, D. Bonnet, P. Ferraris, A. McAfee, Digital transformation: a roadmap for billion-dollar organizations. MIT Center Digit. Bus. Capgemini Consult. 32–64 (2011)
2. E.A. Isaev, N.L. Korovkina, M.S. Tabakova, Evaluation of the readiness of a company's IT department for digital business transformation. Bus. Inform. **2**(44), 55–64 (2018). https://doi.org/10.17323/1998-0663.2018.2.55.64
3. M. Gill, S. Vanboskirk, *The Digital Maturity Model 4.0 Benchmarks: Digital Business Transformation Playbook*. Retrieved from https://forrester.nitrodigital.com/pdf/Forrester-sDigitalMaturityModel4.0.pdf (2016)
4. D. Gunsberg, B. Callow, B. Ryan, J. Suthers, P.A. Baker, J. Richardson, Applying an organisational agility maturity model. J. Organ. Change Manag. **31**(6), 1315–1343 (2018). Retrieved from https://doi.org/10.1108/JOCM-10-2017-0398
5. T. Aguiar, S.B. Gomes, P. Cunha, M. Silva, Digital transformation capability maturity model framework, in *2019 IEEE 23rd International Enterprise Distributed Object Computing Conference (EDOC)* (2019), pp. 51–57
6. S. Berghaus, A. Back, *Stages in Digital Business Transformation: Results of an Empirical Maturity Study* (2016). MCIS 2016 Proc 22. http://aisel.aisnet.org/mcis2016/22
7. J. Bumann, M.K. Peter, Action fields of digital transformation—a review and comparative analysis of digital transformation maturity models and frameworks, in *Digitalisierung und andere Innovationsformen im Management. Innovation und Unternehmertum, Band 2. Edition Gesowip, Basel*, ed. by M. Aeschbacher, K. Hinkelmann, A. Verkuil (2019), pp. 13–40
8. R. Teichert, Digital transformation maturity: a systematic review of literature. Acta Universitatis Agriculturae et Silviculturae Mendelianae Brunensis **67**(6), 1673–1687 (2019)
9. C. Anderson, W. Ellerby, *Digital Maturity Model (DMM): Achieving Digital Maturity to Drive Growth* (2016). Retrieved from www.deloitte.com
10. R. Alvarez, M.R. Martins, M. Silva, Applying the maturity model concept to the servitization process of consumer durables companies in Brazil. J. Manuf. Technol. Manag. **26**, 1086–1106 (2015). https://doi.org/10.1108/JMTM-04-2014-0050
11. O. Lorenzo, Modelos De Madurez Digital: ¿En Qué Consisten Y Qué Podemos Aprender De Ellos? Digital Maturity Models: What Are They and What Can We Learn from Them? Boletin De Estudios Economicos, Vol. LXXI (February), 573–590 (2016b). Retrieved from https://search.proquest.com/openview/937dd70ffd989152b0dde42abed3c418/1?pqorigsite=gscholar&cbl=1536340
12. S. Chanias, T. Hess, *How Digital are We? Maturity Models for the Assessment of a Company's Status in the Digital Transformation* (2016)
13. A. Schumacher, S. Erol, W. Sihn, A maturity model for assessing industry 4.0 readiness and maturity of manufacturing enterprises. Proc CIRP **52**, 161–166 (2016). https://doi.org/10.1016/j.procir.2016.07.040
14. S.R. Hamidi, A.A. Aziz, S. Shuhidan, A. Aziz, M. Mokhsin, *SMEs Maturity Model Assessment of IR4.0 Digital Transformation* (2018)
15. M. Newman, *Digital Maturity Model (DMM): A Blueprint for Digital Transformation* (2017). Retrieved from www.tmforum.org
16. G.C. Kane, D. Palmer, A. Nguyen-Phillips, D. Kiron, N. Buckley, *Achieving digital maturity, 15329194* (Massachusetts Institute of Technology, Cambridge, MA, Cambridge), pp-1–32 (2017)
17. D. Suarez, *Transformación Digital*, Datta Business Innovation (2019), pp. 146–156
18. M.D. Tuya, M.L.D. Tuya, Creating value through information and knowledge flow: lessons from the public sector applied to the private sector. J. Creating Value **5**(2), 210–221 (2019). https://doi.org/10.1177/2394964319851951
19. A. De Carolis, M. Macchi, E. Negri, S. Terzi, A maturity model for assessing the digital readiness of manufacturing companies, in *Advances in Production Management Systems. The Path to Intelligent, Collaborative and Sustainable Manufacturing. APMS 2017. IFIP Advances*

in Information and Communication Technology ed. by H. Lödding, R. Riedel, K.D. Thoben, G. von Cieminski, D. Kiritsis, vol. 513 (Springer, Cham, 2017). https://doi.org/10.1007/978-3-319-66923-6_2
20. H. Gimpel, S. Hosseini, R.X.R. Huber, L. Probst, M. Röglinger, U. Faisst, Structuring digital transformation: a framework of action fields and its application at ZEISS. J. Inf. Technol. Theory Appl. **19**(1), 3 (2018)
21. L. Canetta, B. Andrea, E. Montini, Development of a digitalization maturity model for the manufacturing sector, in *2018 IEEE International Conference on Engineering, Technology and Innovation (ICE/ITMC)* (2018), pp. 1–7
22. D. Schallmo, K. Lang, D. Hasler, K. Ehmig-Klassen, C. Williams, *An Approach for a Digital Maturity Model for SMEs based on Their Requirements* (2020)
23. CMMI Institute, Obtenido de CMMI Institute (2018). https://cmmiinstitute.com/
24. I.V. Aslanova, A.I. Kulichkina, *Digital Maturity: Definition and Model* (2020). https://doi.org/10.2991/aebmr.k.200502.073
25. N. Brown, I. Brown, From digital business strategy to digital transformation—how: a systematic literature review, in *Proceedings of the South African Institute of Computer Scientists and Information Technologists 2019 (SAICSIT '19)*. Association for Computing Machinery, New York, NY, USA, Article 13 (2019), pp. 1–8. https://doi.org/10.1145/3351108.3351122
26. S.R. Carreiro, J.L. Martinho, An industry 4.0 maturity model proposal. J. Manuf. Technol. Manag. (2019). https://doi.org/10.1108/JMTM-09-2018-0284
27. A. De Carolis, M. Macchi, E. Negri, S. Terzi, Guiding manufacturing companies towards digitalization a methodology for supporting manufacturing companies in defining their digitalization roadmap, in *2017 International Conference on Engineering, Technology and Innovation (ICE/ITMC)*, (2017), pp. 487–495. https://doi.org/10.1109/ICE.2017.8279925

Chapter 65
Analysis of the Productivity of Pair Programming Versus Solo Programming in a University Classroom

Ricardo Naranjo Sánchez and Cristhian Villegas Vizhñay

Abstract The present study explains the development of an experimental process about the comparison of programming styles, which was objective to measure productivity in software creation applying two different styles: Pair Programming and Solo Programming. There was a search for information by different relevant online databases about experimental studies that cover the comparison in the use of known agile methodologies. For each study, it was verified that the address experiments about the implementation of programming styles for software development. The results show that experimentation in this area, in most studies, generates significant results in software development by comparing of the productivity of both programming styles, to determinate the feasibility that programming in pairs can have on individual programming and validate that it is applicable in companies dedicated to software development. In addition, this study provides suggestions for futures experimentation where more detailed parameters and longer execution times are established to obtain different results on the implementation of programming styles belonging to the extreme programming.

65.1 Introduction

Software engineering is the application of a systematic approach to knowledge, oriented to the methods, design, implementation, operation, testing, documentation and maintenance of software [1].

The introduction of different agile methodologies in software development has become relevant in the industry, becoming successful in several countries [2–5]. These agile methodologies were born with the need to accelerate development processes in companies that urgently needed systems to maintain their competitiveness. It does not yet have the necessary evidence to support most of the current practices applicable to software development [6]. In the essence, this is due to the

R. N. Sánchez (✉) · C. V. Vizhñay
Universidad Politécnica Salesiana, Campus Centenario, Robles 107 y Chambers, Guayaquil, Ecuador
e-mail: rnaranjo@ups.edu.ec

Á. Rocha et al. (eds.), *Communication, Smart Technologies and Innovation for Society*, Smart Innovation, Systems and Technologies 252,
https://doi.org/10.1007/978-981-16-4126-8_65

lack of a solid theoretical foundation. One way to obtain this scientific knowledge is through experimentation.

Software development is carried out under a guiding approach for the developer, providing ease for organization and collaborative work to satisfy the requirements of end users [7, 8].

In the programming subjects of the universities in general, it is usual to work individually on the exercises and practices of the class [9], due to the physical space of the work spaces of the computer laboratories and the number of students enrolled in the subject [10].

Extreme programming created by Kent Beck [11] in the late 90's, it is one of the best known agile approaches [12], refers to pair programming as a practice that provides benefits. Several empirical studies have been carried out that highlight the effects of the implementation of this practice [13–17].

It is proposed to carry out the design, execution and analysis of a software development experiment, with the main objective of observing the influence that the application of two different programming styles has on productivity. For this, there will be a group of experimentation subjects made up of students of the Software Engineering subject of the Universidad Politécnica Salesiana Guayaquil (UPS-G), where the following question is posed: Are there differences in productivity when applying the styles of pair programming and solo programming?

For this process, two programming styles were used: Pair Programming (PP) and Solo Programming (SP). In PP, two programmers design, code and test the software together on a computer [2], whereas in SP, one person takes care of all development work.

65.2 Related Work

Among the studies that have been carried out analyzing PP versus SP, it was observed in some that the task in pairs showed a shorter duration in completing a task than doing it individually; in the same way, other benefits are reported [10, 16, 18, 19].

Some studies showed mixed results, [18] used 15 professionals who were distributed in five pairs and five individuals, who programmed a script for databases, where the values given by the experiment reflected acceptance of the PP showing a 29% decrease in the time of duration to complete the tasks compared to the SP.

The study of [19] used 16 students for the experimentation that were distributed in five pairs and six individuals, who solved four exercises. The creators of the experiment found no relevant differences between the styles of PP and SP.

The experiment of [16] used five pairs and five individuals, where the creators of the experiment showed acceptance of the PP, reporting a 52% decrease in the time of completion of the task in contrast to the SP.

The study by [10] used 94 students as study subjects divided into 54 pairs and 40 individuals, developing the same software under the same conditions. The results reflect that the students who worked in pairs had better results and perceived easier work, unlike the students who programmed individually, being able to detect errors more quickly.

In the study of [20], 24 students were used who experienced three work modalities during normal class sessions. The results reflected that pair programming was more productive and accepted by students, unlike individual programming and Mob (collective) programming.

The studies showed diverse results: some reported benefits with different quantitative results, others did not find significant differences; that is why it will be positive to carry out another experiment in order to have a greater appreciation.

65.3 Methodology

The planning or experimental software design is the fundamental part prior to the execution and analysis of the experiment; in this phase, the experiment that was carried out was defined and designed. In the experiment, the following were determined hypotheses, selection of variables, parameters, design of the experiment, selection of experimental subjects, collection and analysis procedure.

In the experiment carried out, the productivity factor and the Pair Programming and Solo Programming styles were considered as factor treatments to determine the effect of these styles on software development productivity.

This section details the process of the design of the experiment, which according to the response variable "productivity" and the styles of PP and SP, followed a random blocking design. The random assignment of 30 experimental subjects into two groups was designed, each group corresponding to one treatment.

65.3.1 Objectives

Determine if the programming style used produces an effect on the productivity of the subjects in the development of low complexity software.

65.3.2 Hypothesis

For the statement of the hypotheses, both null and alternative, the notation described by Jedlitschka in [21], where for each objective the null hypothesis is denoted as H_0 and their corresponding alternative hypotheses like H_1.

In this experiment, there was only one null hypothesis and one alternative hypothesis, therefore:

- H_0 = "There is no difference in the productivity obtained by the development of low complexity software with the use of two different styles, Pair Programming and Solo Programming."
- H_1 = "If there is a difference in the productivity obtained by the development of low complexity software with the use of two different programming styles, Pair Programming and Solo Programming."

65.3.3 Variables

In this experiment, a set of factors that may be relevant to the experiment were considered, i.e., they may have an impact on the results of the experiment, being classified as will be seen below in variables and parameters.

The variables are those that represent the cause and effect of the study:

- *Dependent variable*: Within the present experiment, there is only one response variable that has been defined as "Productivity."
- *Independent variable*: For the evaluation of productivity, "Programming Style" was taken into account as an independent variable, which has two levels: Pair Programming and Solo Programming.
- *Blocking Variable*: The creation of a software that was developed in a session of 1 h was determined.

65.3.4 Parameters

The parameters are the control variables that should be kept constant in the test of the experiment, or random failing that, to avoid that the results of the study may be biased:

- *Time for the execution of the experiment*: To control the influence of this parameter, the defined time for the experiment session is 1 h. In the event that the exercise is completed in a shorter time, the time at which it was completed will be indicated.
- *Programming language and tool*: As a control strategy for this aspect, it has been defined that the development of the exercise will be carried out in the Python programming language using the Google Collaboratory (Colab) cloud tool for the PP and SP styles.

65.3.5 Design of the Experiment

In this experiment, it was proposed to use a random blocking design, doing a statistical analysis and organizing two experimental groups, group 1 made up of 10 couples and group 2 made up of 10 subjects, to obtain the results of the experiment. In addition,

Table 65.1 Configuration of the experimental design

Session	Exercise	Treatment	Time
1	A	Pair programming	1 h
		Solo programming	

Table 65.2 Distribution of participants in the session

Session	Number of subjects (Individual)	Number of subjects (Pair)	Total subjects
1	10	10 * 2	10 + 20 = 30

the development of low complexity software is contemplated in an estimated time of 1 h of experimentation. Only one problem was posed, the requirements of which were specified to the experiment participants at the beginning of the session.

For the configuration of the experiment, the following distribution was proposed in the following Table 65.1:

For this proposed design, in order to maintain an equitable distribution in both experimental groups, in relation to the number of participants, with a time limit of 60 min (1 h), the following characteristics were proposed:

- For the session, the experimental subjects were randomly distributed into two work groups, one to apply Pair Programming and the other to apply Solo Programming.
- There was 1 exercise during the session, for the subjects of both work groups to develop it Table 65.2.

It was sought that during the execution of the experiment, all the subjects were in equal technical and environmental conditions, for which the realization of the same was defined in one of the computer laboratories of the UPS-G, where all the subjects have access to desktop computers with the same technical characteristics. In this way, the influence on the response variable that may derive from a technical problem was controlled.

65.3.6 Participants

The participants are students from the Universidad Politécnica Salesiana, who have basic knowledge of the Python programming language and were distributed and organized in development teams for the session, which was configured by PP or SP.

65.4 Results

65.4.1 Statistic Analysis

To carry out the analysis of this experiment, we used the Student's T test, in order to evaluate whether the experimental groups (two) show significant differences in the result. This test was performed using SPSS statistics software.

When entering the data of the experiment in the SPSS statistics software, corresponding to the experimental subjects, it gave us the first results to verify that the entered data are correct, and there is no lack of information, taking into account that for the first group, "Pair Programming" was 10 couples, and for the second group, "Solo Programming" was 10 subjects. This statement can be seen in greater detail in Table 65.3.

In Table 65.4, you can see a descriptive table where some statistical calculations of both programming styles are specified, Pair Programming (1) and Solo Programming (2). The referential value in the aforementioned table is the value of the average which indicates the numerical difference between both styles, having the pair programming an average of 37,3850 and the individual programming an average of 36,4100. The difference between the values is 0.975, where apparently pair programming turned out to have an advantage over the other programming style.

Table 65.3 Case processing summary

Functionality	Programming style	Valid		Missing		Total	
		N	Percent (%)	*N*	Percent (%)	*N*	Percent (%)
	Pair	10	100	0	0.0	10	100
	Solo	10	100	0	0.0	10	100

Table 65.4 Descriptive table of the statistical analysis of pair programming and solo programming

	Programming style			Statistic	Std. error
Functionality	Pair	Mean		37,3850	8,68,561
		95% confidence interval for mean	Lower bound	17,7368	
			Upper bound	57,0332	
	Solo	Mean		36,4100	7,46,623
		95% confidence interval for mean	Lower bound	19,5202	
			Upper bound	53,2998	

In the following section, the steps used to carry out the comparison of the experiment and obtain the result of the response variable "Productivity" are related.

- Step 1: Drafting hypotheses.
  - Hypothesis: Productivity in group 1 "PP" is higher than that of group 2 "SP."

    H_1 = Exists a significant difference in the mean results of both groups.
    H_0 = Does not exist a significant difference in the mean results of both groups.

- Step 2: Determine alpha (α).

  Our experiment had a quantitative analysis with a confidence of 95%; therefore, the percentage of error that the experiment will have is determined:

$$\text{Alpha}(\alpha) = 5\% = 0.05$$

- Step 3: Reading the test value (*P*-Value).
  - Normality: The random variable (VA) must follow a normal distribution in the two groups. The "Shapiro Wilk" test was used for a sample size < **30**, where the values of the column were taken as a reference "Statistic." To determine the normal distribution of the VA, it was determined:

    P-Value $\geq \alpha$ Accept H_0 = The data come from a normal distribution.
    P-Value $< \alpha$ Accept H_1 = The data does not come from a normal distribution Tables 65.5 and 65.6.

  Conclusion: The productivity variable in the two groups behave normally.

  - Equality of variance: Check the equality of variance using Levene's test.

    P-Value $\geq \alpha$ Accept H_0 = The variances are equal.
    P-Value $< \alpha$ Accept H_1 = There is a significant difference between the variances Tables 65.7 and 65.8.

Table 65.5 Shapiro–Wilk test of normality

		Kolmogorov-Smirnov*			Shapiro–Wilk		
Functionality	Programing style	Statistic	df	Sig	Statistic	df	Sig
	Pair	0.209	10	0.200*	0.890	10	0.168
	Solo	0.194	10	0.200*	0.951	10	0.676

Table 65.6 Comparison of values between the normality test and the percentage of error

Equality of variance		
P-Value (Group 1) = 0.890	>	$\alpha = 0.05$
P-Value (Group 2) = 0.951	>	$\alpha = 0.05$

Table 65.7 Levene's test

		F	Sig
Functionality	Equal variances assumed	0.058	0.813
	Equal variances not assumed		

Table 65.8 Comparison of values between the equality of variance and the percentage of error

Equality of variance		
P-value (Group 1) = 0.813	>	*A* = 0.05

Table 65.9 Analysis of the student's T test for equality of means

		F	Sig	*t*	d*f*	Sig. (2-tailed)
Functionality	Equal variances assumed	0.058	0.813	0.085	18	0.933
	Equal variances not assumed			0.085	17.603	0.933

Table 65.10 Comparison of T student test values and the percentage of error

Student's t test		
P-Value (Group 1) = 0.813	>	$\alpha = 0.05$

Conclusion: The equality of variance in both groups is the same.

- Step 4: Statistical resolution.
  - Calculate test value (*P*-Value)

    If the obtained probability *P*-Value $\leq \alpha$, reject H_0.
    If the obtained probability *P*-Value $> \alpha$, no reject H_0.

Table 65.9 showed a bilateral significance (Sig. 2-tailed) with equal values for both programming styles, so the value of the equal variances in Table 65.7 was taken as a reference comparison with the error rate shown in Table 65.10.

SPSS analysis conclusion: There is no significant difference between the mean results of both groups.

65.5 Discussion

Once the experiment was finished, the indications given at the beginning of the session were taken into account to evaluate the programming carried out by the working groups and verify that the code delivered is correct and functional. The results of the experiment showed that pair programming may be the most appropriate approach for its implementation since it presents a difference of 0.975 in the mean as analyzed in Table 65.4, over individual programming. The perception with the naked eye seems to be feasible with the first method, but applying the statistical analysis by means of the SPSS statistics software when entering the variables, more precise results can be evidenced by means of the Student's T test for independent samples that allowed us to determine whether the productivity is significant in the types of programming applied in the experiment.

Our results are similar to those of [2, 19, 22] not finding differences regarding the type of programming applied (Pair Programming and Solo Programming). Unlike other studies such as those of [10, 20] where their results reflected that partner programming is the method accepted by students and is a style that can be applied to a new teaching modality in universities.

It must be considered that the study was carried out with a small sample of participants and a time limit of 1 h, so it will be necessary, for a better interpretation of the results of this experiment, to carry out an investigation that is more in-depth and that allows to verify these findings.

65.6 Conclusion

The experimental process is a complex process in which a great diversity of factors intervene that must be considered in the planning phase, which represents a demand for collaborative work and coordination between the actors in the process, for which a constant communication by different synchronous or asynchronous technological means.

The students who were evaluated in the two programming modalities studied in this experiment showed similar results, with which we can conclude that there is no significant difference in the productivity variable. Pair programming provides benefits by having a second person to be able to solve problems during the execution of a specific activity, but individual programming limits continuous interaction with other people and can mean poor performance if the resolution of a problem is not clear certain exercise.

It is necessary to continue conducting experiments, where more detailed parameters are established and deepened that allow obtaining a different result on the implementation of these programming styles belonging to extreme programming as part of Agile Methodologies applied to software development.

References

1. I. Xplore, International Standard ISO/IEC/IEEEXplore, **I** (2017).

2. O.S. Gomez, A.A. Aguileta, R.A. Aguilar, J.P. Ucan, R.H. Rosero, K. Cortes-Verdin, An empirical study on the impact of an IDE tool support in the pair and solo programming. IEEE Access **5**, 9175–9187 (2017). https://doi.org/10.1109/ACCESS.2017.2701339
3. O. Salo, P. Abrahamsson, Agile methods in European embedded software development organisations: a survey on the actual use and usefulness of extreme programming and scrum. IET Softw. **2**(1), 58–64 (2008). https://doi.org/10.1049/iet-sen:20070038
4. G.S. Matharu, A. Mishra, H. Singh, P. Upadhyay, Empirical study of agile software development methodologies. ACM SIGSOFT Softw. Eng. Notes **40**(1), 1–6 (2015). https://doi.org/10.1145/2693208.2693233
5. P. Rodríguez, J. Markkula, M. Oivo, K. Turula, Survey on agile and lean usage in Finnish software industry. Int. Symp. Empir. Softw. Eng. Meas. 139–148 (2012). https://doi.org/10.1145/2372251.2372275.
6. N. Juristo, A.M. Moreno, S. Vegas, A survey on testing technique empirical studies: How limited is our knowledge. ISESE 2002—Proceedings, 2002 Int. Symp. Empir. Softw. Eng. 161–172 (2002). https://doi.org/10.1109/ISESE.2002.1166935
7. B. Choudhary, S.K. Rakesh, An approach using agile method for software development. 2016 1st Int. Conf. Innov. Challenges Cyber Secur. ICICCS 2016 (ICICCS), 155–158 (2016). https://doi.org/10.1109/ICICCS.2016.7542304
8. F. Nurullah, G. Wang, E.R. Kaburuan, A.N. Fajar, The collaboration of DevOps automation and SOA to accelerate software development culture. 1st 2018 Indones. Assoc. Pattern Recognit. Int. Conf. Ina. 2018—Proc. 262–266 (2019). https://doi.org/10.1109/INAPR.2018.8627022
9. K. Umapathy, A.D. Ritzhaupt, A meta-analysis of pair-programming in computer programming courses. ACM Trans. Comput. Educ. **17**(4), 1–13 (2017). https://doi.org/10.1145/2996201
10. R. Ventura, R. Hernández, J. Antonio, H. Izaguirre, A.L. Mendoza, D. Bravo Herrera, Comparación de la programación individual y por pares en los cursos universitarios. Temas Cienc. y Tecnol. **21** (61), 11–22 (2017)
11. K. Beck, *Extreme Programming Explained, Second Edition* (2004)
12. M.M. Müller, Two controlled experiments concerning the comparison of pair programming to peer review. J. Syst. Softw. **78**(2), 166–179 (2005). https://doi.org/10.1016/j.jss.2004.12.019
13. M.A. Domino, R.W. Collins, A.R. Hevner, Controlled experimentation on adaptations of pair programming. Inf. Technol. Manag. **8**(4), 297–312 (2007). https://doi.org/10.1007/s10799-007-0016-8
14. G. Canfora, A. Cimitile, F. Garcia, M. Piattini, C.A. Visaggio, Evaluating performances of pair designing in industry. J. Syst. Softw. **80**(8), 1317–1327 (2007). https://doi.org/10.1016/j.jss.2006.11.004
15. T. Bipp, A. Lepper, D. Schmedding, Pair programming in software development teams—an empirical study of its benefits. Inf. Softw. Technol. **50**(3), 231–240 (2008). https://doi.org/10.1016/j.infsof.2007.05.006
16. K.M. Lui, K.C.C. Chan, J. Nosek, The effect of pairs in program design tasks. IEEE Trans. Softw. Eng. **34**(2), 197–211 (2008). https://doi.org/10.1109/TSE.2007.70755
17. S. Xu, V. Rajlich, Empirical validation of test-driven pair programming in game development. Proc.—5th IEEE/ACIS Int. Conf. Comput. Info. Sci., ICIS 2006. Conjunction with 1st IEEE/ACIS, Int. Work. Component-Based Softw. Eng., Softw. Arch. Reuse, COMSAR 2006 **2006**, 500–505 (2006). https://doi.org/10.1109/ICIS-COMSAR.2006.34
18. J.T. Nosek, The case for collaborative programming. Commun. ACM **41**(3), 105–108 (1998). https://doi.org/10.1145/272287.272333
19. J. Nawrocki, A. Wojciechowski, Experimental evaluation of pair programming. Eur. Softw. Control Metrics 99–101 (2001) [Online]. Available: http://citeseerx.ist.psu.edu/viewdoc/summary?doi=10.1.1.19.1689.
20. V.R. Hern, S. Armando, G. Moya, "Ramón Ventura Roque Hernández* | Sergio Armando Guerra Moya** | Adán López Mendoza***," (2020), pp. 39–55
21. A. Jedlitschka, A. Tn, *Reporting Guidelines for Controlled Experiments in Software Engineering Dietmar Pfahl* (2005), pp. 95–104

22. A.A. Aguileta, O.S. Gómez, Estudio de calidad y eficiencia de un enfoque de desarrollo software secuencial con programadores solos y en pareja. Ingeniare. Rev. Chil. Ing. **27**(2), 304–318 (2019). https://doi.org/10.4067/s0718-33052019000200304

Chapter 66
Quality Evaluation of a Spring Cloud Microservices Architecture Implementation

Antonio Quiña-Mera, David Flores Landeta, Xavier Mauricio Rea-Peñafiel, and Cathy Guevara-Vega

Abstract In the practice of software engineering, it is a challenge to design highly scalable efficient software architectures for the development of quality web applications. In this sense, several proposals have emerged, such as the microservice-oriented architecture and the Spring Cloud development tools that have several characteristics that aim to respond to these demands. For this reason, we base this research on the Design Science Research (DSR) approach to answer the question, Does the implementation of the Spring Cloud architecture work? The method used to answer this question consisted of designing and implementing a software artifact (Spring Cloud implementation) and then validating it using the quality assessment in use according to ISO/IEC 25022. The evaluation results showed that the Spring Cloud implementation reached an efficiency of 83.04%, the satisfaction of use 78.68% of the expected value, both equivalent to a satisfactory degree. Therefore, the study concludes that the Spring Cloud implementation works to develop reliable, efficient, functional, and user-friendly software.

66.1 Introduction

In software engineering (SE) practice, there are features in web application development that make it challenging to design highly scalable architectures, uniform resource identification and manipulation, and optimal caching [1]. Because of the above and considering the dynamic nature of health and sports activities, we consider it essential that software development be considered a cloud-oriented architectural style through services, which is highly maintainable and validatable through quality standards [2].

A. Quiña-Mera (✉) · D. Flores Landeta · X. M. Rea-Peñafiel · C. Guevara-Vega
Department of Software Engineering in Applied Sciences Faculty, Universidad Técnica del Norte, 17 de Julio Avenue, 100105 Ibarra, Ecuador
e-mail: aquina@utn.edu.ec

A. Quiña-Mera · C. Guevara-Vega
Research Group E-CIER, 17 de Julio Avenue, 100105 Ibarra, Ecuador

Á. Rocha et al. (eds.), *Communication, Smart Technologies and Innovation for Society*, Smart Innovation, Systems and Technologies 252,
https://doi.org/10.1007/978-981-16-4126-8_66

Institutions or organizations specialized in sports areas choose to manage their information with new systems that streamline their information to comply with laws and regulations that benefit athletes, contribute to performance, and obtain the best results in the short, medium, and long term [3]. It is necessary to define the software development processes and manage them through an adequate software development model not to compromise the quality of the software product, as it happens in some modern IT companies [4]. In governmental institutions that manage development in sports, there is the participation of people who excel in different disciplines such as soccer, athletics, swimming. However, accidents inside and outside the sports scenarios are common [5], so an efficient registration and medical follow-up of athletes is required.

According to the above, we base the study with the Design Science Research (DSR) approach that helps evaluate the criteria of value and usefulness of the software product; therefore, we question whether the implementation of Spring Cloud architecture work? The research aims to answer the research question by evaluating the quality in use of the implementation of Spring Cloud components in an application using ISO/IEC 25022. The implementation of the application consisted of automating the registration and medical follow-up of athletes in the areas of Psychology and Physiotherapy of the Sports Federation of Imbabura-Ecuador (FDI-E).

The rest of the paper is structured as follows: Sect. 66.2 Research design: we establish the research activities based on DSR, theoretical foundation, and artifact design and build (software implementation). Section 66.3 Results: Evaluation of quality in use results of the software artifact. Section 66.4 Discussion: discussion of the research. Section 66.5 Conclusions and future work.

66.2 Research Design

The study design was established based on the DSR guidelines [6], see Table 66.1.

Table 66.1 Research design

Activity	Components	Paper section
Problem diagnosis	Problem; objective Population and sample	Introduction Research design
Theoretical foundation	Arquitectura de Spring Cloud architecture; scrum framework; ISO/IEC 25022	Research design
Artifact design and built: software implementation	Requirements; design (process and architecture); development and deployment; workshop and satisfaction survey	Research design
Evaluation of the software artifact	Evaluation of the quality of use	Results

66.2.1 Population and Sample

The work team consisted of a scrum master (project leader) and two FDI-E stakeholders for the diagnostic and requirements elicitation activities. We evaluated the software product (artifact) using the metrics of the ISO/IEC 25022 standard. To obtain the components described in the metrics, we used a workshop (eight practical tasks) of using the software artifact and a system usability scale (SUS) survey applied to 20 users from General Medicine, Physiotherapy, Psychology, and support students at FDI-E. The total population was 23 people, composed of 3 members from the software development team, and 20 users become complex as it grows.

66.2.2 Theoretical Foundation

Spring Cloud architecture. It is a microservices-oriented architecture based on Spring Boot that provides components such as libraries, tools, and common patterns for distributed systems. Among the prominent components are configuration management, service discovery, automatic switches, intelligent routing, micro-proxy, a control bus, single-use tokens, global locks, leadership election, session distribution, and cluster state [7, 8]. In a microservices architecture, each microservice is an independent codebase, unlike monolithic architecture, which is made up of a single code base that becomes complex as it grows [9]. Complexity in monolithic systems appears as new services are added, limiting their scalability and performance; consequently, it forces companies to make new investments in software products [9]. Figure 66.1 shows the Spring Cloud microservices architecture composed of modularized applications (Service Register, Config Server, distributed tracing, and API Gateway) to interact and complement functionalities in the development of "Spring Boot Apps" projects, which are then deployed in the cloud [7].

SCRUM Framework. It is a set of best practices for working collaboratively to get the best possible outcome from the software development project [11]. These practices support each other to establish an approach for highly productive work teams [12].

ISO/IEC 25022 standard. It defines quality in use measures [10] for the characteristics defined in ISO/IEC 25010 [13]. To define the quality in use model, the level of importance of the characteristics described in ISO/IEC 25022 was determined. The efficiency metrics evaluate resources used concerning the accuracy and completeness with the use of the software; in some instances, it could be compared with experts' efficiency. The characteristics represent the relative performance of the number of resources used under certain conditions [14].

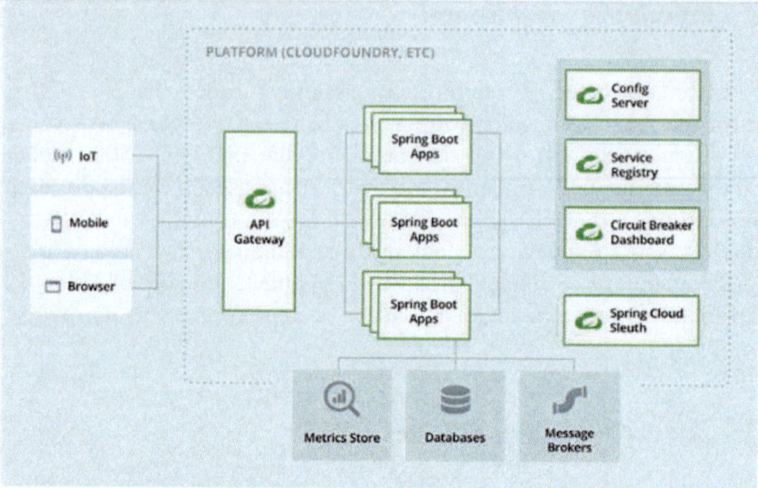

Fig. 66.1 General Spring Cloud architecture [10]

66.2.3 *Software Artifact Design and Implementation*

In this activity, the following tasks are established: (a) definition of requirements [15], (b) application design, and (c) implementation and deployment of the application.

a. **Requirement's definition**. In this phase, the requirements to automate the registration and medical supervision in Psychology (Socio/Economic) and Physiotherapy of FDI-E. Athletes of the FDI-E. The requirements were represented by user stories written in one or two sentences using the user's natural language without elaborating many formal documents and without requiring much time to manage them. We collected ten user stories and recorded them in a stack called Product Backlog, which defines software development requirements. As a result of the Sprint Backlog review, several artifacts could be obtained, such as process diagrams, use cases, and the application's database model.
b. **Application design**. We designed the application's architecture based on microservices architectures' style with Spring Cloud tools to implement patterns that provide resilience, reliability, and coordination of microservices, see Fig. 66.2.

Table 66.2 shows the Spring Cloud components and tools used in the design.

c. **Implementation and deployment of the application**. During the development of the application, the processes and documents used by the FDI-E medical staff to carry out physiotherapeutic activities and psychological sessions were analyzed. We used the scrum framework in an iterative-incremental process; the iteration called Sprint lasted four weeks. The structure of each Sprint was: a planning meeting that detailed the technical work tasks of the user stories

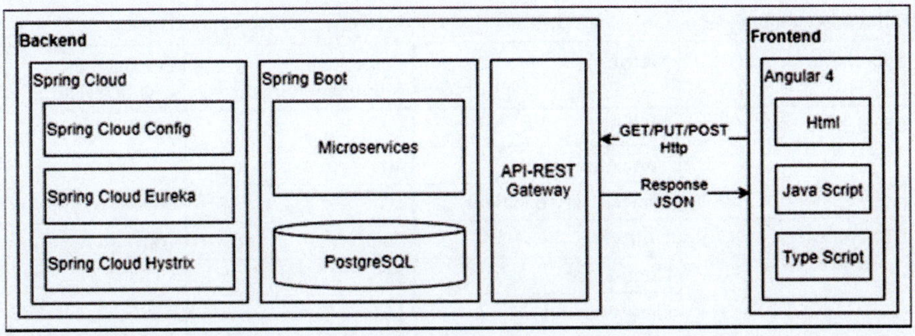

Fig. 66.2 Spring Cloud application's architecture

Table 66.2 Spring Cloud components and tools

Component	Tool	Characteristics	Description
Spring Cloud Config	Spring Cloud Config	Distributed/versioned configuration	Manages external configurations backed by a file hosted in a Git repository, the configurations can be used by other software
Spring Cloud Netflix	Spring Cloud Eureka Spring Cloud Hystrix	Microservices registration and discovery service	Integration with various Netflix OSS (open-source software) components such as circuit breakers, microservice logging, load balancing, and routing

(Sprint backlog); the development of the increment of the software product; an increment review meeting; and a retrospective meeting to perform a continuous improvement of the project execution. Table 66.3 shows the summary of the

Table 66.3 Summary of development and implementation activities

Phase	Sprint (duration)	Deliverables
Pre-game	Sprint 0 (40 h)	Requirements: 10 user stories; product backlog Design (architecture): process diagram; software architecture; initial database diagram
Game	Sprint 1 (40 h) Sprint 2 (40 h)	Software Product Increment 1 Software Product Increment 2
Post-game	Sprint 3 (40 h)	Satisfaction survey Deployment of the software product delivery-receipt Act

Table 66.4 Quality in use metrics

Sub-characteristics	Metric	Percentage Sub-characteristic	Percentage Characteristic
Efficiency	Task time	50	100
	Time efficiency	40	
	Unnecessary behavior	10	
Utility	Utility	50	100
	Confidence	35	
	Comfort	15	

activities and deliverables obtained in each phase of the project implementation and development.

66.3 Results

After the design and implementation of the software artifact, we will validate the artifact using the in-use quality assessment based on ISO/IEC 25022 through the following activities: (1) Definition of the quality model in use; (2) Measurement of the quality model in use; and (3) Evaluation of the quality model in use.

66.3.1 *Definition of the Quality in Use Model*

The product owner and the scrum master of the project defined the quality model by assigning the importance level and weighting of each quality in use characteristic, sub-characteristic, and attribute using efficiency, and satisfaction metrics, see Table 66.4.

66.3.2 *Measurement of the Quality in Use Model*

We measured the quality characteristics through a workshop and a satisfaction survey on using the software implemented in the FDI-E.

To measure the workshop, we used efficiency metrics based on the ISO/IEC 25022 standard. The workshop contains three objectives and eight specific tasks according to the software's functionality, as shown in Table 66.5.

The information collected by the workshop includes:

- Start time. The current time at which it started will be taken for each task, always starting from the start screen.

Table 66.5 Proposed workshop

Workshop goal	To measure the efficiency of the registration and medical supervision software around Psychology and Physical Therapy			
Characteristics	Efficiency		Sub-characteristics	Efficiency
Method	Measuring user performance			
Metric	Metric description	Functional objectives	Task No.	Task description
Task time	The time required to successfully complete a task	Enter relevant patient information	1	Edit patient information
			2	Enter a new medical benefit with two types of benefits that the patient will have
			3	Enter a new medical discount
			4	Enter new suggested medical diagnosis for the patient
Time efficiency	The efficiency with which users achieve their goals more time at which the use of software	Supervision of a patient in the Physiotherapy area	5	Enter a new physiotherapeutic diagnosis with session numbers (5)
			6	Enter three physiotherapeutic check-ups
			7	Edit physiotherapeutic control number two
Unnecessary behavior	The proportion of actions performed by the user that was not necessary to accomplish the task	Running a report from a start date to an end date	8	Perform a physiotherapeutic report from 01/01/2020 to 02/29/2020

- End time. It is the time when the task was finished.
- Total minutes. The total minutes are the difference between the end time and the start time; this will help tabulate the information collected.
- Number of actions that were not necessary to accomplish the task. The number of actions required to perform the task by the software expert was established, and the difference with the total number of actions performed by the user was obtained.

- Number of actions performed by the user. The number of actions performed by the user through the evaluator's close observation to perform the indicated task was counted.

Furthermore, we use a satisfaction survey to measure satisfaction metrics for using the Spring Cloud implementation. We structured the survey with questions based on system usability scale (SUS) with a Likert scale of 5 responses [16].

66.3.3 Evaluation of the Quality in Use Model

In this section, we will evaluate the quality model. Table 66.6 shows the quality model's evaluation defined in Table 66.4 compared to the measurement results obtained in the workshop and satisfaction survey.

The result of the evaluation of the model shows on the one hand that the efficiency characteristics reached 83.04% and satisfaction 78.68% of the expected value, both equivalent to a satisfactory degree. On the other hand, it shows that the results of the metrics obtained are classified as favorable, except for the "unnecessary behavior" metric, classified as unsatisfactory.

The overall result of the evaluation shows that the implemented application obtained an acceptable score for its quality of use, and satisfactorily complying with most of the parameters provided by the ISO/IEC 25022 standard.

Table 66.6 Evaluation of the quality in use model

Sub-characteristics	Metric	% SC	Scale 1–10	Result (%)	Sum (%)	Scoring level	LS
Efficiency	Task time	50	8.83	44.17	83.04	Exceeds requirements	S
	Time efficiency	40	8.72	34.89		Objective range	S
	Unnecessary behavior	10	3.98	3.98		Minimally acceptable	U
Satisfaction	Utility	50	8.50	42.50	78.68	Objective range	S
	Confidence	35	6.50	22.75		Objective range	S
	Comfort	15	8.95	13.43		Exceeds requirements	S

SC Percentage sub-characteristic; *LS* Level of satisfaction; *S* Satisfactory; *U* Unsatisfactory

66.4 Discussion

Liu et al. [17] indicate in their study that the microservices architecture's flexibility have allowed software applications to be diverse domains. They even determine the use of more advanced techniques and frameworks to maintain an ecosystem of microservices versioning and dynamic rules of self-configuration of the entire infrastructure's scaling [18]. In this sense, we were able to corroborate their study by implementing registration and medical monitoring of athletes in the FDI-E with Spring Cloud. Obtaining quantitative and qualitative results by integrating tools (Spring Cloud Eureka and Hystrix) that help monitor real-time API microservices' behavior when they are invoked from user interfaces. We complement the application monitoring with the implementation of the JavaMelody dependency, which monitors Java applications, displaying statistical results such as memory usage, CPU performance, HTTP sessions, JDBC connections, HTTP errors, SQL errors, and Spring errors. During the study, we found some limitations when implementing Spring Cloud with the Spring Cloud Hystrix and JavaMelody interfaces. We noticed errors coming from the basic Spring Boot application when mapping the database with Java Persistence API (JPA) generates infinite JSON loops of parent–child calls and vice versa; the solution to this problem is necessary the @JsonIgnore, @JsonIgnoreProperties, or fetch=FetchType.EAGER statements depending on the case.

66.5 Conclusions and Future Work

We conducted this study based on the Design Science Research (DSR) approach to answer the question Does the Spring Cloud architecture implementation work? The method to answer this question consisted of establishing a conceptual basis for designing and implementing a software artifact (Spring Cloud implementation) and then validating this artifact. We performed the validation of the artifact by evaluating the quality in use based on the ISO/IEC 25022 standard. We implemented Spring Cloud in the practical case of automating the registration and medical supervision of athletes in the areas of Psychology and Physiotherapy of the Sports Federation of Imbabura—Ecuador (FDI-E).

The evaluation of the quality in use of the Spring Cloud implementation shows that the efficiency characteristics reached 83.04%, the satisfaction reached 78.68% of the expected value, both equivalent to a satisfactory degree, in summary, the implementation obtained a score of acceptable satisfactorily fulfilling most of the parameters provided by the ISO/IEC 25022 standard. We conclude and answer the research question by saying that the Spring Cloud implementation works to develop reliable, efficient, helpful, and comfortable use software.

As future work, we propose to continue generating more empirical evidence of the use of Spring Cloud and using the internal and external quality assessment of ISO/IEC standards to validate software artifacts.

References

1. M. Athanasopoulos, K. Kontogiannis, Extracting REST resource models from procedure-oriented service interfaces. J. Syst. Softw. **100**, 149–166 (2015)
2. C.D. Farinango, J.S. Benavides, J.D. Cerón, D.M. López, R.E. Álvarez, Human-centered design of a personal health record system for metabolic syndrome management based on the ISO 9241-210:2010 standard. J. Multidiscip. Healthc. **11**, 21–37 (2018)
3. Gobierno del Ecuador, *Ley del deporte educación física y recreación* (Ecuador, 2015), p. 3
4. J. Yu, Research process on software development model, IOP Conf. Ser. Mater. Sci. Eng. **394**(3) (2018)
5. Federación deportiva de Imbabura, Centro Médico de Federación Deportiva de Imbabura atiende a la ciudadanía en general (2020)
6. A. Hevner, S. Chatterjee, Design science research in information systems, in *Integrated Series in Information Systems*, 39th edn. (Springer, Boston, 2010), pp. 9–22
7. Pivotal, Spring Cloud (2020)
8. Spring.io, Spring Cloud
9. J. Lewis, M. Fowler, Microservices (2014)
10. R. Maddila, Microservices with Spring Boot. Microservice architectures are the 'new… | by Ramesh Maddila | Medium (2020)
11. J.A. Quiña-Mera, L. Chamorro Andrade, J. Montaluisa Yugla, D. Chicaiza Angamarca, C.P. Guevara-Vega, Improving software project management by applying agile methodologies: a case study, in *International Conference on Applied Technologies—ICAT 2020*, 2020.
12. SCRUM.org, The Scrum Guide (2020)
13. ISO/IEC, ISO—ISO/IEC 25010:2011—Systems and software engineering—Systems and software Quality Requirements and Evaluation (SQuaRE)—System and software quality models (2011)
14. ISO/IEC, *Systems and Software Engineering—Systems and Software Quality Requirements and Evaluation (SQuaRE)—Measurement of Quality in Use*, 1st edn., vol. 1 (International Organization for Standardization, Geneva, 2016)
15. C.P. Guevara-Vega, E.D. Guzmán-Chamorro, V.A. Guevara-Vega, A.V.B. Andrade, J.A. Quiña-Mera, Functional requirement management automation and the impact on software projects: case study in Ecuador. Adv. Intell. Syst. Comput. **918**, 317–324 (2019)
16. ISO/IEC, ISO—ISO/IEC 14598-1:1999—Information technology—Software product evaluation—Part 1: General overview (1999)
17. L. Liu, X. He, Z. Tu, Z. Wang, MV4MS: a spring cloud based framework for the co-deployment of multi-version microservices, in *Proceedings—2020 IEEE 13th International Conference on Services Computing, SCC 2020*, 2020, pp. 194–201
18. P. Jamshidi, A. Sharifloo, C. Pahl, H. Arabnejad, A. Metzger, G. Estrada, Fuzzy self-learning controllers for elasticity management in dynamic cloud architectures, in *Proceedings—2016 12th International ACM SIGSOFT Conference on Quality of Software Architectures, QoSA 2016*, 2016, pp. 70–79

Chapter 67
Vibrations Associated with the Quality of Combustion Derived from Induced Failures in a Spark Ignition Engine

Diego Díaz Sinche, Jairo Castillo Calderón, and Rubén Carrión Jaura

Abstract The present investigation aims to analyze engine vibrations and HC emissions when causing faults not detected by the on-board diagnostic system (OBD II) in the air induction system and the ignition system of a gasoline vehicle. First, experimental measurements were performed with a vibration analyzer and a unidirectional accelerometer mounted on the engine block. In parallel, hydrocarbon (HC) measurements were performed to contrast the influence of combustion with the simulated failures. Secondly, signal processing was performed by applying the fast Fourier transform to a set of blocks superimposed on an interval of 512 points, in order to force the periodicity of the signal and thus have the most information related to the degree of severity of the failures. The results obtained indicate that the failure of a spark plug causes higher vibrations in the engine, and this is associated with a higher amount of HC amounting to 188 ppm for the use of fuel with a RON of 87.

67.1 Introduction

The vibrations in the engine have a great influence on the emissions and performance of the vehicle. Currently, researchers are focused on the measurement of vibrations for the study of combustion and inspection and maintenance routines. For this, they use methods to determine the rotational speed from a toothed wheel that is synchronized with the rotation of an end of the crankshaft [1]. One of the main drawbacks of this method is the use of a sampling frequency that is too high to have a good precision of the measurement [2]. Employing an indirect method associated with the vibration signal produces good results. Such a method evaluates a particular spectrum through the acquisition of data in the time domain and its analysis in the frequency domain. With this, it is possible to overcome the disadvantages mentioned in medium power motors [1].

D. D. Sinche (✉) · J. C. Calderón · R. C. Jaura
Facultad de la Energía, las Industrias y los Recursos Naturales no Renovables, Universidad Nacional de Loja, Ciudad Universitaria Guillermo Falconí, Loja, Ecuador
e-mail: diego.v.diaz@unl.edu.ec

Á. Rocha et al. (eds.), *Communication, Smart Technologies and Innovation for Society*, Smart Innovation, Systems and Technologies 252,
https://doi.org/10.1007/978-981-16-4126-8_67

The analysis of signals in the frequency domain has been used extensively in the field of mechanical fault diagnosis as a visual tool. There are works such as that of Bing Li and Shuang-shan Mi, which have carried out this analysis to classify the vibration signal [3]. The use of a pressure sensor that allows detecting the state of combustion can be prohibitive due to the cost of implementation, for which, according to Yoonwoo Lee and Seunghyun Lee, they replace this sensor with an accelerometer, correlating the vibration in the engine block caused by combustion. It was established that vibrations in the frequency range of 2.5–8 kHz are associated with the vibration of the crankshaft, 1–3 kHz corresponds to a stroke of the piston, and between 0.1 and 8 kHz, there was the combustion frequency. Finally, the appropriate range for the correlation with the state of combustion, and the rate of heat release occurs in the range of 2 kHz [4]. Consequently, the accelerometers used to measure disturbances as non-intrusive diagnostic techniques are not required to withstand very high temperatures and pressures, are relatively inexpensive, and do not require structural modifications to the engine. This technique is one of the most important in a variety of investigations, for which it will be provided with a technique of analysis of the signal by blocks to force the signal to be periodic [11, 17–19].

Due to the great information that surface vibration can contain in an internal combustion engine, a variety of signal processing methods can be used, and one of the most common is frequency domain analysis, according to Shaobo. Ji and Xin Lan used the Laplace S transform method to obtain frequency information related to combustion pressure and surface vibration [5]. We also have other tools such as convolution methods [6], spectrum analysis [7], and models that use neural networks [8].

Noise and vibrations are important factors that affect the engine; generally, during combustion, we have the presence of noise and vibrations [18]. The spark plug is of great importance in the efficient operation of the engine, and a failure in this component increases the values of polluting emissions, noise, and vibrations [20]. Wear of this component for the period of useful life tends to increase the distance between electrodes, being an optimum operating value of 0.8 mm, and a failure in the ignition system with 0.9 mm, which causes an increase in the noise and vibrations at low revs [9]. Moosavian and Najafi, perform an analysis of vibrations with accelerometers to determine spark plug failures, for a normal and an abnormal operating condition with the use of neural networks [10]. This is related to the current work in the determination of the amplitude of the spectra that is related to the diagnosis of defects in the engine spark plugs, determining that the vibration noises are too loud and the defects in the spark plug become unrecognizable, for which noise elimination processes must be carried out previously.

The main objective of the research work is to study the influence that exists between engine vibration and combustion associated with failures in engine auxiliary elements, which are not identified by the diagnostic system onboard the vehicle, such as the air filter and spark plug electrode light. For which, a methodology is applied based on the development of an algorithm that allows obtaining the analog signal of the vibration, and with signal processing techniques and applying the FFT, the frequency spectrum associated with the simulated faults is obtained. One of the advantages of the proposed methodology is, on the one hand, the use of superimposed blocks that allow to reconstruct the signal in the frequency domain without any loss of spectrum and, on the other hand, the measurement of the vibration signal can be obtained regardless of the number of cylinders of the engine. This is achieved with the use of a transducer mounted on the block, taking advantage of the transfer of resonance from the cylinder to the surface of the engine [23].

The method used to acquire the engine block vibration signal requires a one-way accelerometer. Due to the advantages mentioned, the proposed method is suitable for studying a correlation between combustion and vibration. In engines, there is a great variety of sources of vibration typical of combustion, piston stroke, intake and exhaust valves, and gear operation among others [10]. Based on this, engine noise can be classified as a source of mechanical noise, combustion noise, and accessory noise [4, 21].

The vibration analysis is performed in the frequency domain with a total of 4096 samples, where the low frequency is related to the maximum pressure and the intermediate and high frequencies with the maximum pressure rate [22]. According to the study by Yoonwoo Lee, it was evidenced that the vibrational frequency that seeks a correlation between combustion and vibration has a presence in the range of 0.5–3.5 kHz [4], which will be the area to which we give greater importance in the research. Another important factor that must be considered in the event of non-regular operation is the decrease in performance, and increased vibration associated with the increase in hydrocarbons [20].

67.2 Materials and Methods

67.2.1 Faults Generated in the Engine

The influence of vibration and combustion associated with two types of failures of the auxiliary systems of the engine equipped with four indirect injection cylinders, with a variable distribution system as indicated in Table 67.1, was studied. Accelerometer for recording the vibration signal in the motor housing with a measurement frequency of 4096 samples/second. The normal operation of the engine was considered as a reference system and faults that are not interpreted by the OBD II system were

Table 67.1 Technical specifications

Hyundai accent vehicle engine specifications	
Displacement	1368cc Kappa MPI CVVT—Dual
Valve number per cylinder	16 DOHC
Power	99 HP @ 6000 rpm
Torque	133.4 Nm @ 4000 rpm
Number of cylinders	4

simulated. Such failures are common due to careless preventive maintenance such as the air filter and the spark plug.

In the case of the air filter, it is considered a failure due to the retention of particulate material in the filter material of 6 g. If this pneumatic resistance is eliminated in the air induction circuit, the useful life of the engine is significantly reduced and this leads to a reduction in combustion efficiency. On the other hand, if the air inlet is blocked, the volumetric efficiency decreases which leads to higher fuel consumption [12]. The second fault corresponds to a spark plug of an electrode with a light equal to 1.3 mm, which simulates the wear of the electrodes for a long time of engine operation. This involves a higher energy consumption of the electrical system, affecting the time of spark, engine efficiency, fuel consumption, vibrations, noise, and polluting emissions [13].

67.2.2 *Experimental Setup*

Measurements were made using Ecopais fuel with a RON of 87. Figure 67.1 illustrates, on the one hand, the ADASH VA4Pro vibration analyzer is used for signal acquisition, and on the other hand, an active unidirectional ADASH AC102-1A magnetic-based transducer that allows analysis from the maximum frequency between 3000 and 5000 Hz. The objective is to find the fault that causes the greatest vibration and is not recognized by OBD II.

It should be considered that the loss of resolution is associated with the maximum value of the signal and the range of the converter. Thus, in the case of the ADASH VA4Pro equipment with four channels based on piezoelectric integrated circuits for alternating and direct current, which has 24 bits the A/D converter, and has 2^{24} divisions in the signal range, for an input voltage of $\pm$ 12 Vpp for AC, allows detecting voltage changes of $12/2^{24}$ mV. Finally, the measurement equipment defines a high-pass filter at the input of the signal.

The signal processing in the time domain is sampled over a period of one second which requires the processing of the data digitally and analysis in the frequency

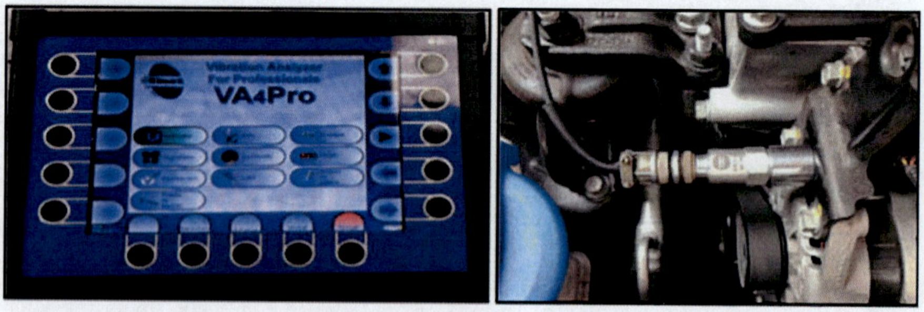

Fig. 67.1 Adash VA4Pro measurement equipment and Adash AC102-1A accelerometer

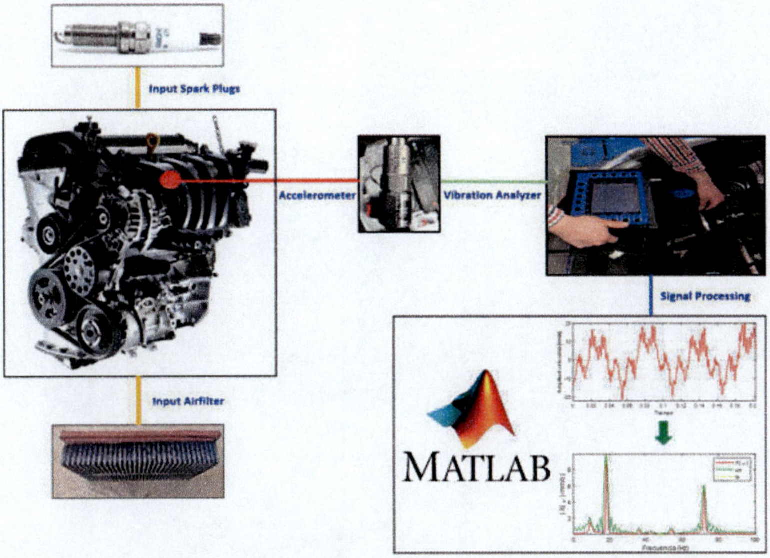

Fig. 67.2 Schematic of the experimental setup

domain. In data collection, a measurement protocol was established based on the ISO 5348: 1998 Mechanical vibration and shock—Mechanical mounting of accelerometers standard, for which eight measurements were established, including replicas. In the experiment, single-point measurements were developed to study the effect of vibrations, and the other magnitudes were obtained with integration and derivation operations as observed in the experimental configuration scheme of Fig. 67.2. In addition to this, the fundamental measurement established for disturbances is made in units of velocity, that is, in mm/s.

67.2.3 Data Processing

For the collection of experimental data, the method based on vibrations and spectrum analysis is used, according to the protocol established by the research group for data collection as detailed in Fig. 67.3. The four records were made in a frequency range between 10 and 1000 Hz for one second with one repetition, at a rotational speed of the motor of 2000 rpm and several samples of 4096 points. The treatment of the data is carried out based on an average, of which only up to 1024 points were considered for the calculation of the FFT. In each block of 1024 points, a Hanning window was applied before the calculation of the FFT; with this, we forced so that the signal is periodic within the selected block and the leakage error is reduced.

Next, the expression for the calculation of the points that are applied to a certain window is defined.

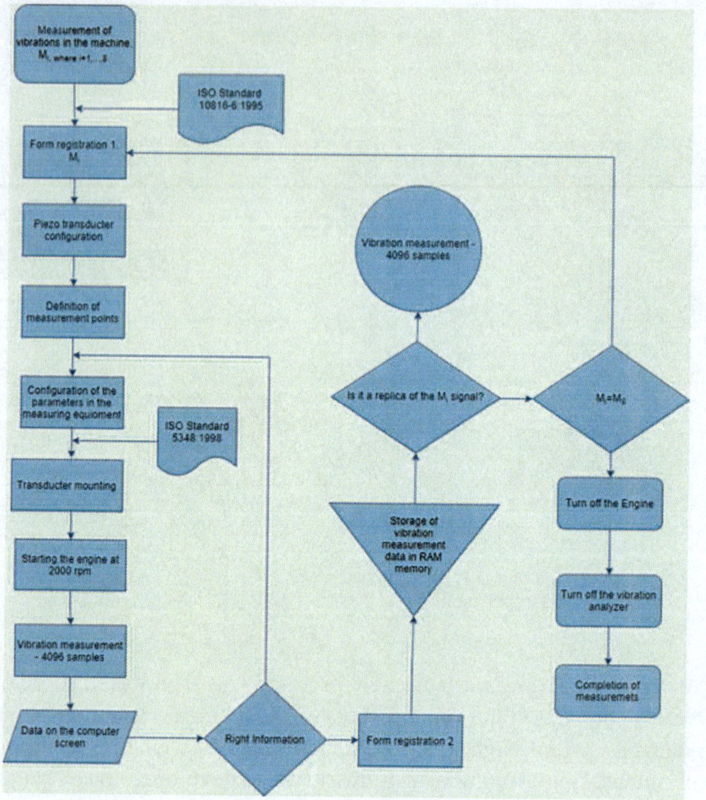

Fig. 67.3 Protocol for sampling the vibration signal in the engine housing

$$V_i = \frac{1}{2}\left(1 - \cos\left(\frac{2 \cdot \pi(i - 1)}{N - 1}\right)\right) \tag{67.1}$$

where N is the number of points established in the FFT calculation and the subscript "i" allows the sweep from 1 to N points. The window obtained for the analysis is indicated in Fig. 67.4. Since the window cancels the temporal signal at the beginning and end of the established block of 1024 points, the blocks have been overlapped to perform the analysis by 50%. This leads to the second analysis starting at 512 points as indicated in Table 67.2.

The application of the window will mean a reduction in the energy associated with the signal; therefore, it is necessary to multiply the temporal data by a window factor (f_v) that compensates for said reduction, I feel in this case $f_v = 1.633$ as seen in Fig. 67.3. The result in each block will be a set of complex numbers with the same number of data, of which, according to the Nyquist theorem, only the first 512 data will be necessary.

Fig. 67.4 Hanning

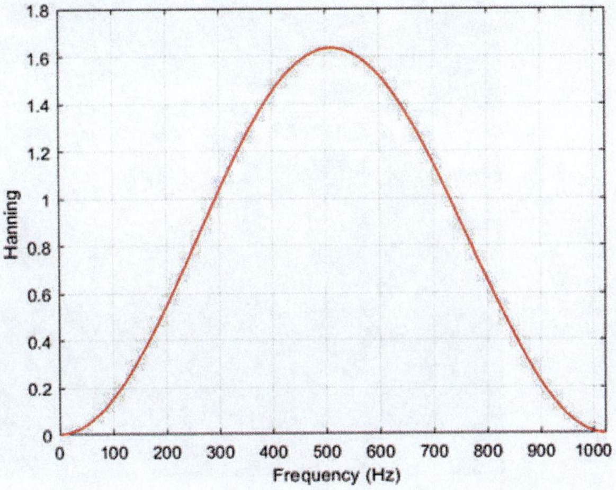

Table 67.2 Block overlap

Block	1	2	3	4	5	6	7
Points	0–1024	512–1536	1024–2048	1036–2560	2048–3072	2561–3584	3072–4096

Each frequency value associated with each component of the FFT module is obtained knowing that the first number corresponds to the continuous value of the signal ($\omega = 0$ Hz), and the following correspond to frequencies equal to $\Delta\omega$ until it reaches half the sampling frequency of 512 Hz. For the analysis, the resolution of frequency $\Delta\omega = 1$ Hz and is equal to the inverse of the time duration of the analyzed data block of 1024 points. In the data processing, only the module of the first 512 points was considered, and also, due to the scaling that MATLAB performs for the calculation of the FFT, its result is multiplied by two and divided by the number of points in the block. The maximum frequency that can be analyzed will be half the sampling frequency to avoid aliasing errors, for which a low pass filter was used to eliminate all frequencies from the signal starting from the maximum frequency. It should be noted that the application of this filter presents a progressive attenuation and for this reason, the Fourier analysis near this area presents incorrect data. After calculating the set of signal spectra, the signal blocks were averaged. Figure 67.4 shows the result in logarithmic scale, the average for the four measurements made in components of speed and acceleration.

To quantify the severity of vibration and the effect of combustion, experimental measurements of the pollutant gases from the engine were carried out. For this, a gas analyzer was used that was connected to the outlet of the exhaust pipe as shown in Fig. 67.5. During the vibration measurements, the gas measurement was developed in parallel, considering the value of hydrocarbons. For each test, the experimental unit is subjected to a measurement protocol defined in the gas regulations NTE INEN 2204 (HC < 200 ppm) [24] (Fig. 67.6).

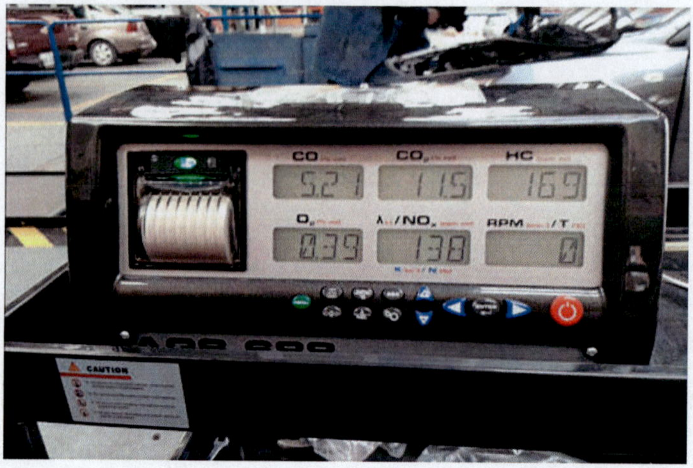

Fig. 67.5 Gas analyzer connected to the exhaust outlet of the experimental unit

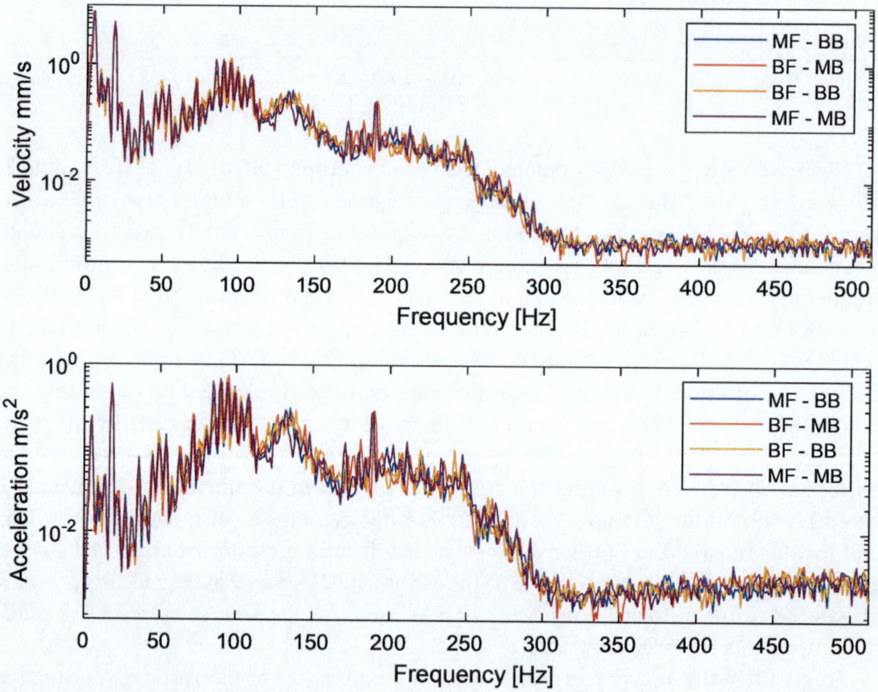

Fig. 67.6 Frequency spectrum in the magnitude of speed and acceleration associated with the experimental tests, where MF corresponds to a damaged filter, BF to a good filter, MB to a spark plug with a 1.3 mm gap, and BB to a spark plug with a 0.8 mm gap

67.3 Results and Discussion

The measurements were made with the normal operating temperature of the engine; eight measurements were taken including a replica in each experiment for a fixed speed corresponding to a static test established by the NTE INEN 2204. The first test consists of simulating an associated engine failure to a defective filter and spark plug in good condition called MF-BB. The second test simulates the failure of a spark plug and a good condition of the filter that is established as BF-MB. The third test consists of an optimal working condition of the filter and sparks plug designated as BF-BB, and finally, the last test corresponds to a malfunction of the spark plug and filter chosen as MF-MB. For each test described above, a replica was carried out with the same procedure, to ensure that the data collection is subject to the same source of the variability described in Table 67.3. In Fig. 67.7, it can be seen, the repetitions carried out in each measurement. Because the aftershock comes from a previous experimental run, commonly scattered include sources of variability caused by environmental factors that change over time, such as engine temperature.

The signals in the frequency domain were analyzed by applying the FFT. With the signals obtained, a comparison was made to identify differences in the frequency components of the signal with different combinations of failure during motor operation. In the averaged spectra, it is possible to identify the rotational speed of the motor from the recorded amplitude; this is illustrated in Fig. 67.8. In the first block, we have an air filter failure simulation compared with an optimal motor operation, which identifies that at a constant speed of rotation of the motor during the experimental tests of 2000 rpm, the associated frequency is 33.33 Hz, and if we consider that the detonation occurs in the compression phase, we have that the amplitude is $0.5692\ \text{m/s}^2$ at a frequency close to 18 Hz and has harmonics up to seven times its turning frequency.

The energy associated with a certain defect appears distributed in the lateral bands of the spectrum because of the small oscillations in the motor's working speed. In the range of frequencies between 94 and 100 Hz, we have the highest vibration amplitudes that correspond to harmonics of the motor rotation. In these circumstances,

Table 67.3 Process variables

Run order	Fuel (RON)	Air filter (g)	Spark gap (mm)	Experiment
1	87	0	1.0	BF-BB
2	87	0	1.3	BF-MB
3	87	6	1.0	MF-BB
4	87	6	1.3	MF-MB
5	87	0	1.0	BF-BB
6	87	0	1.3	BF-MB
7	87	6	1.0	MF-BB
8	87	6	1.3	MF-MB

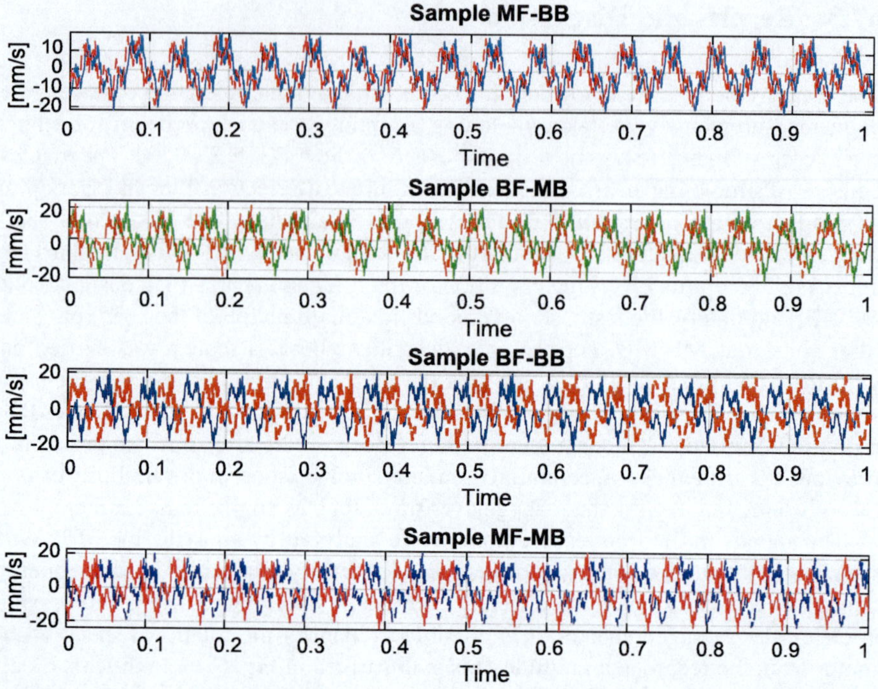

Fig. 67.7 Time domain signal analysis

there may be an advance or delay of the harmonics about the base signal due to the closed-loop control carried out by the electronic control unit (ECU) to reduce the threshold of the vibration allowed with the signal sent from the sensor, knocking due to filter failure. With a clogged air filter, the ECU adjusts a higher amount of fuel, thereby causing increased pollutant emissions [14].

From the analysis, we have that a greater effect on the engine during its operation is the failure of the spark plugs because it generates a peak greater than 0.7456 m/s^2 at a frequency of 95 Hz. For a simulation of filter and spark plug failures, we have a greater presence of sidebands at the peak frequency. This greatly influences the vibrations in the motor to be longer for the same instant of time. Smaller peaks can be seen that can be related to natural frequency responses of the support structure. Spikes can arise from some motor imbalance caused by mounting brackets and are accentuated by simulated motor failures [15].

Considering an analysis of the spectrum in speed units of Fig. 67.9, we have an invalid result for very low frequencies in this case below 5 Hz because of high noise in the speed units of the accelerometer signal. The 18 Hz peak corresponding to the combustion process in each cycle can also be identified by repeatability in all measurements. In the speed measurements, they do not indicate any discrepancy between defective and normal conditions, except for the range close to 100 Hz where there are small differences [15].

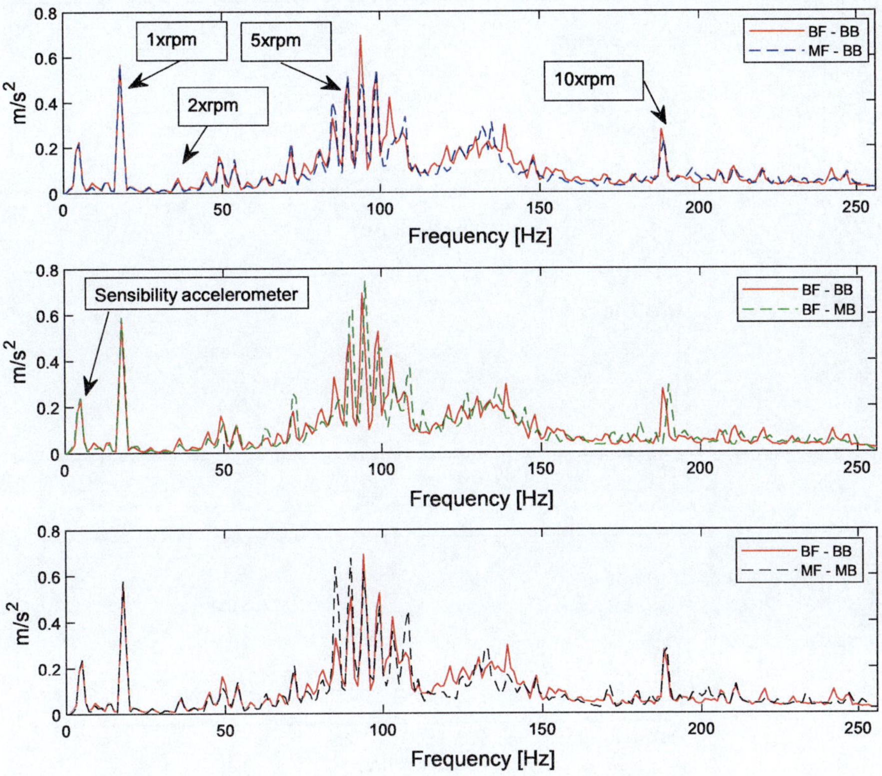

Fig. 67.8 Frequency spectrum in units of acceleration

In the readings of the spectrum of the vibrations in a great part of the range of frequencies of analysis, they gave a similarity of the behavior, but the analysis of the RMS and the crest factor of the vibration measured in units of acceleration, velocity, and displacement between 10 and 1000 Hz, identify a promising difference. The state of vibration of the motor can be determined in a general way according to the ISO 10816-6 standard for alternative machines, the value obtained in none of the cases does the vibrations indicate that there may be serious damage to the machine, according to the results obtained in Table 67.4. It is worth mentioning that during the experimental tests, measurements of the engine exhaust gases were taken, with hydrocarbons (HC) being the most relevant in the investigation to associate the simulation of the failure with the state of combustion.

Finally, we have that a greater distance between the electrodes of the iridium spark plug increases the vibrations and this leads to an increase in the HC [20]. A high value of HC implies bad combustion and increases fuel consumption. It can be said that one of the most important points to consider is the state of the spark plugs due to the greater severity it presents in the performance of the engine [16]. However, a filter with high retention of particulate matter ceases to fulfill its function and causes

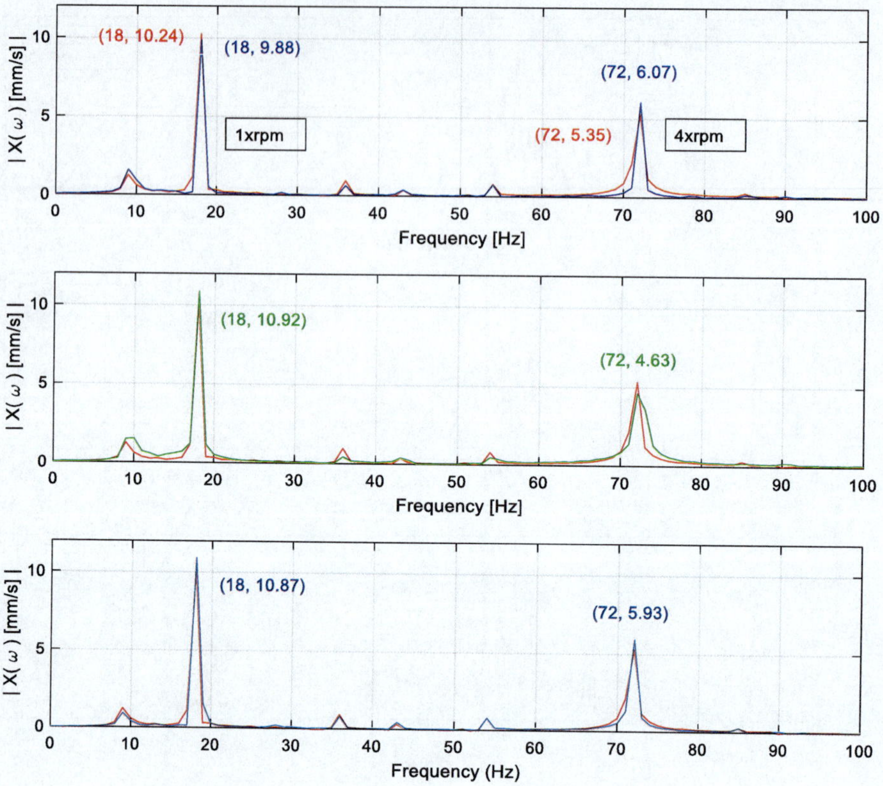

Fig. 67.9 Frequency spectrum in units of velocity

Table 67.4 Vibration analysis in units of velocity, acceleration, and displacement

Experiment	MF-BB	BF-MB	BF-BB	MF-MB
Rate RMS from 10 to 1000 Hz (mm/s)	0.24289	0.25797	0.24958	0.25801
Maximum speed value (mm/s)	6.924	7.6686	7.3506	7.6256
Speed crest factor (mm/s)	28.5063	29.7262	29.4521	29.5557
Acceleration RMS from 10 to 1000 Hz (m/s^2)	0.58116	0.64058	0.62164	0.66435
10–1000 Hz offset RMS (μm)	1.08029	1.8562	1.8545	1.8628
Hydrocarbons (HC) (ppm)	183	188	148	115

premature wear on the engine. Gültekin, N., Mayda, M., come to determine that as the space of the spark plug increases, the number of revolutions, vibration, and noise increases proportionally, a result similar to the one obtained in the present investigation. In addition, the value of exhaust emissions (HC) increases at distances of 0.7, 0.8, and 1.0 mm between spark plugs and decreases to distances 0.9 mm, thus, obtaining a significant reduction in pollutant emissions. In the investigation, two calibrations were proposed, one of 1.0 mm and the other of 1.3 mm, and in the second case, it is seen that the emissions with this spark plug distance reduce the amount of HC in the exhaust emissions, giving an effect equal of the spark plug gap of 0.9 mm [20].

67.4 Conclusions

In the present investigation, an analysis of vibrations in the frequency domain was carried out using an algorithm based on the calculation of the FFT and a block overlap technique, which forces the temporal signal to be periodic within the selected block. The development of this algorithm makes it possible to analyze the influence of combustion associated with engine vibration, for which faults were simulated in the air induction system and the ignition system that is not detected by OBD II. From the analysis of the results, a defect in the spark plug related to an increase in the light between electrodes causes a greater vibration peak, and this effect is reduced in harmonics higher than the combustion cycle of the engine. This is related to the level of hydrocarbons recorded in parallel in the vibration measurements. The effect of the simulated faults does not indicate a great difference for a 188-ppm state. On the other hand, a fault in the filter and the spark plugs also causes sidebands in the spectrum and attenuates the frequency peaks in higher ranges.

References

1. H. Lin, K. Ding, A new method for measuring engine rotational speed based on the vibration and discrete spectrum correction technique. Measurement **46**, 2056–2064 (2013)
2. C.D. Dicenzo, B. Szabados, N.K. Sinna, Digital measurement of angular velocity for instrumentation and control. IEEE Trans. Ind. Electron. Control Instrum. IECI **23**(1), 83–86 (1976)
3. B. Li, S.-S. Mi, P.-Y. Liu, Z.-J. Wang, Classification of time-frequency representations using improved morphological pattern spectrum for engine fault diagnosis. Sound Vib. **332**, 3329–3337 (2013)
4. Y. Lee, S. Lee, H. Choi, et al., Analysis of vibration on an engine block caused by combustion in a diesel engine. Int. J. Autom. Technol. **20**, 187–195 (2019)
5. S. Ji, X. Lan, J. Lian, H. Wang, M. Li, Y. Cheng, W. Yin, Combustion parameter estimation for ICE from surface vibration using frequency spectrum analysis. Measurement **128**, 485–494 (2018)
6. C. Mobley, Non-intrusive in-cylinder pressure measurement of internal combustion engines. SAE Technical Paper 1999-01-0544 (1999)
7. R. Villarino, J. Böhme, Peak pressure position estimation from structure-borne sound. SAE Technical Paper 2005-01-0040 (2005)
8. R. Johnsson, Cylinder pressure reconstruction based on complex radial basis function networks from vibration and speed signals. Mech. Syst. Signal Process. **20**, 1923–1940 (2006)
9. Z. Özçelik, N. Gültekin, Effect of iridium spark plug gap on emission, noise, vibration of an internal combustion engine. Energy Appl. **6**, 44–48 (2019)

10. S.A. Moosavian, G. Najafi, B. Ghobadian, S.M. Jafari, B. Sakhaei, M. Khazaee, Fault diagnosis in engine spark plug by vibration analysis using neural network. Engine Res. **28**, 21–29 (2013)
11. Z.g. Zhang, E. Jiaqiang, G. Zhang, A novel dense spectrum correction algorithm for extracting vibration signals in internal combustion engine and its application. J. Cent. South Univ. **19**, 2810–2815 (2012)
12. Y. Huang, E.C.Y. Ng, Y. Yam, C.K.C. Lee, N.C. Surawski, W. Mok et al., Impact of potential engine malfunctions on fuel consumption and gaseous emissions of a euro VI diesel truck. Energy Convers. Manag. **184**, 521–529 (2019)
13. T. Badawy, X. Bao, H. Xu, Impact of spark plug gap on flame kernel propagation and engine performance. Appl. Energy **191**, 311–327 (2017)
14. N.J. Bugli, Automotive Engine Air Cleaners—Performance Trends. SAE Technical Paper (2001)
15. J.A. Grajales, H. Quintero, J. Lopez Lopez, C. Romero, E. Henao, O. Cardona, Engine diagnosis based on vibration analysis using different fuel blends. Diagnostyka **18**, 27–36 (2018)
16. I. Gravalos, S. Loutridis, D. Moshou, T. Gialamas, D. Kateris, Z. Tsiropoulos, P. Xyradakis, Detection of fuel type on a spark ignition engine from engine vibration behaviour. Appl. Therm. Eng. **54**, 171–175 (2013)
17. G. DeBotton, Vibration signature analysis as a fault detection method for SI engines. SAE Trans. **107**(1998), 1–6 (2021)
18. K. Jafarian, M. Mobin, R. Jafari-Marandi, E. Rabiei, Misfire and valve clearance faults detection in the combustion engines based on a multi-sensor vibration signal monitoring. Measurement 527–536 (2018)
19. J.-D. Wu, C.-Q. Chuang, Fault diagnosis of internal combustion engines using visual dot patterns of acoustic and vibration signals. NDT E Int. **38**(8), 605–614 (2005)
20. N. Gültekin, M. Mayda, Investigation of effects of spark plug gap on vibration, noise and HC emission in a gasoline engine. Int. J. Energy Appl. Technol. **5**, 56–60 (2018)
21. S. Klinchaeam, P. Nivesrangsan, M. Lokitsangthong, Condition monitoring of a small four-stroke petrol engine using vibration signals. KMITL Sci. Technol. J. **9**, 9–17 (2009)
22. X. Zhao, Y. Cheng, L. Wang, S. Ji, Real time identification of the internal combustion engine combustion parameters based on the vibration velocity signal. J. Sound Vib. **390**, 205–217 (2017)
23. L. Barelli, G. Bidini, C. Buratti, R. Mariani, Diagnosis of internal combustion engine through vibration and acoustic pressure non-intrusive measurements. Appl. Therm. Eng. **29**(8–9), 1707–1713 (2009)
24. NTE INEN 2204:2016: Gestión Ambiental. Aire. Vehículos Automotores. Límites Permitidos de Emisiones Producidos por Fuentes Móviles Terrestres que Utilizan Gasolina, Quito-Ecuador (2016)

Chapter 68
Modified First Stage of French Vertical Flow Constructed Wetlands for Wastewater Treatment in Highlands: Start-Up of the System

María B. Arévalo-Durazno, Jorge A. García Zumalacarregui, Andrea Narváez, and Andrés Alvarado

Abstract Worldwide, decentralized sanitation systems represent an appropriate, sustainable, simpler, and cost-effective option for communities where conventional wastewater treatment technologies have failed by not considering the inherent characteristics of those societies. In this research, the start-up phase of a cost-effective decentralized sanitation technology is presented: The modified first stage of French vertical flow constructed wetlands for raw wastewater treatment. In the three pilot units comprising the modified first stage, *Lolium perenne* was the vegetation planted and showed great adaptability to the climatic conditions of the highlands and wastewater characteristics from a combined sewerage. In order to evaluate the performance of the system, in situ parameters were measured and physical chemical analyses were performed from samples of the three units. Efficiencies over 45% were obtained for the removal of organic matter and total suspended solids. The pilots demonstrated that the first stage of French vertical flow constructed wetlands is robust and reliable from their start-up period and treating hydraulic overloads.

M. B. Arévalo-Durazno (✉)
Facultad de Ciencia y Tecnología, Universidad del Azuay, Cuenca, Ecuador
e-mail: barevalo@uazuay.edu.ec

J. A. García Zumalacarregui · A. Narváez · A. Alvarado
Departamento de Recursos Hídricos y Ciencias Ambientales, Universidad de Cuenca, Cuenca, Ecuador

J. A. García Zumalacarregui
Facultad de Ciencias Agropecuarias, Universidad de Cuenca, Av. 12 de Octubre y Don Bosco, 010220 Cuenca, Ecuador

A. Alvarado
Facultad de Ingeniería, Universidad de Cuenca, Av. 12 de Abril s/n, 010107 Cuenca, Ecuador

Á. Rocha et al. (eds.), *Communication, Smart Technologies and Innovation for Society*,
Smart Innovation, Systems and Technologies 252,
https://doi.org/10.1007/978-981-16-4126-8_68

68.1 Introduction

Rapid population growth and uncontrolled urban development have provoked the direct or poorly treated discharge of large amounts of wastewater into receiving bodies, causing severe environmental and health risks. It is estimated that until 2017 only 45% of the global population had access to a safe managed sanitation service [1, 2]. The situation is more difficult for rural or peri-urban areas where a central sewer network is not feasible from a technical and economical point of view. In such areas, decentralized systems, that include natural wastewater treatment processes, have been considered a sustainable option, due to their conceptual simplicity and low operation and maintenance costs [3–5].

Among the decentralized sewerage systems, French vertical flow constructed wetlands (VFCW) have been widely used [6]. Their special feature of treating raw wastewater (without any primary treatment) reduces the total area and costs needed for the treatment. Moreover, the sludge management of VFCWs is easier in comparison to dealing with the sludge from primary treatment, which has represented a major challenge in the implementation of sustainable and low-cost sanitation [7, 8].

The classical VFCWs configuration consists of 2 stages of filters. The first stage is comprised of 3 filters and the second stage by 2 filters. In the first stage of vertical flow constructed wetlands (FS-VFCW), each filter receives the raw wastewater during the feeding period, which lasts between 3 and 4 days, before a resting period of twice the feeding time [7]. During the feeding period, the raw wastewater is introduced in pulses or batches that vary from 6 to 24 per day [6]. The sizing of the system is based on an acceptable organic load expressed as a surface unity per person equivalent (PE). For the FS-VFCW is recommended a total area of 1.2 m^2 per PE divided over 3 filters, that is, each filter operates with an organic load of < 300 g COD/(m^2 d), < 150 g SS/(m^2 d), and < 25–30 g TKN/(m^2 d) and a hydraulic loading rate (HL) of 0.37 m^3/(m^2 d) [7].

The treatment of unsettled wastewater in FS-VFCW results in the formation of a sludge layer at the surface of the bed. This deposit layer improves the treatment performance in many processes, such as filtration, water retention, water distribution, ammonia adsorption, and biological activity [9]. The characteristics, rate of formation, and mineralization of the deposit layer are strongly dependent on the age of the FS-VFCW and influenced by local conditions [10]. During the first months of operation of the FS-VFCW, the absence of the deposit layer can affect the efficiency of the treatment, since the age of the plant has a positive effect on reducing the concentration of pollutants in the effluent [11].

In France, the VFCW are commonly cultivated with Phragmites Australis [12], while in warmer climates as in Brazil, the system is planted with Tifton-85 (Cynodon dactylon Pers) [13]. Lolium perenne performs well in cold climates, adapts easily, and continues to proliferate even at extremely cold periods of the year, and also presents a higher removal rate than other species under the same conditions [14]. Among the most relevant characteristics of the selected crop is its capacity for uniform distribution on the filter surface.

The design of FS-VFCW is quite empirical, based on the knowledge gained over years of laboratory studies and full-scale models. They have been successfully applied in France and in tropical regions [9, 13]. The recommended design values [7] arise from the treatment of raw wastewater coming from a separate sewerage system (concentrated wastewater). There is still uncertainty about the performance of this kind of system treating diluted wastewater, and operating in the high altitudes of the tropical region, where there is less seasonal climatic variation, more constant year-round radiation, and relatively low mean temperature (around 15 °C). Some changes may exist in the operation of the system and the formation and role of the deposit layer.

The performance in the start-up period of the French VFCW is quite unknown. Moreover, the starting-up using diluted influents is uncommon due to the high hydraulic loads needed for biomass growth. Most publications report the results from mature systems which have already efficiencies over or close to the ones expected in the design. A better understanding in the behavior of the VFCW since the first days of operation is needed. This will allow to take into consideration important in site aspects during design and planning that will have a positive effect in the future performance of the system.

This paper presents the first results of the design and the start-up of a modified FS-VFCW pilot plant located after a preliminary treatment process of a municipal wastewater treatment plant (WWTP). The effluent from a combined sewer system, which contains wastewater and stormwater, is being used as the influent to the system, due to this, the pollutants present in the sewage are diluted. The feeding of the system began immediately after the crops were planted, and the monitoring for characterization of both influent and effluent, was carried out from the first day of feeding. The aim of this research is to obtain a better understanding of the performance of a modified FS-VFCW treating highly diluted raw wastewater in the Andean region of Ecuador during the start-up phase of the system.

68.2 Methods

68.2.1 Study Site and Pilot-Scale Units

The experimental plant for wastewater treatment consists of three units of the modified first stage vertical flow constructed wetlands of the French System. The plant is located at Ucubamba WWTP in Cuenca (2°52′15.1″ S, 78°56′30.8″ W), at the south of Ecuador in the Andean region, at 2500 m.a.s.l. and with an annual average temperature of 14 °C.

The plant was designed with three pilot-scale units (W1, W2, W3) of 9.81 m^2 (filter surface area) each, with length to width ratio of 1.3:1 (Fig. 68.1). The depth of each unit was 1 m and the medium depth of 0.7 m. From the top to the bottom, the filter media used was: a 30 cm layer of small gravel (ϕ 2–10 mm), 20 cm of

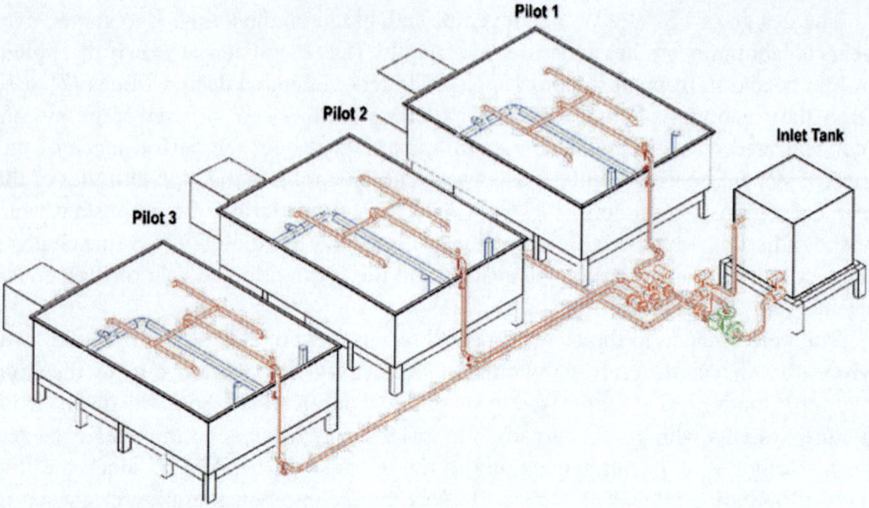

Fig. 68.1 Schematic overview of the pilot-scale units

middle gravel (ϕ 10–20 mm), and 20 cm of large gravel (ϕ 20–60 mm). The feeding system was constituted by ϕ 50 mm PVC pipes placed above the top layer with four outlets to facilitate the even distribution of the influent over the whole surface. And for the drainage system, two perforated ϕ 110 mm PVC pipes were positioned at the bottom and connected to the surface for aeration. The vegetation planted was *Lolium perenne.*

The experimental plant treats the raw wastewater with the same characteristics as the one generated by Cuenca city. After the coarse screen and grit removal of the WWTP, wastewater is derived to an inlet tank, where a submerged pump controlled by a programmable logic controller (PLC) feeds the wetlands.

68.2.2 Operation Mode, Sampling, Monitoring, and Testing

For the start-up of the system, the first objective was to achieve the establishment of the Lolium perenne in each unit. The strategy was to feed the units the time until it could be seen that the plants will survive without wastewater. We checked the plants to become green and then moved to the next unit.

The three units received the same HL of 0.936 $m^3/(m^2\ d)$ and were fed intermittently with a volume of 0.382 m^3 for 2.5 min each hour (24 batches a day), resulting in an instantaneous hydraulic loading rate of 0.93 $m^3/(m^2\ h)$.

In order to evaluate the performance of the system in the start-up phase, in situ parameters and physical chemical analyses were performed in samples collected in the influent and effluent of the three units, during the first month of operation.

Dissolved oxygen (DO) and pH were measured in four samples a day, in situ using a WTW multi 3420 probe. For chemical oxygen demand (COD), biochemical oxygen demand BOD_5, and total suspended solids (TSS) grab point samples were taken, at the same hour of the day during the whole sampling campaign, in the inlet and outlet pipes of the wetlands. COD and TSS were analyzed following the *Standard Methods for the Examination of Water and Wastewater* [15]. BOD_5 was analyzed using a respirometric system Lovibond BD 600. COD was measured in all the samples taken, while BOD_5 was measured only in 14 samples and TSS in 5 samples.

The comparative statistical analysis between COD removal efficiencies in the three units, the ANOVA test was performed using R software.

68.3 Results and Discussion

68.3.1 Start-Up of the Treatment System and Plant Growth

The pilot units started their normal operation in September 2020. One sampling campaign has been conducted for the start-up period during September–October 2020. Before the units could operate under normal conditions, several mechanical and hydraulic issues were overcome by trial-and-error activities and by using a programmable logic controller (PLC) for the calibration of the feeding regime of the pilots. The pilots were planted with Lolium perenne on their first day of feeding. In day 1, W1 was planted, and it took 8 days until the plants were established on the bed. Thus, on day 9, W1 was put to rest and W2 was planted with the same species and starting the feeding period. It took 13 days to W2 to get an established population of Lolium perenne on its bed. On day 22, W3 was planted and loaded with the wastewater. In this unit, after 6 days the plants were already green. Thus, on day 28, W1 was again fed. During this period, it could be seen the rapid adaptation of the Lolium perenne to the climatic conditions of the zone and the wastewater characteristics of the influent. Besides, this plant demonstrated that could resist extended resting periods.

Figure 68.2 shows the evolution of the plants in W1, W2 and W3 since the first day *Lolium perenne* was planted in each unit. As it can be observed in the middle figures, the feeding cycle in each wetland finished when the plants were green enough to assure that they won´t die in the resting period. The figures on the right confirm the last fact.

In W1, the wetland that rested the longest for 19 days, the plants continued to grow and populate the whole bed. According to Ren [14], *Lolium perenne* is a cool-season perennial gramineous terrestrial lawn plant that has the great advantage of being evergreen in cold period and showed that it could improve wastewater treatment efficiencies in cold climates. In this study, *L. perenne* demonstrated that it is a good choice for constructed wetlands in highlands with extended resting periods.

Fig. 68.2 Evolution of *Lolium perenne* in, W1 (first row), W2 (second row), and W3 (third row). From left to right: first day of feeding, final day of feeding, final day of resting periods

68.3.2 Influent Characterization

The influent of the pilots is the sewage of Cuenca city that enters Ucubamba WWTP. The wastewater feeds the pilots after it has passed a coarse screen and the grit removal (raw wastewater). Cuenca's sewerage is a combined system. The quality characteristics of the flux are presented in Table 68.1. The high variability in the influent is a consequence of the climatic conditions of Cuenca (dry and rainy) that change significantly from day to day, and because of the grab point samples taken.

As it can be seen, the mean values of COD and BOD_5 concentrations are rather low which shows the diluted nature of the wastewater because of the type of sewage

Table 68.1 Quality characteristics of the influent to the pilots

Parameter	Measure unit	Mean value	Standard deviation	Coeff. of variation
pH	–	8.5	0.8	0.1
COD	mg/L	255	233	0.91
BOD5	mg/L	105	102	0.97
TSS	mg/L	184	239	1.3

system. This is especially important because it has an implication for the hydraulic loading rates that are applied in the wetlands. According to the first design guidelines for the classical French system, the hydraulic loading rate of 0.37 m/d is applied on the filter in operation for separate networks which results in an organic load of 300 g COD/(m^2 d) [7]. In this case, the HL applied was 2.5 times greater and resulted in an average OL of 217 g COD/(m^2 d). The start-up of the system with hydraulic overloads represents a big challenge because the diluted influent could wash out the biomass that starts to grow on the granular media. In the next section, performance of the pilots, it can be seen that this atypical start showed successful results.

The mean value of pH confirms the domestic origin of the wastewater.

68.3.3 Performance of the Pilots

The performance of the pilots has been evaluated in terms of COD, BOD5, and TSS removal. Additionally, the increase of OD in the effluent was evaluated.

Figure 68.3 illustrates the concentrations of COD, BOD_5, TSS, and DO in the influent and effluent of the three units. For COD, BOD_5, and TSS, it can be seen that the effluent concentrations are rather low and stable, contrary to the influent ones. This fact shows the robustness and reliability of the system. Lombard-Latune et al. [16] compared different treatment technologies and concluded that French VFCW are the most reliable technology; they complied with the treatment targets the 90–95% of the time. It can be observed that these kinds of wetlands are robust from the very beginning, despite the fact that they don´t have yet all the advantages of a mature system. An important aspect was that during the monitoring days it was evidenced that the pilots can cope with the great variability of the effluent quality, responding with adequate removals of organic matter and total suspended solids.

COD and BOD_5 concentrations in the effluent are all below 200 mg/L and 100 mg/L, respectively, which are the regulatory thresholds in Ecuador for the discharge of treated wastewater to a body of freshwater [17]. For TSS, only in W1, there is an effluent concentration above the regulatory limit of 130 mg/L due to an unusual high influent concentration registered.

The DO concentrations increase from influent to effluent showing the aerobic conditions of the wetlands, which is the desired condition for FS-VFCW. Median concentrations in the three units are above 4 mg/L but below 6 mg/L. It is difficult to compare these results with other studies because DO in water varies depending on the altitude and temperature, and those conditions in this research are unique. In Brazil, for completely different climatic conditions: annual average temperature of 21.8 °C at 760 m.a.s.l. similar concentrations were found in the effluent of a mature FS-VFCW by García Zumalacarregui [13]. In the same system, de Andrade Moraes [18] analyzed the influence of the deposit layer in DO concentrations depending on the cycle day and found median values around 4 mg/L in the first day of feeding in a wetland with a 7.2 cm sludge layer height and on day 4 in a wetland with a 0.5 cm sludge layer height. When the sludge layer is not formed yet, the aeration

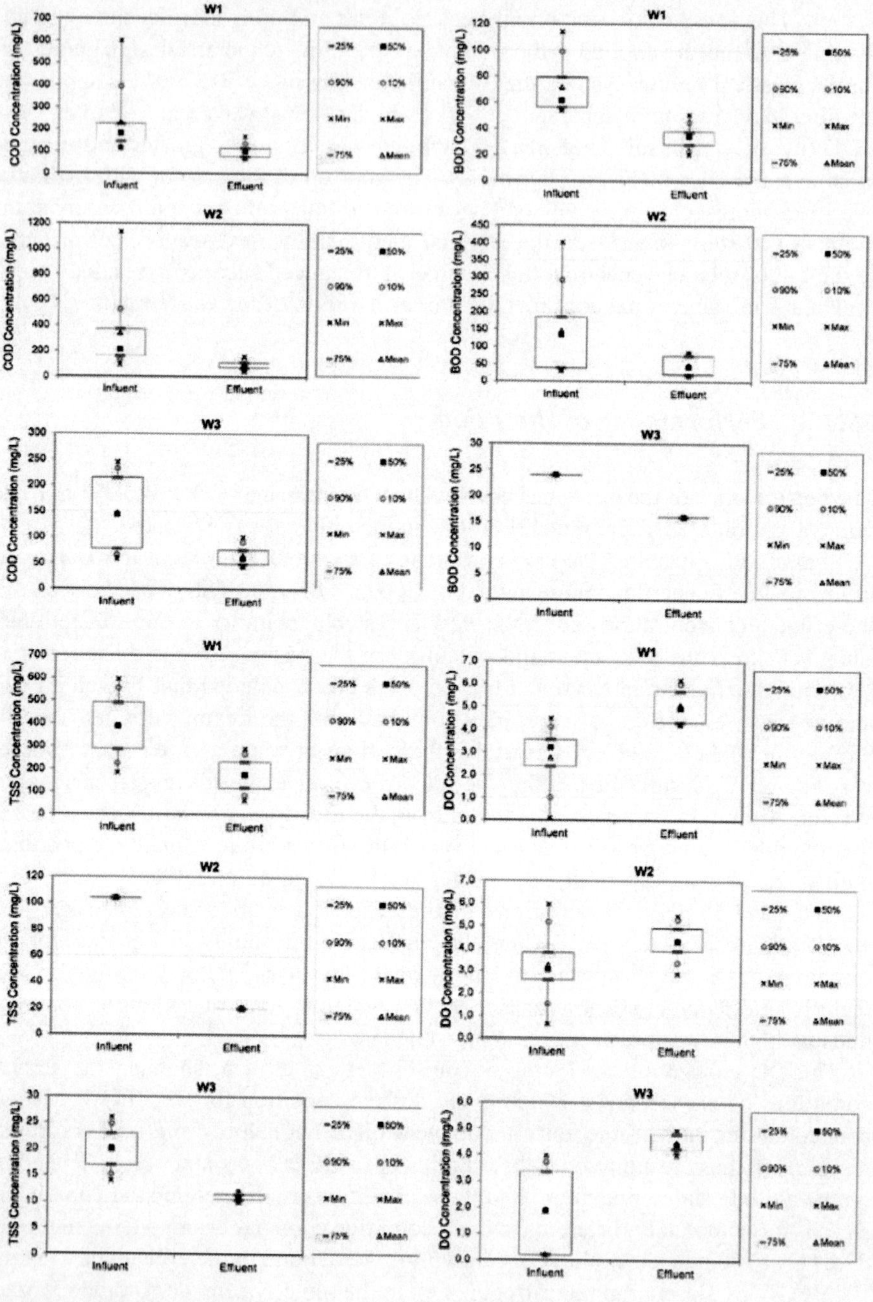

Fig. 68.3 COD, BOD_5, TSS, and DO concentrations in the influent and in the effluent of the three pilots during the start-up period

Table 68.2 Average removal efficiencies of COD, BOD_5, and TSS in the pilots

Parameter	W1 (%)	W2 (%)	W3 (%)
Chemical oxygen demand	47.4	62.2	46.8
Biochemical oxygen demand	51.3	56.1	33.3
Total suspended solids	58.2	80.8	41.2

Table 68.3 Results of the ANOVA test for COD removal efficiencies in the three pilots

Source of variation	SS	df	MS	*F*	*P*-value	F crit
Between groups	1159.458	2	579.729076	1.54581655	0.2401378	3.5545571
Within groups	6750.557	18	375.030968			
Total	7910.015	20	239			

capacity of the system is higher and its possible to have concentrations higher than 4 mg/L during the first feeding days. Thus, it is expected that in the pilots of this research, after some months of operation the aeration capacity will drop and the OD concentrations will decrease.

Regarding the removal efficiencies of COD, BOD_5, and TSS, Table 68.2 displays the average values. Most of the efficiencies are above 45% for the three parameters analyzed showing good performance of the pilots in their start-up phase. Efficiencies over 80% are reported from FS-VFCW in mature systems around the world. The main difference of the mentioned values is the lack of the sludge deposit layer in the pilots of this study, which is considered a key factor for improving the treatment efficiency [9]. Paing et al. [11] concluded that the age of the wetlands has a positive effect on performance, the concentrations of the pollutants in the effluent decrease when the wetlands get older.

Comparing the efficiencies of the three units, a one-way analysis of variance (ANOVA) at the 5% significance level (Table 68.3) indicated that based on the samples; it cannot be concluded that the mean COD removal efficiencies are significantly different between the three VFCWs. This fact also shows the robustness of the system because three beds that operate in the same conditions achieve similar efficiencies.

68.4 Conclusions

The results of the start-up period of FS-VFCW indicated that even though the units don't have yet the characteristics which make a mature system efficient, removal efficiencies above 45% can be obtained in terms of COD, BOD5, and TSS which is comparable and even higher than classic anaerobic units widely used in the region for decentralized systems. The pilots demonstrated that from the first day of operation, the French system is robust and reliable despite the high variability of the influent and can cope with hydraulic overloads. Finally, the plant *Lolium perenne* demonstrated that it is a good choice for vertical flow constructed wetlands in highlands with intermittent feeding and extended resting periods. Considering the results obtained in this study, it is expected that the system will significantly improve its performance in the future, reaching even higher removal efficiencies of the pollutants.

Funding This research was developed within the framework of the project "Innovación y experimentación para el estudio de sistemas sostenibles de tratamiento de efluentes domésticos en altura" funded by the Research Directorate of the University of Cuenca (DIUC) under the XVth call for research proposals. This manuscript is an outcome of the Doctoral Program in Water Resources by the first author, offered by Universidad de Cuenca, Escuela Politécnica Nacional, and Universidad Técnica Particular de Loja.

References

1. WHO, Sanitation (2019), https://www.who.int/news-room/fact-sheets/detail/sanitation. Accessed 16 Mar 2021
2. M.A. Massoud, A. Tarhini, J.A. Nasr, Decentralized approaches to wastewater treatment and management: applicability in developing countries. J. Environ. Manag. **90**, 652–659 (2009)
3. T.A. Nanninga, I. Bisschops, E. López, J.L. Martínez-Ruiz, D. Murillo, L. Essl, M. Starkl, Discussion on sustainable water technologies for peri-urban areas of Mexico City: balancing urbanization and environmental conservation. Water **4**, 739–758 (2012)
4. N.K. Singh, A.A. Kazmi, M. Starkl, A review on full-scale decentralized wastewater treatment systems: techno-economical approach. Water Sci. Technol. **71**, 468–478 (2015)
5. M. von Sperling, Comparison of simple, small, full-scale sewage treatment systems in Brazil: UASB–maturation ponds–coarse filter; UASB–horizontal subsurface-flow wetland; vertical-flow wetland (first stage of French system). Water Sci. Technol. **71**, 329–337 (2015)
6. M. Andrade, J.A. García, C.M. Trein, V.V.M. Ferreira, M. von Sperling, Outflow dynamics in a French system of vertical wetlands operating with an extended feeding cycle. Water Sci. Technol. **79**, 699–708 (2019)
7. P. Molle, A. Liénard, C. Boutin, G. Merlin, A. Iwema, How to treat raw sewage with constructed wetlands: an overview of the French systems. Water Sci. Technol. **51**, 11–21 (2005)
8. L.F. Calderón-Vallejo, C.F. Andrade, E.S. Manjate, C.A. Madera-Parra, M. von Sperling, Performance of a system with full- and pilot-scale sludge drying reed bed units treating septic tank sludge in Brazil. Water Sci. Technol. **71**, 1751–1759 (2015)
9. P. Molle, French vertical flow constructed wetlands: a need of a better understanding of the role of the deposit layer. Water Sci. Technol. **69**, 106–112 (2014)
10. M. Kania, M. Gautier, A. Imig, P. Michel, R. Gourdon, Comparative characterization of surface sludge deposits from fourteen French Vertical Flow Constructed Wetlands sewage treatment plants using biological, chemical and thermal indices. Sci. Total Environ. **647**, 464–473 (2019)
11. J. Paing, A. Guilbert, V. Gagnon, F. Chazarenc, Effect of climate, wastewater composition, loading rates, system age and design on performances of French vertical flow constructed wetlands: a survey based on 169 full scale systems. Ecol. Eng. **80**, 46–52 (2015)
12. J. Paing, J. Voisin, Vertical flow constructed wetlands for municipal wastewater and septage treatment in French rural area. Water Sci. Technol. **51**, 145–155 (2005)
13. J.A. García Zumalacarregui, M. von Sperling, Performance of the first stage of the French system of vertical flow constructed wetlands with only two units in parallel: influence of pulse time and instantaneous hydraulic loading rate. Water Sci. Technol. **78**, 848–859 (2018)

14. Y. Ren, Y. Liu, J. Sun, H. Lu, L. Yang, C. Chen, Y. Han, *Lolium perenne* as the cultivation plant in hydroponic Ditch and constructed wetland to improve wastewater treatment efficiency in a cold region. Wetlands **36**, 659–665 (2016)
15. APHA/AWWA/WEF, *Standard Methods for the Examination of Water and Wastewater*, 22nd edn. (Washington, DC, USA, 2012)
16. R. Lombard-Latune, F. Leriquier, C. Oucacha, L. Pelus, G. Lacombe, B. Le Guennec, P. Molle, Performance and reliability comparison of French vertical flow treatment wetlands with other decentralized wastewater treatment technologies in tropical climates. Water Sci. Technol. **82**, 1701–1709 (2020)
17. TULSMA, Norma de Calidad Ambiental y de descarga de efluentes: Recurso Agua. Anexo 2, Libro VI (Ministerio del Ambiente, Quito, Ecuador, 2015)
18. M.A. de Andrade Moraes, J.A. García Zumalacarregui, C.M. Trein, M. von Sperling, Dynamics of the behaviour of a vertical wetland (French system) operating in warm-climate conditions, evaluated by means of variables continuously measured *in situ*. Water Sci. Technol. **82**, 954–966 (2020)